HANDBOOK OF BUILDING MATERIALS FOR FIRE PROTECTION

HANDBOOK OF BUILDING MATERIALS FOR FIRE PROTECTION

Charles A. Harper Editor-in-Chief
Technology Seminars, Inc., Lutherville, MD 21094

McGRAW-HILL
New York Chicago San Francisco Lisbon London Madrid
Mexico City Milan New Delhi San Juan Seoul
Singapore Sydney Toronto

The **McGraw·Hill** Companies

Cataloging-in-Publication Data is on file with the Library of Congress

Copyright © 2004 by The McGraw-Hill Companies, Inc. All rights reserved. Printed in the United States of America. Except as permitted under the United States Copyright Act of 1976, no part of this publication may be reproduced or distributed in any form or by any means, or stored in a data base or retrieval system, without the prior written permission of the publisher.

1 2 3 4 5 6 7 8 9 0   DOC/DOC   0 9 8 7 6 5 4 3

ISBN 0-07-138891-5

The sponsoring editor for this book was Kenneth P. McCombs and the production supervisor was Sherri Souffrance. It was set in Times Roman by Achorn Graphic Services, Inc. The art director for the cover was Handel Low.

Printed and bound by RR Donnelley.

McGraw-Hill books are available at special quantity discounts to use as premiums and sales promotions, or for use in corporate training programs. For more information, please write to the Director of Special Sales, McGraw-Hill Professional, Two Penn Plaza, New York, NY 10121-2298. Or contact your local bookstore.

 This book is printed on recycled, acid-free paper containing a minimum of 50% recycled, de-inked fiber.

Information contained in this work has been obtained by The McGraw-Hill Companies, Inc. ("McGraw-Hill") from sources believed to be reliable. However, neither McGraw-Hill nor its authors guarantee the accuracy or completeness of any information published herein and neither McGraw-Hill nor its authors shall be responsible for any errors, omissions, or damages arising out of use of this information. This work is published with the understanding that McGraw-Hill and its authors are supplying information but are not attempting to render engineering or other professional services. If such services are required, the assistance of an appropriate professional should be sought.

CONTENTS

Contributors xi
Preface xiii

Chapter 1 Fundamentals of the Fire Hazards of Materials Dr. Frederick W. Mowrer 1.1

1.0 Introduction *1.1*
1.1 Fundamentals of Combustion *1.2*
   1.1.1 Stoichiometry *1.3*
   1.1.2 Thermochemistry *1.7*
   1.1.3 Flame Temperatures *1.12*
1.2 Gases, Mists, and Dusts *1.16*
   1.2.1 Flammability Limits for Gases and Vapors *1.16*
   1.2.2 Ignition Energy *1.19*
   1.2.3 Flame Speed *1.20*
   1.2.4 Ignition of Mists and Dusts *1.20*
1.3 Liquids *1.20*
   1.3.1 Fire Point *1.21*
   1.3.2 Vapor Pressure *1.22*
   1.3.3 Flammability Limits of Liquids *1.23*
   1.3.4 Liquid Mixtures *1.23*
   1.3.5 Burning Rate *1.24*
1.4 Solids *1.26*
   1.4.1 Flaming Ignition of Solid Materials *1.29*
   1.4.2 Critical Heat Flux and Effective Ignition Temperature *1.35*
   1.4.3 Pyrolysis and Burning Rates *1.37*
   1.4.4 Flame Spread *1.38*
   1.4.5 Self-Heating and Smoldering Combustion *1.41*
   1.4.6 Effects of Fire Retardants *1.42*
1.5 Smoke Production *1.42*
   1.5.1 Vision Obscuration *1.43*
   1.5.2 Toxicity of Combustion Products *1.45*
   1.5.3 Nonthermal Damage *1.46*
1.6 Enclosure Effects *1.46*
1.7 Summary *1.48*
1.8 References *1.48*

Chapter 2 Materials Specifications, Standards, and Testing Dr. Archibald Tewarson, Dr. Wai Chin, and Dr. Richard Shuford 2.1

2.1 Introduction *2.1*
2.2 Materials Characteristics *2.3*
2.3 Softening and Melting Behaviors of Materials *2.4*
2.4 Vaporization, Decomposition, and Charring Behaviors of Materials *2.5*
2.5 Ignition Behavior of Materials *2.8*

2.6 Flame Spread, Fire Growth, and Burning Behavior of Materials   *2.13*
2.7 Popular Standard Test Methods for the Burning Behavior of Materials   *2.15*
    2.7.1 ASTM D 5865 and ISO 1716: Test Method for Gross Heat of Complete Combustion   *2.16*
    2.7.2 ASTM E 136 and ISO 1182: Standard Test Method for Behavior of Materials in a Vertical Tube Furnace at 750   *2.16*
    2.7.3 ASTM E 906, ASTM E 2058, and ASTM E 1354 and ISO 5660: Standard Test Methods for Release Rates of Material Vapors, Heat, and Chemical Compounds   *2.17*
    2.7.4 ASTM E 119: Standard Test Methods for Fire Tests of Building Construction and Materials—The Fire Endurance Test   *2.24*
    2.7.5 ASTM E 1529: Standard Test Methods for Determining Effects of Large Hydrocarbon Pool Fires on Structural Members and Assemblies   *2.25*
2.8 Popular Standard Test Methods for Flame Spread and Fire Growth   *2.26*
    2.8.1 prEN ISO and FDIS 11925-2: Reaction to Fire Tests for Building Products—Part 2: Ignitability when Subjected to Direct Impingement of Flame   *2.26*
    2.8.2 UL 94: Standard Test Methodology for Flammability of Plastic Materials for Parts in Devices and Appliances   *2.27*
    2.8.3 ASTM D 2863 (ISO 4589): Test Methodology for Limited Oxygen Index   *2.28*
    2.8.4 ASTM E 162 (D 3675): Standard Test Method for Surface Flammability Using a Radiant Energy Source   *2.30*
    2.8.5 ASTM E 1321 (ISO 5658): Standard Test Method for Determining Material Ignition and Flame Spread Properties (LIFT)   *2.32*
    2.8.6 ASTM E 648 (ISO 9239-1): Standard Test Method for Critical Radiant Flux of Floor-Covering Systems Using a Radiant Heat Energy Source   *2.34*
    2.8.7 ASTM E 84: Standard Test Method for Surface Burning Characteristics of Building Materials   *2.36*
    2.8.8 FM Global Approval Class 4910 (NFPA 318): Standard Test Methods for Clean Room Materials for the Semiconductor Industry   *2.38*
    2.8.9 ASTM E 603: Standard Guide for Room Fire Experiments   *2.39*
Nomenclature   *2.49*
References   *2.51*

Chapter 3 Plastics and Rubber *Dr. Richard E. Lyon* **3.1**

3.1 Introduction   *3.1*
3.2 Polymeric Materials   *3.1*
    3.2.1 Monomers, Polymers, and Copolymers   *3.1*
    3.2.2 Polymer Architectures   *3.3*
    3.2.3 Commercial Materials   *3.7*
    3.2.4 Thermodynamic Quantities   *3.8*
3.3 The Burning Process   *3.15*
    3.3.1 The Fire Triangle   *3.15*
    3.3.2 Chemical Changes During Burning   *3.16*
3.4 Fire Behavior of Plastics   *3.28*
    3.4.1 Ignition   *3.28*
    3.4.2 Steady Burning   *3.34*
    3.4.3 Unsteady Burning   *3.40*
Definition of Terms   *3.44*
References   *3.46*

Chapter 4 Flame Retardants for Plastics *Dr. Elisabeth S. Papazoglou* **4.1**

4.1 Introduction   *4.1*
4.2 Overview of the Flame Retardants Industry   *4.2*
4.3 Mechanisms of Flame Retardancy   *4.5*

4.4 Classes of Commercial Flame Retardants   *4.7*
    4.4.1 Hydrated Minerals   *4.7*
    4.4.2 Halogenated Materials   *4.9*
    4.4.3 Antimony Trioxide   *4.21*
    4.4.4 Phosphorus Additives   *4.23*
    4.4.5 Intumescent Flame-Retardant Systems   *4.36*
4.5 Polymer Families—Selection of Flame Retardant   *4.36*
    4.5.1 Polypropylene   *4.38*
    4.5.2 Polyethylene   *4.44*
    4.5.3 Styrenics   *4.44*
    4.5.4 ABS   *4.48*
    4.5.5 Polycarbonate   *4.49*
    4.5.6 PC/ABS Blends   *4.50*
    4.5.7 Nylon   *4.58*
    4.5.8 Thermoplastic Polyesters   *4.66*
    4.5.9 Polyvinyl Chloride   *4.69*
    4.5.10 Thermosets   *4.72*
    4.5.11 Elastomers/Rubber   *4.73*
4.6 Nanocomposites   *4.76*
    4.6.1 Layered Silicates   *4.76*
    4.6.2 Polymer Nanocomposite Structures   *4.77*
    4.6.3 Preparation Methods   *4.77*
    4.6.4 Flame-Retardant Properties of Nanocomposites   *4.78*
    4.6.5 Mechanism of Flame Retardancy in Nanocomposites   *4.80*
References   *4.84*

Chapter 5 Fibers and Fabrics *Dr. Debbie J. Guckert, Susan L. Lovasic, and Dr. Roger F. Parry*   5.1

5.1 The Role of Fabrics in Fire Protection   *5.1*
5.2 Fibers and Their Properties   *5.1*
    5.2.1 Materials from Which Fibers Are Formed   *5.1*
    5.2.2 Forms of Fibers Available   *5.1*
    5.2.3 Fiber Properties   *5.2*
    5.2.4 Flammability Characteristics   *5.2*
    5.2.5 Flame Retardants   *5.5*
5.3 Fabric Types   *5.7*
5.4 Flammability of Fabrics   *5.7*
    5.4.1 Characteristics of Burning Fabrics   *5.7*
    5.4.2 Thermal Performance Tests for Fabrics   *5.8*
    5.4.3 Performance Standards   *5.19*
5.5 Applications—Protective Clothing   *5.23*
    5.5.1 Burn Injuries   *5.24*
    5.5.2 Flash Fires   *5.25*
    5.5.3 Structural and Wildlands Fires   *5.26*
    5.5.4 Electric Arcs   *5.32*
    5.5.5 Molten Metals   *5.35*
    5.5.6 Soiling and Cleaning of Protective Clothing   *5.36*
5.6 Applications—Furnishings   *5.36*
    5.6.1 Structures   *5.37*
    5.6.2 Transportation   *5.45*
5.7 Challenges for the Future   *5.47*
Acknowledgments   *5.48*
References   *5.48*
Appendix   *5.51*

Chapter 6 Structural Materials *Dr. Nestor R. Iwankiw, Jesse J. Beitel, and Richard G. Gewain* **6.1**

6.1 Structural Materials Used in Construction   *6.1*
    6.1.1 Introduction   *6.1*
    6.1.2 Construction Materials   *6.1*
6.2 Development of Fire Resistance Testing   *6.2*
    6.2.1 Historical Fire Events   *6.2*
    6.2.2 Early Fire Resistance Test Procedures   *6.3*
    6.2.3 Standard Fire Resistance Tests   *6.4*
    6.2.4 Test Equipment   *6.7*
    6.2.5 Failure Criteria   *6.7*
    6.2.6 Special Hazard Resistance Tests ("High-Rise" Curves)   *6.10*
6.3 Structural Materials and Fire   *6.11*
    6.3.1 Reaction of Structural Materials to Fire   *6.11*
6.4 Protection of Structural Materials from Fire   *6.22*
    6.4.1 Fire-Resistive Materials   *6.23*
    6.4.2 Fire-Resistive Systems   *6.28*
6.5 Determination of Fire Resistance by Testing   *6.32*
    6.5.1 Steel Construction   *6.32*
    6.5.2 Concrete Construction   *6.42*
6.6 Determination of Fire Resistance by Calculation   *6.43*
    6.6.1 Steel Construction   *6.44*
    6.6.2 Concrete Construction   *6.60*
6.7 Application of Fire Resistance Ratings   *6.61*
References   *6.61*

Chapter 7 Wood and Wood Products *Dr. Marc Janssens, and Dr. Bradford Douglas* **7.1**

7.1 Units   *7.1*
7.2 Introduction   *7.1*
    7.2.1 Wood   *7.1*
    7.2.2 Forestry   *7.1*
    7.2.3 Wood and Carbon   *7.2*
    7.2.4 Wood and Fire   *7.2*
7.3 Wood as a Construction Material (U.S. Customary Units)   *7.2*
    7.3.1 Sawn Timber   *7.2*
    7.3.2 Panel Products   *7.3*
    7.3.3 Engineered Wood Products   *7.4*
    7.3.4 Other Materials   *7.5*
7.4 Physical and Chemical Characteristics (S.I. Units)   *7.6*
    7.4.1 Botanical Categories   *7.6*
    7.4.2 Physical Structure   *7.6*
    7.4.3 Moisture   *7.7*
    7.4.4 Chemical Composition   *7.7*
    7.4.5 Thermal Decomposition and Pyrolysis   *7.8*
    7.4.6 Fire-Retardant Treatments   *7.9*
7.5 Thermal Properties (S.I. Units)   *7.10*
    7.5.1 Wood and Char   *7.10*
    7.5.2 Other Materials   *7.23*
7.6 Mechanical Properties (U.S. Customary Units)   *7.24*
    7.6.1 Properties at Normal Temperatures   *7.24*
    7.6.2 Properties at Elevated Temperatures   *7.25*
7.7 Reaction to Fire (S.I. Units)   *7.27*
    7.7.1 Ignition   *7.27*
    7.7.2 Heat Release and Charring Rate   *7.29*

7.7.3 Products of Combustion   7.35
7.7.4 Surface Flame Spread   7.36
7.8 Fire Resistance (U.S. Customary Units)   7.40
7.8.1 Exposed Wood Members   7.40
7.8.2 Protected Wood-Frame Construction   7.48
References   7.52

Chapter 8 Liquids and Chemicals *Dr. A. Tewarson and Dr. G. Marlair*   8.1

8.1 Introduction   8.1
   8.1.1 Accidents Involving Fluids Stored in Warehouses   8.2
   8.1.2 Accidents Involving Release of Fluids Contained in Vessels, Tanks, and Pipes   8.4
8.2 Properties Associated with the Ignition, Combustion, and Flame Spread Behaviors of Fluids   8.5
8.3 Vaporization and Boiling Characteristics of Fluids   8.5
8.4 Ignition Characteristics of Fluids   8.10
   8.4.1 Flash Points of Fluids   8.11
   8.4.2 Autoignition Temperature of Fluids   8.12
   8.4.3 Hazard Classification of Fluids Based on Ignition Resistance   8.13
8.5 Flammability Characteristics of Fluids   8.16
8.6 Combustion Characteristics of Fluids   8.18
   8.6.1 Release Rate of Fluid Vapors in Pool Fires   8.20
   8.6.2 Heat Release Rate   8.24
   8.6.3 Release Rates of Products   8.35
8.7 Smoke Point   8.37
Nomenclature   8.38
References   8.40

Chapter 9 Materials in Military Applications *Usman Sorathia*   9.1

9.1 Introduction   9.1
9.2 Composites in Military Applications   9.2
9.3 Polymer Composites   9.6
   9.3.1 Conventional and Advanced Matrix Resins   9.7
   9.3.2 Fabrication Techniques   9.9
   9.3.3 Cost of Composite Systems   9.10
9.4 Fire Threat   9.11
9.5 Polymer Composites and Fire   9.12
9.6 Fire Requirements and Regulations for Polymer Composites   9.13
   9.6.1 Infrastructure and Fire Regulations   9.16
   9.6.2 Ground Transportation and Fire Regulations   9.19
   9.6.3 Air Transportation and Fire Regulations   9.19
   9.6.4 Commercial Marine Transportation and Fire Regulations   9.20
   9.6.5 Military Use of Composites and Fire Regulations   9.21
9.7 Fire Performance and Test Methods for Composites   9.21
   9.7.1 Surface Flammability   9.22
   9.7.2 Smoke and Combustion Gas Generation   9.22
   9.7.3 Fire Growth   9.24
   9.7.4 Fire Resistance   9.37
   9.7.5 Structural Integrity Under Fire   9.51
   9.7.6 Passive Fire Protection   9.60
   9.7.7 Active Fire Protection   9.76
References   9.76

Index   I.1

CONTRIBUTORS

Jesse J. Beitel   *Hughes Associates, Inc., Baltimore, Maryland* (CHAP. 6)

Wai K. Chin   *Army Research Laboratory, Aberdeen, Maryland* (CHAP. 2)

Bradford Douglas   *American Forest and Paper Association, Washington, D.C.* (CHAP. 7)

Richard G. Gewain   *Hughes Associates, Inc., Baltimore, Maryland* (CHAP. 6)

Debbie J. Guckert   *E.I. duPont de Nemours, Richmond, Virginia* (CHAP. 5)

Nestor R. Iwankiw   *Hughes Associates, Inc., Chicago, Illinois* (CHAP. 6)

Susan L. Lovasic   *E.I. duPont de Nemours, Richmond, Virginia* (CHAP. 5)

Marc L. Janssens   *Southwest Research Institute, San Antonio, Texas* (CHAP. 7)

Richard E. Lyon   *Federal Aviation Administration, Atlantic City International Airport, New Jersey* (CHAP. 3)

Frederick W. Mowrer   *University of Maryland, College Park, Maryland* (CHAP. 1)

Guy Marlair   *INERIS, Verneuil-Halatte, France* (CHAP. 8)

Elisabeth S. Papazoglou   *Great Lakes Polymers, West Lafayette, Indiana* (CHAP. 4)

Roger F. Parry   *E.I. duPont de Nemours, Richmond, Virginia* (CHAP. 5)

Richard Shuford   *Army Research Laboratory, Aberdeen, Maryland* (CHAP. 2)

Archibald Tewarson   *Factory Mutual Research, Norwood, Massachusetts* (CHAPS. 2 AND 8)

Usman A. Sorathia   *Naval Surface Warfare Center, West Bethesda, Maryland* (CHAP. 9)

PREFACE

While always important, the broad field of fire protection has, in recent years, appropriately received ever increasing attention. Higher concentrations of people and buildings, wider use of materials in processing, more critical and costly equipment and systems, all contribute to the need for greater understanding and control of fire protection in materials, systems, and fabrication and processing operations. Fortunately, both academia and business have risen to meet the challenge. Many universities now have outstanding degree courses in fire protection, and an increasing number of businesses are including fire protection specialists in their organizations. All of this is becoming increasingly unified through excellent professional associations.

Since, in one way or another, materials are the source of fire and fire hazards, it is appropriate that a broad-ranging book be provided for those having interests or needs in the use of materials in analysis, design, fabrication, and processing. This new *Handbook of Building Materials for Fire Protection* is the first major book devoted completely to materials. As such, it will be invaluable to all of those in this field, and to all others having fire and safety concerns. I feel honored to serve as Editor for this book, and to have had the opportunity to work with the group of truly outstanding people who are the chapter authors for the book. A look at the list of contributors on page xi shows clearly that this is a group of well known and highly respected people in the field of fire protection. Their contributions to this field are invaluable, and their stature is unequaled. The information, data, and guidelines provided in their chapters will be a source of great importance to all of the readers of this sourcebook.

The organization and coverage of materials in this book is well suited to reader convenience. The first chapter, by Dr. Frederick Mowrer of the University of Maryland, provides excellent explanations of all of the fundamentals of fire hazards of materials, including flammability, smoke, etc. Next is a most thorough chapter covering the all-important subject of materials specifications, standards, and testing. The lead author is the well-known and widely respected Dr. Archibald Tewarson of Factory Mutual Research, and his excellent co-authors are Dr. Richard Shuford and Dr. Wai Chin of the Army Research Laboratory. Following this are two important chapters on plastics, the first by Dr. Richard Lyon of the FAA, and the second on flame retardants for plastics by Dr. Elisabeth Papazoglou of Great Lakes Polymers. Then comes a chapter on the critical materials area of fibers and fabrics, authored by the outstanding DuPont team of Dr. Debbie Guckert, Dr. Roger Parry, and Susan Lovasic. Following this are two chapters in the most important area of construction materials, specifically, structural materials by the highly respected Hughes Associates team of Jesse Beitel, Richard Gewain, and Dr. Nestor Iwankiw; and wood and wood products by Dr. Marc Janssens of Southwest Research Institute and Dr. Bradford Douglas of the American Forest and Paper Association. The next chapter covers the materials area so important in materials processing, namely, liquids and chemicals, authored by the well-known team of Dr. Archibald Tewarson and Dr. Guy Marlair. The final chapter covers the all-encompassing and broad area of materials systems in military equipment, authored by the virtual spokesman for this area, Usman Sorathia. A review of the above will readily convince any reader of this book that it would be difficult to match this author team and the breadth of materials covered. It has indeed been an honor and a pleasure to work with this group in producing this major contribution to the field of fire protection. I feel that this book will be an invaluable addition to the bookshelves of any person with any interest in fire protection.

Charles A. Harper

ABOUT THE EDITOR-IN-CHIEF

Charles A. Harper, formerly Manager of Materials Engineering and Technologies for Westinghouse Electric Corporation, is now President of Technology Seminars, an organization devoted to presenting educational seminars to industry on important modern materials technology areas. Mr. Harper is also a series editor for the McGraw-Hill materials science and engineering titles. He is a chemical engineering graduate of The Johns Hopkins University, where he also served as adjunct professor.

CHAPTER 1
FUNDAMENTALS OF THE FIRE HAZARDS OF MATERIALS

Frederick W. Mowrer, PhD, PE, FSFPE
Department of Fire Protection Engineering
University of Maryland

1.0 INTRODUCTION

The fire hazards associated with materials, products, and assemblies used in buildings and other structures have long been a subject of concern and regulation. The hazards of flammable gases and flammable and combustible liquids are widely recognized and regulated with respect to their production, transportation, and utilization. Historically, combustible materials used as part of a building's construction have been regulated more so than the furnishings and contents brought into buildings, but in recent years the fire hazards and risks associated with furnishings and contents have come under increased scrutiny as the contribution of these products to fires becomes more widely recognized.

During the past century, a large number of fire test methods were developed and adopted for regulatory purposes. Many of these fire test methods evaluate only one or a few of the relevant fire hazard characteristics of a product or of a component in a product. The results of these fire test methods are often cast in the form of derived indices that are convenient for regulatory purposes but may bear only a tenuous relationship with the actual fire hazards represented by products in end use. Consequently, some products have been approved for use despite having objectionable fire hazard characteristics.

Over the past decade, there has been an international movement toward the development of performance-based building fire safety analysis methods and the adoption of performance-based building fire safety regulations. At the heart of this movement is the specification of fire scenarios based on expected fires and analysis of the expected conditions resulting from these fires. Consequently, the movement toward performance-based fire safety analysis, design, and regulation demands a better understanding of the fire hazards of materials and the dynamics of building fires than traditional prescriptive approaches to fire safety have required.

Evaluation of the fire hazards of materials is complicated, because so many variables can influence the process. These variables include material properties and configurations, environmental conditions, and enclosure effects. Consequently, the fire hazards associated with different materials depend not only on their chemical and physical properties, but also on their applications. For example, textile materials applied to walls and ceilings pose fire hazards and risks different from the same materials used as floor coverings; a stack of folded newspapers will burn much differently than the same quantity of loosely packed shredded newsprint; a Christmas tree fire will cause more severe conditions in a family room than in a hotel ballroom. Methods are needed to evaluate the fire hazards of materials, products, and assemblies under a full range of anticipated use conditions.

Ultimately, the following issues should be addressed when assessing the fire hazards of materials, products, and assemblies:

- How easy is it to ignite the product?
  - Is it prone to self-heating and, if so, under what circumstances might it self-heat to ignition?
  - Under what circumstances can it be ignited by different ignition sources with different intensities and exposure durations?

- How fast does fire grow and spread on the product once ignited?
  - Under what circumstances will it propagate fire versus burning out locally?
- How big does the fire become and how long does it burn?
  - How much heat will be released as a function of time?
- How much smoke is produced as the product burns?
  - What are the products of combustion?

Once these issues are addressed, the consequences of a fire involving the material, product, or assembly in a particular application can be considered:

- What fire conditions will result from burning of the material, product, or assembly?
  - What temperatures and heat fluxes will develop in different locations?
  - What smoke concentrations will occur in different locations?
  - Will these fire conditions have an influence on the burning characteristics of the material?
  - Will other materials, products, or assemblies be ignited, and, if so, how will they burn once ignited?
- What will be the consequences of these fire conditions?
  - On the structure?
  - On people within the structure?
  - On contents, furnishings, and equipment within the structure?
  - On the environment and on nearby structures?

Because the answers to these questions depend on environmental variables as well as material properties and configurations, it would be desirable to identify and measure fundamental material properties that could be used with appropriate analytical or computational models to evaluate the expected performance of a material, product, or assembly under a full range of potential conditions of use. While significant progress has been made toward this objective in recent years, the current state-of-the-art does not yet permit comprehensive analyses to be performed. This is one reason why reliance is still placed on traditional, index-based, pass-fail fire test methods for regulatory purposes. With improved understanding of enclosure fire dynamics and material flammability, this situation is changing.

In this chapter, fundamental aspects of the fire hazards of materials are addressed. Basic combustion issues are introduced, followed by a discussion of the fire and flammability hazards associated with gaseous, liquid, and solid materials. Smoke production is then discussed along with methods that are used to characterize conditions in a smoke cloud. Finally, enclosure effects on the dynamics of building fires are introduced. In subsequent chapters, these concepts are applied to specific classes of materials and applications.

1.1 FUNDAMENTALS OF COMBUSTION

Fires are a form of combustion, involving the exothermic reaction of a fuel with an oxidizer that yields combustion products and energy in the form of heat. In general, this can be expressed as:

$$\text{Fuel} + \text{oxidizer} \rightarrow \text{products} + \text{heat}$$

Combustion reactions associated with fire include both *flaming combustion* and *smoldering combustion*. Flaming combustion involves the reaction of fuel and oxidizer in the gas phase, while smoldering combustion involves a reaction at the surface of a condensed phase solid. Examples of smoldering combustion include a cigar ember and charcoal briquettes on a barbecue grill. Flaming combustion associated with liquid and solid fuels requires the evaporation or *pyrolysis* of fuel molecules before the combustion reaction occurs in the gas phase, where the term pyrolysis refers to the decomposition of a compound caused by heat.

Flames can be distinguished as either premixed or diffusion, laminar or turbulent. In a *premixed flame*, the fuel and oxidizer are mixed together before entering the reaction zone; in a *diffusion flame*, the reaction zone, represented by the flame, occurs at the interface where the separate fuel and oxidizer streams meet. The flame on a gas-stove burner is an example of a premixed flame, while a candle flame and a bonfire are examples of diffusion flames. The candle flame, with its smooth, unchanging shape, is an example of a laminar diffusion flame, while the bonfire, with its jumping and flickering flames, is an example of a turbulent diffusion flame.

Most fires of hazardous proportions involve turbulent diffusion flames, but there are notable exceptions. A fire involving a gas leak that mixes with air before igniting will propagate as a premixed flame once ignited. Once this mixture has burned, continued burning of gas issuing from the source of the leak will be in the form of a diffusion flame. Similarly, a fire involving a flammable liquid that evolves sufficient flammable gases at its surface to form an ignitable mixture with air will initially propagate across the liquid surface as a premixed flame, but then will sustain as a diffusion flame rising above the liquid pool.

1.1.1 Stoichiometry

The term *stoichiometry* is used to describe the quantitative relationship between reactants and products in a chemical reaction. A *stoichiometric reaction* is one in which there are no excess reactants in the product stream. As noted by Strehlow [1], the practical purpose of stoichiometry is to determine exactly how much air is needed to completely oxidize a fuel to the products carbon dioxide, water vapor, nitrogen, and sulfur dioxide. Typically, the stoichiometry of a reaction is expressed on a molar basis. Consider, for example, a fuel that contains carbon, hydrogen, oxygen, nitrogen, and sulfur (CHONS) in arbitrary proportions. The balanced stoichiometric relationship for this fuel would be:

$$C_u H_v O_w N_x S_y + \left(u + \frac{v}{4} - \frac{w}{2} + y\right)\left(O_2 + \frac{X^\circ_{N_2}}{X^\circ_{O_2}} N_2\right)$$

$$\rightarrow u CO_2 + \frac{v}{2} H_2O + y SO_2 + \left(\frac{X^\circ_{N_2}}{X^\circ_{O_2}}\left(u + \frac{v}{4} - \frac{w}{2} + y\right) + \frac{x}{2}\right) N_2 \quad (1.1)$$

In a practical stoichiometric calculation with air as the oxidizer, the standard atmospheric mole fraction of nitrogen is taken as $X^\circ_{N_2} = 0.79$, while that of air is taken as $X^\circ_{O_2} = 0.21$, with other, minor components of air, including argon, carbon dioxide, and water vapor, neglected. Thus, the ratio of nitrogen to oxygen in air is normally considered to be $X^\circ_{N_2}/X^\circ_{O_2} = 0.79/0.21 = 3.76$.

From Eq. (1.1), the number of moles of oxygen required to completely react with 1 mole of fuel can be determined as:

$$\left(\frac{n_{O_2}}{n_f}\right)_{stoich} = \frac{\left(u + \frac{v}{4} - \frac{w}{2} + y\right)}{1} = \left(u + \frac{v}{4} - \frac{w}{2} + y\right) \quad (1.2)$$

A stoichiometric reaction can also be written on a mass basis rather than on a mole basis. As noted by Drysdale [2], a global stoichiometric reaction can be written on a mass basis as:

$$1 \text{ kg fuel} + r \text{ kg air} \rightarrow (1 + r) \text{kg products} \quad (1.3)$$

where r represents the *air stoichiometric ratio*, which is the stoichiometrically required mass of air needed to completely react with a unit mass of fuel. Sometimes the symbol s is used instead of r to represent the air stoichiometric ratio, so the reader must be careful to understand the terminology and nomenclature used in different publications. The *oxygen stoichiometric ratio* can be determined from the air stoichiometric ratio based on the mass fraction of oxygen in air, which under normal conditions is $X^\circ_{O_2,\infty} = 0.233$:

$$r_{O_2} = Y^\circ_{O_2,\infty} \cdot r_{air} \tag{1.4}$$

The oxygen stoichiometric ratio can be calculated from Eq. (1.2):

$$r_{O_2} = \left(\frac{m_{O_2}}{m_f}\right)_{stoich} = \frac{n_{O_2} \cdot MW_{O_2}}{n_f \cdot MW_f} = \frac{\left(u + \dfrac{v}{4} - \dfrac{w}{2} + y\right) \cdot MW_{O_2}}{1 \cdot (u \cdot MW_C + v \cdot MW_H + w \cdot MW_O + x \cdot MW_N + y \cdot MW_S)} \tag{1.5}$$

It is important to recognize that the molecular weight of a diatomic oxygen molecule ($MW_{O_2} \approx 32$) is twice that of an oxygen atom ($MW_O \approx 16$). Note that all the elements in the denominator of Eq. (1.5) are monatomic, while the oxygen in the air is diatomic. Oxygen stoichiometric ratios are provided for selected fuels in Table 1.1; Babrauskas [3] provides a more extensive tabulation of oxygen stoichiometric ratios for different fuels.

Once the oxygen stoichiometric ratio is calculated from Eq. (1.5), the air stoichiometric ratio is readily calculated from Eq. (1.4). For air under normal conditions, the air stoichiometric ratio is 4.292 (i.e., 1/0.233) times the oxygen stoichiometric ratio.

Example: Determine the oxygen and air stoichiometric ratios for the combustion of propane.

Solution: First, write the stoichiometric reaction of propane:

$$C_3H_8 + 5(O_2 + 3.76\ N_2) \rightarrow 3\ CO_2 + 4\ H_2O + 18.8\ N_2$$

Next, calculate the oxygen stoichiometric ratio:

$$r_{O_2} = \left(\frac{n_{O_2} \cdot MW_{O_2}}{n_f \cdot MW_f}\right)_{stoich} = \frac{5 \cdot 32}{1 \cdot ((3 \cdot 12.01) + (8 \cdot 1.01))} = \frac{160}{44.1} = 3.63\ \text{g } O_2/\text{g fuel}$$

Finally, calculate the air stoichiometric ratio from the oxygen stoichiometric ratio:

$$r_{air} = \frac{r_{O_2}}{Y^\circ_{O_2,\infty}} = \frac{3.63\ \text{g } O_2/\text{g fuel}}{0.233\ \text{g } O_2/\text{g air}} = 15.57\ \text{g air/g fuel}$$

This calculation indicates that the complete combustion of 1 g of propane will consume 3.63 g of oxygen, which is the amount of oxygen in 15.57 g of air under normal conditions.

The stoichiometric, or ideal, yields of combustion products can be determined on a mole basis or on a mass basis in a manner similar to the determination of the oxygen stoichiometric ratio. For a fuel with the composition shown in Eq. (1.1), the ideal yields of each of the combustion products are provided in Table 1.2. Note that the mass yields are simply the molar yields multiplied by the ratio of the species molecular weight to the fuel molecular weight:

$$y_i = x_i \cdot \frac{MW_i}{MW_f} \tag{1.6}$$

Fuels and oxidizers are rarely mixed in exact stoichiometric proportions even in well-controlled systems such as internal combustion engines, let alone in uncontrolled fires. Internal combustion engines are typically regulated with a slight excess of air for emission control, while fires may entrain too much or too little air for complete combustion depending on the ventilation characteris-

TABLE 1.1 Fuel and Oxygen Heats of Combustion for Selected Fuels at 25°C [3]

Fuel	Formula	Oxygen stoichiometric ratio (r_{O_2}) (g O_2/g fuel)	Net fuel heat of combustion (kJ/mol fuel)	Net fuel heat of combustion (kJ/g fuel)	Oxygen heat of combustion (kJ/g O_2)
Alkanes	C_nH_{2n+2}				
Methane	CH_4	4.000	802.48	50.03	12.51
Ethane	C_2H_6	3.725	1428.02	47.49	12.75
Propane	C_3H_8	3.629	2044.01	46.36	12.78
Butane	C_4H_{10}	3.579	2657.25	45.72	12.77
Pentane	C_5H_{12}	3.548	3245.31	44.98	12.68
Hexane	C_6H_{14}	3.528	3855.25	44.74	12.68
Heptane	C_7H_{16}	3.513	4464.91	44.56	12.68
Octane	C_8H_{18}	3.502	5075.94	44.44	12.69
Nonane	C_9H_{20}	3.493	5685.32	44.33	12.69
Decane	$C_{10}H_{22}$	3.486	6294.47	44.24	12.69
Alkenes	C_nH_{2n}				
Ethylene	C_2H_2	3.422	1323.12	47.17	13.78
Propene	C_3H_6	3.422	1926.84	45.79	13.38
Butene	C_4H_8	3.422	2541.89	45.31	13.24
Pentene	C_5H_{10}	3.422	3130.60	44.64	13.04
Hexene	C_6H_{12}	3.422	3740.07	44.44	12.99
Heptene	C_7H_{14}	3.422	4350.36	44.31	12.95
Octene	C_8H_{16}	3.422	4659.68	44.20	12.92
Alkynes	C_nH_{2n-2}				
Acetylene	C_2H_2	3.072	1255.65	48.22	15.7
Propyne	C_3H_4	3.195	1849.57	46.17	14.45
Alcohols	$C_nH_{2n+1}OH$				
Methanol	CH_3OH	1.500	638.88	19.94	13.29
Ethanol	C_2H_5OH	2.084	1235.14	26.81	12.87
Propanol	C_3H_7OH	2.396	1843.56	30.68	12.81
Miscellaneous		(listed alphabetically by name)			
Acetone	C_3H_6O	2.204	1658.76	28.56	12.96
Carbon monoxide	CO	0.571	282.90	10.10	17.69
Cellulose	$C_6H_{10}O_5$	1.184	2613.70	16.12	13.61
Methyl ethyl ketone	C_4H_8O	2.441	2268.27	31.46	12.89
Methy methacrylate	$C_5H_8O_2$	2.078	2563.82	25.61	12.33
Styrene	C_8H_8	3.073	4219.75	40.52	13.19
Toluene	C_7H_8	3.126	3733.11	40.52	12.97
Vinyl chloride	C_2H_3Cl	1.408	1053.75	16.86	11.97
Xylene	C_8H_{10}	3.165	4333.45	40.82	12.90

tics of the enclosures where they occur. For mixtures with compositions other than stoichiometric, it is convenient and useful to normalize the actual mixture composition to the stoichiometric mixture composition. Two dimensionless quantities are commonly used for this purpose; these are the *equivalence ratio* and the *mixture fraction*.

The *fuel equivalence ratio* is defined as the actual fuel to air ratio normalized by the stoichiometric fuel to air ratio:

$$\Phi = \frac{(n_f/n_{air})_{actual}}{(n_f/n_{air})_{stoich}} = \frac{(m_f/m_{air})_{actual}}{(m_f/m_{air})_{stoich}} = r \cdot (m_f/m_{air})_{actual} \quad (1.7)$$

TABLE 1.2 Ideal Yields of Combustion Products for a Stoichiometric Reaction

Species I	Molar yield, x_i (n_i/n_f)	Mass yield, y_i (m_i/m_f)
CO_2	u	$u \cdot \dfrac{MW_{CO_2}}{MW_f}$
H_2O	$\dfrac{v}{2}$	$\dfrac{v}{2} \cdot \dfrac{MW_{H_2O}}{MW_f}$
SO_2	y	$y \cdot \dfrac{MW_{SO_2}}{MW_f}$
N_2	$\left(\dfrac{X^o_{N_2}}{X^o_{O_2}}\right)\left(u + \dfrac{v}{4} - \dfrac{w}{2} + y\right) + \dfrac{x}{2}$	$\left[\left(\dfrac{X^o_{N_2}}{X^o_{O_2}}\right)\left(u + \dfrac{v}{4} - \dfrac{w}{2} + y\right) + \dfrac{x}{2}\right] \cdot \dfrac{MW_{N_2}}{MW_f}$

As demonstrated by Eq. (1.7), an equivalence ratio of less than unity has excess air and is fuel lean, an equivalence ratio of exactly unity represents stoichiometric conditions and an equivalence ratio of greater than unity is fuel rich and air lean. The mass-based global reaction represented by Eq. (1.3) can be rewritten for mixtures other than stoichiometric in terms of the equivalence ratio as:

$$1 \text{ kg fuel} + \frac{r}{\Phi} \text{ kg air} \rightarrow \left(1 + \frac{r}{\Phi}\right) \text{ kg products} \tag{1.8}$$

For mixtures with equivalence ratios of less than unity, excess air will be carried through the reaction and appear in the products of combustion. For mixtures with equivalence ratios of greater than unity, there is insufficient air to fully react with the fuel. Consequently, unburned fuel will be carried through the reaction and appear in the products of combustion. As an idealization, it can be assumed that only products of complete combustion are generated and only pure fuel goes unreacted if there is insufficient oxygen available. With this idealization, the fraction of fuel burned can be expressed as:

$$X_f = \frac{m_{f,\text{burn}}}{m_{f,\text{avail}}} = \text{MIN}\left(1, \frac{1}{\Phi}\right) \tag{1.9}$$

For fuel-rich situations where the equivalence ratio is greater than unity, the products of combustion would be produced in the same proportion as the fraction of fuel consumed, i.e., as $1/\Phi$. Based on this idealized approximation, the relative mass of oxygen in the product stream would be:

$$X_{O_2} = \frac{m_{O_2,\text{prod}}}{m_{O_2,\text{avail}}} = \text{MAX}(0, 1 - \Phi) \tag{1.10}$$

Normalized ideal yields for combustion products, oxygen, and fuel are illustrated in Fig. 1.1 as a function of equivalence ratio. For combustion products, the normalized ideal yields represent the expected yields in the exhaust stream relative to the ideal yields provided in Table 1.2. For oxygen, the normalized yield represents the expected concentration of oxygen in the exhaust stream relative to the ambient concentration of oxygen. For fuel, the normalized yield represents the expected concentration of fuel in the exhaust stream relative to the concentration of fuel entering the reaction.

In reality, products of partial combustion, including carbon monoxide, soot, and a range of unburned hydrocarbons, are likely to be generated as the equivalence ratio of a mixture approaches and exceeds unity. Consequently, the ideal yields shown in Fig. 1.1 should be considered as an idealization against which actual product yields can be compared rather than as an accurate relationship for product yields, particularly for equivalence ratios greater than unity.

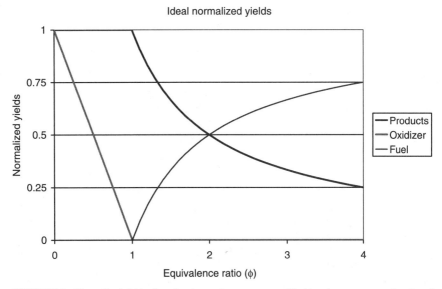

FIGURE 1.1 Normalized yields of combustion products, oxygen, and fuel in exhaust stream as a function of equivalence ratio.

As an alternative to the equivalence ratio concept, the mixture fraction concept normalizes the combustion reaction on the basis of the fraction of a mixture that starts as fuel. The mixture fraction is a conserved quantity representing the fraction of material at a given point in space that originated as fuel [4]. The mixture fraction Z is defined as:

$$Z = \frac{r_{O_2} Y_f - (Y_{O_2} - Y_{O_2,\infty})}{r_{O_2} Y_f^I + Y_{O_2,\infty}} \quad (1.11)$$

The oxygen stoichiometric ratio, r_{O_2}, is defined in Eq. (1.5). The fuel mass fraction is represented as Y_f, with the symbol Y_f^I representing the fuel mass fraction in the fuel stream. The oxygen mass fraction is represented as Y_{O_2}, with the symbol $Y_{O_2,\infty}$ representing the ambient oxygen concentration, normally 0.233. The mixture fraction varies from $Z = 0$ in regions containing no fuel, where the oxygen concentration is at its ambient value, to $Z = 1$ in regions containing only fuel. The mixture fraction concept is used in the combustion submodel of some computer-based fire models [4], but is not as widely used as the equivalence ratio for presenting yield data.

1.1.2 Thermochemistry

The energy released by a fire, and particularly the rate at which this energy is released, to a large extent governs the hazards associated with fires. *Thermochemistry* addresses the amount of energy released in exothermic reactions, including combustion, as well as the amount of energy absorbed in endothermic reactions. In general, the energy release or absorption that accompanies a chemical reaction is known as the *heat of reaction* because this change in energy typically manifests as a temperature change within the system, although it should be known more accurately as the enthalpy of reaction. It is important to note that thermochemistry does not address the rate at which energy is released or the final composition of a reaction. For gaseous mixtures of fuels and oxidizers, these are the subjects of chemical kinetics and equilibrium, while for liquid and solid fuels, heat transfer and diffusion are also important.

Some basic thermodynamic concepts are relevant to this discussion. First, as noted by Drysdale [5], it is appropriate to limit the discussion to gases because flaming combustion occurs in the gas phase. At temperatures of interest for fire applications, it is reasonable to assume ideal gas behavior, which is expressed as:

$$PV = nRT \tag{1.12}$$

where P is the absolute pressure, V is the volume of the mixture, n is the number of moles within the mixture, R is the ideal gas constant, and T is the absolute temperature of the mixture.

For a mixture of ideal gases, Dalton's law applies. Dalton's law states that the total pressure of a mixture of ideal gases is the sum of the partial pressures for each of the components:

$$P = \sum_{i \, \text{species}} P_i \tag{1.13}$$

Each component fills the entire volume of the mixture and has the same temperature as the mixture, i.e., the ideal gases are perfectly mixed.

The first law of thermodynamics deals with the conservation of energy, including the conversion of energy from one form to another. For fire applications, it is generally reasonable to ignore changes in kinetic energy and potential energy of a system relative to changes in the internal energy. Consequently, the change in internal energy can be expressed as:

$$\Delta E = \Delta U = U_2 - U_1 = Q - W \tag{1.14}$$

where U is the internal energy of the system, Q is the heat transferred to the system, and W is the work done by the system. As noted by Glassman [6], the change in internal energy of a system depends only on the initial and final state of the system, typically expressed in terms of its temperature and pressure, and is independent of the means by which the state is attained.

Fires, as opposed to explosions, typically occur under conditions of essentially constant pressure. Consequently, the work done as a result of the expansion of the fire gases can be taken into account as:

$$W = P(V_2 - V_1) \tag{1.15}$$

This expression for the work term can be substituted into the energy equation and expressed as:

$$Q = (U_2 + PV_2) - (U_1 + PV_1) = H_2 - H_1 \tag{1.16}$$

where H is the enthalpy, defined as $H \equiv U + PV$. For an ideal gas at constant pressure, changes in enthalpy are related to changes in temperature as:

$$dH = C_p dT \tag{1.17}$$

where c_p is the heat capacity at constant pressure for the system.

In a chemically reactive system, the *heat of reaction* is defined as the difference between the heat of formation of the products and the heat of formation of the reactants. Mathematically, the heat of reaction at a reference temperature, T, can be expressed as:

$$\Delta H_r = \sum_{i \, \text{products}} n_i (\Delta H_f^\circ)_{T,i} - \sum_{j \, \text{reactants}} n_j (\Delta H_f^\circ)_{T,j} \tag{1.18}$$

Tables of standard heats of formation for many substances have been compiled, typically at a standard reference temperature of 298.15 K (25°C). The best known and most comprehensive of these compilations are the JANAF tables [7]. By definition, the heats of formation for the elements in their standard states are arbitrarily assigned a value of zero for convenience. Standard heats of formation at a reference temperature of 298.15 K (25°C) are provided in Table 1.3 for some substances commonly involved in combustion reactions. Substances are listed in order of descending heat of formation, from most positive to most negative. This order ranks compounds from the least stable to

TABLE 1.3 Standard Heats of Formation at a Reference Temperature of 298.15 K (25°C) [8]

Name	Chemical symbol	State	ΔH_f° (kJ/mol)	Molecular weight (g/mol)
Carbon	C	Vapor	716.68	12.01
Nitrogen atom	N	Gas	472.68	14.01
Oxygen atom	O	Gas	249.18	16.00
Acetylene	C_2H_2	Gas	226.73	26.04
Hydrogen atom	H	Gas	218.00	1.01
Ozone	O_3	Gas	142.67	48.00
Hydrogen cyanide	HCN	Gas	135.14	27.03
Nitric oxide	NO	Gas	90.29	30.01
Benzene	C_6H_6	Gas	82.93	78.11
Benzene	C_6H_6	Liquid	48.95	78.11
Ethylene	C_2H_4	Gas	52.47	28.05
Hydrazine	N_2H_4	Liquid	50.63	32.05
Hydroxyl radical	OH	Gas	38.99	17.01
Oxygen	O_2	Gas	0.00	32.00
Nitrogen	N_2	Gas	0.00	28.01
Hydrogen	H_2	Gas	0.00	2.02
Carbon	C	Solid	0.00	12.01
Chlorine	Cl_2	Gas	0.00	70.91
Ammonia	NH_3	Gas	−45.94	17.03
Ethylene oxide	C_2H_4O	Gas	−52.64	44.05
Methane	CH_4	Gas	−74.87	16.04
Ethane	C_2H_6	Gas	−83.80	30.07
Hydrogen chloride	HCl	Gas	−92.31	36.46
Propane	C_3H_8	Gas	−104.70	44.10
Carbon monoxide	CO	Gas	−110.53	28.01
Butane	C_4H_{10}	Gas	−127.10	58.12
Pentane	C_5H_{12}	Gas	−146.8	72.15
Hexane	C_6H_{14}	Gas	−167.2	86.18
Pentane	C_5H_{12}	Liquid	−173.5	72.15
Heptane	C_7H_{16}	Gas	−187.8	100.20
Hexane	C_6H_{14}	Liquid	−198.7	86.18
Methanol	CH_3OH	Gas	−201.10	32.04
Heptane	C_7H_{16}	Liquid	−224.4	100.20
Methanol	CH_3OH	Liquid	−238.40	32.04
Water	H_2O	Gas	−241.83	18.02
Octane	C_8H_{18}	Liquid	−250.30	114.23
Water	H_2O	Liquid	−285.83	18.02
Sulfur dioxide	SO_2	Gas	−296.81	64.06
Dodecane	$C_{12}H_{26}$	Liquid	−352.1	170.33
Carbon dioxide	CO_2	Gas	−393.51	44.01
Sulfur trioxide	SO_3	Gas	−395.77	80.06

the most stable. Values for the standard heats of formation listed in Table 1.3 were obtained from the July 2001 release of NIST Standard Reference Database Number 69 [8], which is available online (http://webbook.nist.gov/chemistry/) at the time this chapter is written.

At temperatures different from the standard reference temperature, the heat of formation of a pure substance can be calculated as:

$$H(T) = (\Delta H_f^\circ)_{T_{ref}} + \int_{T_{ref}}^{T} c_p \, dT = (\Delta H_f^\circ)_{T_{ref}} + (H_T - H_{T_{ref}}) \quad (1.19)$$

Specific heats (c_p) and enthalpy changes ($H_T - H_{T_{ref}}$) are tabulated as a function of temperature in the JANAF Tables as well as in other tabulations of thermochemical properties. The NIST Standard Reference Database Number 69 provides either tables of specific heats as a function of temperature or coefficients for polynomial curve fits for specific heats as a function of temperature, depending on the substance. The polynomial curve fits have the form of the Shomate equation:

$$C_p^\circ = A + B(T/1000) + C(T/1000)^2 + D(T/1000)^3 + E/(T/1000)^2 \, (\text{J/mol.K}) \quad (1.20)$$

This polynomial equation can be substituted into Eq. (1.17) and integrated to yield:

$$H_T^\circ - H_{298.15}^\circ = A(T/1000) + B(T/1000)^2/2 + C(T/1000)^3/3 + D(T/1000)^4/4 - E/(T/1000) + F$$

$$(1.21)$$

Values for the polynomial coefficients A, B, C, D, E, and F are provided in Table 1.4 for some substances commonly involved in combustion reactions.

For the case of a stoichiometric combustion reaction, the heat of reaction is known as the *heat of combustion* of a fuel. As noted by Strehlow [1], the heat of combustion of any CHONS fuel is the heat released by the complete oxidation of the fuel at a temperature near room temperature in accordance with the stoichiometry of Eq. (1.1). Heats of combustion for selected fuels are tabulated in Table 1.1. By convention, the heats of combustion are listed as positive values although the heats of reaction are actually negative, indicating that the products of combustion are at a lower energy level, and hence more stable, than the reactants.

Two heats of combustion are associated with fuels. These are known as the *gross heat of combustion* and the *net heat of combustion*, or as the *high heating value* and the *low heating value*. The gross heat of combustion is based on water in the products of combustion being in the liquid phase, while the net heat of combustion is based on water in the products of combustion being in the vapor phase. The net heat of combustion is less than the gross heat of combustion by an amount equal to the latent heat of vaporization of the water produced in the combustion reaction, which from Table 1.3 is seen to be 44 kJ/mol or 2.44 kJ/g of water.

Example: Calculate the gross and net heats of combustion for gaseous hexane at a temperature of 25°C and a pressure of 1 atm (101,325 Pa).

Solution: First, write the stoichiometric reaction for hexane in air:

$$C_6H_{14} + 9.5(O_2 + 3.76 N_2) \rightarrow 6 CO_2 + 7 H_2O + 35.72 N_2$$

TABLE 1.4 Polynomial Coefficients for Use with Equations 1.20 and 1.21 [8]

Substance	Nitrogen	Oxygen	Water		Carbon dioxide	
Formula	N_2	O_2	H_2O		CO_2	
Temp. range (K)	298–6000	298–6000	500–1700	1700–6000	298–1200	1200–6000
A	26.09200	29.65200	30.09200	41.96426	24.99735	58.16639
B	8.218801	6.137261	6.832514	8.622053	55.18696	2.720074
C	−1.976141	−1.186521	6.793435	−1.499780	−33.69137	−0.492289
D	0.159274	0.095780	−2.534480	0.098119	7.948387	0.038844
E	0.044434	−0.219663	0.082139	−11.15764	−0.136638	−6.447293
F	−7.989230	−9.861391	−250.8810	−272.1797	−403.6075	−425.9186

Next, calculate the heat of reaction as the difference between the heats of formation of the products less the heats of formation of the reactants under standard conditions. Values listed in Table 1.2 can be used for this purpose. For the case of the gross heat of combustion, with water as a liquid, the solution is:

$$\Delta H_r = [(6 \cdot (-393.53)) + (7 \cdot (-285.83)) + (35.72 \cdot 0.00)]$$
$$- [(1 \cdot (-167.2) + (9.5 \cdot 0.00) + (35.72 \cdot 0.00)]$$
$$= -4{,}194.79 \text{ kJ/mol} \div 86.18 \text{ g/mol} = -48.67 \text{ kJ/g} = -\Delta H_{c,\text{gross}}$$

The net heat of combustion would be calculated similarly, but with a different value for the heat of formation of the gaseous water molecules:

$$\Delta H_r = [(6 \cdot (-393.53)) + (7 \cdot (-241.83)) + (35.72 \cdot 0.00)]$$
$$- [(1 \cdot (-167.2) + (9.5 \cdot 0.00) + (35.72 \cdot 0.00)]$$
$$= -3{,}886.79 \text{ kJ/mol} \div 86.18 \text{ g/mol} = -45.10 \text{ kJ/g} = -\Delta H_{c,\text{net}}$$

Alternatively, the latent heat of evaporation of the water produced by the reaction can be calculated and subtracted from the gross heat of combustion to yield the same answer:

$$\Delta H_{c,\text{net}} = 4{,}194.79 \text{ kJ/mol} - (7 \text{ mol H}_2\text{O} \cdot 44 \text{ kJ/mol H}_2\text{O}) = 3{,}886.79 \text{ kJ/mol}$$

While heats of combustion are normally associated with fuels, it has been recognized for more than 80 years that the amount of heat released per unit of oxygen consumed from the atmosphere is nearly constant for most hydrocarbon- or cellulose-based fuels of practical interest for fire applications [9]. Huggett [10] further developed this concept for fire applications more than 20 years ago when the first calorimeters based on the concept of oxygen consumption calorimetry were being developed [11].

This concept is demonstrated in Table 1.1, where *oxygen heats of combustion* ($\Delta H_c/r_{O_2}$) are tabulated along with fuel heats of combustion. A value of 13.1 kJ/g O_2 is typically cited as an average value for the oxygen heat of combustion. This value is generally used in standard tests based on oxygen consumption calorimetry in the absence of more specific information for a particular fuel. As evident from Table 1.1, this value of 13.1 kJ/g O_2 is within approximately 5 percent of the actual values for the oxygen heat of combustion for most of the fuels listed.

Example: Determine the oxygen heat of combustion for hexane from the previous example.

Solution: From the previous example, the fuel net heat of combustion was determined to be 45.10 kJ/g fuel. The oxygen stoichiometric ratio is calculated as:

$$r_{O_2} = \frac{9.5 \cdot 32}{86.18} = 3.53$$

From these values, the oxygen heat of combustion is calculated to be:

$$\frac{\Delta H_c}{r_{O_2}} = \frac{45.10 \text{ kJ/g fuel}}{3.53 \text{ g } O_2/\text{g fuel}} = 12.78 \text{ kJ/g } O_2$$

This value is about 2.5 percent lower than the average value of 13.1 kJ/g O_2 typically cited.

The approximate constancy of the oxygen heat of combustion is widely used for experimental and computational purposes. Combined with the development of real-time electronic oxygen sensors approximately 25 years ago, the principle of oxygen consumption calorimetry has permitted the development of open calorimeters with capacities ranging from watts up to tens of megawatts. With this relationship between heat release and oxygen consumption, it is not necessary to know the composition or the heats of combustion of the materials or products that are burning to obtain a reasonably accurate heat release rate for engineering purposes. These devices rely on the accurate measurement of flow rates and oxygen concentrations in exhaust streams rather than on the impractical measurement of sensible heat released by a fire. For computational purposes, the oxygen heat of combustion permits the ready calculation of oxygen consumption in fires regardless of the fuel and establishes a "ventilation limit" on the rate of heat release in an enclosure fire:

$$\dot{Q}_{VL} = \dot{m}_{O_2} \frac{\Delta H_c}{r_{O_2}} = \dot{m}_{\text{air}} \frac{\Delta H_c}{r_{\text{air}}} \qquad (1.22)$$

where $\dot{m}_{air}$ is the rate of air flow into the enclosure and the *air heat of combustion* ($\Delta H_c/r_{air}$) has a nearly constant value of approximately 3.0 MJ/kg air (i.e., 13.1 MJ/kg O_2 × 0.233 kg O_2/kg$_{air}$).

1.1.3 Flame Temperatures

The temperatures that will be achieved by a burning mixture of gases will depend on the composition of the gases as well as on the heat losses to the boundaries of the system. The *adiabatic flame temperature* is the maximum temperature that a mixture of gases will achieve; the term *adiabatic* indicates that there will be no heat losses from the gases to the system boundaries. All of the heat released in an adiabatic reaction will act to increase the enthalpy and consequently the temperature of the products, which will include excess air or excess fuel if the mixture is not stoichiometric.

The adiabatic flame temperature can be calculated in several ways. All the methods discussed here presuppose knowledge of the composition of the products, which may not be known if dissociation or incomplete combustion are significant. Iterative computer-based methods can determine the composition of the products as well as the adiabatic flame temperature if kinetic parameters and rate constants for all the reactions of interest are specified. In general, the adiabatic flame temperature can be expressed implicitly as the temperature that solves the following enthalpy equation:

$$H_{products} = H_{reactants} \tag{1.23}$$

where $H_{products}$ is the total enthalpy of the products and $H_{reactants}$ is the total enthalpy of the reactants:

$$H_{products} = \sum_{i\,products} [n_i((\Delta H_f^\circ)_{298\,15} + (H_i)_{T_{ad}} - (H_i)_{298\,15})]$$

$$= \sum_{i\,products} \left[n_i \left((\Delta H_f^\circ)_{298\,15} + \int_{298\,15}^{T_{ad}} c_{pi} dT \right) \right] \tag{1.24}$$

$$H_{reactants} = \sum_{j\,reactants} [n_j((\Delta H_f^\circ)_{298.15} + (H_j)_{T_{ad}} - (H_j)_{298.15})]$$

$$= \sum_{j\,reactants} \left[n_j \left((\Delta H_f^\circ)_{298.15} + \int_{298.15}^{T_i} c_{pi} dT \right) \right] \tag{1.25}$$

This general solution permits the reactants to be at an initial temperature T, different from each other as well as different from the standard reference temperature of 25°C. This form of the solution is suitable for use with the JANAF tables, which tabulate standard heats of formation $(\Delta H_f^\circ)_{298.15}$, as well as enthalpy changes as a function of temperature $(H_T - H_{298.15})$ for a wide range of species involved in combustion reactions.

Use of the JANAF tables generally requires an iterative solution for the adiabatic flame temperature. First, an adiabatic flame temperature must be guessed, then the total enthalpy of the products at that temperature must be calculated and compared with the total enthalpy of the reactants. If the total enthalpies of the products and reactants are different from each other, new temperatures must be selected until the solution converges.

For most fire applications, the reactants can be assumed to be at 25°C with little loss of accuracy. Provided there is sufficient oxygen to react with the fuel (i.e., $\Phi \leq 1$), the adiabatic flame temperature can be calculated explicitly as follows:

$$H_{products} - H_{reactants} = \sum_{i\,products} \left[n_i \left((\Delta H_f^\circ)_{298.15} + \int_{298.15}^{T_{ad}} c_{pi} dT \right) \right] - \sum_{j\,reactants} [n_j((\Delta H_f^\circ)_{298.15})]$$

$$= \left(\sum_{i\,products} [n_i((\Delta H_f^\circ)_{298.15})] - \sum_{j\,reactants} [n_j((\Delta H_f^\circ)_{298.15})] \right) \tag{1.26}$$

$$+ \sum_{i\,products} \left[n_i \left(\int_{298.15}^{T_{ad}} c_{pi} dT \right) \right] = -\Delta H_c + \sum_{i\,products} \left[n_i \left(\int_{298.15}^{T_{ad}} c_{pi} dT \right) \right] = 0$$

This relationship can also be expressed more simply as:

$$\sum_{i\,products}\left[n_i\left(\int_{298.15}^{T_{ad}} c_{pi}\,dT\right)\right] = \sum_{i\,products}[n_i((H_i)_{T_{ad}} - (H_i)_{298.15})] = \Delta H_c \quad (1.27)$$

An equation of the form of Eq. (1.20) can be substituted into the integral for the temperature dependence of the specific heat to yield an equation of the form of Eq. (1.21) for the heat capacity for each product. Because of the nonlinear relationships expressed by Eqs. (1.20) and (1.21), iterative numerical solutions will still be necessary, but such solutions will converge rapidly. As a simple closed-form alternative, an appropriate average specific heat can be selected to remove this term from the integral. With this approximation, the solution for the adiabatic flame temperature becomes:

$$T_{ad} = T_o + \frac{\Delta H_c}{\sum_{i\,products}(n_i \bar{c}_{pi})} \text{ (molar basis)} = T_o + \frac{\Delta H_c}{\sum_{i\,products}(m_i \bar{c}_{pi})} \text{ (mass basis)} \quad (1.28)$$

Example: Use the JANAF Tables to estimate the adiabatic flame temperature for a stoichiometric mixture of propane in air initially at 25°C. Neglect dissociation for this estimate.

Solution: From a previous example, the stoichiometric reaction of propane in air is:

$$C_3H_8 + 5(O_2 + 3.76N_2) \rightarrow 3CO_2 + 4H_2O + 18.8N_2$$

The total enthalpy of the reactants at 25°C is the sum of their enthalpies of formation:

$$H_{reactants} = \sum_{j\,reactants}[n_j((\Delta H_f^\circ)_{298.15})] = (1 \cdot -104.7) + 5 \cdot 0.00) + (18.8 \cdot 0.00) = -104.7 \text{ kJ}$$

The total enthalpy of the products at the adiabatic flame temperature is:

$$H_{products} = \sum_{i\,products}[n_i((\Delta H_f^\circ)_{298.15} + (H_i)_{T_{ad}} - (H_i)_{298.15})]$$

$$= [3 \cdot (-393.51 + (H_i)_{T_{ad}} - (H_i)_{298.15})]_{CO_2}$$
$$+ 4 \cdot (-241.83 + (H_i)_{T_{ad}} - (H_i)_{298.15})]_{H_2O}$$
$$+ [18.8 \cdot (0.00 + (H_i)_{T_{ad}} - (H_i)_{298.15})]_{N_2}$$

Equating the total enthalpies of the products and reactants and rearranging terms yields:

$$-104.70 + 1180.53 + 967.32 = 2043.15$$
$$= [3 \cdot ((H_i)_{T_{ad}} - (H_i)_{298.15})]_{CO_2} + [4 \cdot ((H_i)_{T_{ad}} - (H_i)_{298.15})]_{H_2O}$$
$$+ [18.8 \cdot ((H_i)_{T_{ad}} - (H_i)_{298.15})]_{N_2}$$

As a first guess, select an adiabatic flame temperature of 2400 K. For this temperature, the enthalpies for the three products, taken from the online version of NIST Standard Reference Database Number 69, July 2001 [8], are:

Product	n_i (mol)	$(H_i)_T - (H_i)_{298.15}$ (kJ/mol)	$n_i[(H_i)_T - (H_i)_{298.15}]$ kJ
CO_2	3	115.8	347.4
H_2O	4	93.74	375.0
N_2	18.8	70.50	1325.4
Total			2047.8

This value of 2047.8 kJ for the enthalpy of the products is very close to the value of 2043.15 kJ for the enthalpy of the reactants, indicating that the adiabatic flame temperature would be within a few degrees of 2400 K. Further iterations are not necessary unless a closer approximation is needed. Because the JANAF Tables are tabulated in increments of 100 K, further refinement would require interpolation of the tabulated values. It is noted that the solution could have been simplified by starting with the heat of combustion of propane (2043.15 kJ/mol) rather than recalculating it based on the heats of formation of the products and the reactants.

The actual adiabatic flame temperature for a near-stoichiometric mixture will be lower than the value calculated here because the products will partially dissociate into a number of atomic, molecular, and free radical species at such high temperatures. Thus, the assumption to ignore dissociation was not a good one in this case. Several computer programs are available for the calculation of chemical equilibrium conditions in combustion reactions, including the STANJAN program [12] developed at Stanford University and the CHEMKIN program suite [13] developed at Sandia National Laboratories and currently distributed by Reaction Design, Inc.

Example: Estimate the adiabatic flame temperature for a propane/air mixture with an equivalence ratio of 0.5 initially at 25°C.

Solution: For this case, there is twice as much air as required for complete combustion so the reaction equation becomes:

$$C_3H_8 + 10(O_2 + 3.76N_2) \rightarrow 3CO_2 + 4H_2O + 37.6N_2 + 5O_2$$

As in the previous example, the heat of combustion of the propane will be 2043.15 kJ/mol.

As a first guess, select an adiabatic flame temperature of 1600 K. For this temperature, the enthalpies for the four products, taken from the online version of NIST Standard Reference Database Number 69, July 2001 [8], are:

Product	n_i (mol)	$(H_i)_T - (H_i)_{298.15}$ (kJ/mol)	$n_i[(H_i)_T - (H_i)_{298.15}]$ kJ
CO_2	3	67.57	202.71
H_2O	4	52.91	211.64
N_2	37.6	41.81	1572.06
O_2	5	44.12	220.60
Total			2207.01

The total enthalpy of the products exceeds the total enthalpy of the reactants, so the first guess of the flame temperature is too high. As a second guess, select an adiabatic flame temperature of 1500 K. For this temperature, the enthalpies of the four products are:

Product	n_i (mol)	$(H_i)_T - (H_i)_{298.15}$ (kJ/mol)	$n_i[(H_i)_T - (H_i)_{298.15}]$ kJ
CO_2	3	61.71	185.13
H_2O	4	48.15	192.60
N_2	37.6	38.34	1441.58
O_2	5	40.46	202.30
Total			2021.61

With this guess, the total enthalpy of the products is much closer to the correct value, but is a bit low. A more accurate answer can now be determined by linear interpolation between these two estimates:

$$T_{ad} = 1{,}500 + \frac{2{,}043.15 - 2{,}021.61}{2{,}207.01 - 2{,}021.61} \cdot 100 = 1{,}511.6 K$$

This answer is expected to be much closer to the actual adiabatic flame temperature for this mixture than the previous example because dissociation is not as significant at this substantially lower temperature. The temperature is much lower for this case because additional air is being heated up and carried through the reaction. This additional air serves as "thermal ballast," much as any other diluent with similar heat capacity would.

From the previous examples, it is apparent that nitrogen from the air is the largest constituent in the product stream for fuels burned in air. Nitrogen typically constitutes more than 70 percent of the products on a molar basis and about 65 percent of the thermal capacity of the products. For over-ventilated fires, these ratios become even higher because excess air is present. Because of this, it is relatively common in fire applications to consider the products of combustion to have the thermal properties of air, i.e.:

$$\Delta H_c = \left(1 + \frac{n_{air}}{n_f \Phi}\right)((H_{air})_T - (H_{air})_{298.15}) \quad (1.29)$$

or, alternatively,

$$\Delta T_{ad} \approx \frac{\Delta H_c}{\left(1 + \frac{n_{air}}{n_f \Phi}\right)(\bar{c}_{p,air})} \quad (1.30)$$

where $\dfrac{n_{air}}{n_f} = \dfrac{n_{O_2}}{n_f} \dfrac{n_{air}}{n_{O_2}} \ll \dfrac{n_{O_2}}{0.21 \cdot n_f}$

Example: Estimate the adiabatic flame temperature for a propane/air mixture at an equivalence ratio of 0.5 at an initial temperature of 25°C, using the properties of air.

Solution: As in the previous examples, the heat of combustion of propane is 2043.15 kJ/mol. For propane, the number of moles of oxygen per mole of fuel is 5. Therefore, it is necessary to determine the enthalpy change such that:

$$((H_{air})_T - (H_{air})_{298.15}) = \frac{\Delta H_c}{\left(1 + \frac{n_{air}}{n_f \Phi}\right)} = \frac{2043.15 \text{ kJ/mol } C_3H_8}{\left(1 + \frac{5}{(0.21 \cdot 0.5)}\right) \text{mol air/mol } C_3H_8} = 42.02 \text{ kJ/mol air}$$

From the JANAF Table for air (Appendix B1 of Strehlow [1]), the enthalpy of formation at 1500 K is 38.75 kJ/mol, while that at 1600 K is 42.27 kJ/mol. Linear interpolation between these values yields:

$$T_{ad} = 1,500 + \frac{42.02 - 38.75}{42.27 - 38.75} \cdot 100 = 1,592.9 K$$

This value is 81.3 K higher than the value calculated using the properties of the actual products of combustion in the previous example. However, this difference is less than 7 percent of the adiabatic flame temperature rise, which might be considered an acceptable error for some applications given the simplicity of this method.

Alternatively, estimate the average heat capacity of air to be approximately 33 J/mol K over the temperature range of interest and calculate the adiabatic temperature rise as follows:

$$\Delta T_{ad} \approx \frac{\Delta H_c}{\left(1 + \frac{n_{air}}{n_f \Phi}\right)(\bar{c}_{p,air})} = \frac{2043.15}{\left(1 + \frac{5}{0.21 \cdot 0.5}\right)(0.033)} = 1273.44 K$$

$$T_{ad} = T_o + \Delta T_{ad} = 298.15 + 1273.44 = 1571.59 K$$

This value is still somewhat higher than the value calculated based on the actual products of combustion, but is relatively close to the correct answer. The adiabatic temperatures based on the heat capacity of air are somewhat high because the specific heat of carbon dioxide is much higher than those of the oxygen and nitrogen in air on a mass as well as on a molar basis.

The actual temperatures associated with flames in real fires are lower than the adiabatic flame temperatures calculated above for several reasons, including:

- Heat losses due to radiation from the luminous flames; these losses typically represent 20 to 40 percent or more of the heat released in a turbulent diffusion flame [14, 15]
- Heat losses due to heat transfer with enclosure boundaries or other solid surfaces within an enclosure
- Entrainment of excess air into the flame
- Incomplete combustion due to flame quenching caused by the effects identified above

In a fire, the equivalence ratio of the mixture changes with time and location for at least two reasons: (1) the heat release rate of the fire changes with time, and (2) additional air is entrained in the fire plume rising above the fire source as a function of height. This contrasts with internal combustion engines, furnaces, and other designed combustors, where fuel-air mixtures are carefully regulated by design to achieve desired equivalence ratios.

In fire applications, calculation of the adiabatic flame temperature of a fuel-oxidizer-diluent mixture has value primarily as a means to estimate whether a mixture is likely to be flammable rather than as a method to accurately calculate actual flame temperatures. This critical adiabatic flame temperature concept is discussed in Sec. 1.2.2.

1.2 GASES, MISTS, AND DUSTS

Flammable gases, mists composed of liquid droplets, and dust clouds composed of solid particulates all pose a serious and immediate fire and potential explosion hazard if released and dispersed into the atmosphere in sufficient concentrations to form an ignitable mixture. Once ignited, such mixtures can propagate a flame with considerable speed and, if confined, can generate pressures sufficient to damage enclosing structures if appropriate emergency pressure-venting measures are not taken. Under some conditions, propagating flames can accelerate to the speed of sound and transition from a deflagration to a detonation.

Because of the dangers associated with the release of flammable gases, mists, and dusts into the atmosphere, systems conveying such materials and appliances utilizing such materials are typically designed with special precautions to reduce the potential for an accidental discharge into the atmosphere or to permit rapid discovery if an accidental discharge does occur. For example, an odorizing agent, mercaptan, is typically added to natural gas and liquified petroleum gas in trace amounts to permit detection of these otherwise odorless flammable gases at very low concentrations through the sense of smell. In areas where flammable concentrations of gases, mists, or dusts are expected to occur, designs to minimize potential ignition sources may be employed, including the use of "explosion-proof" or "intrinsically safe" electrical fixtures.

1.2.1 Flammability Limits for Gases and Vapors

Not all mixtures of flammable gases in air or in other oxidizers can propagate a flame. Only mixtures at concentrations within the flammability limits will normally propagate a flame, where the *flammability limits* are the range of concentrations, typically expressed in terms of volume or mole percent in air, which will propagate a flame indefinitely. The flammability limits are bounded by the *lower flammability limit* and the *upper flammability limit*. Mixtures below the lower flammability limit are called *fuel lean,* while mixtures above the upper flammability limit are called *fuel rich.* For many gases, the range of the flammability limits tends to increase with temperature. This concept is illustrated qualitatively in Fig. 1.2 [16]. While these effects will generally be relatively small over the range of typical ambient temperatures, it is important to recognize that tabulated flammability lim-

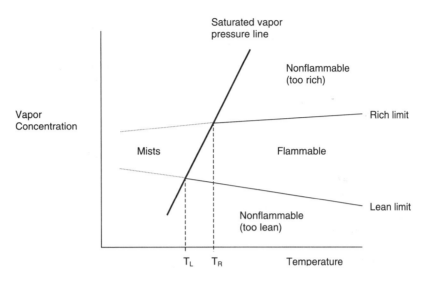

FIGURE 1.2 Qualitative illustration of flammability limits as a function of temperature [16].

its have been determined at specific temperatures and pressures, which should be specified if different from standard conditions.

The primary apparatus used to determine flammability limits was developed at the U.S. Bureau of Mines in the 1950s. This apparatus consists of a vertical tube 1.5 m (5 ft) long and 50 mm (2 in) in diameter with a spark or small flame ignitor located at the bottom end. Homogenous mixtures of gas-air or gas-air-diluent mixtures are introduced into the tube, then the ignitor is activated. A gas mixture is deemed flammable if the flame propagates at least halfway up the tube, a distance of 75 cm (2.5 ft). The lower and upper flammability limits are determined experimentally in this apparatus by bracketing the greatest and least concentrations where flame propagation occurs at a given temperature and pressure, i.e.:

$$LFL_{T,P} = 1/2(C_{L,f} + C_{L,nf})_{T,P} \qquad (1.31)$$

$$UFL_{T,P} = 1/2(C_{U,f} + C_{U,nf})_{T,P} \qquad (1.32)$$

where $C_{L,f}$ is the least concentration at the lower limit that is flammable, $C_{L,nf}$ is the greatest concentration at the lower limit that is not flammable, $C_{U,f}$ is the least concentration at the upper limit that is flammable, and $C_{U,nf}$ is the greatest concentration at the upper limit that is not flammable at the referenced temperature and pressure. Flammability limits for some representative fuels are provided in Table 1.5. Zabetakis [16] provides a comprehensive review and tabulation of flammability limits of gases and vapors in air and in air-diluent mixtures at atmospheric pressure.

Flammable gases will not always be uniformly mixed in a gas-air or gas-air-diluent mixture. Concentration gradients may exist, particularly near the source of a flammable gas or for flammable gases that have molecular weights much different from the air or air-diluent mixture. Bcause of these concentration gradients, some parts of a gas mixture may be within the flammability limits even if the mixture is outside the flammable range on average. For this reason, it is desirable to maintain concentrations of flammable gases well outside the flammability limits. For example, in industrial dryers used to dry parts being cleaned or coated with flammable finishes, it is common practice to provide ventilation at a rate sufficient to maintain the concentration of flammable vapors at less than 25 percent of the lower flammability limit to reduce the potential that regions within the flammable range will occur.

TABLE 1.5 Stoichiometric Concentrations and Flammability Limits for Representative Fuels [16]

Fuel	Formula	Stoichiometric concentration in air (Vol. %)	Lower flammability limit (LFL) (% Stoich.)	Upper flammability limit (UFL) (% Stoich.)	LFL energy concentration in air at STP (kJ/m^3)
Alkanes					
Methane	CH_4	9.5	52.6	157.9	1790.1
Ethane	C_2H_6	5.7	52.6	217.5	1911.3
Propane	C_3H_8	4.0	52.5	237.5	1915.5
n-Butane	C_4H_{10}	3.1	58.1	271.0	2134.0
n-Pentane	C_5H_{12}	2.6	54.7	300.0	2027.1
n-Hexane	C_6H_{14}	2.2	55.5	336.4	2064.3
n-Heptane	C_7H_{16}	1.9	56.0	352.6	2091.6
n-Octane	C_8H_{18}	1.7	57.5	—	2151.6
n-Nonane	C_9H_{20}	1.5	57.5	—	2156.7
n-Decane	$C_{10}H_{22}$	1.3	56.1	430.8	2106.8
Alkenes					
Ethylene	C_2H_4	6.5	41.2	553.8	1594.4
Propene	C_3H_6	4.4	53.8	250.0	2063.7
Butene	C_4H_8	3.4	47.3	294.1	1815.2
Pentene	C_5H_{10}	2.7	51.9	—	1956.0
Hexene	C_6H_{12}	2.3	—	—	—
Alkynes					
Acetylene	C_2H_2	7.8	32.3	1282.1	1400.5
Alcohols					
Methanol	CH_3OH	12.2	48	408	1910.3
Ethanol	C_2H_5OH	6.5	50	292	1818.9
Propanol	C_3H_7OH	4.5	49	300	1810.1
Miscellaneous					
Acetone	C_3H_6O	5.0	59	233	1924.5
Carbon monoxide	CO	29.5	34	676	1577.7
Toluene	C_7H_8	2.3	43	322	1999.1

1.2.1.1 Lower Flammability Limit

The lower flammability limit is generally of more interest than the upper flammability limit for fire hazard analyses. Where practical, it is desirable to maintain concentrations of flammable gases below the lower flammability limit. While concentrations above the upper flammability limit will not propagate a flame, the interface between such concentrations and air in the atmosphere will necessarily include a region within the flammability limits and therefore pose some risk of ignition that will not exist for concentrations below the lower flammability limit. Several methods have been developed to characterize the lower flammability limits. These methods are discussed in this section.

Zabetakis [16] notes that the lower flammability limit for the paraffin hydrocarbon series, also known as the alkanes and as the saturated hydrocarbons, falls within the range of 45 to 50 g/m^3 for most of the series when expressed as a mass concentration. Drysdale [2] notes that the heats of combustion for the alkanes are approximately the same when expressed on a mass basis, with a value of 44 to 45 kJ/g and, consequently, the lower flammability limit for the alkanes can also be expressed as a critical energy density, with a value of approximately 2100 kJ/m^3. Drysdale notes that this critical energy density concept may be more generally applicable than just to the alkanes. Energy densities at the lower flammability limit are provided for selected fuels in Table 1.5. These values tend to corroborate the concept of the critical energy density for a wide range of fuels, with a few notable exceptions, such as hydrogen, acetylene, and carbon monoxide, which tend to be highly reactive.

The lower flammability limit can also be expressed in terms of the equivalence ratio of a gas-air mixture. Zabetakis [16] notes that the lower flammability limit has a relatively constant value of $\Phi_{LFL} \approx 0.55$ for most of the paraffin hydrocarbons. As shown in Table 1.5, this approximation is generally valid for the alkane series of hydrocarbons.

The concept of a critical adiabatic flame temperature at the lower flammability limit has been observed to have great utility [17]. This concept suggests that the adiabatic flame temperature at the lower flammability limit is approximately 1600 K ($\pm$150 K) for a wide range of fuel-oxidizer-diluent mixtures over a wide range of initial temperatures, with a few notable exceptions. The exceptions include hydrogen at 980 K, carbon monoxide at 1300 K, and acetylene at 1280 K. As noted previously, these lower critical adiabatic flame temperatures are an indication of the higher reactivity of these substances. The value of the critical adiabatic flame temperature concept is that it permits evaluation of the flammability of a wide range of fuel-oxidizer-diluent mixtures over a wide range of initial temperatures. For example, Beyler [17] describes the application of this concept to evaluation of the potential for ignition of a smoke layer composed of combustion products, oxygen, and unburned fuel at elevated temperatures in a room fire.

The critical adiabatic flame temperature concept can be applied to a mixture of flammable gases to evaluate the flammability of the mixture. For mixtures of gases in air, LeChatelier's rule can also be used to evaluate the flammability of the mixture and is easier to calculate than the adiabatic flame temperature. LeChatelier's rule indicates that a mixture of flammable gases will exceed the lower flammability limit if the following relationship holds true [17]:

$$\sum_i (C_i/LFL_i) \geq 1 \qquad (1.33)$$

One of the important implications of this relationship is the influence a small quantity of a more volatile flammable liquid can have on the flammability of a large quantity of a less volatile liquid. For example, a small quantity of gasoline contaminating a large tank of diesel fuel can cause the vapor space within the diesel fuel tank to be raised above the lower flammability limit. This issue is discussed further in Sec. 1.3.

1.2.1.2 Upper Flammability Limit

Upper flammability limits are expressed in terms of percent of stoichiometric concentration in Table 1.5. On a volume basis, the upper flammability limit tends to decrease with increasing molecular weight, but on a mass concentration basis, the upper limit tends to increase with increasing molecular weight. Unlike the lower flammability limit, the upper flammability limit cannot be characterized in terms of a constant mass concentration, critical adiabatic flame temperature, or critical energy density at the upper limit. For most applications, the goal will be to maintain flammable concentrations below the lower flammability limit, but for some applications typically involving highly volatile liquids, the goal will be to maintain flammable concentrations well above the upper flammable limit to avoid the potential for accidental ignition.

1.2.2 Ignition Energy

A mixture of gases within the flammability limits can be ignited by a concentrated energy source, such as an electrical arc, a mechanical sparks or a pilot flame. Such a mixture might also ignite spontaneously if the bulk temperature of the mixture is raised to the *autoignition temperature* of the mixture. The minimum ignition energy of a concentrated source is generally very low, on the order of 0.1 to 1 mJ, with the lower energy levels associated with near-stoichiometric mixtures [16]. Such energy levels can be achieved by the discharge of an electrostatic arc, so special precautions are needed to prevent potential ignition sources in areas where flammable mixtures of gases are anticipated.

Autoignition temperatures are reported in the literature [see, e.g., Refs. 16, 18], but it is important to recognize that autoignition temperatures are scale-dependent. In general, the rate of energy generation within an isothermal mixture of gases will be proportional to the volume of the gases, while the rate of heat losses from the mixture will be proportional to the surface area of the boundaries. For example, autoignition temperatures for gas mixtures are commonly measured in a spherical vessel, so the volume of gases and the rate of energy generation are proportional to the cube of the sphere radius while the surface area of the vessel and the rate of heat losses are proportional to the square of the sphere radius. At its simplest, autoignition will occur when the rate of heat generation in the mixture volume exceeds the rate of heat losses to the vessel surface, leading to a runaway reaction.

This discussion suggests that there should be an inverse relationship between vessel size and autoignition temperature. More importantly, it suggests that reported autoignition temperatures should not be considered universal values, but rather as device-dependent values. For example, Drysdale [2] presents data from Setchkin [19] demonstrating the inverse relationship between reactor volume and autoignition temperature. Drysdale also presents several classical theories of thermal explosion, including those of Semenov [20] and Frank-Kamenetskii [21].

1.2.3 Flame Speed

Once a mixture of gases within the flammable range ignites, the flame will propagate away from the point of ignition through the mixture at a rate that depends primarily on the rate of heat transfer from the flame to unburned gases ahead of the flame. The concept of a fundamental flame speed has been developed and analyzed. Drysdale [2] reviews the theory behind the concept.

For near-stoichiometric mixtures of flammable gases in air, fundamental flame speeds in the range of approximately 0.3 to 0.5 m/s are typical. These represent values that are associated with laminar flame propagation through a quiescent mixture of gases. Actual flame speeds can accelerate to values orders of magnitude higher than the fundamental flame speed, particularly where obstructions exist to cause turbulence and force acceleration of the flow through a restriction. The potential for transition from subsonic propagation, i.e., deflagration, to supersonic propagation, i.e., detonation, has been recognized where such obstructed flow paths exist.

1.2.4 Ignition of Mists and Dusts

Mists of liquid fuels and dust clouds of solid particulates are addressed in this section because finely dispersed mists and dusts behave much like flammable vapors with respect to their potential for ignition. This is due to the large surface area to volume ratio of small droplets and particles. Such droplets and particles are readily vaporized with relatively little heat input, permitting them to behave much like vapors when dispersed in the atmosphere.

Mists and dust clouds have minimum concentrations at which they will ignite and propagate a flame much like the lower flammability limit for gases. Fig. 1.2, from Zabetakis [16], shows that mists are expected to have limit concentrations analogous to vapors, particularly for small droplets. Schwab [22] addresses the factors influencing the explosibility of dust clouds and summarizes the explosion characteristics of various dusts based on a compilation of data reported in a series of U.S. Bureau of Mines reports. Minimum explosion concentrations for the dusts he includes in his summary are in the range of 15 to 180 g/m^3, comparable to the 48 g/m^3 associated with flammable vapors at the lower flammability limit.

1.3 LIQUIDS

Liquid fuels are separated into two categories, flammable and combustible, for purposes of transportation regulation. These same categories are widely used in other applications as well to distinguish the relative ignition hazards of flammable and combustible liquids. Within the United States, a flammable liquid is defined as a liquid having a *flash point* below 37.8°C (100°F) and having a vapor pressure not exceeding 276 kPa (40 psia) at 37.8°C (100°F), while a combustible liquid is a liquid with a flash point at or above 37.8°C (100°F). The flash point is the minimum temperature at which a liquid gives off vapors in sufficient concentrations to form an ignitable mixture with air near the liquid surface. It is the temperature denoted as T_L in Fig. 1.2.

Flammable and combustible liquids are subdivided to further distinguish their ignitability hazards. Flammable liquids are designated as Class I liquids and are subdivided as follows:

- Class I-A liquids include those having a flash point below 23°C (73°F) and having a boiling point below 38°C (100°F).
- Class I-B liquids include those having a flash point below 23°C (73°F) and having a boiling point at or above 38°C (100°F).
- Class I-C liquids include those having a flash point at or above 23°C (73°F) and below 38°C (100°F).

Thus, Class 1-A and Class 1-B liquids would be expected to release a flammable concentration of gases over a full range of normal room temperatures, while Class 1-C liquids would be expected to release a flammable concentration of gases only under relatively warm room temperatures.

Combustible liquids are designated as Class II or III and are subdivided as follows:

- Class II liquids are those having a flash point at or above 38°C (100°F) and below 60°C (140°F).
- Class III-A liquids are those having a flash point at or above 60°C (140°F) and below 93°C (200°F).
- Class III-B liquids are those having a flash point at or above 93°C (200°F).

This classification system is intended to reflect that flammable liquids might be expected to routinely release an ignitable mixture of vapors at a range of normal ambient temperatures, while Class II combustible liquids would be expected to release an ignitable mixture of vapors only at relatively high, and therefore relatively unusual, ambient temperatures. Class III combustible liquids would virtually never reach their flash points under ambient conditions, but could form an ignitable mixture if heated to temperatures above their flash points.

While a liquid fuel will not release an ignitable mixture of vapors at temperatures below its flash point, such a liquid can be ignited and burn if dispersed on a porous medium. A hurricane lamp is a classic example of this. Under these conditions, liquid fuel becomes trapped in the porous medium, preventing the liquid from circulating away from the heat source, as it would in a liquid pool. With the application of heat, as from a match, the liquid temperature is increased locally to above its flash point, permitting ignition. Once ignited, liquid fuel "wicks" through the porous medium by capillary action to sustain the fire. This same process permits combustible liquids, such as kerosene, to be used as effective accelerants if dispersed on fibrous materials such as carpeting.

To a large extent, the hazards associated with the ignition of liquid fuels are the same as those associated with gases because ignition occurs in the vapor phase. Consequently, the discussions of flammability limits and autoignition temperatures provided in the subsection on gases also apply to liquids. Once ignition occurs, however, the burning rate of liquids will be governed largely by the heat feedback to the liquid surface. This is one way that the hazards of liquids are different from those of gases.

1.3.1 Fire Point

The *fire point* is the lowest temperature of a liquid at which vapors are released fast enough to support continuous combustion at the liquid surface. The fire point is typically a few degrees higher than

TABLE 1.6 Flash Points and Fire Points for Selected Liquid Fuels [2]

Fuel	Formula	Flash point (°C)	Fire point (°C)
Alkanes			
n-Hexane	C_6H_{14}	−22	
n-Heptane	C_7H_{16}	−4	
n-Octane	C_8H_{18}	13	
n-Nonane	C_9H_{20}	31	
n-Decane	$C_{10}H_{22}$	44	61
Alkenes			
Ethylene	C_2H_4	−121	
Propene	C_3H_6	−108	
Butene	C_4H_8	−80	
Pentene	C_5H_{10}	−18*	
Alcohols			
Methanol	CH_3OH	12	13
Ethanol	C_2H_5OH	13	18
Propanol	C_3H_7OH	15	26
Miscellaneous			
Acetone	C_3H_6O	−18	
Toluene	C_7H_8	4	

* Open-cup; all other flash points are closed-cup.

the flash point because at the flash point sufficient vapors are being released from the liquid surface to support a flash of flame across the surface, but not to support continued combustion once these vapors flash. At the fire point, the rate of vapor evolution is sufficient to support continuous combustion. From a practical standpoint, the flash point is normally used to characterize the relative volatility of a liquid rather than the fire point. Flash points and fire points for selected liquid fuels are provided in Table 1.6.

1.3.2 Vapor Pressure

The potential for development of a flammable concentration of fuel vapors in the vapor space above a liquid fuel under ambient conditions depends primarily on the vapor pressure exerted by the fuel. This vapor pressure provides a measure of the volatility of a fuel. Vapors escape from a liquid surface exposed to the atmosphere even in the absence of a localized energy source because of energy associated with the temperature of the atmosphere. The higher the temperature, the higher the pressure exerted by the vapors will be and the more rapid the vaporization will be.

An equilibrium vapor pressure develops in closed systems, such as a storage container or a fuel tank. At equilibrium, the rate of fuel evaporation is offset by the rate of fuel condensation. This equilibrium vapor pressure is a strong function of temperature. An equilibrium vapor pressure does not develop for open systems, as the evaporating fuel dissipates in the atmosphere and is not offset by fuel condensation. The equilibrium vapor pressure can be considered as the upper limit for the vapor pressure that will exist above a liquid surface in an open system.

The concentration of fuel vapors above a liquid surface can be evaluated to determine if the concentration is within the flammability limits. The temperature-dependent vapor pressures of pure liquids can be calculated according to the Clapeyron-Clausius equation [23] as:

$$\log_{10} p^\circ = (-0.2185 E/T) + F \qquad (1.34)$$

TABLE 1.7 Vapor Pressure Constants for Some Organic Compounds [23]

Compound	Formula	E	F	Temperature range (°C)
n-Pentane	n-C_5H_{12}	6595.1	7.4897	−77 to 191
n-Hexane	n-C_6H_{14}	7627.2	7.7171	−54 to 209
Cyclohexane	c-C_6H_{12}	7830.9	7.6621	−45 to 257
n-Octane	n-C_8H_{18}	9221.0	7.8940	−14 to 281
iso-Octane	C_8H_{18}	8548.0	7.9349	−36 to 99
n-Decane	n-$C_{10}H_{22}$	10,912.0	8.2481	17 to 173
n-Dodecane	n-$C_{12}H_{26}$	11,857.7	8.1510	48 to 346
Methanol	CH_3OH	8978.8	8.6398	−44 to 224
Ethanol	C_2H_5OH	9673.9	8.8274	−31 to 242
n-Propanol	n-C_3H_7OH	10,421.1	8.9373	−15 to 250
Acetone	$(CH_3)_2CO$	7641.5	7.9040	−59 to 214
Methyl ethyl ketone	$CH_3CO \cdot CH_2CH_3$	8149.5	7.9593	−48 to 80
Benzene	C_6H_6	8146.5	7.8337	−37 to 290
Toluene	$C_6H_5CH_3$	8580.5	7.7194	−28 to 31

where E and F are constants, T is the temperature in Kelvin, and $p°$ is the equilibrium vapor pressure in millimeters Hg. Values for E and F are provided in Table 1.7 for selected liquid fuels [23].

1.3.3 Flammability Limits of Liquids

The flammability limits for selected organic liquids are tabulated in Table 1.5 along with those for flammable gases. To a large extent, flammable gases are distinguished from pure flammable liquids only because flammable gases are above their boiling point under ambient conditions, while flammable liquids are below their boiling point under ambient conditions. Thus, the discussion regarding flammability limits presented above for gases also applies for liquids.

When expressed on a volume or molar basis, the lower flammability limit for liquid vapors tends to be lower than the lower flammability limit for gaseous fuels. As discussed in Sec. 1.2.1 and illustrated in Table 1.5, however, when expressed on a mass concentration or critical energy density basis, the lower flammability limit for liquid vapors is virtually the same as for gaseous fuels. This is because of the higher molecular weight but relatively constant heats of combustion on a mass basis of liquid fuels in comparison with gaseous fuels.

1.3.4 Liquid Mixtures

Raoult's law can be used to evaluate the flammability limits of liquid mixtures that can be approximated as "ideal solutions" [2]. Raoult's law states that the vapor pressures associated with each component in a multicomponent mixture will be:

$$p_i = x_i p_i° \qquad (1.35)$$

where $x_i (\equiv n_i / \Sigma_i n_i)$ is the mole fraction of component i in the liquid mixture and $p_i°$ is the equilibrium vapor pressure of component i. Perhaps the most significant aspect of this relationship with respect to fire hazard is the effect a small quantity of a highly volatile liquid can have when mixed with a large quantity of a liquid with lower volatility. A classic example of this is a diesel fuel tank contaminated with a small quantity of gasoline. A tank containing only diesel fuel will be below the lower flammability limit, while a tank containing only gasoline will be above the upper flammability limit under a wide range of ambient conditions [24]. If a small quantity of gasoline is added to a tank of diesel fuel, as is sometimes done to help prevent waxing of the diesel fuel in cold weather, then the vapor space in the diesel tank may be within the flammable range and therefore more susceptible to ignition.

1.3.5 Burning Rate

Once a flammable liquid ignites, its burning rate will depend on either the rate of liquid flow, e.g., from a leaking pipe or vessel, or by the area and thickness of the pool formed by the liquid. If a liquid flows and forms an unconfined pool, the area of the pool formed by the flowing liquid will be governed by the flow rate and the burning rate per unit area of the fuel. Such an unconfined pool will continue to grow until a balance develops between the flow rate and the burning rate. For a confined pool, such as within a tank or diked area, the burning rate will depend on the surface area of the pool within the confinement.

The burning rate of a liquid pool fire is governed by heat feedback to the fuel surface. The steady mass-burning rate per unit area of pool surface $\dot{m}''_f$, can be expressed as [11]:

$$\dot{m}''_f = \frac{\dot{q}''_{net}}{\Delta H_g} \qquad (1.36)$$

where $\dot{q}''_{net}$ is the net heat flux at the fuel surface and ΔH_g is the heat of gasification of the liquid fuel. The heat of gasification is composed of the heat of vaporization ΔH_v and the sensible heat needed to raise the temperature of the liquid from its initial temperature to its boiling point:

$$\Delta H_g = \Delta H_v + c(T_b - T_o) \qquad (1.37)$$

The net heat flux at the fuel surface is the difference between the various heat fluxes entering and leaving the fuel surface:

$$\dot{q}''_{net} = \dot{q}''_{f,c} + \dot{q}''_{f,r} + \dot{q}''_{ext} - \dot{q}''_{rr} - \dot{q}''_k \qquad (1.38)$$

where $\dot{q}''_{f,c}$ is the convective flame heat flux to the surface, $\dot{q}''_{f,r}$ is the radiative flame heat flux to the surface, $\dot{q}''_{ext}$ is the external heat flux, e.g., from enclosure boundaries, to the fuel surface, $\dot{q}''_{rr}$ is the heat flux reradiated from the fuel surface, and $\dot{q}''_k$ is the net effect of conduction losses from the fuel surface into the pool and its bounding surfaces. For thin films of liquid on massive substrates, such as unconfined spills on concrete floors, these conduction losses can dominate the heat-transfer processes and consequently the burning rate, while conduction losses to boundaries will be minor for relatively large, deep pools.

Convection dominates the heat transfer back to the fuel surface for very small pool diameters, on the order of 10 cm or less. For larger fires, which are generally of more interest for fire hazard analyses, radiation dominates the heat transfer back to the fuel surface. Based on a grey-gas model for flame radiation, the radiative heat flux to the fuel surface can be expressed as:

$$\dot{q}''_{f,r} = \sigma T_f^4 [1 - \exp(-\kappa L_m)] \qquad (1.39)$$

where σ is the Stefan-Boltzmann constant (5.67×10^{-11} kW/m²(K⁴)), T_f is the mean absolute flame temperature (K), κ is the flame absorption coefficient (m⁻¹) and L_m is the mean beam length for the flame (m). The mean beam length depends on the height and shape of the flame, which in turn depends on the pool diameter; a scale factor β is typically used to correlate the unknown, mean beam length with the known pool diameter:

$$L_m = \beta D \qquad (1.40)$$

Eq. (1.40) can be substituted into Eq. (1.39), which in turn can be substituted into Eq. (1.36) to yield an expression for the mass-burning rate for pool fires [25]:

$$\dot{m}''_f = \dot{m}''_{f,\infty}[1 - \exp(-\kappa \beta D)] \qquad (1.41)$$

where $\dot{m}''_{f,\infty}$ is the asymptotic mass-burning rate for large fire diameters. The asymptotic mass-burning rate can be expressed as:

$$\dot{m}''_{f,\infty} = \frac{\sigma T_f^4}{\Delta H_g} \quad (1.42)$$

Eq. (1.42) is a form of Eq. (1.36) with the assumption that the net heat flux to the surface of a large diameter pool can be represented in terms of the blackbody radiation from the optically thick flames back to the pool surface. Based on measurements of mass-burning rates and flame temperatures, it should be recognized that Eq. (1.42) would overestimate the actual asymptotic mass-burning rate by a factor of approximately 4. This significant difference has been attributed to the blocking of incident radiation by vapors rising from the pool surface [26].

Babrauskas [25] tabulates values for pool burning rate parameters for a range of liquid fuels; these data are incorporated in Table 1.8 along with other properties of interest for liquid pool burning. Once the fuel mass-burning rate is determined, the heat release rate of a pool fire can be calculated as:

$$\dot{Q}_f = \dot{m}''_f A_f \Delta H_c \chi_{eff} \quad (1.43)$$

where A_f is the surface area of the fuel, ΔH_c is the heat of combustion of the fuel, and χ_{eff} is a combustion efficiency factor. The product $\Delta H_c \chi_{eff}$ is sometimes called the *effective heat of combustion*. Tewarson [15] has tabulated effective heats of combustion for a wide range of fuels based on bench-scale measurements.

TABLE 1.8 Data for Pool Burning Rate Calculations [25]

Material	Boiling point (°C)	Density (kg/m³)	Heat of combustion (MJ/kg)	Heat of vaporization (kJ/kg)	Asymptotic burning rate (kg/s/m²)	$\kappa\beta$ (m⁻¹)
Simple organic fuels						
Butane	0	573	45.7	386	0.078	2.7
Hexane	69	650	44.7	365	0.074	1.9
Heptane	98	675	44.6	365	0.101	1.1
Benzene	80	874	40.1	432	0.085	2.7
Xylene	139	870	40.8	343	0.090	1.4
Acetone	56	791	25.8	521	0.041	1.9
Alcohols						
Methanol	64	796	20.0	1101	0.017	∞
Ethanol	78	794	26.8	837	0.015	∞
Petroleum products						
Gasoline	These products are blends of many components so they do not have a specific boiling point	740	43.7	These products are blends of many components so they do not have a specific heat of vaporization	0.055	2.1
Kerosine		820	43.2		0.039	3.5
JP-4		760	43.5		0.051	3.6
JP-5		810	43.0		0.054	1.6
Transformer oil		760	46.4		0.039	0.7
Fuel oil, heavy		940–1000	39.7		0.035	1.7
Crude oil		830–880	42.5–42.7		0.022–0.045	2.8

TABLE 1.9 Combustibility Ratios for Selected Fuels [15]

Fuel	Combustibility ratio $\Delta H_c/\Delta H_g$
Red oak (solid)	2.96
PVC (granular)	6.66
Nylon (granular)	13.10
PMMA (granular)	15.46
Methanol (liquid)	16.50
Polypropylene (granular)	21.37
Polystyrene (granular)	23.04
Polyethylene (granular)	24.84
Styrene (liquid)	63.30
Heptane (liquid)	92.83

For large-diameter pools burning at the asymptotic mass-burning rate, the net heat flux at the fuel surface can be estimated by inverting Eq. (1.36) and substituting the asymptotic burning rates and heats of vaporization from Table 1.8. As noted by Friedman [26], based on this type of analysis, the flames above a large pool fire impose a net heat flux of about 30 kW/m² on the liquid surface. For typical flame temperatures of approximately 900°C, the blackbody radiation would be approximately 120 kW/m². This demonstrates that Eq. (1.42) tends to overestimate the asymptotic burning rate by approximately a factor of 4.

1.3.5.1 Combustibility Ratio

Substitution of Eq. (1.36) into Eq. (1.43) yields:

$$\dot{Q}_f'' = \dot{q}_{net}'' \cdot \chi_{eff} \cdot \frac{\Delta H_c}{\Delta H_g} \tag{1.44}$$

Eq. (1.44) indicates that the heat release rate per unit area $\dot{Q}_f''$ of a fuel depends on the ratio between the fuel heat of combustion and the fuel heat of gasification. This *combustibility ratio* represents the ratio between the energy released by complete combustion of a unit mass of fuel and the energy required to evaporate or pyrolyze a unit mass of fuel. Tewarson [15] has referred to this ratio as the *heat release parameter*. Some representative values for the combustibility ratio or heat release parameter are provided for selected fuels in Table 1.9.

The combustibility ratio can be considered as a measure of the relative volatility of a fuel, with higher combustibility ratios representing more volatile fuels. The inverse of the combustibility ratio represents the fraction of energy released by complete combustion needed to continue the steady burning of a material. As indicated in Eq. (1.44), the combustibility ratio is the material property governing the heat release rate of a condensed fuel, while the net heat flux to the fuel surface is the environmental variable influencing the heat release rate of a condensed fuel.

1.4 SOLIDS

The majority of potential fuels that will contribute to fires in the built environment are solid materials. Solid materials have several potential fire hazards that should be considered, including the potential for self-heating to ignition, the propensity to smolder, the ease of ignition, and the rate of flame spread and heat release once ignited. The fire hazards of solid combustible materials are particularly complex because they depend on several factors, including the physical form, orientation, and chemical properties of the material and the environmental conditions to which the material is subjected. Surface heat flux and atmospheric oxygen concentration are the two most important environmental factors.

Flaming ignition and combustion at the surface of a solid involves both chemical and physical processes. Similar to liquids, solid materials must gasify before they can burn in a flame at the surface of the material. Unlike liquids, however, solid combustibles do not exert a significant vapor pressure under ambient conditions. Solids must be heated to gasify. The gasification of solid materials generally involves the thermally induced decomposition of complex molecules in a process known as pyrolysis. These combustible gases must be released from the surface rapidly enough to form an ignitable mixture with air at the surface of the material. Because these vapors are constantly being whisked away from the heated surface by convective currents arising from the heating process or from other forces, they must be constantly replaced by new vapors from the surface to form an ignitable mixture with air. As the gasification rate increases with continued heating, a flammable concentration of gases may form near the surface.

Even when a flammable concentration of fuel vapors exists near a solid surface, *piloted ignition* will not occur unless an ignition source is present where a flammable concentration of gases exists. This is analogous to the flash point of a liquid. Sustained burning at the solid surface will not occur unless vapors continue to evolve at a rate sufficient to maintain combustion following ignition. This is analogous to the fire point of a liquid. Ignition can occur in the absence of a pilot ignition source if the gases are hot enough to ignite spontaneously. This is analogous to the autoignition temperature of a flammable gas or liquid and is known as *unpiloted ignition*.

The pyrolysis rate of a solid surface varies strongly with temperature. The pyrolysis rate is typically represented in terms of an exponential Arrhenius expression. While detailed models of the pyrolysis process have been and continue to be developed, for engineering purposes it is usually appropriate to neglect the details of this process and to use an effective ignition temperature as an indication of ignition. Based on this engineering approximation, a solid material is assumed to be inert below its effective ignition temperature and to begin burning when the surface is heated to the effective ignition temperature. This treatment is generally satisfactory for thick materials with a virtually unlimited supply of fuel, but may not be adequate for thin materials because the material may slowly and fully pyrolyze before ignition occurs. The inert approximation does not address this situation.

Once ignition occurs, the burning rate of a solid can be treated in the same way as a liquid for engineering purposes, with Eq. (1.36) used to calculate the mass-burning rate of the solid. In order to apply Eq. (1.36) to solid materials, the net heat flux to the fuel surface must be evaluated in accordance with Eq. (1.37) and an effective heat of gasification must be associated with the solid material. For solids, the *effective heat of gasification* is similar to the thermodynamic heat of gasification of a pure liquid; it represents the quantity of heat that must be absorbed by the material to gasify a unit mass of the material. Some representative effective heats of gasification for solid materials are provided in Table 1.10 along with the heats of gasification for some pure liquid fuels for compari-

TABLE 1.10 Effective Heats of Gasification for Selected Solid Materials [15]

Material	Effective heat of gasification (kJ/g)	Material	Effective heat of gasification (kJ/g)
Polyethylene (solid)	2.32	Hexane (liquid)	0.50
Polycarbonate (solid)	2.07	Heptane (liquid)	0.55
Polypropylene (solid)	2.03	Octane (liquid)	0.60
Douglas fir wood	1.82	Decane (liquid)	0.69
Polystyrene (solid)	1.76	Hexadecane (liquid)	0.92
PMMA (solid)	1.62		

son purposes. Note that the effective heats of gasification for solids tend to be higher than the heats of gasification of liquid fuels.

Solid materials can be characterized in several ways. One important distinction is between materials that are char forming and those that are not. The char layer that forms on the surface of char-forming materials as they burn acts to reduce the rate of heat transfer to the interior of the material where pyrolysis is occurring and consequently reduces the rate of pyrolysis and burning over time. This reduction in heat transfer is due to both the insulating characteristics of the char layer as well as reradiation from the char surface, which can heat up to temperatures well above the pyrolysis temperature of the material. For materials that do not form a char layer, the surface temperature will be at or near the pyrolysis temperature and the net heat flux to the pyrolysis zone will be the net heat flux to the surface of the material. This distinction between materials that form a char layer and those that do not is illustrated qualitatively in Fig. 1.3a and 1.3b, which are intended to schematically illustrate differences in heat and mass transfer between the two types of materials.

Materials that do not char tend to burn at higher rates than those that do under similar exposure conditions. This is due to the higher net heat flux at the pyrolysis front for materials that do not char. For example, Fig. 1.4 shows the heat release rate per unit area for red oak and for PMMA under similar exposure conditions. Red oak is a char-forming material, while PMMA is not. Among solid materials that do not char, the class of plastic materials known as thermoplastics is prevalent. These materials tend to soften and melt under fire exposure conditions and burn much like a liquid pool fire when ignited. The receding and dripping behavior exhibited by these materials when used in vertical and suspended applications complicates the fire hazard analysis of these materials.

Another important distinction between solid materials is based on the thickness of the material. A distinction is made between *thermally thin* and *thermally thick* solids. A thermally thin material is thin enough to neglect temperature gradients through the thickness of the material. The temperature of a thermally thin material is assumed to be uniform through the thickness of the material. In contrast, a thermally thick material is typically treated as a semi-infinite solid for heat-transfer analysis. Implications of these distinctions are discussed in the following subsections.

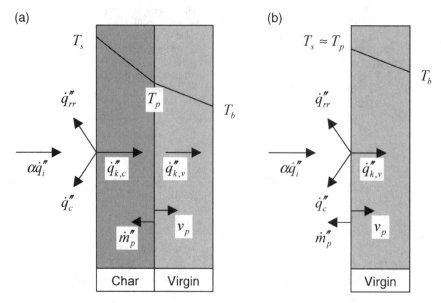

FIGURE 1.3 (a) Schematic diagram of heat-transfer processes for char-forming materials. (b) Schematic diagram of heat-transfer processes for materials that do not char.

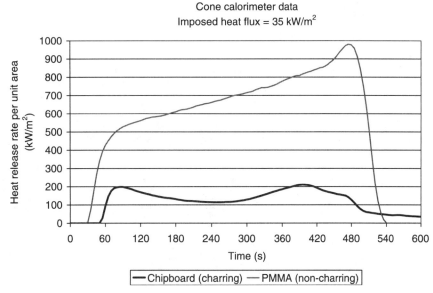

FIGURE 1.4 Representative heat release rates per unit area for a charring and a non-charring material in the cone calorimeter under similar exposure conditions.

1.4.1 Flaming Ignition of Solid Materials

The flaming ignition of solid materials requires the evolution of sufficient combustible vapors at the surface of the material to form a flammable mixture with air; an ignition source must also be present. The concept of an effective ignition temperature is used to evaluate the thermal conditions when such a flammable mixture will form. With this approximation, the ignition process can be considered in terms of thermally thin and thermally thick theories of heat transfer.

1.4.1.1 Ignition of Thermally Thin Materials

A thermally thin material is a material with negligible internal resistance to heat flow, such that temperature gradients through the material are much smaller than those at the surface of the material. The Biot number is used to evaluate the ratio between surface and internal resistances to heat flow. If the Biot number evaluates to a value much less than unity, then a material can be considered as thermally thin, i.e.:

$$Bi \equiv \frac{h_t \delta}{k} = 1.0 \qquad (1.45)$$

The parameter h_t represents the surface total heat-transfer coefficient (kW/m²·K), k is the conductivity of the solid material (kW/m·K), and δ represents the characteristic thickness of the material (m). In general, the characteristic thickness is defined as the volume of the solid divided by the surface area through which heat is transferred; for an infinite flat plate with both sides exposed, the characteristic thickness is simply the plate half-thickness $L/2$, while for an infinite plate with an insulated back the characteristic thickness would be the plate thickness L. If the Biot number evaluates to a value less than 0.1, then errors associated with the thermally thin analysis are known to be less than 5 percent [27]. The accuracy will increase for smaller values of the Biot number and will decrease for larger values.

A number of cases amenable to analytical solution are developed to illustrate the behavior of thermally thin materials in response to constant imposed heat fluxes. All cases consider a thermally thin material of thickness L, conductivity k, specific heat c_p, and density ρ, that is subjected to a constant imposed heat flux of magnitude $\alpha \dot{q}''_i$ on one surface. Here, α represents the grey-body absorptivity of the surface and $\dot{q}''_i$ represents the intensity of the incident radiant heat flux (kW/m²).

Case 1 considers the response of a material that absorbs the imposed heat flux without any losses due to conduction, convection, or reradiation from either the front or back surface; this is the adiabatic case. Case 2 considers a material that is subjected to the imposed heat flux on its front surface and has a convective-radiative boundary condition on the front surface and a perfectly insulated back surface. Case 2 would be the idealized representation of a thin film or fabric covering a highly insulating material. Examples of products that might be expected to approach Case 2 behavior would include a thin vapor barrier over thermal insulation or upholstery fabric over polyurethane foam padding. Case 3 considers a material that is subjected to the imposed heat flux on its front surface and has a convective-radiative boundary condition on both the front and back surfaces. Examples of Case 3 behavior would include free-hanging textiles or thin metal walls or bulkheads. The different thermally thin cases are illustrated in Fig. 1.5a to Fig. 1.5c.

For Case 1, the energy balance at the surface can be expressed as:

$$\rho c L \frac{dT}{dt} = \alpha \dot{q}''_i \tag{1.46}$$

For a constant imposed heat flux, Eq. (1.46) can be integrated to yield the following solution:

$$\Delta T_s = \frac{\alpha \dot{q}''_i t}{\rho c L} \tag{1.47}$$

The time to ignition can be determined for the Case 1 solution if the ignition temperature is known:

$$t_{ig} = \frac{\rho c L \Delta T_{ig}}{\alpha \dot{q}''_i} \tag{1.48}$$

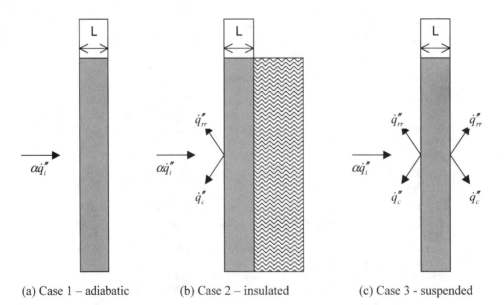

(a) Case 1 – adiabatic (b) Case 2 – insulated (c) Case 3 - suspended

FIGURE 1.5 Schematic representation of thermally thin cases.

Generally speaking, Eqs. (1.46) through (1.48) will only provide an accurate assessment of the rate of heating and the resulting temperature rise in a thermally thin solid at very short times, before the temperature of the material has increased significantly. As the temperature of a material increases above ambient, convective and reradiative heat losses from the surface of the material become increasingly significant. Ultimately, these losses become equal to the rate of heat input, at which point steady conditions would occur.

In order to evaluate the response of a thermally thin material more accurately, the boundary conditions for Case 2 are considered next. For this case, heat losses occur from the exposed face of a material, with the back face considered to be perfectly insulated. Heat losses occur from the exposed face due to both convection and reradiation:

$$\dot{q}''_{loss} = \dot{q}''_c + \dot{q}''_{rr} = h_c(T_s - T_o) + \varepsilon\sigma(T_s^4 - T_o^4) \equiv h_t(T_s - T_o) \quad (1.49)$$

Convective heat transfer is treated in terms of simple Newtonian cooling, with an appropriate convective heat-transfer coefficient h_c. For the free convection conditions typical of many surface heating scenarios in fire, a convective coefficient of $h_c = 0.015$ kW/m$^2 \cdot$ K has been found to be reasonably accurate [28]. In Eq. (1.49), the nonlinear reradiative surface heat losses are linearized to permit an analytical solution of the differential equation. This is done as follows:

$$\dot{q}''_{rr} = \varepsilon\sigma(T_s^4 - T_o^4) \equiv h_r(T_s - T_o) \quad (1.50)$$

Eq. (1.50) can then be solved for the linearized radiative heat-transfer coefficient:

$$h_r = \frac{\varepsilon\sigma(T_s^4 - T_o^4)}{(T_s - T_o)} = \varepsilon\sigma(T_s^3 + T_s^2 T_o + T_s T_o^2 + T_o^3) \quad (1.51)$$

Eq. (1.51) demonstrates that the linearized radiative heat-transfer coefficient is actually a strongly nonlinear function of surface temperature T_s. For engineering purposes, an appropriate constant average value for this parameter must be determined to take advantage of the analytical solution that is possible with this linearized assumption. With this linearized reradiation term, the energy balance at the material surface for Case 2 becomes:

$$\rho c L \frac{dT}{dt} = \alpha \dot{q}''_i - h_t \Delta T \quad (1.52)$$

Eq. (1.52) can be nondimensionalized by defining a characteristic temperature rise ΔT_c and a characteristic time t_c as:

$$\Delta T_c \equiv \frac{\alpha \dot{q}''_i}{h_t} \quad (1.53)$$

$$t_c \equiv \frac{\rho c L}{h_t} \quad (1.54)$$

For Case 2, the characteristic temperature rise represents the maximum temperature rise of a material with perfect back face insulation under steady-state conditions, assuming heat is being lost from the front face by convection and reradiation at the same rate it is being absorbed from the incident heat flux. By substituting Eqs. (1.53) and (1.54) into Eq. (1.52), the nondimensional form of Eq. (1.52) becomes:

$$\frac{d\Delta T_s}{\Delta T_c - \Delta T_s} = \frac{dt}{t_c} \quad (1.55)$$

Eq. (1.55) can be integrated with appropriate limits and expressed in nondimensional terms as:

$$\frac{\Delta T_s}{\Delta T_c} = 1 - \exp\left(-\frac{t}{t_c}\right) \quad (1.56)$$

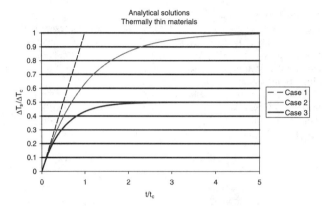

FIGURE 1.6 Analytical solutions for thermally thin solids exposed to constant incident heat fluxes.

For Case 3, the energy balance at the material surface is slightly different from Case 2 because heat is lost from both the front and back faces:

$$\rho c L \frac{dT_s}{dt} = \alpha \dot{q}_i'' - 2h_t \Delta T_s \qquad (1.57)$$

By substituting Eqs. (1.53) and (1.54) into Eq. (1.57), the nondimensional version of Eq. (1.57) becomes:

$$\frac{d\Delta T_s}{\Delta T_c - 2\Delta T_s} = \frac{dt}{t_c} \qquad (1.58)$$

Eq. (1.57) can be integrated with appropriate limits and expressed nondimensionally as:

$$\frac{\Delta T_s}{\Delta T_c} = \frac{1}{2}\left[1 - \exp\left(-\frac{2t}{t_c}\right)\right] \qquad (1.59)$$

Eqs. (1.53) and (1.54) can also be substituted into Eq. (1.46) to nondimensionalize the Case 1 scenario. With these substitutions for Case 1, Eq. (1.46) can be expressed nondimensionally as:

$$\frac{d\Delta T_s}{\Delta T_c} = \frac{dt}{t_c} \qquad (1.60)$$

Eq. (1.60) can be integrated for Case 1, with appropriate limits to yield:

$$\frac{\Delta T_s}{\Delta T_c} = \frac{t}{t_c} \qquad (1.61)$$

The nondimensional solutions for the three thermally thin cases are illustrated in Fig. 1.6. Note that all three solutions have the same slope at the origin. This occurs because there are no heat losses from the surface until the material starts to heat up for Cases 2 and 3. Also, note that the asymptotic temperature rise for Case 3 is exactly one-half the asymptotic temperature rise for Case 2. This difference occurs because heat is being lost from both faces for Case 3 in comparison with only one face for Case 2.

The characteristic temperature rise and the characteristic time can be redefined for Case 3 to be more representative of the Case 3 scenario. For Case 3, the characteristic temperature rise and the characteristic time can be redefined as:

$$\Delta T_c \equiv \frac{\alpha \dot{q}_i''}{2h_t} \quad (1.62)$$

$$t_c = \frac{\rho c L}{2h_t} \quad (1.63)$$

When the characteristic temperature rise and the characteristic time are defined in this way for Case 3, Eq. (1.56) becomes the nondimensional solution for Case 3 as well as for Case 2. These definitions are preferable for Case 3 from the standpoint that the characteristic temperature rise now represents the asymptotic temperature rise of the material rather than twice the asymptotic temperature rise.

These analytical solutions only apply for situations where the imposed heat flux at the surface remains constant. They also are restricted by the assumption of a constant, total heat-transfer coefficient and constant material properties. For other boundary conditions, such as variable imposed heat fluxes or heat-transfer coefficients, or for variable temperature-dependent material properties, numerical solution of the governing energy equation is generally necessary. Despite the limitations of the analytical solutions, they are useful for elucidating the environmental and material parameters governing the heating and ignition of thermally thin solid materials. The environmental parameters include the imposed heat flux at the surface and the total heat-transfer coefficient, while the material properties include the material thermal absorptivity and the product of $\rho c \delta$, where δ is the characteristic thickness of the material. For planar materials, this product represents the mass per unit area, sometimes called the surface density, times the specific heat of the material; it is the heat capacity per unit area of the material.

1.4.1.2 Ignition of Thermally Thick Materials

A thermally thick material is a material that is sufficiently thick to be treated as a semi-infinite solid for purposes of heat-transfer analysis. A number of cases can be considered for thermally thick materials based on the boundary conditions at the surface of the material. For all cases, constant material properties are assumed. Case 1 considers a constant net heat flux at the surface, while Case 2 considers a constant incident heat flux with convective and reradiative cooling of the surface. Case 3 represents the asymptotic long-time solution for Case 2. As will be demonstrated, Case 1 represents the short-term solution for Case 2.

For a constant imposed heat flux at the surface of a semi-infinite solid, the change in surface temperature with time for Case 1 can be expressed in dimensional terms as:

$$\Delta T_s = \alpha \dot{q}_i'' \sqrt{\frac{4t}{\pi k \rho c}} \quad (1.64)$$

where $\alpha \dot{q}_i''$ represents the incident heat flux absorbed at the surface and the product $k\rho c$ is the *thermal inertia* of the material [$(kW/m^2 \cdot K)^2 \times s$]. For a given heat flux at the surface, the thermal inertia is the material property that governs the rate of surface temperature rise and consequently the time to ignition. The lower the thermal inertia is, the more quickly the surface of a material will heat up and ignite. For many solid materials, the conductivity of the material is approximately proportional to its density, so the thermal inertia of a material is a strong function of the bulk density of a material, with low-density materials heating up more quickly than high-density materials. Thus, the physical form of a material is an important factor in terms of the potential for ignition and flame spread.

For Case 2, the surface temperature rise above ambient with time can be expressed as:

$$\Delta T_s = \frac{\alpha \dot{q}_i''}{h_t} \left[1 - \exp\left(\frac{h_t^2 t}{k\rho c}\right) \cdot \mathrm{erfc}\left(\sqrt{\frac{h_t^2 t}{k\rho c}}\right) \right] \quad (1.65)$$

The thermally thick solutions can be nondimensionalized by defining an appropriate characteristic temperature rise and an appropriate characteristic time. The characteristic temperature rise is the same as for the thermally thin material:

$$\Delta T_c \equiv \frac{\alpha \dot{q}_i''}{h_t} \quad (1.53)$$

The characteristic time for the thermally thick cases is different from the thermally thin cases and can be expressed as:

$$t_c \equiv \frac{k\rho c}{h_t^2} \quad (1.66)$$

When these definitions are substituted into the dimensional equations, the nondimensional solution for the thermally thick Case 1 becomes:

$$\frac{\Delta T_s}{\Delta T_c} = \sqrt{\frac{4t}{\pi t_c}} \quad (1.67)$$

The nondimensional solution for the thermally thick Case 2 becomes:

$$\frac{\Delta T_s}{\Delta T_c} = [1 - \exp(t/t_c) \cdot erfc(\sqrt{t/t_c})] \quad (1.68)$$

The Case 2 solution involves the complementary error function, which is defined as:

$$erfc(x) = 1 - erf(x) = 1 - \frac{2}{\sqrt{\pi}} \int_0^x e^{-\eta^2} d\eta \quad (1.69)$$

Because the complementary error function does not permit a closed-form solution, series or tabulated solutions are typically employed for the solution of Eq. (1.68). As demonstrated by Long et al. [29], Taylor series expansions of Eq. (1.68) around the two limits of $t/t_c \to 0$ and $t/t_c \to \infty$ do yield closed-form solutions. The Case 1 solution, represented by Eq. (1.67), is the short-time limit solution for Eq. (1.68), while the long-term limit solution, called Case 3 here, can be represented nondimensionally as:

$$\frac{\Delta T_s}{\Delta T_c} = 1 - \left[\frac{\pi t}{t_c}\right]^{-1/2} \quad (1.70)$$

These nondimensional solutions for thermally thick solids are illustrated in Fig. 1.7. Note that as $t/t_c \to 0$, the Case 2 solution converges to the Case 1 solution, while as $t/t_c \to \infty$, the Case 2 solution converges to the Case 3 solution as expected from the Taylor series limit solutions. However, it is also significant to note that the limit solution for short times (Case 1) diverges from the exact solution (Case 2) by 10 percent by the time $t/t_c = 0.013$, and this difference continues to grow over time. Thus, the short-time solution (Case 1), represented by Eqs. (1.64) and (1.67), loses accuracy relatively quickly with time. Physically, this is due to the rapidly increasing convective and reradiative heat losses from the exposed surface as it heats up. The Case 1 solution does not consider these losses.

Similarly, the long-time solution (Case 3) is not within 10 percent of the exact solution (Case 2) until $t/t_c > 2.0$ and is not within 1 percent of the exact solution until $t/t_c > 10.0$. Thus, the closed form solutions are less than 90 percent accurate within the time range of $0.013 < t/t_c < 2.0$. Unfortunately, this is the period of interest for the ignition of many solid materials. The exact solution represented by Eq. (1.68) can be used, but it requires iterative solution.

The thermally thick solutions presented above are based on a constant imposed heat flux and on constant material properties. For scenarios with variable heat fluxes or variable temperature-dependent material properties, numerical solutions are generally necessary. Nonetheless, these idealized solutions permit the relevant variables influencing the heating and ignition of thermally thick solids to be identified. The thermal inertia, representing the product of $k\rho c$, is the primary ma-

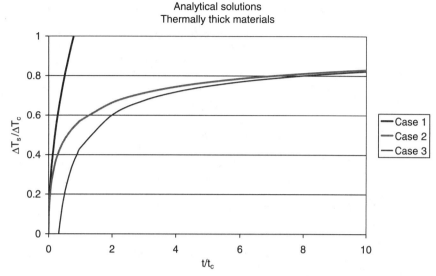

FIGURE 1.7 Analytical solutions for thermally thick solids exposed to constant incident heat fluxes.

terial property influencing the time to ignition, while the imposed heat flux is the primary environmental variable influencing the time to ignition of thermally thick solids.

1.4.2 Critical Heat Flux and Effective Ignition Temperature

The *critical heat flux* of a solid material is the minimum external heat flux that will cause evolution of sufficient vapors at the fuel surface to permit ignition of the material. Critical heat fluxes can be evaluated under conditions of both piloted and unpiloted ignition. Critical heat fluxes for piloted ignition tend to be lower than for unpiloted ignition. Some representative critical heat fluxes under piloted conditions are provided in Table 1.11. Little data is available for unpiloted ignition, but unpiloted critical heat fluxes for wood products tend to be two to three times higher than piloted critical heat fluxes for the same products. It is typically assumed that some type of ignition source is

TABLE 1.11 Critical Heat Fluxes for Selected Solid Materials Under Piloted Conditions [15]

Material	Piloted critical heat flux (kW/m^2)
Red oak (solid)	10
Douglas fir (solid)	10
PMMA (solid)	11
Nylon (solid)	15
Polyethylene (solid)	15
Polycarbonate (solid)	15
Polypropylene (solid)	15

likely to be present when a material is subjected to fire heat fluxes, so it is generally conservative to use piloted critical heat fluxes for fire hazard analyses.

From a practical standpoint, the potential for ignition at a particular external heat flux is evaluated for a fixed period, such as 15 or 20 min, depending on the test standard. Thus, reported critical heat fluxes may not represent the absolute minimum heat flux capable of igniting a surface under longer-term exposures.

Effective ignition temperatures are derived from critical heat flux data. A quasi-steady energy balance at the surface of a material can be expressed as:

$$\dot{q}''_{crit} = \left(\sum_i h_i\right) \cdot (T_{ig} - T_o) \quad (1.71)$$

where T_{ig} represents the effective ignition temperature and $\Sigma_i h_i$ represents the sum of all the relevant heat-transfer coefficients for convection, reradiation, and conduction appropriate for the boundary conditions for the geometry being evaluated. Eq. (1.71) can be inverted to evaluate the ignition temperature rise above ambient:

$$\Delta T_{ig} = \frac{\dot{q}''_{crit}}{\left(\sum_i h_i\right)} \quad (1.72)$$

For thermally thin materials, the heat-transfer coefficients will include convection and reradiation from the exposed face if the back face is insulated (Case 2) and from both the front and back face if the thermally thin material is exposed on both faces (Case 3). Thus, for these cases, the ignition temperature rise is the same as the respective characteristic temperature rise described earlier. Characteristic (ignition) temperature rise is illustrated as a function of imposed (critical) heat flux for the thermally thin cases in Fig. 1.8; these curves are based on a black surface ($\varepsilon = 1$) and an ambient temperature of 293 K. The characteristic temperature rise is the same as the ignition temperature rise for the thermally thin cases because there are no conduction losses within the material.

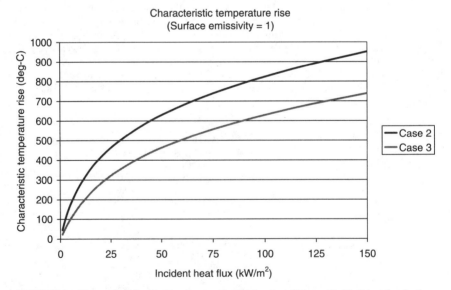

FIGURE 1.8 Characteristic (ignition) temperature rise as a function of imposed (critical) heat flux for thermally thin materials. The Case 2 solution also represents the upper limit solution for thermally thick materials with perfectly insulated back faces.

TABLE 1.12 Effective Ignition Temperatures for Selected Thermally Thick Materials Based on Negligible Conduction Losses [30]

Material	Effective ignition temperature (°C)
PMMA	278
Hardboard	298
Fiberboard, low density	330
Fiber insulation board	355
Douglas fir particleboard	382
Plywood, plain	390
Polyurethane foam, flexible	390
Polycarbonate (solid)	528

For thermally thick solids, the heat-transfer coefficients will include convection and reradiation from the exposed surface as well as conduction into the material. When conduction losses are negligibly small, the ignition temperature rise for a thermally thick material is the same as for the Case 2 thermally thin material. Thus, the characteristic temperature rise for Case 2 shown in Fig. 1.8 also represents the upper limit for the effective ignition temperature of a thermally thick solid.

Much of the ignition temperature data that has been reported in the literature for thermally thick materials (see, e.g., Ref. 30) assumes that the back surface of a material is perfectly insulated, such that the conduction losses into the material will become negligibly small for long exposure periods. Any errors in this assumption will tend to cause the estimation of ignition temperatures that are too high, with the magnitude of the error depending on the magnitude of the conduction losses relative to the convection and reradiation losses. If the magnitude of the conduction losses can be determined, then the actual ignition temperature can be estimated from the reported ignition temperature as:

$$\frac{\Delta T_{ig}}{\Delta T_c} = \frac{(h_c + h_r)}{(h_c + h_r + h_k)} = 1 - \frac{h_k}{\sum_i h_i} \quad (1.73)$$

For example, if conduction into the surface constitutes 10 percent of the total heat-transfer term, then the actual ignition temperature rise, ΔT_{ig}, would be 90 percent of the reported temperature rise ΔT_c, based on negligible conduction losses. With this caveat, representative ignition temperatures based on the negligible conduction loss assumption are provided in Table 1.12 [30,31].

1.4.3 Pyrolysis and Burning Rates

Once the surface of a solid material has reached its ignition temperature, the surface will begin to pyrolyze and burn. The mass pyrolysis rate of a solid is treated in a manner analogous to the steady mass-burning rate of a liquid, i.e.:

$$\dot{m}''_p = \frac{(\dot{q}''_{net})_p}{\Delta H_g} \quad (1.74)$$

where $(\dot{q}''_{net})_p$ represents the net heat flux at the pyrolysis front. For materials that do not char, the pyrolysis front will remain at the fuel surface, as illustrated in Fig. 1.3b, with the fuel surface receding as the material pyrolyzes. For materials that form a char layer, the pyrolysis front will penetrate into the interior of the material; the net heat flux at the pyrolysis front will be the difference between the rate of heat conduction through the char layer to the pyrolysis front and the rate of heat conduction through the virgin material from the pyrolysis front, as illustrated in Fig. 1.3a. Char layers that crack and form fissures will behave in a more complicated way, but the basic concepts will remain the same.

Once the mass pyrolysis rate is determined, the heat release rate per unit area associated with this pyrolysis rate can be calculated as:

$$\dot{Q}''_f = \dot{m}''_p \Delta H_c \chi_{\text{eff}} = \dot{q}''_{\text{net}} \chi_{\text{eff}} \frac{\Delta H_c}{\Delta H_g} \quad (1.75)$$

Eq. (1.75) shows the relevance of the combustibility ratio to the heat release rate of a solid material, just as it was demonstrated in Sec. 1.3.5.1 for liquid fuels. Combustibility ratios for selected solid fuels are provided in Table 1.9 along with those for selected liquid fuels.

1.4.4 Flame Spread

Flame spread on the surface of a solid material can be considered as a sequence of ignitions, with the rate of flame spread governed by how quickly new elements of the fuel surface are raised to the ignition temperature of the material by the heat flux imposed by the advancing flame and any external sources. Flame spread will not occur if a burning element does not burn long enough to cause the ignition of an adjacent element; localized burnout will occur under these conditions. In this respect, the potential for flame spread can be considered as a race between the ignition of new fuel elements and the burnout of elements that have already ignited and are burning.

Fundamentally, the rate of flame spread v_{fs} can be expressed as the rate at which the pyrolysis front advances along the surface of a material. In differential form, this can be expressed as [30]:

$$v_{fs} = \frac{dx_p}{dt} \quad (1.76)$$

Two modes of flame spread on solid materials are typically considered: *opposed flow* and *wind aided*, also known as *concurrent flow*. Opposed-flow flame spread refers to situations where the flame is spreading in the direction opposite the direction of airflow, while wind-aided flame spread refers to situations where the flame is spreading in the same direction as the airflow. The airflow direction is normally induced by the fire itself. Examples of opposed-flow flame spread include flame spread down or laterally on a vertical surface and radial flame spread on a floor. Examples of wind-aided flame spread include upward flame spread on a vertical surface and flame spread beneath a ceiling.

The hazards associated with wind-aided flame spread tend to be more severe than those associated with opposed-flow flame spread because larger areas of the material surface are exposed to heat fluxes intense enough to cause ignition. Because wind-aided flame spread generally represents the more severe hazard, analysis of wind-aided flame spread is presented here. Quintiere and Harkleroad [31] describe a practical test procedure for evaluating the parameters needed to describe opposed-flow flame spread on thick materials burning in air. This test procedure is embodied in ASTM Standard E1321 [28]. These references should be consulted for more information on opposed-flow flame spread.

A widely used model for wind-aided flame spread was first postulated by Quintiere [32] and has been modified and applied by many others [see, e.g., Refs. 33–36]. The physical basis for this model is illustrated in Fig. 1.9. The model can be applied to situations where burnout is significant as well as to situations where burnout is not significant. The case where burnout is not significant is considered first.

As conceived, the flame spread model assumes that a length of fuel, denoted as x_p, is pyrolyzing and burning with a known heat release rate per unit area, $\dot{Q}''$. The flame from this burning length of fuel extends beyond the pyrolysis front and imposes a steady and uniform heat flux, $\dot{q}''_f$, at the fuel

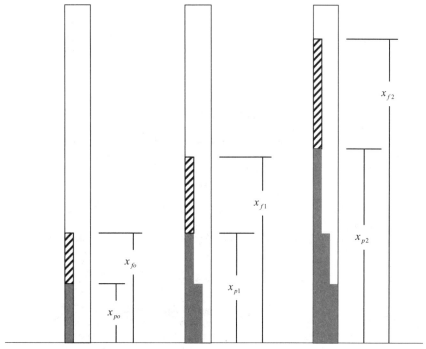

FIGURE 1.9 Conceptual illustration of the concurrent-flow flame spread process on a vertical surface.

surface over a length denoted as x_f, as illustrated in Fig. 1.9. This heat flux causes ignition of the region $(x_f - x_p)$ after a period denoted as t_{ig}, which will depend on the magnitude of the heat flux from the flame and the thermal properties of the fuel surface. The flame then propagates in stepwise fashion after every time interval of t_{ig}, with the previous flame length becoming the new pyrolysis length at each time step.

Based on this model, a material that does not burn out will always propagate a flame if (1) the flame length is longer than the pyrolysis length, i.e., $x_f > x_p$ and (2) the flame heat flux is greater than the critical heat flux for the material, i.e., $\dot{q}''_f > \dot{q}''_{crit}$. The flame length relative to the pyrolysis length depends on the heat release rate of the fuel. For wall fires, flame lengths have been correlated with the heat release rate per unit width of wall $\dot{Q}'$ as:

$$x_f = K_f [\dot{Q}']^n \tag{1.77}$$

where K_f is an appropriate flame length correlating coefficient and n is an appropriate power, generally with a value between $1/2$ and 1. Delichatsios et al. [37] suggest values of 0.052 m$^{5/3}$/kW$^{2/3}$ for K_f and $2/3$ for n. In order to determine when the flame length exceeds the pyrolysis length, it is necessary to convert the heat release rate per unit length of surface to the heat release rate per unit area of surface:

$$\dot{Q}' = \dot{Q}'' \cdot x_p \tag{1.78}$$

Eq. (1.78) is then substituted into Eq. (1.77) and both sides are divided by x_p to yield:

$$\frac{x_f}{x_p} = \frac{K_f[\dot{Q}'' \cdot x_p]^n}{x_p} = K_f[\dot{Q}'']^n x_p^{n-1} \tag{1.79}$$

If Eq. (1.79) evaluates to a value greater than unity, then the flame length will be greater than the pyrolysis length, and flame spread would be expected based on the model for a material that does not burn out. For example, if the values for K_f and n suggested by Delichatsios et al. are substituted into Eq. (1.79), then the flame length will exceed the pyrolysis length when:

$$\dot{Q}'' > 84.3\sqrt{x_p} \qquad (1.80)$$

where the units would be kW/m² for the heat release rate per unit area and m for the pyrolysis length. This relationship has the counterintuitive effect of requiring higher heat release rates per unit area for longer pyrolysis lengths, suggesting that a fire may grow to a point and then stop growing because the flame no longer extends beyond the pyrolysis zone.

Cleary and Quintiere [33] have suggested that the flame length correlation expressed by Eq. (1.77) can be linearized with a correlating coefficient of $K_f = 0.01$ m²/kW and a power of $n = 1$. With this approximation, the flame length is directly proportional to the pyrolysis length, i.e., $x_f = K_f \dot{Q}'' x_p$. This relationship can then be substituted into Eq. (1.76) to yield:

$$v_{fs} = \frac{dx_p}{dt} \approx \frac{x_f - x_p}{t_{ig}} = \frac{x_p(K_f \dot{Q}'' - 1)}{t_{ig}} \qquad (1.81)$$

Eq. (1.81) can be integrated with appropriate limits to yield an expression for the flame spread rate:

$$\frac{x_p}{x_{po}} = \exp\left(\frac{(K_f \dot{Q}'' - 1)t}{t_{ig}}\right) \qquad (1.82)$$

Eq. (1.82) suggests that wind-aided flame spread will accelerate exponentially provided that $K_f \dot{Q}'' > 1$ and will decay to extinction otherwise. For the value of $K_f = 0.01$ m²/kW suggested by Cleary and Quintiere, this would indicate that a heat release rate per unit area of at least 100 kW/m² would be needed for flame spread to occur.

For materials where burnout is significant, an additional criterion for flame spread exists. The burning duration of the material must exceed the ignition time, i.e., $t_b > t_{ig}$. Mowrer [38] applied this concept to painted gypsum wallboard and deduced that a critical heat flux exists for materials where burnout is significant. The burning duration of a material with finite combustible thickness can be approximated as:

$$t_b \approx \frac{Q''}{\dot{Q}''} = \frac{m'' \Delta H_c \chi_{eff}}{\dot{m}'' \Delta H_c \chi_{eff}} = \frac{\rho \delta}{\dot{q}''_{net,b} / \Delta H_g} \qquad (1.83)$$

where $\rho \delta$ is the combustible mass per unit area, ΔH_g is the effective heat of gasification associated with this combustible mass and $\dot{q}''_{net,b}$ is the net heat flux at the fuel surface associated with the burning material. In general, this net heat flux is the difference between the incident heat fluxes being imposed on the surface and the convective, reradiative and conduction losses from the surface.

The time to ignition of a thermally thick material exposed to a constant, net incident heat flux at the surface can be approximated as:

$$t_{ig} \approx \frac{\pi}{4} k\rho c \left[\frac{\Delta T_{ig}}{\dot{q}''_{net,ig}}\right]^2 \qquad (1.84)$$

The net heat flux associated with Eq. (1.84) is slightly different from that associated with Eq. (1.83) because it incorporates conduction losses directly. If it is assumed that the net heat flux terms in Eqs. (1.83) and (1.84) are proportional to each other, i.e., $\dot{q}''_{net,b} = X\dot{q}''_{net,ig}$, then the ratio between the burning duration and the ignition time can be expressed as:

$$\frac{t_b}{t_{ig}} \approx \frac{\rho \delta \Delta H_g / \dot{q}''_{net,b}}{\frac{\pi}{4} k\rho c \left[\frac{\Delta T_{ig}}{\dot{q}''_{net,ig}}\right]^2} = \left(\frac{4\rho \delta \Delta H_g}{\pi k\rho c \Delta T_{ig}^2}\right) \left(\frac{\dot{q}''_{net,ig}}{\chi}\right) \qquad (1.85)$$

For flame spread to occur according to this model, Eq. (1.85) must evaluate to a value greater than unity, suggesting that the net heat flux for ignition must be:

$$\dot{q}''_{net,ig} > \chi\left(\frac{\pi k\rho c\Delta T_{ig}^2}{4\rho\delta\Delta H_g}\right) \tag{1.86}$$

Because of the simplifications and assumptions associated with this derivation, it is not likely to yield accurate quantitative results. The main purpose of this discussion has been to demonstrate the expected existence of a critical heat flux for flame spread on materials where burnout is significant and the parameters influencing the magnitude of this critical heat flux. This relationship shows that this critical heat flux for flame spread is expected to increase with the thermal inertia of the material and to decrease as the combustible mass per unit area increases. For example, as additional layers of paint are added to a surface, the critical heat flux for flame spread will decrease and the potential for flame spread will increase.

The larger implication of this discussion is that certain thin materials may exhibit desirable behavior in some fire tests by burning out, but exhibit undesirable behavior in the field due to differences in the imposed heat fluxes between the fire test and the field conditions.

1.4.5 Self-Heating and Smoldering Combustion

Sec. 1.4 has described hazards associated with the flaming ignition and burning of solid materials. Some solid materials also have the potential for self-heating to ignition or the propensity to smolder, which represent different, but no less important aspects of the fire hazards of solid materials.

Analysis of the self-heating of solid materials is analogous to the theory of thermal explosion used to analyze the autoignition temperature of gas mixtures. Gray [39] presents a discussion of the Semenov [20] and Frank-Kamenetskii [21] theories of thermal ignition applied to self-heating to ignition of solid materials. The potential for self-heating to ignition of a solid, sometimes called *spontaneous combustion,* involves a balance between the rate of heat generation within the solid material and the rate of heat losses from the surface of the material.

The phenomenon of self-heating to ignition is commonly associated with relatively large piles of porous materials, which can favor the internal generation of heat while restricting heat losses from the surface. Some materials have a recognized propensity for self-heating, while others do not. Many of the materials with a recognized tendency for self-heating based on experience are identified in the NFPA *Fire Protection Handbook* [40], but, as noted in the NFPA Handbook, the absence of a material from the list does not mean it cannot self-heat.

Materials with a propensity for self-heating will only self-heat to ignition under certain conditions of restricted heat loss. A classic example of a material that can self-heat to ignition is linseed oil dispersed on a cotton rag. If such rags are crumpled up and thrown in a corner, thus restricting heat losses, there is a relatively high likelihood that they will self-heat to ignition. If the same rags are hung on a clothesline to dry, there is virtually no chance that they will self-heat to ignition. Another example is low-density wood fiberboard, which will not typically self-heat to ignition when stacked in small quantities at room temperature, but has been known to self-heat to ignition when stored in large quantities (e.g., railroad boxcar) at moderately elevated ambient temperatures.

Drysdale [2] notes that there appear to be two main factors associated with the potential for a bulk solid to self-heat to ignition: (1) the material must be sufficiently porous to allow oxygen to permeate throughout the mass and (2) the material must yield a rigid char when undergoing thermal decomposition. One of the distinguishing features of a material that has self-heated to ignition is evidence of burning at the core of the material rather than at the surface. In some cases, evidence of such internal burning is destroyed by the subsequent fire initiated by the self-heating, but in many cases such evidence of internal heating remains after the fire is extinguished and may be used by investigators to help identify the cause of the fire as self-heating.

Materials that have self-heated to ignition will generally smolder within the core of the porous material. Smoldering is an exothermic combustion reaction that occurs between a solid fuel and a gaseous oxidizer at the surface of the fuel. The smoldering reaction will propagate outward toward the surface of the material at a rate governed by the rate of oxygen diffusion into the material, which in turn will depend on the permeability of the material. Smolder velocities on the order of approximately 10^{-4} m/s are typical, so the heat release rate associated with smoldering combustion is generally many orders of magnitude lower than that associated with flaming fires. But, as noted by Ohlemiller [41], smoldering can still constitute a serious fire hazard for two reasons: (1) smoldering provides a pathway to flaming combustion that can be initiated by heat sources too weak to initiate flaming combustion directly, and (2) smoldering typically yields a substantially higher conversion of a fuel to toxic products such as carbon monoxide than a well-ventilated flame does. For these reasons, the potential for a material to smolder should be addressed.

Materials susceptible to smoldering combustion tend to be porous materials with high surface area to volume ratios and pathways for oxygen permeation. At the same time, these porous materials tend to be relatively good insulators, trapping heat near the reaction zone. In this respect, materials with a tendency to smolder are similar to materials subject to self-heating to ignition. Many materials with the potential for self-heating to ignition also demonstrate a propensity to smolder. Ohlemiller [41] provides a qualitative review of the practical aspects of smoldering combustion as well as a comprehensive bibliography of the relevant literature.

1.4.6 Effects of Fire Retardants

Fire retardants are sometimes added to solid materials to change their flammability characteristics. Fire retardants may simply be inert inorganic filler materials, they may be hydrated materials, or they may be chemical reactive materials, such as halogenated compounds. The influence of these fire retardants can be considered qualitatively within the context of the discussion presented above. Further details on the effects of various fire retardants are presented in subsequent chapters.

Inert inorganic filler materials simply serve to replace organic materials within a solid composite material. These inorganic materials absorb heat along with the organic materials, thus contributing to the effective heat of gasification of the composite material, but they do not contribute to the effective heat of combustion of the composite material. The net effect of this will be a decrease in the combustibility ratio of the composite material and a consequent decrease in the pyrolysis rate. Some inorganic filler materials may also promote char formation, further reducing the pyrolysis rate of the composite material.

Hydrated filler materials, such as alumina trihydrate ($Al_2O_3 \cdot 3H_2O$), serve much the same function of inert filler materials from the standpoint that they contribute to the effective heat of gasification without contributing to the effective heat of combustion of the composite material. Hydrated filler materials are particularly effective as heat absorbers due to the high heat of vaporization of the water of hydration of the bound water molecules in the hydrated materials. When this water of hydration is vaporized along with the organic pyrolyzates as the composite material is heated, the water vapor molecules act to dilute the concentration of the fuel vapors at the fuel surface, thus decreasing the flammability of these vapors. Thus, hydrated materials influence flammability in both the solid and vapor phases.

The performance of fire retardants that act through chemical reaction is more complex than that of the simple or hydrated inorganic fillers. Some chemical fire retardants act in the solid phase, promoting char formation, while others act in the gas phase, inhibiting combustion of the flammable gases released by the solid. As noted by Friedman [26], however, such fire-retarded materials can still burn vigorously in a fully developed fire.

1.5 SMOKE PRODUCTION

Smoke can be defined [42] as the airborne solid and liquid particulates and gases evolved when a material undergoes pyrolysis or combustion, together with the quantity of air that is entrained or otherwise mixed into the mass of combustion products. Based on this definition, smoke is not only the actual product of combustion, but also includes the volume of air contaminated by the combustion products. As more air is mixed into the smoke, e.g., due to entrainment in a fire plume, the volume of smoke increases while the concentrations of combustion products decrease. Smoke production represents a significant aspect of the fire hazards of materials for several reasons, including (1) vision obscuration, (2) toxicity of combustion products, and (3) nonthermal damage to structures, equipment, and stored commodities.

1.5.1 Vision Obscuration

Smoke tends to obscure vision at concentrations much lower than those required to cause toxic or incapacitating effects. For building occupants, the obscuration caused by smoke is significant primarily in terms of its effects on visibility and wayfinding. People immersed in a smoke cloud may become disoriented and unable to see exit signs or other visual cues to help them evacuate a building. Evacuation speeds tend to decrease as smoke concentrations increase, causing occupants to be immersed in a smoke cloud for longer periods. For firefighters, the obscuration caused by smoke can make it difficult for them to find the source of a fire as well as cause disorientation if they lose their way within a building.

Smoke obscuration is typically considered in terms of Bouguer's law:

$$\frac{I_\lambda}{I_\lambda^o} = \exp(-\sigma_s \rho Y_s L) = \exp(-kL) \tag{1.87}$$

where I_λ is the light intensity transmitted through path length L, I_λ^o is the light intensity at the source, σ_s is the mass specific extinction coefficient (m²/g$_s$), ρY_s is the mass concentration of smoke (g$_s$/m³), and L is the path length through the smoke (m).

The product $\sigma_s \rho Y_s$ is normally referred to as the extinction coefficient k. As noted by Mulholland and Croarkin [43], the general utility of this approach is based on the hypothesis that σ_s is nearly universal for postflame smoke production from overventilated fires. As demonstrated by Mulholland and Croarkin, the value of σ_s is constant for a wide range of fuels, with a value of 8.7 ± 1.1 (95 percent confidence interval) m²/g$_s$. From this, smoke mass concentrations can be inferred from light extinction measurements, a useful experimental consequence. A further consequence is that smoke yield factors can be deduced from fuel mass loss rates made in conjunction with light extinction measurements. Previously, smoke yield factors were determined from filter collection and gravimetric measurements, a much more tedious process that could only produce integrated average values, not real-time transient results.

While useful and theoretically accurate, this fundamental approach requires the determination and knowledge of a smoke yield factor. An alternative approach that has been used more widely in engineering applications is the mass optical density, which is related to the specific extinction coefficient. The mass optical density is defined as:

$$D_m = \frac{DV}{m_f} \tag{1.88}$$

where D_m is the mass optical density (m²/g$_f$), which represents the obscuration potential per unit mass of fuel released, D is the optical density of the smoke cloud (m^{-1}), V is the volume into which the smoke is dispersed (m³), and m_f is the mass of fuel released (g). Some representative values for mass optical densities of selected fuels are provided in Table 1.13. Eq. (1.88) can be rearranged to solve for the optical density of a cloud of smoke in terms of the mass optical density and the equivalent mass fraction of fuel in the cloud:

$$D = \frac{D_m m_f}{V} = D_m \rho Y_f \tag{1.89}$$

where Y_f is the equivalent mass fraction of fuel in the mixture ($\equiv \phi/\phi + r \approx \phi/r$).

TABLE 1.13 Mass Optical Densities for Selected Fuels Under Small-Scale Well-Ventilated Flaming and Pyrolysis Conditions [15, 26, 45]

Material	$D_m(m^2/g_f)$ Flaming	$D_m(m^2/g_f)$ Pyrolysis
Plywood	0.017	0.29
Douglas fir	—	0.28
PMMA	0.07–0.11	0.15
Nylon	0.23	
Polyethylene	0.29	
Polypropylene	0.53	

The optical density is the base 10 equivalent of the extinction coefficient; it is related to obscuration as:

$$\frac{I}{I_o} = 10^{-D_m \rho Y_f L} = 10^{-DL} \tag{1.90}$$

The extinction coefficient k is related to the optical density D as $k = 2.303D$. Consequently, the yield of smoke can be expressed as the ratio between the smoke mass fraction and the fuel mass fraction:

$$f_s = \frac{Y_s}{Y_f} = \frac{2.303 D_m}{\sigma_s} \tag{1.91}$$

To the extent that the specific extinction coefficient does have a relatively constant value independent of the fuel source, this relationship can be used to estimate the smoke mass yield factor based on widely reported mass optical density data. For example, many hydrocarbon-based fuels have mass optical densities of approximately 0.3 m²/g. Assuming a value for σ_s of 8.7 m²/g as suggested by Mulholland and Croakin would result in a calculated smoke mass yield factor of approximately 7.9 percent for these fuels.

Finally, yet another method for characterizing the smoke production potential of different fuels has been used, particularly in conjunction with the cone calorimeter [44]. This method makes use of the specific extinction area σ_a, which is the Naperian base equivalent of the mass optical density:

$$\sigma_a = \frac{kV}{m_f} = 2.303 D_m \tag{1.92}$$

$$k = \sigma_a \rho Y_f \tag{1.93}$$

$$f_s = \frac{Y_s}{Y_f} = \frac{\sigma_s}{\sigma_s} \tag{1.94}$$

Keeping all these different parameters straight requires considerable effort because of the similar units for the specific extinction coefficient (m²/g_s), the mass optical density (m²/g_f), and the specific extinction area (m²/g_f).

Once the optical density of a volume of smoke is determined, one of several calculations can be used to estimate visibility through the smoke. The simplest approach is to simply suggest that the visibility distance L_{vis} varies inversely with the optical density of a smoke cloud:

$$L_{vis} = \frac{C}{D} \tag{1.95}$$

The constant C depends on the lighting conditions; Mulholland [45] quotes an extensive study by Jin [46] and indicates that the visibility of light-emitting signs was found to be two to four times greater than the visibility of light-reflecting signs. Mulholland correlates Jin's data as $kS = 8$ ($DS = 3.5$) for light-emitting signs and $kS = 3$ ($DS = 1.3$) for light-reflecting signs, where k is the extinction coefficient of the smoke and S is the vision distance through the smoke. Jin's data were obtained with human subjects viewing smoke through glass, eliminating irritant effects of smoke as a factor.

For scenarios where people are expected to be immersed in smoke, the ability to see through the smoke is a function of the irritancy of the smoke as well as its optical properties. Many types of smoke can be highly irritating at relatively low concentrations. This irritancy can cause eyes to water, reducing vision, or even cause people to keep their eyes shut, effectively blinding them. Jin and Yamada [47] studied these effects by subjecting people to wood smoke with and without goggles to protect their eyes from the irritant effects of the smoke. As noted by Mulholland [45], they found that visual acuity without goggles decreased markedly for smoke with an extinction coefficient greater than 0.25 m^{-1}.

1.5.2 Toxicity of Combustion Products

The combustion products produced by fires can cause injury or death if people are subjected to high enough concentrations for sufficient periods. The product of the concentration by the exposure duration is known as the *dose*. In general, there is an inverse relationship between the concentration and the exposure period required to cause incapacitation or death. The dose required to cause incapacitation or death depends on the toxicity of the individual components of a smoke cloud, on potential synergistic effects between components, as well as on the sensitivity of the victim.

Carbon monoxide is recognized as the primary toxicant produced in building fires. Because of the dominant role played by carbon monoxide in building fire–related deaths, issues related to its production in building fires are discussed in the next subsection. Purser [48] provides a comprehensive review of the current state of knowledge regarding the toxicity of different combustion products, including carbon monoxide and many others.

1.5.2.1 Carbon Monoxide

Carbon monoxide is a product of incomplete combustion of carbonaceous fuels. It is produced in large quantities in some building fires primarily due to an imbalance between the production of combustible vapors and the availability of oxygen to completely burn the fuel. The production and subsequent transport of carbon monoxide during building fires depend on several variables. Key variables included the combustion mode, whether smoldering or flaming, and ventilation, particularly for flaming fires. Other factors include temperature and suppression effects as well as fuel type and geometry.

The concentration of carbon monoxide in the environment depends on its rate of production in the fire. The rate of carbon monoxide generated in building fires is typically addressed in terms of two factors: a yield factor and a fuel mass loss rate. This is represented as:

$$\dot{m}_{CO} = f_{CO} \cdot \dot{m}_f \tag{1.96}$$

In flaming fires, the yield of carbon monoxide has been observed to be a function of several variables, including the fire ventilation, represented in terms of a global equivalence ratio, the temperature of the smoke, and the mixing of fresh air into the smoke [49]. For engineering purposes, Gottuk and Lattimer [50] suggest that the yield of carbon monoxide in flaming fires can be expressed as a function of the global equivalence ratio for a compartment:

$$f_{CO} = \begin{matrix} 0 \\ 0.2\phi - 0.1 \\ 0.2 \end{matrix} \quad \text{for} \quad \begin{matrix} \phi < 0.5 \\ 0.5 \leq \phi \leq 1.5 \\ \phi > 1.5 \end{matrix} \tag{1.97}$$

This relationship suggests that the yield of carbon monoxide will be insignificant if an enclosure is provided with twice the ventilation required for complete combustion of the fuel being released.

It also suggests that the yield of carbon monoxide will increase linearly with the global equivalence ratio up to a maximum yield of 0.2 g CO per gram of fuel released when the global equivalence ratio reaches a value of 1.5 and remains at this level for further increases in the global equivalence ratio. This does not mean that the production of CO will remain constant, but that the yield of CO per unit mass of fuel released will remain approximately constant. Because the equivalence ratio is related directly to the fuel release, this means that the quantity of CO produced will continue to increase as the fire becomes further underventilated, i.e., as the equivalence ratio continues to increase.

The underventilation of flaming fires in enclosures is typically associated with the phenomenon of flashover. Flashover is a transitional phase of enclosure fires during which the intensity of a fire escalates from burning one or a few objects within the enclosure to involvement of virtually all exposed combustible surfaces. As a consequence of this escalation in the burning rate, enclosure fires typically transition from overventilated to underventilated because of flashover. Because the production of carbon monoxide will increase by orders of magnitude due to this transition, the prevention of flashover is frequently one of the primary objectives of an effective fire-protection strategy. Enclosure effects are addressed in more detail in Sec. 1.6.

1.5.3 Nonthermal Damage

The term *nonthermal damage* is used to distinguish damage caused by smoke contamination from damage caused by direct exposure to the heat released by a fire. Nonthermal damage includes the effects of discoloration, odor, corrosion, changes in electrical conductivity, and other effects caused by the deposition of smoke on solid surfaces. In many fires, it is quite common for the monetary losses associated with nonthermal damage to be many times greater than those associated with direct thermal damage. This is particularly true in buildings housing sensitive electronic equipment, such as telecommunications centers or clean rooms, or in buildings where sensitive products are being stored, such as pharmaceuticals or foodstuffs. Discussion of the full range of nonthermal damage that can be caused by fire is beyond the scope of this chapter. Reference 15 should be consulted for more information on nonthermal damage.

1.6 ENCLOSURE EFFECTS

During the very early stages of a building fire, the fire behaves much as it would outside. There is little if any interaction between the fire and its surroundings. This situation changes rapidly, however, as the fire and smoke begin to interact with the enclosure shortly after ignition. Enclosure fires can go through a sequence of four stages, as illustrated schematically in Fig. 1.10, which Mowrer [51] has identified as: (1) fire plume/ceiling jet period, (2) enclosure smoke-filling period, (3) preflashover ventilated period, and (4) postflashover ventilated period. Not all enclosure fires go through all stages. In fact, successful fire protection strategies will typically prevent flashover, which represents the transition from the third to the fourth stage.

During the first stage, the hot gases generated by a fire rise in a buoyant plume, entraining fresh air as they rise to the ceiling. When the plume impinges on the ceiling, the hot buoyant gases can rise no further, so they begin to spread out radially beneath the ceiling. Once this ceiling jet is confined by the walls of the enclosure, the layer of smoke that is forming beneath the ceiling will begin to bank down from the ceiling and start to fill the room with smoke. This begins the second stage of enclosure fires.

As this smoke layer banks further down from the ceiling, the fire plume entrains less fresh air as it rises to the ceiling. Instead, the plume begins to recirculate the smoke that is already present above the smoke layer interface while injecting new smoke into the smoke layer from below. As less and less fresh air is entrained into the plume, the temperature of the smoke layer increases along with the concentrations of combustion products within the smoke layer. In a closed room, the second stage of enclosure fires will continue until the entire room is filled with smoke. Eventually, such a fire is likely to smother itself as it uses up the available oxygen within the room, much as a candle within an inverted jar will go out. The fire may not go completely out, but it is likely to burn at a greatly

Stage 1. Fire plume/ceiling jet period

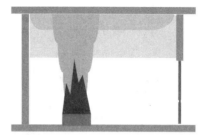
Stage 2. Enclosure smoke filling period

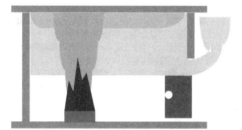
Stage 3. Preflashover vented period

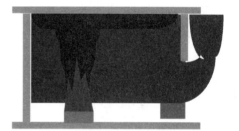
Stage 4. Postflashover vented period

FIGURE 1.10 Schematic illustration of the stages of enclosure fires.

reduced rate until oxygen is reintroduced into the room, e.g., through a broken window or an opened door. At this point, the fire may rapidly reinvigorate. Under some circumstances, backdraft may occur [52].

If there are door or window openings between the fire enclosure and adjacent spaces, smoke will begin to flow from the fire enclosure to these adjacent spaces once the smoke layer descends to the top of these vents. Hydrostatic pressure differences arise as a result of the temperature difference between the hot gases in the fire room and the cool gases in the adjacent spaces. These pressure differences induce the flow of hot gases from the fire enclosure to the adjacent spaces and the flow of cool air from the adjacent spaces into the fire room. This is the third stage of an enclosure fire.

The rate of airflow into the ventilated enclosure will be governed by the size, shape, and locations of openings between the fire enclosure and adjacent spaces as well as by any mechanical ventilation of the enclosure. A balance will develop between the rate of air inflow to the enclosure and the rate of smoke outflow from the enclosure. The smoke layer will descend to the elevation where this balance between air inflow and smoke outflow occurs.

As the fire continues to develop, conditions within the enclosure will continue to change. The temperature of the smoke layer will continue to increase along with the temperature of the walls and ceiling in contact with the smoke layer. As these temperatures continue to increase, the hot smoke layer and the walls and ceiling will start to radiate significant heat fluxes back to objects located within the room. Eventually, this heat flux can reach a level where objects in the lower layer begin to pyrolyze and spontaneously burst into flame. This transition is known as flashover and it begins the fourth and final stage of enclosure fires, the postflashover stage.

In relatively small enclosures typical of residential and office rooms, flashover has been observed to occur when the smoke layer reaches a temperature of approximately 600°C. When the smoke layer reaches this temperature, it imposes a radiant heat flux at floor level of approximately 25 kW/m^2, which is sufficient to ignite a range of ordinary combustibles in a relatively short period. Consequently, a smoke layer temperature of 600°C or a heat flux at floor level of 25 kW/m^2 are widely used to indicate the likelihood that flashover will occur.

With virtually every combustible fuel surface ignited during the transition to flashover, the rate of burning within the fire enclosure typically becomes ventilation limited during the postflashover stage. This means that the rate of heat release within the enclosure is governed by the rate at which air can flow into the enclosure, not by the rate at which the fuel surfaces are pyrolyzing. Because insufficient air is entering the enclosure to burn all the fuel that is being pyrolyzed, flames typically extend out the ventilation openings during the postflashover stage. Because the postflashover fire is underventilated (i.e., the global equivalence ratio is greater than unity) and the fuel release rate is high, substantial quantities of carbon monoxide and other products of incomplete combustion are typically generated in postflashover fires. This carbon monoxide can then travel throughout a building, threatening occupants far from the actual fire enclosure.

A number of enclosure fire models have been developed to simulate the complex interactions of heat and mass transfer that occur during enclosure fires. Zone models divide the fire enclosure, and in multiroom models the adjacent spaces as well, into two primary control volumes, the smoke layer and the lower layer, then apply conservation equations for mass, species, and energy to each control volume. Different submodels are used to describe the fire and the transfer of heat and mass across the boundaries of these control volumes. Field or CFD models divide the computational domain into thousands of much smaller control volumes, then solve the conservation equations for mass, species, energy, and momentum for each cell. Quintiere [53] describes the bases for zone models, while Cox and Kumar [54] describe the bases for CFD models.

1.7 SUMMARY

In this chapter, the primary material parameters and environmental variables influencing the fire hazards of materials have been introduced and discussed. Considerable progress has been made over the past three decades to more quantitatively evaluate the fire hazards of materials and to model the consequences of fires in buildings. It is now possible to derive effective material flammability properties from bench-scale flammability tests and to use this data in enclosure fire models to evaluate the growth and development of fires in enclosures and the spread of heat and smoke throughout buildings. This progress has led to the emergence of performance-based analysis, design, and regulation of building fire safety as a practical approach.

While considerable progress has been made in evaluating the fire hazards of materials, continued research is needed to advance the state of the art and the knowledge base. The large number of variables influencing the fire hazards of materials, including enclosure effects, complicates evaluation of these hazards. The problem is further compounded by the critical nature of fire spread, where small perturbations in input conditions can have large consequences on outcomes. Consequently, while the analysis methods presented here and in subsequent chapters allow better estimation of the fire hazards of materials, extreme care must be exercised in their application. In particular, users must recognize when a calculation is producing results near a critical limit, where a small variation associated with uncertainties in input parameters might yield a large difference in outcome.

This chapter has introduced many of the fundamental aspects of the fire hazards of materials in terms of both chemical and physical properties. In subsequent chapters, many of the topics introduced in this chapter are expanded and applied to specific materials, products, and applications.

REFERENCES

1. Strehlow, R.A., *Combustion Fundamentals,* McGraw-Hill, Inc., New York, 1984.
2. Drysdale, D., *An Introduction to Fire Dynamics,* 2nd edition, John Wiley and Sons, Chichester, 1999.
3. Babrauskas, V., Tables and Charts, Appendix A, *Fire Protection Handbook,* 18th edition, National Fire Protection Association, Quincy, MA, 1997.

4. McGrattan, K.B., Baum H.R., Rehm, R.G., Hamins, A., Forney, G.P., Floyd, J.E., and Hostikka, S., Fire Dynamics Simulator (Version 2)—Technical Reference Guide, *NISTIR 6783,* National Institute of Standards and Technology, Gaithersburg, MD, November 2001.
5. Drysdale, D.D., "Thermochemistry," *SFPE Handbook of Fire Protection Engineering,* 3rd edition, National Fire Protection Association, Quincy, MA, 2002.
6. Glassman, I., *Combustion,* Academic Press, Inc., New York, 1977.
7. Chase Jr., M.W., Davies C.A., Davies Jr., J.R., Fulrip, D.J., McDonald, R.A., and Syverud, A.N., *JANAF Thermochemical Tables,* 3rd edition, *Journal of Physical and Chemical Reference Data,* 14 (Supplement 1), 1985.
8. NIST Standard Reference Database Number 69, National Institute of Standards and Technology, Gaithersburg, MD, July 2001.
9. Thornton, W.M., "Relation of Oxygen to the Heat of Combustion of Organic Compounds," *Philosophical Magazine and Journal of Science,* 33: 196–203, 1917.
10. Huggett, C., "Estimation of Rate of Heat Release by Means of Oxygen Consumption Measurements," *Fire and Materials,* 4 (2): 61–65, June 1980.
11. Janssens, M., "Calorimetry," *SFPE Handbook of Fire Protection Engineering,* 3rd edition, National Fire Protection Association, Quincy, MA, 2002.
12. STANJAN reference.
13. Kee, R.J., Rupley, F.M., Meeks, E., and Miller, J.A., CHEMKIN-III: A Fortran Chemical Kinetics Package for the Analysis of Gas Phase Chemical and Plasma Kinetics, Report No. UC-405/SAND96-8216, Sandia National Laboratories, Livermore, CA, 1996.
14. Markstein, G.H., "Relationship between Smoke Point and Radiant Emission from Buoyant Turbulent and Laminar Diffusion Flames," *Twentieth Symposium (International) on Combustion,* The Combustion Institute, Pittsburgh, PA, 1985, pp. 1055–1061.
15. Tewarson, A., "Generation of Heat and Chemical Compounds in Fires," *SFPE Handbook of Fire Protection Engineering,* 3rd edition, National Fire Protection Association, Quincy, MA, 2002.
16. Zabetakis, M.G., "Flammability Characteristics of Combustible Gases and Vapors," *Bulletin 627,* Bureau of Mines, U.S. Department of the Interior, Washington, DC, 1965.
17. Beyler, C.L., "Flammability Limits of Premixed and Diffusion Flames," *SFPE Handbook of Fire Protection Engineering,* 3rd edition, National Fire Protection Association, Quincy, MA, 2002.
18. National Fire Protection Association, *Fire Protection Guide to Hazardous Materials,* National Fire Protection Association, Quincy, MA, 2001.
19. Setchkin, N.P., "Self-ignition temperatures of combustible liquids," *Journal of Research, National Bureau of Standards,* 53: 49–66, 1954.
20. Semenov, N.N., "Theories of combustion processes," *Zeitschrift für Physikalische Chemie,* 48: 571–582, 1928.
21. Frank-Kamenetskii, D.A., "Temperature distribution in reaction vessel and stationary theory of thermal explosion," *Journal of Physical Chemistry (USSR),* 13: 738–755, 1939.
22. Schwab, R.F., "Dusts," *Fire Protection Handbook,* 18th edition, National Fire Protection Association, Quincy, MA, 1997.
23. Lide, R.A., *CRC Handbook of Chemistry and Physics,* 83rd edition, CRC Press, Boca Raton, FL, 2002.
24. Mowrer, F.W., Milke, J.A., and Clarke, R.M., "Heavy Truck Fuel System Fire Safety Study," DOT HS 807 484, U.S. Department of Transportation, National Highway Traffic Safety Administration, September 1989.
25. Babrauskas, V., "Burning Rates," *SFPE Handbook of Fire Protection Engineering,* 3rd edition, National Fire Protection Association, Quincy, MA, 2002.
26. Friedman, R., *Principles of Fire Protection Chemistry and Physics,* 3rd edition, National Fire Protection Association, Quincy, MA, 1998.
27. Kreith, F.A., and Bohn, M.S., *Principles of Heat Transfer,* 6th edition, Brooks/Cole Publishing Co., 2000.
28. ASTM E1321-97a, *Standard Test Method for Determining Material Ignition and Flame Spread Properties,* ASTM International, West Conshohocken, PA, 2002.
29. Long Jr., R.T., Torero, J.L., Quintiere, J.G., and Fernandez-Pello, A.C., "Scale and Transport Considerations on Piloted Ignition of PMMA," *Fire Safety Science—Proceedings of the 6th International Symposium,* International Association for Fire Safety Science, 2000.

30. Quintiere, J.G., "Surface Flame Spread," *SFPE Handbook of Fire Protection Engineering*, 3rd edition, National Fire Protection Association, Quincy, MA, 2002.
31. Quintiere, J.G., and Harkleroad, M., "New Concepts for Measuring Flame Spread Properties," *Fire Safety: Science and Engineering, ASTM STP 882*, American Society for Testing and Materials, West Conshohocken, PA, 1985.
32. Quintiere, J.G., "A Simulation Model for Fire Growth on Materials Subject to a Room-Corner Test," *Fire Safety Journal*, 20: 313–339, 1993.
33. Cleary, T.G., and Quintiere, J.G., "A Framework for Utilizing Fire Property Tests," *Fire Safety Science, Proceedings of the 3rd International Symposium*, Elsevier Applied Science, London, 1991.
34. Mowrer, F.W., and Williamson, R.B., "Flame Spread Evaluation for Thin Interior Finish Materials," *Fire Safety Science, Proceedings of the 3rd International Symposium*, Elsevier Applied Science, London, 1991.
35. Grant, G.B., and Drysdale, D.D., "Numerical Modeling of Early Fire Spread in Warehouse Fires," *Fire Safety Journal*, 24: 247–278, 1995.
36. McGraw Jr., J.R., and Mowrer, F.W., "Flammability and Dehydration of Painted Gypsum Wallboard Subjected to Fire Heat Fluxes," *Fire Safety Science—Proceedings of the 6th International Symposium*, International Association for Fire Safety Science, 2000, pp. 1003–1014.
37. Delichatsios, M.A., Panagiotou, T., and Kiley, F., "The Use of Time to Ignition Data for Characterising the Thermal Inertia and the Minimum (Critical) Heat Flux for Ignition or Pyrolysis," *Combustion and Flame*, 84: 323–332, 1991.
38. Mowrer, F.W., "Flammability of Oil-based Painted Gypsum Wallboard Subjected to Fire Heat Fluxes," to appear in *Fire and Materials*, 2002.
39. Gray, B., "Spontaneous Combustion and Self-Heating," *SFPE Handbook of Fire Protection Engineering*, 3rd edition, National Fire Protection Association, Quincy, MA, 2002.
40. National Fire Protection Association, "Table A-10. Materials Subject to Spontaneous Heating," *Fire Protection Handbook*, 18th edition, National Fire Protection Association, Quincy, MA, 1997.
41. Ohlemiller, T.J., "Smoldering Combustion," *SFPE Handbook of Fire Protection Engineering*, 3rd edition, National Fire Protection Association, Quincy, MA, 2002.
42. NFPA 92B, *Guide for Smoke Management Systems in Malls, Atria, and Large Areas*, National Fire Protection Association, Quincy, MA, 2000.
43. Mulholland, G.W., and Croakin, C., "Specific Extinction Coefficient of Flame Generated Smoke," *Fire and Materials*, 24 (5): 227–230, 2000.
44. Babrauskas, V., "The Cone Calorimeter," *SFPE Handbook of Fire Protection Engineering*, 3rd edition, National Fire Protection Association, Quincy, MA, 2002.
45. Mulholland, G.W., "Smoke Production and Properties," *SFPE Handbook of Fire Protection Engineering*, 3rd edition, National Fire Protection Association, Quincy, MA, 2002.
46. Jin, T., "Visibility and Human Behavior in Fire Smoke," *SFPE Handbook of Fire Protection Engineering*, 3rd edition, National Fire Protection Association, Quincy, MA, 2002.
47. Jin, T., and Yamada, T., "Irritating Effects of Fire Smoke on Visibility," *Fire Science and Technology*, 5 (1): 79–89, 1985.
48. Purser, D.A., "Toxicity Assessment of Combustion Products," *SFPE Handbook of Fire Protection Engineering*, 3rd edition, National Fire Protection Association, Quincy, MA, 2002.
49. Pitts, W.M., "The Global Equivalence Ratio Concept and the Formation Mechanisms of Carbon Monoxide in Enclosure Fires," *Progress in Energy and Combustion Science*, 21: 197–237, 1995.
50. Gottuk, D.T., and Lattimer, B.Y., "Effects of Combustion Conditions on Species Production," *SFPE Handbook of Fire Protection Engineering*, 3rd Edition, National Fire Protection Association, Quincy, MA, 2002.
51. Mowrer, F.W., "Enclosure Smoke Filling Revisited," *Fire Safety Journal*, 33: 93–114, 1999.
52. Fleischmann, C.M., Pagni, P.J., and Williamson, R.B., "Quantitative Backdraft Experiments," *Fire Safety Science—Proceedings of the 4th International Symposium*, International Association for Fire Safety Science, Boston, MA, 1994, pp. 337–348.
53. Quintiere, J.G., "Compartment Fire Modeling," *SFPE Handbook of Fire Protection Engineering*, 3rd edition, National Fire Protection Association, Quincy, MA, 2002.
54. Cox, G., and Kumar, S., "Modeling Enclosure Fires Using CFD," *SFPE Handbook of Fire Protection Engineering*, 3rd edition, National Fire Protection Association, Quincy, MA, 2002.

CHAPTER 2
MATERIALS SPECIFICATIONS, STANDARDS, AND TESTING

Archibald Tewarson
FM Global Research, 1151 Boston Providence Turnpike, Norwood, MA 02062

Wai Chin and Richard Shuford
U.S. Army Research Laboratory, Aberdeen Proving Ground, Aberdeen, MD 31005

2.1 INTRODUCTION

Uncontrolled fires present hazards to life and property due to release of smoke, toxic and corrosive compounds (nonthermal hazard), and release of heat* (thermal hazard) in all fire stages (preignition, ignition, fire growth and flame spread, steady state, and decay). The nonthermal and thermal hazards are created primarily due to the use of products made of combustible materials (both natural and synthetic). These products are used in a variety of ways in residential, private, government, industrial, transportation, and manufacturing applications. Consequently, numerous fire scenarios need to be considered for testing of products. As a simplification, two types of standard methods have, therefore, been developed for testing of products:

- Test methods to comply with specific regulations or voluntary agreements: These types of test methods are usually larger than laboratory-scale tests that are included in the *prescriptive-based fire codes*.† Generally, products are tested under a defined fire condition in their end-use configurations.
- Small-scale standard test methods: These types of test methods have been developed based on qualitative experience as well as on an understanding of fire stages and associated hazards. In the tests, relatively simple measurements are made for various fire properties of the materials at each stage of the fire and associated hazards. These types of standard test methods are useful for the *performance-based fire codes* that are being considered to augment or replace the prescriptive-based fire codes‡ [1–3].

Both types of standard test methods for products in their end-use configurations and materials used for the construction of products are promulgated by various national and international standards organizations and government and private agencies, including [4, 5]:

- Australia (*Standards Australia*, SA)
- Canada (*Canadian General Standards Board*, CGSB)
- Europe (*International Electrotechnical Commission*, IEC; *European Committee for Electrotechnical Standardization*, CENELEC; *European Committee for Standardization*, CEN, *International Standards Organization*, ISO)

---
*Heat is released in three forms: conductive (hot surfaces), convective (hot gases), and radiative (flame). The heat release rate is commonly identified with the fire intensity.
†The codes reflect expectations for the level of fire protection.
‡An example of the prescriptive-based code for passive fire protection is the specified fire-resistance rating for an interior wall, whereas for the performance-based code, it would be a prediction for the desired passive fire protection based on the engineering standards, practices, tools, and methodologies.

- Finland (*Finnish Standards Association*, SFS)
- France (*Association Europeene des Constructeurs de Materiel Aerospatial*, AECMA; *Association Francaise de Normalisation*, AFNOR)
- Germany (*Deutsches Institut fur Normung*, DIN)
- India (*Indian Standards Institution*, ISI)
- Israel (*Standards Institution of Israel*, SII)
- Italy (*Ente Nazionale Italiano di Unifacazione*, UNI)
- Japan (*Japanese Standards Association*, JSA)
- Korea (*Korean Standards Association*, KSA)
- New Zealand (*Standards New Zealand*, SNZ)
- Nordic countries (Nordtest: Denmark, Finland, Greenland, Iceland, Norway, and Sweden)
- People's Republic of China (*China Standards Information Center*, CSIC)
- Russia (*Gosudarstvennye Standarty State Standard*, GOST)
- South Africa (*South African Bureau of Standards*, SABS)
- Taiwan, Republic of China (*Bureau of Standards, Metrology, and Inspection*, BSMI)
- United Kingdom (*British Standards Institution*, BSI; *Civil Aviation Authority*, CAA)
- United States (examples of government agencies: *Department of Transportation*, DOT; military, MIL; *National Aeronautics and Space Administration*, NASA; examples of private agencies: *American National Standards Institute*, ANSI, *American Society for Testing and Materials*, ASTM; *Building Officials & Code Administrators International Inc.*, BOCA; *Electronic Industries Alliance*, EIA; *FM Approvals*; *Institute of Electrical and Electronics Engineers*, IEEE; *National Fire Protection Association*, NFPA; *Underwriters Laboratories*, UL).

Each national and international standards organization, both government and private industries from each country listed above, as well as others, use their own standard test methods for the evaluation of the products and materials. Consequently, there are literally thousands of standard test methods used on a worldwide basis [5–9]. The national and international standards organizations list their test methods in standard catalogues, such as the CEN [10], FM Approvals [11], UL [12], ISO [13], ASTM [14], and many others.

Because there are thousands of standard testing methods in use today, products accepted in one country may be unacceptable in another, creating confusion and serious problems for the manufacturers and fire safety regulator. Vigorous efforts are thus being made, especially in Europe, to harmonize the standard test methods.* Recently, the European Commission's, *Single Burning Item (SBI) and Reaction to Fire Classification* [10] is an example of harmonizing hundreds of European standard testing methods for building products into a single standard test method. The SBI test method (EN 13823) for testing the fire safety of construction products will be widely used by the manufacturers to allow for the affixing of C marking that will indicate compliance with the Essential Requirements of the Union Directive 89/106/EEC. In addition, new regulations, Euroclasses,† and test methods designated EN ISO are in a process of being introduced that will be used throughout Europe [15, 16].

Further harmonization is expected as many regulatory agencies are considering augmenting or replacing the prescriptive-based fire codes (currently in use) by the performance-based fire codes. In the performance-based fire codes, engineering methods are used that utilize data for the fire properties as inputs for the assessment of the fire performance of buildings and products [1–3]. The data for the fire properties can be obtained from many standard test methods currently in use worldwide by modifying the test

---

*ISO, IEC, Nordtest, CEN, and the U.S. Federal Aviation Administration's (FAA) standards criteria are internationally acceptable for regulations.

†There are seven main Euroclasses for building materials for walls, ceiling, and floors: A1, A2, B, C, D, E, and F [15, 16]. A1 and A2 represent different degrees of limited combustibility. B to E represent products that may go to flashover in a room within certain times [15, 16]. F means that no performance is determined [15, 16]. Thus, there are seven classes for linings and seven class for floor coverings [15, 16]. There are additional classes of smoke and any occurrence of burning droplets [15, 16].

procedures and data acquisition methodology. Because fire properties will be measured quantitatively, the standard test methods will be automatically harmonized worldwide and the assessment for the fire resistance of materials and products will become reliable because it will be subject to quantitative verification.

2.2 MATERIALS CHARACTERISTICS

Various products are constructed from natural and synthetic materials containing carbon, hydrogen, oxygen, nitrogen, sulfur, and halogen atoms that are attached to each other by a variety of chemical bonds in the structure. The synthetic materials are identified as thermoplastics,* elastomers,† and thermosets‡ [17, 18]. Wood and cotton are examples of the natural materials, and polyethylene, polypropylene, polystyrene, nylon, polyvinyl chloride, and organic matrix composites are examples of the synthetic materials.

On exposure to heat, thermoplastics and elastomers typically soften and melt without any significant charring. The burning of thermoplastics and elastomers is accompanied by dripping of burning droplets that collect at the bottom and burn as liquid pool fires. Liquid pool fires are one of the most dangerous stages in a fire. Natural materials and thermosets generally do not soften and melt but decompose and form varying amounts of char on exposure to heat. Nonhalogenated and moderately halogenated thermoplastics and elastomers and natural materials have low fire resistance and are identified as *ordinary materials.* Some of the thermoplastics and elastomers and most of the thermosets, identified as *high-temperature materials,* and *highly halogenated thermoplastics and elastomers,*§ have high fire resistance.

Because of the low fire resistance, ordinary materials require fire-retardant treatments to increase their fire resistance when used in the construction of various products [19, 20]. Products constructed from fire-retarded ordinary materials satisfy the standard test method requirements for residential applications, where protection from only small fires is required, but not the standard test method requirements for public and industrial (government and private) occupancies. For these occupancies, protection from large fires is required and products are constructed from materials with a high degree of fire retardancy or from high-temperature materials and highly halogenated thermoplastics and elastomers that have high fire resistance.

The fire resistance of the materials and products is identified in terms of the ease or difficulty with which materials or products undergo a transformation through the following processes when exposed to heat:

- softening and melting
- decomposition, vaporization, and charring
- ignition
- flame spread and fire growth
- release of heat
- release of smoke, toxic, and corrosive compounds

Thus, standard test methods have been designed to assess the ease or difficulty with which materials and products undergo the above processes. In some test methods, the assessment is made only visu-

---

Thermoplastics are linear or branched materials that can be melted and remelted repeatedly upon the application of heat. They can be molded and remolded into virtually any shape [17, 18].

†*Elastomers* are chemically or physically cross-linked rubbery materials that can easily be stretched to high extensions and rapidly recover their original dimensions. Thermoplastic elastomers can be melted and remelted repeatedly and molded and remolded [17, 18].

‡*Thermosets* are rigid materials having a short network in which chain motion is greatly restricted by a high degree of cross-linking. They are intractable once formed and degrade and char rather than melt upon the application of heat [17, 18].

§Highly halogenated materials, such as polytetrafluoroethylene (Teflon), may form drops and pools of molten material on heat exposure. However, due to high halogen content, it is hard to ignite and burn the drops and pools of the molten material. Thus, these materials do not require fire-retardant treatments and are accepted without the active fire protection in products designed and assembled for various industrial (government and private) applications.

ally, while in others, quantitative measurements are made. As discussed above, for harmonization and reliable assessment of fire resistance of materials, it is necessary to use quantitative measurements, rather than visual observations.

2.3 SOFTENING AND MELTING BEHAVIORS OF MATERIALS

On exposure to heat, thermoplastics and elastomers soften, melt, and flow away from the heat source and drip as burning drops, igniting other materials in close proximity or collecting at the bottom and burning as a liquid pool fire (one of the most hazardous conditions in a fire).

The softening, melting, and flow properties of materials are characterized by the glass transition (T_{gl}) and melting (T_m) temperatures and *melt flow index* (MFI) [17, 18, 21, 22]. The values of these parameters are listed in Table 2.1, where data are taken from Refs. 17, 18, 21, and 22. The glass tran-

TABLE 2.1 Glass Transition Temperature, Melting Temperature, and Melt Flow Index of Materials [17, 18, 21, 22][a]

Material	Temperature (°C)		Melt flow index (g/10 min)
	Glass transition	Melting	
Ordinary			
Polyethylene low density, PE-LD	−125	105–110	1.4
PE-high density, PE-HD		130–135	2.2
Polypropylene, PP (atactic)	−20	160–165	21.5
Polyvinylacetate, PVAC	28	103–106	
Polyethyleneterephthalate, PET	69	250	
Polyvinyl chloride, PVC	81	189	
Polyvinylalcohol, PVAL	85		
PP (isotactic)	100		
Polystyrene, PS	100		9.0
Polymethylmethacrylate, PMMA	100–120	130; 160	2.1, 6.2
High temperature			
Polyetherketone, PEK	119–225		
Polyetheretherketone, PEEK		340	
Polyethersulfone, PES	190		
Highly halogenated			
Perfluoro-alkoxyalkane, PFA	75	300–310	
TFE,HFP,VDF fluoropolymer 200		115–125	20
TFE,HFP,VDF fluoropolymer 400		150–160	10
TFE,HFP,VDF fluoropolymer 500		165–180	10
Polyvinylideneufloride, PVDF	−35	160–170	
Ethylenechlorotrifluoroethylene, ECTFE		240	
Ethylenetetrafluoroethylene, ETFE		245–267	
Perfluoroethylene-propylene, FEP		260–270	
MFA		280–290	
Tetrafluoroethylene, TFE	−130	327	

[a] HFP: hexafluoropropylene; VDF: vinylidene fluoride; MFA: copolymer of TFE and perfluoromethyl vinyl ether (PMVE: Hyon®).

sition and melting temperatures of ordinary materials are lower than those for the high-temperature and highly halogenated materials. Ordinary materials, such as polypropylene, burn with much higher-intensity fires releasing large amounts of products of complete and incomplete combustion compared to the fires of the high-temperature and highly halogenated materials.

Pool fires are the most dangerous stage of a fire and govern the fire intensity. For example, in intermediate-scale tests in 100- and 500-kW calorimeters for automobile parts made of synthetic materials, pool fires strongly affected their burning behaviors irrespective of their size, shape, and configuration [23]. The observation was consistent with the material melt flow characteristics of polycarbonate, polycarbonate, polyethyleneterephthalate, and a sheet molding compound used for the construction of the parts [24].

Currently, there are no standard test methods for quantifying the resistance to softening and melting. There are only qualitative standard test methods, such as the prEN ISO 11925-2 [10, 13] and the UL 94 [12], where the ignition of cotton, placed under the specimen, is used as a criterion of ease of the dripping of hot molten material. Test methods used by the manufacturers (such as nonfire standards) to supply the data for the softening and melting of the materials [24] could be adopted as a standard test method for the assessment of fire resistance of materials. In addition, softening, melting, and pooling characteristics of materials should be considered in the assessment of fire hazards.

2.4 VAPORIZATION, DECOMPOSITION, AND CHARRING BEHAVIORS OF MATERIALS

On exposure to heat, materials vaporize, decompose, and char. The vaporization, decomposition, and charring characteristics of materials depend on their thermal stability (chain rigidity and strong interchain forces), characterized by the vaporization T_v and decomposition temperatures T_d, heat losses due to surface reradiation, conduction, and convection $\dot{q}''_{loss}$, heat of gasification (ΔH_g), and the char yield. All these parameters affect release rate of material vapors and thus the flame spread and fire growth is given by [25–27]:

$$\dot{m}'' = \dot{q}''_n / \Delta H_g \tag{2.1}$$

where $\dot{m}''$ is the release rate of the material vapors per unit surface area of the material [g/(m²·s)] and $\dot{q}''_n$ is the net heat flux to the surface per unit surface area of the material (kW/m²) defined as:

$$\dot{q}''_n = \dot{q}''_e + \dot{q}''_f - \dot{q}''_{loss} \tag{2.2}$$

where $\dot{q}''_e$ is the external heat flux per unit surface area of the material (kW/m²), $\dot{q}''_f$ is the flame heat flux per unit surface area of the material (kW/m²), $\dot{q}''_{loss}$ is in kW/m², and ΔH_g is in kJ/g and is expressed as:

For thermoplastics and elastomers:

$$\Delta H_g = \int_{T_a}^{T_m} c_{p,s} dT + \Delta H_m + \int_{T_m}^{T_v} c_{p,l} dT + \Delta H_v \tag{2.3}$$

For thermosets:

$$\Delta H_g = \int_{T_a}^{T_d} c_{p,s} dT + \Delta H_d \tag{2.4}$$

where ΔH_m is the heat of melting of the material (kJ/g), ΔH_v is the heat of vaporization of the material (kJ/g), ΔH_d is the heat of decomposition of the material (kJ/g), T_a is ambient temperature (°C), T_m, T_v, and T_d are the melting, vaporization, and decomposition temperatures of the material (°C), respectively, and $c_{p,s}$ and $c_{p,l}$ are the heat capacities of the original solid and molten material [kJ/(g·K)], respectively. The heat of gasification combines the effects of important individual components into a single parameter and relates it to the release rate of material vapors and thus to the flame spread and fire growth.

The values of ΔH_g, $\dot{q}''_{loss}$, and char yields have been measured and reported in the literature using the following standard tests:

- ASTM E 1354 and ISO 5660 (cone calorimeter) [13, 14]
- ASTM E 2058 (fire propagation apparatus) [14]

The ΔH_g value can also be measured by nonfire standard (thermal analysis) tests [24]. The properties discussed above are needed as inputs for the performance-based fire codes [3]. T_v and T_d values,

TABLE 2.2 Vaporization/Decomposition Temperature, Limiting Oxygen Index and UL 94 Ratings for Polymeric Materials [28]

Material	T_v/T_d (°C)	Char yield (%)	UL 94
Ordinary (horizontal burning, HB)			
Poly(α-methylstyrene)	341	0	HB
Polyoxymethylene (POM)	361	0	HB
Polystyrene (PS)	364	0	HB
Polymethylmethacrylate (PMMA)	398	2	HB
Polyurethane elastomer (PU)	422	3	HB
Polydimethylsiloxane (PDMS)	444	0	HB
Poly(acrylonitrile-butadiene-styrene) (ABS)	444	0	HB
Polyethyleneterephthalate (PET)	474	13	HB
Polyphthalamide	488	3	HB
Polyamide 6 (PA6)-Nylon	497	1	HB
Polyethylene (PE)	505	0	HB
High temperature (vertical burning, V)			
Cyanate ester of Bisphenol-A (BCE)	470	33	V-1
Phenolic Triazine Cyanate Ester (PT)	480	62	V-0
Polyethylenenaphthalate (PEN)	495	24	V-2
Polysulfone (PSF)	537	30	V-1
Polycarbonate (PC)	546	25	V-2
Liquid Crystal Polyester	564	38	V-0
Polypromellitimide (PI)	567	70	V-0
Polyetherimide (PEI)	575	52	V-0
Polyphenylenesulfide (PPS)	578	45	V-0
Polypara(benzoyl)phenylene	602	66	V-0
Polyetheretherketone (PEEK)	606	50	V-0
Polyphenylsulfone (PPSF)	606	44	V-0
Polyetherketone (PEK)	614	56	V-0
Polyetherketoneketone (PEKK)	619	62	V-0
Polyamideimide (PAI)	628	55	V-0
Polyaramide (Kevlar®)	628	43	V-0
Polybenzimidazole (PBI)	630	70	V-0
Polyparaphenylene	652	75	V-0
Polybenzobisoxazole (PBO)	789	75	V-0
Polyvinylchloride (PVC)	270	11	V-0
Polyvinylidenefluoride (PVDF)	320–375	37	V-0
Polychlorotrifluoroethylene (PCTFE)	380	0	V-0
Fluorinated Cyanate Ester	583	44	V-0
Polytetrafluoroethylene (PTFE)	612	0	V-0

TABLE 2.3 Heat Losses and Heat of Gasification of Materials [25–27]

Material	$\dot{q}''_{loss}$ (kW/m²)	ΔH_g (kJ/g) DSC	ASTM E 2058
Ordinary			
Filter paper	10		3.6
Corrugated paper	10		2.2
Douglas fir wood	10		1.8
Plywood/fire retarded (FR)	10		1.0
Polypropylene, PP	15	2.0	2.0
Polyethylene, PE, low density	15	1.9	1.8
PE-high density	15	2.2	2.3
Polyoxymethylene, POM	13	2.4	2.4
Polymethylmethacrylate, PMMA	11	1.6	1.6
Nylon 6,6	15		2.4
Polyisoprene	10		2.0
Acrylonitrile-butadiene-styrene, ABS	10		3.2
Styrene-butadiene	10		2.7
Polystyrene, PS foams	10–13		1.3–1.9
PS granular	13	1.8	1.7
Polyurethane, PU, foams-flexible	16–19	1.4	1.2–2.7
PU foams-rigid	14–22		1.2–5.3
Polyisocyanurate, PIU, foams	14–37		1.2–6.4
Polyesters/glass fiber	10–15		1.4–6.4
PE foams	12		1.4–1.7
High temperature			
Polycarbonate, PC	11		2.1
Phenolic foam	20		1.6
Phenolic foam/FR	20		3.7
Phenolic/glass fiber	20		7.3
Phenolic-aromatic polyamide	15		7.8
Halogenated			
PE/25 % chlorine (Cl)	12		2.1
PE/36 % Cl	12		3.0
PE/48 % Cl	10		3.1
Polyvinylchloride, PVC, rigid	15		2.5
PVC plasticized	10		1.7
Ethylene-tetrafluoroethylene, ETFE	27		0.9
Perfluoroethylene-propylene, FEP	38		2.4
Ethylene-tetrafluoroethylene, ETFE	48		0.8–1.8
Perfluoro-alkoxyalkane, PFA	37		1.0

although not measured in the ASTM E 1354, ISO 5660, and ASTM E 2058 standard test methods, can be measured by nonfire standard test methods [24], which can easily be adopted as fire standard test methods.

Examples of the useful data that are available in the literature [25–28] for vaporization and decomposition temperatures, char yields, surface reradiation loss, and heat of gasification for mate-

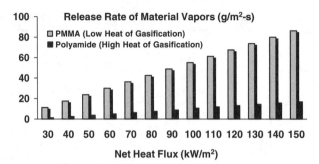

FIGURE 2.1 Release rate of material vapors versus net heat flux, calculated from the heat of gasification and heat loss values given in Table 2.3 for a low heat of gasification material (PMMA) and a high heat of gasification material (polyamide).

rials are listed in Tables 2.2 and 2.3. Fig. 2.1 shows examples of the effectiveness of the heat of gasification in increasing the fire resistance of materials. In the figure, release rates of material vapors at various heat flux values are calculated for a material with low heat of gasification (*polymethylmethacrylate*, PMMA) and for a material with high heat of gasification (polyamide) using data from Table 2.3. Polyamide is expected to contribute about one-sixteenth the amount of combustible vapors to the fire compared with PMMA.

2.5 IGNITION BEHAVIOR OF MATERIALS

The ignition resistance of a polymeric material is expressed in terms of the delay in igniting a material when exposed to heat. The ignition delay depends on the magnitude of the heat exposure and the thickness of the material d relative to the thermal penetration depth δ defined as [29–32]:

$$\delta = \sqrt{\alpha t_{ig}} = \sqrt{(k/\rho c)t_{ig}} \tag{2.5}$$

where α is the thermal diffusivity of the polymer (mm/s), t_{ig} is the time to ignition (s), k is the thermal conductivity of the polymer [kW/(m·K)], ρ is the density of the polymer (g/m³), and c is the heat capacity of the polymer [kJ/(g·K)]. A material behaves as a thermally thin material when $\delta > d$ and as a thermally thick material when $\delta < d$. The times to ignition under thermally thin and thick material behaviors satisfy the following relationships [25, 26, 29–32]:

Thermally thin material behavior:

$$1/t_{ig} = (\dot{q}''_e - \dot{q}''_{loss})/\Delta T_{ig} d\rho c \tag{2.6}$$

Thermally thick material behavior:

$$1/t_{ig}^{1/2} = (\dot{q}''_e - \dot{q}''_{loss})/(\Delta T_{ig}\sqrt{k\rho c}) \tag{2.7}$$

where ΔT_{ig} is the ignition temperature above ambient (K), and $d\rho$ is commonly defined as the areal density of the thin material (g/m²). The term $\Delta T_{ig} d\rho c$ is defined as the *thermal response parameter* (TRP) of a material behaving as a thermally thin material [(kW·s)/m²] and the term $\Delta T_{ig}\sqrt{k\rho c}$ is defined as the TRP of a material behaving as a thermally thick material [(kW·s^{1/2})/m²]. The TRP values represent the ignition resistance of the materials.

The thermally thin and thick material behaviors are shown in Figures 2.2 and 2.3, respectively, where linear relationships are found between $1/t_{ig}$ and $\dot{q}''_e$ and between $1/t_{ig}^{1/2}$ and $\dot{q}''_e$, respectively

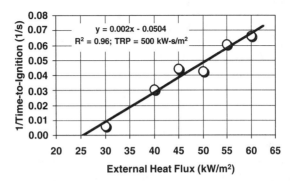

FIGURE 2.2 Relationship between time to ignition for a thermally thin material and external heat flux. Data in the figure are for a physically thin fabric (<3 mm thick) wrapped around a 100-mm² and 10-mm-thick ceramic insulation block measured in the ASTM E 2048 apparatus.

[Eqs. (2.6) and (2.7), respectively]. The inverse of the slope provides the TRP values for thermally thin and thick material behaviors, respectively. The minimum flux at which there is no ignition represents the $\dot{q}''_{loss}$ value. When convective and conductive heat losses are negligibly small under quiescent airflow condition, $\dot{q}''_{loss} \approx \dot{q}''_{cr}$ defined as *critical heat flux* (CHF), at or below which there is no ignition. CHF value represents the ignition resistance of a material and is related to the T_{ig} value.*

For the assessment of fire hazards and protection requirements using the performance-based fire codes, the engineering methods need effective thermal inertia ($k\rho c$). The $k\rho c$ values for materials can be derived from the TRP values with ignition temperature either estimated from the CHF value or measured directly.

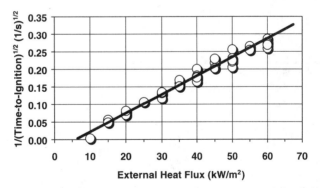

FIGURE 2.3 Relationship between time to ignition and external heat flux for a thermally thick material. Data in the figure are for a physically thick (10-mm-thick) and 100-mm² slab of PMMA measured in the ASTM E 2048 apparatus.

---

*T_{ig} (°C) ≈ $[(q''_{cr})^{0.25} \times 364] - 273$, assuming heat losses to be mainly due to reradiation under quiescent airflow condition and the surface acting as a black body with an ambient temperature of 20°C.

Standard test methods have been developed for examining the ignition resistance of materials. Some test methods provide qualitative data, while others provide partial or complete quantitative data for the ignition resistance of materials. The following are examples of the common standard test methods used for examining the ignition resistance of materials:

- ISO 871 (T_{ig}, hot oven) [13]
- ASTM D 1929 (T_{flash}: flash ignition temperature; T_{ig}: spontaneous) [14]
- ASTM E 1352 (qualitative—cigarette ignition of upholstered furniture) [14]
- ASTM E 1353 (qualitative—cigarette ignition resistance of components of upholstered furniture) [14]
- ASTM F 1358 (qualitative—effects of flame impingement on materials used in protective clothing not designed primarily for flame resistance) [14]
- ASTM C 1485 (CHF value of exposed attic floor insulation using an electric radiant heat energy source) [14]
- ASTM E 648 (CHF value of floor covering systems using a radiant heat energy source) [14]
- ASTM E 1321 and ISO 5658 (CHF and TRP values of materials) [13, 14]
- ASTM E 1354 and ISO 5660 (CHF and TRP values of materials) [13, 14]
- ASTM D 1929 (T_{ig} values for plastics) [14]
- ASTM E 2058 (CHF and TRP values of materials) [14]

Tests performed in the apparatus specified in three standards listed above, i.e., ASTM E 1321 and ISO 5658 (*lateral ignition and flame spread test,* LIFT, apparatus), ASTM E 1354 and ISO 5660 (cone calorimeter), and ASTM E 2058 (fire propagation apparatus) provide complete sets of fire properties for the assessment of ignition resistance of materials. These apparatuses also provide data in a format that is useful for the engineering methods in the performance-based fire codes.

Examples of the data for CHF and TRP from the tests performed according to ASTM E 1354 and ISO 5660 (cone calorimeter) and ASTM E 2058 (fire propagation apparatus) are listed in Table 2.4, where data are taken from Refs. 25, 26, and 33–37. Materials with high ignition resistance have high CHF and TRP values.

TABLE 2.4 Critical Heat Flux and Thermal Response Parameter from Cone Calorimeter (ASTM E 1354) and the Fire Propagation Apparatus (ASTM E 2058) [25, 26, 33–37]

	ASTM E 2058		ASTM E 1354	
Material[a]	CHF (kW/m^2)	TRP (kW-s$^{1/2}$/m^2)	CHF (kW/m^2)	TRP (kW-s$^{1/2}$/m^2)
	Ordinary			
Tissue paper	10	95		
Newspaper	10	108		
Wood (red oak)	10	134		
Corrugated paper	10	152		
Wood (Douglas fir, Fire retarded, FR)	10	251		222
Wool				232
Polyethylene (PE)	15	454		526
PE				364
Cross linked polyethylene (XLPE)				385

TABLE 2.4 (*Continued*)

Material[a]	ASTM E 2058		ASTM E 1354	
	CHF (kW/m^2)	TRP (kW-s$^{1/2}$/m^2)	CHF (kW/m^2)	TRP (kW-s$^{1/2}$/m^2)
Ordinary				
Polypropylene (PP)-1	15	288		291
PP-2	15	323		377
PP-3	10	277		
PP-4	15	333		
PP-5				556
PP/glass fiber				377
Polymethylmethacrylate, PMMA	10	274		222
Polyoxymethylene, POM	10	250		357
Polystyrene, PS	20	146		556
PS-fire retarded (FR)				667
PS foam			20	168
PS foam-FR			20	221
Nylon				333
Nylon 6	20	154		379
Polybutyleneterephthalate, PBT				588
Polyethyleneterephthalate, PET	10	174		435
Poly(acrylonitrile-butadiene-styrene), ABS				317
ABS-FR				556
ABS-PVC				357
Vinyl thermoplastic elastomer			20	294
Polyurethane foam, PU				76
Thermoplastic PU-FR				500
EPDM/styrene-acrylonitrile (SAN)				417
Isophthalic polyester				296
Polyvinyl ester				263
Epoxy				457
Acrylic paneling, FR				233
High temperature				
Polysulfone (PSF)	30	469		
Polyetheretherketone (PEEK)	30	550		
Polycarbonate (PC)-1	20	357		370
PC-2	20	434	20	455
PC-3	30	455		
PC-4	30	455		
Polyphenylene oxide, PPO-polystyrene (PS)				455
Halogenated				
Polyvinylchloride, PVC, flexible-1	10	215		
PVC, flexible-2	10	263		
PVC, flexible-3				244
PVC flexible-4 (LOI 25%)				285
PVC flexible-5				333
PVC-fire retarded (FR) flexible-1				222
PVC-FR flexible-2				263
PVC-FR flexible-3 (LOI 34%)				345

TABLE 2.4 (*Continued*)

Material[a]	ASTM E 2058		ASTM E 1354	
	CHF (kW/m^2)	TRP (kW-s$^{1/2}$/m^2)	CHF (kW/m^2)	TRP (kW-s$^{1/2}$/m^2)
Halogenated				
PVC-FR flexible-4 (LOI 30%)				397
PVC-FR flexible-5 (LOI 28%)				401
PVC rigid-1				357
PVC rigid-2				385
PVC rigid-3 (LOI 50%)			25	388
PVC rigid-4			25	390
PVC rigid-5				417
Chlorinated PVC, CPVC-1	40	435		
CPVC-2			25	1111
ABS-PVC, flexible	19	73		
Ethylenechlorotrifluoroethylene, ECTFE	38	450		
Polychlorotrifluoroethylene, PCTFE	30	460		
Ethylene-tetrafluoroethylene copolymer, ETFE	25	481		
Polyvinylidenefluoride, PVDF	40	506		
Polytetrafluoroethylene, PTFE	50	654		
Perfluoroethylene-propylene copolymer, FEP	50	680		
Nylon/glass fiber				359
Fiber based composites				
Polyester/glass fiber (30%)			20	256
Isophthalic polyester/glass fiber (77%)				426
Polyvinyl ester/glass fiber-1		10		312
Polyvinyl ester/glass fiber-2		10		429
Polyvinyl ester/glass fiber (69%)-3		15		444
Epoxy/glass fiber-1				288
Epoxy/glass fiber-2				334
Epoxy/glass fiber (69%)-3	10	410		388
Epoxy/glass fiber-4	10	400		397
Epoxy/glass fiber-5	10	420		433
Epoxy/glass fiber-6				512
Epoxy/glass fiber-7				517
Epoxy/glass fiber-8				555
Epoxy/glass fiber-9				592
Epoxy/glass fiber-10	15	667		665
Epoxy/graphite fiber-1			25	484
Epoxy/graphite fiber-2				493
Epoxy/graphite fiber-3	24	667		554
Cyanate ester/glass fiber				302
Cyanate ester/graphite fiber	20	1000		
Phenolic/glass fiber	20	610		
Phenolic/kevlar fiber	15	403		
Acrylic/glass fiber				180
Polyphenylenesulfide/glass fiber	20	909		
Epoxy/phenolic/glass fiber	20	1250		
Polyphenyleneoxide, PPO/glass fiber				435
Polyphenylenesulfide, PPS/glass fiber-1			25	588
PPS/glass fiber-2			25	623

TABLE 2.4 (*Continued*)

Material[a]	ASTM E 2058		ASTM E 1354	
	CHF (kW/m^2)	TRP (kW-s$^{1/2}$/m^2)	CHF (kW/m^2)	TRP (kW-s$^{1/2}$/m^2)
Fiber based composites				
PPS/graphite fiber-1			25	330
PPS/graphite fiber-2			25	510
Polyarylsulfone/graphite fiber			25	360
Polyethersulfone/graphite fiber			25	352
Polyetheretherketone, PEEK/glass fiber (30%)-1			20	301
PEEK/graphite fiber-2				514
Polyetherketoneketone, PEKK/glass fiber			25	710
Bismaleimide, BMI/graphite fiber-1				513
BMI/graphite fiber-2			25	515
BMI/graphite fiber-3				608
BMI/graphite fiber-4			25	605
Phenolic/glass fiber-1			25	382
Phenolic/glass fiber-2			25	409
Phenolic/glass fiber-3			25	641
Phenolic/glass fiber (45%)-4				683
Phenolic/glass fiber-5			25	728
Phenolic/glass fiber-6			25	738
Phenolic/glass fiber-7			25	765
Phenolic/glass fiber-8			25	998
Phenolic/graphite fiber-1			25	398
Phenolic/graphite fiber-2			25	684
Phenolic/graphite fiber-3			25	982
Phenolic/PE fiber				267
Phenolic/aramid fiber				278
Polyimide/glass fiber			25	844

[a] Generic materials marked by numbers 1 to 9 have different compositions as well as in some cases different manufacturers. Materials are identified in the references from where the data were taken.

2.6 FLAME SPREAD, FIRE GROWTH, AND BURNING BEHAVIOR OF MATERIALS

This is one of the most critical stages in a fire and is primarily responsible for creating fire hazards because of the release of heat, smoke, and toxic and corrosive compounds. As a result, numerous small-, intermediate-, and large-scale standard test methods have been developed to assess the flame spread, fire growth, and peak (steady) burning behaviors of materials and products both qualitatively and quantitatively.

Flame spread and fire growth can be considered as an advancing ignition front in which the leading edge of the flame acts as the source of heat* and the source of ignition [29–31]. It can occur on horizontal, inclined, and vertical surfaces, parallel or opposite to the airflow direction (upward, concurrent, downward, or lateral flame spread). One of the following fire behaviors may be observed during flame spread and fire growth process:

- Nonpropagating fire behavior: There is no flame spread beyond the ignition zone.
- Decelerating fire behavior: Flame spread rate† beyond the ignition zone decreases with time and spread stops before covering the entire surface of the material or the product.

---
*Flames can heat the surface ahead of the flame front in many ways. These depend on the mode of fire spread, orientation, wind, and nature of material or product [31].
†Flame spread rate is the velocity at which the ignition front travels over the surface.

- Propagating fire behavior: Flame spreads beyond the ignition zone until the entire surface of the material or the product is involved on fire.
- Accelerating fire behavior: Flame spread rate beyond the ignition zone increases rapidly covering the entire surface of the material or the product, the flames extend far beyond the surface of the material or product in a relatively short time.

For spreading fires, the leading edge of the flame transfers heat ahead of the zone. As a result, surface temperature increases and reaches the ignition temperature of the material or the product (satisfying the CHF value) and maintains the temperature until the vapors generated from the material or the product ignite (satisfying the TRP value). For solids, the most significant heat transfer rate is at the surface over a length x_f [31]. The flame spread rate is expressed as [31]:

$$V = x_f/t_{ig} \tag{2.8}$$

where V is the flame spread rate (mm/s), x_f is in mm, and the time to ignition in seconds is given in Eqs. (2.6) and (2.7) for thermally thin and thick material behaviors, respectively. Thus, from Eqs. (2.6), (2.7), and (2.8), the flame spread rates for thermally thin and thick material behaviors can be expressed as [31]:

Thermally thin material behavior:

$$V = x_f(\dot{q}''_e + \dot{q}''_f - \dot{q}''_{loss})/\Delta T_{ig}d\rho c \tag{2.9}$$

where $\dot{q}''_f$ is the flame heat flux transferred ahead of the flame front (kW/m²) and is dependent on the heat release rate.

Thermally thick material behavior:

$$V^{1/2} = x_f^{1/2}(\dot{q}''_e + \dot{q}''_f - \dot{q}''_{loss})/\Delta T_{ig}\sqrt{k\rho c} \tag{2.10}$$

The $\dot{q}''_e$, $\dot{q}''_f$, and $\dot{q}''_{loss}$ values depend on variety of conditions, such as the generic nature, shape, size, and arrangement of the materials and products, airflow rate and direction, proximity to enclosure walls and ceiling, and others. Furthermore, $\dot{q}''_f$ depends on the heat release rate and relationships have been developed between this and the flame spread rate under various conditions [25, 26, 29–32].

After the flame has spread to the entire exposed surface, the material or the product burns at its peak intensity, with burning intensity decreasing as the material or product is consumed. The peak burning intensity is also defined as steady state burning. The generation rate of material vapors at the peak burning intensity is expressed by Eq. (2.2) and the release rates of heat and various chemical compounds are expressed as follows [25, 26]:

$$\dot{Q}''_{ch} = \Delta H_{ch}\dot{m}'' = (\Delta H_{ch}/\Delta H_g)(\dot{q}''_e + \dot{q}''_f - \dot{q}''_{loss}) \tag{2.11}$$

$$G''_j = y_j\dot{m}'' = (y_j/\Delta H_g)(\dot{q}''_e + \dot{q}''_f - \dot{q}''_{loss}) \tag{2.12}$$

where $\dot{Q}''_{ch}$ is the chemical heat release rate* per unit surface area of the material or the product (kW/m²), G''_j is the release rate of compound j per unit surface area of the material or products (g/m²s), ΔH_{ch} is the chemical heat of combustion† (kJ/g), and y_j is the yield of compound j (g/g).

The following are the most common flame spread and fire growth characteristics utilized for the measurements in the standard test methods:

- extent and rate of flame spread on horizontal, vertical, or inclined surfaces
- melting, dripping, and ignition of materials near the sample by the hot burning molten droplets
- minimum heat flux or surface temperature for flame spread
- minimum oxygen concentration for flame spread

---

*Chemical heat release rate consists of convective heat release rate plus the radiative heat release rate if there are no losses.
†Chemical heat of combustion consists of convective heat of combustion plus the radiative heat of combustion.

- heat release rate during flame spread, growth, and steady burning
- release rates of material vapors, smoke, CO, CO_2, and hydrocarbons during flame spread, growth, and steady burning

The following are the most common characteristics of the burning intensity of a fire that are utilized for the measurements in the standard test methods:

- heat release rate
- release rates of material vapors, smoke, CO, CO_2, and hydrocarbons

The standard test methods specify apparatuses to be used and type of measurements that need to be made for characterizing flame spread, fire growth, and burning intensity of materials and products. Some of the standard test methods specify small-scale tests, while others specify intermediate- and large-scale tests. In most of the standard test methods, only a limited number of flame spread, fire growth, and burning characteristics are quantified, some of which are included in the prescriptive-based fire codes and utilized as the acceptance criteria of the materials and products by various regulatory agencies. There are, however, a limited number of standard test methods that specify apparatus capable of providing quantitative data for the fire properties of materials and products, which are utilized in the performance-based fire codes for the assessment of hazards and protection from fires.

In almost all the small-scale standard test methods, different apparatus are specified for the characterization of the flame spread and fire growth behavior and for the characterization of the burning behavior of the materials. In the intermediate- and large-scale standard test methods, however, single tests are specified to characterize flame spread and fire growth, as well as the burning behaviors of the materials and products.

There are numerous standard test methods available worldwide on the subjects of characterization of flame spread and fire growth and burning intensity of materials and products. However, there are many common features in these worldwide standard test methods and thus it is possible to describe them in a generalized manner. Furthermore, some of the test methods are becoming popular all over the world as they provide quantitative information for the fire properties of materials and products. With increased frequency of use of these popular standard test methods on a worldwide basis, there will be automatic harmonization of the standard test methods. Data measured according to these standardized test methods will be reliable, as they will be subject to quantitative verification.

2.7 POPULAR STANDARD TEST METHODS FOR THE BURNING BEHAVIOR OF MATERIALS

The burning behavior of materials is examined by measuring the release rates of material vapors, heat, and chemical compounds (including smoke) in the apparatus specified in the standard test methods. From these measurements, the following fire properties are derived:

- heat of gasification and heat losses (Sec. 2.4, Table 2.3)
- chemical, convective, and radiative heats of combustion (ratio of the summation of the heat release rate to the summation of the release rate of material vapors)
- yields of various chemical compounds (ratio of the summation of the release rate of each compound to the summation of the release rate of material vapors)
- combustion efficiency (ratio of the heat of combustion to the net heat of complete combustion)
- generation efficiency of chemical compounds (ratio of the yield of a compound to the maximum possible stoichiometric yield of the compound based on the elemental composition of the material)

The heat of complete combustion is measured according to ASTM D 5865 and ISO 1716 test methods [13, 14]. The release rates of material vapors, heat, and various chemical compounds (including smoke) are measured according to ASTM E 906 (*Ohio State University heat release rate,* OSU-HRR, apparatus), ASTM E 2058 (fire propagation apparatus), and ASTM E 1354 and ISO 5660 (cone calorimeter) [13, 14]. Smoke released in flaming and nonflaming fires of materials is also characterized by following these standard test methods as well by the ASTM E 662 (smoke density chamber) [14].

2.7.1 ASTM D 5865 and ISO 1716: Test Method for Gross Heat of Complete Combustion [13, 14]

This standard test method specifies the use of small-scale test apparatus to quantify gross heat of complete combustion under controlled conditions. Detailed description of the apparatus and its sketch are included in the standard. The specimen weighing 0.8 to 1.2 g and contained in an open platinum, quartz, or base metal alloy crucible is burned in 100% oxygen in an oxygen bomb calorimeter. The gross heat of complete combustion is computed from the temperature before, during, and after the combustion of the specimen with proper allowance for thermochemical and heat transfer corrections. The *gross heat of complete combustion* is defined as the quantity of energy released when a unit mass of specimen is burned in a constant volume enclosure, with combustion products being gaseous, including water.

The gross heat of complete combustion is used to determine the *net heat of complete combustion,** which is defined as the quantity of energy released when a unit mass of specimen is burned at constant pressure, with all the combustion products, including water, being gaseous. Maxwell [38] has listed both gross and net heat of complete combustion of gaseous and liquid hydrocarbons, alcohols, glycols and glycerols, ethers, aldehydes, and ketones. Maxwell's data show that the net heat of complete combustion for these compounds $\approx$ 0.9274 × gross heat of complete combustion of the compounds with a standard deviation of 0.0438. The net heat of complete combustion for solid and foamed materials are listed in Refs. 25 and 26.

In Europe, the gross heat of complete combustion (*gross calorific potential*, PCS), measured by following the ISO 1716 standard test method, is used for the classification of reaction to fire performance for construction products (prEN 13501-1) [10]:

- construction products excluding floorings:
  - Class A1: PCS $\leq$ 1.4 to 2.0 MJ/kg.
  - Class A2: PCS $\leq$ 3.0 to 4.0 MJ/kg.
- floorings:
  - Class A1$_{fl}$: PCS $\leq$ 1.4 to 2.0 MJ/kg.
  - Class A2$_{fl}$: PCS $\leq$ 3.0 to 4.0 MJ/kg.

This standard test method incorporates the fundamental principles for the energy associated with the complete combustion of materials and thus is independent of fire scenarios [39]. The gross and net heat of complete combustion of materials are used in the performance-based fire codes for the assessment of fire hazards associated with the use of products and protection needs.

2.7.2 ASTM E 136 and ISO 1182: Standard Test Method for Behavior of Materials in a Vertical Tube Furnace at 750°C [13, 14]

This standard test method specifies the use of a small-scale apparatus to assess the noncombustibility behavior of building construction materials under the test conditions. Detailed description of the

---

*If the percentage of hydrogen atoms in the sample is known: net heat of complete combustion (kJ/g) = gross heat of complete combustion (kJ/g) − 0.2122 × mass percent of hydrogen atoms, where heats of combustion are in kJ/g [14]. If the percentage of hydrogen atoms is not known: net heat of complete combustion in kJ/g = 10.025 + (0.7195) × gross heat of combustion in kJ/g [14].

apparatus and its sketch are included in the standard. The standard test apparatus consists of two concentric, vertical refractory tubes, 76 and 102 mm (3 and 4 in) inside diameter and 210 to 250 mm (8.5 to 10 in) in length. Electric heating coils outside the larger tube are used to apply heat. A controlled flow of air is admitted tangentially near the top of the annular space between the tubes and passes to the bottom of the inner tube. The top of the inner tube is covered. Temperatures are measured by thermocouples at the center: (1) between the two concentric tubes, (2) close to specimen location, and (3) at the sample surface.

Test specimens are used in granular or powdered form contained in a 38 × 38 × 51-mm holder. The specimen in the holder is placed in the center of the inside vertical refractory tube after the temperature at the specimen location is maintained at 750 ± 5.5°C for 15 min. The test is continued until all the temperatures have reached their maximum values. Visual observations are made throughout the test on the specimen behavior, combustion intensity, smoke formation, melting, charring, etc. The specimen is weighed before and after the test. The data measured in the test are used to assess the following specimen behaviors:

- weight loss, $\Delta m \leq 50$ percent
- surface and interior temperature, $\Delta T \leq 30°C$
- there is either no flaming, i.e., flaming duration $t_f = 0$, or no flaming after the first 20 s, $t_f \leq 20$ s.

In Europe, data from ISO 1182 are used for the classification of reaction to fire performance for construction products (prEN 13501-1) [10]:

- construction products excluding floorings:
  - Class A1: $\Delta T \leq 30°C$ and $\Delta m \leq 50$ percent and $t_f = 0$
  - Class A2: $\Delta T \leq 30°C$ and $\Delta m \leq 50$ percent and $t_f \leq 20$ s
- floorings:
  - Class A1$_{fl}$: $\Delta T \leq 30°C$ and $\Delta m \leq 50$ percent and $t_f = 0$.
  - Class A2$_{fl}$: $\Delta T \leq 30°C$ and $\Delta m \leq 50$ percent and $t_f \leq 20$ s.

This standard test method incorporates the fundamental behavior of materials associated with the resistance to ignition and combustion up to 750°C (about 60 kW/m^2) and thus is independent of fire scenarios [39]. The test is capable of providing quantitative data for the performance-based fire codes for the assessment of fire hazards associated with the use of products and protection needs.

2.7.3 ASTM E 906, ASTM E 2058, and ASTM E 1354 and ISO 5660: Standard Test Methods for Release Rates of Material Vapors, Heat, and Chemical Compounds [13, 14]

These standard test methods specify the use of small-scale apparatus to quantify the fire properties of materials. The apparatus specified are the following:

- ASTM E 906 (the OSU-HRR apparatus)
- ASTM E 2058 (the fire propagation apparatus)
- ASTM E 1354 and ISO 5660 (cone calorimeter)

Detailed description of these apparatus and their sketches are included in their respective standards. A summary of the design, capacity, and types of measurements made in these apparatus are listed in Table 2.5. Examples of the data measured in the cone calorimeter (ASTM E 1354) and reported in Refs. 36, 37, and 40 are listed in Tables 2.6 and 2.7. Examples of the data measured in the fire propagation apparatus (ASTM E 2058) and reported in Refs. 25, 26, 34, 35, and 41 are listed in Tables 2.8 and 2.9.

TABLE 2.5 ASTM Test Apparatuses Used for the Measurements of Release Rates of Material Vapors, Heat and Chemical Compounds Including Smoke [14]

Design/test conditions	ASTM apparatuses		
	E 906 (OSU-HRR)	E 2058 (Fire propagation)	ASTM E 1354 (Cone)
Airflow	Co-flow	Co-flow/natural	Natural
Oxygen concentration (%)	21	0–60	21
Co-flow airflow velocity (m/s)	0.49	0–0.146	NA
External heaters	Silicone carbide	Tungsten-quartz	Electrical coils
External heat flux (kW/m^2)	0–100	0–65	0–100
Sampling duct flow (m^3/s)	0.04	0.035–0.364	0.012–0.035
Sample (mm)-horizontal	110 × 150	100 × 100	100 × 100
Sample (mm)-vertical	150 × 150	100 × 600	100 × 100
Ignition source	Pilot flame	Pilot flame	Spark plug
Ventilation controlled	No	Yes	No
Flame radiation simulation by O_2	No	Yes	No
Heat release rate capacity (kW)	8	50	8
Ignition-time	Yes	Yes	Yes
Release rate of vapors	No	Yes	Yes
Release rate of chemical compounds	Yes	Yes	Yes
Light obscuration by smoke	Yes	Yes	Yes
Gas phase corrosion	No	Yes	No
Fire propagation	No	Yes	No
Chemical heat release rate	Yes	Yes	Yes
Convective heat release rate	Yes	Yes	No
Radiative heat release rate	No	Yes	No
Flame extinction-water, Halon and alternates	No	Yes	No

Data in Tables 2.6 and 2.8 show that for ordinary materials (such as thermoplastics, which melt easily), heat release rates are very high in the range predicted for the liquid pool fires. In addition, data in these tables suggest that burning behaviors of generically similar materials under comparable test conditions in the cone calorimeter and fire propagation apparatus are very similar, for example:

- Heat release rates at 50 kW/m^2 for pool fires of polyethylene, polypropylene, nylon 6, and acrylonitrile-butadiene-styrene are in the range of 1133 to 1304 kW/m^2 given in Table 2.6 (cone calorimeter) and 1004 to 1341 kW/m^2 given in Table 2.8 (fire propagation apparatus).
- Heat release rates for thermoplastics with glass fibers and charring-type thermoplastics and high-temperature and halogenated materials given in Tables 2.6 (cone calorimeter) and 2.8 (fire propagation apparatus);
- Heat of combustion and yields of compounds given in Tables 2.7 (cone calorimeter) and 2.9 (fire propagation apparatus).

The apparatus specified in the ASTM E 906, ASTM E 1354, and ASTM E 2058 and ISO 5660 standard test methods have been developed to provide fire property data. Some of these data are used by the prescriptive-based as well as in the performance-based fire codes for the fire hazard analyses and protection needs for residential, private, government, and industrial occupancies, transport, manufacturing, and others.

TABLE 2.6 Peak Heat Release Rate Measured in the ASTM E 1354 (Cone Calorimeter) [36, 37, 40]

	Peak chemical heat release rate (kW/m^2)								
	External heat flux (kW/m^2)								$\Delta H_{ch}/\Delta H_g$
Material[a]	20	25	30	40	50	70	75	100	(kJ/kJ)[b]
			Ordinary						
High density polyethylene, HDPE	453		866	944	1133				21
Polyethylene, PE	913			1408		2735			37
Polypropylene, PP	377		693	1095	1304				32
Polypropylene, PP	1170			1509		2421			25
PP/glass fiber (1082)		187			361		484	432	6
Polystyrene, PS	723			1101		1555			17
Nylon	517			1313		2019			30
Nylon 6	593		802	863	1272				21
Nylon/glass fiber (1077)		67			96		116	135	1
Polyoxymethylene, POM	290			360		566			6
Polymethylmethacrylate, PMMA	409			665		988			12
Polybutyleneterephthalate, PBT	850			1313		1984			23
Acrylonitrile-butadiene-styrene, ABS-1	683		947	994	1147				14
ABS-2	614			944		1311			12
ABS-FR	224			402		419			4
ABS-PVC	224			291		409			4
Vinyl thermoplastic elastomer	19			77		120			2
Polyurethane, PU foam	290			710		1221			19
EPDM/Styrene acrylonitrile, SAN	737			956		1215			10
Polyester/glass fiber (30%)	NI			167	231				6
Isophthalic polyester	582		861	985	985				20
Isophthalic polyester/glass fiber (77%)	173		170	205	198				2
Polyvinyl ester	341		471	534	755				13
Polyvinyl ester/glass fiber (69%)-1	251		230	253	222				2
Polyvinyl ester/glass fiber-2		75			119		139	166	1
Polyvinyl ester/glass fiber-3		377					499	557	2
Epoxy	392		453	560	706				11
Epoxy/glass fiber-1	164		161	172	202				2
Epoxy/glass fiber-2		159			294		191	335	2
Epoxy/glass fiber-3		81			181		182	229	2
Epoxy/glass fiber-4					40		246	232	2
Epoxy/glass fiber-5		231			266		271	489	3
Epoxy/glass fiber-6		230			213		300	279	1
Epoxy/glass fiber-7		175			196		262	284	2
Epoxy/glass fiber-8		20			93		141	202	2
Epoxy/glass fiber-9		39			178		217	232	2
Epoxy/glass fiber-10		118			114		144	173	1
Epoxy/graphite fiber-1		NI					197	241	2
Epoxy/graphite fiber-2		164			189		242	242	2
Epoxy/graphite fiber-3		105			171		244	202	3
Cyanate ester/glass fiber		121			130		196	226	2
Kydex Acrylic paneling, FR	117			176		242			3
			High temperature						
Polycarbonate, PC-1	16			429		342			21
PC-2	144			420		535			14
Cross linked polyethylene (XLPE)	88			192		268			5
Polyphenylene oxide, PPO-polystyrene (PS)	219			265		301			2
PPO/glass fiber	154			276		386			6

TABLE 2.6 (*Continued*)

Material[a]	Peak chemical heat release rate (kW/m²) External heat flux (kW/m²)								$\Delta H_{ch}/\Delta H_g$ (kJ/kJ)[b]
	20	25	30	40	50	70	75	100	
High temperature									
Polyphenylenesulfide, PPS/glass fiber-1		NI			52		71	183	3
PPS/graphite fiber-2		NI					60	80	2
PPS/glass fiber-3		NI			48		88	150	2
PPS/graphite fiber-1		NI			94		66	126	1
Polyarylsulfone/graphite fiber		NI			24		47	60	1
Polyethersulfone/graphite fiber		NI			11		41	65	0.3
Polyetheretherketone, PEEK/glass fiber (30%)-1	NI			35	109				7
PEEK/graphite fiber-2					14		54	85	1
Polyetherketoneketone, PEKK/glass fiber		NI			21		45	74	1
Bismaleimide, BMI/graphite fiber-1		160					213	270	1
Bismaleimide, BMI/graphite fiber-2		128			176		245	285	2
Bismaleimide, BMI/graphite fiber-3		NI					172	168	(1)
Bismaleimide, BMI/graphite fiber-4		NI			74		91	146	1
Phenolic/glass fiber-1					165				
Phenolic/glass fiber-2		NI			66		102	122	1
Phenolic/glass fiber-3		NI			66		120	163	2
Phenolic/glass fiber-4		NI			47		57	96	1
Phenolic/glass fiber-5		NI			81		97	133	1
Phenolic/glass fiber-6		NI			82		76	80	(1)
Phenolic/glass fiber-7		NI			190		115	141	1
Phenolic/glass fiber-8		NI			132		56	68	1
Phenolic/graphite fiber-1		NI					159	196	2
Phenolic/graphite fiber-2		NI			177		183	189	(1)
Phenolic/graphite fiber-3		NI			71		87	101	1
Phenolic/PE fiber		NI			98		141	234	3
Phenolic/aramid fiber		NI			51		93	104	1
Phenolic insulating foam				17	19		29		1
Polyimide/glass fiber		NI			40		78	85	1
Wood									
Douglas fir	237			221		196			(−)
Hemlock	233		218	236	243				(−)
Textiles									
Wool	212		261	307	286				5
Acrylic fiber	300		358	346	343				6
Halogenated									
PVC flexible-3 (LOI 25%)	126		148	240	250				5
PVC-FR (Sb₂O₃) flexible-4 (LOI 30%)	89		137	189	185				5
PVC-FR (triaryl phosphate) flexible-5 (LOI 34%)	96		150	185	176				5
PVC rigid-1	40			175		191			3
PVC rigid-2	75			111		126			2
PVC rigid-3	102			183		190			2
PVC rigid-1 (LOI 50%)	NI		90	107	155				3
PVC rigid-2	NI		101	137	157				3
Chlorinated PVC (CPVC)	25			84		93			1

[a] Generic materials marked by numbers 1 to 10 have different compositions as well as in some cases different manufacturers. Materials are identified in the references from where the data were taken.

[b] From the slopes of the heat release rate versus external heat flux linear relationship (Eq. 11).

TABLE 2.7 Average Effective (Chemical) Heat of Combustion and Smoke Yield Calculated from the Data Measured in the ASTM E 1354 Cone Calorimeter and Reported in Refs. 36, 37, and 40

Materials[a]	ΔH_{ch} (MJ/kg)	y_{sm} (g/g)[b]
Ordinary		
High density polyethylene, HDPE	40.0	0.035
Polyethylene, PE	43.4	0.027
Polypropylene, PP	44.0	0.046
Polypropylene, PP	42.6	0.043
PP/glass fiber	NR	0.105
Polystyrene, PS	35.8	0.085
PS-FR	13.8	0.144
PS foam	27.7	0.128
PS foam-FR	26.7	0.136
Nylon	27.9	0.025
Nylon 6	28.8	0.011
Nylon/glass fiber	NR	0.089
Polyoxymethylene, POM	13.4	0.002
Polymethylmethacrylate, PMMA	24.2	0.010
Polybutyleneterephthalate, PBT	20.9	0.066
Polyethyleneterephthalate, PET	14.3	0.050
Acrylonitrile-butadiene-styrene, ABS	30.0	0.105
ABS	29.4	0.066
ABS-FR	11.7	0.132
ABS-PVC	17.6	0.124
Vinyl thermoplastic elastomer	6.4	0.056
Polyurethane, PU foam	18.4	0.054
Thermoplastic PU-FR	19.6	0.068
EPDM/Styrene acrylonitrile, SAN	29.0	0.116
Polyester/glass fiber (30%)	16.0	0.049
Isophthalic polyester	23.3	0.080
Isophthalic polyester/glass fiber (77%)	27.0	0.032
Polyvinyl ester	22.0	0.076
Polyvinyl ester/glass fiber (69%)-1	26.0	0.079
Polyvinyl ester/glass fiber-2	NR	0.164
Polyvinyl ester/glass fiber-3	NR	0.128
Epoxy	25.0	0.106
Epoxy/glass fiber (69%)-1	27.5	0.056
Epoxy/glass fiber-2	NR	0.142
Epoxy/glass fiber-3	NR	0.207
Epoxy/glass fiber-4	NR	0.058
Epoxy/glass fiber-5	NR	0.113
Epoxy/glass fiber-6	NR	0.115
Epoxy/glass fiber-7	NR	0.143
Epoxy/glass fiber-8	NR	0.149
Epoxy/glass fiber-9	NR	0.058
Epoxy/glass fiber-10	NR	0.086
Epoxy/graphite fiber-1	NR	0.082
Epoxy/graphite fiber-2	NR	0.049
Cyanate ester/glass fiber	NR	0.103
Acrylic/glass fiber	17.5	0.016
Kydex Acrylic paneling, FR	10.2	0.095

TABLE 2.7 (*Continued*)

Materials[a]	ΔH_{ch} (MJ/kg)	y_{sm} (g/g)[b]
High-temperature		
Polycarbonate, PC-1	21.9	0.098
PC-2	22.6	0.087
Cross linked polyethylene (XLPE)	23.8	0.026
Polyphenylene oxide, PPO-polystyrene (PS)	23.1	0.162
PPO/glass fiber	25.4	0.133
Polyphenylenesulfide, PPS/glass fiber-1	NR	0.063
PPS/graphite fiber-1	NR	0.075
PPS/glass fiber-2	NR	0.075
PPS/graphite fiber-2	NR	0.058
Polyarylsulfone/graphite fiber	NR	0.019
Polyethersulfone/graphite fiber	NR	0.014
Polyetheretherketone, PEEK/glass fiber (30%)	20.5	0.042
PEEK/graphite fiber	NR	0.025
Polyetherketoneketone, PEKK/glass fiber	NR	0.058
Bismaleimide, BMI/graphite fiber-1	NR	0.077
Bismaleimide, BMI/graphite fiber-2	NR	0.096
Bismaleimide, BMI/graphite fiber-3	NR	0.095
Bismaleimide, BMI/graphite fiber-4	NR	0.033
Phenolic/glass fiber (45%)-1	22.0	0.026
Phenolic/glass fiber-2	NR	0.008
Phenolic/glass fiber-3	NR	0.037
Phenolic/glass fiber-4	NR	0.032
Phenolic/glass fiber-5	NR	0.031
Phenolic/glass fiber-6	NR	0.031
Phenolic/glass fiber-7	NR	0.015
Phenolic/glass fiber-8	NR	0.009
Phenolic/graphite fiber-1	NR	0.039
Phenolic/graphite fiber-2	NR	0.041
Phenolic/graphite fiber-3	NR	0.021
Phenolic/PE fiber	NR	0.054
Phenolic/aramid fiber	NR	0.024
Phenolic insulating foam	10.0	0.026
Polyimide/glass fiber	NR	0.014
Wood		
Douglas fir	14.7	0.010
Hemlock	13.3	0.015
Textiles		
Wool	19.5	0.017
Acrylic fiber	27.5	0.038

TABLE 2.7 (*Continued*)

Materials[a]	ΔH_{ch} (MJ/kg)	y_{sm} (g/g)[b]
Halogenated		
PVC flexible-3 (LOI 25%)	11.3	0.099
PVC-FR (Sb_2O_3) flexible-4 (LOI 30%)	10.3	0.078
PVC-FR (triaryl phosphate) flexible-5 (LOI 34%)	10.8	0.098
PVC rigid-1	8.9	0.103
PVC rigid-2	10.8	0.112
PVC rigid-3	12.7	0.103
PVC rigid-1 (LOI 50%)	7.7	0.098
PVC rigid-2	8.3	0.076
Chlorinated PVC (CPVC)	5.8	0.003

[a] Generic materials marked by numbers 1 to 10 have different compositions as well as in some cases different manufacturers. Materials are identified in the references from where the data were taken.
[b] y_{sm}(g/g) = 0.0994 × (average extinction area) × 10^{-3} [25, 26].

TABLE 2.8 Peak Release Rates of Heat and Compounds from the Combustion of Materials in the ASTM E 2058 Fire Propagation Apparatus[a] [25, 26, 34, 35, 41]

	Release rates				
	Compounds (g/m^2-s)				Heat (kW/m^2)
Materials[b]	CO	CO_2	HC[c]	Smoke	
Nylon 6	0.40	22.5	<0.01	0.66	301
Polyvinylchloride, PVC-1	1.42	37.3	0.21	3.08	527
PVC-2	1.05	15.6	0.12	1.64	219
Polypropylene, PP-1	1.00	66.2	0.08	2.09	926
PP-2	1.95	78.8	0.36	2.86	1110
PP-3	3.52	88.3	1.17	3.35	1254
PP-4	1.43	77.0	0.16	2.35	1078
PP-5	0.78	54.0	0.08	1.73	755
PP-6	1.64	71.4	0.23	2.40	1004
Polyethylene, PE	2.34	91.4	0.57	2.04	1296
High density polyethylene, HDPE	3.95	93.1	1.40	2.52	1341
Ethylene-propylene-diene rubber copolymers, EPDM	0.60	18.0	0.01	0.91	242
Polystyrene, PS	0.64	28.1	0.07	2.34	381
Polyethyleneterephthalate, PET	0.27	9.42	0.02	0.44	125
Acrylonitrile-butadiene-styrene, ABS-PVC	0.60	11.5	0.03	1.08	158
Polycarbonate, PC-1	1.26	44.7	0.08	3.29	486
PC-2	1.74	51.2	0.21	4.75	559
Natural rubber	0.79	29.2	0.05	3.34	396
Cotton/polyester	0.53	36.3	0.03	1.51	488
Sheet molding compound	0.61	25.5	0.03	2.26	345

[a] Combustion in normal air at 50 kW/m^2 of external heat flux in the ASTM E 2058 apparatus.
[b] Generic materials marked by numbers 1 to 6 have different compositions as well as in some cases different manufacturers. Materials are identified in the references from where the data were taken.
[c] HC-total hydrocarbons.

TABLE 2.9 Average Heat of Combustion and Yields of Products from the Data Measured in the ASTM E 2058 Fire Propagation Apparatus [25, 26, 34, 35, 41]

Material	Composition	y_j(g/g) CO	CO_2	HC^a	Smoke	ΔH_{ch} (kJ/g)
Ordinary						
Polyethylene, PE	CH_2	0.024	2.76	0.007	0.060	38.4
Polypropylene, PP	CH_2	0.024	2.79	0.006	0.059	38.6
Polystyrene, PS	CH	0.060	2.33	0.014	0.164	27.0
Polystyrene foam	$CH_{1.1}$	0.061	2.32	0.015	0.194	25.5
Wood	$CH_{1.7}O_{0.73}$	0.004	1.30	0.001	0.015	12.6
Polyoxymethylene, POM	$CH_{2.0}O$	0.001	1.40	0.001	0.001	14.4
Polymethylmethacrylate, PMMA	$CH_{1.6}O_{0.40}$	0.010	2.12	0.001	0.022	24.2
Polyester	$CH_{1.4}O_{0.22}$	0.075	1.61	0.025	0.188	20.1
Nylon	$CH_{1.8}O_{0.17}N_{0.17}$	0.038	2.06	0.016	0.075	27.1
Flexible polyurethane foams	$CH_{1.8}O_{0.32}N_{0.06}$	0.028	1.53	0.004	0.070	17.6
Rigid polyurethane foams	$CH_{1.1}O_{0.21}N_{0.10}$	0.036	1.43	0.003	0.118	16.4
High temperature						
Polyetheretherketone, PEEK	$CH_{0.63}O_{0.16}$	0.029	1.60	0.001	0.008	17.0
Polysulfone, PSO	$CH_{0.81}O_{0.15}S_{0.04}$	0.034	1.80	0.001	0.020	20.0
Polyethersulfone, PES		0.040	1.50	0.001	0.021	16.7
Polyetherimide, PEI	$CH_{0.65}O_{0.16}N_{0.05}$	0.026	2.00	0.001	0.014	20.7
Polycarbonate, PC	$CH_{0.88}O_{0.19}$	0.054	1.50	0.001	0.112	16.7
Halogenated						
PE + 25% Cl	$CH_{1.9}Cl_{0.13}$	0.042	1.71	0.016	0.115	22.6
PE + 36% Cl	$CH_{1.8}Cl_{0.22}$	0.051	0.83	0.017	0.139	10.6
PE + 48% Cl	$CH_{1.7}Cl_{0.36}$	0.049	0.59	0.015	0.134	5.7
Polyvinylchloride, PVC	$CH_{1.5}Cl_{0.50}$	0.063	0.46	0.023	0.172	7.7
Chlorinated PVC	$CH_{1.3}Cl_{0.70}$	0.052	0.48	0.001	0.043	6.0
Polyvinylidenefluoride, PVDF	CHF	0.055	0.53	0.001	0.037	5.4
Polyethylenetetrafluoroethylene, ETFE	CHF	0.035	0.78	0.001	0.028	7.3
Polyethylenechlorotrifluoroethylene, ECTFE	$CHCl_{0.25}F_{0.75}$	0.095	0.41	0.001	0.038	4.5
Polytetrafluoroethylene, TFE	CF_2	0.092	0.38	0.001	0.003	2.8
Perfluoroalkoxy, PFA	$CF_{1.6}$	0.099	0.42	0.001	0.002	1.8
Polyfluorinated ethylene propylene, FEP	$CF_{1.8}$	0.116	0.25	0.001	0.003	1.0

[a] HC-total gaseous hydrocarbon.

2.7.4 ASTM E 119: Standard Test Methods for Fire Tests of Building Construction and Materials—The Fire Endurance Test [14]

This standard test method specifies use of a large-scale furnace for testing of walls, columns, floors, and other building members, under high fire exposure conditions. Detailed description of the apparatus and its sketch are included in the standard. Fire resistance is expressed in terms of time to reach the critical point, that is, $1/2$, 2, 6 h, and other ratings of building materials and assemblies as they are exposed to heat. The building materials and assemblies are exposed to heat in a natural gas or propane fueled furnace with the temperature increasing as follows:

5 minutes	538°C	10 minutes	704°C	30 minutes	843°C
1 hour	927°C	2 hour	1010°C	4 hour	1093°C
≥ 8 hour	1260°C				

The standard test method has been designed to test the following building materials and assemblies in the furnace:*

- Bearing and nonbearing walls and partitions: The area exposed to fire is ≥ 9 m² (100 ft²) with neither dimension less than 2.7 m (9 ft).
- Columns: The length of the column exposed to fire is ≥ 2.7 m (9 ft).
- Protection for structural steel columns: The length of the protected column is ≥ 2.4 m (8 ft) held in a vertical orientation. The column is exposed to heat on all sides.
- Floors and roofs: The area exposed to fire is ≥ 16 m² (180 ft²) with neither dimension ≥ 3.7 m (12 ft).
- Loaded restrained and unrestrained beams: The length of the beam exposed to fire is ≥ 3.7 m (12 ft) and tested in a horizontal position.
- Protection for solid structural steel beams and girders: The length of beam or girder exposed to the fire is ≥ 3.7 m (12 ft) tested in a horizontal position.
- Protective members in walls, partition, floor, or roof assemblies: The sizes used are the same as above for the respective specimens.

Various criteria are used for the acceptance of the specimens:

- Sustains itself or with the applied load without passage of flame or gases hot enough to ignite cotton waste or the polymer-based hose assembly for a period equal to that for which classification is desired.
- There is no opening that projects water from the stream beyond the unexposed surface during the time of water stream test.
- Rise in the temperature on the unexposed surface remains ≤ 139°C above its initial temperature.
- Transmission of heat through the protection during the period of fire exposure for which classification is desired maintains the average steel temperature of ≤ 538°C (measured temperature ≤ 649°C).
- For steel structural members (beams, open-web steel joists, etc.) spaced more than 1.2 m (4 ft), the average temperature of steel needed is ≤ 593°C (measured temperature ≤ 704°C) during the classification period.

Although this is a large-scale standard test method with intense heat-exposure conditions in a furnace, it does not capture the mode of heat transfer and other processes, especially radiative heat transfer from hot soot particles in the flame and aerodynamic conditions present in fires burning in the open, i.e., exposure of structural members and assemblies to large hydrocarbon pool fires.

2.7.5 ASTM E 1529: Standard Test Methods for Determining Effects of Large Hydrocarbon Pool Fires† on Structural Members and Assemblies [14]

The standard test method specifies a large-scale test similar to ASTM E 119, except that exposure of specimens consists of rapidly increasing heat flux. Detailed description of the apparatus and its

---

*As needed, load is applied to the specimens throughout the test to simulate a maximum load condition in their end-use application.
†A large pool fire is defined as that resulting from hundreds (or thousands) of gallons of liquid hydrocarbon fuel burning over a large area (several hundred to a thousand square meters) with relatively unrestricted airflow and release of chemical compounds. A range of temperatures, velocities, heat fluxes, and chemical conditions exists and varies dramatically with time and spatial location.

sketch are included in the standard. In this test, the specimen surface is exposed to an average heat flux of 158 ± 8 kW/m^2 attained within the first 5 min and maintained for the duration of the test. The temperature of the environment reaches ≥ 815°C after the first 3 min of the test and remains between 1010 and 1180°C at all times after the first 5 min. This standard test method is used to determine the response of columns, girders, beams or structural members, and fire-containment walls or either homogeneous or complete construction exposed to a rapidly increasing heat flux. In this standard test method, the control of both heat flux and temperature is specified, as compared with ASTM E 119, where only the temperature is specified.

Performance is defined as the period during which structural members or assemblies will continue to perform their intended function when subjected to fire exposure. The results are reported in terms of time increments, such as $^1/_2$, $^3/_4$, 1, 1$^1/_2$ h, and others.

The tests are performed in a manner similar to that in the ASTM E 119, except for the heat flux and temperature profiles. For example, in this standard test method, a heat flux exposure of 158 kW/m^2 to the specimen surface is specified within the first 5-min of the test. In ASTM E 119, a heat flux exposure of 35 kW/m^2 at 5 min and 118 kW/m^2 at 60 min to the specimen surface is specified.

Testing of structural members and assemblies exposed to simulated exposure to large, free-burning (outdoors), fluid-hydrocarbon-fueled pool fires is needed for the design of facilities for the hydrocarbon processing industry (oil refineries, petrochemical plants, offshore oil production platforms, and others) and chemical plants. In the future, the testing may also be used in the design of high-rise buildings because of the extreme terrorist act that occurred in New York City on September 11, 2001. There was a complete collapse of the World Trade Center Towers due to exposure to very hot pool fires from the large spillage of aviation gasoline [42].

2.8 POPULAR STANDARD TEST METHODS FOR FLAME SPREAD AND FIRE GROWTH

In the standard test methods, specifications are made for visual observations of the movement of flame and char during the test and measurements of the surface temperature and release rates of material vapors, heat, and chemical compounds, including smoke. Both small- and large-scale flame spread and fire growth tests are performed using materials and products. The following are some of the popular standard test methods for characterizing flame spread and fire growth behaviors of materials and products.

2.8.1 prEN ISO and FDIS 11925-2: Reaction to Fire Tests for Building Products—Part 2: Ignitability when Subjected to Direct Impingement of Flame [10,13]

This standard test method specifies the use of small-scale apparatus for testing. Detailed description of the apparatus and its sketch are included in the standard. The apparatus consists of a stainless steel 800-mm-high, 700-mm-long, and 400-mm-wide chamber with an exhaust duct attached at the top of the chamber. In the test, a 250-mm-long and 180-mm-wide specimen with thickness ≤ 60 mm is used. The specimen is placed between two halves of a U-shaped stainless steel frame holder held together by screws or clamps. Each arm of the holder is 15 mm wide and 5 mm thick. The total length and width of the holder are 370 and 110 mm, respectively, with an 80-mm-wide open mouth. The frame with the sample hangs vertically inside a stainless steel chamber.

The vertically hanging holder can move closer to or away from a 45° propane gas burner (similar to a bunsen burner). A 100 × 50 × 10-mm deep aluminum foil tray containing filter paper is placed beneath the specimen holder and replaced between the tests.

The flame from the burner is applied for 15 or 30 s and the burner is retracted smoothly. The location of flame application depends on the shape and construction of the specimens. For the 15-s flame application, the test duration is 20 s after flame application. For the 30-s flame application, the test duration is 60 s after flame application. The following observations are made in the test:

- ignition of the specimen
- flame spread F_s up to 150 mm and time taken
- presence of flaming droplets
- ignition of the filter paper below the specimen

In Europe, data from ISO 11925-2 are used for the classification of reaction to fire performance for construction products (prEN 13501-1) [10]:

- construction products excluding floorings:
  - Class B: $F_s \leq 150$ mm within 60 s for 30-s exposure
  - Class C: $F_s \leq 150$ mm within 60 s for 30-s exposure
  - Class D: $F_s \leq 150$ mm within 60 s for 30-s exposure
  - Class E: $F_s \leq 150$ mm within 20 s for 15-s exposure
- floorings:
  - Class B_{fl}: $F_s \leq 150$ mm within 20 s for 15-s exposure
  - Class C_{fl}: $F_s \leq 150$ mm within 20 s for 15-s exposure
  - Class D_{fl}: $F_s \leq 150$ mm within 20 s for 15-s exposure
  - Class E_{fl}: $F_s \leq 150$ mm within 20 s for 15-s exposure

This standard test method cannot predict the fire behavior of materials and products other than those used in the test. Thus, it is useful only to screen the materials and products for their resistance to flame spread and burning under the test conditions. The use of this standard test method can lead to erroneous ranking of their end-use application conditions.

2.8.2 UL 94: Standard Test Methodology for Flammability of Plastic Materials for Parts in Devices and Appliances [12]

This standard test method specifies the use of a small-scale apparatus for testing and is similar to prEN ISO and FDIS 11925-2 tests. Detailed description of the apparatus and its sketch are included in the standard. In the test, both horizontal (HB) and vertical burning (V) behaviors of 127-mm (5-in)-long, 13-mm (0.5-in)-wide, and up to 13-mm (0.5-in)-thick material samples are examined. The horizontal burning test is performed for 94HB classification of materials.

In the horizontal burning test, the sample is placed on top of a wire gauge and ignited by a 30-s exposure to a bunsen burner at one end. The material is classified as 94HB if over the entire length of the sample (76 mm, or 3.0 in), the flame spread rate is: (1) less than 38 mm/min for 3- to 13-mm-thick sample and (2) less than 76 mm/min or the extent of flame spread is less than 102 mm (4.0 in) for less than 3-mm-thick sample.

In the vertical burning test, the bottom edge of the sample is ignited by a 5-s exposure to a bunsen burner with a 5-s delay and repeated five times until the sample ignites. The materials are classified as 94V-0, 94V-1, or 94V-2 based on the flaming combustion time after removal of the test flame, total flaming combustion time after 10 test flame applications, flaming and glowing, and dripping. The criteria for classifying the materials as 94V-0, 94V-1, and 94V-2 are listed in Table 2.10.

The relative resistance of materials to flame spread and burning according to UL 94 is HB < V-2 < V-1 < V-0. Examples of the UL 94 classification of materials are listed in Table 2.2 along with their vaporization and decomposition temperatures and char yields, where data are taken from Ref. 28. All the ordinary materials listed in the table, which generally have low fire resistance, are classified as HB. Most of the high-temperature and halogenated polymers listed in the table, which generally have high fire resistance, are classified as V-0.

As intended, the test criteria are applicable to materials used for the construction of small parts in metallic devices and appliances exposed to small ignition sources. The test criteria were not developed for use in large devices and appliances that are made entirely of the materials and exposed to high-intensity ignition sources. The test was not developed for predicting the fire behavior of mate-

TABLE 2.10 The UL 94V-0, 94V-1, and 94V-2 Material Classification Criteria [12]

	Classification		
Criterion	94V-0	94V-1	94V-2
A. Flaming combustion time after removal of the test flame (s)	≤10	≤30	≤30
B. Total flaming combustion time after 10 test flame applications for each set of five specimens (s)	≤50	≤250	≤250
C. Burning with flaming or glowing combustion up to the holding clamp	No	No	No
D. Dripping flaming particles that ignite the dry absorbent surgical cotton located 12-in (305-mm) below the test specimen	None	None	Yes
E. Glowing combustion persisting for more than 30 seconds after the second removal of the test flame (s)	None	≤60	≤60

rials and products expected in actual fires, but rather to screen them out in terms of their resistance to fire spread and burning.*

2.8.3 ASTM D 2863 (ISO 4589): Test Methodology for Limited Oxygen Index [14]

This standard test method specifies the use of a small-scale apparatus for testing. Detailed description of the apparatus and its sketch are included in the standard. In the test, a 70- to 150-mm (2.8- to 5.9-in)-long, 6.5-mm (0.26-in)-wide, and 3-mm (0.12-in)-thick vertical sheet of a material is placed inside a glass cylinder, with gas flowing in an upward direction. The test is performed at various oxygen concentrations under ambient temperature, with sample ignited at the top, to determine the minimum oxygen concentration at or below which there is no downward flame spread, which is defined as the *limited oxygen index* (LOI) of the material.

The decrease in the oxygen concentration decreases the flame spread rate by decreasing the flame heat flux transferred ahead of the flame front [Eq. (2.10)] [26, 27, 43], such as shown in Fig. 2.4 where data are taken from Ref. 44. Near the LOI value, the flame front is not able to supply the heat

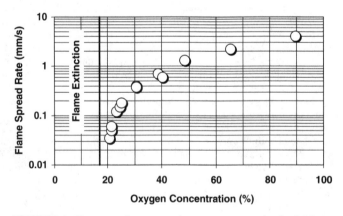

FIGURE 2.4 Flame spread rate versus the oxygen concentration for the downward fire spread over PMMA sheet with an LOI value of 17 percent. Data are taken from Ref. 44.

---

*The major limitation with this standard test method is that it is very difficult to assess the fire behavior of products for heat exposure and environmental conditions and shape, size, and arrangements of the products other than those used in the test.

TABLE 2.11 Ambient Temperature Limited Oxygen Index (LOI) Values for Materials [26, 28, 45]

Material	LOI	Material	LOI
Ordinary		Polypara(benzoyl)phenylene	41
		Polybenzimidazole (PBI)	42
Polyoxymethylene	15	Polyphenylenesulfide (PPS)	44
Cotton	16	Polyamideimide (PAI)	45
Cellulose acetate	17	Polyetherimide (PEI)	47
Natural rubber foam	17	Polyparaphenylene	55
Polypropylene	17	Polybenzobisoxazole (PBO)	56
Polymethylmethacrylate	17	Composites	
Polyurethane foam	17		
Polyethylene	18	Polyethylene/Al_2O_3(50%)	20
Polystyrene	18	ABS/glass fiber (20%)	22
Polyacrylonitrile	18	Epoxy/glass fiber (65%)	38
ABS	18	Epoxy/glass fiber (65%)-300°C	16
Poly(α-methylstyrene)	18	Epoxy/graphite fiber (1092)	33
Filter paper	18	Polyester/glass fiber (70%)	20
Rayon	19	Polyester/glass fiber (70%)-300 °C	28
Polyisoprene	19	Phenolic/glass fiber (80%)	53
Epoxy	20	Phenolic/glass fiber (80%)-100°C	98
Polyethyleneterephthalate (PET)	21	Phenolic/Kevlar® (80%)	28
Nylon 6	21	Phenolic/Kevlar® (80%)-300 °C	26
Polyester fabric	21	PPS/glass fiber (1069)	64
Plywood	23	PEEK/glass fiber (1086)	58
Silicone rubber (RTV, etc)	23	PAS/graphite (1081)	66
Wool	24	BMI/graphite fiber (1097)	55
Nylon 6,6	24–29	BMI/graphite fiber (1098)	60
Neoprene rubber	26	BMI/glass fiber (1097)	65
Silicone grease	26	Halogenated	
Polyethylenephthalate (PEN)	32		
High temperature		Fluorinated Cyanate Ester	40
		Neoprene	40
Polycarbonate	26	Fluorosilicone grease	31–68
Nomex®	29	Fluorocarbon rubber	41–61
Polydimethylsiloxane (PDMS)	30	Polyvinylidenefluoride	43–65
Polysulfone	31	PVC (rigid)	50
Polyvinyl ester/glass fiber (1031)	34	PVC (chlorinated)	45–60
Polyetherketoneketone (PEKK)	35	Polyvinylidenechloride (Saran®)	60
Polyimide (Kapton®)	37	Chlorotrifluoroethylene lubricants	67–75
Polypromellitimide (PI)	37	Fluorocarbon (FEP/PFA) tubing	77–100
Polyaramide (Kevlar®)	38	Polytrichloroethylene	95
Polyphenylsulfone (PPSF)	38	Polytrichlorofluorethylene	95
Polyetherketone (PEK)	40		
Polyetherketoneketone (PEKK)	40		

flux required to satisfy the CHF and TRP values of the materials and thus the flame is extinguished. This is shown in Fig. 2.4 for PMMA with an LOI value of 17 percent.

Thus, materials with higher LOI values require higher flame heat flux because of high CHF and TRP values or higher resistance to ignition (and thus to flame spread). Examples of LOI values measured at the ambient temperature for various materials are listed in Table 2.11 taken from Refs. 26, 28, and 45. The LOI values are arranged based on the generic nature of the materials.

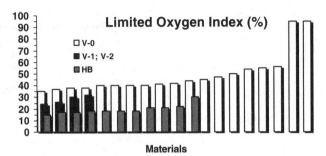

FIGURE 2.5 Relationship between LOI values and UL 94 classification of materials. Data are taken from Refs. 26, 28, and 45. UL classification of materials is listed in Table 2.2 and LOI values are listed in Table 2.11.

The LOI values and UL 94 classification of materials are interrelated as shown in Fig. 2.5. The LOI values for V-0 materials are ≥ 35, whereas the LOI values are <30 for materials classified as V-1, V-2, and HB. As intended, the standard test method is applicable to materials with small surface areas, such as in small parts of metallic devices and appliances exposed to small ignition sources, similar to the UL 94 test.

This standard test method has not been developed to predict the fire behavior of materials expected in actual fires, but rather to screen materials for low and high resistance to fire propagation.* For the majority of high-temperature and highly halogenated materials, the LOI values are ≥ 40. These polymers have high resistance to ignition and combustion, as well as flame spread, independent of fire size and ignition source strength [25, 26, 34, 46].

2.8.4 ASTM E 162 (D 3675): Standard Test Method for Surface Flammability Using a Radiant Energy Source [14]

This standard test method specifies a small-scale apparatus. Detailed description of the apparatus and its sketch are included in the standard. In this test method, a vertical sample 460-mm (18-in) × 150-mm (6-in) wide and up to 25-mm (1-in) thick is used. The sample is exposed to a temperature of 670 ± 4°C at the top from a 300-mm (18-in) × 300-mm (12-in) inclined radiant heater with the top of the heater closest to and the bottom farthest away from the sample surface. The sample is ignited at the top and flame spreads in the downward direction.

In the test, measurements are made for the arrival time of flame at each of the 75-mm (3-in) marks on the sample holder and the maximum temperature rise of the stack thermocouples. The test is completed when the flame reaches the full length of the sample or after an exposure time of 15-min, whichever occurs earlier (provided the maximum temperature of the stack thermocouples is reached). *Flame spread index* I_s is calculated from the measured data, defined as the product of flame spread factor F_s and the heat evolution factor Q.

Many materials and products have been tested using this standard test method. Examples of the data are listed in Table 2.12, where data are taken from Refs. 37 and 47. The I_s values vary from 0 to 2220, suggesting large variations in the fire spread behavior of materials.

Many regulations and codes specify the I_s value as an acceptance criterion of materials and products. For example, for structural composites inside naval submarines [37] and for passenger

---

*The major limitation with this standard test method is that it is sensitive to ambient temperature. The LOI values decrease with increase in the ambient temperature. Thus, it is very difficult to assess the fire behavior of products for heat exposure and environmental conditions other than those used in the test. The effects of temperature on LOI values have been examined in various studies that are listed in Refs. 25, 26, 35, and 59.

TABLE 2.12 Flame Spread Indices for Materials from ASTM E 162 Test [37, 47]

Material[a]	Thickness (mm)	Flame spread index, I_s
Polyurethane polyether rigid foam		2220
Polyurethane polyether flexible foam		1490
Polyurethane polyester rigid foam-1, FR		1440
Polyurethane polyester flexible foam, FR		1000
Polyurethane polyester rigid foam-2, FR		880
Acrylic, FR	3.2	376
Polystyrene	1.7	355
Polyester/glass fiber (21%)	1.6	239
1087 Vinyl ester/glass fiber		156
Plywood, FR, exterior	6.4	143
Polystyrene, rigid foam		114
Phenolic, laminate	1.6	107
Red oak	19.1	99
Polyester-FR/glass fiber (27%)	2.4	66
Phenolic/polyethylene fiber		48
Epoxy/glass fiber-1		43
Phenolic/aramid fiber		30
Vinyl ester/glass fiber		27
Epoxy/glass fiber-2		23
Phenolic/graphite fiber-1		20
Bismaleimide/graphite fiber-1		17
Polystyrene, rigid foam, FR		13
Bismaleimide/graphite fiber-2		13
Nylon/glass fiber		13
Epoxy/glass fiber-3		12
Bismaleimide/graphite fiber-3		12
Epoxy/glass fiber-4		11
Epoxy/glass fiber-5		11
Polyurethane polyether flexible foam, FR		10
Polyvinylchloride, PVC	3.7	10
Polyarylsulfone/graphite fiber		9
Polyphenylenesulfide/glass fiber-1		8
Polyphenylenesulfide/glass fiber-2		7
Phenolic/glass fiber-1		6
Phenolic/graphite fiber-2		6
Phenolic/glass fiber-2		5
Phenolic/glass fiber-3		4
Phenolic/glass fiber-4		4
Phenolic/glass fiber-5		4
Phenolic/glass fiber-6		4
PVC, FR	3.7	3
Bismaleimide/graphite fiber-4		3
Phenolic/graphite fiber-3		3
Polyphenylenesulfide/glass fiber-3		3
Polyphenylenesulfide/glass fiber-4		3
Polyetheretherketone/graphite fiber		3
Polyetheretherketone/glass fiber		3
Polyimide/glass fiber		2
Phenolic/glass fiber-7		1
Asbestos cement board	4.8	0

[a] Generic materials marked by numbers 1 to 7 have different compositions as well as in some cases different manufacturers. Materials are identified in the references from where the data were taken.

cars and locomotive cabs [48,49], the following I_s values are specified for the acceptance of the materials:

- $I_s < 20$ for structural composites inside naval submarines
- $I_s \leq 25$ for cushions, mattresses, and vehicle components made of flexible cellular foams for passenger cars and locomotive cabs and thermal and acoustic insulation for buses and vans
- $I_s \leq 35$ for all vehicle components in passenger cars and locomotive cabs and for seating frame, seating shroud, panel walls, ceiling, partition, windscreen, HVAC ducting, light diffuser, and exterior shells in buses and vans
- $I_s \leq 100$ for polymers for passenger cars and locomotive cabs that are used for their optically transparent properties

The above listed criteria for the I_s values (<20) suggest that structural composites for inside naval submarines are expected to have high resistance to flame spread and heat release if exposed to heat flux values similar to those used in the ASTM E 162. In addition, materials used in passenger cars, locomotive cabs, buses, and vans with I_s values of ≤ 25 and ≤ 35 are expected to have relatively higher resistance to fire spread and heat release rate compared with ordinary materials with I_s values of ≤ 100 under low-heat-exposure conditions.

This test method has not been developed to predict the fire behavior of materials and products expected in actual fires, but rather to screen them for low and high resistance to fire spread under conditions of lower-intensity heat exposure.*

2.8.5 ASTM E 1321 (ISO 5658): Standard Test Method for Determining Material Ignition and Flame Spread Properties (LIFT) [13, 14]

This standard test method specifies the use of an intermediate-scale test apparatus to determine the material properties related to piloted ignition of a vertically oriented sample under a constant and uniform heat flux and to lateral flame spread on a vertical surface due to an externally applied radiant-heat flux. Detailed description of the apparatus and its sketch are included in the standard. For the ignition test, a 155-mm (6-in) square sample is exposed to a nearly uniform heat flux and the time-to-flame attachment is measured.

For the flame spread test, an 800-mm (31-in)-long and 155-mm (6-in)-wide horizontal sample turned vertically on its side is used. The sample is placed in front of a 280-mm (11-in) × 483-mm (19-in) radiant heater with a 15° orientation to the heater such that the sample surface is exposed to decreasing heat flux. The heat flux varies from 31 kW/m² at the top near the pilot flame (50-mm from the top) to 2 kW/m² at the bottom (750-mm from the top). Fig. 2.6 shows examples of the data for the flame spread rate versus the external heat flux measured in the ISO 5658 and reported in Ref. 50. The products in the figure are identified in Table 2.13. As can be noted, the flame spread rate increases with increase in the external heat flux as expected from Eq. (2.10).

The flame spread rate is correlated by the following relationship, where the numerator in Eq. (2.10) is replaced by Φ, defined as the flame-heating parameter (kW²/m³) [14, 30, 31]:

$$V = \Phi/k\rho c(\Delta T_{ig})^2 \qquad (2.13)$$

The test provides data that are used to derive T_{ig}, $k\rho c$, Φ, and the minimum temperature for flame spread $T_{s,min}$, which are listed in Table 2.14 as taken from Ref. 30. Results from the lateral flame spread tests in the ASTM E 1321 apparatus are similar to downward flame spread, except for materials with excessive melting and dripping [14, 30, 31]. Data in Table 2.14 are generally representative of common construction or interior finish materials [30].

---

*The major limitation with this standard test method is that it is very difficult to assess the fire behaviors of products for heat exposure and environmental conditions and shape, size, and arrangements of the products other than those used in the test.

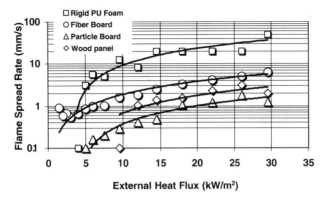

FIGURE 2.6 Fire spread rate versus the external heat flux measured in the ISO 5658 (LIFT). Data are taken from Ref. 50.

The properties derived from the ASTM E 1321 (ISO 5658) provide information about the flame spread characteristics of materials and can serve as an indication of their hazardous characteristics [14, 30, 31]. The test results provide ignition and flame spread properties of materials needed by flame spread and fire growth theories [14, 30, 31]. The analysis may be used to rank materials performance by some set of criteria applied to the correlation; or the analysis may be employed in fire risk growth models to develop a more rational and complete risk assessment for wall materials [14, 30, 31].

$T_{s,min}$ is related to T_{ig} as shown in Fig. 2.7. The correlation between these parameters is included in the figure. Thus, materials with higher ignition temperature (higher ignition resistance) have higher resistance to flame spread (this is to be expected).

The ASTM E 1321 (ISO 5658) test method has been developed to provide pertinent data needed by the prescriptive- and performance-based fire codes for the fire hazard analyses and protection needs for residential, private, government, and industrial occupancies, transport, manufacturing, and others.

TABLE 2.13 Building Products Tested in the ASTM E 1321 (ISO 5658) Standard Test Apparatus [50]

No	Product	Thickness (mm)	Density (kg/m³)
1	Insulating fiberboard	13	250
2	Medium density fiberboard	12	600
3	Particle board	10	750
4	Gypsum plaster board	13	700
5	PVC wall covering on gypsum plaster board	0.70	240
6	Paper wall covering on gypsum plaster board	0.60	200
7	Textile wall covering on gypsum plaster board	0.70	370
8	Textile wall covering on mineral wool	50	100
9	Melamine faced particle board	1.2	810
10	Expanded polystyrene (PS)	50	20
11	Polyurethane rigid (PUR) foam	30	30
12	Wood panel (spruce)	11	530
13	Paper wall covering on particle board	0.60	200

TABLE 2.14 Ignition and Fire Spread Properties Derived from the ASTM E 1321 Test [30]

Material	T_{ig} (°C)	$k\rho c$ (kW²-s/m⁴-K²)	Φ (kW²/m³)	$T_{s,min}$ (°C)	$\Phi/k\rho c$ (m-K²/s)	V (mm/s)
Synthetic						
Polyisocyanurate foam (5.1 cm)	445	0.02	4.9	275	245	36.4
Foam, rigid (2.5 cm)	435	0.03	4.0	215	133	20.3
Polyurethane foam, flexible (2.5 cm)	390	0.32	11.7	120	37	6.2
PMMA Type G (1.3 cm)	378	1.02	14.4	90	14	2.5
PMMA Polycast (1.6 mm)	278	0.73	5.4	120	7	1.8
Polycarbonate (1.5 mm)	528	1.16	14.7	455	13	1.6
Carpets						
Carpet (acrylic)	300	0.42	9.9	165	24	5.3
Carpet #2 (wool, untreated)	435	0.25	7.3	335	29	4.4
Carpet (nylon/wool blend)	412	0.68	11.1	265	16	2.6
Carpet #1 (wool, stock)	465	0.11	1.8	450	16	2.3
Carpet #2 (wool, treated)	455	0.24	0.8	365	3	0.5
Natural						
Plywood, plain (1.3 cm)	390	0.54	12.9	120	24	4.1
Gypsum board, (common) (1.3 mm)	565	0.45	14.4	425	32	3.7
Gypsum board, FR (1.3 cm)	510	0.40	9.2	300	23	3.0
Plywood, plain (6.4 mm)	390	0.46	7.4	170	16	2.7
Fiberglass shingle	445	0.50	9.0	415	18	2.7
Douglas fir particle board (1.3 cm)	382	0.94	12.7	210	14	2.4
Hardboard (3.2 mm)	365	0.88	10.9	40	12	2.3
Hardboard (nitrocellulose paint)	400	0.79	9.8	180	12	2.1
Asphalt shingle	378	0.70	5.3	140	8	1.3
Fiber insulation board	355	0.46	2.2	210	5	0.9
Particle board (1.3 cm stock)	412	0.93	4.2	275	5	0.7
Hardboard (6.4 mm)	298	1.87	4.5	170	2	0.5
Hardboard (gloss paint) (3.4 mm)	400	1.22	3.5	320	3	0.5
Gypsum board, wallpaper (S142M)	412	0.57	0.79	240	1	0.2

2.8.6 ASTM E 648 (ISO 9239-1): Standard Test Method for Critical Radiant Flux of Floor-Covering Systems Using a Radiant Heat Energy Source [13, 14]

This standard test method specifies the use of an intermediate-scale test method, similar in principle to ASTM E 1321 (ISO 5658). Detailed description of the apparatus and its sketch are included in the standard. A 1.0-m (39.4-in)-long and 0.20-m (7.9-in)-wide horizontal sample is exposed to radiant heat flux in the range of 1 to 11 kW/m² from a 30°-inclined radiant panel all contained inside a chamber. The sample surface closer to the radiant heater is exposed to 11 kW/m². The radiant flux decreases as the distance between the sample surface and the radiant heater increases to the lowest value of 1 kW/m².

A pilot flame ignites the sample surface exposed to 11 kW/m², and flame spread is observed until the flame is extinguished at some downstream distance due to the decrease in the radiant flux. The radiant flux at this distance is defined as the *critical radiant flux* (CRF) of the sample:

$$CRF = \dot{q}''_{cr} - \dot{q}''_f(x) \tag{2.14}$$

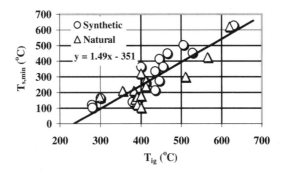

FIGURE 2.7 Relationship between the minimum temperature for fire spread and the ignition temperature of materials. All the data were measured in ASTM E 1321 (LIFT) apparatus and reported in Ref. 30.

where $\dot{q}''_f(x)$ is the flame heat flux at distance x where flame is extinguished (kW/m²). Thus, materials and products for which the radiant fraction of the flame heat flux is higher would have lower CRF values. Materials with a higher radiant fraction of the flame heat flux have lower resistance to flame spread due to higher heat transfer efficiency ahead of the flame front.

This test method was developed as a result of the need for a flammability standard for carpets and rugs to protect the public against fire hazards [51]. Consequently, several carpet systems were tested by this standard [51–53]. Examples of the CRF values for selected materials taken from Ref. 52 are listed in Table 2.15.

In the United States, ASTM E 648 standard test method is specified for the classification of the interior floor finish in buildings in the NFPA 101 Life Safety Code (Table 2.17 lists the interior finish classification limitations) [54]:

- Class I interior floor finish: CRF > 4.5 kW/m²
- Class II interior floor finish: 2.2 kW/m² < CRF < 4.5 kW/m²

TABLE 2.15 Critical Radiant Flux for Carpets from ASTM E 648 [52][a]

Fiber weight (oz)	Style	Fiber type	Yarn	Adhesive	CRF (kW/m²)
28	Cut pile	Nylon 6,6	BCF[b]	Nu Broadlok II[c]	18.5
26	Loop pile	Polypropylene	BCF[b]	Supra STIX 90[c]	2.8
50	Cut pile	Nylon 6,6	Staple	Supra STIX 90[c]	4.6
				Supra STIX 90[d]	4.8
				Supra STIX 90[e]	4.7
28	Cut pile	Nylon 6,6	Staple	Supra STIX 90[c]	3.0
28	Loop pile	Nylon 6,6	BCF[b]	Nu Broadlok II[c]	17.5
50	Cut pile	Wool	Staple	Supra STIX 90[c]	6.4
24	Loop pile	Nylon 6	BCF[b]	Supra STIX 90[c]	3.2
24	Loop pile	Nylon 6	BCF[b]	Supra STIX 90[c]	3.1

[a] Preheat time-2 minutes.
[b] BCF-bulk continuous filament.
[c] Substrate-Sterling Board (high-density inorganic fiber reinforced cement board).
[d] Substrate-Ultra Board.
[e] Ester Board.

In Europe, ISO 9293-1 with test duration of 30 min is specified for the Euroclasses for flooring in prEN 13501-1 [10]:

- Class $A2_{fl}$: CRF ≥ 8 kW/m² and smoke production (s_1 = smoke ≤ 750 percent, minimum; s_2 = not s_1);
- Class B_{fl}: CRF ≥ 8 kW/m² and smoke production (s_1 = smoke ≤ 750 percent, minimum; s_2 = not s_1);
- Class C_{fl}: CRF ≥ 4.5 kW/m² and smoke production (s_1 = smoke ≤ 750 percent, minimum; s_2 = not s_1);
- Class D_{fl}: CRF ≥ 3 kW/m² and smoke production (s_1 = smoke (≤ 750 percent, minimum; s_2 = not s_1).

CRF represents the resistance of a material or product to flame spread. It characterizes the ability of the leading edge of the flame in providing sufficient heat flux ahead of the flame to satisfy the flame spread requirements of the material and the product. The intent of the test method is to separate materials and products with higher flame spread resistance from those with lower resistance. It is not intended to predict the flame spread behaviors of materials and products in actual fires.*

2.8.7 ASTM E 84: Standard Test Method for Surface Burning Characteristics of Building Materials [14]

This standard test method specifies the use of a larger-scale apparatus for testing. It is one of the most widely specified methods. Detailed description of the apparatus and its sketch are included in the standard. In this 10-min test, a 7.3-m (24-ft)-long and 0.51-m (20-in)-wide horizontal sample is used inside a 7.6-m (25-ft)-long, 0.61-m (24-in)-wide sample location and 0.31-m (12-in)-deep tunnel. Two gas burners, located 0.19 m (7 in) below the specimen surface and 0.31 m (12 in) from one end of the tunnel are used as ignition sources. The two burners release 88 kW of heat creating a gas temperature of 900°C near the specimen surface. The flames from the burners cover 1.37 m (4.5 ft) of the length and 0.45 m (17 in) of the width of the sample [a surface area of 0.63 m² (7 ft²)]. Air enters the tunnel 1.4 m (54 in) upstream of the burner at a velocity of 73-m (240-ft)/min. The test conditions are set such that for red oak flooring control material, flame spreads to the end of the 7.3-m (24-ft)-long sample in 5.5 min or a flame spread rate is 22 mm/s.

In the test, measurements are made for the percent light obscuration by smoke flowing through the exhaust duct, gas temperature [7.0 m (23 ft) from the burner], and location of the leading edge of the flame (visual measurement) as functions of time. The measured data are used to calculate the *flame spread index* (FSI) and *smoke developed index* (SDI) from the flame-spread-distance time and percent-light-absorption time areas, respectively. Some typical FSI values are listed in Table 2.16 as taken from Ref. 14.

The NFPA 101 Life Safety Code uses the ASTM E 84 test data for the following classification of building products (Table 2.17 lists the interior finish classification limitations) [54]:

- Class A interior wall and ceiling finish: FSI 0 to 25; SDI 0 to 450
- Class B interior wall and ceiling finish: FSI 26 to 75; SDI 0 to 450
- Class C interior wall and ceiling finish: FSI 76 to 200; SDI 0 to 450

The FSI value represents the resistance to flame spread and decreases with increase in the resistance. The intent of the test method is to separate materials and products with higher flame spread resistance from those with lower resistance. It has not been designed to predict the flame spread behavior of materials and products in actual fires. The test method does not provide any information on the melting behaviors of the materials and products leading to pool fires and thus is limited in its application.†

---

*The major limitation with this standard test method is that it is very difficult to assess the fire behaviors of products for heat exposure and environmental conditions and shape, size, and arrangements of the products other than those used in the test.

†The major limitation with this standard test method is that it is very difficult to assess the fire behaviors of products for heat exposure and environmental conditions and shape, size, and arrangements of the products other than those used in the test.

TABLE 2.16 ASTM E-84 Flame Spread Index for Materials [14]

Material	Flame spread index (FSI)
Plywood, fir, exterior	143
Douglas fir plywood	91
Rigid polyurethane foam	24
Douglas fir plywood/FR	17
Composite panel	17
Type X gypsum board	9
Rigid polystyrene foam	7

TABLE 2.17 Interior Finish Classification Limitations: NFPA 101 Life Safety Code [54]

Occupancy	Exits	Access to exits	Other spaces
Assembly, new >300 occupant load	A	A or B	A or B
Assembly, new ≤300 occupant load	A	A or B	A, B, or C
Assembly, existing >300 occupant load	A	A or B	A or B
Assembly, existing ≤300 occupant load	A	A or B	A, B, or C
Education, new	A	A or B	A or B; C on partitions
Education, existing	A	A or B	A, B, or C
Day care centers-new	A I or II	A I or II	A or B NR
Day care centers-existing	A or B	A or B	A or B
Group day care homes-new	A or B	A or B	A, B or C
Group day care homes-existing	A or B	A, B, or C	A, B or C
Family day care homes	A or B	A, B, or C	A, B or C
Health care-new	A or B	A or B; C on lower portion of corridor wall	A or B; C in small individual rooms
Health care-existing	A or B	A or B	A or B
Detention and correctional-new	A; I	A; I	A, B or C
Detention and correctional-existing	A or B; I or II	A or B; I or II	A, B or C
1- and 2-family dwellings, lodging or rooming houses	A, B or C	A, B or C	A, B or C
Hotels and dormitories-new	A; I or II	A or B; I or II	A, B, or C
Hotels and dormitories-existing	A or B; I or II	A or B; I or II	A, B, or C
Apartment buildings-new	A; I or II	A or B; I or II	A, B, or C
Apartment buildings-existing	A or B; I or II	A or B; I or II	A, B, or C
Mercantile-new	A or B	A or B	A or B
Mercantile-existing, class A or B	A or B	A or B	Ceilings-A or B; walls-A, B, or C
Mercantile-existing, class C	A, B or C	A, B or C	A, B or C
Business and ambulatory health care-new	A or B; I or II	A or B; I or II	A, B, or C
Business and ambulatory health care-existing	A or B	A or B	A, B or C
Industrial	A or B	A, B or C	A, B or C
Storage	A or B	A, B or C	A, B or C

A: Flame spread index (FSI) from ASTM E 84-0-25, Smoke Development Index (SDI) from ASTM E 84-0-450; B: FSI-26-75, SDI-0-450; C: FSI-76-200, SDI-0-450; I: Critical Radiant Flux (CRF) from ASTM E 648 > 4.5 kW/m^2; II: 2.2 < CRF < 4.5 kW/m^2.

2.8.8 FM Global Approval Class 4910 [11] (NFPA 318 [55]): Standard Test Methods for Clean Room Materials for the Semiconductor Industry

This standard test method specifies use of a small-scale apparatus (ASTM E 2058 fire propagation apparatus) for the quantification of flammability of materials used for the construction of products in clean rooms for the semiconductor industry. Detailed description of the apparatus and its sketch are included in the standard.

In this standard test method, a parameter defined as the *fire propagation index* (FPI), which relates to the large-scale vertical flame spread behaviors of materials, is used to evaluate the flame spread behavior of materials [25, 26, 34]. FPI is the equivalent of $V^{1/2}$ in Eq. (2.10), with the numerator replaced based on the relationship between $\dot{q}''_f$ and $\dot{Q}''_{ch}$ [25, 26, 34]:

$$FPI = 1000 \left(\frac{(0.42 \dot{Q}_{ch}/w)^{1/3}}{(\Delta T_{ig} \sqrt{k\rho c})} \right) \quad (2.15)$$

where FPI is in $(m/s^{1/2})/(kW/m)^{2/3}$, $\dot{Q}_{ch}$ is the chemical heat release rate (kW), and w is the specimen width (m).

The standard test method also specifies the use of SDI for the smoke release characteristics of materials during flame spread. SDI is expressed by the following expression, based on the relationship among FPI, $\dot{m}''$, and release rate of smoke [Eq. (2.12)] [25, 26, 34]:

$$SDI = (FPI) y_s \quad (2.16)$$

where SDI is in $(g/g)(m/s^{1/2})/(kW/m)^{2/3}$ and the yield of smoke y_s is in grams per gram (g/g). The standard test method specifies the following criteria for the selection of materials for the construction of products for the clean rooms of the semiconductor industry:

- FPI $\leq$ 6 $(m/s^{1/2})/(kW/m)^{2/3}$
- SDI $\leq$ 0.4 $(g/g)(m/s^{1/2})/(kW/m)^{2/3}$

For quantifying the FPI and SDI values, ignition, combustion, and flame spread tests are performed. Ignition and combustion tests are performed in normal air using square [100 mm (4 in)] or round [100 mm (4 in) diameter] horizontal samples. Ignition tests are performed at various external heat flux values, whereas combustion tests are performed at 50 kW/m² of external heat flux. Flame spread tests are performed using 100-mm (4-in)-wide, 305-m-high (12-in) and 3-mm (0.1-in) to 25-mm (1-in)-thick vertical specimens. In the flame spread tests, large-scale flame radiative heat flux is simulated by using 40% oxygen concentration [43].

The flame radiative heat flux increases with oxygen concentration due to an increase in the flame temperature and soot concentration and a decrease in the diffusion flame height (or reduction in the soot residence time in the flame), creating an ideal path for enhanced flame radiation [43]. The increase in flame radiative heat flux with increase in oxygen concentration is reflected in the increase in the flame spread rate, as shown in Fig. 2.8 with data taken from Ref. 44.

Ignition tests are performed to obtain the $\Delta T_{ig}\sqrt{k\rho c}$ values for the calculation of FPI values from Eq. (2.15). Combustion tests are performed to obtain the y_s value to calculate SDI value from Eq. (2.16). Flame spread tests are performed to obtain $\dot{Q}_{ch}$ values for the calculation of FPI values from Eq. (2.15).

FPI and SDI values have been determined for variety of materials, electrical cables, and conveyor belts [25, 26, 34, 56–61], examples of which are listed in Table 2.18. Visual observations made during the flame spread tests in the ASTM E 2058 and in the large-scale parallel-panel tests indicate that:

- For FPI $\leq$ 6, flames are close to extinction conditions and flame spread is limited to the ignition zone (area where surface is exposed to 50 kW/m² of external heat flux).
- For 6 < FPI $\leq$ 10, flame spread is decelerating and stops short of the sample length.

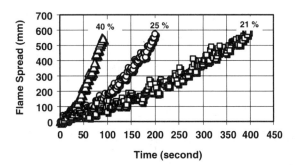

FIGURE 2.8 Upward flame spread versus time on a 600-mm-high and 25-mm-diameter vertical cylinder of PMMA measured in the larger version of the ASTM E 2058 apparatus. Data are taken from a study on the subject [44].

- For $10 < \text{FPI} \leq 20$, there is flame spread beyond the ignition zone and the rate increases with the FPI value.
- For $\text{FPI} > 20$, flame spread beyond the ignition zone is rapid.

There are, however, some cases where there is no flame spread or decelerating flame spread in the large-scale parallel-panel test for materials with FPI values >6 but ≤ 9 $(m/s^{1/2})/(kW/m)^{2/3}$, such as indicated by the data in Table 2.19 taken from Refs. 25, 26, 34, 56, and 57. Thus, for such cases, the standard test method specifies the use of the larger-scale parallel-panel test. The materials are accepted if there is no flame spread beyond the ignition zone in the large-scale parallel-panel test with FPI values ≤ 9 $(m/s^{1/2})/(kW/m)^2$.

The intent of this standard test method is to provide guidance for the selection of materials that can be used in clean room products of the semiconductor industry without active fire protection. The use of active fire protection agents, especially ordinary water, is deleterious for the manufacture of the chips due to contamination. Only those materials that do not have flame spread beyond the ignition zone, under simulated large-scale flame radiative heat flux conditions in the ASTM E 2058 apparatus or in the large-scale parallel-panel test, satisfy the requirements for their use without the active fire protection. The standard test method has not been designed to predict the fire spread behavior of materials and products under a variety of fire conditions that may be present in other occupancies.

2.8.9 ASTM E 603: Standard Guide for Room Fire Experiments [14]

One major reason for performing room fire tests is to learn about various fire stages in the room so that results of standard fire test methods can be related to the performance of the products in full-scale room fires. In addition, some of the tests or their reduced versions are used for the acceptance of building products as they are specified in the prescriptive-based fire codes.

The ASTM E 603 is a guide written to assist in conducting full-scale compartment fire tests dealing with any or all stages of fire in a compartment. Whether it is a single- or multiroom test, observations can be made from ignition to flashover or beyond full-room involvement. Examples of the full-scale room fire tests are:

- FM Global Approval Class No. 4880 for Building Wall and Ceiling Panels and Coatings and Interior Finish Materials [11]
- ISO 9705: Full-Scale Fire Test for Surface Products [13]
- EN 13823: SBI [10, 15, 16].

TABLE 2.18 Flame Spread and Smoke Release Characteristics of Materials [25, 26, 34, 56–61]

Flame spread		Smoke release		
Material[a]	FPI	Material[a]	SDI	Quality
Ordinary				
Polystyrene	34	Polystyrene	5.6	Copious, black
PVC-PVC cable	36	PVC-PVC cable	4.1	Copious, black
Polypropylene (PP)	32	PE-PVC cable	3.8	Copious, black
PE-PVC cable	28	Polybutyleneterephthalate	2.2	Copious, black
Polybutyleneterephthalate	32	Fire retarded-polypropylene	2.1	Copious, black
Polymethylmethacrylate	31	Silicone-PVC cable	2.0	Copious, black
Fire retarded-polypropylene	30	Polypropylene (PP)	1.8	Very large, black
Silicone-PVC cable	17	Acrylonitrile-butadiene-styrene	0.80	Small, black
Polyoxymethylene	15	Polyester/glass fibers (70%)	0.68–0.91	
Wood slab	14	Polymethylmethacrylate	0.62	Small, light grayish
Polyester/glass fibers (70%)	10–13	Polyoxymethylene	0.03	Very small, grayish-white
Acrylonitrile-butadiene-styrene	8	Wood slab	0.20	Very small, grayish-white
High temperature				
Polyetherimide, PEI	8	Polycarbonate	2.1	Copious, black
Phenolic/Kevlar fibers (84%)	8	Polyphenyleneoxide	1.6	Very large, black
Epoxy/glass fibers (65–76%)	5–11	Epoxy/glass fibers (65–76%)	0.61–2.1	
Epoxy/graphite (71%)	5	Epoxy/graphite (71%)	0.54	
Highly modified PP	4–5	Cyanate/graphite (73%)	0.41	
Highly modified PVC	1–4	Polyetheretherketone, PEEK-1	0.40	Very small, grayish-white
Phenolic/glass fibers (80%)	3	Phenolic/Kevlar fibers (84%)	0.33	
Cyanate/graphite (73%)	4	PPS/glass fiber (84%)	0.29	
PPS/glass fiber (84%)	3	Highly modified PP	0.19–0.40	
Epoxy/phenolic/glass fibers (82%)	2	Epoxy/phenolic/glass fibers (82%)	0.18	
Polycarbonate	14	Polysulfone, PSO	0.18	
Polyphenyleneoxide	9	Polyetherimide, PEI	0.15	Very small, grayish-white
Polysulfone, PSO	9	Polyethersulfone, PES	0.15	
Polyetherimide, PEI	8	Phenolic/glass fibers (80%)	0.07	
Polyethersulfone, PES	7	Phenol-formaldehyde	0.06	
Polyetheretherketone, PEEK-1	6	Highly modified PVC	0.03–0.29	
Phenol-formaldehyde	5	PEEK-2	0.03	Very small, grayish-white
PEEK-2	4			
Halogenated				
PE-25% chlorine	15	PE-25% chlorine	1.7	Very large, black
PVC (flexible)	16	PVC (flexible)	1.6	Very large, black
PE-36% Cl	11	PE-36% Cl	1.5	Large, black
PE-48% Cl	8	PE-48% Cl	1.4	Large, black
ETFE (Tefzel®)	7	ETFE (Tefzel®)	0.18	
PVC-D (rigid)	7	PVC-D (rigid)	0.70	Small, grayish
PVC-E (rigid)	6	PVC-E (rigid)	0.30	Very small, grayish-white
PVC-F (rigid)	4	PVC-F (rigid)	0.30	Very small, grayish-white
Polyvinylidenefluoride	4	ECTFE (Halar®)	0.15	
PTFE, Teflon®	4	Polyvinylidenefluoride	0.12	Very small, grayish-white
PVDF (Kynar®)	4	PVDF (Kynar®)	0.12	
ECTFE (Halar®)	4	TFE (Teflon®)	0.01	
TFE (Teflon®)	4	PFA (Teflon®)	0.01	
PFA (Teflon®)	2	FEP (Teflon®)	0.01	
FEP (Teflon®)	3	CPVC (Corzan®)	0.01	
CPVC (Corzan®)	1	TFE, (Teflon®)	0.01	Very small, grayish-white

[a] Generic materials marked by A to F or by numbers 1 to 2 have different compositions as well as in some cases different manufacturers. Materials are identified in the references from where the data were taken.

TABLE 2.19 Fire Propagation Index and Mode and Extent of Flame Spread in Large Scale Tests [25, 26, 34, 56, 57]

Large-scale test	Material/product[a]	FPI[b]	Fire spread mode[c]	Fire spread beyond the ignition zone (% of total height)[c]
4.9-m long × 0.61-m wide parallel Marinite® sheets covered with electrical cables	PVC-PVDF	7	Decelerating	2
	XLPE-EVA	7	Decelerating	45
	XLPE-Neoprene	9	Decelerating	14
	XLPO-XLPO	9	Decelerating	45
	PE-PVC	20	Accelerating	100
2.4-m high × 0.61-m wide parallel slabs of materials	PVDF	4	None	0
	PVC-1	4	None	0
	PVC-2	6	None	0
	ETFE	7	Decelerating	14
	PVC-3	8	Decelerating	12
	FR-PP	30	Accelerating	100
	PMMA	31	Accelerating	100
	PP	32	Accelerating	100

[a] PVC: polyvinylchloride; PVDF: polyvinylidenefluoride; XLPE: cross-linked polyethylene; EVA: ethylvinylacetate; XLPO: cross-linked polyolefin; PE: polyethylene; ETFE: ethylenetrifluoroethylene; PP: polypropylene; FR: fire retarded; PMMA: polymethylmethacrylate. PVCs marked by numbers 1 to 3 have different compositions as well as manufacturers. They are identified in the references from where the data were taken.
[b] From ASTM E 2058.
[c] Determined from the measurements for the ignition zone length (in depth burning of the sample), flame spread length (thin pyrolyzed surface layer), and total height of the sample.

2.8.9.1 FM Global Approval Class No. 4880: Test for Building Wall and Ceiling Panels and Coatings and Interior Finish Materials [11]

This standard test method specifies use of a large-scale test, identified as the "25-ft Corner Test," to evaluate flame spread characteristics of building walls and ceiling panels and coatings. Detailed description of the apparatus and its sketch are included in the standard.

The test is performed in 7.6-m (25-ft)-high, 15.2-m (50-ft)-long, and 11.6-m (38-ft)-wide walls and ceiling forming a corner of a building. The products tested are typically panels with a metal skin over the insulation core material. The panels installed on the walls and ceiling are subjected to a growing exposure fire at the base of the corner. The growing exposure fire consists of burning 340-kg (750 lb), 1.2-m (4-ft) × 1.2-m (4-ft) oak crib pallets, stacked 1.5 m (5 ft) high with a peak heat release rate of about 3 MW.

In the test, measurements are made for the surface temperatures (at 100 equidistant locations on the walls and ceiling) and the length of flame on the walls under the ceiling (visually). After the test, visual measurements are made for the flame spread by the extent of charring on the walls and ceiling. The product is considered to have failed the test if within 15 min either of the following occur:

- flame spread on the wall and ceiling extends to the limits of the structure.
- flame extends outside the limits of the structure through the ceiling smoke layer.

The fire environment within the "25-ft Corner Test" has been characterized by heat flux and temperature measurements [62]. It has been shown that the flame spread boundary (measured visually by the extent of surface charring) is very close to the CHF boundary for the material, very similar to the flame spread behavior in the ASTM E 1321 (ISO 5658) and consistent with Eq. (2.10), that is, $\dot{q}''_f \leq \dot{q}''_{loss} \approx \dot{q}''_{cr}$. A good correlation has been developed between the extent of flame spread and the ratio of the convective heat release rate to $\Delta T_{ig} \sqrt{k\rho c}$, measured in the ASTM E 2058 apparatus [Eq. (2.10) with numerator replaced by the convective heat release rate for the 15-min test] [62, 63].

This test has been instrumental in encouraging the development of other large- and intermediate-scale standard corner tests, such as ISO 9705 [13] and prEN 13823 (SBI) [10].

2.8.9.2 ISO 9705: Room/Corner Test Method for Surface Products [13]

This standard test method specifies the use of a large-scale test to simulate a well-ventilated fire. Detailed description of the apparatus and its sketch are included in the standard. Fire is started at the corner of a 3.6-m-long, 2.4-m-high, and 2.4-m-wide room with a 0.8-m-wide and 2.0-m-high doorway. The walls and ceiling with a total surface area of 32 m^2 (344 ft^2) are covered with the specimen. The ignition source, located in the corner of the room, consists of a propane-fueled 0.17-m^2 sandbox burner set to produce a heat release rate of 100 kW* for the first 10 min. If the flashover does not occur, then the sandbox burner output is increased to produce a heat release rate of 300 kW for another 10 min. The test is ended after 20 min or as soon as the flashover is observed.

A hood attached to a sampling duct is used to capture heat and chemical compounds that are released during the test. In the sampling duct, measurements are made for the gas temperature, concentrations of chemical compounds released in the fire and oxygen, light obscuration by smoke, total flow of the mixture of air and chemical compounds, and heat flux values at various locations in the room. Two parameters are used for ranking the products [15, 16, 64]:

- *Fire growth rate* (FIGRA) index: This is defined as the peak heat release rate in kilowatts during the period from ignition to flashover (excluding the contribution from the ignition source) divided by the time at which the peak occurs (kW/s).
- *Smoke growth rate* (SMOGRA) index: This is defined as the 60-s average of the peak *smoke production rate* (SPR, in m^2/s) divided by the time at which this occurs, and the value is multiplied by 1000 (m^2/s^2). SPR is defined as [ln (I_0 / I) / ℓ] $\dot{V}$, where I/I_0 is the fraction of light transmitted through smoke, ℓ is the optical path length (m), and $\dot{V}$ is the volumetric flow rate of the mixture of smoke and other compounds and air (m^3/s). SPR can also be expressed in terms of grams of smoke released per second as [ln(I_0/I)/ℓ]V ($\lambda\rho_s \times 10^{-6}/\Omega$), where λ is the wavelength of light (0.6328 µm used in the cone) [13, 14], ρ_s is the density of smoke (1.1 × 10^{6}g/m^3) [65], and Ω is the coefficient of particulate extinction (7.0) [65]. Thus, SPR (in m^2/s) multiplied by 0.0994 changes the unit to grams per second (for $\lambda = 0.6328$ µm).

Numerous products have been tested during the last 10 years following the ISO 9705 Room/Corner Test Method [64, 66]. Data including the FIGRA and SMOGRA indices for the building products, taken from Refs. 64 and 66, are listed in Tables 2.20, 2.21, and 2.22.

Under similar burning conditions, the combustion chemistry responsible for release of heat and smoke are conserved and thus release rates of heat and smoke are interrelated:

$$\text{SMOGRA/FIGRA} = y_s/\Delta H_{ch} \qquad (2.17)$$

This interrelationship was found for the data from the ISO 9705 tests [64], as shown in Fig. 2.9. Data in Fig. 2.13 are taken from Table 2.21. The relationship appears reasonable given that data are from large-scale tests where products do not always behave as expected and considering data scatter.

Fire properties of products tested in the ISO 9705 room have been quantified in the ASTM E 1354 and ISO 5660 (cone calorimeter) and ASTM E 1321 and ISO 5658 (LIFT) such as those

---

*A 100-kW diffusion flame is used to simulate a burning large wastepaper basket and a 300-kW diffusion flame is used to simulate a burning small upholstered chair [15, 16].

TABLE 2.20 Data Measured in the ISO 9705 Room/Corner Test [64]

Sample #	Product	Time to (s) Flashover	Peak HRR	Peak SPR	Peak HRR kW	Peak SPR m^2/s	FIGRA Index (kW/s)	SMOGRA Index (m^2/s^2)
M01	Plasterboard		626	626	94	0.4	0.17	0.6
M02	FR polyvinylchloride (PVC)		761	761	129	1.7	0.17	2.2
M03	FR extruded polystyrene (PS) board	96		96	>1000	24.0	9.4	250
M04	PUR foam panel/aluminum faces	41		41	>1000	12.6	22	307
M05	Varnished mass timber, pine	106		106	>1000	13.0	8.5	123
M06	FR Chip board		1200	965	423	15.7	0.35	16
M07	FR polycarbonate (PC) panel (three layers)		265	696	147	1.7	0.55	2.4
M08	Painted plasterboard		625	1200	71	0.5	0.11	0.4
M09	Paper wall covering on plasterboard		640	645	394	1.2	0.62	1.9
M10	PVC wall carpet on plasterboard	675		675	>1000	11.2	1.0	17
M11	Plastic-faced steel sheet on mineral wool		685	685	95	4.8	0.14	7.0
M12	Unvarnished mass timer, spruce	170		170	>1000	4.6	5.3	27
M13	Plasterboard on PS		625	1065	83	0.6	0.13	0.6
M14	Phenolic foam	640		640	>1000	7.0	1.1	11
M15	Intumescent coat on particle board	700		700	>1000	25.0	1.0	36
M16	Melamine faced medium density foam board	150		135	>1000	6.9	6.0	51
M17	PVC water pipes		865	645	75	5.0	0.09	7.8
M18	PVC covered electric cables		605	675	272	6.2	0.45	9.2
M19	Unfaced rockwool		1180	1175	73	0.8	0.06	0.7
M20	Melamine faced particle board	165		165	>1000	8.7	5.5	53
M21	Steel clad PS foam sandwich panel	970		970	>1000	6.8	0.72	7.0
M22	Ordinary particle board	155		155	>1000	12.5	5.8	81
M23	Ordinary plywood (Birch)	160		160	>1000	8.1	5.6	51
M24	Paper wall covering on particle board	165		160	>1000	5.0	5.5	31
M25	Medium density fiberboard	190		175	>1000	8.4	4.7	48
M26	Low density fiberboard	58		0.55	>1000	13.4	16	244
M27	Plasterboard/FR PUR foam core		1105	1180	146	3.7	0.13	3.1
M28	Acoustic mineral fiber tiles		855	645	44	0.5	0.05	0.7
M29	Textile wallpaper on calcium silicate board		1135	1185	648	0.6	0.57	0.5
M30	Paper-faced glass wool	18	1500	25	>1000	5.7	50	228

listed in Table 2.22 taken from Ref. 66. Various studies have been performed for the correlation of the data from the ISO 9705 Room/Corner Test and ASTM E 1354 and ISO 5660 and ASTM E 1321 and ISO 5658 (LIFT) [66–70]. In addition, these data have been used to develop predictive models for the fire behaviors of interior finish materials from their fire properties [71]. The following are examples of flame spread models for the interior finish materials of buildings:

- Ostman-Nussbaum model [72]
- Karlsson and Magnusson model [73]
- Wickstrom-Goransson model [74]

TABLE 2.21 ISO 9705 Room/Corner Test Ranking of Building Products Based on Fire Growth Rate Index [64]

Rank	Product	FIGRA Index kW/s	SMOGRA Index m^2/s^2	Flashover 100 kW	Flashover 300 kW
1	Mineral wool	0.01	0.0	No	No
2	Plasterboard	0.03	0.3	No	No
3 (M28)	Acoustic mineral fiber tiles	0.05	0.7	No	No
4 (M19)	Unfaced rockwool	0.06	0.7	No	No
5	Plastic faced steel sheet on mineral wool-1	0.07	5.6	No	No
6 (M17)	PVC water pipes	0.09	7.8	No	No
7 (M08)	Painted plasterboard	0.11	0.4	No	No
8 (M13)	Plasterboard on polystyrene	0.13	0.6	No	No
9 (M27)	Plasterboard/FR PUR foam core	0.13	3.1	No	No
10 (M11)	Plastic faced steel sheet on mineral wool-2	0.14	7.0	No	No
11 (M01)	Plasterboard	0.15	0.6	No	No
12 (M02)	FR PVC	0.17	2.2	No	No
13	Painted gypsum board	0.20	0.8	No	No
14	Metal faced noncombustible board	0.22	13	No	No
15 (M06)	FR chip board	0.35	16	No	No
16	Painted glass tissue faced glasswool	0.37	1.0	No	No
17	FR particle board-1	0.41	8.8	No	No
18 (M18)	PVC covered electric cables	0.45	9.2	No	No
19 (M07)	FR polycarbonate panel three layered	0.55	2.4	No	No
20 (M29)	Textile wallpaper on calcium silicate board	0.57	0.5	No	No
21 (M09)	Paper wall covering on plasterboard-1	0.62	1.9	No	No
22 (M21)	Steel clad expanded PS sandwich panel	0.72	7.0	No	Yes
23	Glass fabric faced glasswool	0.73	3.7	No	No
24 (M15)	intumescent coat on plasterboard	1.0	36	No	Yes
25 (M10)	PVC wall carpet on plasterboard	1.0	17	No	Yes
26	Textile wall covering/gypsum	1.1	16	No	Yes
27	PVC wall carpet on gypsum	1.1	37	No	Yes
28 (M14)	Phenolic foam	1.1	11	No	Yes
29	Paper wall covering on plasterboard-2	1.1	3.9	No	Yes
30	Textile wall covering on plasterboard	1.1	1.9	No	Yes
31	FR plywood	1.1	8.1	No	Yes
32	FR particle board-2	1.1	20	No	Yes
33	Melamine faced particle board-1	1.9	58	Yes	Expected
34	PUR foam covered with steel sheets	3.6	40	Yes	Expected
35 (M25)	Medium density fiber board-1	4.7	48	Yes	Expected
36 (M12)	Unvarnished mass timber (spruce)	5.3	27	Yes	Expected
37 (M20)	Melamine faced particle board-2	5.5	53	Yes	Expected

TABLE 2.21 (*Continued*)

Rank	Product	FIGRA Index kW/s	SMOGRA Index m^2/s^2	Flashover 100 kW	300 kW
38 (M24)	Paper wall covering on particle board-1	5.5	31	Yes	Expected
39 (M23)	Ordinary plywood (Birch)-1	5.6	51	Yes	Expected
40 (M22)	Ordinary particle board	5.8	81	Yes	Expected
41 (M16)	Melamine faced medium density fiberboard	6.0	51	Yes	Expected
42	Ordinary plywood-2	6.0	55	Yes	Expected
43	Particle board	6.0	60	Yes	Expected
44	Paper wall covering on particle board-2	6.3	61	Yes	Expected
45	Plywood	6.4	48	Yes	Expected
46	Wood panel	6.5	38	Yes	Expected
47	Acrylic glazing	6.6	8.8	Yes	Expected
48	Medium density fiberboard-2	6.7	43	Yes	Expected
49	FR PS 50-mm	6.8	130	Yes	Expected
50 (M05)	Varnished mass timber (pine)	8.5	123	Yes	Expected
51	FR expanded PS 80-mm	8.5	72	Yes	Expected
52 (M03)	FR extruded PS board	9.4	250	Yes	Expected
53	FR PS 20-mm	10	311	Yes	Expected
54	FR expanded PS 40-mm	10	148	Yes	Expected
55	Combustible faced mineral wool	11	20	Yes	Expected
56	FR expanded PS 25-mm	11	100	Yes	Expected
57	Insulating fiberboard	13	61	Yes	Expected
58 (M26)	Low density fiberboard	16	244	Yes	Expected
59	Textile wall covering on mineral wool	16	295	Yes	Expected
60 (M04)	PUR foam panel with aluminum foil faces	22	307	Yes	Expected
61 (M30)	Paper faced glass wool	50	228	Yes	Expected
62	Polyurethane foam	64	2329	Yes	Expected

- Quintiere Room-Corner model [75]
- Qian and Saito Model [76]
- WPI Room/Corner Fire Model [77]
- HAI/Navy Corner Fire Model [78]
- Quintiere-Dillon Room Corner Fire Model [66, 79]

The ISO 9705 Room/Corner Standard Test Method has been adopted by various agencies, such as the *International Maritime Organization* (IMO), for qualifying fire-restricting materials for use as structural components and compartment linings on high-speed watercraft using the following criteria [80]:

- average net heat release rate ≤ 100 kW
- maximum 30-s average net heat release rate ≤ 500 kW

TABLE 2.22 Flashover Time from ISO 9705 Room/Corner Test and Fire Properties of Materials from ASTM E 1354/ISO 5660 and ASTM E 1321 (ISO 5658) [66]

#	Product	ISO 9705 Time to flashover (s)	ASTM E 1321 (ISO 5658)					ASTM E 1354/ISO 5660		
			CHF (kW/m^2)	T_{ig} (s)	$T_{s,min}$ (s)	$k\rho c$ (kW/m^{2}K)^2	Φ (kW^2/m^3)	ΔH_{ch} (kJ/g)	ΔH_g (kJ/g)	E/A (kJ/m^2)
R4.01	FR chipboard	None	25	505	507	4.024	0.0	7.9	4.5	34.2
R4.02	Gypsum	None	26	515	517	0.549	0.0	3.2	4.8	2.2
R4.03	Polyurethane foam/aluminum	41					0.0	18.2		32.9
R4.04	Polyurethane foam/paper		6	250	77	0.199	8.7	18.0	5.5	30.8
R4.05	Expanded polystyrene 40-mm-1	96	7	275	77	1.983	1.2	28.2	4.5	38.7
R4.06	Acrylic	141	4	195	195	2.957		24.0	3.0	89.5
R4.07	FR PVC	None	16	415	352	1.306	0.2	6.8	4.2	16.1
R4.08	Three layered polycarbonate	None	24	495	167	1.472	0.0	21.5	3.6	58.1
R4.09	Mass timber	107	10	330	77	0.530	6.9	15.7	6.5	68.2
R4.10	FR plywood	631	22	480	197	0.105	0.7	10.3	8.8	51.8
R4.11	Plywood	142	8	290	147	0.633	2.2	10.8	3.9	64.6
R4.20	Expanded polystyrene 40-mm-2	87	8	295	77	1.594	4.2	27.8	11.2	33.9
R4.21	Expanded polystyrene 80-mm	None	23	490	77	0.557	7.1	27.9	9.4	25.5

- average smoke production rate ≤ 1.4 m^2/s
- maximum 60-s average smoke production rate ≤ 8.3 m^2/s
- limits on downward flame spread and flaming drops or debris

The intent of this standard test method is to separate materials and products with higher flame spread resistance from those with lower resistance. It has not been designed to predict the flame spread behavior of materials and products under conditions other than used in the test.

The flame spread behavior of products in the ISO 9705 test has not yet been modeled successfully; however, modeling shows promise of success in the future. A successful modeling of the flame

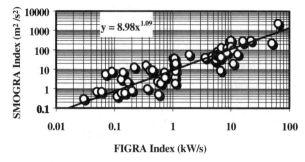

FIGURE 2.9 Relationship between SMOGRA and FIGRA indices for the building products tested according to ISO 9705. Data for the indices are taken from Ref. 64.

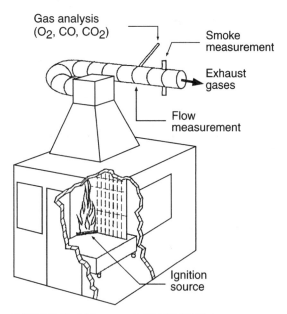

FIGURE 2.10   SBI test setup (prEN 13823) [10].

spread behavior of products in the test is expected to provide tools to predict flame spread behaviors of products for conditions other than those used in the test.

2.8.9.3   prEN 13823: The Single Burning Item [10]

This standard test method is the latest test specification utilizing an intermediate-scale apparatus, which is shown in Fig. 2.10. The apparatus consists of a trolley with two 1.5-m-high, 1.0-m-wide, and 0.5-m-wide vertical noncombustible boards mounted at 90° to each other. The test specimen (wall and ceiling materials) are mounted and fixed onto the noncombustible boards in a manner representative of end use. The ignition source consists of a 31-kW propane right-angled triangular sandbox burner (each side measures 250 wide and 80 mm high) placed at the bottom of the vertical corner. The test is performed inside a 2.4-m-high and 3.0-m² room with top attached to a hood connected to a sampling duct through which heat and chemical compounds released during the fire test are exhausted. Evenly distributed airflow along the floor of the test room is achieved by introducing the air under the floor of the trolley through perforated plates.

In the sampling duct, measurements are made for the gas temperature, concentrations of chemical compounds released in the fire, and oxygen, light obscuration by smoke, and total flow of the mixture of air and chemical compounds. The parameters used for the assessment of fire performance of specimens are the following:

- heat release rate obtained from the measurements for oxygen depletion in the sampling duct
- smoke release from the light obscuration by smoke in the sampling duct
- horizontal flame spread observed visually, i.e., time taken to reach the extreme edge of the main 1.5 × 1.0-m sample panel
- falling molten droplets and particles

The performance of the specimen is evaluated over a period of 20 min. However, the test is terminated earlier if any of the following conditions occur:

- heat release rate >350 kW at any instant or >280 kW over a period of 30 s
- sampling duct temperature >400°C at any instant or >300°C over a period of 30 s
- material falling onto the sandbox burner substantially disturbs the flame of the burner or extinguishes the burner by choking.

The test data are used to obtain the following parameters to rank the fire performance of the specimens:

1. FIGRA index
2. SMOGRA index*
3. THR_{600s}: total heat released within 600 s
4. TSP_{600s}: total smoke released within 600 s
5. LFS: lateral flame spread
6. flaming and nonflaming droplets or particles and ignition of the paper† (prEN ISO 11925-2)

Data measured in the prEN 13823 (SBI) test and reported in Ref. 81 are listed in Table 2.23. The products listed as M1 to M30 are identified in Table 2.21 and are ranked according to the peak heat release rate measured in the prEN 13823 test. Product ranking according to FIGRA index from the ISO 9605 from Table 2.21 is also included in Table 2.23. These two rankings do not agree as expected, as the test conditions are different.

In Europe, data from prEN 13823 are used for the classification of reaction to fire performance for construction products (prEN 13501-1) [10]:

- construction products excluding floorings:
  - Class A2: FIGRA ≤ 120 W/s; LFS < edge of specimen; THR_{600s} ≤ 7.5 MJ; smoke production and melting and burning drops
  - Class B: FIGRA ≤ 120 W/s; LFS < edge of specimen; THR_{600s} ≤ 7.5 MJ; smoke production and melting and burning drops
  - Class C: FIGRA ≤ 250 W/s; LFS < edge of specimen; THR_{600s} ≤ 15 MJ; smoke production and melting and burning drops
  - Class D: FIGRA ≤ 750 W/s; smoke production and melting and burning drops

The use of this standard test method for regulatory purposes is very similar to that of the ASTM E 84 standard test method. The intent of this standard test method is to separate materials and products with higher flame spread resistance from those with lower resistance. It has not been designed to predict the flame spread behavior of materials and products in actual fires.‡

---

*s_1 = SMOGRA ≤ 30 m²/s² and TSP_{600s} ≤ 50 m²; s_2 = SMOGRA ≤ 180 m²/s² and TSP_{600s} ≤ 200 m²; s_3 = neither s_1 nor s_2.

†d_0 = no flaming droplets or particles in prEN 13823 within 600 s; d_1 = no flaming droplets or particles persisting longer than 10 s in prEN 13823 within 600s; d_2 = neither d_0 nor d_1 (ignition of paper in prEN ISO 11925-2 results in a d_2 classification).

‡The major limitation with this standard test method is that it is very difficult to assess the fire behaviors of products for heat exposure and environmental conditions and shape, size, and arrangements of the products other than those used in the test.

MATERIALS SPECIFICATIONS, STANDARDS, AND TESTING   **2.49**

TABLE 2.23 Ranking of Building Products Based on FIGRA (ISO 9705) [64] and Peak Heat Release Rate (prEN 13823: SBI) [81]. Products are Identified in Table 2.21

Rank	ISO 9705		Rank	PrEN 13823 (SBI)		
	FIGRA kW/s	SMOGRA m^2/s^2		Peak heat release rate (kW)	ΔT_{max}	Peak smoke production rate (m^2/s)
M28	0.05	0.7	M19	2.8	36.0	0.12
M19	0.06	0.7	M21	3.8	37.7	0.12
M17	0.09	7.8	M01	4.7	34.5	0.06
M08	0.11	0.4	M13	4.9	33.9	0.12
M13	0.13	0.6	M08	4.9	34.9	0.09
M27	0.13	3.1	M28	5.2	35.4	0.09
M11	0.14	7.0	M27	6.3	36.2	0.14
M01	0.15	0.6	M14	7.3	40.4	0.11
M02	0.17	2.2	M11	7.6	37.0	0.45
M06	0.35	16	M06	8.8	44.2	0.25
M18	0.45	9.2	M15	9.5	43.4	0.11
M07	0.55	2.4	M29	12.9	38.1	0.10
M29	0.57	0.5	M09	13.2	39.2	0.08
M09	0.62	1.9	M02	25.9	45.0	3.56
M21	0.72	7.0	M10	37.4	61.0	0.70
M15	1.0	36	M17	41.5	58.6	4.50
M10	1.0	17	M05	61.1	96.5	0.09
M14	1.1	11	M20	67.8	110.5	0.17
M25	4.7	48	M12	70.1	104.4	0.12
M12	5.3	27	M23	76.0	125.0	0.25
M20	5.5	53	M16	81.1	117.1	0.13
M24	5.5	31	M22	83.6	127.3	0.28
M23	5.6	51	M24	86.1	131.6	0.23
M22	5.8	81	M30	106.0	127.3	0.13
M16	6.0	51	M25	134.2	204.3	0.12
M05	8.5	123	M18	138.3	186.7	1.89
M03	9.4	250	M26	142.6	186.4	0.25
M26	16	244	M04	159.5	192.2	1.75
M04	22	307	M07	330.1	235.9	7.03
M30	50	228	M03	373.9	328.9	6.23

NOMENCLATURE

A	Total exposed surface area (m^2)
c	Heat capacity (kJ/g-K)
CHF	Critical heat flux (kW/m^2)
CRF	Critical radiant flux (kW/m^2)
d	Material thickness (mm)
E	Energy (kJ)
F_s	Flame spread (mm)
FIGRA	Fire growth rate (kW/s)
FPI	Fire Propagation Index $(m/s^{1/2})/)kW/m)^{2/3}$
FSI	Flame spread index
$\dot{G}''_j$	Release rate of compound j per unit surface area of the material $(g/m^2$-s$)$

Symbol	Description
ΔH_{ch}	Effective (chemical) heat of combustion (kJ/g)
ΔH_d	Heat of decomposition (kJ/g)
HRP	Heat release parameter ($\Delta H_{ch}/\Delta H_g$, kJ/kJ)
ΔH_g	Heat of gasification (kJ/g)
ΔH_m	Heat of melting (kJ/g)
ΔH_T	Net heat of complete combustion (kJ/g)
ΔH_v	Heat of vaporization (kJ/g)
HRR	Heat release rate (kW)
I	Intensity of light transmitted through smoke
I_o	Intensity of incident light
k	Thermal conductivity (kW/m-K)
ℓ	Optical path length (m)
LFS	Lateral flame spread (m)
LOI	Limited oxygen index
$\dot{m}''$	Release rate of material vapors per unit surface area of the material (g/m^2-s)
PCS	Gross calorific potential (kJ/g)
$\dot{q}''_e$	External heat flux per unit surface area of the material (kW/m^2)
$\dot{q}''_{cr}$	Critical heat flux per unit surface area of the material (kW/m^2)
$\dot{q}''_f$	Flame heat flux per unit surface area of the material (kW/m^2)
$\dot{q}''_{loss}$	Heat loss per unit surface area of the material (kW/m^2)
$\dot{q}''_n$	Net heat flux per unit surface area of the material (kW/m^2)
$\dot{q}''_{rr}$	Surface reradiation loss per unit surface area of the material (kW/m^2)
$\dot{Q}''_{ch}$	Chemical heat release rate per unit area of the material (kW/m^2)
$\dot{Q}'$	Chemical heat release rate per unit width of the material (kW/m)
SDI	Smoke development index
SMOGRA	Smoke growth rate (m^2/s^2)
SPR	Smoke production rate (m^2/s)
t_f	Duration of sustained flaming (s)
t_{ig}	Time-to-ignition (s)
T_d	Decomposition temperature (°C)
T_{gl}	Glass transition temperature (°C)
ΔT_{ig}	Ignition temperature above ambient (°C)
T_m	Melting temperature (°C)
$T_{s,min}$	Minimum temperature for flame spread (°C)
T_v	Vaporization temperature (°C)
THR	Total heat release (MJ)
TRP	Thermal response parameter (kW-s$^{1/2}$/m^2)
TSP	Total smoke production (m^2)
w	Width (m)
y_j	Yield of product j (g/g)
x_f	Flame heat transfer distance over the surface (mm)

Greek

Symbol	Description
α	Thermal diffusivity (mm/s)
δ	thermal penetration depth (mm)
λ	Wavelength of light (μm)
ρ	density (g/cm^3)
Φ	Flame-heating parameter (kW^2/m^3)
Δm	Weight loss (%)
Ω	Coefficient of particulate extinction

Subscripts

Symbol	Description
a	Ambient
ch	Chemical
d	Decomposition
f	Flame

g	Gasification
ig	Ignition
rr	Reradiation
l	Liquid
loss	Loss
n	net
s	Solid
sm	Smoke
T	Total or complete
v	Vaporization

Superscripts

.	Per unit of time (1/s)
′	Per unit of width (1/m)
″	Per unit of area (1/m²)

REFERENCES

1. Beck, V., "Performance-Based Fire Engineering Design and Its Application in Australia," *Fire Safety Science, Fifth International Symposium,* International Association for Fire Safety Science, Japan, 1997.
2. Meacham, B.J., "Concepts of a Performance-Based Building Regulatory System for the United States," *Fifth International Symposium,* International Association for Fire Safety Science, Japan, 1997.
3. Hall, J.R. (Editor), "ASTM's Role in Performance-Based Fire Codes and Standards," *ASTM STP 1377,* American Society for Testing and Materials, West Conshohocken, PA, 1999.
4. Worldwide Standards Service for Windows, HIS, Englewood, CO, 2002.
5. Troitzsch, J., *International Plastics Flammability Handbook—Principles, Regulations, Testing and Approval,* Macmillan Publishing, New York, 1983.
6. Landrock, A.H., *Handbook of Plastics Flammability and Combustion Toxicology—Principles, Materials, Testing, Safety, and Smoke Inhalation Effects,* Noyes Publications, Park Ridge, NJ, 1983.
7. *Flammability Testing of Building Materials—An International Survey,* Document No. TH 42126, British Standards Institution, London, UK, 2000.
8. Makower, A.D., *Fire Tests—Buildings Products, and Materials,* British Standards Institution, London, UK.
9. Hilado, C.J., *Flammability Test Methods Handbook,* Technomic Publication, Westport, CT, 1973.
10. European Committee for Standardization (CEN) (http://www.cenorm.be/).
11. FM Approvals (http://www.fmglobal.com/research_standard_testing/product_certification/approval_standards.html).
12. UL Standards (http://ulstandardsinfonet.ul.com/catalog/).
13. ISO Standards (http://www.iso.ch/iso/en/isoonline.frontpage).
14. ASTM Standards (http://www.astm.org).
15. Sundstrom, B., "European Classification of Building Products," *Interflam'99, 8th International, Fire Science & Engineering Conference, Volume 2,* Interscience Communications, London, UK, 1999.
16. Sundstrom, B., and Christian, S.D., "What Are the New Regulations, Euroclasses, and Test Methods Shortly to be Used throughout Europe," *Conference Papers, Fire and Materials,* Interscience Communications, London, UK, 1999.
17. Mark, J.E. (Editor), *Physical Properties of Polymers Handbook,* American Institute of Physics, Woodbury, NY, 1996.
18. Harper, C.A. (Editor), *Handbook of Materials for Product Design,* 3rd edition, McGraw-Hill Handbooks, McGraw-Hill, New York, 2001.
19. Lyons, J.W., *The Chemistry and Uses of Fire Retardants,* John Wiley and Sons, New York, 1970.
20. Papazoglou, E., "Flame Retardants for Plastics" (chapter 4 of this handbook).

21. Domininghaus, H., *Plastics for Engineers—Materials, Properties, Applications,* Hanser Publishers, New York, 1988.
22. Scheirs, J. (Editor), "Modern Fluoropolymers," *Wiley Series in Polymer Science High Performance Polymers for Diverse Applications,* John Wiley and Sons, New York, 2000.
23. Ohlemiller, T.J., and Shields, J.R., *Burning Behavior of Selected Automotive Parts from a Minivan,* Docket No. NHTSA-1998-3588-26 (www.nhtsa.dot.gov).
24. Abu-Isa, I.A., Cummings, D.R., and LaDue, D., *Thermal Properties of Automotive Polymers I. Thermal Gravimetric Analysis and Differential Scanning Calorimetry of Selected Parts from a Dodge Caravan,* Report R&D 8775, General Motors Research and Development Center, Warren, MI, June 1998.
25. Tewarson, A., "Generation of Heat and Chemical Compounds in Fires," *SFPE Handbook of Fire Protection Engineering,* 3rd edition, National Fire Protection Association Press, Quincy, MA, 2002.
26. Tewarson, A., "Flammability," *Physical Properties of Polymers Handbook,* American Institute of Physics, Woodbury, NY, 1996.
27. Tewarson, A., and Pion, R.F., "Flammability of Plastics. I: Burning Intensity," *Combustion and Flame,* 26: 85–103, 1976.
28. Lyon, R.E., *Solid-State Thermochemistry of Flaming Combustion,* Technical Report DOT/FAA/AR-99/56, Federal Aviation Administration, Airport and Aircraft Safety, Research and Development Division, William J. Hughes Technical Center, Atlantic City, NJ, July 1999.
29. Fernandez-Pello, A.C., and Hirano, T., "Controlling Mechanisms of Flame Spread," *Combustion Science & Technology,* 32: 1–31, 1983.
30. Quintiere, J.G., "Surface Flame Spread," *SFPE Handbook of Fire Protection Engineering,* 3rd edition, National Fire Protection Association Press, Quincy, MA, 2002.
31. Quintiere, J.G., *Principles of Fire Behavior,* Delmar Publishers, New York, 1998.
32. Drysdale, D., *An Introduction to Fire Dynamics,* John Wiley and Sons, New York, 1985.
33. Tewarson, A., Abu-Isa, I.A., Cummings, D.R., and LaDue, D.E., "Characterization of the Ignition Behavior of Polymers Commonly used in the Automotive Industry," *Fire Safety Science, Sixth International Symposium,* International Association for Fire Safety Science, France, 1999. U.S. National Highway Traffic Safety Administration Docket No. NHTSA-1998-3588-71, December 13, 1999.
34. Tewarson, A., Khan, M.M., Wu, P.K.S., and Bill, R.G., "Flammability of Clean Room Polymeric Materials for the Semiconductor Industry," *Journal of Fire and Materials,* 25: 31–42, 2001.
35. Tewarson, A., *Fire Hardening Assessment (FHA) Technology for Composite Systems,* Technical Report ARL-CR-178 (under contract DAAL01-93-M-S403), Army Research Laboratory, Aberdeen Proving Ground, MD, November 1994.
36. Hirschler, M.M., "(a) Heat Release from Plastic Materials," Babrauskas, V., and Grayson, S.J. (Editors), Heat Release in Fires, E & FN Spon, Chapman & Hall, New York, 1992, pp. 375–422.
37. Sorathia, U., and Beck, C., "Fire-Screening Results of Polymers and Composites," *Improved Fire- and Smoke-Resistant Materials for Commercial Aircraft Interiors A Proceedings,* Committee on Fire- and Smoke-Resistant Materials, National Materials Advisory Board, Commission on Engineering and Technical Systems, National Research Council, Publication NMAB-477-2, National Academy Press, Washington, DC, 1995.
38. Maxwell, J.B., *Data Book on Hydrocarbons Application to Process Engineering,* D. Van Nostrand, New York, 1950.
39. Clarke, F.B., "Issues Associated with Combustibility Classification: Alternate Test Concepts," *Fire Safety Science, Proceedings of the Fifth International Symposium,* International Association for Fire Safety Science, Japan, 1997.
40. Scudamore, M.J., Briggs, P.J., and Prager, F.H., "Cone Calorimetry—A Review of Tests Carried Out on Plastics for the Association of Plastic Manufacturers in Europe," *Fire and Materials,* 15: 65–84, 1991.
41. Tewarson, A., *A Study of the Flammability of Plastics in Vehicle Components and Parts,* Technical Report J.I. OB1R7.RC, U.S. National Highway Traffic Safety Administration Docket No. NHTSA-1998-3588-1, July 17, 1998.
42. Rehm, R.G., Pitts, W.M., Baum, H.R., Evans, D.D., Prasad, K., McGrattan, K.B., and Forney, G.P., "Initial Model for Fires in the World Trade Center Towers," *Invited Lecture #1, 7th International Symposium on Fire Safety Science,* Worcester Polytechnic Institute, Worcester, MA, June 16–21, 2002.

43. Tewarson, A., Lee, J.L., and Pion, R.F., "The Influence of Oxygen Concentration on Fuel Parameters for Fire Modeling," *Eighteenth Symposium (International) on Combustion,* Combustion Institute, Pittsburgh, PA, 1981.
44. Tewarson, A., and Ogden, S.D., "Fire Behavior of Polymethylmethacrylate," *Combustion and Flame,* 89: 237–259, 1992.
45. NFPA 53, 1999 edition, Recommended Practice on Materials, Equipment, and Systems Used in Oxygen-Enriched Atmospheres, National Fire Protection Association, Quincy, MA, 1999.
46. Tewarson, A., *Thermo-Physical and Fire Properties of Automotive Polymers,* Technical Report J.I.OB1R7.RC, FM Global Research, Norwood, MA, June 2002.
47. Hilado, C.J., *Flammability Handbook for Plastics,* Technomic Publishing, Stanford, CT, 1969.
48. "Test Procedures and Performance Criteria for the Flammability and Smoke Emission Characteristics of Materials Used in Passenger Cars and Locomotive Cabs," *Federal Register, Rules and Regulations,* 64 (91), May 12, 1999.
49. Department of Transportation, Federal Transit Administration, Docket 90-A, "Recommended Fire Safety Practices for Transit Bus and Van Materials Selection," *Federal Register,* 58 (201), October 20, 1993.
50. Sundstrom, B., *Results and Analysis of Building Materials Tested According to ISO and IMO Spread of Flame Tests,* Technical Report SP-RAPP 1984:36, National Testing Institute, Fire Technology, Boras, Sweden, 1984.
51. Benjamin, I.A., and Adams, C.H., "The Flooring Radiant Panel Test and Proposed Criteria," *Fire Journal,* 70 (2): 63–70, March 1976.
52. Davis, S., Lawson, J.R., and Parker, W.J., *Examination of the Variability of the ASTM E 648 Standard with Respect to Carpets,* Technical Report NISTIR 89-4191, National Institute of Standards and Technology, Gaithersburg, MD, October 1989.
53. Tu, K., and Davis, S., *Flame Spread of Carpet Systems Involved in Room Fires,* Technical Report NBSIR 76-1013, National Bureau of Standards, Gaithersburg, MD, 1976.
54. NFPA 101 Life Safety Code, National Fire Codes—A Compilation of NFPA Codes, Standards, Recommended Practices and Guides, Volume 5, National Fire Protection Association, Quincy, MA, 2000.
55. NFPA 318 Standard for the Protection of Cleanrooms, National Fire Codes, Volume 6, National Fire Protection Association, Quincy, MA, 2000.
56. Tewarson, A., Bill, R.G., Alpert, R.L., Braga, A., DeGiorgio, V., and Smith, G., *Flammability of Clean Room Materials,* J.I. OYOE6.RC, FM Global Research, Norwood, MA, 1999.
57. Tewarson, A., and Khan, M.M., *Electrical Cables—Evaluation of Fire Propagation Behavior and Development of Small-Scale Test Protocol,* Technical Report J.I. 0M2E1.RC, FM Global Research, Norwood, MA, January 1989.
58. Khan, M.M., *Classification of Conveyor Belts Using a Fire Propagation Index,* Technical Report J.I. OT1E2.RC, FM Global Research, Norwood, MA, June 1991.
59. Tewarson, A., and Macaione, D., "Polymers and Composites—An Examination of Fire Spread and Generation of Heat and Fire Products," *Journal of Fire Sciences,* 11: 421–441, 1993.
60. Tewarson, A., *Fire Hardening Assessment (FHA) Technology for Composite Systems,* Technical Report ARL-CR-178, Army Research Laboratory, Aberdeen Proving Ground, MD, 1994.
61. Apte, V.B., Bilger, R.W., Tewarson, A., Browning, G.J., Pearson, R.D., Fidler, A., *An Evaluation of Flammability Test Methods for Conveyor Belts,* Grant Report ACARP-C5033, Workcover NSW, Londonderry Occupational Safety Center, Londonderry, NSW, Australia, December 1997.
62. Newman, J.S., and Tewarson, A., "Flame Spread Behavior of Char-Forming Wall/Ceiling Insulation," *Fire Safety Science, Third International Symposium,* International Association for Fire Safety Science, Elsevier Applied Science, New York, 1991.
63. Newman, J.S., "Integrated Approach to Flammability Evaluation of Polyurethane Wall/Ceiling Materials," *Polyurethanes World Congress,* Society of the Plastics Industry, Washington, DC, 1993.
64. Sundstrom, B., Hees, P.V., and Thureson, P., *Results and Analysis from Fire Tests of Building Products in ISO 9705, the Room/Corner Test; the SBI Research Program,* Technical Report SP-RAPP, 1998:11, Swedish National Testing and Research Institute, Fire Technology, Boras, Sweden, 1998.
65. Newman, J.S., and Steciak, J., "Characterization of Particulates from Diffusion Flames," *Combustion and Flame,* 67: 55–64, 1987.

66. Dillon, S.E., *Analysis of the ISO 9705 Room/Corner Test: Simulation, Correlations, and Heat Flux Measurements,* NIST-GCR-98-756, National Institute of Standards and Technology, Gaithersburg, MD, August 1998.
67. Babrauskas, V., and Wickstrom, U., "The Rational Development of Bench-Scale Fire Tests for Full-Scale Fire Prediction," *Fire Safety Science, Proceedings of the Second International Symposium,* International Association for Fire Safety Science, Hemisphere Publishing, New York, 1989.
68. Kokkala, M.A., "Sensitivity of the Room/Corner Test to Variations in the Test System and Product Properties," *Fire and Materials,* 17: 217–224, 1993.
69. Richardson, L.R., Cornelissen, A.A., and Dubois, R.P., "Physical Modeling Fire Performance of Wall-Covering Materials," *Fire and Materials,* 17: 293–303, 1993.
70. Ostman, B.A.L., and Tsantaridis, L.D., "Correlation between Cone Calorimeter Data and Time to Flashover in the Room Fire Test," *Fire and Materials,* 18: 205–209, 1994.
71. Beyler, C.L., Hunt, S.P., Lattimer, B.Y., Iqbal, N., Lautenberger, C., Dembsey, N., Barnett, J., Janssens, M., Dillon, S., and Grenier, A., *Prediction of ISO 9705 Room/Corner Test Results,* Report No. CG-D-22-99, U.S. Coast Guard Research and Development Center, Groton, CT, November 1999.
72. Ostman, B.A.L., and Nussbaum, R.M., "Correlation between Small-Scale Rate of Heat Release and Full-Scale Room Flashover for Surface Linings," *Fire Safety Science, Proceedings of the Second International Symposium,* International Association for Fire Safety Science, Hemisphere Publishing, New York, 1989.
73. Karlsson, B., and Magnusson, S.R., "Combustible Wall Lining Materials: Numerical Simulation of Room Fire Growth and the Outline of a Reliability Based Classification Procedure," *Fire Safety Science, Proceedings of the Third International Symposium,* International Association for Fire Safety Science, Elsevier Applied Science, New York, 1991.
74. Wickstrom, U., and Goransson, U., "Fuel-Scale/Bench-Scale Correlations of Wall and Ceiling Linings," *Fire and Materials,* 16: 15–22, 1992.
75. Quintiere, J.G., "A Simulation Model for Fire Growth on Materials Subject to a Room-Corner Test," *Fire Safety Journal,* 20: 313–339, 1993.
76. Qian, C., and Saito, K., "An Empirical Model for Upward Flame Spread Over Vertical Flat and Corner Walls," *Fire Safety Science, Proceedings of the Fifth International Symposium,* pp. 285–296, International Association for Fire Safety Science, Japan, 1997.
77. Wright, M.T., *Flame Spread on Composite Materials for Use in High Speed Crafts,* Master of Science thesis, Worcester Polytechnic Institute, Worcester, MA, 1999.
78. Lattimer, B.Y., Hunt, S.P., Sorathia, U., Blum, M., Garciek, T., MacFarland, M., Lee, A., and Long, G., *Development of a Model for Predicting Fire Growth in a Combustible Corner,* NSWCCD-64-TR-199, U.S. Navy, Naval Surface Warfare Center—Carderock Division, Bethesda, MD, September 1999.
79. Dillon, S.E., Quintiere, J.G., and Kim, W.H., "Discussion of a Model and Correlation for the ISO 9705 Room-Corner Test," *Fire Safety Science, Proceedings of the Sixth International Symposium,* International Association for Fire Safety Science, France, 2000.
80. Grenier, A.T., Janssens, M.L., and Dillon, S.E., "Predicting Fire Performance of Interior Finish Materials in the ISO 9795 Room/Corner Test," *Conference Papers, Fire and Materials 2001, 7th International Conference,* San Francisco, Interscience Communications, London, UK, January 2001.
81. Smith, D.A., and Shaw, K., "Evolution of the Single Burning Item Test," *Flame Retardant '98,* London, UK, Interscience Communications, London, UK, February 1990.

CHAPTER 3
PLASTICS AND RUBBER

Richard E. Lyon
Fire Safety Branch AAR-440
Federal Aviation Administration
William J. Hughes Technical Center
Atlantic City International Airport, NJ 08405

3.1 INTRODUCTION

Plastics represent a large and growing fraction of the fire load in public and residential environments, yet relatively little is known about the factors that govern their fire behavior. This is due in large part to the variety of plastics in use, the large number of flammability tests, and the lack of a consensus opinion on what standardized fire test method(s) of fire response best describes the fire hazard. Moreover, the most widely used plastics are those that are least expensive and these tend to be the most flammable. Fig. 3.1 shows the fire hazard (see heat release capacity, Sec. 3.3) versus the truckload price of commercial plastics and elastomers. It is seen that fire hazard and cost span over two orders of magnitude, but the commodity and engineering plastics costing less than $10 per pound comprise over 95 percent of plastics in use and these vary by about a factor of 10 in flammability and price. Specialty plastics costing over $10 per pound are typically heat and chemical resistant (e.g., polymers with aromatic backbones and fluoroplastics) and these tend to also be of low flammability. This chapter examines passive fire protection from a materials engineering perspective. The goal is to develop an understanding of the relationship between the fire behavior of plastics and their properties and identify flammability parameters that can be measured, tabulated, and used to predict fire hazard. Several books have reviewed the flammability parameters of solids [1–17], liquids, and gases [18–20] in relation to their fire behavior.

3.2 POLYMERIC MATERIALS

Plastics and elastomers are commercial products based on polymers (long-chain synthetic organic molecules) that are formulated to obtain specific properties for a particular application. Polymers may be blended together and/or mixed with additives, fillers, or reinforcements to reduce cost, improve heat and light resistance, increase flame retardance, stiffness, toughness, or myriad other physical, chemical, and aesthetic properties. Thus, tens of thousands of commercial products (plastics and elastomers) are derived from a few dozen polymers, with the overwhelming majority being the commodity plastics derived from hydrocarbon monomers continuously obtained from petrochemical feedstocks (i.e., polyolefins and styrenics). The following is a brief introduction to polymers and their chemistry. The interested reader should consult the many excellent texts on polymer science and engineering for more detail.

3.2.1 Monomers, Polymers, and Copolymers

Monomers are reactive liquids or gases that are the building blocks of polymers. Polymers in turn comprise the major component of commercial plastics and elastomers. A single polymer molecule is produced when thousands of liquid or gaseous monomers link together through controlled chemical

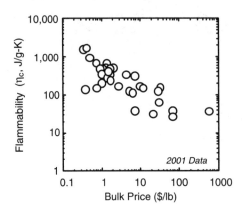

FIGURE 3.1 Flammability (heat release capacity) of plastics versus cost.

reactions called polymerization to produce a long chain. The molecular weight of the polymer increases as additional monomers are added to the chain, with a corresponding increase in boiling point so that the physical state of the reaction mixture changes from a gaseous or liquid monomer to a viscous oil, and finally to a solid. This physical process is reversed in a fire when the chemical bonds in the polymer chain are broken by heat and the polymer reverts back to an oil, liquid, and finally a gas that can mix with oxygen in the flame and undergo combustion (see Fig. 3.6). Thus, the chemical structure of the polymer is closely related to the amount of heat liberated by combustion (see Table 3.3). A detailed description of polymer synthetic chemistry is beyond the scope of this chapter, but a few examples are shown in Fig. 3.2. Generally, polymer molecules are formed when one or more types of monomers add together to form a long chain with practical molar masses ranging from about 50,000 to several million grams per mole. By comparison, the molar mass of automotive gasoline (e.g., octane) is about 100 g/mol. If the monomers react to form a chain without producing any by-products, the polymerization is termed addition, and the chain grows from one end as monomers are sequentially added. Addition polymerization of a single monomer produces a homopolymer such as polyethylene from ethylene gas in Fig. 3.2, while more than one monomer yields a copolymer such as *ethylene-propylene rubber* (EPR), which is an elastomer at room temperature. All of the vinyl polymers and copolymers and most of those containing "ene" in their chemical name (except PBT, PET, PPE, and PPO) in Table 3.1 are addition polymers, as is PA6.

If a small molecule is eliminated during the polymerization, e.g., water is eliminated in the ethylene glycol–terephthalic acid reaction to make *poly(ethyleneterephthalate)* (PET) in Fig. 3.2, then the polymerization is called a condensation polymerization. Condensation polymerization accounts for about half of the polymers in Table 3.1. Engineering plastics (PBT, PET, PPE, PPO, nylons, polysulfones) and many low-cost thermosets (phenolics, aminos, ureas) are condensation polymers.

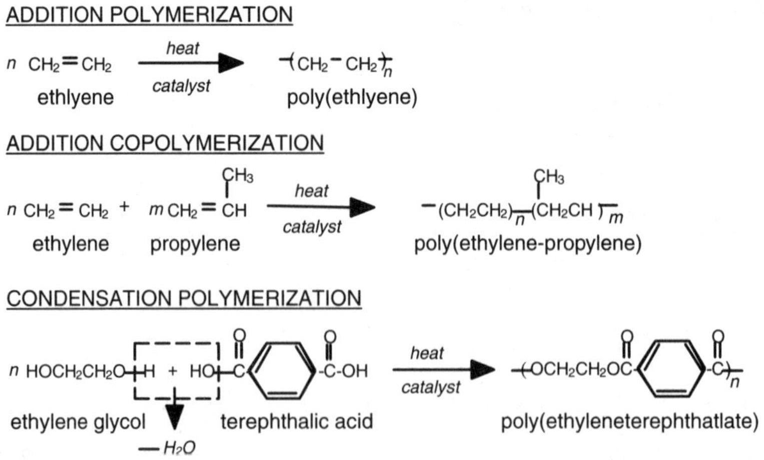

FIGURE 3.2 Examples of plastics made by addition (PE, EPR) and condensation (PET) polymerization.

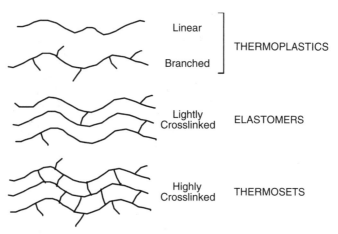

FIGURE 3.3 Molecular architectures for linear, branched, lightly cross-linked, and highly cross-linked plastics and elastomers.

Condensation polymerization involves at least two separate monomers that react together with the elimination of a small molecule that must be continuously removed from the polymerization mixture to achieve high molar mass (thermoplastics) or good structural properties (thermosets).

3.2.2 Polymer Architectures

Molecular. The monomers used to make polymers can have two or more reactive ends or functional groups, $f = 2, 3,$ or 4 (typically). Linear polymer chains result if there are two reactive groups ($f = 2$), and linear chains with occasional intramolecular branches or intermolecular cross-links are produced if the average functionality is between 2 and 3 ($f = 2$ to 3). Linear and branched polymer chains can flow when heated and these are called thermoplastics. Lightly cross-linked polymers cannot flow but can be stretched to several times their initial length with instantaneous or delayed recovery depending on whether the polymer is above or below its glass transition temperature, respectively. If the monomers have an average functionality $f > 3$ the result is a highly cross-linked polymer network with a large number of intermolecular chemical bonds. These polymer networks cannot flow when heated and are called thermoset polymers. Fig. 3.3 shows a schematic diagram of these basic molecular architectures. The implication for fire safety of these two types of polymers is that thermoplastics can melt and drip at, or prior to, ignition if they do not char first, and the flaming drips can spread the fire. For this reason the most common flammability test rates plastics for self-extinguishing tendency as well as the propensity to form flaming drips [21]. Thermoset polymers thermally degrade to volatile fuel without dripping and so limit the fire to their own surface.

Supramolecular. Fig. 3.4 shows schematic diagrams of the two basic types of large-scale supramolecular structure of polymers: amorphous and (semi)crystalline. If the polymer chains are linear and the repeat unit (monomer sequence) is asymmetric or highly branched, the polymer chains in bulk are disordered (amorphous), and if there are no fillers or contaminants to scatter visible light, then these materials are usually clear [e.g., Lucite/Plexiglas polymethyl methacrylate, Lexan polycarbonate, flexible PVC, or silicone rubber]. Amorphous polymers have only a single thermal transition corresponding to a second-order thermodynamic transition known as the glass transition temperature T_g. Below the glass transition temperature, the amorphous polymer is a rigid solid, while above T_g, it is a rubber or highly viscous liquid depending on whether it is cross-linked or not. Above the glass transition temperature, there is a 10^6 reduction in stiffness and a change in the slope of density ρ, heat capacity c, and thermal conductivity κ versus temperature. Fig. 3.5 is a schematic plot of dynamic

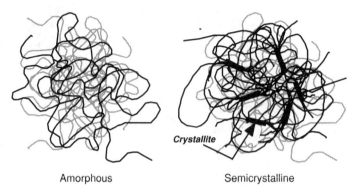

FIGURE 3.4 Amorphous and semicrystalline polymer morphologies.

modulus (stiffness) versus reduced temperature T/T_g showing the dramatic change in stiffness between the glassy state and the rubbery or fluid state. The thermal properties κ, ρ, and c are plotted in reduced form in Fig. 3.5 by normalizing each property p by its value at the glass transition temperature, that is, $p_i(T)/p_i(T_g) = 1$ at $T = T_g$. Fig. 3.5 shows the qualitative changes in κ, ρ, and c with temperature.

If the monomer sequence is fairly regular and symmetric the polymer chain can crystallize into ordered domains known as crystallites that are dispersed in the amorphous (disordered) polymer as illustrated schematically in Fig. 3.4. At the melting temperature T_m, the crystallites melt and the

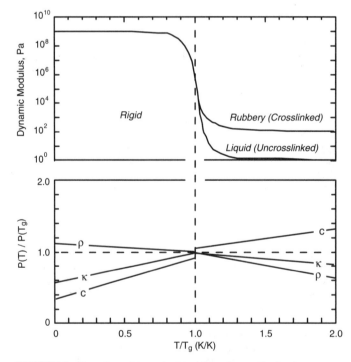

FIGURE 3.5 Dynamic modulus and reduced thermal properties $P = \kappa, \rho, c$ versus reduced temperature (T/T_g). Slope of κ, ρ, c changes at the glass transition temperature $T = T_g$.

TABLE 3.1 Plastics and Elastomers: Nomenclature, Glass Transition Temperature (T_g), and Melting Temperature (T_m)

Polymer (Common or Trade Name)	Abbreviation	T_g (K)	T_m (K)
Thermoplastics			
Acrylonitrile-butadiene-styrene	ABS	373	—
Cellulose acetate	CA	503	—
Cellulose acetate butyrate	CAB	413	—
Cellulose acetate propionate	CAP	463	—
Cellulose nitrate	CN	—	—
Cellulose proprionate	CP	—	—
Polychlorotrifluoroethylene	CTFE	373	493
Polyethylene-acrylic acid salt (ionomer)	EAA	—	358
Polyethylenechlorotrifluoroethylene	ECTFE	—	513
Epoxy (PHENOXY-A)	EP	373	—
Epoxy Novolac (PHENOXY-N)	EPN	438	—
Polyethylene-tetrafluoroethylene (TEFZEL)	ETFE	—	543
Ethylene vinyl acetate	EVA	—	378
Fluorinated ethylene propylene	FEP	331	548
Poly(styrene-butadiene)	HIPS	373	—
Poly(p-phenyleneisophthalamide)	KEVLAR	—	820
Polyarylate (liquid crystalline)	LCP	—	603
Poly(m-phenyleneisophthalamide)	NOMEX	—	680
Polytrifluoroethylene	P3FE	304	—
Polyamide 11	PA11	—	475
Polyamide 12	PA12	—	458
Polyamide 6	PA6	313	533
Polyamide 6/10	PA610	—	493
Polyamide 6/12	PA612	—	480
Polyamide 6/6	PA66	323	533
Polyaramidearylester	PAE	—	—
Polyaryletherketone	PAEK	453	—
Polyamideimide (TORLON)	PAI	548	—
Polyacrylonitrile	PAN	368	408
Polyarylate	PAR	463	—
Poly1-butene	PB	249	400
Polybenzimidazole	PBI	698	—
Poly(p-phenylenebenzobisoxazole)	PBO	>900	—
Polybutyleneterephthalate	PBT	313	510
Polycarbonate of bisphenol-A	PC	423	—
Polycarbonate/ABS blend	PC/ABS	398	—
Polyethylene (high density)	PE HD	195	408
Polyethylene (low density)	PE LD	148	373
Polyethylene (medium density)	PE MD	195	396
Polyethylene (crosslinked)	PE XL	195	396
Polyetheretherketone	PEEK	419	607
Polyetherimide (ULTEM)	PEI	486	—
Polyetherketoneketone	PEKK	430	578
Polyethylmethacrylate	PEMA	338	—
Polyethylenenaphtalate	PEN	—	533
Polyethyleneoxide	PEO	213	308
Polyethersulfone (RADEL-A)	PESU	495	—
Polyethyleneterephthalate	PET	342	528
Poly(tetrafluoroethylene-perfluoroether)	PFA	—	583
Polyimide	PI	610	—
Polymethylmethacrylate	PMMA	387	—
Poly(4-methyl-1-pentene)	PMP	303	505
Poly(α-methyl)styrene	PMS	441	—
Polyoxymethylene	POM	204	453
Polypropylene	PP	253	444
Polyphthalamide (AMODEL)	PPA	393	583
Polyphenyleneether	PPE	358	535
Poly(2,6-dimethylphenyleneoxide)	PPO	482	548
Polypropyleneoxide	PPOX	198	—

TABLE 3.1 (*Continued*)

Polymer (Common or Trade Name)	Abbreviation	T_g (K)	T_m (K)
Thermoplastics			
Polyphenylenesulfide	PPS	361	560
Polyphenylsulfone (RADEL-R)	PPSU	492	—
Polystyrene	PS	373	—
Polysulfone	PSU	459	—
Polytetrafluoroethylene	PTFE	240	600
Polytetramethyleneoxide	PTMO	190	320
Polyvinylacetate	PVAC	304	—
Polyvinylbutyral	PVB	324	—
Polyvinylchloride (plasticized/flexible)	PVC (flex)	248	—
Polyvinylchloride (rigid)	PVC (rigid)	354	—
Polyvinylchloride (chlorinated)	CPVC	376	—
Polyvinylidenechloride	PVDC	255	468
Polyvinylidenefluoride	PVDF	233	532
Polyvinylfluoride	PVF	253	503
Polyvinylcarbazole	PVK	423	—
Polyvinylalcohol	PVOH	358	523
Poly(benzoyl-1,4-phenylene) (POLY-X)	PX	433	—
Poly(styrene-acrylonitrile)	SAN	393	—
Elastomers			
Polybutadiene	BDR	175	—
Polyisobutylene (butyl rubber)	BR	214	—
Polyethylene (chlorinated)	CPE	261	—
Polychloroprene (Neoprene)	CR	233	—
Chlorosulfonated polyethylene	CSPE	274	—
Ethylene-propylene-diene	EPDM	224	—
Poly(vinylidenefluouride-hexafluoropropylene)	FKM	255	—
Polypropyleneoxide-allyglycidylether	GPO	198	—
Nitrile-butadiene (Buna-N)	NBR	243	—
Polyisoprene (natural)	NR	203	—
Polyurethane rubber	PUR	223	—
Styrene-butadiene rubber	SBR	240	—
Polydimethylsiloxane (silicone)	SIR	146	—
Thermosets			
Bismaleimide	BMI	573	—
Benzoxazine of bisphenol-A/aniline	BZA	423	—
Cyanate ester of hexafluorobisphenol-A	CEF	546	—
Cyanate ester of bisphenol-A	CEA	543	—
Cyanate ester of bisphenol-E	CEE	548	—
Cyanate ester of bisphenol-M	CEM	528	—
Cyanate ester of tetramethylbisphenol-F	CET	525	—
Diallylphthalate	DAP	423	—
Epoxy	EP	393	—
Melamine formaldehyde	MF	—	—
Phenol formaldehyde	PF	443	—
Polyimide	PI	623	—
Cyanate Ester from Novolac (phenolic triazine)	PT	375	—
PU (isocyanurate/rigid)	PU	—	—
Silicone resin	SI	473	—
Urea formaldehyde	UF	—	—
Unsaturated polyester	UPT	330	—
Vinylester	VE	373	—

entire polymer becomes amorphous and can flow. Because the melting temperature of the crystallites is above the glass transition temperature (typically $T_m/T_g \approx$ 1.3 to 2.0 K/K), crystallinity raises the flow temperature of the plastic and makes it more rigid. However, crystallinity does not prevent flaming drips as the melting temperature is usually much lower than the ignition temperature (compare Tables 3.1 and 3.6). Crystallinity does not exceed 90 to 95 percent in bulk polymers, with 20 to 80 percent being typical, because the polymer chains are too long to pack into an orderly crystal lattice without leaving some dangling ends that segregate into disordered (amorphous) domains. Crystallites usually are of sufficient size to scatter visible light so that natural/unfilled semicrystalline plastics are translucent or white. Semicrystalline polymers of commercial importance include polyethylene, polypropylene, PET, polytetrafluoroethylene, and the polyamides (nylons).

3.2.3 Commercial Materials

Table 3.1 lists some plastics and elastomers for which a reasonably complete set of fire and thermal properties were available. Abbreviations conform to the recommended International Standards Organization (ISO) 1043-1 (thermoplastics and thermosets) and ASTM D1418 (elastomers) designations. The following definitions apply to the commercial plastics and elastomers in this chapter:

Thermoplastic. A linear or branched polymeric solid that flows with the application of heat and pressure at the glass transition temperature (amorphous) or the crystalline melting temperature (semicrystalline), whichever is higher. Different thermoplastics can be blended together in the molten state to obtain new compositions, called alloys, with improved toughness (PC/ABS, HIPS), better high-temperature properties (PS/PPO), or better flame retardancy (PVC/PMMA). Reinforced thermoplastic grades typically contain chopped fiberglass or carbon fibers at 30 to 40 percent by weight to increase strength and stiffness. Continuous sheet and profile are made by extrusion, and individual parts and shapes by injection molding, rotational molding, etc.

Elastomer. A lightly cross-linked linear polymer that is above its glass transition temperature at room temperature (i.e., is rubbery). Elastomers exhibit high extensibility (>100 percent strain) and complete, instantaneous recovery. Cross-linking can be by permanent chemical bonds (thermoset), which form in a process called vulcanization, or by thermally labile glassy or ionic domains that can flow with the application of heat (thermoplastic elastomer). Commercial elastomers are typically compounded with oils, fillers, extenders, and particulate reinforcement (carbon black, fumed silica). Vulcanized elastomers (e.g., tires) are cured in closed heated molds, while thermoplastic elastomers can be extruded, compression molded, or injection molded.

Thermoset. A rigid polymer made from two or more multifunctional monomers. Polymerization to a highly cross-linked network gives the final form (typically in a mold) that will not flow with application of heat or pressure. Thermoset polymers degrade thermally rather than flow because the intermolecular bonds are permanent chemical ones. Thermosets are typically brittle and commercial formulations are usually compounded with chopped fiberglass or mineral fillers to improve strength and reduce cost.

The generic fire property data tabulated in this chapter for plastics and elastomers are averages of values within sources and between sources (typically 1 to 3) for each material unless the values differed by more than about 20 percent, in which case the range is specified. No attempt was made to establish the composition of commercial products reported in the literature and nominal values are used throughout. The tabulated fire and thermal properties are thus representative of the average of commercial formulations. Polymeric materials listed by name (e.g., polyethylene terephthalate/PET) are assumed to be natural (unmodified) polymers, copolymers, and blends containing at most a few weight percent of stabilizers and processing aids. Flame-retardant grades are designated by the suffix -FR which usually refers to an additive level sufficient to achieve a self-extinguishing rating in a bunsen burner test of ignition resistance, e.g., Underwriters Laboratory test for flammability of plastic materials (UL 94) [21]. Flame-retardant formulations are proprietary but can include inert fillers such as alumina trihydrate (ATH) and flame-retardant chemicals [7, 9, 10, 13, 17]. Thermoplastics, thermosets, and elastomers reinforced with chopped glass fibers are designated by the suffix -G. Reinforcement level is 30 to 40 percent by weight unless otherwise noted. Filled grades designated by the suffix -M contain mineral fillers such as talc, calcium carbonate, etc., at unspecified levels.

3.2.4 Thermodynamic Quantities

Thermal Properties. The rate at which heat is transported and stored in polymers in a flame or fire is of fundamental importance because these processes determine the time to ignition and burning rate. There are no good theories to predict the thermal conductivity κ (W/m·K), heat capacity c (kJ/kg·K), or density ρ (kg/m³) of condensed phases (e.g., solid or molten polymers) from chemical structure, but empirical structure-property correlations have been developed that allow calculation of thermal properties from additive atomic [29] or chemical group [30] contributions if the chemical structure of the plastic is known. Table 3.2 lists generic thermophysical properties at 298 K gathered from the literature [22–33, 35–39] for a number of common thermoplastics, thermoset resins, elastomers, and fiberglass-reinforced plastics. Entries are individual values, averages of values from different sources, or averages of a range of values from a single source, and therefore represent in most cases a generic property value with an uncertainty of about 10 to 20 percent. Empirical structure-property correlations [29, 30] were used to calculate thermal properties of several polymers at 298 K from their chemical structure when these could not be found in the literature. The general trend of κ, ρ, and c with temperature is shown in reduced form in Fig. 3.5 relative to the values of these properties at the glass transition temperature.

Thermal conductivity increases with degree of crystallinity and the temperature dependence of the thermal conductivity of polymers varies widely in the literature [31–33]. However, a rough approximation of temperature dependence of the thermal conductivity relative to its value at the glass transition temperature $\kappa(T_g)$ is [29, 30]

$$\kappa \approx \kappa(T_g)\left[\frac{T}{T_g}\right]^{0.22} \quad (T < T_g) \tag{3.1a}$$

$$\kappa \approx \kappa(T_g)\left\{1.2 - 0.2\left[\frac{T}{T_g}\right]\right\} \quad (T > T_g) \tag{3.1b}$$

The relationship between density and temperature can be expressed (neglecting the abrupt change on melting of semicrystalline polymers) to a first approximation [30]

$$\frac{1}{\rho} = \frac{1}{\rho_0} + B(T - T_0) \tag{3.2}$$

where $\rho = \rho(T)$ is the density at temperature T, ρ_0 is the density at temperature $T_0 = 298$ K, and $B = 5 \pm 2 \times 10^{-7}$ m³/(kg·K) is the volume thermal expansivity per unit mass. Neglecting crystalline melting, the temperature dependence of the heat capacity can be approximated [29, 30]

$$c = (c_0 + \Delta c)(0.64 + 1.2 \times 10^{-3}T) \approx \frac{3}{4}c_0(1 + 1.6 \times 10^{-3}T) \tag{3.3}$$

where $c = c(T)$ in units of kJ/kg·K is the heat capacity at temperature T, c_0 is the heat capacity at standard temperature $T_0 = 298$ K, and Δc is the change in heat capacity at the glass transition temperature.

The product $\kappa\rho c$ is a quantity called the thermal inertia that emerges from the transient heat transfer analysis of ignition time [see Eq. (3.52)]. The individual temperature dependence of κ, ρ, and c revealed by Eqs. (3.1) through (3.3) and experimental data for about a dozen plastics [22–39] suggest that the product of these terms (i.e., the thermal inertia) should have the approximate temperature dependence:

$$\kappa\rho c(T) \approx \kappa_0\rho_0 c_0 T/T_0 = (\kappa\rho c)_0 T/T_0 \tag{3.4}$$

where κ_0, ρ_0, c_0 are the room temperature (T_0) values listed in Table 3.2. Another thermal parameter that emerges from unsteady heat transfer analyses [see Eqs. (3.52) and (3.58)] is the thermal diffusivity $\alpha = \kappa/\rho c$. Thermal diffusivities of polymers at T_0 reported in the literature [26–33] or calculated from κ, ρ, and c are listed in the last column of Table 3.2. Thermal diffusivity generally decreases with temperature according to the approximate relationship derived from experimental data [32, 33]

$$\alpha(T) = \alpha_0 T_0/T$$

Heat of Combustion (HOC). At constant pressure and when no nonmechanical work is done, the heat (Q, q) and enthalpy (H, h) of a process are equal. The flaming combustion of polymers at atmospheric pressure satisfies these conditions. The high-pressure adiabatic combustion of a polymer in a bomb calorimeter satisfies these conditions approximately, since the fractional pressure change is small. Consequently, the terms heat and enthalpy are used interchangeably in polymer combustion. Heats of combustion of organic macromolecules can be calculated from the oxygen consumed in the combustion reaction [40–45]. Oxygen consumption is, in fact, the basis for most modern bench- and full-scale measurements of heat release in fires [41, 42]. The principle of oxygen consumption derives from the observation that for a wide range of organic compounds, including polymers, the heat of complete combustion per mole of oxygen consumed is a constant E that is independent of the composition of the polymer. Mathematically,

$$E = h_{c,p}^o \left[\frac{nM}{n_{O_2} M_{O_2}} \right] = \frac{h_{c,p}^o}{r_O} = 13.1 \pm 0.7 \text{ kJ/g O}_2 \quad (3.5)$$

where $h_{c,p}^o$ is the net heat of complete combustion of the polymer solid with all products in their gaseous state, n and M are the number of moles and molecular weight of the molecule or polymer repeat unit, respectively, n_{O_2} is the number of moles of O_2 consumed in the balanced thermochemical equation, and $M_{O_2} = 32$ g/mol is the molecular weight of diatomic oxygen. In Eq. (3.5), the quantity $r_O = [n_{O_2} M_{O_2}/nM]$ is the oxygen-to-fuel mass ratio.

To illustrate the thermochemical calculation of the net HOC we use as an example poly(methylmethacrylate) (PMMA), which has the chemical structure

$$-\!\!\!\left[\text{CH}_2\text{-}\underset{\underset{\underset{\text{CH}_3}{|}}{\overset{\overset{|}{\text{O}}}{\underset{|}{\text{C}=\text{O}}}}}{\overset{\overset{\text{CH}_3}{|}}{\text{C}}}\right]_{\!\!n}\!\!- \quad \text{Poly(methylmethacrylate)}$$

The methylmethacrylate repeat unit shown in brackets has the atomic composition $C_5H_8O_2$ so the balanced chemical equation for complete combustion is

$$C_5H_8O_2 + 6 \text{ O}_2 \rightarrow 5 \text{ CO}_2 + 4 \text{ H}_2O$$

Thus, 6 moles of O_2 are required to completely convert 1 mole of PMMA repeat unit to carbon dioxide and water. Inverting Eq. (3.5)

$$h_{c,p}^o = E\left[\frac{n_{O_2} M_{O_2}}{nM}\right] = \frac{(13.1 \text{ kJ/g O}_2)(6 \text{ mol O}_2)(32 \text{ g O}/m_2\text{ol O}_2)}{(1 \text{ mol PMMA})(100 \text{ g/mol PMMA})} = 25.15 \text{ kJ/g}$$

Table 3.3 lists net heats of complete combustion for plastics and elastomers obtained from the literature [39–41]. Values in parentheses were calculated from the elemental composition as illustrated above.

Heat of Gasification. In principle, the heat (enthalpy) of gasification is the difference between the enthalpy of the solid in the initial state and the enthalpy of the volatile thermal-decomposition products

TABLE 3.2 Thermal Properties of Plastics

Polymer	κ W/m·K	ρ kg/m^3	c_p kJ/kg·K	α m^2/s × 10^7
ABS	0.26	1050	1.50	1.65
BDR	0.22	970	1.96	1.16
BR	0.13	920	1.96	0.72
CA	0.25	1250	1.67	1.20
CAB	0.25	1200	1.46	1.43
CAP	0.25	1205	1.46	1.42
CE	0.19	1230	1.11	1.39
CN	0.23	1375	1.46	1.15
CP	0.20	1300	1.46	1.05
CPVC	0.48	1540	0.78	4.00
CR	0.19	1418	1.12	1.20
CTFE	0.23	1670	0.92	1.50
DAP	0.21	1350	1.32	1.18
DAP-G	0.42	1800	1.69	1.38
EAA	0.26	945	1.62	1.70
ECTFE	0.16	1690	1.17	0.81
EP	0.19	1200	1.7	1.12
EPDM	0.20	930	2.0	1.08
EP-G	0.42	1800	1.60	1.46
EPN	0.19	1210	1.26	1.25
ETFE	0.24	1700	1.0	0.66
EVA	0.34	930	1.37	2.67
FEP	0.25	2150	1.17	0.99
HIPS	0.22	1045	1.4	1.54
LCP	0.20	1350	1.20	1.24
MF	0.25	1250	1.67	1.20
MF-G	0.44	1750	1.67	1.51
NBR	0.25	1345	1.33	1.40
NR	0.14	920	1.55	0.98
P3FE	0.31	1830	1.08	1.41
PA11	0.28	1120	1.74	1.44
PA11-G	0.37	1350	1.76	1.56
PA12	0.25	1010	1.69	1.46
PA6	0.24	1130	1.55	1.37
PA610	0.23	1100	1.51	1.38
PA612	0.22	1080	1.59	1.28
PA66	0.23	1140	1.57	1.29
PA6-G	0.22	1380	1.34	1.19
PAEK	0.30	1300	1.02	2.27
PAI	0.24	1420	1.00	1.69
PAN	0.26	1150	1.30	1.74
PAR	0.18	1210	1.20	1.24
PB	0.22	920	2.09	1.14
PBI	0.41	1300	0.93	3.40
PBT	0.22	1350	1.61	1.01
PC	0.20	1200	1.22	1.36
PC-G	0.21	1430	1.10	1.34
PE (HD)	0.43	959	2.00	2.24
PE (LD)	0.38	925	1.55	2.65
PE (MD)	0.40	929	1.70	2.53

TABLE 3.2 (*Continued*)

Polymer	κ W/m·K	ρ kg/m³	c_p kJ/kg·K	α m²/s $\times 10^7$
PEEK	0.20	1310	1.70	0.90
PEI	0.23	1270	1.22	1.48
PEKK	0.22	1280	1.00	1.72
PEMA	0.18	1130	1.47	1.08
PEO	0.21	1130	2.01	0.90
PESU	0.18	1400	1.12	1.15
PET	0.20	1345	1.15	1.29
PET-G	0.29	1700	1.20	1.42
PF	0.25	1300	1.42	1.35
PFA	0.25	2150	1.0	1.16
PF-G	0.40	1850	1.26	1.72
PI	0.11	1395	1.10	0.72
PI-TS	0.21	1400	1.13	1.33
PMMA	0.20	1175	1.40	1.19
PMP	0.17	834	1.73	1.18
PMS	0.20	1020	1.28	1.53
POM	0.23	1420	1.37	1.18
PP	0.15	880	1.88	0.89
PPA	0.15	1170	1.40	0.92
PPE	0.23	1100	1.19	1.76
PPO	0.16	1100	1.25	1.16
PPO-G	0.17	1320	1.31	0.98
PPS	0.29	1300	1.02	2.19
PPSU	0.18	1320	1.01	1.35
PS	0.14	1045	1.25	1.04
PS-G	0.13	1290	1.05	0.96
PSU	0.26	1240	1.11	1.89
PTFE	0.25	2150	1.05	1.11
PU	0.21	1265	1.67	0.99
PUR	0.19	1100	1.76	0.98
PVAC	0.16	1190	1.33	1.03
PVC (flex)	0.17	1255	1.38	0.98
PVC (rigid)	0.19	1415	0.98	1.34
PVDC	0.13	1700	1.07	0.91
PVDF	0.13	1760	1.12	0.68
PVF	0.13	1475	1.30	0.72
PVK	0.16	1265	1.23	1.02
PVOH	0.20	1350	1.55	0.96
PX	0.32	1220	1.3	2.02
SAN	0.15	1070	1.38	1.02
SBR	0.17	1100	1.88	0.82
SI-G	0.30	1900	1.17	1.35
SIR	0.23	970	1.59	1.49
UF	0.25	1250	1.55	1.29
UPT	0.17	1230	1.30	1.06
UPT-G	0.42	1650	1.05	1.85
VE	0.25	1105	1.30	1.74

TABLE 3.3 Net Heats of Complete Combustion and Chemical Formulae of Plastics (Calculated Values in Parentheses. Averages Indicated by ±1 Standard Deviation)

Polymer	Chemical formula	Net heat of complete combustion MJ/kg
ABS	$C_{15}H_{17}N$	36.0 ± 3.0
BMI	$C_{21}H_{14}O_4N_2$	(26.3)
BR	C_4H_8	42.7
BZA	$C_{31}H_{30}O_2N_2$	33.5
CA	$C_{12}H_{16}O_8$	17.8
CAB	$C_{12}H_{18}O_7$	22.3
CAP	$C_{13}H_{18}O_8$	(18.7)
CEA	$C_{17}H_{14}O_2N_2$	28.8
CEE	$C_{16}H_{12}O_2N_2$	28.4
CEF	$C_{16}H_{12}O_2N_2$	18.3
CEM	$C_{26}H_{24}O_2N_2$	33.1
CEN	$C_{24}H_{15}O_3N_3$	28.8 ± 1.4
CET	$C_{19}H_{18}O_2N_2$	30.0
CN	$C_{12}H_{17}O_{16}N_3$	10.5 ± 3.1
CP	$C_{15}H_{22}O_8$	(21.0)
CPE (25% Cl)	$C_{10}H_{19}Cl$	31.6
CPE (36% Cl)	C_4H_7Cl	26.3
CPE (48% Cl)	$C_8H_{18}Cl_3$	20.6
CPVC	$CHCl$	12.8
CR	C_4H_5Cl	18.6 ± 8.9
CSPE	$C_{282}H_{493}Cl_{71}SO_2$	26.7
CTFE	C_2ClF_3	5.5 ± 3.5
DAP	$C_7H_7O_2$	26.2
EAA	C_5H_8O	(32.4)
ECTFE	$C_4H_4F_3Cl$	13.6 ± 1.9
EP	$C_{21}H_{24}O_4$	32.0 ± 0.8
EPDM	C_5H_{10}	38.5
EPN	$C_{20}H_{11}O$	29.7
ETFE	$C_4H_4F_4$	12.6
EVA	C_5H_9O	(33.3)
FEP	C_5F_{10}	7.7 ± 4.0
FKM	$C_5H_2F_8$	12.5 ± 2.5
HIPS	$C_{14}H_{15}$	42.5
KEVLAR	$C_{14}H_{10}O_2N_2$	(27.3)
LCP	$C_{39}H_{22}O_{10}$	25.8
MF	$C_6H_9N_6$	18.5
NBR	$C_{10}H_{14}N$	33.1 ± 0.4
NOMEX	$C_{14}H_{10}O_2N_2$	26.5 ± 1.2
NR	C_5H_8	42.3
P3FE	C_2HF_3	(11.9)
PA11	$C_{11}H_{21}ON$	34.5
PA12	$C_{12}H_{23}ON$	(36.7)
PA6	$C_6H_{11}ON$	28.8 ± 1.1
PA610	$C_{16}H_{30}O_2N_2$	(33.4)
PA612	$C_{18}H_{34}O_2N_2$	(34.5)
PA66	$C_{12}H_{22}O_2N_2$	30.6 ± 1.8
PAEK	$C_{13}H_8O_2$	30.2
PAI	$C_{15}H_8O_3N_2$	24.3
PAN	C_3H_3N	31.0
PAR	$C_{23}H_{18}O_4$	(29.9)
PB	C_4H_9	43.4
PBD	C_4H_6	42.8

TABLE 3.3 (*Continued*)

Polymer	Chemical formula	Net heat of complete combustion MJ/kg
PBI	$C_{20}H_{12}N_4$	21.4
PBO	$C_{14}H_6O_2N_2$	28.6
PBT	$C_{12}H_{12}O_4$	26.7
PC	$C_{16}H_{14}O_3$	30.4 ± 0.8
PC/ABS	$C_{45}H_{43}O_6N$	(32.4)
PE (HD)	C_2H_4	43.8 ± 0.7
PE (LD)	C_2H_4	(44.8)
PE (MD)	C_2H_4	(44.8)
PEEK	$C_{19}H_{12}O_3$	30.7 ± 0.6
PEI	$C_{37}H_{24}O_6N_2$	29.0 ± 1.0
PEKK	$C_{20}H_{12}O_3$	30.3
PEMA	$C_6H_{10}O_2$	(27.6)
PEN	$C_{14}H_{10}O_4$	(25.2)
PEO	C_2H_4O	24.7
PESU	$C_{12}H_8O_3S$	24.9 ± 0.4
PET	$C_{10}H_8O_4$	22.2 ± 1.4
PF	C_7H_5O	28.6
PFA	C_5OF_{10}	5.0
PI	$C_{22}H_{10}O_5N_2$	25.4
PMMA	$C_5H_8O_2$	25.0 ± 0.1
PMP	C_6H_{12}	43.4
PMS	C_9H_{10}	40.4
POM	CH_2O	15.7 ± 0.2
PP	C_3H_6	43.1 ± 0.4
PPA	$C_{14}H_{19}O_2N_2$	(30.1)
PPE	C_6H_4O	(29.6)
PPO	C_8H_8O	32.9 ± 0.3
PPOX	C_3H_6O	28.9
PPS	C_6H_4S	28.3 ± 0.7
PPSU	$C_{24}H_{16}O_4S$	27.2
PS	C_8H_8	40.5 ± 1.3
PSU	$C_{27}H_{22}O_4S$	29.2 ± 0.3
PTFE	C_2F_4	6.0 ± 0.7
PTMO	C_4H_8O	31.9
PU	$C_6H_8O_2N$	24.3 ± 2.1
PUR	$C_{80}H_{120}O_2N$	26.3 ± 2.5
PVAC	$C_4H_6O_2$	21.5
PVB	$C_8H_{14}O_2$	30.7
PVC (flex)	$C_{26}H_{39}O_2Cl$	24.7 ± 3.5
PVC (rigid)	C_2H_3Cl	16.7 ± 0.4
PVDC	$C_2H_2Cl_2$	13.1 ± 4.9
PVDF	$C_2H_2F_2$	13.7 ± 0.6
PVF	C_2H_3F	20.3
PVK	$C_{14}H_{11}N$	(36.4)
PVOH	C_2H_4O	22.2 ± 1.2
PX	$C_{13}H_8O$	37.4
SAN	$C_{27}H_{27}N$	(38.8)
SBR	$C_{10}H_{13}$	42.0
SI	$C_{12}H_{10}O_3Si_2$	(24.4)
SIR	C_2H_6OSi	17.1 ± 3.0
UF	$C_3H_6O_2N_2$	20.8 ± 8.7
UPT	$C_{12}H_{13}O_3$	24.4 ± 5.8
VE	$C_{29}H_{36}O_8$	(27.8)

at the pyrolysis temperature. Thus, the heat of gasification is expected to be a thermodynamic quantity comprised of the sum of the enthalpies required to bring the polymer from the solid state at the initial (room) temperature T_0 and pressure P_0 (1 atm) to the gaseous state at the pyrolysis temperature and pressure T_p and P_0, respectively. If the stored heat on a molar basis is ΔH_s, the enthalpy of fusion (melting) for semicrystalline polymers is ΔH_f, the bond dissociation enthalpy is ΔH_d, and the enthalpy of vaporization of the decomposition products is ΔH_v, then the molar heat of gasification is

$$\Delta H_g = \Delta H_s + \Delta H_f + \Delta H_d + \Delta H_v \qquad (3.6)$$

Table 3.4 illustrates the magnitude of these enthalpic terms on a mass basis for amorphous poly(methylmethacrylate), polystyrene, and semicrystalline polyethylene. Values in joules per gram (J/g) are obtained by dividing the molar heat by the molecular weight of the gaseous decomposition products M_g. The stored heat Δh_s was obtained by numerical integration of heat capacity versus temperature [35] from ambient to the dissociation temperature. Unfortunately, experimental data for c versus T for polymers is scarce, but a reasonable approximation for the stored heat is obtained by integrating the analytic expression for the heat capacity [Eq. (3.2)] between room temperature (T_0) and the onset degradation temperature (T_d)

$$\Delta h_s = \int_{T_0}^{T_d} c(T)\,dT = \frac{3}{4}c_0\left\{(T_d - T_0) + 0.8 \times 10^{-3}\frac{T_d^2}{T_0^2}\right\} \approx \frac{3}{4}c_0(T_d - T_0) \qquad (3.7)$$

where c_0 and T_d are calculable from the polymer chemical structure using empirical molar group contributions [29, 30]. The dissociation (bond-breaking) enthalpy Δh_d is assumed to be equal to the heat of polymerization but opposite in sign for these polymers that thermally degrade by random or end-chain scission [34] (see Table 3.5). The degradation product for polyethylene is assumed to be a tetramer (i.e., octane with $M_g = 112$ g/mol) for the purpose of calculating the heats of dissociation and vaporization on a mass basis for this polymer, and the degree of polyethylene crystallinity is taken to be 90 percent. All other enthalpies in Table 3.4 were obtained from handbooks [35] using monomer molecular weights M to convert the energies to a mass basis. The values for h_g in the second to last row were obtained by summing the individual enthalpies according to Eq. (3.6) for each polymer.

In practice, the heat of gasification per unit mass of solid h_g is rarely calculated because detailed and reliable thermodynamic data for the polymer and its decomposition products are generally unavailable except for the most common polymers. Direct laboratory measurement of h_g using differential thermal analysis and differential scanning calorimetry have been reported, but h_g is usually measured in a constant heat flux gasification device or fire calorimeter. In these experiments a plot of mass loss rate per unit surface area (mass flux) versus external heat flux has slope $1/L_g$ where

$$L_g = \frac{h_g}{1 - \mu} \qquad (3.8)$$

TABLE 3.4 Components of the Heat of Gasification of PMMA, PS, and PE

Polymer	PMMA	PS	PE
Monomer MW (g/mole)	100	104	28
Fuel MW (g/mole)	100	104	112
Δh_s (J/g)	740	813	803
Δh_f (J/g)	amorphous	amorphous	243
Δh_d (J/g)	550	644	910
Δh_v (J/g)	375	387	345
$h_g = \Sigma \Delta h_i$ (J/g)	1665	1850	2301
h_g (measured) J/g	1700	1800	2200

is the heat absorbed per unit mass of volatile fuel produced and µ is the nonfuel fraction (char or inert filler). The last row in Table 3.4 lists the average of h_g values for these noncharring polymers (see Table 3.11). Agreement is seen to be quite good between experimental values and thermochemical calculations of h_g. Table 3.11 in the section on "Steady Burning" contains L_g values for about 75 plastics, thermosets, and elastomers.

3.3 THE BURNING PROCESS

3.3.1 The Fire Triangle

Strictly speaking, solid polymers do not burn. Rather, it is their volatile thermal decomposition products that burn in the gas phase when mixed with oxygen and ignited. Ignition occurs when the concentration of volatile fuel gases reaches the lower flammability limit for the particular fuel-air mixture. Polymers do not burn in the condensed state because of the low solubility and diffusivity of oxygen and the low oxidation rate at the decomposition temperature. In fact, thermal degradation of the surface layer of polymer in the presence of a heat source is thought to occur in a reducing, rather than an oxidizing, environment. Low-molecular-weight volatile organic compounds are produced that mix with atmospheric oxygen above the polymer surface to form a flammable mixture that, when ignited, combusts, producing a luminous flame. The surface temperature of the burning plastic cannot greatly exceed its thermal decomposition temperature until all of the volatile fuel is depleted because until this occurs excess thermal energy is consumed by vaporization (mass transfer) of the volatile fuel rather than being stored in the solid as a temperature rise. The surface temperature of plastics at ignition, also called the fire point temperature, should therefore be close to the thermal degradation temperature (see Table 3.6). At these temperatures, the thermal degradation reactions at the plastic surface are faster than the rate at which heat is absorbed. Consequently, it is the latter process (i.e., heat transfer) that governs the burning rate, heat release rate, and smoke evolution during flaming combustion. The chemical structure of the plastic or elastomer determines the thermal stability (ignition temperature), fuel fraction, potential HOC of the fuel gases, and the products of combustion.

Fig. 3.6 illustrates the three coupled processes required for flaming combustion: (1) heating of the polymer, (2) thermal decomposition of the solid polymer to gaseous fuel, and (3) ignition and combustion of the fuel gases in air. An ignition source or thermal feedback of radiant energy from the flame supplies heat to the polymer surface that causes thermolysis of primary chemical bonds in the polymer molecules. Evaporation of the low-molar-mass degradation products and the reaction of these with air (oxygen) in the combustion zone above the surface releases heat and produces carbon dioxide, water, and incomplete combustion products such as carbon monoxide, mineral acids, unburned hydrocarbons, and soot. In order to resist burning, the fire cycle must be broken at one or more places.

Several comprehensive texts have been written on the chemistry and physics of gas phase combustion [18–20]. In contrast, combustible solids (with the exception of wood) have received relatively little attention. The remainder of this chapter examines the flaming combustion of solids, specifically plastics, from a phenomenological perspective. Recent developments in the metrology and modeling of fire and its impact on materials provide the basis for relating polymer ignition

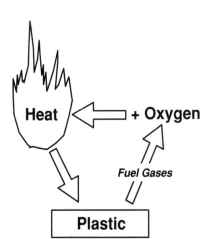

FIGURE 3.6 The fire triangle. Heating of the plastic generates volatile thermal degradation products (fuel gases) that mix with air forming a combustible mixture. Ignition of the combustible mixture releases heat that continues the burning process.

and burning to measurable, macroscopic flammability parameters. Connecting these macroscopic flammability parameters to the kinetics and thermodynamics of the fuel-generation process provides a thermochemical basis for the solid-state processes of flaming combustion.

3.3.2 Chemical Changes during Burning

The elementary fuel-generation step of a solid in a fire is thermal degradation [46–63]. Typically, it is the fraction and rate of production of volatile fuel at fire temperatures and the HOC of this fuel that determine the flammability of plastics and elastomers. Short-term thermal stability and reduced fuel fraction (increased char yield) are achieved by eliminating hydrogen atoms from the polymer molecule so that recombination of carbon radicals to form char during thermal degradation is kinetically favored over hydrogen abstraction/termination reactions that produce volatile fuel fragments. A low HOC is observed when heteroatoms (e.g., halogens, nitrogen, phosphorus, sulfur, silicon, boron, and oxygen) replace carbon and hydrogen in the polymer molecule. Heteroatoms form stable gas phase combustion products that are either low in fuel value (i.e., N_2, SO_2, hydrogen halides) or thermally stable solid oxides (i.e., SiO_2, P_2O_5, B_2O_3) that precipitate onto the polymer surface and act as mass- and thermal-diffusion barriers.

Thermal Decomposition of the Solid. The basic thermal degradation mechanism leading to volatile fuel generation in polymers involves primary and secondary decomposition events. The primary decomposition step can be main-, end-, or side-chain scission of the polymer [5, 46–48]. Subsequent thermal degradation reactions depend largely on the chemical structure of the polymer but typically proceed by hydrogen transfer to α- or β-carbons, nitrogen or oxygen, intramolecular exchange (cyclization), side-chain reactions, small-molecule (SO_2, CO_2, S_2) elimination, molecular rearrangement, and/or unzipping to monomer [5, 46–48, 51]. Unzipping or depolymerization of vinyl polymers is characterized by a kinetic chain length or "zip length," which is the average number of monomer units produced by a decomposing radical before the radical is deactivated by termination or transfer. Mathematically, the zip length is the ratio of the rate constants for initiation to termination. Aromatic backbone polymers such as polycarbonate, polyimide, and polyphenyleneoxide tend to decompose in varying degrees to a carbonaceous char residue through a complex set of reactions involving cross-linking and bond scission [7]. A generally applicable, detailed mechanism for thermal degradation of aromatic polymers is unlikely.

The enthalpy of the solid → gas phase change has been related to the global activation energy for pyrolysis E_a measured in a laboratory *thermogravimetric analyzer* (TGA) [34, 52]. In particular, the average molecular weight of the decomposition products M_g is related to the heat of gasification per unit mass of solid h_g

$$h_g = \frac{E_a}{M_g} \tag{3.9}$$

In this case, the average molar mass of the decomposition products M_g and the molar mass of the monomer or repeat unit M should be in the ratio

$$\frac{M_g}{M} = \frac{E_a}{M h_g} \tag{3.10}$$

Polymers that pyrolyze to monomer by end-chain scission (depolymerize/unzip) at near-quantitative yield such as PMMA, polyoxymethylene, and polystyrene should have M_g equal to the monomer molar mass M, that is, $M_g/M \approx 1$. Polymers such as polyethylene and polypropylene that decompose by main-chain scission (cracking) to multimonomer fragments have $M_g/M > 1$. In contrast, polymers with complex molecular structures and high molar mass repeat units ($M \geq 200$ g/mol) such as nylon, cellulose, or polycarbonate degrade by random scission, cyclization, small-molecule splitting, or chain stripping of pendant groups (e.g., polyvinylchloride) and yield primarily low-molar-mass species (water, carbon dioxide, alkanes, mineral acids) relative to the starting monomer so that $M_g/M < 1$. Table 3.5 shows fuel/monomer molar mass ratios M_g/M calculated as E_a/Mh_g according to

TABLE 3.5 Heats of Gasification, Pyrolysis Activation Energy, Char Yield, and Calculated Molecular Weight of Decomposition Products for Some Polymers

Polymer	M (g/mol)	L_g (kJ/g)	μ (g/g)	h_g (kJ/g)	E_a (kJ/mol)	M_g/M	Pyrolysis products
				Chain cracking			
PP	42	1.9	0	1.9	243	3.0	C_2–C_{90} saturated and unsaturated
PE	28	2.2	0	2.2	264	4.3	hydrocarbons
				Unzipping			
PS	104	1.8	0	1.8	230	1.2	40–60% monomer
PMMA	100	1.7	0	1.7	160	0.94	100% monomer
POM	30	2.4	0	2.4	84	1.2	100% monomer
				Intramolecular scission			
PA 66	226	2.1	0	2.1	160	0.3	H_2O, CO_2, C_5 HC's
PVC	62	2.7	0.1	2.4	110	0.7	HCl, benzene, toluene
Cellulose	162	3.2	0.2	2.6	200	0.5	H_2O, CO_2, CO
PT	131	5.0	0.6	2.0	178	0.3	Complex mixture of low mo-
PC	254	2.4	0.3	1.7	200	<1	lecular weight products
PEI	592	3.5	0.5	1.8	≈275	<1	
PPS	108	3.8	0.5	1.9	≈275	<1	
PEEK	288	3.4	0.5	1.7	≈275	<1	
PAI	356	4.8	0.6	1.9	≈275	<1	
PX	180	6.4	0.7	1.9	≈275	<1	

Eq. (3.10) for some of the commercial polymers listed in Tables 3.1 to 3.4. Global pyrolysis activation energies for the thermally stable engineering plastics listed in the last four rows of Table 3.5 are estimated to be in the range $E_a = 275 \pm 25$ kJ/mol [30, 35, 47, 49]. Qualitative agreement is observed between the modes of pyrolysis (end-chain scission, random scission, chain stripping) and the calculated fragment molecular weight using Eq. (3.10), suggesting that the global pyrolysis activation energy determined from mass loss rate experiments is the molar enthalpy of pyrolysis of the degradation products. Surprisingly, the heat of gasification per unit mass of solid $h_g = (1-\mu)L_g$ remains constant at about 2.0 kJ/g over this broad range of thermal stability and decomposition modes.

Phenomenological schemes that account for some or all of the pyrolysis products of combustible solids (gas, tar, primary char, secondary char, secondary gas) have been proposed [46–63] wherein the decomposition steps occur sequentially (series), simultaneously (parallel), or in some combination of series/parallel steps. All of the models predict rate-dependent peak decomposition temperatures. A simple solid-state fuel-generation model that shows reasonable agreement with thermal analysis data [50, 52], numerical models of fire behavior [63], and experimental data [63] is

$$\text{Polymer Solid, } P \underset{k_r}{\overset{k_i}{\rightleftarrows}} \text{Reactive Intermediate, } I^* \begin{array}{c} \xrightarrow{k_g} \text{Fuel Gases, } G (\uparrow) \\ \xrightarrow{k_c} \text{Char, } C \end{array}$$

in which the thermal degradation of polymer mass P is assumed to occur in a single step involving rapid equilibrium between the polymer and an active intermediate I^* that simultaneously produces gas G and char C. Fig. 3.7 shows data [50, 52] for a variety of pure, unfilled polymers plotted as the char yield measured after flaming combustion in a fire calorimeter versus the char residue at 900 ± 100°C for the same material after anaerobic pyrolysis in a TGA at a heating rate of about 10 K/min.

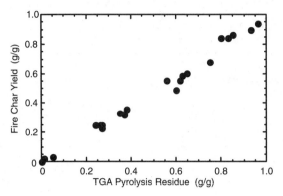

FIGURE 3.7 Char yield of plastics after burning versus anaerobic pyrolysis residue in TGA.

It is seen that the char yield of a material in a fire is essentially equal to its residual mass fraction after pyrolysis in an oxygen-free environment at temperatures representative of the char temperature in a fire. Although oxidative degradation products have been identified at the surface of noncharring olefinic polymers after flaming combustion, the data in Fig. 3.10 suggest that oxidation reactions in the solid during flaming combustion are not important to the overall fuel fraction as evidenced by the close agreement between fire char yield and anaerobic pyrolysis residue.

The phenomenological decomposition scheme above can be solved for the instantaneous fuel and char fractions in terms of the mass of polymer (P), intermediate (I^*), gas (G), and char (C) as follows. If k_i is the rate constant for initiation and k_r, k_g, and k_c are the rate constants for termination by recombination (k_r), hydrogen transfer to gaseous species (k_g), and cross-linking to char (k_c), respectively, then neglecting solid-state oxidation, the thermal decomposition reactions are [50, 52, 62]

$$P \underset{k_r}{\overset{k_i}{\rightleftarrows}} I^* \quad \text{(rapid equilibrium)} \tag{3.11}$$

$$I^* \xrightarrow{k_g} G \quad \text{(slow)} \tag{3.12}$$

$$I^* \xrightarrow{k_c} C \quad \text{(slow)} \tag{3.13}$$

and the system of rate equations for the species at time t is

$$\frac{dP}{dt} = -k_i P + k_r I^* \tag{3.14}$$

$$\frac{dI^*}{dt} = -k_i P - (k_r + k_g + k_c) I^* \tag{3.15}$$

$$\frac{dG}{dt} = k_g I^* \tag{3.16}$$

$$\frac{dC}{dt} = k_c I^* \tag{3.17}$$

According to the stationary-state hypothesis, $dI^*/dt \approx 0$, so that Eq. (3.15) provides the useful result

$$I^* = \left[\frac{k_i}{k_r + k_g + k_c}\right] P = KP$$

where $K = k_i/(k_r + k_g + k_c)$ is the pseudo-equilibrium constant for the polymer dissociation reaction. As the ratio of initiation to termination rate constants, K represents the kinetic chain length for degradation by depolymerization. Substituting $I^* = KP$ into Eqs. (3.14), (3.16), and (3.17),

$$\frac{dP}{dt} = -[k_i - Kk_r]P \tag{3.18}$$

$$\frac{dG}{dt} = k_g KP \tag{3.19}$$

$$\frac{dC}{dt} = k_c KP \tag{3.20}$$

With $I^* \ll P, G, C$, the total mass balance in terms of the initial mass m_0 is

$$m_0 = P + G + C + I^* \approx P + G + C \tag{3.21}$$

From Eqs. (3.18) to (3.21) with $dm_0/dt = 0$

$$\frac{dP}{dt} = -\frac{dC}{dt} - \frac{dG}{dt} = -[k_i - Kk_r]P \tag{3.22}$$

The sensible mass of the sample as measured, for example, in a TGA or fire calorimeter test is

$$m = P + C + I^* \approx P + C$$

and with Eqs. (3.18) to (3.20)

$$\frac{dm}{dt} = \frac{dP}{dt} + \frac{dC}{dt} = -\frac{dG}{dt} = -Kk_g P \tag{3.23}$$

Eq. (3.22) can be solved immediately for P in the isothermal case with initial condition $P = P_0 = m_0$ at $t = 0$,

$$P = m_0 \exp(-[k_i - Kk_r]t) \tag{3.24}$$

Substituting the isothermal result for P into Eq. (3.23) and separating variables

$$\int_{m_0}^{m} dm' = -\int_{0}^{t} Kk_g m_0 \exp(-k_p t)\, dt \tag{3.25}$$

where k_p in the exponential of the integrand on the right-hand side of Eq. (3.25) is the overall rate constant for pyrolysis and is assumed to have the Arrhenius form

$$k_p = k_i - Kk_r = K(k_g + k_c) = A \exp\left(-\frac{E_a}{RT}\right) \tag{3.26}$$

in terms of the global activation energy E_a and frequency factor A for pyrolysis. The isothermal solution of Eq. (3.25) is

$$\frac{m(t)}{m_0} = 1 - \left[\frac{k_g}{k_g + k_c}\right](1 - e^{-k_p t}) \tag{3.27a}$$

or

$$\frac{m(t)}{m_0} = Y_c(T) + [1 - Y_c(T)]e^{-k_p t} \tag{3.27b}$$

Eqs. (3.27a) and (3.27b) show that the mass fraction $m(t)/m_0$ decreases exponentially with time and approaches an equilibrium value at a particular temperature as $t \to \infty$

$$Y_c(T) = \frac{m(\infty)}{m_0} = \frac{k_c}{k_g + k_c} \tag{3.28}$$

where $Y_c(T)$ is the equilibrium residual mass fraction or char yield at temperature T in terms of the rate constants for gas and char formation. Eq. (3.28) predicts a finite char yield at infinite time if $k_c > 0$ and zero char if $k_c = 0$.

The physical significance of a temperature-dependent, equilibrium char yield as the ratio of rate constants for gas and char formation is consistent with the use of group contributions for the char-forming tendency of polymers developed by Van Krevelen [30, 46] (see the following section). If k_g and k_c have Arrhenius forms, Eq. (3.28) can be written

$$Y_c(T) = \left[1 + \frac{A_g}{A_c}\exp[-(E_g - E_c)/RT]\right]^{-1} \quad (3.29)$$

where E_c, E_g and A_c, A_g are the activation energies and frequency factors for char and gas formation, respectively. The crossover temperature T_{cr} is defined as the temperature at which the rates of gasification and cross-linking are equal, i.e., when $k_g = k_c$,

$$T_{cr} = \frac{(E_g - E_c)}{R \ln[A_g/A_c]} \quad (3.30)$$

It follows from Eq. (3.30) that the crossover condition, $k_g = k_c$, corresponds to the equilibrium residual mass fraction, $Y_c(T_{cr}) = 0.50$. If $Y_c(T)$ is the char yield at a temperature above the major mass loss transition temperature or the char yield is independent of temperature (e.g., an inert filler), then $Y_c(T) = \mu$ = constant and Eq. (3.27) is the solution for the isothermal mass loss history of a filled polymer with a nonvolatile mass fraction μ satisfying the rate law

$$\frac{dm}{dt} = -k_p(m - \mu m_0) \quad (3.31)$$

although Eq. (3.31) was not assumed a priori in the present derivation.

Charring. Char is the carbonaceous solid that remains after flaming combustion of the polymer. The char yield is the mass fraction of char based on the original weight of material. Charring competes with the termination reactions that generate volatile species and so reduces the amount of fuel in a fire. In addition, char acts as a heat and mass transfer barrier that lowers the flaming heat release rate. Fig. 3.7 demonstrated that the char yield in a fire is roughly equal to the anaerobic pyrolysis residue at high (flame) temperatures. Thus, char formation takes place in the solid state where oxidation reactions are slow compared to polymer dissociation and gas/char formation. The equivalence between the char yield and pyrolysis residue of a material permits a molecular interpretation of this important material fire parameter using the large volume of published thermogravimetric data and its correlation with chemical structure [30, 46].

Pyrolysis/char residue has the character of a thermodynamic quantity because it depends only on temperature and the composition of the material through the enthalpy barriers to gas and char formation, E_g, E_c, in Eq. (3.29). More precisely, char yield is a statistical thermodynamic concept wherein the total free energy of the char system at a particular (reference) temperature is the sum of the individual group contributions. Van Krevelen [30, 46] has devised a method for calculating the pyrolysis residue ($\approx$ char yield) of a polymer from its chemical composition and the observation that the char-forming tendency of different groups is additive and roughly proportional to the aromatic (i.e., nonhydrogen) character of the group. The char yield is calculated by summing the char-forming tendency per mole of carbon of the chemical groups, $C_{FT,i}$ and dividing by the molecular weight of the repeat unit

$$Y_c = \frac{C_{FT}}{M} \times M_{\text{carbon}} \times 100 = \frac{\sum_{i=1}^{N} n_i C_{FT,i}}{\sum_{i=1}^{N} n_i M_i} \times 1200 \quad (3.32)$$

The $C_{FT,i}$ is the amount of char per structural unit measured at 850°C divided by 12 (the atomic weight of carbon), i.e., the statistical amount of carbon equivalents in the char per structural unit of polymer. Negative corrections are made for aliphatic groups containing hydrogen atoms in proximity to char-forming groups because of the possibility for disproportionation and subsequent volatilization of chain-terminated fragments that are no longer capable of cross-linking. The method is empirical and relatively simple to use and good agreement is obtained with the measured pyrolysis residues (see Table 3.7). The char yield of polymers under anaerobic conditions is thus well described using the additive molar contributions of the individual groups comprising the polymer.

Kinetic Heat Release Rate. The previous results apply to the isothermal (constant temperature) case but processes of interest in fire and flammability are nonisothermal, e.g., thermogravimetric analyses at constant heating rate or fuel generation in the pyrolysis zone of a burning polymer. To calculate the instantaneous mass fraction $m(t)/m_0$ during a constant heating rate experiment where $dT/dt = $ constant $= \beta$, begin by eliminating P between Eqs. (3.29) and (3.31) and integrating

$$\int_{m_0}^{m} dm' = (1 - Y_c) \int_{P_0}^{P} dP' \qquad (3.33)$$

or since $P_0 = m_0$,

$$\frac{m(T)}{m_0} = Y_c(T) + [1 - Y_c(T)] \frac{P(T)}{P_0} \qquad (3.34)$$

For nonisothermal conditions $P(T)/P_0$ in Eq. (3.34) is obtained from Eq. (3.22)

$$\int_{P_0}^{P} \frac{dP'}{P'} = -\int_{0}^{t} k_p dt' = -\frac{A}{\beta} \int_{T_0}^{T} \exp\left(-\frac{E_a}{RT}\right) dT' \qquad (3.35)$$

where the constant heating rate $\beta = dT/dt$ transforms the variable of integration from time t to temperature T, and A and E_a are the global frequency factor and activation energy of pyrolysis, respectively.

The right-hand side of Eq. (3.35) is the exponential integral, which has no closed-form solution. However, a good (± 2 percent) approximation for the exponential integral over the range of activation energies and temperatures encountered in thermal analysis and combustion is [64]

$$-\frac{A}{\beta} \int_{T_0}^{T} \exp\left(-\frac{E_a}{RT}\right) dT' \approx \frac{-ART^2}{\beta(E_a + 2RT)} \exp\left(-\frac{E_a}{RT}\right) = \frac{-k_p RT^2}{\beta(E_a + 2RT)} \qquad (3.36)$$

Defining

$$y = \frac{k_p RT^2}{\beta(E_a + 2RT)} \qquad (3.37)$$

the solution of Eq. (3.35) takes the form

$$\frac{P(T)}{P_0} = e^{-y} \qquad (3.38)$$

Substituting Eq. (3.38) into Eq. (3.34), the residual mass fraction in a constant heating rate experiment is

$$\frac{m(T)}{m_0} = Y_c(T) + [1 - Y_c(T)]e^{-y} \qquad (3.39)$$

which is the same form as the isothermal solution Eq. (3.27). Eqs. (3.37) and (3.39) show that the mass fraction is a function only of temperature and heating rate for a given set of material properties. Eq. (3.39) provides a good fit to data for residual mass fraction versus temperature [50, 52] such as that shown in Fig. 3.8A for PMMA and PAI. The fractional mass loss rate during a linear temperature ramp is obtained by differentiating Eq. (3.39) with respect to time,

$$\frac{-1}{m_0}\frac{dm(T)}{dt} = (1 - Y_c(T))\frac{de^{-y}}{dt} + (1 - e^{-y})\frac{dY_c(T)}{dt}$$

$$= (1 - Y_c(T))k_p(T)e^{-y} + \beta Y_c(T)(1 - Y_c(T))\frac{E_g - E_c}{RT^2}(1 - e^{-y}) \qquad (3.40)$$

Because the rate of change of $Y_c(T)$ is small compared to the fractional mass loss rate at pyrolysis [50, 52], a good approximation is $Y_c(T) = \mu = $ constant so that $dY_c/dt = 0$ and Eq. (3.40) simplifies to

$$\frac{-1}{m_0}\frac{dm(T)}{dt} = (1 - \mu)k_p e^{-y} \qquad (3.41)$$

Eq. (3.41) describes the fractional mass loss rate versus temperature at constant heating rate such as that shown in Fig. 3.8B for PMMA and PAI. The maximum value of the fractional mass loss rate (e.g., the peak heights in Fig. 3.8B) can be found by differentiating Eq. (3.41) with respect to time and setting this second derivative of the residual mass fraction equal to zero,

$$-\frac{d^2}{dt^2}\left(\frac{m(T)}{m_0}\right) = \beta(1-\mu)\frac{d}{dT}[k_p e^{-y}] = (1-\mu)k_p e^{-y}\left(\frac{\beta E_a}{RT^2} - k_p\right) = 0 \qquad (3.42)$$

Eq. (3.42) has two roots: the trivial case $\mu = 1$ and

$$k_p(\max) = \frac{\beta E_a}{RT_p^2} \qquad (3.43)$$

where T_p is the temperature at maximum mass loss rate during the course of the linear heating history. Fig. 3.8 shows TGA data at a constant heating rate of 10 K/min for two plastics of widely differing thermal stability: polymethylmethacrylate (PMMA) and polyamideimide (PAI). The onset of thermal degradation (mass loss) is seen as a knee in the mass fraction versus temperature curves (Fig. 3.8A). The onset degradation temperature T_d corresponds roughly to the temperature at which 5 percent of the pyrolyzable mass (initial mass minus char mass) is lost and values of $T_d = 350$ and 495 for PMMA and PAI, respectively, are shown in Fig. 3.8A. The residual mass at the end of the experiment is the pyrolysis residue. For pure polymers, the pyrolysis residue is the carbonaceous char fraction. For filled polymers, this pyrolysis residue will contain the inert filler in addition to the char (if any).

The time derivative of the mass fraction at each temperature in Fig. 3.8A is plotted in Fig. 3.8B. The temperature at the peak mass loss rate is T_p in Eq. (3.43) and this is seen to be 375° and 605°C for PMMA and PAI, respectively. The peak mass loss rate temperature corresponds roughly to the temperature at which 50 percent of the pyrolyzable mass is lost.

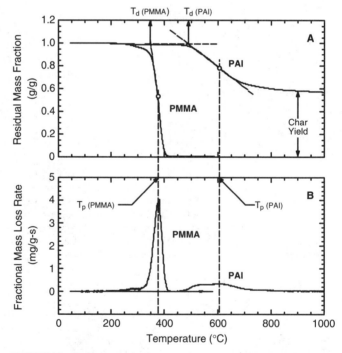

FIGURE 3.8 Residual mass fraction (A) and mass loss rate (B) of PMMA and PAI versus temperature at a heating rate of 10 K/min in nitrogen illustrating method used to obtain T_d and T_p from thermogravimetric data.

An analytic result for the peak fractional mass loss rate in a constant heating rate experiment is obtained by substituting Eq. (3.43) into Eq. (3.41)

$$\left.\frac{-1}{m_0}\frac{dm}{dt}\right|_{max} = (1-\mu)\frac{\beta E_a}{e^r RT_p^2} \quad (3.44)$$

where the exponent r of the natural number e in the denominator has the value

$$r = \left[1 + \frac{2RT_p}{E_a}\right]^{-1} \quad (3.45)$$

For the usual case where $E_a \gg 2RT_p$ [58–62], Eq. (3.44) simplifies to

$$\left.\frac{-1}{m_O}\frac{dm}{dt}\right|_{max} \approx (1-\mu)\frac{\beta E_a}{eRT_p^2} \quad (3.46)$$

The temperature at peak mass loss rate T_p is obtained from the root E_a/RT_p of Eq. (3.49) written in the form

$$\ln\left[\frac{E_a}{RT_p}\right]^2 + \left[\frac{E_a}{RT_p}\right] + \ln\left[\frac{\beta R}{AE_a}\right] = 0 \quad (3.47)$$

Table 3.6 lists onset degradation temperatures (T_d) and maximum pyrolysis rate temperatures (T_p) for common plastics and elastomers obtained in a TGA at a heating rate of 10 K/min. The variability in decomposition temperatures of a plastic measured on different TGA instruments is about ±5°C. Real differences in decomposition temperatures for plastics from different sources are about ±10°C as seen by comparing PMMA decomposition temperatures in Fig. 3.8 and the average values T_d = 354 ± 8°C and T_p = 383 ± 9°C for eight samples of PMMA reported in Table 3.6. Also listed in Table 3.6 are the experimental values of the surface temperature at piloted ignition for the same [65–67] or similar [1–4, 68–71] plastics.

Eq. (3.47) shows that the peak mass loss temperature T_p increases with heating rate [50, 52]. There is general agreement [50, 52] between Eq. (3.44) and experimental data for plastics over a wide range of heating rates. By way of example, Eq. (3.44) predicts for PMMA with E_a = 160 kJ/mol [30], μ = 0, and T_p = 375°C (648 K) a peak mass loss rate at 10 K/min of (0.167 K/s)(160 kJ/mol)/($e^{0.94}$)(8.314 J/mol·K)(648 K)^2 ≈ 3 mg/g·s, which is in reasonable agreement with the value 3.7 mg/g·s in Fig. 3.8B.

The maximum specific heat release rate of the plastic is obtained by multiplying the peak kinetic mass loss rate [Eq. (3.46)] by the HOC of the pyrolysis gases. If h_c^o is the HOC of the pyrolysis gases, the maximum value of the specific heat release rate is [50, 72–74]

$$\dot{Q}_c^{max}(W/kg) = \left.\frac{-h_c^o}{m_O}\frac{dm}{dt}\right|_{max} = \frac{\beta h_c^o(1-\mu)E_a}{eRT_p^2} = \frac{\beta h_{c,s}^o E_a}{eRT_p^2} \quad (3.48)$$

where $h_{c,s}^o$ is the HOC of the pyrolysis gases per unit mass of original solid, which is related to the HOC of the polymer $h_{c,p}^o$ (see Table 3.3) and its char $h_{c,\mu}^o$ as

$$h_c^o = \frac{h_{c,p}^o - \mu h_{c,\mu}^o}{1-\mu} = \frac{h_{c,s}^o}{1-\mu} \quad (3.49)$$

Fig. 3.9 contains data for the specific heat release rate of plastics measured at a heating rate of 258 K/min (4.3 K/s) in a pyrolysis-combustion flow calorimeter [73, 74]. It is not immediately obvious that the specific heat release rate has any intrinsic value as a predictor of fire behavior, and much theoretical and experimental work is ongoing [72–74] to develop this relationship because of the ease of measuring specific heat release rate in the laboratory using small samples (milligrams) and the good correlation between this quantity and the ignition resistance and burning rate of plastics [50, 52, 72–74]. A rate-independent material flammability parameter emerges from this analysis when the maximum specific heat release rate $\dot{Q}_c^{max}$ (Eq. (3.48)] is normalized for heating rate [72]

$$\eta_c = \frac{\dot{Q}_c^{max}}{\beta} = \frac{h_{c,s}^o E_a}{eRT_p^2} \quad (3.50)$$

TABLE 3.6 Decomposition and Ignition Temperatures of Plastics (Average Values ±10°C)

Polymer	ISO/ASTM Abbreviation	T_d °C	T_p °C	T_{ign} °C
	Thermoplastics			
Acrylonitrile-butadiene-styrene	ABS	390	461	394
ABS FR	ABS-FR	—	—	420
Polybutadiene	BDR	388	401	378
Polyisobutylene (butyl rubber)	BR	340	395	330
Cellulose Acetate	CA	250	310	348
Cyanate Ester (typical)	CE	448	468	468
Polyethylene (chlorinated)	CPE	448	476	—
Polyvinylchloride (chlorinated)	CPVC	—	—	643
Polychloroprene rubber	CR	345	375	406
Polychlorotriuoroethylene	CTFE	364	405	580
Poly(ethylene-chlorotrifluoroethylene)	ECTFE	445	465	613
Phenoxy-A	EP	—	350	444
Epoxy (EP)	EP	427	462	427
Poly(ethylene-tetrafluoroethylene)	ETFE	400	520	540
Polyethylenevinylacetate	EVA	448	473	—
Fluorinated ethylene propylene	FEP	—	—	630
Poly(styrene-butadiene)	HIPS	327	430	413
Poly(styrene-butadiene) FR	HIPS-FR	—	—	380
Poly(p-phenyleneterephthalamide)	KEVLAR	474	527	—
Polyarylate (liquid crystalline)	LCP	514	529	—
Melamine formaldehyde	MF	350	375	350
Polyisoprene (natural rubber)	NR	301	352	297
Polytrifluoroethylene	P3FE	400	405	—
Polyamide 12	PA12	448	473	—
Polyamide 6	PA6	424	454	432
Polyamide 610	PA610	440	460	—
Polyamide 612	PA612	444	468	—
Polyamide 66	PA66	411	448	456
Polyamide 6 (glass reinforced)	PA6-G	434	472	390
Polyamideimide	PAI	485	605	526
Polyacrylamide	PAM	369	390	—
Polyacrylonitrile	PAN	293	296	460
Polyarylate (amorphous)	PAR	469	487	—
Polybutene	PB	—	390	—
Polybenzimidazole	PBI	584	618	—
Polybutylmethacrylate	PBMA	261	292	—
Polybenzobisoxazole	PBO	742	789	—
Polybutyleneterephthalate	PBT	382	407	382
Polybutyleneterephthalate	PBT-G	386	415	360
Polycarbonate	PC	476	550	500
Polycarbonate/ABS (70/30)	PC/ABS	421	475	440
Polycarbonate (glass reinforced)	PC-G	478	502	420
Polycaprolactone	PCL	392	411	—
Polyethylene (high density)	PE HD	411	469	380
Polyethylene (low density)	PE LD	399	453	377
Polyethylacrylate	PEA	373	404	—
Polyethylene-acrylic acid salt	PEAA	452	474	—
Polyetheretherketone	PEEK	570	600	570
Polyetherimide	PEI	527	555	528

TABLE 3.6 (*Continued*)

Polymer	ISO/ASTM Abbreviation	T_d °C	T_p °C	T_{ign} °C
	Thermoplastics			
Polyetherketone (e.g., KADEL)	PEK	528	590	—
Polyetherketoneketone	PEKK	569	596	—
Polyethylmethacrylate	PEMA	246	362	—
Polyethylenenaphthalate	PEN	455	495	479
Polyethyleneoxide	PEO	373	386	—
Polyethersulfone	PESU	533	572	502
Polyethyleneterephthalate	PET	392	426	407
Phenol formaldehyde	PF	256	329	429
Polytetrafluoroethylene-perfluoroether	PFA	—	578	—
Phenol formaldehyde	PF-G	—	—	580
Polymethylmethacrylate	PMMA	354	383	317
Poly(4-methyl-1-pentene)	PMP	—	377	—
Poly(α-methyl)styrene	PMS	298	333	—
Poly(α-methylstyrene)	PMS	250	314	—
Polyoxymethylene	POM	323	361	344
Polypropylene	PP	354	424	367
Polypropylene (isotactic)	PP (iso)	434	458	—
Polyphthalamide (AMODEL)	PPA	447	488	—
Polyphenyleneether	PPE	—	418	426
Poly(2,6-dimethylphenyleneoxide)	PPO	441	450	418
Polypropyleneoxide	PPOX	292	343	—
Polyphenylenesulfide	PPS	504	545	575
Polyphenylsulfone	PPSU	557	590	575
Polystyrene	PS	319	421	356
Polysulfone	PSU	481	545	510
Polytetrafluoroethylene	PTFE	543	587	630
Polytetramethyleneoxide	PTMO	—	352	—
PU (isocyanurate/rigid)	PU	271	422	378
Polyetherurethane rubber	PUR	324	417	356
Polyvinylacetate	PVAC	319	340	—
Polyvinylbutyral*	PVB	333	373	—
Polyvinylchloride (50% DOP)	PVC (ex)	249	307	318
Polyvinylchloride (rigid)	PVC (rigid)	273	285	395
Polyvinylchloride/polyvinylacetate blend	PVC/PVAC	255	275	—
Polyvinylidenechloride	PVDC	225	280	—
Polyvinylidenefluoride	PVDF	438	487	643
Polyvinylfluoride	PVF	361	435	476
Polyvinylcarbazole	PVK	356	426	—
Polyvinylalcohol	PVOH	298	322	—
Polyvinylpyridine	PVP	385	408	—
Polypara(benzoyl)phenylene	PX	476	602	—
Poly(styrene-acrylonitrile)	SAN	389	412	368
Phenylsilsesquioxane (silicone) resin	SI	475	541	—
Silicone rubber	SIR	456	644	407
Poly(stryene-maleic anhydride)	SMA	337	388	—
Polyimide thermoplastic	TPI	523	585	600
Polyurethane thermoplastic	TPU	314	337	271
Unsaturated polyester	UPT	330	375	380
Unsaturated polyester	UPT-G	—	—	395

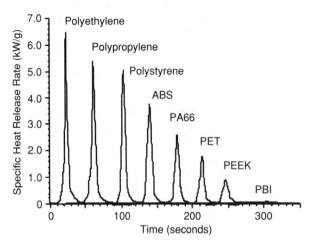

FIGURE 3.9 Specific heat release rate histories for some of the polymers in Table 3.7 (horizontally shifted for clarity). Dividing the maximum value (peak height) by the heating rate in the test ($\beta = 4.3$ K/s) gives the heat release capacity listed in Table 3.7.

The flammability parameter η_c has the units and significance of a heat [release] capacity (J/g·K) when the linear heating rate is β(K/s) and it contains only thermochemical properties of the material and the fundamental constants e, R. The heat release capacity η_c is a molecular-level flammability parameter that is the potential heat release per degree of temperature rise at the surface of a burning plastic. Table 3.7 contains ranked η_c values (± 10 percent) for commercial plastics and elastomers along with the measured HOC of the fuel gases $h^o_{c,s}$ and char yield μ [74].

TABLE 3.7 Heat Release Capacity, Heat of Combustion of Fuel Gases, and Char Yield of Plastics and Elastomers

Polymer	Abbreviation	HR capacity (J/g·K)	Total HR (kJ/g)	Char (%)
Polyethylene (low density)	PE LD	1676	41.6	0
Polypropylene	PP	1571	41.4	0
Epoxy (aliphatic amine cure)	EPA	1100	27	6
Polyisobutylene	BR	1002	44.4	0
Polystyrene	PS	927	38.8	0
Polystyrene (Isotactic)	PS (iso)	880	39.9	0
Polyhexamethylene sebacamide	PA610	878	35.7	0
Poly-2-vinylnaphthalene	PVN	834	39.0	0
Polyvinylbutyral	PVB	806	26.9	0.1
Polylaurolactam	PA12	743	33.2	0
Poly α-methylstyrene	PMS	730	35.5	0
Polyhexamethylene dodecanediamide	PA612	707	30.8	0
Acrylonitrile-butadiene-styrene	ABS	669	36.6	0
Phenoxy-A	EP	657	26.0	3.9
Polyethyleneoxide	PEO	652	21.6	1.7
Polyhexamethanyleneadipamide	PA66	615	27.4	0
Polyphthalamide	PPA	575	32.0	0
Polyphenyleneether	PPE	553	22.4	23
Polyvinylalcohol (>99%)	PVOH	533	21.6	3.3
Polcaprolactone	PCL	526	24.4	0

TABLE 3.7 (*Continued*)

Polymer	Abbreviation	HR capacity (J/g·K)	Total HR (kJ/g)	Char (%)
Polymethylmethacrylate	PMMA	514	24.3	0
Dicyclopentadienyl bisphenol cyanate ester	CED	493	20.1	27.1
Polycaprolactam	PA6	487	28.7	0
Polybutyleneterephthalate	PBT	474	20.3	1.5
Polyethylmethacrylate	PEMA	470	26.4	0
Polymethylmethacrylate	PMMA	461	23.2	0
Polyepichlorohydrin	ECR	443	13.4	4.8
Poly-n-butylmethacrylate	PBMA (n)	412	31.5	0
Poly-2,6-dimethyl-1,4-phenyleneoxide	PPO	409	20.0	25.5
Polyisobutylmethacrylate	PBMA (iso)	406	31.3	0
Polyethylmethacrylate	PEMA	380	26.8	0
Polyarylate	PAR	360	18.0	27
Polycarbonate of bisphenol-A	PC	359	16.3	21.7
Polysulfone of bisphenol-A	PSU	345	19.4	28.1
Polyethyleneterephthalate	PET	332	15.3	5.1
Bisphenol E cyanate ester	CEE	316	14.7	41.9
Polyvinylacetate	PVAC	313	19.2	1.2
Polyvinylidenefluoride	PVDF	311	9.7	7
Polyethylenenaphthylate	PEN	309	16.8	18.2
Poly(p-phenyleneterephthalamide)	KEVLAR	302	14.8	36.1
Bisphenol A cyanate ester	CEA	283	17.6	36.3
Tetramethylbisphenol F cyanate ester	CET	280	17.4	35.4
Poly(styrene-maleicanhydride)	SMAH	279	23.3	2.2
Epoxy novolac/Phenoxy-N	EPN	246	18.9	15.9
Polynorbornene	PN	240	21.3	6
Bisphenol-M cyanate ester	CEM	239	22.5	26.4
Polyethylenetetrafluoroethylene	ETFE	198	10.8	0
Polychloroprene	CR	188	16.1	12.9
Polyoxymethylene	POM	169	14.0	0
Polyacrylic Acid	PAA	165	12.5	6.1
Poly-1,4-phenylenesulfide	PPS	165	17.1	41.6
Liquid crystalline polyarylate	LCP	164	11.1	40.6
Polyetheretherketone	PEEK	155	12.4	46.5
Polyphenylsulphone	PPSU	153	11.3	38.4
Polyvinylchloride	PVC (rigid)	138	11.3	15.3
Polyetherketone	PEK	124	10.8	52.9
Novolac cyanate ester	CEN	122	9.9	51.9
Polyetherimide	PEI	121	11.8	49.2
Poly-1,4-phenyleneethersulfone	PESU	115	11.2	29.3
Polyacrylamide	PAK	104	13.3	8.3
Polyetherketoneketone	PEKK	96	8.7	60.7
Phenylsilsequioxane resin (toughened)	SI	77	11.7	73.1
Poly(m-phenylene isophthalamide)	NOMEX	52	11.7	48.4
Poly-p-phenylenebenzobisoxazole	PBO	42	5.4	69.5
LaRC-1A polyimide	PI	38	6.7	57
Polybenzimidazole	PBI	36	8.6	67.5
Polytetrafluoroethylene	PTFE	35	3.7	0
Polyamideimide	PAI	33	7.1	53.6
Hexafluorobisphenol-A cyanate ester	CEF	32	2.3	55.2
Thermoplastic polyimide	TPI	25	6.6	51.9
LaRC-CP2 polyimide	PI	14	3.4	57
LaRC-CP1 polyimide	PI	13	2.9	52

3.4 FIRE BEHAVIOR OF PLASTICS

The continuum-level treatment of the fire behavior of plastics that follows disregards the discrete (molecular) structure of matter so that the temperature distribution and, more importantly, its derivatives, are continuous throughout the material. In addition, the material is assumed to have identical thermal properties at all points (homogeneous) and in all directions (isotropic). The concept of a continuous medium allows fluxes to be defined at a point, e.g., a surface in one-dimensional space. Chemical reactions in the solid (pyrolysis) and flame (combustion) are assumed to occur so rapidly that the burning rate is determined solely by the heat transfer rate. Differential [75–78] and integral [79, 80] condensed-phase burning models have been developed from the continuum perspective with coupled heat and mass transfer for both charring and noncharring polymers. All of these models must be solved numerically for the transient (time-dependent) mass loss and heat release rates.

In the following simplified treatment of ignition and burning, the material response of a semi-infinite solid is assumed to be amenable to analysis by unsteady and steady heat transfer, respectively, at a constant surface heat flux. These simplified energy balances allow for the development of algebraic (scaling) relationships between the thermal properties of a plastic and its fire response, but ignore many important details such as transient behavior (see Fig. 3.13) that can only be captured through detailed numerical analyses.

3.4.1 Ignition

Ignition of plastics is a complicated phenomenon because the finite-rate solid-state thermochemistry (pyrolysis) is coupled to the gas phase chemistry (combustion) through the heat feedback from the flame (see Fig. 3.6). Ignition criteria for liquids and gaseous fuel/air mixtures are well established [3, 18–20, 81] because only the thermodynamic (equilibrium) state of the system need be considered. In particular, the reaction of gaseous fuels with air will be self-sustaining if the volumetric energy (heat) release of the equilibrium mixture is above a minimum (critical) value [81]. Sustained ignition of liquids and solids is complicated by the fact that there is dynamic coupling between the gas phase combustion and condensed-phase fuel-generating reactions because energy must be supplied to raise the temperature of the condensed phase to the fire point [3, 82] to generate combustible gases. The coupled, time-dependent nature of condensed-phase flaming combustion gives rise to a variety of proposed criteria for piloted ignition of solids [3, 82–84], but these can be roughly divided into thermal (solid state) and chemical (gas phase) criteria. Examples of thermal criteria for piloted ignition are a critical radiant heat flux and an ignition temperature. A piloted ignition temperature corresponds to a temperature at which the solid plastic decomposes to volatile fuel at a rate sufficient to maintain a flammable mixture at the igniter. Fig. 3.10 is a plot of ignition temperature versus gasification temperature of liquid and solid fuels. Plotted in Fig. 3.10 on the vertical axis are the piloted ignition and fire point temperatures of liquid and solid [1–4, 65–71] fuels, respectively, versus the mean thermal decomposition temperature of plastics [$(T_d + T_p)/2$ from Table 3.6], and the open cup flash point temperature of liquid hydrocarbons [81]. It is seen that the thermal decomposition temperature of plastics measured in laboratory thermogravimetric analysis at heating rates in the vicinity of 10 K/min give reasonable predictions of piloted ignition temperatures in standard ignition tests [85] and surface temperature measurements at piloted ignition [65–71].

Eq. (3.47) and experimental data [50] show that the decomposition temperature of polymers increases with heating rate, and there is some evidence that surface temperatures at ignition show a corresponding increase with radiant heating intensity [50]. Fig. 3.11 is a plot of measured surface temperatures at piloted ignition [67, 68] over a range of external heat fluxes for various plastics showing that the effect is small for these plastics over this range of heat flux.

Chemical criteria for solid ignition include a boundary layer reaction rate [82] and a critical pyrolyzate mass flux [3, 84], both of which are equivalent to establishing a lower flammability limit at the ignition source for a fixed test geometry and ventilation rate. Table 3.8 shows mass fluxes measured at ignition [67] and extinction [71, 88] for a number of plastics. Also listed are the effective HOCs h_c^{eff} EHOC of the fuel gases and the product of the mass flux and EHOC at ignition. It is seen that the heat release rate at ignition/extinction is relatively independent of the type of plastic.

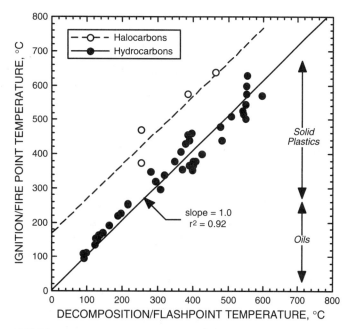

FIGURE 3.10 Ignition/fire point temperature versus decomposition/flash point temperature for solids/liquids.

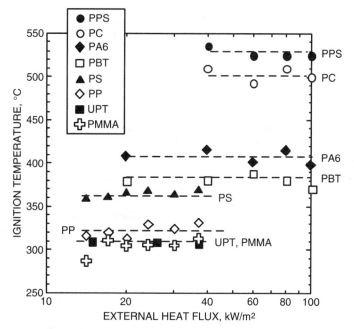

FIGURE 3.11 Ignition temperature versus external heat flux for PPS, PC, PA6, PBT [67], as well as PP, UPT, and PMMA [68].

TABLE 3.8 Effective Heat of Combustion (EHOC), Mass Loss Rate (MLR), and Heat Release Rate (HRR) of Polymers at Incipient Burning (Extinction and Ignition)

Material	HOC kJ/g	MLR g/m²-s	HRR kW/m²
At extinction			
POM	14.4	4.5	65
PMMA	24.0	3.2	77
PE	38.4	2.5	96
CPE	13.6	7.0	95
PP	38.5	2.7	104
PS	27.0	4.0	108
PUR (foam)	17.4	5.9	101
PU (foam)	13.2	7.7	102
Extinction average:		4.7 ± 2.0	94 ± 15
At ignition			
PMMA	24.8	4.4	109
EP	20.4	4.4	90
PA6	29.8	3.0	89
PBT	21.7	3.4	74
PC	21.2	3.4	72
PPS	23.5	3.6	85
PEN	22.9	2.7	62
PPA	24.2	3.1	75
PEEK	21.3	3.3	70
PESU	22.4	3.7	83
PPSU	23.8	4.3	102
Ignition average:		3.6 ± 0.6	83 ± 14

Thus, a chemical criterion is probably sufficient for ignition to occur but a critical surface temperature near the thermal decomposition temperature (see Table 3.6) is necessary to begin the fuel-generation process. Prior to ignition, the temperature history of a semi-infinite thickness of solid plastic is described by the one-dimensional energy equation for unsteady heat conduction with no internal heat generation and constant κ

$$\rho c \frac{\partial T}{\partial t} - \rho c v \frac{\partial T}{\partial x} = \kappa \frac{\partial^2 T}{\partial x^2} \tag{3.51}$$

where T is the temperature at location x in the solid polymer and $\alpha = \kappa/\rho c$ is the polymer thermal diffusivity in terms of its thermal conductivity κ, density ρ, and heat capacity c (see Table 3.2); v is the regression velocity of the burning surface. During the preheat phase prior to ignition, there is no surface regression, so $v = 0$ and Eq. (3.51) reduces to

$$\frac{\partial^2 T}{\partial x^2} - \frac{1}{\alpha} \frac{\partial T}{\partial t} = 0 \tag{3.52}$$

The solution of Eq. (3.52) for the ignition time t_{ign} of a thermally thick sample with constant α and net heat flux $\dot{q}_{net}$ at the surface $x = 0$ is [89]

$$t_{ign} = \frac{\pi}{4} \kappa \rho c \left[\frac{T_{ign} - T_0}{\dot{q}_{net}} \right]^2 \tag{3.53}$$

where T_{ign} is the (piloted) ignition temperature and T_o is the ambient initial temperature. If the sample thickness b is less than a millimeter or so, ignition occurs at time

$$t_{ign} = \rho bc \left[\frac{T_{ign} - T_o}{\dot{q}_{net}} \right] \tag{3.54}$$

Eqs. (3.53) and (3.54) define a time to ignition that is determined by the net heat flux and the ignition (decomposition) temperature, sample thickness, and thermal and transport properties of the material κ, ρ, c. The net heat flux at the surface, $\dot{q}_{net} = \dot{q}_{ext} - \dot{q}_{rerad} - \dot{q}_{conv} - \dot{q}_{cond}$ is the heat influx from an external source $\dot{q}_{ext}$ minus the heat losses by reradiation $\dot{q}_{rerad}$ and convection $\dot{q}_{conv}$ to the cooler environment, and conduction into the solid $\dot{q}_{cond}$, respectively. For high heat fluxes and/or thermally thick samples, substituting the net heat flux at incipient (pre)ignition into Eq. (3.53) and rearranging gives

$$\frac{1}{\sqrt{t_{ign}}} = \frac{\dot{q}_{ext} - \dot{q}_{crit}}{TRP} = \frac{\dot{q}_{ext} - CHF}{TRP} \tag{3.55}$$

where

$$TRP = \sqrt{\pi \kappa \rho c / 4} (T_{ign} - T_\infty) \tag{3.56}$$

is a quantity known as the thermal response parameter (TRP) [88, 90] and

$$CHF = \dot{q}_{rerad} + \dot{q}_{conv} + \dot{q}_{cond} \cong \sigma(T_{ign}^4 - T_0^4) + h(T_{ign} - T_0) \tag{3.57}$$

is the critical heat flux for ignition. Eq. (3.55) suggests that CHF can be obtained experimentally as the $\dot{q}_{ext}$ intercept at $1/\sqrt{t_{ign}} = 0$ from a linear plot of $1/\sqrt{t_{ign}}$ versus external heat flux. However, the assumption of a semi-infinite solid breaks down at the long times/low heat fluxes near the critical condition when the sample temperature approaches the surface temperature and Eq. (3.53) no longer applies. Critical heat fluxes are best obtained by bracketing procedures and/or by measuring the external heat flux at which the flame spread rate asymptotically approaches zero in a gradient heat flux experiment [91]. Fig. 3.12 shows experimental data [66] for time to ignition at various heat fluxes for polycarbonate (PC) and the graphical procedures used to obtain TRP and CHF. Table 3.9 is a listing of TRPs reported for these plastics or calculated from heat flux and time-to-ignition data [65–71, 90–108], averaged for multiple values, along with the measured and calculated CHF. Calculated values of CHF were obtained from Eq. (3.57) with $h = 15$ W/m²/K [91, 92] with $T_{ign} = (T_d + T_p)/2$. The agreement between measured and calculated CHF is within the variation in CHF from different sources.

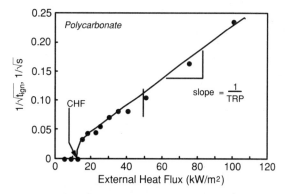

FIGURE 3.12 Reciprocal square root of time to ignition versus external heat flux for polycarbonate showing graphical method for determining CHF and TRP. (Data from Ref. 66.)

TABLE 3.9 Thermal Response Parameters (TRP) and Critical Heat Fluxes (CHF) for Ignition

Polymer	TRP kW·s$^{1/2}$m^{-2}	Critical heat flux (CHF) kW/m^2	
		(Measured)	(Calculated)
ABS	365	9–15	19
ABS FR	330	13	19
BR	211	19	16
CE (typical)	534	27	22
CPVC	591	40	—
CR	245	20–37	17
CTFE	460	30	16
ECTFE	410	38–74	43
EP	425	20	13
EP-G	462	10–15	13
ETFE	478	17–27	32
FEP	682	38–50	47
HIPS	420	—	15
HIPS-FR	351	—	15
LCP-G (30%)	—	32	30
LCP-M (45%)	—	22	30
MF	324	25	14
NBR	308	26	—
NR	294	17	11
P3FE	504	—	17
PA6	461	15–20	20
PA66	352	15–21	20
PA6-G (10%)	303	—	22
PA6-G (20%)	315	—	22
PA6-G (30%)	318	—	22
PA6-G (5%)	371	—	22
PAI	378	40–50	33
PBI	—	≈60	41
PBT	520	20	16
PBT-G (10%)	317	—	17
PBT-G (20%)	308	—	17
PBT-G (30%)	325	—	17
PBT-G (5%)	381	—	17
PC	455	15–20	29
PC/ABS	344	—	21
PC/ABS-FR	391	—	—
PC-G (10%)	383	—	26
PC-G (20%)	362	—	26
PC-G (30%)	373	—	26
PC-G (5%)	402	—	26
PE HD	343	15	21
PE LD	454	—	19
PE-XL	442	—	—
PE-XL/FR	581	—	—
PEEK	623	30–40	39
PEEK-G	301	—	—
PEI	435	25–40	32
PEN	545	24	24

TABLE 3.9 (*Continued*)

Polymer	TRP kW·s$^{1/2}$m^{-2}	Critical heat flux (CHF) kW/m^2 (Measured)	(Calculated)
PESU	360	19–30	34
PESU-G	258	—	—
PET	403	10–19	18
PF	537	15–26	9
PF-G	610	20	9
PFA	787	—	38
PMMA	274	6–23	12
POM	269	13	12
PP	193–336	15	21
PPA-G	—	29	23
PPA-G/FR	—	15	23
PPE	323	—	15
PP	415	15	16
PP-FR	310	10	—
PPO	342	19	16
PPS	395	35–38	37
PPS-G (5%)	450	—	37
PPS-G (10%)	468	—	37
PPS-G (20%)	490	—	37
PPS-G (30%)	466	—	37
PPSU	512	32–35	33
PS	355	6–13	10
PSU	424	26	24
PTFE	654	50	34
PUR	347	23	10
PVC (rigid)	410	15–28	7
PVC (flex)	174	21	9
PVDF	609	30–50	26
PVF	303	—	15
PX	626	—	28
SBR	198	10–15	—
SIR	429	34	23
TPI	—	36–50	32
UPT	343	—	10
UPT-G	483	10–15	12
UPT-M	752	—	—
VE	285	—	—
VE-G	443	—	—

Table 3.10 is a list of thermal inertia values. Values in the first column of Table 3.10 were calculated from Table 3.2 as the product of room temperature values, that is, $\kappa_0\rho_0 c_0$ (298 K) = $(\kappa\rho c)_0$. The second column lists the thermal inertia at ignition calculated as $\kappa\rho c = (\kappa\rho c)_0 T_{ign}/T_0$ according to Eq. (3.4). The last column in Table 3.10 lists experimental values for $\kappa\rho c$ extracted from the TRP in Table 3.9 using Eq. (3.56) and T_{ign} reported in Table 3.6. It is seen that the approximation $\kappa\rho c(T_{ign}) \approx (\kappa\rho c)_0 T_{ign}/T_0$ gives qualitative (± 25 percent) agreement with experimental values.

TABLE 3.10 Thermal Inertia: Measured and Calculated (All Values in Units $kW^2\text{-s-m}^{-4}\text{-K}^{-2}$)

Polymer	$\kappa\rho c(T_0)$	$(\kappa\rho c)_0 T_{ign}/T_0$	$\kappa\rho c$ (Fire data)
ABS	0.41	0.92	1.1
EP	1.2	1.9	1.6
FEP	0.63	1.9	1.6
MF	0.52	1.1	1.3
EP	0.39	0.91	1.6
PPO	0.22	0.51	0.77
HIPS	0.31	0.73	1.5
PA6	0.42	1.0	1.4
PA66	0.41	1.0	0.50
PAI	0.34	0.91	0.72
BR	0.42	0.91	1.3
PBT	0.48	1.1	1.1
PC	0.29	0.76	0.96
CR	0.30	0.69	0.75
SIR	0.35	0.81	1.8
PET	0.59	1.3	1.4
PEEK	0.45	1.3	0.68
PEI	0.36	0.96	0.95
PESU	0.28	0.73	0.72
PUR	0.37	0.78	1.6
PE MD	0.63	1.3	1.4
ETFE	0.41	1.1	0.58
PE HD	0.82	1.8	3.9
PEAA	0.40	1.1	1.0
PEN	0.0	0.0	2.1
TPI	0.17	0.49	1.1
BR	0.23	0.47	0.61
NR	0.20	0.38	1.5
PMMA	0.33	0.65	1.1
POM	0.45	0.93	0.90
PPE	0.30	0.71	0.82
PPS	0.38	1.1	0.55
PPO	0.29	0.68	0.77
PPSU	0.24	0.68	0.68
PP	0.25	0.53	0.91
PS	0.18	0.39	0.74
PS-FRP	0.18	0.37	3.6
PSU	0.36	0.94	1.3
PTFE	0.56	1.7	0.85
PVC	0.26	0.59	0.35
PVF	0.25	0.63	0.51
SI	0.67	1.5	1.6
SBR	0.35	0.78	0.58
UPT	0.73	1.6	2.2

3.4.2 Steady Burning

Once sustainable ignition has occurred, steady, one-dimensional burning of the polymer is assumed. Steady burning at a constant surface heat flux is treated as a stationary state by choosing a coordinate system that is fixed to the surface and moving at the recession velocity v. If there is no internal heat generation or absorption, the one-dimensional heat conduction equation [Eq. (3.51)] applies. Because semicrystalline polymers absorb the heat of fusion during melting at temperatures below the decomposition temperature, Eq. (3.51) is only approximate for these materials. Under steady-state conditions, $dT(x)/dt = 0$ so that Eq. (3.51) becomes, for steady burning [50, 52]

$$\frac{d^2T}{dx^2} + \frac{v}{\alpha}\frac{dT}{dx} = 0 \qquad (3.58)$$

The general solution of Eq. (3.58) for a material with constant thermal diffusivity is

$$T(x) = c_1 + c_2\exp[-vx/\alpha] \qquad (3.59)$$

Two boundary conditions are needed to evaluate the constants of integration c_1 and c_2 in Eq. (3.59). Conservation of energy at the pyrolysis front $x = 0$ gives

$$\kappa\frac{dT(x)}{dx}\bigg|_{x=0} = -\dot{q}_{net} + \rho v\Delta h_v = -c_2\frac{\kappa v}{\alpha} \qquad (3.60)$$

from which $c_2 = (\alpha/\kappa v) - (\Delta h_v/c)$ with Δh_v the latent heat of vaporization of the pyrolysis products and the net heat flux at the surface ($x = 0$)

$$\dot{q}_{net} = \dot{q}_{ext} + \dot{q}_{flame} - \dot{q}_{rerad} - \dot{q}_{cond} \qquad (3.61)$$
$$= \dot{q}_{ext} + \dot{q}_{flame} - \dot{q}_{loss}$$

Eq. (3.61) defines the net heat flux into the surface under conditions of flaming combustion $\dot{q}_{net}$ as the difference between the heat flux entering the surface from an external radiant energy source $\dot{q}_{ext}$ and/or surface flame $\dot{q}_{flame}$, and the heat losses $\dot{q}_{loss}$ due to surface reradiation, convection, and conduction into the solid.

On the rear face of the semi-infinite slab ($x = \infty$) specify $dT/dx = 0$ or, equivalently, $T(\infty) = T_o = c_1$ where T_o is the backside (ambient) temperature. The final temperature distribution during steady-state burning of a semi-infinite thickness of combustible plastic is

$$T(x) = T_O + \left[\frac{\dot{q}_{net}}{\rho cv} - \frac{\Delta h_v}{c}\right]\exp\left[-\frac{v}{\alpha}x\right] \qquad (3.62)$$

The steady burning velocity of the surface $x = 0$ at temperature $T(0) = T_p$ from Eq. (3.62) is

$$v = \frac{1}{\rho}\frac{\dot{q}_{net}}{(c(T_p - T_O) + \Delta h_v)} = \frac{1}{\rho}\frac{\dot{q}_{net}}{h_g} \qquad (3.63)$$

where the total heat of gasification h_g per unit original mass of polymer is [see Eq. (3.6)],

$$h_g = (\Delta h_s + \Delta h_f + \Delta h_d) + \Delta h_v \approx c(T_p - T_O) + \Delta h_v. \qquad (3.64)$$

Eqs. (3.62) and (3.63) allow the steady-state temperature distribution in the burning solid polymer to be expressed

$$T(x) - T_O = (T_p - T_O)\exp\left(-\frac{c\,\dot{q}_{net}}{\kappa\,h_g}x\right)$$

which is in general agreement with experimental data for the temperature gradient in steadily gasifying PMMA slabs [50].

Conservation of mass for a virgin polymer of density ρ that pyrolyzes to an inert fraction or char residue μ gives [50, 52]

$$\rho v = \frac{\dot{m}_g}{1 - \mu} \qquad (3.65)$$

where $\dot{m}_g$ is the mass loss rate of pyrolysis gases per unit surface area. Defining a heat of gasification per unit mass of volatiles [Eq. (3.8)]

$$L_g = \frac{h_g}{1 - \mu}$$

and combining Eqs. (3.8) and (3.65)

$$\dot{m}_g = \frac{\dot{q}_{net}}{h_g/(1 - \mu)} = \frac{\dot{q}_{net}}{L_g} \qquad (3.66)$$

shows that the heat of gasification per unit mass of solid polymer h_g can be determined from the reciprocal slope of a plot of areal mass loss rate versus external heat flux if the char yield is measured after the test, since from Eqs. (3.61) and (3.66)

$$\dot{m}_g = \frac{\dot{q}_{ext}}{L_g} - \left(\frac{\dot{q}_{flame} - \dot{q}_{loss}}{L_g}\right) \qquad (3.67)$$

Multiplying Eq. (3.67) by the net heat of complete combustion of the volatile polymer decomposition products h_c^0 and the gas phase combustion efficiency χ gives the usual result for the average heat release rate of a burning specimen [3, 71].

$$\dot{q}_c = \chi h_c^0 \dot{m}_g = \chi(1-\mu)\frac{h_c^0}{h_g}\dot{q}_{net} = \text{HRP}\,\dot{q}_{net} \qquad (3.68)$$

The dimensionless material sensitivity to external heat flux in Eq. (3.68)

$$\text{HRP} = \chi(1-\mu)\frac{h_c^0}{h_g} = \chi\frac{h_c^0}{L_g} \qquad (3.69)$$

is called the heat release parameter [102]. Fire calorimetry is used to obtain HRP as the slope of heat release rate versus external heat flux or as the ratio h_c^{eff}/L_g from individual measurements. Tewarson [71] has reported HRP values for many common polymers and composites and has used this fire parameter for predicting the fire propagation tendency and heat release rate of materials [90, 103, 104]. Table 3.11 lists values for HRP, the effective heat of flaming combustion HOC = χh_c^0, and the heat of gasification L_g for plastics, elastomers, flame-retarded (-FR) plastics, and fiberglass reinforced plastics (-G) as reported in or calculated from data in the literature [65–72, 88, 90, 91, 93–108].

Fig. 3.13 shows typical heat release rate histories for thermally thick and thin polymers that gasify completely or form a char during burning. It is apparent that none of these heat release rate histories shows a constant (steady-state) value of heat release rate over the burning interval as presumed in the derivation of Eq. (3.68). Thus, the interpretation of time-varying heat release rate histories in terms of Eq. (3.68) is a subject of active research [109, 110] that attempts to account for phase transitions (e.g., melting), time-varying temperature gradients, finite sample thickness, and char formation during burning. The left-hand curves in Fig. 3.13 are characteristic of noncharring ($\mu = 0$) plastics of different thicknesses. The heat release histories for the charring plastics on the

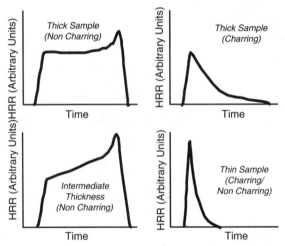

FIGURE 3.13 Typical heat release rate (HRR) histories for thick and thin samples of charring and noncharring plastic.

right-hand side of Fig. 3.13 show the typical peak in heat release rate soon after ignition followed by a depression in the heat release rate as the char layer increases in thickness. The growing char layer insulates the underlying plastic from the surface heat flux so that the net heat flux at the in-depth pyrolysis front decreases with time. Char can also act as a mass diffusion barrier to the volatile fuel. Charring polymers can be linear, branched, or cross-linked thermoplastics, elastomers, or thermosets having amorphous or semicrystalline morphologies.

In all cases the time integral (area under) the heat release rate history per unit mass of plastic consumed by burning is the effective heat of flaming combustion (EHOC). The EHOC is determined primarily by the combustion chemistry in the flame and the ventilation rate. Combustion efficiency decreases when halogens are present in the polymer molecule or gas phase active flame-retardant chemicals are added, when soot or smoke is produced in large yield, or when there is insufficient oxygen for complete conversion of the organic fuel to carbon dioxide and water. Flaming combustion efficiency appears to be relatively independent of the charring tendency of a polymer.

Fig. 3.14 shows fire calorimetry data for the average heat release rate of plastics versus their char yield after burning or pyrolysis. The linear dependence of heat release rate on char yield predicted by Eq. (3.68) is not observed in Fig. 3.14, indicating that char formation contributes more to flammability reduction than simply reducing the fuel fraction, e.g., acts as a barrier to the transfer of mass and heat during burning as illustrated schematically in Fig. 3.13.

Fig. 3.15 shows experimental heat release rate histories for 6-mm-thick samples of high-density polyethylene (ultra high molecular weight) at external heat fluxes of 35, 50, 65, and 80 kW/m². At no time during the heat release rate history of polyethylene is steady (time independent) burning observed. Fig. 3.16 is a plot of data for the peak heat release rate of PA6 versus external heat flux from two different sources [67, 99] showing typical variation. The slope of the linear fit of data in Fig. 3.16 is the heat release parameter HRP, and the intercept is the heat release rate at zero external heat flux HRR_0.

The significance of HRR_0 as a flammability parameter is seen by combining Eqs. (3.61) and (3.68) and separating the heat release rate into unforced and forced components

$$\dot{q}_c = \left[\left(\chi(1-\mu)\frac{h_c^0}{h_g}\right)(\dot{q}_{\text{flame}} - \dot{q}_{\text{loss}})\right] + \left[\chi(1-\mu)\frac{h_c^0}{h_g}\right]\dot{q}_{\text{ext}} \qquad (3.70)$$

or

$$\dot{q}_c = HRR_0 + HRP\,\dot{q}_{\text{ext}} \qquad (3.71)$$

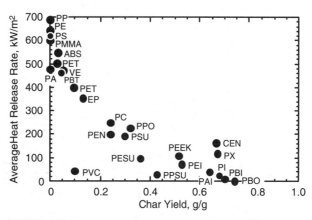

FIGURE 3.14 Average heat release rates versus pyrolysis residue/char yield of plastics.

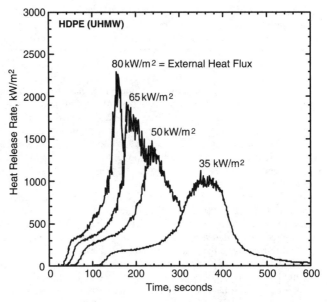

FIGURE 3.15 Heat release rate histories of polyethylene (high density/ultra high molecular weight) at 35, 50, 65, and 80 kW/m² external heat flux showing effect on time to ignition and peak/average heat release rates.

Eqs. (3.70) and (3.71) show that the heat release rate of the plastic is comprised of an intrinsic heat release rate HRR_0 in unforced flaming combustion and an extrinsic heat release rate HRP $\dot{q}_{ext}$ in forced flaming combustion. Because HRR_0 is primarily driven by the flame heat flux, it will be a function of the size and ventilation rate in the fire test and therefore is expected to be somewhat apparatus dependent and reflective of a particular combustion environment.

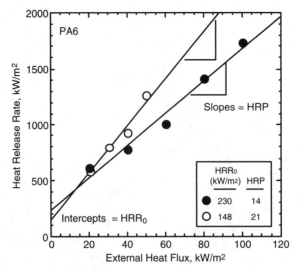

FIGURE 3.16 Peak heat release rate versus external heat flux for PA6 showing graphical procedure for determining HRP and HRR_0 as the slope and intercept, respectively. (Solid circles, Ref. 99; open circles, Ref. 67.)

The first term on the right-hand side of Eq. (3.71), HRR_0, has the units and significance of an ideal [101] or intrinsic [111] heat release rate of the material burning under ambient (unforced) conditions and has the functional form

$$HRR_0 = \left(\chi(1-\mu)\frac{h_c^0}{h_g}\right)(\dot{q}_{flame} - \dot{q}_{loss}) = HRP(\dot{q}_{flame} - \dot{q}_{loss}) \quad (3.72)$$

Increasing the oxygen concentration in the combustion atmosphere increases the flame heat flux [112] and thus increases HRR_0 [107, 111], since HRP and $\dot{q}_{loss}$ depend primarily on the properties of the solid.

TABLE 3.11 Effective Heats of Combustion (EHOC), Heats of Gasification (L_g), and Heat Release Parameters (HRP) of Plastics and Elastomers

Plastic	EHOC, (MJ/kg)	L_g (MJ/kg)	HRP
ABS	29.0	2.3	13
ABS-FR	10.2	2.5	4
CEA	25.9	4.0	7
CEE	25.1	5.1	5
CEF	16.9	3.0	6
CEM	28.9	3.0	10
CEN	20.6	5.0	4
CET	25.9	3.5	7
CPE (25% Cl)	22.6	2.1	11
CPE (36% Cl)	10.6	2.8	4
CPE (48% Cl)	7.2	3.3	2
CPVC	3.9	2.0	2
CR	17.6	2.0	9
CTFE	6.5	0.7	9
ECTFE	4.6	1.5	3
EP	20.4	1.6	13
EPDM	29.2	1.9	15
EP-G	21.1	2.3	9
ETFE	7.3	1.1	6
EVA	30.8	—	—
FEP	1.3	1.5	2
HIPS	28.1	2.0	14
HIPS-FR	10.3	2.1	5
LCP	14.8	—	—
NR	30.2	2.0	15
PA6	29.8	1.5	20
PA66	28.2	2.1	18
PAI	19.3	4.8	4
PBI	22.0	5.5	4
PBT	21.7	1.4	16
PBT-FR	13.7	2.3	6
PC	21.2	2.4	9
PC/ABS	22.4	2.0	11
PC/ABS-FR	17.7	2.5	7
PC-FR	10.4	3.5	3
PE HD	40.3	2.2	18
PE LD	40.3	1.9	21
PEEK	21.3	3.4	6
PEEK-G	20.5	5.8	4

TABLE 3.11 (*Continued*)

Plastic	EHOC, (MJ/kg)	L_g (MJ/kg)	HRP
PEI	21.8	3.5	6
PEN	22.9	2.5	5
PESU	22.4	5.6	4
PESU-G	16.0	5.3	3
PET	18.0	1.4	13
PE-XL	31.4	6.3	5
PE-XL/FR	20.1	5.2	4
PF	16.3	5.4	3
PFA	2.2	0.9	2
PMMA	24.8	1.7	14
POM	14.4	2.4	6
PP	41.9	1.9	22
PPA	24.2	1.4	17
PPO/PS	22.9	1.5	15
PPO-G	25.4	8.5	3
PPS	23.5	6.3	4
PPSU	23.8	5.6	4
PS	27.9	1.8	16
PS-FR	13.8	2.0	7
PSU	20.4	2.5	8
PTFE	4.6	2.3	2
PU	16.3	3.3	5
PUR	24.0	1.3	18
PVC (flexible)	11.3	2.3	5
PVC (rigid)	9.3	2.7	3
PVC/PMMA	10.0	4.0	3
PVDF	3.8	1.9	2
PVF	4.1	1.0	4
PX	20.0	5.0	4
SBR	31.5	2.3	18
SIR	21.7	2.7	8
TPI	12.0	2.4	5
TPU	23.5	2.3	10
UPT	23.4	3.0	8
UPT-FR	15.0	4.2	4
VE	22.0	1.7	13

3.4.3 Unsteady Burning

The heat release rate in forced flaming combustion is the single best indicator of the fire hazard of a material in an enclosure [113], yet none of the tens of billions of pounds [114] of flame-retardant plastics sold worldwide each year in consumer electronics, electrical equipment, building and construction, home furnishings, automobiles, and public ground transportation are required to pass a heat release rate test. Instead, a flame test of ignition resistance [21] is the only fire hazard assessment required of these plastics [12]. Ignition resistance (flammability) tests rank materials according to the duration or extent of burning after removal of a small-flame (e.g., a bunsen burner) ignition source. The Underwriters Laboratory [21] test for flammability of plastics rates materials according to their tendency to burn at a particular rate in a horizontal configuration (UL 94 HB) or self-

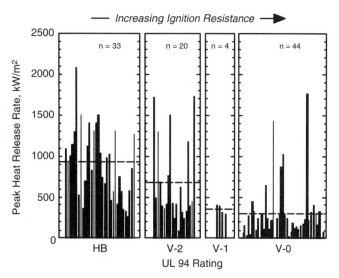

FIGURE 3.17 Peak heat release rates of 106 plastics at external heat flux of 50 ± 10 kW/m² grouped according to ignition resistance rating in the UL 94 test.

extinguish in a specified period of time (10 or 30 s) in a vertical orientation (UL 94 V-0/1/2) after being ignited from below by a bunsen burner. Another widely used flammability rating is the limiting oxygen index (LOI) [115], which is the oxygen concentration in a metered gas stream flowing past a thin (2 to 3 mm) downward-burning bar of material that results in flame extinguishment at 180 s. The tendency of a material to cease combustion or self-extinguish after removal of the ignition source is termed ignition resistance. Empirical correlation of bench-scale [116, 117] and full-scale [118–125] fire behavior with standardized [21, 115] ignition resistance (flammability) tests indicates a general trend of improved fire safety for flame-resistant and fire-retardant plastics [125]. However, a formal relationship between fire behavior and flammability, and between the flammability tests themselves, remains obscure.

Fig. 3.17 shows the peak heat release rates of 106 plastics, plastic blends, and composites measured at an external heat flux of 40, 50, or 60 kW/m² (typically 50 kW/m²). Peak heat release rates are grouped according to their flammability rating in the Underwriters Laboratory test (UL 94) for flammability of plastics [21]. Average peak heat release rates for the n samples in each category are indicated by the horizontal dashed lines in Fig. 3.5, and these are 926 ± 427, 686 ± 513, 359 ± 54, and 294 ± 340 kW/m² for HB, V-2, V-1, and V-0 ratings, respectively. Although the average value of the peak heat release rate decreases as ignition resistance increases, the range and variation of heat release rates in Fig. 3.17 for any particular UL 94 rating precludes the use of a single heat release rate measurement as an indicator of flammability. More important in context of fire safety is the observation that ignition resistance is a poor predictor of fire hazard, i.e., heat release rate [113].

The poor agreement between heat release rate in forced flaming combustion and the self-extinguishing tendency of plastics in the absence of external heating can be understood by consideration of Eq. (3.70). After a solid combustible material is ignited and the ignition source is removed, flaming will cease if the rate of heat released by the burning sample is insufficient to drive the fuel generation (evaporation or pyrolysis) at a rate that will maintain a flammable mixture in the gas phase. As the lower limit of flammability of diffusion flames shows a strong dependence on the fuel composition and concentration [126], a critical mass flux criterion for persistent ignition of plastics would not be expected to broadly apply because the volatile fuel species (pyrolyzate) can vary significantly between polymers [51] so that the lower flammability limit would be material dependent.

Table 3.8 indicates that a critical value of the heat release rate is a better indicator of incipient burning (ignition or extinction). During the ignition phase of the UL 94 or limiting oxygen index test when the bunsen burner is in contact with the plastic specimen [Eq. (3.71)]

$$HRR_0 + HRP\dot{q}_{ext} - \dot{q}_c = 0 \qquad (3.73)$$

If a critical heat release criterion extends to these flammability tests and incipient burning satisfies

$$HRR_0 + HRP\dot{q}_{ext} - \dot{q}_c^* \geq 0 \qquad (3.74)$$

where $\dot{q}_c^* = 90$ kW/m^2 (see Table 3.8) is the critical heat release rate for sustained burning, then, after removal of the bunsen burner flame when $\dot{q}_{ext}^* = 0$, burning will only continue if

$$HRR_0 \geq \dot{q}_c^* = 90 \; kW/m^2 \qquad (3.75)$$

Eq. (3.74) is a general criterion for sustained, piloted ignition of solids analogous to the fire point equation [3], and Eq. (3.75) is an energy balance criterion for self-sustained burning of plastics. Thus, self-extinguishing behavior in flammability tests after removal of the bunsen burner is expected for HRR$_0$ less than about 90 kW/m^2. Table 3.12 compares HRR$_0$ obtained by extrapolation from forced flaming combustion (see Fig. 3.16), measured for self-sustaining combustion under ambient conditions [128], and calculated from the flame heat flux and heat loss measurements [101]. Table 3.12 shows reasonable agreement between HRR$_0$ obtained in these different ways for a variety of common plastics.

Table 3.13 lists ranked values for HRR$_0$ of thermoplastics, thermosets, and elastomers obtained from standardized fire calorimetry tests as the heat release rate intercept at zero external heat flux. Included in Table 3.13 are fire retardant (-FR) and glass reinforced (-G) versions of these materials along with a generic UL 94 rating for all of the materials. It is seen that unsteady burning or self-extinguishing behavior in the UL 94 test (V rating) is observed when HRR$_0$ falls below about 90 kW/m^2, in general agreement with the critical heat release rate at incipient burning (ignition/extinction) deduced from Table 3.8. Toward the bottom of Table 3.13, the HRR$_0$ assume negative values because halogen-containing/flame-retarded (FR) plastics burn with a low flame heat flux and heat-resistant/thermally stable polymers have large surface heat losses. In either case, the term $\dot{q}_{flame} - \dot{q}_{loss}$ in HRR$_0$ could (and apparently does) assume negative values as per Eq. (3.70).

TABLE 3.12 Intrinsic Heat Release Rates HRR$_0$ Obtained by Extrapolation from Forced Flaming Combustion Compared to Direct Measurements [128] and Calculation [101] in Self Sustaining Combustion

Polymer	HRR$_0$, kW/m^2		
	Extrapolated (Table 13)	Measured	Calculated
PET	424 ± 168	353	—
PS	410 ± 66	—	240
PP	369 ± 79	415	202
ABS	359 ± 66	162	—
EVA	—	254	—
PA12	—	245	—
PA66	240 ± 59	—	—
PMMA	217 ± 47	180	240
PA6	187 ± 55	150	—
PE	145 ± 93	75–180	126
POM	162 ± 30	—	148
PC	89 ± 32	5	−200
PEN	57 ± 13	15	—
PF	—	—	16

TABLE 3.13 Intrinsic Heat Release Rate (HRR$_0$) and Flammability (UL 94 Rating) of Plastics Ranked by HRR$_0$

Polymer	HRR$_0$ (kW/m^2)	UL 94 Rating
HIPS	510 ± 77	HB
PP	369 ± 79	HB
PET	424 ± 168	HB
PS	410 ± 66	HB
ABS	359 ± 66	HB
PBT	341 ± 106	HB
UPT	261 ± 105	HB
PC/ABS	259 ± 43	HB
PA66	240 ± 59	HB
PMMA	217 ± 47	HB
PS-FR	205 ± 27	V2
PPO/PS	192 ± 22	HB
PA6	187 ± 55	HB
PC/ABS-FR	178 ± 36	V1
VE	169 ± 44	HB
PESF	168 ± 23	V1
HIPS-FR	164 ± 30	V2
POM	162 ± 30	HB
EP	160 ± 46	HB
PE	145 ± 93	HB
PBT-FR	141 ± 130	V2
ABS-FR	117 ± 33	V2
PVC (flex)	91 ± 19	V2
SIR-M	90 ± 13	V0
PX	88 ± 18	V0
PC	89 ± 32	V2
PEN	57 ± 13	V2
ETFE	44 ± 31	V0
PVC (rigid)	9 ± 25	V0
UPT-FR	−31 ± 10	V0
CPVC	−34 ± 9	V0
PE-XL/FR	−38 ± 28	V0
PAI	−64 ± 16	V0
PASF	−83 ± 25	V0
PTFE	−84 ± 9	V0
PESF-G	−89 ± 92	V0
PEEK	−94 ± 20	V0
PEI	−113 ± 19	V0
ECTFE	−127 ± 6	V0
PPS	−147 ± 30	V0
PBI	−150 ± 36	V0
PC-FR	−191 ± 51	V0
TPI	−201 ± 39	V0
PEEK-G	−261 ± 52	V0

The critical heat flux for burning CHF$_b$ is obtained from Eq. (3.68) when $\dot{q}_c = \dot{q}_c^* \approx 90$ kW/m^2 as per Table 3.8

$$\text{CHF}_b = \dot{q}_{\text{net}}^{\text{crit}} = (\dot{q}_{\text{ext}} + \dot{q}_{\text{flame}} - \dot{q}_{\text{loss}}) = \frac{\dot{q}_c^*}{\text{HRP}} \approx \frac{90 \ kW/m^2}{\text{HRP}} \qquad (3.76)$$

TABLE 3.14 Critical Heat Fluxes for Ignition (Measured) and Burning (Calculated)

	Critical heat flux, kW/m²	
Polymer	Ignition	Burning
ABS	9–15	7
CPVC	40	44
ETFE	17–27	15
FEP	38–50	44
PA6	15–20	5
PA66	15–21	5
PAI	40–50	22
PBI	≈60	22
PBT	20	6
PC	15–20	10
PE	15	4
PEEK	30–40	15
PEI	25–40	15
PEN	24	18
PESU	19–30	22
PET	10–19	7
PMMA	6–23	6
POM	13	15
PP	15	4
PPS	35–38	22
PPSU	32–35	22
PS	6–13	5
PTFE	43–50	44
PVC (rigid)	15–28	29
PVDF	30–50	44
SBR	10–15	5
TPI	36–50	18

Table 3.14 compares critical heat fluxes for ignition (CHF) in Table 3.9 to the critical heat flux for sustained burning CHF_b calculated from Eq. (3.76) using the reported HRP for each plastic (Table 3.11). Table 3.14 shows that CHF_b is about 10 kW/m² less than CHF on average, but the trend is similar.

DEFINITION OF TERMS

α = thermal diffusivity ($\kappa/\rho c$)

A = global frequency factor for pyrolysis (s⁻¹)

A_g = frequency factor for gas generation (s⁻¹)

A_c = frequency factor for char formation (s⁻¹)

b = sample thickness (m)

β = constant heating rate (K/s)

c_1, c_2 = constants of integration (K)

CHF	=	critical heat flux for piloted ignition (kW/m^2)
CHF$_b$	=	critical heat flux to sustain burning (kW/m^2)
χ	=	gas phase combustion efficiency (dimensionless)
c	=	heat capacity (J/g·K)
e	=	the natural number 2.718 . . .
ε	=	surface emissivity of radiant energy (dimensionless)
E_a	=	global activation energy for pyrolysis (J/mol)
E_g	=	activation energy for gas formation (J/mol)
E_c	=	activation energy for char formation (J/mol)
EHOC	=	heat of combustion
h	=	average surface convective heat transfer coefficient (W/m^2·K)
h_c^o	=	net heat of complete combustion of pyrolysis gases (J/kg)
$h_{c,p}^o$	=	net heat of complete combustion of solid polymer (J/kg)
h_g	=	heat of gasification per unit mass of polymer (J/kg)
η_c	=	heat release capacity (J/g·K)
HRP	=	heat release parameter, $\chi h_c^o / L_g$ (dimensionless)
HRR$_0$	=	heat release rate in unforced flaming combustion (kW/m^2)
I^*	=	reactive intermediate for pyrolysis
k_p	=	global Arrhenius rate constant for pyrolysis (s^{-1})
k_g	=	Arrhenius rate constant for gas generation (s^{-1})
k_c	=	Arrhenius rate constant for char formation (s^{-1})
k_i	=	Arrhenius rate constant for initiation of bond breaking (s^{-1})
k_r	=	Arrhenius rate constant for bond recombination (s^{-1})
κ	=	thermal conductivity (W/m·K)
L_g	=	heat of gasification per unit mass of volatile fuel (kJ/g)
m	=	instantaneous sample mass (kg)
m_o	=	initial sample mass (kg)
$\dot{m}_{max}$	=	peak mass loss rate (kg/s)
μ	=	char yield or pyrolysis residue of polymer at 850°C (g/g)
M_g	=	molecular weight of gaseous decomposition species (g/mol)
M	=	monomer molecular weight (g/mol)
$\dot{Q}_c$	=	kinetic heat release rate (W/kg)
$\dot{q}_c$	=	heat release rate (HRR) in flaming combustion (kW/m^2)
$\dot{q}_{cr}$	=	critical heat flux (CHF) for piloted ignition (kW/m^2)

$\dot{q}_{ext}$ = external heat flux in a fire or test (kW/m²)
$\dot{q}_{net}$ = net heat flux to the surface of a burning sample (W/m²)
$\dot{q}_{flame}$ = flame heat flux (W/m²)
$\dot{q}_{loss}$ = heat losses per unit area of surface (kW/m²)
ρ = density (kg/m³)
r = mass loss rate exponent $(1 + 2RT_p/E_a)^{-1}$ (dimensionless)
R = ideal gas constant (= 8.314 J/mol·K)
σ = Boltzmann radiation constant = 5.7×10^{-8} W/m²·K⁴
S = surface area (m²)
t = time (s)
t_{ign} = time to piloted ignition at a constant heat flux (s)
T_p = temperature at peak pyrolysis rate at constant heating rate
T_s = sample or surface temperature
T_o = ambient (room) temperature (298 K)
T_d = onset temperature of thermal degradation
T_{ign} = surface temperature at piloted ignition
TRP = thermal response parameter (kW·s$^{1/2}$/m²) or (kW·s$^{1/2}$·m^{-2})
v = surface recession velocity/burning rate (m/s)
V = volume (m³)
Y_c = temperature-dependent char yield (dimensionless)

REFERENCES

1. Hilado, C.J., *Flammability Handbook for Plastics,* 5th edition, Technomic Publishing Co., Lancaster, PA, 1998.
2. *The SFPE Handbook of Fire Protection Engineering,* 3rd edition, Society of Fire Protection Engineers, Boston, MA, 2002.
3. Drysdale, D., *An Introduction to Fire Dynamics,* 2nd edition, John Wiley & Sons, New York, 1998.
4. Quintiere, J.G., *Principles of Fire Behavior,* Del Mar Publishers, Albany, NY, 1998.
5. Cullis, C.F., and Hirschler, M.M., *The Combustion of Organic Polymers,* Clarendon Press, Oxford, UK, 1981.
6. Aseeva, R.M., and Zaikov, G.E, *Combustion of Polymer Materials,* Hanser, New York, 1985.
7. Nelson, G.L., ed., *Fire and Polymers: Hazards Identification and Prevention,* ACS Symposium Series 425, American Chemical Society, Washington, DC, 1990.
8. Babrauskas, V., and Grayson, S.J., eds., *Heat Release in Fires,* Elsevier Applied Science, New York, 1992.
9. Le Bras, M., Camino, G., Bourbigot, S., and Delobel, R., *Fire Retardancy of Polymers: The Use of Intumescence,* The Royal Society of Chemistry, Cambridge, UK, 1998.
10. Horrocks, A.R., and Price, D., eds., *Fire Retardant Materials,* Woodhead Publishing, Ltd., Cambridge, UK, 2001.
11. Friedman, R., *Principles of Fire Protection Chemistry and Physics,* National Fire Protection Association, Quincy, MA, 1998.

12. Troitzsch, J., *International Plastics Flammability Handbook,* 2nd edition, Carl Hanser, Munich, Germany, 1990.
13. Grand, A.F, Wilkie, and C.A., eds., *Fire Retardancy of Polymeric Materials,* Marcel Dekker, Inc., NY, 2000.
14. Landrock, A.H., *Handbook of Plastics Flammability and Combustion Toxicology,* Noyes Publications, Park Ridge, NJ, 1983.
15. Fire, F.L., *Combustibility of Plastics,* Van Nostrand Reinhold, New York, 1991.
16. Pal, G., and Macskasy, H., *Plastics: Their Behavior in Fires,* Elsevier, New York, 1991.
17. Nelson, G.L., ed., *Fire and Polymers II: Materials and Tests for Hazard Prevention,* ACS Symposium Series 599, American Chemical Society, Washington, DC, 1995.
18. Glassman, I., *Combustion,* 3rd edition, Academic Press, New York, 1996.
19. Turns., S.R., *An Introduction to Combustion—Concepts and Applications,* McGraw-Hill, New York, 1996.
20. Strehlow, R.A., *Combustion Fundamentals,* McGraw-Hill, New York, 1984.
21. *Flammability of Plastic Materials,* UL 94 Section 2 (Horizontal: HB) and Section 3 (Vertical: V-0/1/2), Underwriters Laboratories Inc., Northbrook, IL, 1991.
22. *Plastics 1980: A Desktop Data Bank,* The International Plastics Selector, Cordura Publications, Inc., La Jolla, CA, 1979.
23. Kennedy, G.F., *Engineering Properties and Applications of Plastics,* John Wiley & Sons, New York, 1957.
24. Gent, A.N., ed., *Engineering with Rubber,* Hanser Publishers, Munich, Germany, 1992.
25. Hartwig, G., *Polymer Properties at Room and Cryogenic Temperatures,* Plenum Press, New York, 1994.
26. Goldsmith, A., Waterman, T.E., and Hirschorn, H.J., eds., *Handbook of Thermophysical Properties of Solid Materials,* MacMillan, New York, 1961.
27. Mark, J.E., ed., *Physical Properties of Polymers Handbook,* American Institute of Physics, Woodbury, NY, 1996.
28. *Modern Plastics Encyclopedia,* McGraw-Hill, New York, 1988.
29. Bicerano, J., *Prediction of Polymer Properties,* 2nd edition, Marcel Dekker, Inc., New York, 1996.
30. Van Krevelen, D.W., *Properties of Polymers,* 3rd edition, Elsevier Scientific, New York, 1990.
31. Thompson, E.V., *Thermal Properties, in Encyclopedia of Polymer Science and Engineering,* Vol. 16, John Wiley & Sons, New York, 1989, pp. 711–747.
32. Touloukian, Y.S., ed., *Thermophysical Properties of High Temperature Solid Materials,* Vol. 6: Intermetallics, Cermets, Polymers and Composite Systems, Part II, MacMillan, New York, 1967.
33. Touloukian, Y.S., Powell, R.W., Ho, C.Y., and Nicolaou, M.C., eds., "Thermal Diffusivity," in *Thermophysical Properties of Matter,* Vol. 10, Plenum Press, New York, 1973.
34. Kishore, K., and Pai Verneker, V.R., "Correlation Between Heats of Depolymerization and Activation Energies in the Degradation of Polymers," *Journal of Polymer Science, Polymer Letters,* 14: 7761–7765, 1976.
35. Brandrup, J., and Immergut, E.H., eds., *Polymer Handbook,* 3rd edition, John Wiley & Sons, New York, 1989.
36. *Engineered Materials Handbook,* Vol. 2, Engineering Plastics. ASM International, Metals Park, OH, 1988.
37. *Plastics Digest,* 17(1): 773–889, 1996.
38. *Engineered Materials Handbook,* Vol. 2, Engineering Plastics, ASM International, Metals Park, OH, 1988.
39. Bhowmik, A.K., and Stephens, H.L., *Handbook of Elastomers,* 2nd edition, Marcel Dekker, Inc. New York, 2001.
40. Thornton, W., "The Role of Oxygen to the Heat of Combustion of Organic Compounds," *Philosophical Magazine and Journal of Science,* 33: 196–205, 1917.
41. Hugget, C., "Estimation of Rate of Heat Release by Means of Oxygen Consumption Measurements," *Fire and Materials,* 4(2): 61–65, 1980.
42. Janssens, M., and Parker, W.J., "Oxygen Consumption Calorimetry," in Babrauskas, V., Grayson, S.J., eds., *Heat Release in Fires,* Elsevier Applied Science, London, 1992, pp. 31–59.
43. Babrauskas, V., "Heat of Combustion and Potential Heat," in Babrauskas, and V., Grayson, S.J., eds., *Heat Release in Fires,* Elsevier Applied Science, London, 1992, pp. 207–223.

44. Walters, R.N., Hackett, S.M., and Lyon, R.E., "Heats of Combustion of High Temperature Polymers," *Fire and Materials,* 24(5): 245–252, 2000.
45. Lowrie, R., *Heat of Combustion and Oxygen Compatibility, Flammability and Sensitivity of Materials in Oxygen-Enriched Atmospheres,* ASTM STP 812, B. Werley, ed., American Society for Testing of Materials, Philadelphia, PA, 1983.
46. Van Krevelen, D.W., "Some Basic Aspects of Flame Resistance of Polymeric Materials," *Polymer,* 16: 615–620, 1975.
47. Wolfs, P.M., Van Krevelen, D.W., and Waterman, H.I., *Brennstoff Chemie,* 40: 155–189, 1959.
48. Montaudo, G., and Puglisi, C., "Thermal Degradation Mechanisms in Condensation Polymers," in Grassie, N., ed., *Developments in Polymer Degradation,* Vol. 7, Elsevier Applied Science, New York, 1977, pp. 35–80.
49. Grassie, N., and Scotney, A., "Activation Energies for the Thermal Degradation of Polymers," in Brandrup, J., Immergut, E.H., eds., *Polymer Handbook,* 2nd edition, Wiley Interscience, New York, 1975, pp. 467–47.
50. Lyon, R.E., "Heat Release Kinetics," *Fire and Materials,* 24(4): 179–186, 2000.
51. Grassie, N., "Products of Thermal Degradation of Polymers," in Brandrup, J., and Immergut, E.H., eds., *Polymer Handbook,* 3rd edition, Wiley Interscience, New York, 1989, pp. 365–397.
52. Lyon, R.E., "Solid State Thermochemistry of Flaming Combustion," in Grand, A.F., and Wilkie, C.A., eds., *Fire Retardancy of Polymeric Materials,* Marcel Dekker, Inc., New York, 2000, pp. 391–44745.
53. Broido, A., and Nelson, M.A., "Char Yield on Pyrolysis of Cellulose," *Combustion and Flame,* 24: 263–268, 1975.
54. DiBlasi, C., "Analysis of Convection and Secondary Reaction Effects Within Porous Solid Fuels Undergoing Pyrolysis," *Combustion Science and Technology,* 90: 315–340, 1993.
55. Shafizadeh, F., "Pyrolysis and Combustion of Cellulosic Materials," *Advances in Carbohydrate Chemistry,* 23: 419–425, 1968.
56. Milosavljevic, I., and Suuberg, E.M., "Cellulose Thermal Decompositon Kinetics: Global Mass Loss Kinetics," *Industrial and Engineering Chemistry Research,* 34(4): 1081–1091, 1995.
57. Lewellen, P.C., Peters, W.A., and Howard, J.B., "Cellulose Pyrolysis Kinetics and Char Formation Mechanism," *Proceedings of the 16th International Symposium on Combustion,* The Combustion Institute, 1976, pp. 1471–1480.
58. Nam, J.D., and Seferis. J.C., "Generalized Composite Degradation Kinetics for Polymeric Systems Under Isothermal and Nonisothermal Conditions," *Journal of Polymer Science: Part B, Polymer Physics,* 30: 455–463, 1992.
59. Day, M., Cooney, J.D., and Wiles. D.M., "A Kinetic Study of the Thermal Decomposition of Poly(Aryl-Ether-Ether-Ketone) in Nitrogen," *Polymer Engineering and Science,* 29: 19–22, 1989.
60. Friedman, H.L., "Kinetics of Thermal Degradation of Char Forming Plastics from Thermogravimetry: Application to a Phenolic Plastic," *Journal of Polymer Science: Part C,* 6: 183–195, 1962.
61. Wall, L.A., "Pyrolysis of Polymers," in *Flammability of Solid Plastics,* Vol. 7, Fire and Flammability Series, Technomic Publishers, Westport, CT, 1974, pp. 323–352.
62. Lyon, R.E., "Pyrolysis Kinetics of Char Forming Polymers," *Polymer Degradation and Stability,* 61: 201–210, 1998.
63. Stags, J.E.J., "Modeling Thermal Degradation of Polymers Using Single-Step First-Order Kinetics," *Fire Safety Journal,* 32: 17–34, 1999.
64. Lyon, R.E., "An Integral Method of Nonisothermal Kinetic Analysis," *Thermochimica Acta,* 297: 117–124, 1997.
65. Gandhi, S., Walters, R.N., and Lyon, R.E., "Cone Calorimeter Study of Cyanate Esters for Aircraft Applications," 27th International Conference on Fire Safety, San Francisco International Airport, CA, Jan. 11–15, 1999.
66. Gandhi, S., and Lyon, R.E., "Ignition and Heat Release Parameters of Engineering Polymers," *PMSE Preprints,* 83, ACS National Meeting, Washington, DC, August 2000.
67. FAA unpublished results.
68. Drysdale, D.D., and Thomson, H.E., "The Ignitability of Flame Retarded Plastics," in Takashi, T., ed., *Proceedings of the 4th International Symposium on Fire Safety Science,* IAFSS, 1994, pp. 195–204.

69. Grand, A.F., "The Use of the Cone Calorimeter to Assess the Effectiveness of Fire Retardant Polymers Under Simulated Real Fire Test Conditions," *Interflam 96,* Interscience Communications, Ltd, London, 1996, pp. 143–152.
70. Tewarson, A., Abu-Isa, I., Cummings, D.R., and LaDue, D.E., "Characterization of the Ignition Behavior of Polymers Commonly Used in the Automotive Industry," *Proceedings 6th International Symposium on Fire Safety Science,* International Association for Fire Safety Science, 2000, pp. 991–1002.
71. Tewarson, A., "Generation of Heat and Chemical Compounds in Fires," in *SFPE Handbook of Fire Protection Engineering,* 3rd edition, Society of Fire Protection Engineers, Boston, MA, Section 3, 2002, pp. 82–161.
72. Lyon, R.E., "Heat Release Capacity: A Molecular Level Fire Response Parameter," 7th International Symposium on Fire Safety Science, Worcester Polytechnic Institute, Worcester, MA, June 16–21, 2002.
73. Lyon, R.E., and Walters, R.N., "A Microscale Combustion Calorimeter," Final Report DOT/FAA/AR-01/117; "Pyrolysis-Combustion Flow Calorimetry," *Journal of Analytical and Applied Pyrolysis* (in press).
74. Walters, R.N., and Lyon, R.E., "Calculating Polymer Flammability from Molar Group Contributions," Final Report DOT/FAA/AR-01/31, September 2001.
75. Parker, W., and Filipczak, R., "Modeling the Heat Release Rate of Aircraft Cabin Panels in the Cone and OSU Calorimeters," *Fire and Materials,* 19: 55–59, 1995.
76. Zhou, Y.Y., Walther, D.C., and Fernandez-Pello, A.C., *Combustion and Flame,* 131: 147–158, 2002.
77. Wichman, I.S., and Atreya, A., "A Simplified Model for the Pyrolysis of Charring Materials," *Combustion and Flame,* 68: 231–247, 1987.
78. Bucsi, A., and Rychly, J., "A Theoretical Approach to Understanding the Connection Between Ignitability and Flammability Parameters of Organic Polymers," *Polymer Degradation and Stability,* 38: 33–40, 1992.
79. Moghtaderi, B., Novozhilov, V., Fletcher, D., and Kent, J.H., "An Integral Model for the Transient Pyrolysis of Solid Materials," *Fire and Materials,* 21: 7–16, 1997.
80. Quintiere, J.G., and Iqbal, N., "An Approximate Integral Model for the Burning Rate of a Thermoplastic-Like Material," *Fire and Materials,* 18: 89–98, 1994.
81. Kanury, A.M., "Flaming Ignition of Liquid Fuels," in *SFPE Handbook of Fire Protection Engineering,* 3rd edition, Society of Fire Protection Engineers, Boston, MA, Section 2, 2002, pp. 188–199.
82. Kanury, A.M., "Flaming Ignition of Solid Fuels," in *The SFPE Handbook of Fire Protection Engineering,* 3rd edition, Society of Fire Protection Engineers, Boston, MA, Section 2, 2002, pp. 229–245.
83. Kanury, A.M., *Introduction to Combustion Phenomena,* Gordon and Breach Science Publishers, New York, 1977.
84. Rasbash, J.D., "Theory in the Evaluation of Fire Properties of Combustible Materials," *Proceedings of the 5th International Fire Protection Seminar,* Karlsruhe, September 1976, pp. 113–130.
85. Standard Test Method for Ignition Properties of Plastics, ASTM D 1929, American Society for Testing and Materials, Philadelphia, PA.
86. Standard Test Method for Heat and Visible Smoke Release Rates for Materials and Products Using an Oxygen Consumption Calorimeter, ASTM E 1354, American Society for Testing and Materials, Philadelphia, PA.
87. Standard Test Method for Measurement of Synthetic Polymer Material Flammability Using a Fire Propagation Apparatus, ASTM E 2058, American Society for Testing and Materials, Philadelphia, PA.
88. Tewarson, A., "Experimental Evaluation of Flammability Parameters of Polymeric Materials," in Lewin, M., and Pearce, E.M., eds., *Flame Retardant Polymeric Materials,* Plenum Press, New York, 1982, pp. 97–153.
89. Carslaw, H.S., and Jaeger, J.C., *Conduction of Heat in Solids,* 2nd edition, Clarendon Press, Oxford, UK, 1976, pp. 50–193.
90. Tewarson, A., "Flammability Parameters of Materials: Ignition, Combustion, and Fire Propagation," *Journal of Fire Sciences,* 12(4): 329–356, 1994.
91. Quintiere, J.G., and Harkelroad, M.T., "New Concepts For Measuring Flame Spread Properties," in Harmathy, T.Z., ed., *Fire Safety Science and Engineering, Special Technical Publication 882,* American Society for Testing and Materials, Philadelphia, PA, 1985, pp. 239–267.
92. Babrauskas, V., "The Cone Calorimeter," in Babrauskas, V., and Grayson, S.J., eds., *Heat Release in Fires,* Elsevier Applied Science, London, UK, 1992, Chapter 4, pp. 87–88.

93. Grand, A.F., "Heat Release Calorimetry Evaluations of Fire Retardant Polymer Systems," *Proceedings of the 42nd International SAMPE Symposium,* 42, 1997, pp. 1062–1070.
94. Tewarson, A., and Chin, W., Shuford, R., An Exploratory Study on High Energy Flux (HEF) Calorimeter to Characterize Flammability of Advanced Engineering Polymers, Phase 1—Ignition and Mass Loss Rate, ARL-TR-2102, October 1999.
95. Gandhi, S., Walters, R.N., and Lyon, R.E., "Fire Performance of Advanced Engineering Thermoplastics," *Proceedings of the 6th International Conference on Fire and Materials,* San Antonio, TX, Feb. 22–23, 1999.
96. Effectiveness of Fire Retardant Chemicals at Elevated Temperatures, Phase I SBIR Final Report, U.S. DOT Contract Number DTRS-57-94C-00172, Omega Point Laboratories, May 1995.
97. Innovative Fire Retardant Polymeric Systems, Phase II SBIR Final Report, U.S. DOT Contract Number DTRS-57-96-00011, Omega Point Laboratories, February 2000.
98. Hirschler, M.M., "Heat Release from Plastic Materials," in Babrauskas, V., and Grayson, S., eds., *Heat Release in Fires,* Elsevier Applied Science, New York, pp. 207–232, 1992.
99. Scudamore, M.J., Briggs, P.J., and Prager, F.H., "Cone Calorimetry—A Review of Tests Carried Out on Plastics for the Association of Plastic Manufacturers in Europe," *Fire and Materials,* 15: 65–84, 1991.
100. Hallman, J.R., *Ignition Characteristics of Plastics and Rubber,* Ph.D. dissertation, University of Oklahoma, Norman, OK, 1971.
101. Tewarson, A., and Pion, R.F., "Flammability of Plastics: I. Burning Intensity," *Combustion and Flame,* 26: 85–103, 1976.
102. Tewarson, A., "Heat Release Rates from Burning Plastics," *Journal of Fire & Flammability,* 8: 115–130, 1977.
103. Tewarson, A., "Flammability Evaluation of Clean Room Polymeric Materials for the Semiconductor Industry," *Fire and Materials,* 25: 31–42, 2001.
104. Tewarson, A., Abu-Isa, I.A., Cummings, D.R., and LaDue, D.E., "Characterization of the Ignition Behavior of Polymers Commonly Used in the Automotive Industry," *Proceedings of the 6th International Symposium on Fire Safety Science,* Poitiers, France, 1999, pp. 991–1002.
105. Gandhi, S., Walters, R.N., and Lyon, R.E., "Heat Release Rates of Engineering Polymers," *Proceedings of the 6th International Conference on Fire and Materials,* San Antonio, TX, February 1999.
106. Buch, R.R., "Rates of Heat Release and Related Fire Parameters for Silicones," *Fire Safety Journal,* 17: 1–12, 1991.
107. Hsieh, F.Y, Motto, S.E., Hirsch, D.B., and Beeson, H.D., "Flammability Testing Using a Controlled Atmosphere Cone Calorimeter," presented at the Eighth International Conference on Fire Safety, Milbrae, CA, 1993.
108. Koo, J., Venumbaka, S., Cassidy, P., Fitch, J., Grand, A., and Burdick, J., "Evaluation of Thermally Resistant Polymers Using Cone Calorimetry," *Proceedings of the 5th International Conference on Fire and Materials,* San Antonio, TX, February 23–24, 1998, pp. 183–193.
109. Staggs, J.E.J, "Discussion of Modeling Idealized Ablative Materials with Particular Reference to Fire Testing," *Fire Safety Journal,* 28: 47–66, 1997.
110. Whiteley, R.H., Elliot, P.J., and Staggs, J.E.J., "Steady-State Analysis of Cone Calorimeter Data," *Flame Retardants '96,* Interscience Communications, Ltd., London, pp. 71–78, 1996.
111. Lyon, R.E., "Ignition Resistance of Plastics," *13th Annual BCC Conference on Flame Retardancy of Polymeric Materials,* Stamford, CT, June 3–5, 2002.
112. Tewarson, A., Lee, J.H., and Pion, R.F., "The Influence of Oxygen Concentration on Fuel Parameters for Fire Modeling," *Proceedings of the Eighteenth Symposium (International) on Combustion,* The Combustion Institute, Pittsburgh, PA, 1981, pp. 563–570.
113. Babrauskas, V., and Peacock, R.D., "Heat Release Rate: The Single Most Important Variable in Fire Hazard," *Fire Safety Journal,* 18: 255–272, 1992.
114. Calculation based on U.S. Department of Commerce data for dollar value of flame retardants sold in U.S. and Society of Plastics Industry database on U.S. Plastics sales by use (NAICS) category assuming annual worldwide consumption of FR plastics is three times that of U.S.
115. Standard Test Method for Measuring the Minimum Oxygen Concentration to Support Candle-Like Combustion of Plastics (Oxygen Index), ASTM D 2863, American Society for Testing of Materials, Philadelphia, PA.

116. Gandhi, P.D., "Comparison of Cone Calorimeter Results with UL94 Classification for Some Plastics," presented at the 5th Annual BCC Conference on Flame Retardancy, Stamford, CT, May 24–26, 1994.
117. Bundy, M., National Institute of Standards and Technology, Gaithersburg, MD, private communication.
118. Hill, R.G., Eklund, T.I., and Sarkos, C.P., Aircraft Interior Panel Test Criteria Derived From Full Scale Fire Tests, Federal Aviation Administration Report DOT/FAA/CT-85/23, September 1985.
119. Sarkos, C.P., Filipczak, R.A.. and Abramowitz, A., A Preliminary Evaluation of an Improved Flammability Test Method for Aircraft Materials, Federal Aviation Administration Report DOT/FAA/CT-84/22, December 1984.
120. Babrauskas, V., Fire Hazard Comparison of Fire-Retarded and Non Fire-Retarded Products, NBS Special Publication, SP749, National Institute of Standards and Technology, USA, 1988.
121. Simonson, M., and DePoortere, M., "The Fire Safety of TV Set Enclosure Material," presented at Fire Retardant Polymers, 7th European Symposium, Lille, France, 1999.
122. Bailey, R., and Blair, G., "Small Scale Laboratory Flammability Testing and Real Automotive Interior Fires," presented at Fire Retardant Chemical Association Meeting, Philadelphia, PA, October 14–16, 2001.
123. Grayson, S., "Fire Performance of Plastics in Car Interiors," *Proceedings of Flame Retardants 2002*, London, UK, February 2002.
124. Le Tallec, Y., Sainratm, A., and Strugeon, A., "The Firestarr Project–Fire Protection of Railway Vehicles," *Proceedings Fire & Materials 2001, 7th International Conference and Exhibition*, San Francisco, CA, January 22–24, 2001.
125. Babrauskas, V., "The Effects of FR Agents on Polymer Performance," in Babrauskas, V., and Grayson, S., eds., *Heat Release In Fires*, Elsevier Applied Science, New York, 1992, pp. 423–446.
126. Burgess, M.J., and Wheeler, R.V., *Journal of the Chemical Society*, 99: 2013, 1911.
127. Lyon, R.E., "Fire and Flammability," Fire & Materials Conference, San Francisco, CA, January 2003.

CHAPTER 4
FLAME RETARDANTS FOR PLASTICS

Dr. Elisabeth S. Papazoglou
Technical Services Manager
Great Lakes Polymer Additives
1802 US Highway 2
West Lafayette, IN 47906

4.1 INTRODUCTION

Flame retardants are used in plastics because they increase the material's resistance to ignition, and once ignition occurs they slow down the rate of flame spread. A combustible plastic material does not become noncombustible by incorporation of a flame-retardant additive. However, the flame-retardant polymer resists ignition for a longer time, takes more time to burn, and generates less heat compared to the unmodified plastic [1]. In a well-established study carried out by NBS in 1988, the benefits of flame-retardant plastics were demonstrated by comparing commercial products with and without flame retardants. Use of flame retardants allows time to react and contain a fire until extinguishing media are available. Traditionally, either halogenated compounds, with antimony trioxide as a synergist, or phosphorus compounds have been used to flame retard thermoplastic materials. Inorganic compounds with high water content such as magnesium hydroxide and *alumina trihydrate* (ATH) are also used. The addition of low-molecular-weight compounds that either act as fillers or plasticize the polymer degrades mechanical properties and appearance. Processing can also become more complicated especially in the production of complex or very thin parts.

The successful use of flame retardants in thermoplastics is a compromise or at best a balancing act. Filler-type additives of optimum particle size must be properly dispersed, and processing temperatures have to be chosen to prevent degradation. Brittleness and poor impact properties can result from nonuniform dispersion of such solid additives. Melt blendable additives of low molecular weight will plasticize high-molecular-weight polymers depending on polarity, size, and concentration in the polymer. Heat distortion temperature is drastically decreased upon incorporation of such plasticizing additives. Light and heat stability can suffer especially with incorporation of halogenated materials.

A significant change in flame-retarding standards regarding the evolution of smoke as an additional requirement is emerging and is being addressed by new materials and formulations. Many traditional flame retardants increase smoke evolution as they suppress flame propagation. New materials are being developed to balance flame-retarding efficacy and smoke generation.

New products based on silicone chemistry are gaining momentum as flame retardants, and nanoclays impart flame retardancy in polymers where they can be successfully dispersed. The technology of nanocomposites has niche applications and developments are under way. Nanoclays are currently mostly used in combination with already existing flame-retardant chemistries to meet commercial flame-retardant specifications and pass tests. However, it is clear that the opportunity exists for such a technology to change the landscape of flame-retardant products in the near future.

This chapter reviews flame-retardant additives in terms of their chemistry and functionality, then in terms of their application in plastics, highlighting the essential effects on mechanical, ther-

TABLE 4.1 Consumption of Flame Retardants by Major Region (millions of dollars)

	1988	1992	1995	1998	Average annual growth rate (1998–2003) percent
United States	470	480	585	630	2.8–3.6
Western Europe	332	559	631	685	3–4
Japan	250	317	348	373	3.8
Other Asia	na	na	>244	>390	5.1
Total	>1,052	>1,356	>1,808	2,078	3.5–4%

mal, and durability performance. Additives are categorized as halogenated, phosphorus, inorganic flame retardants, and new technologies. Antimony trioxide and some other essential synergists are presented to provide a complete picture of the flame-retarding additive packages available today. An overview of the market gives the reader a broad perspective of the reach and prospects of this industry and how it is affected by regulations, environmental pressures, and improvements in the standard of living.

4.2 OVERVIEW OF THE FLAME RETARDANTS INDUSTRY

In 1998, the flame retardants business in the United States, Europe, and Asia was worth more than $2.1 billion and consisted of 1.14 million metric tons of materials. The US market for 2000 was worth $835 million and is expected to reach $969 million by 2003 at an annual growth rate of 5.1 percent (BCC Communications Report 2002 [2]). The halogenated chemicals valued at $329 million in 2000 are expected to reach a value of $403 million by 2003. ATH and antimony trioxide were valued at $381 million in 2000. For the period 1998 to 2003, the global flame retardants market is expected to grow at a rate of 3.5 to 4 percent per year both on volume and value basis. There is global understanding on the value of flame retardants in saving lives, and several initiatives to promote and expand the use of flame retardants are under way. At the same time, however, there is also considerable environmental pressure on various chemicals associated with flame retardancy, such as halo-

TABLE 4.2 Consumption of Flame Retardants by Type and Region—1998 (thousands of metric tons)

	US	Europe	Japan	Other Asia	Quantity	Value (million $)
Brominated compounds	68.3	51.5	47.8	97	264.6	790
Organophosphorus compounds	57.1	71	26.0	19	175.1	435
Chlorinated compounds	18.5	24.7	2.1	20	65.3	116
Alumina trihydrate	259	160	42	9	>470	260
Antimony oxides	28	23	15.5	20	>86.5	327.5
Others	42.7	29.8	10.5	na	>83	>149.5
	474.6	360	143.9	>165	1,144.5	2,078

* Source: Reference 3.

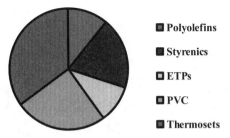

FIGURE 4.1 Flame retardant use by polymer in millions of metric tons.

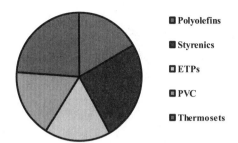

FIGURE 4.2 Flame retardant use by polymer in millions of US dollars.

gens and antimony. This creates a very interesting balance, as there are competing extremes between omitting flame retardants from plastics or proving that the environmental impact of halogens and antimony is not the one depicted by environmental advocates.

Table 4.1 summarizes the consumption of flame retardants by geographic region in terms of dollar value. Table 4.2 categorizes their usage by flame-retardant chemistry—type and volume [3].

Regional differences exist in the use and promotion of flame retardants. The United Kingdom has strict fire standards in home furnishings, whereas the United States does not yet have a uniform standard for home furnishings. Recently, European nations allowed the use of non-flame-retardant TV cabinets versus flame-retardant TV sets required in the United States and Japan. The general trend is that the United States is abiding by regulations such as the UL or other standards and selects the most cost-effective flame retardant. Europe is more influenced by public opinion on environmental concerns, and Japan seems to be the most aggressive in implementing new flame retardants to strike a balance between environmental pressures and fire safety standards.

Growth in the business is almost entirely driven by regulations. New regulations impact growth far more dramatically than growth in end-use markets. The flame-retardant business is an intensely global business. This is not only because the same additive companies participate in all geographic areas, but also because regulations in one geographic area have impact throughout the world. The exports out of Asia and the insistence of global end-users on taking advantage of economies of scale drive the use of the same additive technology globally. Such a technology will meet the most stringent regulation and the most restrictive environmental scrutiny.

Figs. 4.1 and 4.2 depict the flame-retardant use by polymer system as a percentage of the total flame-retardant plastics sales volume (in MT, Fig. 4.1) and value (in United States dollars, Fig. 4.2). The key importance of thermosets and engineering plastics as they affect not only volume but also value is clear from the pie charts.

Figs. 4.3 and 4.4 categorize the flame-retardant plastics use by end application. Both volume (in

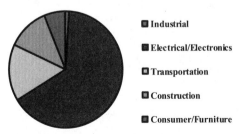

FIGURE 4.3 End markets for flame retardants in millions of metric tons.

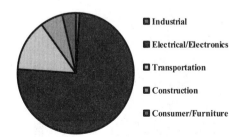

FIGURE 4.4 End markets for flame retardants in millions of US dollars.

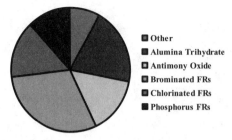

FIGURE 4.5 Use of flame retardants by type in millions of metric tons.

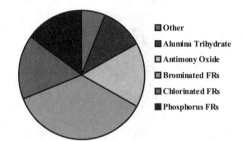

FIGURE 4.6 Use of flame retardants by type in millions of US dollars.

MT, Fig. 4.3) and value (in United States dollars, Fig. 4.4) charts reveal the paramount significance of the electrical and electronics market on the use of flame-retardant additives.

The market is further broken down in terms of flame-retardant chemistries (brominated, phosphorus, ATH, antimony) in Figs. 4.5 and 4.6 for volume (in MT) and value (in United States dollars), respectively. The major producers of flame-retardant additives are summarized in Table 4.3, broken down in terms of additive chemistries.

TABLE 4.3 Global Producers of Flame Retardant Additives

Brominated Compounds
    Ameribrom
    Great Lakes Chemical
    Albemarle Corporation
    Dead Sea Bromine
    Tosoh
Halogenated Phosphorus Compounds
    Akzo Nobel Chemicals
    Albright and Wilson
    Daihachi
Nonhalogenated P-compounds
    Akzo Nobel Chemicals
    Great Lakes Chemical
    Albright and Wilson
    Daihachi
    Ajinomoto
    Clariant
    Bayer
Chlorinated compounds
    Dover Chemical
    Occidental Chemical
    Alumina Trihydrate

Alcan Chemicals Ltd.
Alcoa
Aluchem
J.M. Huber Corporation
Martinswerke Lonza
Nabeltec
Nippon Light Metal Company
Showa Denko
Sumitomo Chemical Co.
VAW
Melamine Compounds
    DSM
Antimony Oxides
    Great Lakes Chemical
    Occidental/Laurel Industries
    Amspec Chemical Corporation
    Nihon Seiko Co.
    Sumitomo Metal Mining
Boron compounds
    Great Lakes Chemical
    U.S. Borax

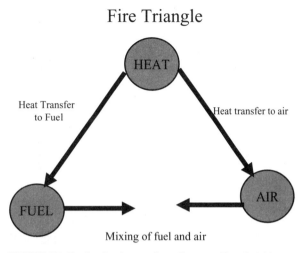

FIGURE 4.7 The fire triangle according to Emmons. (From Ref. 4.)

4.3 MECHANISMS OF FLAME RETARDANCY

To understand the mechanism of action of all flame retardants, it is critical to understand the life cycle of a fire—best depicted by Emmons' fire triangle [4] (Fig. 4.7).

An initial heat source is the beginning of an ignition that further requires fuel and oxygen (through air) to be sustained and to grow.

The polymer provides the fuel in a continuous stream of pyrolysis and decomposition products as the temperature increases [5] (Fig. 4.8).

The ambient atmosphere provides the oxygen necessary to further decompose and create free radicals that propagate combustion from the polymer fragments present in the gas phase. A step-by-step schematic of the pyrolysis-decomposition process is shown in Fig. 4.9.

To stop the fire, this process must be interrupted at one or more stages.

The flame retardants developed inhibit this process according to their mechanism of action in one

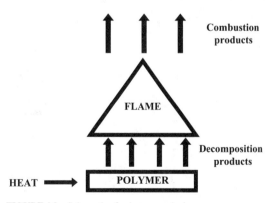

FIGURE 4.8 Schematic of polymer pyrolysis.

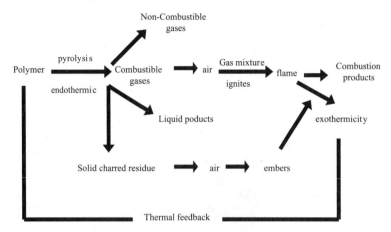

FIGURE 4.9 Diagram of polymer pyrolysis and combustion-decomposition processes.

or more critical stages (Fig. 4.10) [5]. The letters (a), (b), (c), and (d) correspond to the diagram of Fig. 4.10.

a) Gas phase mechanism—flame poisoning

The flame retardants that act in the gas phase are free radical scavengers that bond the oxygen or hydroxy radicals present in the gas phase. Halogenated chemicals act in this fashion by themselves or in combination with antimony trioxide: in this case, the antimony oxyhalides are the active free radical scavengers. Some phosphorus chemicals such as *triphenyl phosphate* (TPP) are known to have some gas phase action, although they seem to participate also in glass-forming, solid phase reactions.

b) Cooling

Flame retardants that decompose in endothermic reactions will cool the combustion environment and therefore slow down reaction pathways. The hydrated minerals alumina trihydrate and mag-

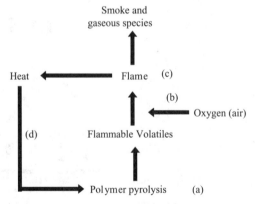

FIGURE 4.10 Stages of polymer combustion.

nesium hydroxide act in such a fashion by releasing water in endothermic decomposition reactions.

c) Dilution of oxygen concentration

The active concentration of oxygen in the gas phase is the driving force for the combustion of carbon radicals and creation of more oxygen and hydroxy radicals that propagate decomposition. The release of inert gases in a fire environment dilutes the active oxygen concentration and decelerates the reaction pathways.

d) Solid phase mechanism

(1) Charring: Silicon- or phosphorus-containing chemicals have the possibility to initiate, at elevated pyrolysis temperatures, cross-linking reactions in the polymer matrix that create an effective barrier to heat transfer and diffusion of gases, thereby providing effective flame retardancy. These reactions resemble the reactions present in the formation of glass and are catalyzed by acidic conditions and certain cations.

(2) Intumescence: The term refers to the formation of a porous carbonaceous char by the combined action of a carbon source, a blowing agent/gas released, and a catalyst necessary to set the reaction in motion. Examples of such systems are combinations of alcohols (carbon source), ammonium compounds (releasing NH_3 as the blowing agent), and phosphorus compounds where the phosphoric acid formed during pyrolysis acts as the reaction catalyst. The strength of the char and its structure are critical in achieving the desired flame-retarding performance [6].

The choice of a flame-retardant additive depends on the best mechanism that can be exploited and put in action for the specific polymer matrix. As an example, charring cannot occur in polypropylene, but is readily possible in polycarbonate and polyphenylene oxide.

4.4 CLASSES OF COMMERCIAL FLAME RETARDANTS

4.4.1 Hydrated Minerals

The acting principle in the use of all hydrated minerals is their decomposition through energy consumption and release of water, resulting in dilution of the combustion gases and cooling of the polymer.

Alumina Trihydrate. Hydrated minerals such as ATH have a high water content and function as flame retardants by liberating their chemically combined water (34.6 percent for ATH), thus cooling the polymer and delaying ignition. The decomposition of ATH is endothermic and follows the reaction:

$$Al(OH)_3 + Energy \longrightarrow Al_2O_3 + H_2O$$

An additional flame-retarding mechanism in action with ATH is that of dilution of combustible gases. This leads to reduced heat release and smoke evolution. The release of water starts at temperatures as low as 205°C for ATH. Table 4.4 shows the water content and decomposition temperature for some hydrated minerals [7].

The main use of ATH in thermoplastics is for wire and cable applications in both polyolefins and *polyvinyl chloride* (PVC). Polyolefin formulations for wire and cable have to pass strict smoke-evolution requirements (such as UL-910), and ATH is critical in meeting cost-effective performance. Because of the high level of ATH needed to meet these strict requirements, load levels of 40 to 60 percent must be used. Sometimes special copolymers have to be used to allow a material with acceptable physical properties to be developed. Particle size selection and surface-coating modifications of ATH are available to improve processability and mechanical properties. Surface modifiers help wetting-out of the ATH particles, improve dispersion, and allow better impact strength and tensile elongation.

TABLE 4.4 Physical Properties of Alumina Trihydrate, Magnesium Hydroxide, and Magnesium Carbonate

	ATH	$Mg(OH)_2$	$Mg_4(CO_3)_3(OH)_2 4H_2O$
Bound water (%)	34.6	31	20.2
Carbon dioxide (%)	—	—	36.1
Minimum decomposition temperature (°C)	205	320	230
Enthalpy of decomposition (cal/g)	−280	−382	−295
Specific gravity (g/cm3)	2.42	2.36	2.24

Table 4.5 depicts the positive effect of surface modification of ATH in maintaining mechanical properties in an *ethylene-vinyl acetate* (EVA) copolymer formulation. Surface modification, however, can have a negative effect on flame-retarding performance and coloring ability.

Magnesium Hydroxide and Magnesium Carbonate. Magnesium hydroxide is another hydrated mineral that is often used in thermoplastic applications instead of ATH. It has a higher decomposition temperature for the release of water (320 versus 205°C), and this makes it more attractive for higher temperature thermoplastics. Combinations of ATH and $Mg(OH)_2$ are sometimes used to achieve release of water and cooling over a range of temperatures, thus reducing heat release and smoke evolution.

$$Mg(OH)_2 + Energy \longrightarrow MgO + H_2O$$

The same comments made for the coating of ATH apply for magnesium hydroxide: Table 4.6 quantifies the improvement in mechanical properties of a polypropylene formulation containing coated magnesium hydroxide, compared to a formulation with uncoated $Mg(OH)_2$. The water resistance and chemical resistance of the coating chemicals used for ATH and $Mg(OH)_2$ are of critical importance as these formulations must pass strict wet electrical testing and other extractability performance tests required by the wire and cable industry.

Magnesium carbonate is the other important hydrated flame-retardant mineral: it decomposes into CO_2 and this makes it a synergistic char former in some systems. Its decomposition starts at 230°C but continues over a broader temperature range.

Fundamental properties of ATH, $Mg(OH)_2$, and magnesium carbonate are described in Table 4.4.

TABLE 4.5 Effect of ATH Coating on Compounded EVA Properties

Test	Uncoated ATH	Amino silane coating	New coating
Tensile strength at break (Mpa)	9.5	12.4	11.5
Elongation at break	130	200	500
Melt flow index 190C/10Kg in g/10 min	1.4	1.5	3.0
Volume resistivity @ 28 days/50C in water, ohm-cm	1E8	1.6E12	1.5E12

TABLE 4.6 Effect of Coated Magnesium Hydroxide on Compounded Polypropylene Properties

Test	Neat resin	Uncoated MgOH	Coated MgOH
Tensile strength at break (Mpa)	23	20.2	12.2
Elongation at break	280	1	224
Melt flow index 230C/5Kg in g · 10 min	7	N/D	6.9
Charpy impact strength	No break	6.4	No break

4.4.2 Halogenated Materials

The acting principle in the use of halogenated materials as flame retardants is the interruption of the radical chain mechanism in the gas phase. The following scheme depicts the reactions occurring:

a. Release of halogen radicals ($X^* = Cl^*$ or Br^*) from the flame-retardant R-X:

$$R - X \longrightarrow R^* + X^*$$

b. Formation of halogen hydroxides (HX)

$$RH + X^* \longrightarrow HX + R^*$$

c. Neutralization of energy-rich radicals

$$HX + H^* \longrightarrow H_2 + X^*$$

$$HX + OH^* \longrightarrow H_2O + X^*$$

Chlorine-Containing Flame Retardants

Chlorinated Paraffins. Chlorinated paraffins contain a maximum of 75 to 76 percent chlorine and are manufactured via chlorination of a paraffin hydrocarbon. They represent the lowest cost halogenated flame retardants available in the market. At levels of 40 to 60 percent of chlorine, the products are liquid of varying viscosities, while at chlorine levels of 70 percent they become solids. Thermal degradation occurs via a dehydrochlorination reaction and starts at 180°C. For most thermoplastic applications chlorinated paraffins have poor thermal stability, limiting their use to PVC, polyurethanes, and some special applications in polypropylene and polyethylene. Polyethylene blown film is an application where the use of chlorinated paraffins would be feasible because of lower processing temperatures. Chlorex 760 (Dover Chemical), a high-softening-point resinous chlorinated paraffin (160°C) is claimed to have improved thermal stability, allowing its use in polypropylene and *high-impact polystyrene* (HIPS).

Mixed halogenated paraffins (bromochlorinated compounds) have also been developed to meet more stringent flammability requirements. Such products contain typically 27 percent chlorine and 32 percent bromine.

Dechlorane Plus. Dechlorane Plus is a thermally stable alicyclic chlorine containing flame retardant manufactured by a Diels-Alder reaction of hexachlorocyclopentadiene and cyclooctadiene. It is one of the earliest flame retardants developed and has been in commercial use since the mid-1960s, currently available from Occidental Chemical Corporation (Niagara Falls). It has an outstanding thermal stability (approximately up to its melting point—350°C) and a high chlorine content (65 percent). It is available in various particle sizes to meet the physical property profile required in various applications.

It is mainly used in polyolefins and nylon materials. For V-0 polypropylene, it is used in conjunction with antimony trioxide and zinc borate as synergists, while in nylon, antimony trioxide or ferric oxide can be used as synergists. Ferric oxide imparts a characteristic red color, and it is used only in applications where such color can be tolerated. Table 4.7 gives the typical properties

TABLE 4.7 Properties of Dechlorane Plus

Appearance	White crystalline solid
Chlorine content	65%, Alicyclic
Melting point	350°C
Specific gravity	1.8–2.0 g/cc

TABLE 4.8 Flame Retardant EVA Formulation for Wire and Cable Applications

Formulation (weight %)	1	2
EVA	47.9	47.9
Dechlorane Plus	25	18
ATO	5	9
Clay (translink −37)	20	20
Agerite Resin D (AO)	0.7	0.7
Luperox 500R (Peroxide)	1.4	1.4
Flammability		
Oxygen index (%)	27	27
UL 44 VW-1 flame test	Pass	Pass
Tensile strength (MPa)	12.7	14.0
(psi)	1846	2034
Elongation (%)	298	262

of Dechlorane Plus. Table 4.8 describes the formulations and properties of Dechlorane Plus in EVA [8]. Nylon and polypropylene formulations are discussed in the section devoted to these thermoplastics.

Brominated Materials. There is no universal brominated flame retardant suitable for all plastic resins and applications. Considering the mechanism of action of brominated materials and the wide range of thermoplastics in use, this is readily explainable. Bromine acts in the gas phase and has to be delivered there; it should be made available at the same temperature at which the thermoplastic pyrolysis occurs and the free radical segments diffuse and are available in the gas phase to combine with bromine. Given the different temperatures of pyrolysis of thermoplastic materials [9] (Table 4.9), no single additive can be used across the board. The successful additive must be thermally stable to allow trouble-free processing, and have adequate compatibility with the polymer to achieve good mechanical properties and avoid phase separation (blooming). Table 4.10 lists the most important brominated flame-retardant additives by trade and chemical name as well as identifies the companies making them.

The most structured way to present brominated flame retardants is to categorize them by the basic brominated raw material used in their manufacture. Six types of brominated materials are currently commercially available, and they are derived from different brominated feedstocks.

Tetrabromophthalic Anhydride and Derivatives. Tetrabromophthalic anhydride and its derivative diol are reactive flame retardants mainly used in polyesters and polyurethanes. They are seldom used as blended additives in thermoplastics. The tetrabromophthalate ester (Table 4.10) is used in PVC, and several variations exist in end chains depending on the starting alcohol. This brominated phthalate ester is a flame-retardant plasticizer in PVC (DP-45), and its main use is in high-end wire and cable formulations (Plenum and riser markets).

Tetrabromobenzoate esters are a newer development where the material is a liquid and can be easily incorporated into the urethane or polyester formulation. The products combine excellent thermal stability and good flame-retardant efficacy. Table 4.11 gives some physical properties for products of this chemical family.

Hexabromocyclododecane (HBCD) and Derivatives. Alicyclic brominated hydrocarbons have generally poor thermal stability (less than 200°C). However, they are very effective in flame-retarding expandable and extruded polystyrene foam. Very low concentrations without any antimony trioxide as a synergist are adequate to allow the polystyrene materials to meet a variety of

TABLE 4.9 Ignition Temperature of Various Polymers

Polymer	Flash ignition temperature (°C)	Self ignition temperature (°C)
Polyethylene	341 to 357	349
Polypropylene, fiber		570
Polyvinyl chloride (PVC)	391	454
Polyvinyl chloride acetate	320 to 340	435 to 557
Polyvinylidene chloride (PVDC)	532	532
Polystyrene	345 to 360	488 to 496
Styrene-acrylonitrile (SAN)	366	454
Acrylonitrile-butadiene styrene (ABS)		466
Styrene methyl methacrylate	329	485
Polymethyl methacrylate (PMMA)	280 to 300	450 to 462
Acrylic, fiber		560
Polycarbonate	375 to 467	477 to 580
Nylon	421	424
Nylon 6,6, fiber		532
Polyetherimide	520	535
Polyethersulfone (PES)	560	560
Polytetrafluoroethylene (PTFE)		530
Cellulose nitrate	141	141
Cellulose acetate	305	475
Cellulose triacetate, fiber		540
Ethyl cellulose	291	296
Polyurethane, polyether rigid foam	310	416
Phenolic, glass fiber laminate	520 to 540	571 to 580
Melamine, glass fiber laminate	475 to 500	623 to 645
Polyester, glass fiber laminate	346 to 399	483 to 488
Silicone, glass fiber laminate	490 to 527	550 to 564
Wool	200	
Wood	220 to 264	260 to 416
Cotton	230 to 266	254

building code specifications. HBCD is the most prominent member of the family, and various versions of "stabilized HBCD" are available commercially to allow a wider processing window in polystyrene applications. These materials are stabilized either through addition of proprietary stabilizers, and/or via selection of the most stable isomers. The presence of metals catalyzes HBCD degradation; especially, zinc stearate used in polystyrene as a lubricant can accelerate the decomposition of HBCD and cause HBr evolution at processing temperatures. A special mechanism for flame retardancy in polystyrene is activated by HBCD-chain scission decomposing polystyrene molecular weight and accelerating dripping. Examples are described in the section devoted to polystyrene.

Tetrabromobisphenol A (TBBA) and Derivatives. TBBA (or simply Tetra) is still the highest volume flame retardant in use today and has been in that position for the last 8 to 10 years. It is used in epoxy resins for printed circuit boards and reacts with the epoxy resin components. This allows high thermal stability and excellent electrical properties for circuit boards. Thermosets constitute the highest volume flame-retardant plastics used (Fig. 4.1) and a very significant portion of the total value of flame-retardant plastics (Table 4.2). Intense efforts to replace TBBA in epoxy circuit boards with nonhalogen alternatives have produced interesting chemical compounds and patents, but no real alternative has yet been discovered.

TABLE 4.10 Halogenated Flame Retardants. Manufacturers-Chemical Structure-Applications

COMPANY	TRADE NAME	STRUCTURE	CAS No.	ABS	HIPS	NYLON	POLYURETHANE	POLYESTERS	POLYOLEFINS	POLYSTYRENE	POLYCARBONATE	PVC	ELASTOMERS	UNSATURATED P	EPOXY RESINS	ADHESIVES & COATINGS	TEXTILES	TPE
Akzo / Great Lakes	Fyrol PBR / DE-60F Special	(structure: brominated diphenyl ether with phosphate, 75%/25%, R=Isopropyl, where x+y=5)	Proprietary				X					X	X	X	X	X	X	
Great Lakes / Shanghai Huazhuan	PHT4 / Shanghai PHT-4	(structure: tetrabromophthalate dimethyl ester)	632-79-1											X				
Albemarle / Great Lakes	Saytex RB-79 / PHT4 Diol	(structure: tetrabromophthalate diol)	77098-07-8				X						X					
Great Lakes / Laurel Industries / Oxychem	DP-45 / Pyronil 45	(structure: tetrabromophthalate diester)	26040-51-7				X	X		X	X	X	X					
Dead Sea Bromine	FR-1808	(structure: brominated indane, where x+y=8)	155613-93-7	X	X													

4.12

TABLE 4.10 *Continued*

COMPANY	TRADE NAME	STRUCTURE	CAS No.	ABS	HIPS	NYLON	POLYURETHANE	POLYESTERS	POLYOLEFINS	POLYSTYRENE	POLYCARBONATE	PVC	ELASTOMERS	UNSATURATED P	EPOXY RESINS	ADHESIVES & COATINGS	TEXTILES	TPE
Albemarle Dead Sea Bromine Great Lakes	Saytex 102E DE-83R FR1210	(decabromodiphenyl ether structure)	1163-19-5	x	x			x	x		x		x	x	x	x	x	
Albemarle Albemarle (x-Ferro) Great Lakes	Saytex 7010 Pyrocheck 68PB PBS 64, PBS 64HW	(brominated polystyrene structure) where: n>1 and x can be 1,2,3,4 or 5	?			x		x										
Albemarle	Saytex 8010	(structure)	Proprietary	x	x	x		x	x	x	x		x	x				
Albemarle, Dead Sea Bromine Great Lakes	Saytex HBCD GLCC CD-75	(hexabromocyclododecane structure)	3194-55-6		x			x	x									
Albemarle Great Lakes	Saytex HBCD-SF Saytex BC70 HS SP-75	(hexabromocyclododecane structure)	3194-55-6	x	x	x		x	x									

TABLE 4.10 *Continued*

COMPANY	TRADE NAME	STRUCTURE	CAS No.	ABS	HIPS	NYLON	POLYURETHANE	POLYESTERS	POLYOLEFINS	POLYSTYRENE	POLYCARBONATE	PVC	ELASTOMERS	UNSATURATED P	EPOXY RESINS	ADHESIVES & COATINGS	TEXTILES	TPE
Great Lakes Teijin Chemicals	BC58 Fireguard 8500		71342-77-3	X	X			X			X		X					
Great Lakes Teijin Chemicals	BC58 Fireguard 7500		94334-64-2															
Albemarle, Dead Sea Bromine, Great Lakes, Teijin Chemicals United Phosphorus Ltd.	Saytex RB100/ CP-2000 FR-1524 (TBBPA) BA-59P Fireguard 2000 TBBA B14		79-94-7	X	X			X			X X	X X			X X	X X		
Albemarle Great Lakes	Saytex 120 DE83R		58936-66-5					X	X X	X X								
Albemarle	Saytex BT-93, BT-93W		32588-76-4					X	X X	X X X	X		X	X X X	X X			

4.14

TABLE 4.10 *Continued*

COMPANY	TRADE NAME	STRUCTURE	CAS No.	ABS	HIPS	NYLON	POLYURETHANE	POLYESTERS	POLYOLEFINS	POLYSTYRENE	POLYCARBONATE	PVC	ELASTOMERS	UNSATURATED P	EPOXY RESINS	ADHESIVES & COATINGS	TEXTILES	TPE
Albemarle Dead Sea Bromine	Saytex HP 7010		88497-56-7			X		X										
Albemarle, Great Lakes Teijin Chemicals	Saytex HP-800 PE-68 Fireguard 3100		21850-44-2		X				X	X			X					
Dead Sea Bromine Great Lakes	FR-1208 79 DE		32536-52-0	X	X								X		X	X		
Dead Sea Bromine Great Lakes	FR 613 PH-73		118-79-6									X						
Dead Sea Bromine	FR513		3296-90-0				X											
Dead Sea Bromine	FR 522		36483-57-5				X											

TABLE 4.10 *Continued*

COMPANY	TRADE NAME	STRUCTURE	CAS No.	ABS	HIPS	NYLON	POLYURETHANE	POLYESTERS	POLYOLEFINS	POLYSTYRENE	POLYCARBONATE	PVC	ELASTOMERS	UNSATURATED P	EPOXY RESINS	ADHESIVES & COATINGS	TEXTILES	TPE
Dead Sea Bromine	FR 1025 FR1025M - monomer	Br-substituted benzyl acrylate ester $-[CH-CH_2]_n-$	59447-57-3			X		X		X								
Oxychem	Dechlorane Plus 515, 25, 35	Dodecachloro structure	13526-88-9	X			X	X	X	X	X	X		X			X	X
Unitex Chemical	Uniplex FRP 44-57	Halogenated phthalate ester	Proprietary									X						
Unitex Chemical	Uniplex FRP-45	Di-2-ethylhexyl tetrabromophthalate	26040-51-7									X						
Unitex Chemical	Uniplex BAP-370	Brominated alkyl phosphate	Proprietary									X						
DaiNippon Inc	EP-16	Brominated Epoxy oligomer	68928-70-1				X	X					X					

TABLE 4.10 *Continued*

COMPANY	TRADE NAME	STRUCTURE	CAS No.	ABS	HIPS	NYLON	POLYURETHANE	POLYESTERS	POLYOLEFINS	POLYSTYRENE	POLYCARBONATE	PVC	ELASTOMERS	UNSATURATED P	EPOXY RESINS	ADHESIVES & COATINGS	TEXTILES	TPE
Dead Sea Bromine	F-2200, F-2001, F-2016, F-2300, F-2300H, F-2400 E, F2400	Brominated Epoxy oligomer	68928-70-1	X				X					X					
Marubishi Yoka	Non-Nen 52								X									

TABLE 4.11 Properties of the Phthalate Ester Family of Additives

Properties	Tetrabromophthalic anhydride (PHT4)	Diol based on PHT4
Appearance	Crystalline powder-light tan	Viscous liquid–light brown
Bromine content (%)	68.2	46
Melting range (°C)	274–277	Liquid
TGA data		
5% weight loss (°C)	229	128
10% weight loss (°C)	242	166
50% weight loss (°C)	277	319
95% weight loss (°C)	297	380
Specific gravity (g/cc)	2.9	1.9

Derivatives of tetrabromobisphenol A include:

1. The bis(2,3-dibromopropyl ether) of tetrabromobisphenol A (PE-68 from GLCC or FR-720 from DSB), is an especially interesting molecule because of the presence of both aliphatic and aromatic bromine on the same molecule. PE-68 is the most effective flame retardant in polypropylene: One only needs 1.5 percent of PE-68 to pass the glow wire test. UL-94 V2 performance can be achieved with 3 percent PE-68 and 1.5 percent antimony trioxide as a synergist. The only disadvantage of PE-68 is its poor compatibility with polypropylene and therefore its tendency to plate out during molding and bloom upon aging.

    A variant of PE-68 based on tetrabromobisphenol S is also available from Japan (Nonen-52, Table 4.10). It has a similar profile to PE-68 and the two are mostly used interchangeably. Claims of less blooming of Nonen-52 compared to PE-68 in certain polypropylene formulations exist, but no general benefit can be seen.

2. Brominated oligomers. A series of materials based on brominated polycarbonate derive from tetrabromobisphenol A and are available at 52 and 58 percent bromine content. The higher brominated material is end-capped with tribromophenol, and sometimes it can have a lower thermal-stability profile. Main uses include polycarbonate and PBT applications for both filled and unfilled resins. Thermoset applications are also possible because the materials possess excellent thermal stability and high molecular weight.

3. Brominated epoxy oligomers (BEOs). BEOs are available in various molecular weights and are being used in thermoset resins or acrylonitrile-butadiene-styrene (ABS). Table 4.12 shows a list of various molecular weight brominated epoxies, along with their bromine content and major uses.

TABLE 4.12 Properties of Brominated Epoxy Oligomers

Commercial name	Molecular weight	Bromine content	Application
F-2001	Less than 1000	500 gr/eq	Thermosets reactive
F-2016	1600	50	ABS, HIPS, ETP
F-2300	3600	51	PBT, ABS, HIPS, PC/ABS
F-2400	50,000	53	PBT, PET, PC/ABS, ABS polyamides, TPU, HIPS
F-3014	1400	60 /Tribromophenol end capped	HIPS, ABS
F-3020	2000	56 /Tribromophenol end capped	HIPS, ABS
F-3516	1600	54 (partially capped)	HIPS, ABS

The brominated epoxy oligomer (EE = 500 g/eq) is recommended for use as a flame-retardant additive for various thermoset plastics such as epoxy, phenolic, and unsaturated polyester resins.

Original mechanical and physical properties are well maintained in the presence of the BEO. Its good UV stability and high styrene solubility make it a good choice for translucent applications in unsaturated polyesters.

Higher molecular weight oligomers (MW from 1600 to 3600) are available and are recommended for use in ABS, HIPS, and *engineering thermoplastics* (ETPs). A BEO with an average MW of 1600 is best for ABS because of its good combination of nonbloom, UV stability, and good thermal stability. Brominated epoxy polymers are also available (MW = 50,000) and can work best in high-temperature thermoplastics such as *polybutylene terephthalate* (PBT), *polyethylene terephthalate* (PET), and polyamides.

Tribromophenol end-capped epoxy oligomers are also available and overcome the problem of metal adhesion sometimes seen during lengthy molding cycles with traditional brominated epoxies.

Brominated Diphenyl Oxides (DPOs). The fully brominated compound decabromodiphenyl oxide is a high melting solid (>300°C) and is the major flame retardant used for HIPS. It is also used in polypropylene and polyethylene because of its cost-effectiveness and its ability to work well in filled systems. However, in polyolefins it tends to bloom at high concentrations and has very poor light-stability performance. Decabromodiphenyl oxide (deca) is the brominated organic compound that comes closest to being of use and effective in most polymer systems. Low-cost polyester formulations (filled PBT) can be formulated for UL-94 V-0 performance with decabromodiphenyl oxide. Thermoset materials can also be successfully formulated with deca. Antimony trioxide is always used as a synergist in deca formulations, and the preferred ratio is three parts of deca to one part of antimony trioxide (3:1 ratio).

Octabromodiphenyl oxide is a blend of brominated diphenyl oxides with an average bromine content of eight bromines. This product has a melting range of 70 to 150°C and is being used to flame retard ABS, along with antimony trioxide as a synergist. Because of rather poor UV stability and some environmental pressure on DPO-containing molecules, its use in ABS is declining. It has mostly been replaced by brominated epoxy oligomers because of their good balance of heat stability, UV stability, bloom resistance, and flame-retardant efficiency.

The lowest brominated diphenyl oxide available commercially is pentabromodiphenyl oxide, a very high viscosity liquid. Its commercial applications revolve around blends with phosphate esters to flame retard flexible polyurethane foams. Such products are DE-60F special, DE-61 from GLLC, or Fyrol PBr from Akzo. No antimony is needed in these applications. Its liquid form and high bromine content also make it a desirable choice in a variety of thermoset applications: epoxy or polyurethane-based resins.

The newest family of highly brominated aromatic compounds for use in styrenics and polyolefins while avoiding the DPO structure are the commercial products Saytex 8010 from Albemarle and Firemaster 2100 from GLCC. DPO-containing molecules are under environmental attack and scrutiny.

Tribromophenol Derivatives. The basic chemical tribromophenol is used as a reactive alcohol in unsaturated polyester applications, especially in pultrusion. The resulting allyl ether, tribromophenyl allyl ether, can be used in polystyrene (beads) or in other thermoplastic applications (PHE-65 by Great Lakes). By reacting tribromophenol with ethylene dibromide one can obtain a bis(tribromophenoxy) ethane; the commercial name for the product is FF-680. This compound, which has a melting point of approximately 225°C, is an excellent flame retardant for ABS in terms of thermal stability and efficacy. Its only drawback is bloom at high temperatures and dark colors.

The reaction product of tribromophenol with phosphorus trichloride is tri-dibromophenyl phosphate, a useful compound containing both phosphorus and bromine in the same molecule. This product (Reoflam PB-460) has been shown to have synergistic effects in oxygen-containing polymers such as polycarbonate or PET [10–13].

Tris(tribromophenyl) triazine (FR 245 from Dead Sea Bromine Group) contains 67 percent bromine, and the combination of aromatic bromine and cyanurate provides high flame-retarding efficacy and good thermal stability. Major uses of FR-245 are in HIPS and ABS, where it combines good UV stability, impact, and flow properties with a strong nonblooming performance.

Dibromostyrene Derivatives or Brominate Polystyrenes. Brominated polymers based basically on variants of polystyrene are available for use in high-temperature engineering resin applications.

The bulk of the commercially available materials are based on brominating polystyrene (Saytex 7010, PyroCheck 68PB, or FR-803P). Another approach (GLCC) is the polymerization of dibromostyrene, which yields a family of brominated polystyrene molecules with tailored molecular weight and bromine content. With this technology, it is possible to produce brominated polystyrenes with reactive end groups to allow better compatibility with polyamides and polyesters.

Grafted copolymers of dibromostyrene with polypropylene are also available (GPP-36 and GPP-39) and can be used in polypropylene applications where superior thermal stability is of importance (fiber and film applications) [14].

Brominated Alcohol Derivatives. Tribromoneopentyl alcohol (FR-513 from DSB) is a reactive flame retardant containing approximately 73 percent aliphatic bromine. It combines high bromine content with exceptional stability and is particularly suitable in applications where thermal, hydrolytic, and light stability are required. Its major uses include that of a reaction intermediate for the manufacture of brominated phosphate esters (PB-370) or as a reactive flame retardant for polyurethanes. Its high solubility in urethanes allows reaction of the single hydroxy functionality to form pendant urethane groups. Effective combination of physical properties and flammability performance can be achieved by the use of mixtures of FR-513 with difunctional brominated glycol (FR-522). The difunctional neopentyl glycol (dibromoneopentyl glycol-FR522 from DSB) contains 60 percent aliphatic bromine and is mainly used in thermoset polyester resins. Resins formulated with FR-522 are claimed to have high chemical and fire resistance, minimal thermal discoloration, and excellent light stability. The glycol is also suitable for use in rigid polyurethane because of its high bromine content and reactivity. As CFC-free foams systems are designed to meet more stringent standards of flame retardancy, FR-522 is a very useful and effective additive. The reaction product of tris-tribromoneopentyl alcohol with phosphorus oxychloride yields tribromoneopentyl phosphate ester, a very effective aliphatic brominated flame retardant (PB-370). The molecule is especially suited for polypropylene molding and polypropylene film and fiber applications. It is melt blendable, melts at 180°C, and has excellent UV stability. It has the best thermal stability of the commercially available aliphatic brominated materials and is the easiest to stabilize under UV light exposure (Table 4.13) [15]. Its flame-retarding efficacy is not as good as that of other brominated compounds (e.g., it is less effective in polypropylene than PE-68); however, it has better blooming performance than PE-68.

Specialty Brominated Compounds. Pentabromobenzyl acrylate (FR-1025) is a brominated flame-retardant monomer containing 71 percent bromine and comes in powder form. It is polymerizable via reactive extrusion processes and can participate in homo- and copolymerization reactions. UL-94 V-0 can be achieved with this acrylate in PBT, polycarbonate, and Nylon 6,6. It offers good thermal stability and excellent shelf life. It acts as a processing aid and improves the compatibility of the polymer matrix with the glass reinforcement.

TABLE 4.13 Color Change Upon UV Exposure in PP and Flame Retarded PP

	Color change ΔE PP control	Color change ΔE PP/aliphatic Br	Color change ΔE PP with aromatic Br
400 hrs	2.45	5.8	14.6
650 hrs	3.25	5.95	15.7
900 hrs	5.90	6.60	17.8
1150 hrs	2.3	7.60	20.1

Octabromo-phenyl-trimethylindan (FR-1808) is a 73 percent bromine-containing flame retardant very versatile in its range of usage. It can be used in styrenics and engineering thermoplastics and imparts good flow and impact properties along with effective flame retardance.

4.4.3 Antimony Trioxide

Antimony trioxide is a fine white powder with a melting point of 656°C, a specific gravity of 5.7, and an antimony metal content of approximately 83 percent. Fine powder of 1.5-μm average particle size is recommended for thermoplastic applications, because of the improved impact properties compared to larger particle size antimony grades. It has a high tinting strength and is satisfactory for white/opaque or pastel applications. A coarser powder of 2.5- to 3-μm size has low tinting strength and is used in PVC, PE film and sheet applications, and, generally, in thermoplastic products where dark tones are desired.

Concerning flame-retarding efficacy, antimony trioxide is synergistic with most halogenated compounds, allowing a decreased amount of halogenated compound to be used to achieve equivalent flame-retardant performance compared to using the halogenated compound on its own. Antimony oxide is nonvolatile, but antimony oxyhalide (SbOX) and antimony trihalide (SbX_3), which are formed in the condensed phase by reaction with the halogenated flame retardants, are volatile and facilitate transfer of the halogen and antimony into the gas phase where they function as radical scavengers [4]. Fig. 4.11 shows the reactions and the temperatures of formation of these antimony-halogen compounds. The temperature range from 250 to 600°C is covered and a continuous stream of radical scavengers is provided to quench the flame. There are some cases where antimony is not recommended as a synergist for the halogenated flame retardant, such as the following:

- Antimony produces smoke, and therefore its content must be minimized for smoke-sensitive applications. Usually, zinc borate or other zinc compounds can be used to reduce smoke.
- Antimony produces afterglow, and its level must be optimized in applications where the UL code or another similar code specifies time of afterglow in the flame-retarding tests. Again, both zinc borate and phosphorus additives tend to reduce afterglow.
- Antimony trioxide degrades the molecular weight and hence the mechanical properties of polycarbonate.
- Antimony trioxide interferes with the chain-scission mechanism of HBCD in polystyrene.
- In PET applications: antimony trioxide acts as a depolymerization catalyst. Sodium antimonate is used in its place to synergize the halogenated flame retardants.

$$Sb_2O_3 + 2RX \xrightarrow{250°C} 2SbOX + H_2O$$

$$5SbOX \xrightarrow{245-280°C} 2Sb_4O_5X_2 + SbX_3$$

$$4Sb_4O_5X_2 \xrightarrow{410-475°C} 5Sb_3O_4X + SbX_3$$

$$3Sb_3O_4X \xrightarrow{475-565°C} 4Sb_2O_3 + SbX_3$$

FIGURE 4.11 Antimony trioxide–halogen synergism: a schematic of key reactions.

TABLE 4.14 Optimum Ratio of Brominated Flame Retardant to Antimony Trioxide

Product	% Br	FR/Antimony Ratio
BA-59P	58.8	2.8:1
PE-68	67.7	2.43:1
PE-68 aliphatic	33.85	4.86:1
BC-52	51.3	3.21:1
BC-58	58.7	2.8:1
CD-75P	74.4	None
FF-680	70	2.35:1
DE-83R	83.3	1.97:1
DE-79	79.8	2.06:1
DE-71	70.8	2.32:1
PDBS 80	59	2.79:1
DP-45	45	3.66:1

Low-dusting, surface-treated antimony powders are available to allow easier handling. Antimony dusts readily and breathing its dust should be avoided. Light coatings of wetting agents such as mineral oil, diisododecyl phthalate, chlorinated paraffin, or other liquid are added to substantially reduce the dusting. Concentrates in thermoplastic carriers are also very popular because they eliminate powder dust and exposure issues. Polyethylene, polypropylene, polystyrene, and EVA concentrates are available in pellet form at active contents of approximately 80 to 90 percent antimony trioxide. Special colloidal dispersions (Azub, GLCC) are available to meet highly demanding applications.

Interference with the antimony halogen reaction will affect the flame retardancy of the polymer. Metal cations from color pigments or "inert" fillers such as calcium carbonate may lead to the formation of stable metal halides, rendering the halogen unavailable for reaction with the antimony oxide. This results in neither halogen nor antimony being transported into the gas phase [16, 17]. Silicones have also been known to interfere with the FR mechanism of halogen-antimony.

Colloidal antimony pentoxide is available in aqueous dispersion and in powder form. It is less tinting than the trioxide and is of great value in film and fiber applications. It is not clear if it has better efficacy than antimony trioxide. Studies suggest that it is marginally better or equivalent, calculated on antimony content. The pentoxide is considerably more expensive than the trioxide.

Table 4.14 summarizes the optimum ratio of brominated product to antimony trioxide for a variety of commercially available flame retardants.

Sodium antimonate is used in PET applications because the trioxide can act as a depolymerization catalyst. Particle size (2 to 5 µm depending on grade) and color are closely controlled to be compatible with the thermoplastic resin.

The issue of phosphorus and antimony interaction has been studied extensively [18, 19]. It is, however, dependent on the resin system utilized and the other additives present in the formulation. Fig. 4.12 gives an example of P-Sb interaction in a PVC plastisol formulation. Table 4.15 also shows the effect on *oxygen index* (OI) of a PVC formulation containing both P and Sb. For the concentration ranges studied, no antagonism could be detected by the oxygen index test. The antagonism can be seen sometimes as a lowering in OI when both P and Sb are present. However, once a critical load level is achieved this antagonism disappears. More discussion is provided in the PVC section.

The effect of the average particle size of antimony trioxide on the physical properties of HIPS [20] and ABS [21] is shown in Tables 4.16 and 4.17.

Decabromodiphenyl oxide and four different antimony trioxide samples were used for the HIPS study (Table 4.16). The smaller the particle size of antimony trioxide, the more effective it is as a flame retardant. Impact properties are also better with the smaller particle size antimony trioxide. However, this formulation had the highest tinting strength and therefore the highest whiteness. Table 4.17 summarizes the data for ABS flame retarded with tetrabromobisphenol A and various-sized

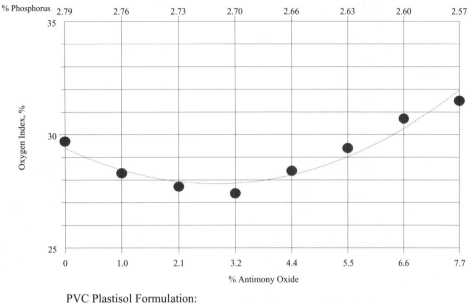

FIGURE 4.12 Antimony-phosphorus interactions in a PVC plastisol formulation.

antimony trioxides. Again, smallest particle size is optimum for flame-retarding efficacy and impact performance.

4.4.4 Phosphorus Additives

The acting principle in the use of most organophosphorus or inorganic phosphorus additives is the formation of a solid surface layer promoted by cross-linking reactions. Certain phosphorus additives such as TPP have been shown to have gas phase activity as well [22].

$$(NH_4PO_3)_n \longrightarrow (HPO_3)_n + n\ NH_3 \text{ (at } T > 250°C)$$

$$P \longrightarrow P_4O_{10} \longrightarrow (HPO_3)_n + H_2O$$

TABLE 4.15 Effect of Antimony and Phosphorus on the Oxygen Index of Flexible PVC

	PHR							
Plasticizer DOP	45	30	30	30	45	30	30	30
Alkyl diaryl phosphate	—	15	—	—	—	15	—	—
TAP	—	—	15	—	—	—	15	—
Chlorinated paraffin	—	—	—	15	—	—	—	15
Antimony trioxide	—	—	—	—	3	3	3	3
Oxygen Index	26.3	28.2	26.2	27.5	30.4	28.6	30.3	31.7

TABLE 4.16 Effect of Antimony Trioxide Particle Size on Physical Properties and Flammability in HIPS

Components	1	2	3	4	5	6
HIPS	100	87.4	83	83	83	83
DBDPO	—	12.6	12	12	12	12
ATO–0.3 microns	—	—	5	—	—	—
ATO–1.0 microns	—	—	—	5	—	—
ATO–2.5 microns	—	—	—	—	5	—
ATO–10.0 microns	—	—	—	—	—	5
Flammability						
Oxygen index (%)	18.3	19.3	25.7	25.2	24.9	24.2
UL-94 @ 1/8″	B	B	V-0	V-0	V-0	V-0
Physical properties						
Izod impact (ft-lb/in)	5.23	2.59	2.07	1.99	1.98	1.77
Tensile strength, psi	7678	6562	6498	6128	5919	5760
Elongation (%)	36.2	23.8	22.9	21.0	18.2	15.7
Flex strength, psi	7132	6665	6546	6584	6456	6343
Whiteness index	Std	24.65	30.71	29.3	29.0	27.5

Decabromodiphenyl oxide is the flame retardant used.

Formation of a protective layer occurs by the production of polyphosphoric acid and carbonization via the release of water

$$(HPO_3)_n + C_x(H_2O)_m \longrightarrow [\text{``C''}]_x + (HOP_3)_n \cdot mH_2O$$

The classes of phosphorus compounds most commonly used as flame retardants are as follows (Table 4.18):

(a) Chlorophosphates. *Trisdichloropropyl phosphate* (TDCP) is a neutral chloroalkyl phosphate ester with good thermal and hydrolytic stability. It contains 7.2 percent phosphorus and 49 percent chlorine and is a liquid at room temperature. It is water insoluble and is compatible with a broad range of polymeric systems to provide the required flame retardancy. It is recommended for use in both polyether- and polyester-based polyurethane foams. It is soluble in alcohols, ketones, chlorinated hydrocarbons, butyl acetate, toluene, and the lower glycol ethers. However, the scorch resistance of TDCP-containing foams is not as good as the scorch performance achieved with blends of brominated materials with phosphate esters.

TABLE 4.17 Effect of Antimony Trioxide Particle Size on Physical Properties and Flammability in ABS

Properties	ABS No FR	FR + ATO-2.5 micron	FR + ATO-1.5 micron	FR + ATO-0.3 micron
Tensile strength, psi	5200	4850	4800	4925
Tensile modulus × 10^5, Ksi	4.8	4.8	4.7	4.8
Flex strength, psi	9750	9025	9050	9275
Melt flow (g/10min)	3	8	10	8.8
DTUL (°C)	73	64	63	65
Izod impact (ft-lb/in)	6.06	1.83	2.15	2.96
Flammability				
UL-94 @ 1/8″	B	V-2	V-0	V-0
UL-94 @ 1/16″	B	V-2	V-2	V-2

TBBA is the flame retardant used.

TABLE 4.18 Non-Halogenated Flame Retardants. Manufacturers-Chemical Structure-Applications

COMPANY	TRADE NAME	STRUCTURE/COMPOSITION	CAS No.	ABS	PC	PC/ABS	POLYAMIDE	POLYURETHANE	POLYESTERS	POLYOLEFIN	POLYSTYRENE	PPO/HIPS	PVC	TPE	PHENOLIC RESINS	ADHESIVE/COATINGS	TEXTILES
Akzo	Akzo 701S	56%, 27% X=Br or Cl, 7% X=Br or Cl						X									
Akzo / Great Lakes / Rhodia	Phosflex 21P, 31P,41P,11P / Reofos 50, 65 / Antiblaze 519,521,524	[HOCH$_2$CH$_2$CH$_2$OCH$_2$CH$_2$CH$_2$O]$_3$—P ; x+y=3, x=0-3 ; Isopropylated triaryl phosphates	28108-99-8 (Phosflex 31P)					X				X	X	X	X	X	X
Akzo / Solutia / Bayer	Phosflex 362 / Santicizer 141 / Disflamoll DPO	2-ethyl hexyl diphenyl phosphate	1241-94-7										X			X	X

TABLE 4.18 *Continued*

COMPANY	TRADE NAME	STRUCTURE/COMPOSITION	CAS No.	ABS	PC	PC/ABS	POLYAMIDE	POLYURETHANE	POLYESTERS	POLYOLEFIN	POLYSTYRENE	PPO/HIPS	PVC	TPE	PHENOLIC RESINS	ADHESIVE/ COATINGS	TEXTILES
Akzo	Phosflex 370	blend of alkyl diaryl and triaryl phosphate esters											x			x	x
Akzo Solutia	Phosflex 390 Santicizer 148	Isodecyl diphenyl phosphate	29761-21-5								x		x				
Akzo	Phosflex 61B, Phosflex 71B	butylated triphenyl phosphate ester	56803-37-3										x	x			

TABLE 4.18 *Continued*

COMPANY	TRADE NAME	STRUCTURE/COMPOSITION	CAS No.	ABS	PC	PC/ABS	POLYAMIDE	POLYURETHANE	POLYESTERS	POLYOLEFIN	POLYSTYRENE	PPO/HIPS	PVC	TPE	PHENOLIC RESINS	ADHESIVE/ COATINGS	TEXTILES
Akzo Akzo Great Lakes Daihachi	Fyroflex RDP, Fyroflex RDP-B Reofos RDP CR 733, CR 735	Resorcinol Bis-(Diphenyl Phosphate)	57583-54-7		x	x	x	x	x		x		x				
Akzo Albemarle Great Lakes	Fyroflex BDP NCENDX P-30 Reofos BAPP	Bisphenol A Bis-(Diphenyl Phosphate)	181028-79-5		x	x						x	x				
Akzo Akzo Akzo Great Lakes Rhodia	Phosflex Lindol, Phosflex 179C, Phosflex 179 EG Disflammol TCP Reofos TCP Antiblaze TCP	Tricresyl Phosphate	1330-78-5					x	x				x	x		x	x
Akzo Great Lakes Rhodia	Phosflex 179A Reofos TXP, Antiblaze TXP	Trixylyl Phosphate	68952-33-0 and 25155-23-1		x				x				x	x		x	x
Akzo	Phosflex 90, 90S	Xylyl diphenyl phosphate							x				x	x		x	x

TABLE 4.18 Continued

COMPANY	TRADE NAME	STRUCTURE/COMPOSITION	CAS No.	ABS	PC	PC/ABS	POLYAMIDE	POLYURETHANE	POLYESTERS	POLYOLEFIN	POLYSTYRENE	PPO/HIPS	PVC	TPE	PHENOLIC RESINS	ADHESIVE/COATINGS	TEXTILES
Akzo, Bayer, Great Lakes, Rhodia	Phosflex TPP, Disflammol TP, Reofos TPP, Albrite TPP	Triphenyl phosphate	115-86-6			x						x	x		x	x	
Akzo, Great Lakes, Rhodia	Phosflex 4, Reofos TBP, Amgard TBPO4	Tributyl phosphate	126-71-6										x			x	x
Akzo, Daihachi, Great Lakes, Rhodia	Phosflex T-BEP, TBXP, Reomol KP-140, Amgard TBPE	Tributoxyethyl Phosphate	78-51-3									x	x	x		x	x
Akzo, Bayer, Great Lakes	Phosflex 112, Disflamol DPK, Santicizer 140, Reofos CDP	cresyl diphenyl phosphate	26444-49-5			x							x			x	
Ferro	Santicizer 142	octyl diphenyl phosphate											x				

TABLE 4.18 Continued

COMPANY	TRADE NAME	STRUCTURE/COMPOSITION	CAS No.	ABS	PC	PC/ABS	POLYAMIDE	POLYURETHANE	POLYESTERS	POLYOLEFIN	POLYSTYRENE	PPO/HIPS	PVC	TPE	PHENOLIC RESINS	ADHESIVE/ COATINGS	TEXTILES
Ferro	Santicizer 144	iso octyl diphenyl phospate											X				
Ferro	Santicizer 147	nonyl diphenyl phosphate											X				
Ferro	Santicizer 154	t-butylphenyl diphenyl phosphate	56803-37-3										X				
Ferro	Santicizer 2148	dodecyl diphenyl phosphate											X				
Daihachi Bayer Albright & Wilson	TOP Disflamoll TOF Amgard TOF	tri (2-ethyl hexyl phosphate)	78-42-2										X				
Daihachi	CR 741C	bisphenol A cresyl phosphate				X							X				

TABLE 4.18 *Continued*

COMPANY	TRADE NAME	STRUCTURE/COMPOSITION	CAS No.	ABS	PC	PC/ABS	POLYAMIDE	POLYURETHANE	POLYESTERS	POLYOLEFIN	POLYSTYRENE	PPO/HIPS	PVC	TPE	PHENOLIC RESINS	ADHESIVE/ COATINGS	TEXTILES
Unitex Chemical	Uniplex FRX 44-94S	Nitrogen and phosphorus based	Proprietary					X		X				X		X	
Rhodia	Antiblaze 1045	neutral cyclic diphosphate ester	Proprietary				X										
Akzo	Antiblaze 75																
Rhodia	Fyrol DMMP / Antiblase DMMP	dimethyl methylphosphonate	756-79-6					X	X								
Akzo	Fyrol 6	diethyl N,N bis[2-hydroxyethyl] aminomethylphosphonat	2781-11-5				X	X									
Rhodia	Antiblaze NR-25	neutral cyclic diphosphate ester															
Rhodia	Albrite PA-75	75% Phenyl acid phosphate in butanol															
Akzo	Fyrol 51	oligomeric phosphonate															X
CHLORINATED PHOSPHATES																	
Akzo	Fyrol FR-2, Fyrol 38, Antiblaze 195, Antiblaze																
Rhodia	TDCP(Amgard 610)	tri[B,B'dichlorisopropyl] phosphate	13674-87-8					X									
Akzo	Fyrol CEF																
Bayer	Disflamoll TCA	tri[B, chlorethyl] phosphate	115-96-8										X				
Akzo	Fyrol PCF																
Rhodia	Antiblaze 80 (TMCP)	tri[2, chloroisopropyl] phosphate	6145-73-9					X	X	X							
Akzo	Fyrol 25	oligomeric chloralkyl phosphate/phosphonate								X					X		
Akzo	Fyrol 99	oligomeric chloralkyl phosphate	109640-81-5				X										
Rhodia	Antiblaze 78	chlorinated phosphonate ester						X									
Rhodia	Antiblaze 100 (Amgard V-6)	chlorinated diphosphate ester	38051-10-4					X								X	
Rhodia	Antiblaze 125	chlorinated p hosphorus ester						X								X	X

(b) Ammonium Polyphosphate (APP). APP, a polymeric flame retardant is used in intumescent coatings and paints. In plastics, if formulated as the only flame-retardant additive in the formulation, it mostly has applications in thermoset resins such as epoxy and unsaturated polyester resins. APP has very good thermal stability and can be used in high-temperature applications. In polyurethane rigid foams, it also finds use when low smoke density and resistance to migration are required.

Microencapsulation with melamine produces a useful product for polyurethane foams (e.g., Exolit AP 462). Combined with pentaerythritol as a synergist, or another polyol, it can be used in polypropylene or polyethylene. These "fully formulated ammonium polyphosphate" (e.g., Exolit AP 750) compounds are a nonhalogen alternative for polyolefins, although the water retention of APP compounds is a serious detriment. Surface treatments have been developed to address this disadvantage and are being tested by the market. APP is also suitable for use in hot melt adhesives or coating applications.

(c) Phosphate esters. Phosphate esters are classified into three types, mainly, triaryl, alkyldiaryl, and trialkyl phosphates. Triaryl phosphates are further classified as natural, if they are produced from naturally derived alcohols (cresols and xylenols), or synthetic, if they are produced from synthetic alcohols (isopropyl phenol and *t*-butyl phenol). Monophosphates have been used mainly in PVC as flame-retardant plasticizers and up to the 1970s in PPO/HIPS blends. Bisphosphates (such as *resorcinol diphosphate* [RDP] and *bisphenol A diphosphate* [BAPP or BADP]) are mainly used in PC/ABS blends. These bisphosphates cause significantly less stress cracking in polycarbonate compared to TPP (a monophosphate) and their usage has expanded significantly in the last 5 years.

Phosphate esters are not very stable hydrolytically. RDP is less stable than BAPP. Phosphine oxides are much more stable thermally and hydrolytically than phosphate esters. The P-C bond is very stable and can have unique applications in nylons and polyesters.

(d) Phosphine Oxides and Derivatives. Phosphine oxide diols can be polymerized into polyesters [PBT, PET], polycarbonates, epoxy resins, and polyurethanes.

An alkyl bishydroxymethyl phosphine oxide may be in use today for epoxy circuit printed boards. Phosphine oxides are expensive and this significantly limits their application.

(e) Red Phosphorus. Red phosphorus can be a very effective flame retardant for plastics: it is effective at low levels and excellent mechanical properties of the matrix are maintained. For electrical and electronic components, red phosphorus could be the ideal flame retardant because of its good electrical properties—low surface conductivity and high tracking resistance. However, handling and processing safety concerns limit its use. Special retardation and stabilization by microencapsulation is necessary to produce a useful commercial additive. Resin concentrates of red phosphorus are also available in various resins (polyamide6,6, phenolic and epoxy resins, polyurethanes).The level of red phosphorus required to achieve UL-94 V-0 in various resins is given in Table 4.19.

(f) Phosphonates. Organophosphonate esters are used in various thermoplastic and thermoset applications. Products such as Antiblaze V490 are used in laminates and block foams. They are suitable for formulations employing CFC, HCFCs, water, or hydrocarbon blowing agents. Table 4.20 shows the typical properties of Antiblaze V490. Added into two-component foam systems, the phos-

TABLE 4.19 Red Phosphorus Concentration for UL-94 V-0 Rating

Polycarbonate	1.2%
Polyethylene terephthalate	3
Filled phenolic resin	3
Polyamide	7
Polyethylene	10
Polystyrene	15

TABLE 4.20 Properties of Organophosphate Ester (Antiblaze V490)

Appearance	Clear liquid
Viscosity @ 25°C	1.5 cps
Odor	Mild
Refractive index (25°C)	1.412–1.418
Flash point	195°F
Phosphorus content (%)	18.6%
Color	50 APHA
Acidity (mg KOH/g)	1.0 mg KOH/g
Water content	0.05%
Specific gravity (20°C)	1.025

phonate functions as a viscosity depressant, improving the processing characteristics of formulations utilizing water-blown or polyester polyols.

Dimethyl methyl phosphonate (DMMP) is used in rigid foam applications that allow the addition of the flame retardant to the foam components immediately prior to or during the manufacturing process. Its high phosphorus content (25 percent) makes it ideal for rigid polyurethane, PIR, or polyester hybrids using hydrocarbon blowing agents. In unsaturated polyester resins, it can be used as a replacement for inorganic flame retardants. Table 4.21 summarizes its physical properties.

Cyclic phosphonates (such as Antiblaze N for textiles or Antiblaze 1045 for plastic applications) are high-phosphorus-content, glass-type liquid additives (21 percent P; Table 4.22). These cyclic phosphonates can be used in epoxies, unfilled thermoplastic polyesters, polycarbonates, and niche polyamides. Reinforced epoxy composites with good retention of mechanical properties can be achieved with 8 to 15 percent loadings depending on resins and curing agents. Flame-retardant PET can be produced with 3 to 5 percent addition levels.

Table 4.23 shows the physical and thermal properties of PET flame retarded with Antiblaze 1045 compared to the unmodified PET starting material. This additive plasticizes PET by increasing chain mobility in both the melt and solid state. Glass transition temperature is lowered by 6°C and heat distortion temperature by 4°C with the addition of 3.5 percent Antiblaze 1045. Molecular weight reduction in PET occurs because of acid-catalyzed reactions by the additive. In nylons, high elongation can be maintained with this cyclic phosphonate ester: 13 to 15 percent is required in Nylon 6 to achieve UL-94 V-0, and in Nylon 6,6, 17 to 19 percent is necessary.

TABLE 4.21 Properties of Dimethyl Methyl Phosphonate (DMMP)

Appearance	Clear liquid
Viscosity @ 25°C	2.0 cps
Odor	Mild
Refractive index (25°C)	1.410–1.416
Flash point	156°F
Phosphorus content (%)	25%
Solubility in water at 20°C (w/w)%	Miscible
Color	25 APHA
Acidity (mg KOH/g)	1.0 mg KOH/g
Water content	0.05%
Specific gravity (20°C)	1.17

TABLE 4.22 Typical Properties of Cyclic Phosphonate (Antiblaze 1045)

Appearance	Glass type liquid
Phosphorus content	20.8%
Acid number, ng KOH/gm	<2
Density	1.25 g/cc
Flash point, closed cup	350°F
Viscosity:	
25°C	1,500,000 cPs
60°C	6,700 cPs
100°C	2,000 cPs
Solubility:	
Water:	Miscible
Acetone	Miscible
Ethanol	Miscible
Pentane	<5
Toluene	>10
Volatility: 200°C	<1
TGA Loss 250°C	2
300°C	7
350°C	13

(g) Ethylene Diamine Phosphate (EDAP). EDAP is an important additive for achieving halogen-free flame retardancy in polyolefins. It is a neutral compound in aqueous systems and is compatible with coadditives used in curable systems such as urethanes and epoxies. Its major disadvantage is its thermal stability—it decomposes at temperatures above 250°C and in practical applications temperatures have to be kept close to 200°C. Table 4.24 shows the physical properties of EDAP. Table 4.25 shows the properties of formulated PP and PE with EDAP.

(h) Melamine and Derivatives. Liebig first discovered melamine in 1834, but its chemistry was not understood and explored until the 1930s. Its chemical name is 2,4,6-triamino-1,3,5 triazine. It is a white crystalline powder with a melting point of 354°C (670°F) and a density of 1.573 g/cc. At its melting point, melamine vaporizes (sublimes) rather than goes through a traditional melt phase transition. It dissociates endothermically, absorbing a significant amount of heat (470 kcal/mol) and

TABLE 4.23 Physical and Thermal Properties of Flame Retardant PET With a Cyclic Phosphonate (Antiblaze 1040)

Property	PET	FR Pet
Antiblaze 1040 (wt%)	0	3.5
UL-94, 1/16"	V-2	V-0
Break strength (Mpa)	48	42
Elongation (%)	9	7
Notched izod (J/m)	59	48
DTUL (°C)	67	63
Tg (°C)	76	71
Tm (°C)	253	250
Crystallinity (%)	48	45

TABLE 4.24 Physical Properties of Ethylene Diamine Phosphate

Appearance	White, free flowing powder
Specific gravity (g/cc)	1.22 g/cc
Bulk density	0.63 g/cc
Phosphorus content	19.6%
H_3PO_4 content	62% by weight
Melting point	Decomposes above 250°C
Particle size	15 microns
TGA data	
250°C	2% weight loss
300°C	15% weight loss
400°C	30% weight loss

acting as a heat sink in fire situations. Its vaporization during a fire dilutes the fuel gases and oxygen near the combustion source, and this further contributes to flame retardancy. Melamine seldom works on its own as a flame-retardant additive. It is usually mixed with another flame retardant that will promote the formation of char or participates in a gas phase reaction to scavenge free radicals. Its low solubility in water and most other solvents, its UV absorption above 250 nm, and its pH of 8.1 are also desirable attributes that promote its use in plastic applications as a flame retardant. When it is used as a flame retardant in polyurethane foams, it becomes the blowing agent for foam. Commercial producers of melamine are shown in Table 4.3.

Melamine Derivatives. Melamine has a pH of 8.1, making it a weak base and allowing it to form stable salts with most organic and inorganic acids. Many of these salts, such as those with boric acid, phosphoric acid, orthophosphoric acid, polyphosphoric acid, cyanuric acid, and sulfuric acid are either commercially viable flame retardants or have the potential to become such [23]. Melamine cyanurate is in use in thermoplastic polyurethanes, unreinforced polyamides, epoxies, and polyesters. Melamine phosphate is used in conjunction with other organophosphorus compounds in olefins and thermoplastic urethanes. Melamine polyphosphate has recently become available (Melapur MP200 from DSM) and can be used in reinforced polyamides and polyesters, epoxies, and thermoplastic urethanes. Fig. 4.14 shows the UV stability achieved with MP200 versus a brominated flame retardant in Nylon 6,6.

TABLE 4.25 Physical Properties of Polypropylene and Polyethylene Flame Retarded with Ethylene Diamine Phosphate (EDAP)

Polymer		PP	FR PP	LDPE	FR LDPE
EDAP–wt%		None	35	None	35
Property	Test method				
UL-94, 1/16″	UL94	No rating	V-0	No rating	V-0
Density (g/cc)	ASTM D1505	0.9	1.02	0.91	1.04
MFI	ASTM D1238	4.0	1.9	8.0	4.0
Break strength, psi	ASTM D638	2800	2500	1300	1100
Yield strength, psi	ASTM D638	4500	3000	1100	900
Elongation, %	ASTM D638	60	110	200	160
IZOD, notched ft-lb/in	ASTM D256	0.8	1.2	3.1	3.5
DTUL, 66 psi, °C	ASTM D644	108	122	43	56

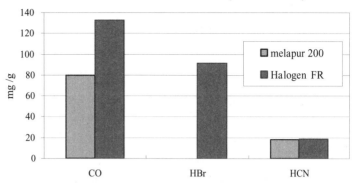

FIGURE 4.13 Comparison of emissions of melamine polyphosphate and a brominated flame retardant in Nylon 6,6. (From Ref. 23.)

Melamine Cyanurate. Melamine cyanurate is a salt of melamine and cyanuric acid. Above 320°C it undergoes endothermic decomposition to melamine and cyanuric acid, acting as a heat sink. The vaporized melamine acts as an inert gas diluting the combustible gases and the oxygen at the combustion zone. The cyanuric acid decomposes the polymer and rapidly removes it from the ignition source. Three distinct mechanisms of flame retardancy are in action with melamine cyanurate.

Melamine Polyphosphate. Melamine polyphosphate is a salt of melamine and polyphosphoric acid. It decomposes at 350°C and the phosphoric acid evolved coats and shields the combustible polymer from the heat source as a glassy surface. Melamine in the gas phase dilutes the oxygen and the combustion gases. The amount of gases evolved during combustion of polymers flame retarded with melamine derivatives is significantly lower than that produced when other types of flame retardants are used. Fig. 4.13 shows a comparison of gases evolved during combustion of Nylon 6,6 with various flame-retardant systems representing different mechanisms. The melamine compound used was Melapur 200—a melamine polyphosphate. The mechanical properties, flammability performance, and electrical properties of Nylon 6 or 6,6 flame retarded with melamine phosphate or melamine polyphosphate are summarized in the section devoted to nylon materials.

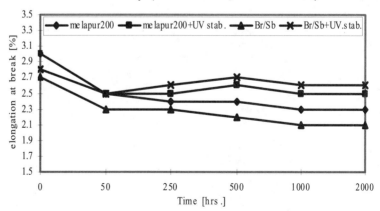

FIGURE 4.14 Comparison of melamine polyphosphate and a brominated flame retardant on the UV stability of Nylon 6,6. (From Ref. 23.)

4.4.5 Intumescent Flame-Retardant Systems

Intumescence is a mechanism of flame retarding a material by causing the material to foam and create an insulating barrier when exposed to heat.

The acting principle during the use of an intumescent material is the formation of a voluminous, insulating protective layer and simultaneous foaming.

A typical intumescent composition consists of

1. Carbon donors—charring agent (e.g., polyalcohols such as starch, pentaerythritol)
2. Acid donors—catalyst (ammonium polyphosphate, usually a P compound)
3. Propellant—blowing agent (melamine or other compound decomposing to yield CO_2 or NH_3).

The phosphorus compounds are expected to catalyze dehydration; however, their true action is multifunctional, as they also become active ingredients of the char structure.

APP-Pentaerythritol System or EDAP-Pentaerythritol System. In these intumescent systems mainly used in polypropylene or special coatings, several key processes occur that result in the formation of a strong char: decomposition of APP or EDAP with release of water and ammonia, phosphorylation of pentaerythritol and thermo-oxidized PP, dehydration, dephosphorylation, cross-linking, carbonization, and formation of a char structure. The blowing agent vaporizes and decomposes to noncombustible gases. The formed char is being swollen by the gases to create a foam structure. Commercial materials based on APP are available as Exolit 750 by Clariant or FR Cross 480 by Budenheim. EDAP fully formulated is available as Amgard NP from Rhodia. A new proprietary system has been recently commercialized by GLCC—Reogard 1000—exploiting similar intumescence-forming reactions through a combination of P-N action.

4.5 POLYMER FAMILIES—SELECTION OF FLAME RETARDANT

Table 4.26 summarizes the mechanism of action of flame retardants used in various polymers. Table 4.9 summarizes the ignition temperatures of various polymers and Table 4.27, the oxygen index characteristics of various thermoplastic and thermoset materials [9]. Table 4.28 shows the improvement in oxygen index achieved with flame retardants incorporated in various polymer systems.

TABLE 4.26 Mechanism of Action of Flame Retardants for Various Polymers

Polymer	Heat sink	Condensed phase	Vapor phase
Polyolens	X	Minor	X
Styrenics			X
Nylon		X	X
PC		X	X
PC/ABS		X	
PBT & PET			X
PPO/HIPS		X	
PVC		X	X
Polyurethane foam	X	X	X
Epoxy			X
Unsat polyester	X		X

TABLE 4.27 Oxygen Index for Some Polymers

Material	Oxygen index range, %
Polyethylene	17.3
Polypropylene	17.0
Polybutadiene	18.3
Chlorinated polyethylene	21.1
Polyvinyl chloride	44
Polyvinyl alcohol	22.5
Polystyrene	17.0
Styrene-acrylonitrile (SAN)	18
Acrylonitrile-butadiene styrene (ABS)	18
Acrylic	17
Polyethylene terephthalate (PET)	20
Polybutelene terephthalate (PBT)	24
Polycarbonate	21
Liquid crystal polymer (LCP)	35
Polyphenylene ether/oxide (PPE/PPO)	24
Nylon 6	23
Nylon 6,6	21
Nylon 6,10	25
Nylon 6,12	25
Polyimide	36.5
Polyamide-imide	43
Polyetherimide	47
Polybenzimidazole	41
Polyacetal	15
Polysulfone	30
Polyethersulfone	37
Polytetrafluoroethylene (PTFE)	95
Polyvinyl fluoride (PVF)	22.6
Polyvinylidene fluoride (PVDF)	43.7
Fluorinated ethylene propylene (FEP)	95
Polychlorotrichloroethylene (CTFE)	83
Ethylene-tetrafluoroethylene (ETFE)	30
Ethylene chlorotrifluoroethylene (ECTFE)	60
Cellulose acetate	16.8
Cellulose butyrate	18.8
Cellulose acetate butyrate	18
Phenolic	18
Epoxy	18
Polyester unsaturated	20
Polybutadiene, rubber	17.1
Styrene-butadiene rubber (SBR)	17–19
Polychloroprene, rubber	26.3
Chlorosulfonated polyethylene, rubber	25.1
Silicone, rubber	26–39
Natural rubber	17.2
Wood	22.4–24.6
Cardboard	24.7
Fiber board, particle board	22.1–24.5
Plywood	25.4

TABLE 4.28 Performance Improvement Achieved with Flame Retardants as Demonstrated by LOI

Base polymer	Commercial abbreviation	Limiting oxygen index of base polymer (LOI–%)	Limiting oxygen index of V-0 grades
Acrylonitrile butadiene styrene	ABS	—	31
Polystyrene	PS	18	26
Polyketone	PK	20	35
Polybutelene terephthalate	PBT	22	29–36
Polyamide	PA	24.5	28
Polyphenylene ether	PPE/PPO	28	37 (PPO/HIPS blend)
Polycarbonate	PC	29	35
Polysulfone	PSU	29.5	36 (V-1)
Polyaryletherketone	PAEK	37	—
Polyethersulfone	PES	38	45

4.5.1 Polypropylene

Halogenated Additives. For electrical and electronic applications (outside the wire and cable market) polypropylene is flame retarded to meet the Underwriters Laboratories (UL-94) specification of V-2 or V-0. Table 4.29 summarizes these specifications. A V-2 product drips and ignites the cotton, while a V-0 product either does not drip or drips but does not ignite the cotton.

For V-2 polypropylene applications the most efficient flame retardant available today is PE-68 or Nonen52 or chemically similar molecules. A load level of 3 to 4 percent is adequate to achieve this rating along with antimony trioxide (1 to 2 percent). The physical properties of polypropylene are not affected significantly because of the low addition levels. PE-68 or its equivalents act as a plasticizer for polypropylene, increasing the melt flow significantly and providing external lubrication during processing. They are not, however, compatible with polypropylene, and tend to phase separate during either molding (plate out) or storage (blooming). This blooming tendency has limited the applications especially in aesthetically critical areas and in dark colors.

Another option for achieving a V-2 rating is PB-370, an aliphatic brominated phosphate ester. Levels of use are 4 to 5 percent with 2 to 3 percent antimony trioxide. Table 4.30 summarizes the flammability and physical properties for various melt flow polypropylenes flame retarded with PB-370. PB-370 is generally more compatible than PE-68 in polypropylene, and blooms only under elevated-temperature aging conditions. Being an aliphatic brominated material, it has the best light stability for polypropylene applications using halogenated flame retardants. Table 4.13 compares the light stability performance versus an aromatic brominated material. Film and fiber applications are a market area where the combined properties of PB-370 make it attractive.

Dripping V-0 polypropylene materials can be formulated with 8 to 10 percent PE-68 or equivalent molecules, and 4 to 5 percent antimony trioxide. The blooming is exacerbated at these high addition levels and several patents exist on incorporation of compatibilizing agents to reduce blooming. No third additive offers a real solution to the problem while maintaining the flame-retarding efficacy of PE-68. As an example, grafted polypropylenes with polar groups, stearates, and esters have been

TABLE 4.29 UL-94 V-0/V-1/V-2 Specifications

Rating	V-0	V-1	V-2
Max individual burn time (sec)	<10	<30	30
Total burn time/5 specimens (sec)	<50	<250	<250
Glow time after second ignition (sec)	<30	<60	<60
Ignites cotton	No	No	Yes

TABLE 4.30 Typical Properties of Injection Molded V-2 Polypropylene with Brominated Phosphate Ester

PP resin (MI = 12)	100	95.5	—	—	—	—
PP resin (MI = 4)	—	—	100	92.5	—	—
PP resin (MI = 30)	—	—	—	—	100	92.5
Brominated phosphate	—	3	—	5	—	5
Antimony trioxide	—	1.5	—	2.5	—	2.5
Flammability						
UL94 @ 1.6 mm	HB	V-2	HB	V-2	HB	V-2
Oxygen index	18	23.6	19	24.5	19	26
Physical properties						
Tensile yield strength (Mpa)	35	33	32.7	32.7	37.5	35.8
Elongation (%)	11	10	11	9	16	10.5
Young's modulus (Mpa)	1,550	1,470	1,452	1,578	1,859	1,872
Flex modulus (Mpa)	2,130	2,050	1,749	1,962	2,190	2,228
Notched izod (J)	24	19	29	16	26	16
Gardner impact (J)	15.7	14.8	30	25	6.3	9.4

proposed and shown to offer some limited help. Blends of PB-370 with PE-68 have also been investigated as a way to balance the properties of V-0 polypropylene.

In order to achieve nondripping V-0 rating, filler type brominated flame retardants are the best. The most cost-effective and traditionally used in both filled and unfilled applications is decabromodiphenyl oxide (DE-83R by GLCC, Saytex 102 by Albemarle). Levels of 20 to 30 percent of deca are needed, depending on the polypropylene melt flow and the levels of fillers added. Antimony trioxide at a ratio of 3:1 is also being used. The particle size of deca and dispersion of both the solid flame retardant and antimony trioxide are critical in obtaining as good physical properties as possible. Addition of talc has been shown to improve heat release and smoke generation in deca formulations. Adequate light/UV stability with deca formulations is very difficult to obtain, regardless of type and level of additives used.

A similar formulation approach is the addition of decabromodiphenyl ethane (Saytex 8010 by Albemarle, also known under the chemical name ethylene bis-pentabromobenzene). Levels of addition are again 20 to 30 percent, and the main advantage is non-DPO-containing formulation. Light stability improves considerably over similar decabromodiphenyl oxide formulations still having the challenges of an aromatic brominated additive. Table 4.31 summarizes V-2 PP formulations of

TABLE 4.31 Formulations and Properties of UL-94 V-2 Unfilled Polypropylene

	Units	DCBDPO	DCBDPE	HBCD
Polypropylene (Profax 6523)	Wt%	89	89	96
FR	Wt%	8	8	3.2
Sb_2O_3	Wt%	3	3	1
Tensile strength at yield	Mpa	20	29.6	—
% Elongation at yield	%	5.5	6.1	—
Izod impact	J/m	42.7	37.4	32
Gardner impact	J/m	205	214	—
HDT	OC	54	53	—
Melt index	G/10 m	5.7	4.8	5.3
300 hrs Xenon arc	DE	15.1	11	—
UL-94	3.2 mm	V-2	V-2	V-2
UL-94	1.6 mm	Fail	V-2	V-2
Oxygen index	% O_2	24.3	25.1	—

TABLE 4.32 Formulations and Properties of UL-94 V-0 Unfilled and Talc Filled Polypropylene

	Units	Virgin PP	DCBDPO	PE-68	HBCD/DCBDPO
Polypropylene (Profax 6523)	Wt%	100	58	90.5	86
FR	Wt%	—	22	9	3/7
Sb_2O_3	Wt%	—	6	3	3
Talc	Wt%	—	14	—	—
Tensile strength at yield	Mpa	32	26	20.7	20
% Elongation at yield	%	11.6	4	11.3	10.4
Izod impact	J/m	32	21.4	16	16
Melt index	G/10 m	4.1	4.6	26	34.3
UL-94	3.2 mm	Burn	V-0	V-0	V-0
UL-94	1.6 mm	Burn	V-0	V-0	V-0
Oxygen index	% O_2	17.8	26.3	28	28.1

unfilled polypropylene with decabromodiphenyl oxide or decabromodiphenyl ethane. Table 4.32 compares V-0 polypropylene formulations achieved with HBCD, deca, and Saytex 8010.

Ethylene bistetrabromophthalimide is also an option in polypropylene, although its efficacy is not as good as that of the decabromodiphenyl oxide or ethane additives. Table 4.33 summarizes the properties of V-0 filled polypropylene with deca, Saytex 8010, or Saytex BT-93. The use of HBCD could present processing stability challenges.

A different approach is possible by using an alicyclic chlorinated flame retardant, such as Dechlorane Plus. Typical levels would be 25 to 30 percent and the ratio of flame retardant to antimony at 2:1. A general concern with all the polypropylene formulations containing a high level of antimony trioxide (more than 10 percent) is the long afterglow times during the UL-94 test. Efforts to substitute part of the antimony trioxide synergist with zinc borate are quite effective in reducing afterglow. Typical formulations have 30 to 50 percent of the total synergist being zinc borate, with the balance still being antimony trioxide. Impact properties can be affected adversely by zinc borate, and a careful balance of flame retardancy with physical properties must always be maintained (Table 4.34).

Nonhalogenated Additives. Although halogenated additives offer the most cost-effective flame retardancy of polypropylene, there are areas where nonhalogen technologies are mandated because

TABLE 4.33 Formulations and Properties of UL-94 V-0 Filled Polypropylene

	Units	DCBDPO	DCBDPE	ETBP
Polypropylene (Profax 6523)	Wt%	58	58	53
FR	Wt%	22	22	27
Sb_2O_3	Wt%	6	6	6
Talc	Wt%	14	14	14
Tensile strength at yield	Mpa	26.2	27.6	26.9
% Elongation at yield	%	4.0	3.1	4.1
Izod impact	J/m	21.4	21.4	21.4
Gardner impact	J/m	120	107	107
HDT	OC	71	67	66
Melt index	G/10 m	4.6	4.1	3.9
300 hrs Xenon arc	DE	15.1	10.0	6.6
Dielectric constant	MHz	2.37	2.40	2.43
UL-94	3.2 mm	V-0	V-0	V-0
UL-94	1.6 mm	V-0	V-0	V-0
Oxygen index	% O_2	26.3	25.8	25.3

TABLE 4.34 FR-Polypropylene with Dechlorane Plus and Various Synergists

Formulation (weight %)	1	2	3
Polypropylene	55	61	62
Dechlorane plus	35	30	28
ATO	4	6	6
Zinc borate	6	4	2
SFR-100 (Silicone FR)	—	1	2
Flammability			
UL-94 @ 3.2mm (1/8″)	V-0	V-0	V-0
@ 1.6mm (1/16″)	V-0	V-0	V-0
Tensile elongation (%)	25	23	40

of either performance requirements or consumer preferences. In the wire and cable arena, smoke generation and corrosion issues present a barrier to the use of halogenated materials. Therefore, alumina trihydrate or magnesium hydroxide are used exclusively in polypropylene wire and cable applications. Levels of addition range from 40 to 60 percent hydrated mineral, creating a rather nonplastic polypropylene with poor impact and elongation. Dispersing agents and specially coated minerals are widely used to improve wetting and dispersion of the filler and allow some improved properties. An example is given in Ref. 7 of an organosilicone additive by Union Carbide/Dow. Good electrical properties can also be achieved with ATH, and this positions it as the largest single use of hydrated minerals.

Another approach to achieve a nonhalogenated polypropylene material is the formation of an intumescent char, via combination of polyol, an acid catalyst, and a blowing agent. The intumescent char insulates the polymer from the heat source and does not allow combustible volatiles to escape; in this way, it has a flame-retarding effect with much reduced heat release and smoke evolution.

Commercially available intumescent materials rely on melamine on nitrogen compounds to produce the blowing agent/gas, on phosphates to play the role of the acid catalyst, and on pentaerythritol or some derivative thereof as a polyol source.

Traditionally, ammonium polyphosphate combined with pentaerythritol (fully formulated compound) has been used to achieve V-0 (nondripping) in unfilled polypropylene. Addition levels of 30 percent are required; the compound is very thermally stable and the additive acts as a filler. The main disadvantage of APP is its poor hydrolytic stability. Sometimes, special compounding must be devised to protect APP from leaching into the extruder water bath. Newer materials have been developed with surface coatings to minimize hydrolysis and their use is being tested by the market.

EDAP combined again with pentaerythritol or other similar compound is another traditionally available material for use in thermoplastics. It is a melt blendable additive compared to the filler-type APP; however, it has limited thermal stability. Improved stability grades are recently available and are being utilized in niche applications.

A new material has been developed for use in nonhalogen PP applications, Reogard 1000. Its details are proprietary, but it is based again on a combination of phosphorus and nitrogen to achieve V-0 at levels of 20 to 25 percent in unfilled polypropylene. It is a melt blendable additive with excellent physical properties and strong FR performance, suited for wire, cable, and building markets. Good hydrolytic stability and electrical properties allow use in wire and cable applications.

It is a white powder additive, which melts at 190 to 200°C, allowing for ease of processing because of its melt blendability in polypropylene. Table 4.35 gives the typical properties for the product showing the low density of the product.

The main technical advantages of Reogard 1000 over the other commercial halogen-free system is its superior water resistance, heat distortion temperature, and impact performance. Compared with the brominated compounds, it offers bloom resistance, nondripping UL-94 V-0, and reduced compound specific gravity, which results in less polymer required to fill a given mold cavity. The superior heat

TABLE 4.35 Typical Properties of Reogard 1000

Typical properties	
Appearance	White powder
Melt range, °C	190–200
Bulk loose density @ 25°C, g/ml	0.6
Bulk packed density @ 25°C, g/ml	0.8

Thermogravimetric analysis (10mg @ 10°C/minute under N_2)			
Weight loss, %	5	10	25
Temperature, °C	274	310	390

Solubility (g/100g solvent @ 20°C)			
Water	1	Toluene	Insoluble
Methylene Chloride	Insoluble	Methyl Ethyl Ketone	1
Methanol	2	Acetone	2
Hexane	<0.5		

distortion temperature, without the need for a talc reinforcing filler, will allow polypropylene compounds flame retarded with Reogard 1000 to be considered for applications where previously only filled grades could be considered. The additional benefit of maintaining the living hinge capability of the polypropylene while meeting the UL-94-V-0 rating allows its use in areas such as complex housings/chassis with snap fit fastenings. Table 4.36 summarizes a comparison of V-0 polypropylene formulations with PE-68, deca, Exolit 750 (fully formulated APP) and Reogard 1000.

TABLE 4.36 Comparative Performance Data of UL-94 V-0 Polypropylene Halogen and Nonhalogen Additives

Formulation	Control	Reogard 1000	Exolit AP-750	PE-68	DE-83R
Polypropylene Profax 6524	100	80	70	87	59
Flame retardant level, %	—	20	30	10	20
Sb_2O_3, TMS HP, %	—	—	—	3	7
Talc filler	—	—	—	—	14
Flammability performance					
UL94 @ 1.6mm	Fail	V-0	V-0	V-0 (Drips)	V-0
Physical properties					
Specific gravity	0.89	0.99	1.06	0.97	1.28
Bloom, 168 hrs @ 70°C	None	None	None	Severe bloom	Mod. bloom
Water absorption, %	<0.1	<0.1	0.4	<0.1	<0.1
Izod impact, unnotched, J/m	>1000	595	425	>1000	390
Izod impact, notched, J/m	48	37	27	32	27
Tensile strength, Mpa	33	30	30	33	28
Elongation @ break, %	450	110	100	300	88
Flexural strength, Mpa	48	48	39	50	
Flexural modulus, GPa	1.4	1.9	1.9	1.2	2.3
HDT @ 0.46 MPa, °C	84	110	93	91	120

Polymer grade: Profax® 6524, MFI = 4, homopolymer.

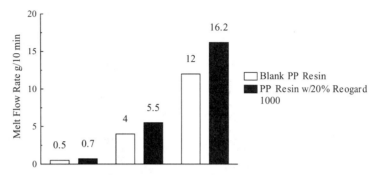

FIGURE 4.15 Melt flow index changes in polypropylene with Reogard 1000.

Reogard 1000 increases the melt flow of the polymer compounds: addition of 20 percent of Reogard 1000 gives approximately a 35 percent increase in the melt flow rate of the base polymer as shown in Fig. 4.15.

Existing halogen-free polyolefin cable formulations based on magnesium hydroxide as a flame-retardant system do give delayed ignition and reduced heat release on the cone calorimeter, mainly as a function of the reduced flammable polymer concentration in the final compounds. Figs. 4.16 and 4.17 give comparative heat release and smoke optical density values, respectively, for a polypropylene compound with Reogard 1000 versus a commercial magnesium-hydroxide-filled polypropylene cable compound and a PVC plenum cable compound [24].

A totally different chemistry has been found to give good flame-retarding properties in polypropylene for fiber applications or V-2 applications. It is based on radical scavenging chemistry by NOR molecules and was developed by Ciba [25]. The additive is used as a synergist with halogen (deca or PB-370) for maximum efficacy.

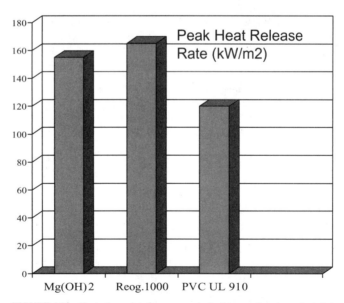

FIGURE 4.16 Heat release data from cone calorimeter experiments conducted at 50 kW/m². Comparison of halogen-free polypropylene formulations versus PVC formulation passing the UL-910 test protocol.

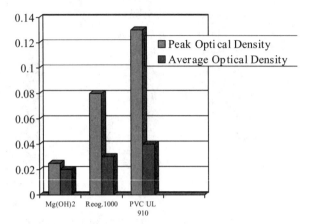

FIGURE 4.17 Smoke evolution from cone calorimeter experiments conducted at 50 kW/m². Comparison of halogen-free polypropylene formulations versus PVC formulation passing the UL-910 test protocol.

4.5.2 Polyethylene

The major application for flame-retardant polyethylene is for electrical and electronic applications, as well as films for packaging/building and construction requirements.

Most of the market is serviced by deca because of its cost performance situation. When light stability and mechanical properties in ultrathin sections become an issue, BT-93 is the product of choice.

Despite the chemical similarity of polypropylene and polyethylene, the degradation and pyrolysis pathways are distinctly different for these two polymers. Therefore, in both stabilization and flame retardancy different chemicals have to be utilized. Pyrolysis temperatures for PE versus PP are very different (Table 4.9), dictating flame retardants that will decompose at different temperatures to be effective. Decabromodiphenyl oxide is a notable exception. This molecule has excellent thermal stability and its range of FR activity spans from polyolefins to styrenics without any compromise. However, it is a solid-filler-type additive and does contain diphenyl oxide. These two characteristics sometimes limit its applications.

The decabromodiphenyl ethane (Saytex 8010, for example) or a proprietary structure from Great Lakes (Firemaster 2100) are viable alternatives in PE for those deca applications where light stability or DPO content are the worry. It depends on the particular formulation at hand and the light stability requirements if Saytex 8010 can meet some of the applications where BT-93 is currently the additive of choice.

Chlorinated paraffins are a low-cost option to flame retard polyethylene; Table 4.37 summarizes such formulations. A comparison of LDPE and HDPE formulations with decabromodiphenyl oxide is shown in Table 4.38.

4.5.3 Styrenics

Crystal Polystyrene. The term is used to describe polystyrene that has not been impact modified. It is almost a misnomer, because polystyrene is an amorphous material with a glass transition temperature of about 100°C.

The big and probably only application for flame-retardant polystyrene is in polystyrene foam for building insulation. Such material can be produced via a reactor process (from expanded polystyrene beads [EPS]) or via a direct extrusion process where a blowing agent is injected into the extruder where thermoplastic polystyrene is fed (XPS foam). The qualifying specification for most applica-

TABLE 4.37 Application of Chlorinated Paraffins in Polyethylene

Components	1	2
Polyethylene	66	72
CP-70 (70% chlorine)	24	—
CP-76 (76% chlorine)	—	20
ATO	10	8
Flammability		
UL-94 @ 3.2mm (1/8″)	V-0	V-0
Oxygen index (%)	24	24.5
Physical properties		
Tensile strength (Mpa)	10.1	10.7
(psi)	1460	1548
Heat distortion (DTUL) 264 psi	40°C	37°C
66 psi	50°C	48°C

tions in the United States is the ASTM E-84 25-ft tunnel test and the competition is rigid polyurethane foam.

In the x-ray photoelectron spectroscopy (XPS) process, HBCD is the material of choice. It is very effective, and concentrations from 1 to 3 percent allow the material to pass the ASTM E-84 specification.

The mechanism of action of HBCD in polystyrene has been the center of research for several years, and it is clear that a chain-scission mechanism of polystyrene occurs because of attack by bromine radicals, along with a less effective traditional bromine gas phase pathway [26]. These synergists accelerate polystyrene chain scission, thus allowing the use of less HBCD to achieve the same flame-retarding performance. Inclusion of antimony is harmful as it disrupts the chain-scission mechanism without being able to provide equally effective gas phase quenching reactions. The chain scission allows the material to drip away from the flame and cool. It gives a V-2 system in the UL characterization and cannot achieve a V-0 because of dripping [27].

The main drawback of HBCD is its thermal stability; as an alicyclic bromine-containing molecule, it lacks the robustness of decabromodiphenyl oxide or any other aromatic bromine-containing

TABLE 4.38 V-0 Polyethylene Formulations

Component	Virgin LDPE	Deca	Dech Plus	Virgin HDPE	Deca
LDPE (%)	100	46.5	39.5	—	—
HDPE (%)	—	—	—	85	57
FR (%)	—	21	25	—	22
ATO (%)	—	7	10	—	6
Talc (%)	—	25	25	15	15
Tensile strength (Mpa)	12.6	11.4	11.4	—	—
Tensile elongation (%)	—	21	12	—	—
Izod impact (J/m)	—	198	59	—	—
Specific gravity (g/cc)	0.924	1.44	1.48	0.954	1.03
UL-94 @ 3.2 mm	Burn	V-0	V-0	Burn	V-0
UL-94 @ 1.6 mm	Burn	V-0	V-1	Burn	V-0
Oxygen index (%)	20	30.6	28.6	20.6	28

TABLE 4.39 Effect of Zinc Stearate on Glow Wire Test in Polystyrene with Stabilized HBCD

HBCD stabilized (%)	3	3
Zinc stearate	0.1	0.0
Glow wire test	Average burn time (sec)	Average burn time (sec)
@ 3mm	30 sec	31 sec
@ 1.5 mm	Fail (>60 sec)	Fail (>30 sec)
@ 1 mm	4 sec	17 sec
Color	Brown streaks	Off-white

material. In formulating a successful flame-retardant XPS application with HBCD, thermal stability is an issue, as well as smoke generation. Zinc has been known to interact negatively with HBCD accelerating its degradation [27] (Table 4.39). A stabilized version of HBCD (SP-75 from Great Lakes) is available to allow a wider processing window in the XPS process and extend the stability of the traditional material. The effect of antimony trioxide on the V-2 performance of formulations containing stabilized HBCD in polystyrene is shown in Table 4.40. It is clear that the chain-scission mechanism is interrupted in the presence of antimony trioxide.

HIPS. Flame-retardant HIPS is a very important material and is used widely in electrical and electronic housings and appliances. The requirement for flammability of TV cabinets is an issue of continuing debate. The United States and Japan elected the UL-94 V-0 standard while Europe allowed V-2 cabinets. However, in 2000, in view of fire statistics and gauging risks associated with TV cabinet fires, Sony and other manufacturers voluntarily adopted the V-0 standard worldwide.

Brominated aromatic flame retardants are the most cost-effective materials for imparting flame retardancy to HIPS. Physical properties are generally maintained; however, UV stability suffers even further (HIPS is not exceptional on its own either!).

The most cost-effective solution in achieving a UL-94 V-0 HIPS material is the use of decabromodiphenyl oxide, along with antimony trioxide as a synergist. The preferred ratio is 3:1 and levels range from 15 to 20 percent depending on the rubber content of HIPS. Such material will not have a good UV stability and will suffer if its use is for housings where exposure to light will be severe. The yellowing of older appliances and copiers/printers is related to the mediocre UV stability of polystyrene combined with the poor stability of decabromodiphenyl oxide.

Ways to address the UV stability question, as well as the environmental attack on decabromodiphenyl oxides (DPOs) exist in the form of alternate materials such as the following:

- Decabromodiphenyl ethane (Saytex 8010) or a proprietary non-DPO-containing material (Firemaster 2100 from GLCC): These materials are as close to drop-in replacements to deca

TABLE 4.40 Effect of ATO on V-2 Performance in HIPS with Stabilized HBCD

HBCD/ATO (%/%)	Average burn time @ 3.2 mm	Average burn time @ 1.6 mm
2/0	2.2 sec (V-2)	1.1 sec (V-2)
1.6/0	2.2 sec (V-2)	5.2 sec (V-2)
1.6/0.53	>30 sec (fail V-2)	>30 sec (fail V-2)
1.6/1.6	>30 sec (fail V-2)	>30 sec (fail V-2)
1/0.33	>30 sec (fail V-2)	>30 sec (fail V-2)
2/0.67	>30 sec (fail V-2)	>30 sec (fail V-2)

TABLE 4.41 Recommended Load Levels for UL-94 V-0 HIPS

	None	Deca	8010	PB-370	SR-245	BT-93
FR (%)	0	12	12	17	12	12
Sb_2O_3	0	4	4	5	4	4
UL-94 @ 1/16″	Burn	V-0	V-0	V-0	V-0	V-0
UL-94 @ 1/8″	Burn	V-0	V-0	V-0	V-0	V-0
LOI	18	28	27	26	26	27

in HIPS as possible. Similar load levels are required for flammability performance and property profiles are the same. These new materials allow better UV stability to be achieved with HIPS, while it is still unclear if a Delta E of 3 to 4 can be achieved with FR HIPS, and if applications away from polyolefins can be won with this new technology.

- Tris tribromoneopentyl phosphate (PB-370 from DSB): This aliphatic brominated phosphate ester gives potentially the best UV stability in a HIPS formulation. Load levels range around 20 percent, but cost is a critical factor as this is a chemical based on expensive and not readily available raw materials.
- SR-245: This material is another way of introducing bromine. It has better UV stability than deca. Exact performance versus the other systems will depend on the application.
- BT-93: Excellent UV stability and good physical properties can be maintained. May require more additive than other systems.

Recommended load levels for UL-94 V-0 HIPS are given in Table 4.41. The physical properties of such products are summarized in Table 4.42.

The addition of a flame retardant reduces the impact strength of the neat resin; however, small addition of an SBS impact modifier can restore the impact strength to its original levels. Table 4.43 demonstrates the efficacy of impact modification at levels as low as 3 percent.

UV Stabilization of Flame-Retardant HIPS. Market requirements vary widely with regard to the desired maximum Delta E color change after 300 h of Xenon arc exposure (ASTM D4459-86 test protocol). Published studies [28] suggest that it is possible to achieve a Delta E of less than 1 with BT-93, while 8010 can approach values of 5 with appropriate stabilization. The choice of HIPS resin is a critical factor in color development, as are the pigments and the additional additives in the formulation.

TABLE 4.42 Typical Physical Properties of UL-94 V-0 HIPS Formulations

FR	None	Deca	Saytex 8010	Saytex BT-93
Yield strength (Mpa)	26	26	26	26
Elongation (%)	1.5	1.5	1.4	1.4
Tensile modulus (Gpa)	2.0	2.2	2.2	2.3
Flex strength	50	51	50	50
Flex modulus (Gpa)	2.0	2.2	2.3	2.2
HDT (C)	75	74	76	77
Izod	3.9	3.7	2.8	2.2
MFI (230/3.8 Kg)	9	13	10	8
Delta E (300 hrs Xenon Arc)	NA	54.9	26.2	6.3
Izod (KJ/m^2)	15	13	8	5

TABLE 4.43 Effect of Impact Modifiers on HIPS UL-94 V-0 Formulations

Izod (KJ/m^2)	Deca	8010	BT-93
No SBS	13	8	5
3% SBS	16	12	10

Table 4.44 shows that the UV stabilities of various HIPS resins vary widely. The addition of BT-93 and antimony trioxide resulted in a color change of 6 in this particular study. Further addition of TiO$_2$ and a hindered amine light stabilizer (HALS) reduced the Delta E to values between 1 and 2.

It is clear that the impact modifier chosen for HIPS, the application color, and the pigment chosen to achieve it will play a critical role in the UV stability of the final formulation. It is very encouraging, however, that materials exist both on the FR and the stabilization side of the industry to address the UV question and position FR HIPS for significant growth.

4.5.4 ABS

Acrylonitrile-butadiene-styrene (ABS) copolymer is an impact modified resin with greater impact strength and higher heat distortion temperature than HIPS (yet not at the performance level of ETPs such as polycarbonate, nylons, and polyesters). Because of the slightly improved physical properties and mostly because of its better surface appearance, ABS commands a higher price compared to HIPS. In flame-retarded applications, HIPS and ABS compete and there is a fair amount of interpolymer substitution occurring. The advantages of ABS are high gloss, ease of molding, high toughness combined with good flexibility, and the ability to be metal plated. However, ABS is still difficult to UV stabilize (benzene polystyrene ring and rubber content), has limited thermal stability compared to ETPs, and has an inherent slightly yellow color.

In the last 5 years, significant market erosion has occurred in FR ABS applications because of consumer preferences for halogen-free materials and thin mold designs requiring increased heat distortion performance. This environment accelerated the conversion of flame-retarded ABS applications (monitors and housings) to PC/ABS materials that can be fairly easily flame retarded with phosphate esters and are halogen free.

The traditional choices for flame-retarding ABS were octabromodiphenyl oxide, tetrabromobisphenol A, bis(tribromophenoxy)ethane (FF680), and brominated epoxy oligomers. However,

TABLE 4.44 UV Stability of Flame Retardant HIPS Formulations

Formulation				
HIPS	100	84	81.4	78.4
BT-93	0	12	12	12
S$_2$O$_3$	0	4	4	4
TiO$_2$	0	0	2	5
HALS	0	0	0.6	0.6
Delta E after 300 hours				
HIPS 1	8.5	6.6	2.0	1.4
HIPS 2	5.7	8.4	3.2	2.4
HIPS 3	11.8	9.0	2.7	1.9
HIPS 4	5.1	7.8	1.6	1.1
HIPS 5	7.5	5.9	1.8	1.3

TABLE 4.45 Comparison of ABS Flame Retarded with Octabromodiphenyl Oxide, Bis(Tribromophenoxy) Ethane, Tetrabromobisphenol A and Brominated Epoxy Oligomer

	Control	FF680	Octa	Tetra	Epoxy oligomer
%FR	0	21.5	18.8	18	21
Sb_2O_3	0	3	3.1	3.6	6.8
Bromine	0	15	15	10.6	11
UL94 @ 1/16″	HB	V-0	V-0	V-0	V-0
OI	18.5	—	—	29.5	—
Notched izod (ft-lb/in)	6.7	3.3	3.7	1.6	2.2
HDT (F)	197	186	193	161	189
UV stability	1.1	3.1	—	9.7	—

environmental pressures and very poor UV stability drove the market away from octabromodiphenyl oxide. Tetrabromobisphenol A formulations still have limited thermal stability and are known to produce "black specs" in the final parts. Most of these formulations have been substituted with brominated epoxy oligomers that offer the best balance of thermal and UV stability. FF680 is very effective in ABS; however, it tends to plate out and the disadvantage is especially limiting in black formulations. It has, however, exceptional thermal stability and excellent FR efficacy. Although not necessarily the most cost-effective options, decabromodiphenyl ethane and ethylene bis(tetrabromophthalimide) can be used to flame retard ABS.

Comparison of ABS with octabromodiphenyl oxide, bis(tribromophenoxy) ethane, tetrabromobisphenol A, and brominated epoxy oligomer is shown in Table 4.45. A comparison of ABS UL-94 V-0 formulations containing BT-93, deca, and 8010 is given in Table 4.46. Decabromodiphenyl oxide destroys the impact properties of ABS, so it is the least desired option.

4.5.5 Polycarbonate

Polycarbonate is an amorphous polymer with significant char-forming characteristics. Its clarity and high-temperature performance position it at the center of the engineered thermoplastics (ETP) marketplace. It can be easily alloyed with ABS or PBT/PET and the resulting materials have achieved healthy growth rates especially in flame-retardant applications.

TABLE 4.46 Properties of UL-94 V-0 ABS

	Control	Tetrabromophthalimide	DCDPE	TetraB-A
Resin	100	78	81.4	75.6
FR	0	18	14.6	20.4
Sb_2O_3	0	4	4	4
% Bromine	0	12.1	12.0	12.1
Physical properties				
Izod impact, 3.2 mm ft-lb/in	3.3	0.6	2.0	1.7
DTUL C	79	82	80	68
MFI, 230C/3800g	5.8	3.3	2.6	14.0
UV stability				
100 hr Xenon Arc	7.9	7.3	21.1	Nd
300 hr Xenon Arc	9.2	9.4	25.5	34.3

Polycarbonate of appropriate molecular weight (melt flow is the industrial measurement used to denote MW) achieves UL-94 V-2 without any addition of flame retardants at a thickness of 1/8 in. To achieve UL-94 V-0 at a variety of thicknesses, the following technologies are available:

- Brominated polycarbonate oligomers (BC-52 and BC-58). These materials are readily compatible with the matrix and affect mechanical properties the least. It is possible to achieve a UL-94 V-0 at 1/16 in with 10 to 15 percent of these materials. Addition of fibrilar PTFE (such as Teflon 6C from Du Pont) as an antidrip agent may be required, although this is not necessary. If PTFE is used less flame retardant is usually necessary to achieve the desired flammability.
- Brominate phosphates. UL-94 V-0 can be achieved at 6 to 8 percent of this material. Transparency can be maintained and physical properties are excellent. No manufacturer is commercial in the United States with this material. However, formulators can achieve similar performance by combining brominated oligomers with triphenyl phosphate or bisphosphates to achieve the P-Br synergy demonstrated to exist with Reoflam PB-460 [10–13].

No antimony is used in polycarbonate because it causes molecular weight degradation of the polymer and a precipitous reduction of mechanical properties.

- Sulfonated salts. Extremely small amounts of sulfonated salts can be used to achieve UL-94 V-0 at 1/8 in, maintaining impact and transparency, and the technology is the subject of several mostly expired patents. At thinner sections, the technology does not work or the impact is lost. Combinations of sulfonated salts with traditional brominated materials could be an interesting exploration.

The challenge in flame-retarded polycarbonate in producing a transparent, high-impact material is the extreme notch sensitivity of the basic polycarbonate matrix. Any amount of a second phase at levels higher than 3 percent results in a significant reduction of notched impact. Therefore, the formulations containing brominated oligomers require the addition of impact modifiers to maintain good Izod, and this compromises the desired transparency.

New silicone additives are available to provide another nonhalogen alternative to V-0 polycarbonate. Table 4.47 gives a comparison of a silicone derivative [29] in polycarbonate versus a phosphate ester, and tetrabromobisphenol A as brominated reference. Technology also exists to incorporate the silicone in the polycarbonate backbone, in a copolymerization-type reaction [30].

4.5.6 PC/ABS Blends

There has been an explosion in the growth of flame-retardant PC/ABS for monitors and other electronic applications from 1990 to 1998 at a yearly growth rate of more than 12 percent. The main reason for this significant growth has been the successful use of nonhalogenated flame retardants in PC/ABS blends, providing good properties and acceptable cost to commercially viable systems. A

TABLE 4.47 Properties of Polycarbonate Resins with Flame Retardants

Properties	PC Resin	PC + TBBA	PC + P-ester	PC + silicone derivative
Izod impact strength [Kg cm/cm]	75	12	4	45
Flexural strength [Kg/cm^2]	920	980	1080	920
Flexural modulus [Kg/cm^2]	23,000	23,200	27,000	22,800
HDT [°C]	134	134	106	133
Melt flow [g/10 min]	22	22	47	22
Flame retardancy UL-94 @ 1/16″	V-2	V-0	V-0	V-0

TABLE 4.48 Recommended Load Levels of Phosphates to Achieve V-0 in PC/ABS and HIPS/PPO Blends

	4:1 PC/ABS	8:1 PC/ABS	m-PPO
Triphenyl phosphate	11	9	13
Resorcinol diphenyl phosphate	11	9	16
Substituted TPP (proprietary)	17	13	18
Bisphenol-A-diphenyl phosphate	14	9	20

host of phosphate esters have been developed to meet the needs of the PC/ABS applications. All phosphate esters significantly lower the heat distortion temperature and impact properties of PC/ABS while increasing melt flow. This behavior has been known as antiplasticization and has been explained in a series of publications [31, 32].

The simplest phosphate ester, triphenyl phosphate, available for many decades for other applications, can be used and is still used for some less demanding PC/ABS applications.

The main disadvantage of TPP and the driving force for developing more advanced additives has been the stress cracking of PC/ABS parts molded from resins containing TPP. TPP is rather volatile and during molding of high-stress areas, it tends to exude (juice) from the plastic part, causing polycarbonate to stress crack. However, TPP with a phosphorus level of around 9 percent is a very efficient flame retardant, and its being a solid offers ease of handling in a variety of compounding environments. Depending on the ratio of polycarbonate to ABS, levels of 8 to 10 percent of TPP are needed to achieve a UL-94 V-0 rating at $1/16$ in. Table 4.48 summarizes the levels needed to achieve UL-94 V-0 for some representative PC/ABS blends for triphenyl phosphate, as well as a series of biphosphates.

The next generation of phosphate esters developed for PC/ABS were bisphosphates in an effort to increase molecular weight, and therefore decrease volatility and mobility/juicing to the surface of the plastic part. Resorcinol diphosphate (RDP) was the first material developed for PC/ABS and it is a liquid additive with 9 percent P content and good efficacy as a flame retardant. Bisphenol A bisphosphate (BAPP or BADP) is another liquid bisphosphate with properties similar to RDP. It is more viscous than RDP and needs to be heated in order to be metered into the extruder. Stress cracking is better with BAPP, probably because of the compatibility of the additive in the polycarbonate matrix.

Some solid monophosphates also exist that could be used in PC/ABS, for example, Daihachi PX200; however, the properties are not as good as with the bisphosphates. Several application patents exist on the use of phosphates in PC/ABS.

From a technical feasibility viewpoint, brominated oligomers as well as brominated phosphate esters can be used to achieve excellent properties in PC/ABS blends. A small amount of brominated PC/ABS is available from compounders. However, the resin suppliers driven by consumer preferences and big OEM demands positioned the nonhalogen PC/ABS materials as environmentally preferred alternatives and offer only nonhalogen PC/ABS resins to the market.

Silicones can also be used in PC/ABS blends and offer a more expensive nonhalogen alternative to phosphate esters. Both phosphorus and silicone work via a glass-forming mechanism that is promoted in the polycarbonate matrix.

Case Study: Flammability Study in PC/ABS and HIPS/PPO Blends with Various Phosphate Esters

UL-94 Analysis. A study was carried out to determine the load level required to achieve V-0 in various PC/ABS and HIPS/PPO blends by statistically analyzing the second burn time during the UL-94 procedure. The additives tested included TPP, substituted TPP, RDP, and bisphenol A diphosphate (BPADP). Fig. 4.18 depicts the chemical structures of these phosphates. The load level recommendations to achieve UL-94 V-0 at $1/16$ inch in 4:1 and 8:1 PC/ABS blends and modified HIPS/PPO (Noryl 730) are given in Table 4.49.

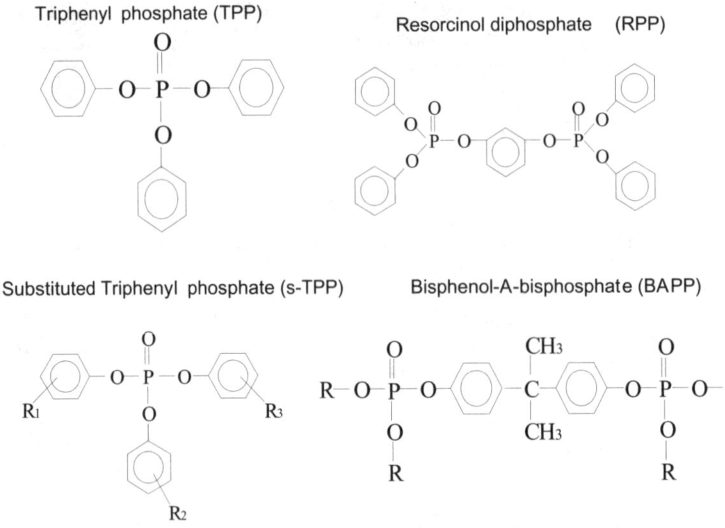

FIGURE 4.18 Structures of phosphate esters used in PC/ABS and HIPS/PPO blends.

The average burn time for the second UL-94 flame application, with 95 percent confidence intervals, is shown as a function of additive concentration: Figs. 4.19 to 4.22 for the 8:1 PC/ABS blends, Figs. 4.23 to 4.26 for the 4:1 PC/ABS blends, and Figs. 4.27 to 4.30 for Noryl 731 depict the data. The critical loading (L_c) is indicated for each system.

All systems were analyzed individually to determine the heuristic onset of V-0 behavior, as well as the L_c. To avoid bias, L_c was first determined as the intersection of the two linear trends observed through statistical analysis of the second burn time average values. Actual additive levels corresponding to the transition point were determined later. A comparison of the values for each system is shown in Table 4.49. The heuristic onset of V-0 behavior corresponds to the L_c value determined

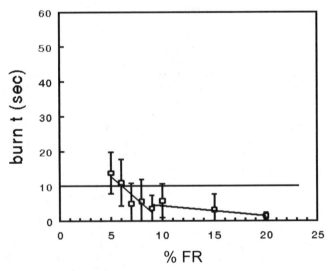

FIGURE 4.19 Second burn times according to the UL-94 test protocol as a function of flame-retardant (TPP) concentration in an 8:1 PC/ABS blend. The bars indicate 95 percent confidence intervals.

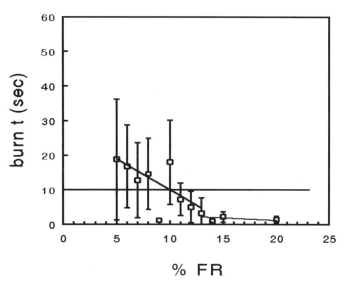

FIGURE 4.20 Second burn times according to the UL-94 test protocol as a function of flame-retardant (substituted TPP) concentration in an 8:1 PC/ABS blend. The bars indicate 95 percent confidence intervals.

statistically. TPP demonstrates the highest degree of flame-retarding efficiency across all resin systems. Resorcinol diphosphate is more effective in PC/ABS than it is in modified HIPS/PPO.

Analysis by Oxygen Index. OI trends, shown as a function of overall phosphorous content in the resin systems, indicate different flame-retardant mechanisms for monophosphates versus oligomeric phosphates. Oligomeric phosphates appear to be more efficient flame retardants as loadings are increased. However, the increased efficiency of oligomerics in the OI results does not translate to

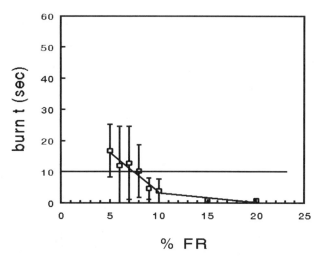

FIGURE 4.21 Second burn times according to the UL-94 test protocol as a function of flame-retardant (RDP) concentration in an 8:1 PC/ABS blend. The bars indicate 95 percent confidence intervals.

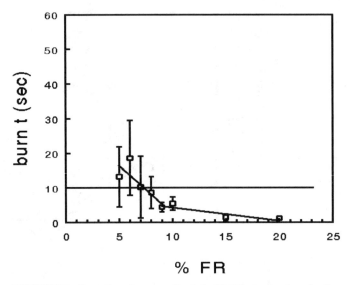

FIGURE 4.22 Second burn times according to the UL-94 test protocol as a function of flame-retardant (BAPP) concentration in an 8:1 PC/ABS blend. The bars indicate 95 percent confidence intervals.

improved UL-94 performance. All phosphates achieved V-0 criteria at equal phosphorous content within the PC/ABS formulations. The oligomeric phosphates were differentiated from monophosphates according to phosphorous content required to achieve UL-94 V-0 in m-PPO.

Several trends were then observed in both PC/ABS systems. First, OI increases linearly with increasing phosphorous content. However, both oligomeric (RDP and BPADP, aka bis) phosphates lie on one line while monomeric phosphates (TPP and substituted 507, aka mono) lie on another line

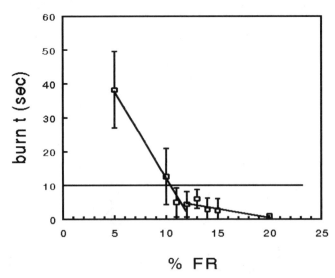

FIGURE 4.23 Second burn times according to the UL-94 test protocol as a function of flame-retardant (TPP) concentration in a 4:1 PC/ABS blend. The bars indicate 95 percent confidence intervals.

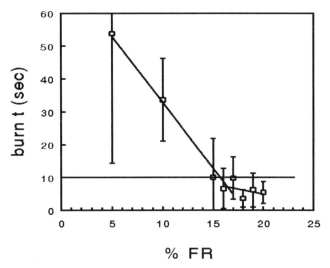

FIGURE 4.24 Second burn times according to the UL-94 test protocol as a function of flame-retardant (substituted TPP) concentration in a 4:1 PC/ABS blend. The bars indicate 95 percent confidence intervals.

of different slope. The oligomeric phosphates appear to be more effective flame retardants because of the higher slope of the line. The difference between oligomeric and monomeric OI trends could suggest different flame-retardant mechanisms for these additive types in PC/ABS systems, although the data contained herein are too limited to clarify mechanistic behavior.

Although oligomeric phosphates appear more effective because OI increases faster than for monomeric phosphates, there is no correlation within these systems between OI and UL-94 V-0 performance. The phosphorous content needed to achieve a V-0 rating falls within a very minor range for all four additive types in each PC/ABS system. All four phosphate additives demonstrate the

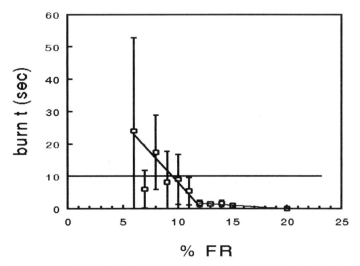

FIGURE 4.25 Second burn times according to the UL-94 test protocol as a function of flame-retardant (RDP) concentration in a 4:1 PC/ABS blend. The bars indicate 95 percent confidence intervals.

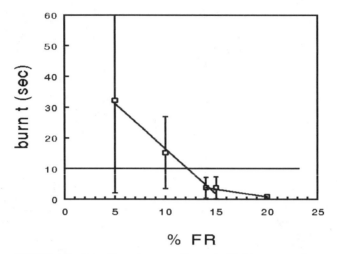

FIGURE 4.26 Second burn times according to the UL-94 test protocol as a function of flame-retardant (BAPP) concentration in a 4:1 PC/ABS blend. The bars indicate 95 percent confidence intervals.

onset of V-0 behavior (described in the previous section) at a phosphorous content of 0.011 to 0.013 percent by weight in the 4:1 PC/ABS blends. In 8:1 PC/ABS systems, the necessary phosphorous content is only 0.008 to 0.01 percent of the formulation.

The OI behavior is very different, however, in the Noryl (m-PPO) resin. Although each additive demonstrates linear OI performance against phosphorous content, there is no distinction between monomeric versus oligomeric phosphates. REOFOS 507 and REOFOS RDP exhibit individual linear trends, while REOFOS TPP and BPADP lie together on one line. On the other hand, the additive types are uniquely differentiated according to oligomeric/monomeric type with respect to phos-

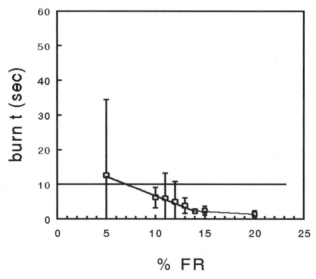

FIGURE 4.27 Second burn times according to the UL-94 test protocol as a function of flame-retardant (TPP) concentration in a HIPS/PPO blend. The bars indicate 95 percent confidence intervals.

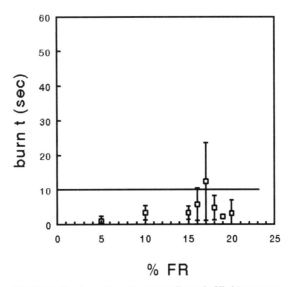

FIGURE 4.28 Second burn times according to the UL-94 test protocol as a function of flame-retardant (substituted TPP) concentration in a HIPS/PPO blend. The bars indicate 95 percent confidence intervals.

phorous content and UL-94 performance. The monomeric phosphates (REOFOS TPP and REOFOS 507) demonstrate the onset of V-0 characteristics at a phosphorous content of 0.012 to 0.013 percent of the Noryl formulations. The oligomeric phosphates (REOFOS RDP and PBADP) require higher phosphorous content to meet V-0 criteria at 0.017 percent P.

Although the relationship between OI and phosphorous content appears linear, the y intercepts are not equal for the respective lines. The reason is that initially the additives worsen the OI perfor-

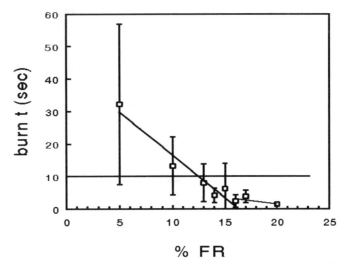

FIGURE 4.29 Second burn times according to the UL-94 test protocol as a function of flame-retardant (RDP) concentration in a HIPS/PPO blend. The bars indicate 95 percent confidence intervals.

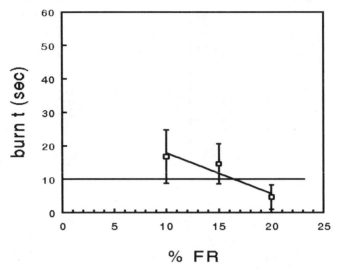

FIGURE 4.30 Second burn times according to the UL-94 test protocol as a function of flame-retardant (BAPP) concentration in a HIPS/PPO blend. The bars indicate 95 percent confidence intervals.

mance against the blank resin systems. Therefore, the observed linearity begins only after a critical level of phosphorous is incorporated into each system. The OI behavior is graphically demonstrated for each system in Figs. 4.31, 4.32, and 4.33 in an 8:1 PC/ABS blend, a 4:1 PC/ABS, and Noryl, respectively.

4.5.7 Nylon

Flame-retardant nylon materials are very important in electronic applications, and a host of excellent additives exist to achieve good combination of physical properties and flame-retarding performance. Table 4.50 characterizes the advantages and disadvantages of the main flame retardants used in polyamides.

All the halogenated flame retardants in nylon require antimony trioxide as a synergist at a ratio of 3:1 (FR/antimony trioxide). Teflon powder at 0.5 percent is usually added to suppress dripping.

Brominated Polystyrenes. Different materials are available to address the needs of regular or high-temperature nylons. The traditional brominated polystyrene (derived from bromination of poly-

TABLE 4.49 Comparison of UL-94 V-0 Load Levels According to the Critical Load Level Method and the Heuristic Arguments

	4:1 PC/ABS		8:1 PC/ABS		Noryl 731 Black	
	L_c	H_{onset}	L_c	H_{onset}	L_c	H_{onset}
REOFOS TPP	11	11	9	9	14	13
REOFOS 507	17	18	13	13	*	18
REOFOS RDP	11	11	10	9	16	16
BPADP	14	14	9	9	20	20

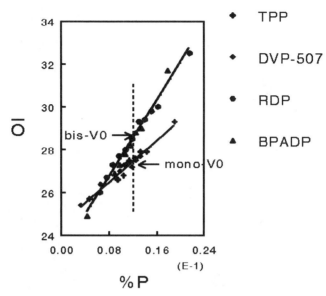

FIGURE 4.31 Oxygen index as a function of phosphorus concentration for TPP, substituted TPP, RDP, and BAPP in 8:1 PC/ABS blends.

styrene) is good for regular nylon applications, but lacks the thermal stability or flow characteristics needed in more demanding polyamide applications. An alternate technology of bromostyrene polymerization yields a host of brominated polystyrene materials of different molecular weight, bromine content, tailored end functional groups, and improved thermal stability.

Table 4.51 summarizes the properties of several of these brominated polystyrenes. PBS refers to polymerized bromostyrene materials and Br-PS refers to bromination of polystyrene. Comparison of

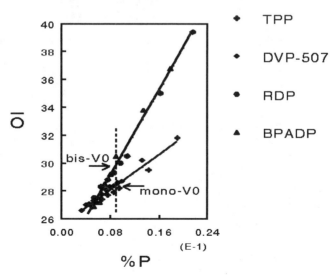

FIGURE 4.32 Oxygen index as a function of phosphorus concentration for TPP, substituted TPP, RDP, and BAPP in 4:1 PC/ABS blends.

TABLE 4.50 Advantages and Disadvantages of the Main Flame Retardants Used in Polyamides

Flame retardant	Advantages	Disadvantages
Halogenated flame retardants		
Brominated styrene polymers	Good thermal stability Excellent efficiency Good flow Good electricals	UV stability Compatibility
Dechlorane Plus	Good electricals Good efficiency	Thermal stability
Decabromodiphenyl oxide	Cost	Limited thermal stability UV stability Physical properties Flow
Decabromodiphenyl ethane	Good thermal stability Good UV stability Non DPO	Flow Cost versus DECA
Nonhalogen additives		
Red phosphorus	Good efficiency Improved arc resistance	Handling safety concerns Limited colors (red . . .)
Melamine salts	Good efficiency	Wet electricals Thermal stability

physical properties of various flame retardants in 30 percent glass-filled Nylon 66 with various brominated polystyrenes is given in Table 4.52. Table 4.53 provides the data for 30 percent glass-filled high-temperature polyamides. Dechlorane Plus finds applications in both filled and unfilled nylons and its performance is summarized in Tables 4.54 and 4.55. The variation of tensile strength as a function of additive concentration in 30 percent glass-reinforced Nylon 6,6 is shown in Fig. 4.34 for various brominated polystyrenes. Tensile elongation is shown in Fig. 4.35, while the molded in color is shown as Fig. 4.36. Brominated polystyrenes such as PDBS-80 and PBS-64 show superior thermal stability. This behavior is further demonstrated with the data for high-temperature polyamides (HTPAs). Figs. 4.37, 4.38, and 4.39 summarize the data on tensile strength, elongation, and notched Izod impact for HTPA.

Unreinforced polyamides can be effectively flame retarded with the use of melamine cyanurate, although some processing difficulties can be experienced because of dispersion and kneading capac-

TABLE 4.51 Properties of Brominated Polystyrenes. PBS Refers to Polymerized Bromostyrene Materials and Br-PS Refers to Bromination of Polystyrene

Property	PBSI (64% Br)	PBSII (64% Br)	PBSIII (60% Br)	Brominated PS
Appearance	Amber pellet	Amber pellet	Amber pellet	—
Bromine (%)	64–65	64–65	59–60	—
Monomer (%)	0.3	0.3	0.3	0
MFI (g/10 min)	5–15	20–35	20–35	—
Tg (°C)	132.7	149.6	144.5	—
Specific gravity	2.0	1.9	1.9	—
Molecular weight	60,000	28,000	60,000	—

TABLE 4.52 Comparison of Physical Properties of Various Brominated Polystyrenes in 30% Glass Filled Nylon 66

Properties	Flame retardant			
	PBSIII	PBSI	PBSII	Br-PS
%FR/%ATO	22/7	20/6.5	20/6.5	19/6.0
UL-94 @ 1/32″ (0.8 mm)	V-0	V-0	V-0	V-0
Tensile strength (Mpa)	134	136	146	123
Tensile elongation (%)	2.5	2.6	2.3	2.5
Izod impact (unnotched) J/m	677	726	811	720
Spiral flow (mm)	1196	1201	1245	1199

TABLE 4.53 Comparison of Physical Properties of Various Brominated Polystyrenes in 30% Glass Filled High Temperature Polyamides

	PBSIII	PBSI	Br-PS
%FR	22	20	19.5
% Antiminy Trioxide	7	6.4	6.3
UL-94 @ 1/32″ (0.8 mm)	V-0	V-0	V-0
Physical properties			
Tensile strength (Mpa)	160	168	150
Tensile elongation (%)	2.1	2.2	1.9
Izod impact (unnotched) J/m	560	651	507

TABLE 4.54 Typical Formulation of Dechlorane Plus in Unfilled Nylon 66

Formulation (weight %)	1	2	3	4	5	6
Nylon 6,6	70	78	70	70	82	85
Dechlorane Plus	20	16	20	20	14	12
ATO	10	2	—	—	—	—
Zinc Borate	—	4	10	5	—	1.5
Zinc Oxide	—	—	—	5	—	—
Iron Oxide	—	—	—	—	4	1.5
Flammability						
UL-94 @ 3.2 mm (1/8″)	V-0	V-0	V-0	V-0	V-0	V-0
@ 1.6 mm (1/16″)	V-0	V-0	V-0	V-0	V-0	V-0
@ 0.8 mm (1/32″)	V-0	V-0	V-0	V-0	V-0	V-0
@ 3.2 mm (1/64″)	V-0	V-0	V-0	V-0	V-0	V-0
Tensile strength (MPa)	58.6	67.1	59.3	56.9	71.6	70.2
(psi)	8609	9744	8603	8264	10,376	10,173
CTI Kc (Volts)	275	450	300	375	275	350

TABLE 4.55 FR-Nylon 6,6 (25% Glass Reinforced) with Dechlorane Plus and Various Synergists

Formulation (weight %)	1	2	3	4
Nylon 6,6	49	48	48	60
Fiberglass	25	25	25	25
Dechlorane Plus	18	18	18	12
ATO	8	—	—	—
Zinc Borate	—	9	—	—
Zinc Oxide	—	—	9	—
Ferric Oxide	—	—	—	3
Flammability				
UL-94 @ 3.2 mm (1/8″)	V-0	V-0	V-0	V-0
@ 1.6 mm (1/16″)	V-0	V-0	V-0	V-0
@ 0.8 mm (1/32″)	V-0	V-1	V-1	V-0
@ 3.2 mm (1/64″)	V-0	V-1	NC	V-0
Tensile strength (MPa)	122.8	121.3	97.2	98.3
(psi)	17,800	17,590	14,080	14,250
CTI Kc (volts)	225	375	325	225

ity issues. The flame retardant required to achieve UL-94 V-0 at 1.6 mm can be summarized as follows:

	Melamine cyanurate
PA 6 and 66 unreinforced	UL-94 V-0 at 6–10% wt
PA 6 and 66 mineral filled	UL-94 V-0 at 13–15% wt
PA 6 and 66 glass reinforced	UL-94 V-2, glow wire 960°C
	CTI > 500 V

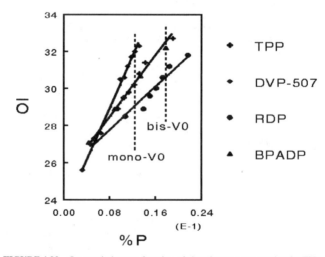

FIGURE 4.33 Oxygen index as a function of phosphorus concentration for TPP, substituted TPP, RDP, and BAPP in HIPS/PPO blends.

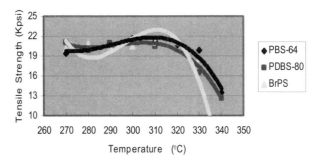

FIGURE 4.34 Tensile strength as a function of molding temperature in 30 percent glass-reinforced Nylon 6,6. Several brominated polystyrenes are compared.

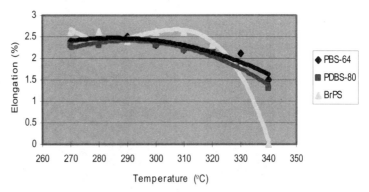

FIGURE 4.35 Tensile elongation as a function of molding temperature in 30 percent glass-reinforced Nylon 6,6. Several brominated polystyrenes are compared.

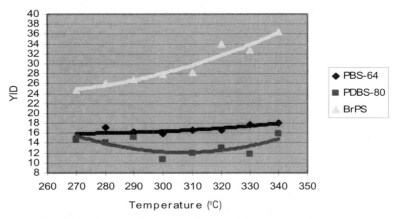

FIGURE 4.36 Molded part color as a function of molding temperature in 30 percent glass-reinforced Nylon 6,6. Several brominated polystyrenes are compared.

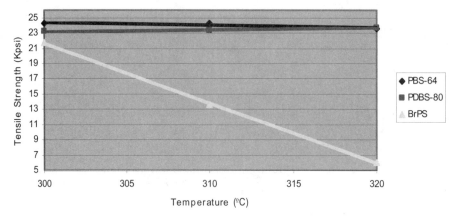

FIGURE 4.37 Tensile strength as a function of molding temperature in 30 percent glass-reinforced HTPA. Several brominated polystyrenes are compared.

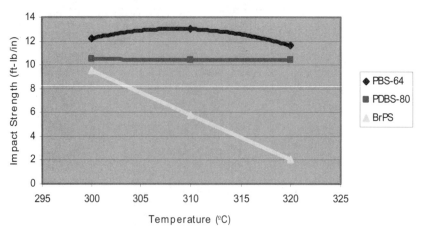

FIGURE 4.38 Tensile elongation as a function of molding temperature in 30 percent glass-reinforced HTPA. Several brominated polystyrenes are compared.

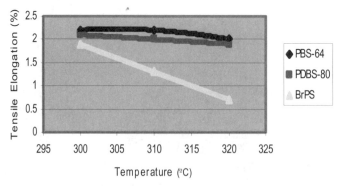

FIGURE 4.39 Impact strength as a function of molding temperature in 30 percent glass-reinforced HTPA. Several brominated polystyrenes are compared.

TABLE 4.56 Flammability, Physical and Electrical Properties of Unreinforced Polyamides: Melamine Compounds versus Halogen/Antimony

	Unreinforced nylon 6 or 66	
Properties	Bromine/Antimony	Melamine Cyanurate (MC25 or MC50)
Specific gravity	1.3	1.18
Tensile strength (Mpa/psi)	75/10,800	80/11,600
Elongation at break (%)	5	5
Flex strength (Mpa/psi)	100/14,000	125/18,100
Flex modulus (Gpa/Ksi)	3/435	3.9/560
Charpy notched impact (KJ/m^2/ft-lb/in)	4/0.8	4/0.8
Charpy notched impact (KJ/m^2/ft-lb/in)	4/0.8	4/0.8
Flammability		
UL-94 @ 1.6 mm	V-0	V-0
UL-94 @ 0.8 mm	NC	V-0
UL-94 @ 0.4 mm	NC	V-0
Smoke evolution (NBS)		
Flaming mode	790	190
Nonflaming mode	390	80
CTI (Volts)	350	>600

TABLE 4.57 Flammability, Physical and Electrical Properties of 25% Glass Reinforced Polyamides: Melamine Compounds versus Halogen/Antimony

	25% Glass Reinforced Nylon 6 or 66	
Properties	Bromine/Antimony	Melamine Polyphosphate (Melapur 200)
Specific gravity	1.54	1.48
Tensile strength (Mpa/psi)	145/21,000	150/21,750
Elongation at break (%)	2.5	2.5
Flex strength (Mpa/psi)	155/22,500	160/23,200
Flex modulus (Gpa/Ksi)	10/1,450	12/1,740
Charpy notched impact (KJ/m^2/ft-lb/in)	7/1.35	7/1.35
Flammability		
UL-94 @ 1.6 mm	V-0	V-0
UL-94 @ 0.8 mm	V-0	V-0
UL-94 @ 0.4 mm	NT	V-0
Smoke evolution (NBS)		
Flaming mode	—	360
Nonflaming mode	—	150
CTI (Volts)	350	350–400

TABLE 4.58 Advantages and Disadvantages of the Most Common Flame Retardants Used in Polyesters

Flame retardant	Advantages	Disadvantages
Brominated styrene polymers	High thermal stability Excellent flow Excellent electricals	Poor UV stability Compatibility
Tetrabromobishenol A Carbonate oligomers	High efficiency Good UV stability Adequate thermal stability	Limitations on electrical properties Limitations on thermal stability
Decabromodiphenyl oxide	Cost	Limited thermal stability UV stability Physical properties
Decabromodiphenyl ethane	Good thermal stability Good UV stability Non DPO	Poor physicals (flow and impact) Cost versus DECA
Brominated epoxy oligomers	Lowest cost alternative Adequate thermal stability	Limitation on physical properties Reduced physicals vs. TBBPA carbonate oligomers

The physical properties and flammability of Nylon 6 flame retarded with melamine cyanurate is shown in Table 4.56. In the reinforced polyamide arena, the recently available melamine polyphosphate (MP200) can be used and its performance is demonstrated in the data of Table 4.57.

4.5.8 Thermoplastic Polyesters

Table 4.58 describes the advantages and disadvantages of the most common flame retardants used in polyesters.

Polybutylene Terephthalate (PBT)

PBT is a crystalline plastic, part of the engineering thermoplastics market (ETPs), with a low heat distortion temperature (DTUL is 125°F at 264 psi). Reinforcement with 30 percent glass increases

TABLE 4.59 Comparison of Physical Properties of Various Flame Retardants in 30% Glass Filled PBT

	Flame retardant			
	PBSIII	PBSI	PBSII	Br-PS
%FR	14	12	13	12
% Antimony Trioxide	4.5	4.0	4.0	4.0
UL-94 @ 1/32″ (0.8mm)	V-0	V-0	V-0	V-0
Physical properties				
Tensile strength (Mpa)	119	122	114	108
Tensile elongation (%)	2.2	2.5	2.3	2.3
Izod impact (unnotched) J/m	619	603	630	576
Spiral flow (mm)	1021	1054	1024	1080

TABLE 4.60 Flame Retardant Unmodified PBT with Brominated Carbonate Oligomers

Components	PBT base resin	PBT with BC-52	PBT with BC-58
Flame retardant/ATO ratio	—	18/3	16/4
UL-94 @ 1/32″	Burns	V-0	V-0
Unnotched Izod impact (ft-lb/in)	>80	7.9	7.0
Gardner impact strength (in-lb)	210	8	6
Heat distortion temperature (°F)	124	146	152
Spiral flow (in)	16.3	12.6	16.2
Flex modulus (Ksi)	327	375	382

Reference: R.C. Nametz, FRCA, 1984.

DTUL to 400°F and allows high-temperature applications. Reinforcement by mineral or mineral/glass filler combinations increases stiffness and toughness. Mineral fillers reduce warpage significantly and improve processing. Thermally stable brominated compounds can all effectively flame retard PBT, and the choice is usually made based on physical properties and processing performance characteristics. In unreinforced PBT 8 to 10 percent bromine from an aromatic brominated compound is required to achieve a UL-94 V-0 at a thickness of $1/16$ in. For thinner sections, more flame retardant is required along with special additives to inhibit dripping. In mineral- and glass-reinforced PBT less flame retardant is needed.

Decabromodiphenyl oxide is the most cost-effective choice and is in use in low-end applications. It does not maintain good physical properties and it blooms to the surface of the PBT resin. Brominate polystyrenes are very effective in unreinforced and glass-reinforced PBT grades and maintain nonblooming performance along with very good physical properties. Table 4.59 shows the flammability and physical properties of PBT flame retarded with various brominated polystyrenes.

Carbonate oligomers (BC-52 and BC-58) are excellent in PBT because they maintain physical properties, processing performance, and nonblooming behavior. Especially, the toughness and impact achieved with BC-52 make this additive very attractive for demanding applications (Table 4.60). High-molecular-weight brominated epoxy oligomers (such as FR 1025) when used in PBT yield materials with excellent color, impact, flow, and thermal stability (Table 4.61).

TABLE 4.61 FR PBT with Brominated Epoxy Oligomers (BEO)–Unfilled and Glass Filled

Components	1	2
PBT	57.6	50
Impact Modifier	18.4	—
Glass fiber	—	30
Brominated epoxy oligomer	17.8	13.5
ATO	6.2	6.5
Flame retardancy		
Oxygen index	30.0	31.5
UL-94 @ 1/8″	V-0	V-0
Physical properties		
Notched Izod impact (ft-lb/in)	3.2	1.6
Tensile strength (psi)	4840	1700
Elongation (%)	162	—
Tensile modulus (Ksi)	190	430
Flex strength (psi)	8150	1650
Flex modulus (Ksi)	70	180

TABLE 4.62 Flame Retardant Mineral Filled PBT

	1	2	3	4
PBT mineral filled	84	84	84	84
Brominated carbonate oligomer	—	12	—	16
Brominated phosphate	—	12	—	16
ATO	4	4	—	—
Oxygen index	31.8	29.7	29.1	31.2
UL-94 @ 1/16″	V-0	V-0	Burns	V-0
Time for flame out (sec)	0	3.7	—	3.1

An aromatic brominated phosphate has also been reported to be used in PBT, and is very effective [31]. The P-Br synergy allows flame retardancy to be achieved in mineral-filled PBT without the use of antimony trioxide (Table 4.62). In glass-filled PBT this brominated phosphate ester improves processability and impact at a use level of 10 percent with 3 percent antimony trioxide. Table 4.62 shows the properties of mineral-filled PBT flame retarded with Reoflam PB-460 (the brominated phosphate ester containing 60 percent bromine and 4 percent phosphorus) or with a brominated carbonate oligomer. Comparison of physical properties of various flame retardants in 30 percent glass-filled PBT is shown in Table 4.63.

In the nonhalogen PBT arena, combinations of melamine cyanurate and phosphorus compounds have shown promise in unreinforced and glass-reinforced PBT. In unreinforced PBT, 5 to 15 percent of melamine cyanurate along with 5 to 10 percent of a phosphate ester (TPP, RDP, or BAPP) will give a UL-94 V-0 compound. In glass-reinforced PBT, a load level of 10 to 15 percent melamine cyanurate gives a UL-94 V-2 rating or a glow wire 960°C rating.

Polyethylene Terephthalate (PET)

Glass-reinforced PET is commercially flame retarded with brominated polystyrene derivatives. As discussed earlier, antimony trioxide cannot be used in PET as a synergist to the halogenated flame retardants because it depolymerizes PET. Sodium antimonate is used as an alternative synergist. The

TABLE 4.63 Flame Retardant Glass Filled PBT

	Control	1	2	3
PBT-30% glass	100	87.5	87.5	86.5
Brominated PS	—	10	—	—
Brominated carbonate oligomer	—	—	10	—
Brominated phosphate	—	—	—	10
ATO	—	2.5	2.5	3.5
Teflon 6C	—	0.5	0.5	0.5
Oxygen index	20.7	26.7	28.5	28.5
UL-94 @ 1/16″	Burns	V-0	V-0	V-0
Time (s)	—	2.6	0	0
Notched izod (ft-lb/in)	1.32	0.8	1.07	1.26
DTUL (°C)	207	205	201	201
Melt flow index (g/10 min)	13.4	16	11.6	19.1
Spiral flow (in)	37	36	26	35
Tensile strength (psi)	17,100	11,600	15,600	15,400
Tensile elongation (%)	3.9	2.3	3.3	4.5
Flex strength (psi)	29,420	20,900	28,350	28,790
Flex modulus (Ksi)	1200	1200	1200	1060

TABLE 4.64 FR PET with Brominated Phosphate—No Need for Antimony Synergist

PET/30% glass	80	82	80
Brominated phosphate	15	18	20
Sodium antimonate	5	—	—
Teflon 6C	0.5	0.5	0.5
Flammability			
Oxygen index	29.4	30.6	36
UL-94 @ 1/16"	V-2	V-2	V-0
Time (s)	5.1	1.2	0.1

TABLE 4.65 Residue Yields Formed in TGA Up to 500°C (N2) 2/1 Polycarbonate/PET

Resin blend (PC/PET)	37%
12% Brominated oligomer	35%
13% Triphenyl phosphate (TPP)	33%
7% Brominated oligomer + 2.5% TPP	31%
6% Brominated phosphate (60/40)	48%

TABLE 4.66 Residue Yields Formed in TGA Up to 500°C (N2) 3/1 Polycarbonate/ABS

12% Brominated oligomer	35%
10% Triphenyl Phosphate (TPP)	57%
9.5% Resorcinol diphosphate (RDP)	59.5%
6% Brominated phosphate (60/40)	54%

use of a brominated phosphate allows PET to be flame retarded without any antimony synergist. Table 4.64 demonstrates this effect.

In the nonhalogen area, it is possible to use a cyclic phosphoanate such as Antiblaze 1040 to flame retard PET. Table 4.65 summarizes the flammability and physical properties.

PC/PET blends can be formulated to achieve UL-94 V-0 properties with excellent physical properties using phosphate esters or combinations of phosphorus and bromine. These materials form a char and prevent flame propagation. Table 4.65 compares char residue yields in a 2:1 PC/PET blend with various flame-retardant additives [33]. Table 4.66 compares similar additives in a 3:1 PC/ABS blend in terms of char residue yield. RDP is the best char former in PC/ABS and a brominated phosphate has the optimum performance in PC/PET.

4.5.9 Polyvinyl Chloride

PVC is one of the most versatile thermoplastic materials regarding processability and range of applications. It can be processed in rigid or plasticized form, but special stabilizers are needed to boost its thermal stability and allow it to survive the processing requirements. Rigid PVC burns with charring and the flame extinguishes immediately upon removal of the ignition source. It has an oxygen index of 44 (Table 4.27). Plasticized PVC, however, because of the high level of combustible plasticizers can continue to burn with a smoky flame. During the decomposition of PVC, hydrogen chloride is eliminated at temperatures between 200 and 300°C. This produces conjugated double bonds in the carbon chain while elimination of water and cyclic reactions form a char residue. Gas phase reactions through the elimination of HCl and charring are the main mechanisms of flame retardancy in PVC.

Aliphatics, aromatics, and condensed aromatics are by-products of PVC pyrolysis and along with HCl contribute significantly to smoke during PVC combustion.

The most stringent flammability requirements in PVC applications arise from the wire and cable industry, especially the plenum-type applications. Specialty plasticizers and smoke suppressants must be combined to achieve flame retardancy for such systems. In such systems, brominated flame retardants such as brominated phthalate esters are used in combination with specialty phosphorus plasticizers. For example, DP-45 and Santicizer 2148 are normally employed to pass the UL-910 test required for plenum applications. The issue of bromine chlorine synergy has been studied in PVC [34].

TABLE 4.67 Effect of Various FR's On the Flammability of PVC (LOI)

	Concentration (phr)	Oxygen index
Control	—	20.8
Antimony trioxide (ATO)	6	25.0
Alumina trihydrate	25	22.6
Zinc borate	25	21.2
Isopropyl phenyl diphenyl phosphate	30	24.2
Isopropyl phenyl diphenyl phosphate	45	25.4

Many additives have been used to cost-effectively flame retard PVC and these include antimony trioxide, alumina trihydrate, zinc borate, chlorinated paraffins, and phosphate esters. Antimony trioxide is very effective in flame-retarding PVC because it acts as a synergist for the halogen contained inherently in the polymer (57 percent chlorine). Phosphate esters are also very effective as they replace part of the combustible plasticizers and act both as flame retardant and plasticizer components of the formulation. Table 4.67 shows the oxygen index of PVC with incorporation of various flame-retardant additives.

Antimony trioxide is usually the most cost-effective option; phosphate esters are the choice when transparency or special smoke requirements must be met. Zinc borate is a very useful coadditive along with antimony trioxide in PVC formulations because it reduces smoke and afterglow—both common side effects of high levels of antimony trioxide. The effect of zinc borate in an ATH/Sb_2O_3 system is shown in Table 4.68.

Phosphate esters are classified into three main groups: triaryl, alkyl diaryl, and trialkyl phosphates. They replace in whole or in part the phthalate or adipate plasticizer in the PVC formulation providing dual functionality. The aromatic phosphate esters are more effective flame retardants as measured by oxygen index, UL-94, or flame spread performance tests. The alkyl groups are detrimental to flammability; however, they improve low-temperature performance and plasticizing efficiency and reduce smoke evolution.

Typical properties of a PVC formulation plasticized with phosphate esters is shown in Table 4.69.

In PVC, several formulations contain both antimony trioxide and phosphate esters. Results are favorable, although some studies point to antagonism [19, 34].

TABLE 4.68 The Effect of Zinc Borate/ATH System on the Heat Release Properties of Flexible PVC (Palatinol 711P at 50 phr Is the Plasticizer)

Additive	1	2	3	4	5
PVC	100	100	100	100	100
ATH (Micral 1500, Solem)	30	30	30	100	50
ATO	15	7.5	—	—	—
Zinc borate	—	7.5	15	15	15
$Mg(OH)_2$	—	—	—	—	50
Stearic acid	—	—	—	1	1
Cone calorimeter results					
5 min average heat release rate (kW/m^2)	171	134	165	115	106
Peak heat release rate (kW/m^2)	273	151	187	133	127
Peak smoke (m^2/g)	1528	1361	1544	628	581
Peak CO (Kg/Kg)	0.36	0.14	0.23	0.13	0.05

Ref: Kelvin Shen–FRCA.

TABLE 4.69 Flame Retardant PVC with Phosphate Esters

Composition	Formulations		
	1	2	3
PVC (parts)	100	100	100
Stabilizers (phr)	6	6	6
Santicizer 148 (phr)	50	—	—
Reofos 65 (phr)	—	50	—
Reofos 3600	—	—	50
Flammability			
Oxygen index	28.5	31.5	28.5
Vertical burn (5903)	0.8	1.1	0.6
After flame (sec)	1.8	1.8	2.6
Char length (in)			
Physical properties			
Hardness, shore A	82	89	82
Modulus (psi)	1510	2090	1603
Low temperature flexibility (°C)	−26	−11	−27
Carbon volatility, 24 hr/90°C	7.3	5.7	5.6

Detailed studies of triaryl phosphate with antimony trioxide in PVC showed this antagonism to occur only in part of the composition range [34, 35]. This antagonistic effect probably is the result of the formation of antimony phosphate, which is very stable and practically inactive as a flame retardant.

An example of PVC electrical sheathing material is given in Table 4.70. A PVC jacketing or insu-

TABLE 4.70 Flame Retardant PVC Electrical Sheathing Material

Components	1	2	3	4	5	6
PVC (D.P. 1,100)	100	100	100	100	100	—
PVC (D.P. 1,300)	—	—	—	—	—	100
Diisodecyl phthalate (DOP)	40	40	40	40	40	40
ATO	20	20	15	15	15	15
Zinc borate	—	—	5	5	10	10
ATH	10	10	20	20	20	20
$Mg(OH)_2$	10	10	—	—	10	10
$CaCO_3$	—	15	—	20	—	—
Mica	15	—	20	—	20	—
Properties (IEEE 383 test)						
Sheath-damaging distance (cm)	90	135	78	123	40	87
Insulation-damaging distance (cm)	55	70	43	62	21	54
Sustained flame time (min)	25	47	12	41	0	18
Pass/fail	Pass	Fail	Pass	Pass	Pass	Pass
State of the remained crust	B	C	A	B	A	B

Reference: Patent assigned to Yakazi Corporation, 1989, by H. Otani et al., Japan Kokai Tokyo Koho, JP 0101,650 from *Use of Zinc Boratein Electrical Applications* by Kelvin Shen and Donald J. Ferm, FRCA.
All formulations contain 5 parts of lead sulfate. D.P. is degree of polymerization.
State of crust classifications:
A = Crust left is round, shrunk, and hard.
B = Crust left is round and shrunk.
C = Crust left is deformed with cracks.

TABLE 4.71 PVC Jacketing or Insulation Compositions Containing Phosphate Plasticizer

Components	1	2	3	4
PVC (D.P. 1,100)	100	100	100	100
Stabilizer	4	4	4	4
Diphenylcresyl phosphate	—	30	30	20
Diphenyloctyl phosphate	—	30	30	20
ATH	—	70	70	70
ATO	—	7	3.5	3.5
Chlorinated paraffin (50% chlorine)	—	—	—	20
Zinc borate	—	—	3.5	3.5
Dioctyl phthalate	60	—	—	—
CaCO$_3$	70	—	—	—
Properties				
Oxygen index (%)	23	37.6	42.4	43
Smoke (ASTM D-2843-70) % extinction	100	99	93	87
DIN 402				
Residual length (%)	0	22.7	27.1	32.4
Flue gas temperature (°C)	>700	191	169	156
Combustion class	—	B1	B1	B1

The stabilizers are: 80% tribasic lead sulfate, 20% of an equal mixture of lead stearate, calcium stearate and paraffin wax. Ref: U.S. Patent 4,246,158 (1981 to Wacker-Chemie by Popp W., and Sedivy J.).

lation composition is shown in Table 4.71. A low-smoke PVC sheet formulation is described in Table 4.72 [37].

4.5.10 Thermosets

Thermoset matrix resins can be based on epoxy, unsaturated polyester, or phenolic chemistries. The resins are usually prepared in two stages, an initially low-MW oligomeric liquid (prepreg—Stage 1) and then its curing during subsequent processing (Stage 2). Reactive strategies for achieving flame retardancy are often employed for thermosets in either the first or the second stage of their preparation.

Epoxy. Epoxy resins are mainly flame retarded with reactive brominated compounds, especially tetrabromobisphenol A. This allows retention of physical and electrical properties and therefore use of such materials for electronic circuit boards, resins for aerospace applications, or as industrial coating and adhesive materials.

TABLE 4.72 Low Smoke PVC Sheet Formulation

Additive (phr)	Relative smoke (%)	HCl produced (mg/g)	Combustion time (sec)	Breakdown voltage (KV/mm)	Tensile strength (Kg/mm^2)
None	100	331	45	40	2.4
Zinc borate (1.5) and ATH (7.5)	53	<5	35	33	1.9
Zinc borate (6) and ATH (40)	42	<5	<1	33	1.9

Additive phosphorus compounds tend to significantly plasticize the matrix and are usually avoided. P-containing molecules used as curing agents or monomers are of value. The reaction of dialkyl or diaryl phosphates on the epoxy groups of the 4,4′-diglycidyl ether of bisphenol A gives a modified resin, which upon curing shows improved flame retardancy compared to its unmodified counterpart, or a resin where the phosphates are included as additives [38].

A true need for a "green" epoxy exists and the area of research in phosphorus-based reactive additives is a very active field. Phosphorus has been incorporated in either the starting precure materials or curing agent or in both [39, 40].

The use of bis(m-aminophenyl)-methylphosphine oxide as a flame-retarding curing agent for epoxies has been extensively studied; however, its commercial use is still not very wide. A dual mechanism of action as both a char former and a gas phase flame retardant has been postulated [41].

Phenolic Resins. The significant aromatic content and high cross-link density of the phenolic resins presents an easier task in achieving flame retardancy in these systems versus the average epoxy-based formulation. Several phosphates such as triphenyl phosphate and resorcinol diphosphate have been shown to effectively flame retard phenolics and now this has been extended to polycyclic phosphonates [42, 43]. Mixtures of halogenated phosphates along with metal hydroxides improve FR; however, the electrical and physical properties cannot be easily optimized [44].

Unsaturated Polyesters. Flame retardancy is achieved in unsaturated polyesters with reactive technology that replaces the traditional monomers with halogenated diols or halogenated dicarboxylic acids.

Tetrabromopthalic anhydride (PHT4), chlorendic anhydride, or tetrabromobisphenol A are used to replace part of the phthalic anhydride, and dibromoneopentyl glycol is used to replace the ethylene or propylene glycol. The use of dibromostyrene as part of the cross-linking monomer mixture has also been used to improve the flame retardance of unsaturated polyesters. In such systems, the use of zinc additives such as zinc hydroxystannate can reduce smoke evolution.

The combination of halogenated resins and alumina trihydrate is used to produce an effective flame-retardant system with improved smoke suppression. A good summary of applications and markets is given in Ref. 45.

The use of phosphate esters along with DMMP has also been known to impart flame retardancy and processability to the unsaturated polyester manufacturing.

Polyurethanes. Polyurethane materials have a diverse range of applications from surface coatings (rigid and flexible) to elastomers (thermoplastic or cured) to foams (flexible or rigid). Flexible polyurethanes constitute the majority of applications and are chemically based on oligomeric polyethers or polyesters (polyols). The urethane linkage formed via the reaction of a hydroxy group with an isocyanate group is a common characteristic of all polyurethane materials.

Polyurethane foams are mostly used in building and construction applications and must pass the ASTM E-84 tunnel test or other similar large-scale flammability requirements. Both flame spread and smoke are important in such tests.

For polyurethane foams, the most commonly used flame retardants include chlorinated phosphate esters (10 to 12 phr) that provide cost-effective performance at a significant disadvantage in thermal resistance (scorch resistance). Table 4.10 describes such products.

It is possible to avoid the scorch in polyurethane foam by incorporating specially formulated phosphorus-bromine products, such as DE-60F or Firemaster 550 from GLCC, a combination of a phosphate ester and a brominated liquid additive. Table 4.73 shows formulations of flexible foam with DE-60F special and other bromine-phosphorus additives for polyurethane foams of various densities. Table 4.74 shows the performance of rigid PU formulated with brominated alcohols [46].

4.5.11 Elastomers/Rubber

The everyday feel of elastomeric materials is a bouncy, stretchy, soft material compared to a more rigid and hard plastic. The official definition of "rubber" according to the *American Society for*

TABLE 4.73 Performance of Various Bromine-Phosphorus Flame Retardant Systems in Polyurethane Formulations

a) Formulations for FMVSS 302–1.8 lb/ft^3 density foam

Components	DE-60F/BZ-54 HP (phr)	FR-38 (phr)
Polyol	100	100
FR	5/5	4
Water	3.3	3.3
Amine	0.48	0.48
Surfactant	1	1
Tin	0.53	0.53
TDI Index	110	110
Scorch resistance	Excellent	Poor

b) Formulations for California TB 117–1.8 lb/ft^3 density foam

Components	FM 550 (phr)	DE-60F special (phr)
Polyol	100	100
FR	9	9
Water	4	4
Amine	0.48	0.48
Surfactant	1	1
Tin	0.52	0.52
TDI index	110	110
Scorch resistance	Excellent	Excellent
Average burn distance (in)		
Initial	3.2	2.6
Heat Aged	4.0	3.6
Rating	Pass	Pass
Smolder (% wt retained)	>99	>99

TABLE 4.74 Performance of Rigid PU Formulated with Brominated Alcohols

Composition (phr)	Tribromoneopentyl alcohol	Dibromoneopentyl glycol
Polyol	80	75
Surfactant	2	2
Catalyst	1.5	1.5
HCFC 141B	25	25
Water	1	1
FR	18	22
DEEP	11	11
Isocyanate	127	138
Cream Time (sec)	19	19
Gel time (sec)	140	68
Tack free time (sec)	203	108
Density, Kg/m3	31	34
Heat resistance (°C)	143	135
Flame retardance	B-2	B-2

Testing and Materials (ASTM) is that of a material capable of recovering quickly from large deformations and being insoluble in a boiling solvent. Therefore, the ability to form a network and manifestation of the rubbery state are essential elements in classifying the material as an elastomer. The network formation in elastomers can be achieved with various processes:

a) chemical cross-linking with a peroxide or sulfur (curing or vulcanization, respectively—rubber formation)
b) Ionic cross-linking—ion cross-links—ionomers
c) Physical linkages that are broken with heat (TPEs with properties similar to rubber but plastics processing capabilities).

Typical applications of flame-retardant elastomers include seals, belts, tubing, wire and cable insulation, roofing, and coated fabrics. Flame-retardant additives can be successfully used for *polystyrene-butadiene rubber* (SBR), *polychloroprene* (CR), *ethylene-propylene rubber* (EPR), *chlorinated polyethyelene* (CPE), EVA, ethylene acrylic copolymer, silicone rubber, styrene block copolymer (SBS), and olefin based thermoplastic elastomers. Several of the flame-retardant additives used in elastomers remain solid at the processing temperatures, and good dispersion of the particles throughout the polymer matrix is critical to maintain good physical properties while achieving the desired ignition resistance. If processing occurs in a batch mixer, as is many times the case in elastomers, the flame retardant should be charged very early and before any plasticizers or processing aids are added. Maximum exposure to shear helps to properly disperse a solid additive, although many times two-pass processing or a flame-retardant predispersion may be required.

Formulations with brominated flame retardants (such as DE-83R, Saytex BT-93, Firemaster 2100, and Saytex 8010) are optimized by including antimony trioxide as a synergist (one part of antimony trioxide for every three parts of brominated additive) [47, 48]. Smoke evolution is further decreased via addition of zinc borate, barium metaborate, zinc oxide, molybdenum oxide, ammonium octamolybdate, ammonium polyphosphate, or zinc stannate. These additives can substitute up to 50 percent of the antimony trioxide in the formulation. Any further reduction in antimony trioxide content negatively impacts flame-retarding efficiency. In coatings or in applications where physical properties must be improved, colloidal antimony pentoxide is available [49].

Many of the fillers typically used in elastomers for reinforcement and cost reduction influence the flammability of the final compound. Clay, talc, and silica work well together with brominated flame retardants. However, carbon black can produce extended afterglow in some formulations. Zinc borate or barium metaborate have been claimed as efficient additives to control afterglow.

Typical elastomer fillers, such as calcium carbonate, alumina trihydrate, magnesium hydroxide, and magnesium carbonate may function as smoke suppressants. However, calcium and magnesium

TABLE 4.75 Summary of Typical Formulations for Elastomers

FR additive	Synergist	Elastomer	Rating
30 phr of BT-93 or DE-83R	10 phr of ATO	Unfilled EPDM	UL-94 V-0
35 phr of DE-83R	17.5 phr of ATO	EVA (18% vinyl acetate)	UL-94 V-0
30 phr of DE-83R	7.5 phr of ATO	EVA (28% vinyl acetate)	UL-94 V-0
15 phr of DE-83R and 40 phr of ATH and 15 phr chlorinated paraffin	5 phr of ATO	SBR	Conveyor belting–heat release and smoke requirement
8 phr BT-93	4 phr ATO	Ionomer (copolymer of elthylene)	UL-94 V2
15 phr DE-83R	5 phr ATO	Styrene block copolymers	UL-94 V-0
21 phr DE-83R	7 phr ATO	Olefinic TPE's	UL-94 V-0
10 phr of DE-83R	0–3 phr of ATO	CPE, CR, silicone rubber	UL-94 V-0

carbonate can interfere with the activities of brominated flame retardants and higher load levels may be required to maintain the same flame retardance. Table 4.75 shows a summary of typical formulations for flame-retardant elastomers.

Chemical modification of poly-1,3 butadiene, to improve flame retardancy was achieved by copolymerization with halogen- and/or phosphorus-containing unsaturated compounds [50].

Phosphorus modification of epoxidized natural rubber with dialkyl or diaryl phosphates has been carried out and these modified 1,4-polydienes yield improved flame-retardant elastomers. Further cross-linking of these P-modified polydienes occurs with methylnadic anhydride [51, 52].

Phosphorus additives such as diethyl phosphonate and halogenated P-compounds such as trichloromethyl phosphonyl dichloride can be added to relatively low-MW poly-1,2 butadienes under radical initiation: the resulting polymers have been used as flame retardants for rubber [53, 54].

Bromination of polybutadienes has also been successful in achieving flame-retardant elastomers because of the gas phase mechanism available in all brominated materials [55].

4.6 NANOCOMPOSITES

Nanocomposites constitute a relatively new development in the area of flame retardancy and can offer significant advantages compared to traditional approaches. Composites in general are structures formed from two or more physically and chemically distinct phases, and the inherent understanding is that their properties are superior to those of the individual components. The component phase morphologies and interfacial properties influence the structure and properties of the composite. Nanocomposites are based on the same principle; however, the scale of phase mixing is at the nanometer scale compared to the microscopic scale (millimeters or micrometers) in conventional composites. This allows nanocomposites to exhibit properties superior to conventionally filled polymers.

The most common nanocomposites are polymer layered silicate structures. Although first reported by Blumstein in 1961, the real commercial interest in the technology did not occur until the 1990s. Key advantages of nanocomposite structures are high modulus and strength, high heat distortion temperature, low gas permeability, improved solvent resistance, and thermal stability. An unexpected benefit of flame retardancy has made this technology the focus of research at NIST and at several academic institutions. The polymer structures based on nanocomposites are usually transparent because of the scale of the particle size compared to the wavelength of scattered light. Their relevance in flame retardancy results from their ability to significantly improve the thermal stability and self-extinguishing characteristics of the polymer matrix where they are incorporated. The best hypothesis for their flame-retarding efficacy is that the scale of mixing allows the layered silicate structure to interfere with the decomposition and pyrolysis mechanism of the polymer matrix, thereby improving flame retardancy [56].

Conventional compounding techniques such as extrusion and casting can be used to fabricate nanocomposites; this makes them very attractive from a cost/performance perspective compared to the traditional composite manufacturing methods. Applications in film and fibers are also possible where they would be prohibited with traditional microscale composites.

4.6.1 Layered Silicates

The most widely used structures to form polymer nanocomposites are layered silicates because of their chemically stable siloxane surfaces, high surface areas, high aspect ratios, and high strengths. Two particular characteristics are exploited for the formation of nanocomposites [57]:

a) The rich intercalation chemistry is used to facilitate exfoliation of silicate nanolayers into individual layers. Consequently, an aspect ratio of 100 to 1000 can be obtained compared to 10 for poorly dispersed particles. Exfoliation is critical because it maximizes interfacial contact between organic and inorganic phases.

b) The ability to modify surface chemistry through ion exchange reactions with organic and inorganic cations.

There are two classes of silicates used in nanocomposites: layered silicates (clay minerals) and phyllosilicates (rock minerals). Clay minerals consist of two structural units: a sheet of silica tetrahedra and two layers of closely packed oxygen or hydroxyl groups in which aluminum, iron, or magnesium atoms are embedded so that each is at the same distance from six oxygens or hydroxyls. The silica tetrahedra are arranged as a hexagonal network where the tips of the tetrahedra all point in the same direction. The same structure of silica tetrahedra is found in phyllosilicates.

Most clay minerals are sandwiches of two structural units, the tetrahedral and the octahedral. The simplest combination is a single layer of silica tetrahedra with an aluminum octahedral layer on top; these 1:1 minerals are of the kaolinite family. The other combination consists of the 2:1 structure (smectite minerals), where an octahedral filling exists between two tetrahedral layers. In smectite minerals, the octahedral sites may be occupied by iron, magnesium, or aluminum. Montmorillonite clay minerals of this group are a very popular choice for nanocomposites because of their small particle size (less than 2 μm) and the ease of polymer diffusion into the particles. Their high aspect ratio (10 to 2000) and high swelling capacity allows efficient intercalation of the polymer. Phyllosilicates include muscovite ($KAl_2(AlSi_3O_{10})(OH)_2$), talc ($Mg_3(Si_4O_{10})(OH)_4$), and mica.

Stacking of the layers leads to a gap between the layers, called the interlayer or gallery. Van der Waals forces hold the gap in place. Isomorphic substitution within the layers generates negative charges that are balanced by cations residing in the interlayer space. The interlayer cations are usually hydrated Na^+ or K^+ and can be easily exchanged with various organic cations such as alkylammonium. This ion exchange modifies the normally hydrophilic silicate surface to organophilic and paves the way for polymer intercalation. The organic cations lower the surface energy of the silicate and improve wetting, with the polymer matrix making the organosilicate more compatible with most commercial plastic materials. The organic cations can also contain specific functional groups to be able to react with the polymer surface and improve adhesion between the inorganic phase and the polymer matrix.

4.6.2 Polymer Nanocomposite Structures

The most important structures of polymer nanocomposite materials are the following:

- Intercalated structures: the monomer or the polymer is sandwiched between silicate layers. The clay layers maintain registry in the system.
- Delaminated or exfoliated: the silica sheets are separated—"exfoliated"—to produce a "sea of polymer with rafts of silicate" [57]. The exfoliated structure can be ordered (silicate layers are unidirectional) or disordered (random dispersion of silicate layers). In such a system, the clay layers have lost registry.
- End-tethered structure: one layer of the silicate or the whole silicate is attached to the end of the polymer chain.

For optimum physical properties, exfoliated structures are better than intercalated ones. However, for flame retardancy, intercalated structures could be adequate to modify the pyrolysis mechanism and induce a solid-phase mechanism of ignition resistance. Further work on the subject shows that optimization depends on the polymer substrate involved [56].

4.6.3 Preparation Methods

Nanocomposites are not formed by the simple mixture and physical proximity of a polymer and a silicate; it is only when the polymer intertwines in the gallery spacings that the benefits of the nanocomposite can be achieved. Kawasumi et al. were able to synthesize nanocomposites utilizing different polymers such as Nylon 6, polyimide, epoxy resin, polystyrene, polycaprolactone, and acrylic. Exfoliation could only be achieved in polymers containing polar functional groups such as amides and imides. This is expected because the clay silicate layers have polar hydroxy groups that can interact only with polymers containing polar functional groups [58].

This is where ion exchange reactions are critical in rendering the silicate surface organophilic and facilitating interaction with a wider class of polymer matrices. Alkyl ammonium cations are normally used to achieve an efficient ion exchange.

Different preparation methods have been successfully employed to form nanocomposites: melt blending and polymerizarion.

Melt Blending. Static mixing or mixing under shear of the silicate with the polymer matrix at a temperature above the polymer softening point allows an annealing process to take place. The polymer chains diffuse from the bulk of the polymer into the galleries between the silicate layers. Giannelis [59] has used the "direct polymer melt" method to prepare an intercalated *polyethylene oxide* (PEO) by heating the polymer and the silicate at 80°C for 6 h. Polystyrene, polyamides, polyesters, polycarbonate, polyphosphazene, and polysiloxane nanonocomposites can be prepared by this method.

Extrusion has also been used successfully to prepare intercalated structures. Gilman et al. [56] have prepared polystyrene layered silicate nanocomposites using extrusion at 150 to 170°C (residence time 24 min). Polypropylene intercalated structures have been prepared by twin-screw extrusion using a maleic anhydride modified polypropylene and modified clay [58, 60]. Twin-screw extrusion compounding is a very efficient process for manufacturing a nanocomposite. A recent review [61] summarized key concepts in the screw design and compounding techniques that need to be in place to produce an exfoliated structure.

Polymerization. This preparation method involves dispersing the clay in the monomer and carrying the polymerization reaction around it. Polystyrene clay nanocomposites can be prepared by polymerizing styrene in the presence of clay [62]. Intercalation of montmorillonite with e-caprolactam yields Nylon 6–clay hybrids [64]. Chemical grafting of polystyrene onto montmorillonite interlayers has also been achieved via addition polymerization reactions [63].

The Toyota research group was the first to observe that exfoliation of layered silicates in Nylon 6 greatly improves the thermal, mechanical, and barrier properties of the polymer. The Nylon 6 layered silicate nanocomposites containing 5 percent of nanoclay by weight exhibited increases in tensile strength by 40 percent, in tensile modulus by 68 percent, in flexural strength by 60 percent, and in flex modulus by 126 percent. The heat distortion temperature exhibited an increase from 65 to 152°C. In situ polymerization of e-caprolactam was first used to prepare the nanocomposites; however, melt blending has been proven to yield similar results [64].

Other polymers containing intercalated silicate structures include unsaturated polyesters where including 1.5 percent by volume of a clay increases the fracture energy from 70 to 138 J/m^2 [65]. The improved thermal stability of these polymers manifests itself also as improved flammability.

Generally, nanocomposite polymers offer

- Improved stiffness without loss of impact strength
- Improved heat distortion temperature
- Improved transparency
- Improved barrier characteristics
- Improved flame retardancy because of the formation of a three-dimensional inorganic network.

4.6.4 Flame-Retardant Properties of Nanocomposites

Blumstein in 1961 was the first to report in the literature nanocomposite structures and in 1965 [66] demonstrated the improved thermal stability of PMMA-layered silicate nanocomposite. These nanocomposites were prepared by free radical polymerization of methyl methacrylate monomer.

TGA studies showed that both linear and cross-linked PMMA intercalated into the Na$^+$ montmorillonite have 40 to 50°C higher decomposition temperatures compared to unmodified PMMA.

Steric factors restricting the thermal motion of the segments sandwiched between the two layers were considered the main mechanisms that contributed to higher thermal stability. Unzipping of the chains can start only when the temperature is high enough to generate a thermal motion adequate to "unlock" the chains from the interlamellar grip.

The original work at the Toyota laboratories showed that a polyamide six-clay nanocomposite containing 5 percent clay exhibits a 40 percent increase in tensile strength, 68 percent increase in tensile modulus, 60 percent increase in flex strength, and 126 percent increase in flex modulus, while the heat distortion temperature increased from 65 to 152°C and the impact strength is lowered only by 10 percent.

The research group at Cornell headed by Giannelis [59, 67, 68] demonstrated similar results for dimethyl siloxane and polyimide nanocomposites. PDMS nanocomposite was prepared by melt intercalation of silanol terminated PDMS into montmorillonite treated with dimethyl ditallow ammonium [68]. PDMS-nanocomposite containing 10 percent by weight of clay exhibits an increase in decomposition temperature of 140°C compared to the unmodified elastomer. Permeability of the nanocomposite structure decreased dramatically and the increased thermal stability was attributed to the hindered diffusion of the cyclic decomposition products from the new structure.

A similar finding occurs in aliphatic polyimide nanocomposites (intercalated and exfoliated structures) [67]. TGA data show higher decomposition temperatures and higher char yields. Montmorillonite and fluorohectorite structures showed identical results when the same nanocomposite structure could be obtained, suggesting that in this particular polymer particle size is not important. Thermal stability as measured by TGA was independent of the cation in the organosilicate.

These polyimide as well as PDMS nanocomposites [59] were exposed to an open flame and stopped burning upon removal of the flame. Still a commercially useful UL-94 V-0 rating could not be obtained. There is definitely flame-retardant efficacy and activity in these structures because the silicate layer acts as a barrier to the diffusion of gaseous products to the flame and shields the polymer from the heat flux. However, the level of inhibition is not yet optimized to the point of yielding commercially acceptable flame-retardant materials. Table 4.76 summarizes the char yield data [67].

Selected Examples of Flame Retardancy in Nanocomposites. The group at the National Institute of Standards and Technology has prepared nanocomposites using montmorillonite and fluorohectorite in polymers as diverse as polypropylene-graft maleic anhydride, polystyrene, Nylon 6, Nylon 12, vinyl ester, and epoxy [69]. Cone calorimetry was used as the investigative tool to explore flame-retardant properties. In all polymers, a common mechanism of flame reduction seems to occur, and the level of efficacy depends on the type of layered silicate used, the level of dispersion achieved, and the level of processing degradation experienced. Table 4.77 summarizes calorimeter data parameters obtained at a low-heat flux rate of 35 kW/m^2 [69].

TABLE 4.76 Char Yield Data

	Char yield % in PEI	Char yield % in PEI-intercalated nanocomposite
At 450C	—	—
After 20 minutes	45	90
After 120 minutes	15	45
At 500C	—	—
After 40 minutes	0	55

TABLE 4.77 Cone Calorimeter Results at 35 KW/m^2

Sample	Peak HRR KW/m^2	Average HRR KW/m^2	Mean Hc	Residue yield (%)
Nylon 6	1010	603	27	1
Nylon 6–delaminated nanocomposite @ 2%	686	390	27	3
Nylon 6–delaminated nanocomposite @ 5%	378	304	27	6
Nylon 12	1710	846	40	0
Nylon 12–delaminated nanocomposite @ 2%	1060	719	40	0
Polystyrene	1120	703	29	0
PS silicate–3% immiscible mixture	1080	715	29	3
PS-intercalated nanocomposite @ 3%	567	444	27	4
PS-DBDPO-antimony/30%	491	318	11	3
PpgMA	2030	861	38	0
PpgMA-intercalated nanocomposite @ 5%	922	651	37	8
Mod-bis-A vinyl ester (A)	879	598	23	0
A-intercalated nanocomposite @ 6%	656	365	20	8
Bis-A/novolac vinyl ester (B)	977	628	21	2
B-intercalated nanocomposite @ 6%	596	352	20	9

HRR = Heat Release Rate.
Hc = Heat of Combustion.

The above results suggest a very significant reduction in the peak and average heat release rate with intercalated nanocomposites. Char yields are not significantly affected and specific heat of combustion, specific extinction area, and CO yield remain unchanged. These results point to a condensed phase mechanism compared to a gas phase mechanism prevalent in halogenated systems. The example with decabromodiphenyl oxide and antimony clearly shows lower heat of combustion and higher CO yield typical of incomplete combustion and a gas phase mechanism of radical quenching.

Bourbigot et al. [70] have shown that pentaerythritol can be replaced by a polyamide-6-montmorillonite nanocomposite in a typical intumescent flame-retardant system. In this system, one third of the ammonium polyphosphate can be removed while maintaining the V-0 rating and significantly reducing heat release rate.

4.6.5 Mechanism of Flame Retardancy in Nanocomposites

Most researchers today agree that the FR mechanism of flame retardancy in nanocomposite-containing polymers is a consequence of high-performance carbonaceous-silicate char buildup on the surface during combustion. During pyrolysis, the nanocomposite structure collapses and the resulting carbonaceous-silicate structure forms a char layer. This layer acts as a thermal barrier, keeping the heat source from reaching the polymer surface, and as a barrier to diffusion of combustible decomposition products into the gas phase. Recent experiments [71] have revealed, via XPS, that the carbon vanishes and the clay accumulates on the surface of a degrading polymer. Residue yields do not improve with nanocomposites; therefore the mechanism of action is not that of retention of the char in the condensed phase.

Gilman et al. studied the effect of an exfoliated versus an intercalated structure on flame retardancy and concluded that no universal optimum exists: for certain polymers, an intercalated structure is best (epoxy and vinyl esters); for others (polystyrene), the exfoliated structure achieves better flame-retarding performance. Processing conditions also affected the flame-retarding efficacy of the polystyrene nanocomposite [56]. Both solvent mixing in toluene and extruder melt blending (170°C under nitrogen or vacuum) produced structures with reduced flammability. However, inclu-

TABLE 4.78 LOI and Char Yield of Kaolin-Containing Nanocomposites

Sample	LOI	Char yield at 600°C (%) (above the clay residue)
Polyvinyl alcohol (PVA)	20.7	3.0
PVA-nanocomposite (5% kaolin-intercalated)	23.7	12.5
Nylon 6	23	0
Nylon 6-nanocomposite (5% kaolin-intercalated)	27.5	6.8
Nylon 6 + 10% kaolin intercalated modified by TPP (1:1)	26.3	—
Polystyrene	18	—
PS-nanocomposite (5% montmorillonite-intercalated)	23	—
PS-7% kaolin-intercalated modified by TPP (1:1)	30	—

sion of air or high extrusion temperatures affected flammability negatively, resulting in a nanocomposite with no improved flammability over the base formulation.

The behavior of polystyrene nanocomposites with different organophilic clays at various load levels was studied by Zhu and Wilkie [62] using bulk polymerization as a preparation method. At a load level as low as 0.1 percent, clay reduced peak heat release rate by 40 percent and improved thermal stability as measured by TGA. Char yields as assessed by cone calorimetry and TGA were unaffected, in agreement with the work of Gilman et al.

Efforts to modify traditional flame retardants with the nanocomposite technology are documented in the work of Lomakin et al. [72] and Ruban et al. [73], who studied PVA, polystyrene, and Nylon 6 with modified TPP. TPP, an effective flame retardant on its own, was intercalated using kaolin, and its effectiveness increased. TPP has been shown to traditionally work in the gas phase, releasing P*, which acts as a radical trap in the gas phase. When TPP is intercalated, the mechanism of degradation of TPP changes to the condensed phase. Char formation is present in intercalated TPP, but is absent in TPP-only combustions. This is evidence of cross-linking and aromatic reactions. Table 4.78 summarizes the combustion properties of polymer nanocomposites containing mineral kaolin.

Combinations of traditional flame retardants with nanocomposites also received early attention in search of a potential synergism or of unique physical properties. PBT is traditionally flame retarded with a combination of brominated polycarbonate and antimony trioxide. General Electric Plastics succeeded in replacing 40 percent of this traditional package by a combination of a PTFE dispersion in styrene-acrylonitrile (50 percent in PTFE) and a montmorillonite clay treated with di(hydrogenated tallow) ammonium salt. The inventors [74] claim that a special synergism exists between PTFE and organomontmorillonite because both these additives are needed to improve flame retardancy.

Another example of a successful combination of traditional flame retardants and nanocomposites is Okada's work in polyethylene [75]. Table 4.79 summarizes his results on peak heat release and

TABLE 4.79 Polyethylene with Various Nanofillers: Peak Heat Release from Cone Data at 50 kW/m² and Elongation

Sample	PHR (kW/m²)	Elongation (%)
Polyethylene (PE)	1327	980
PE + SBAN N-400 (10 phr)	687	900
PE + DCBDO + ATO (10 phr)	1309	830
PE + DCBDO + ATO (15 phr)	1189	720
PE + APP (10 phr)	1272	590
PE + APP (15 phr)	989	490
PE + SBAN N-400 (10 phr) + APP (5 phr)	493	900
PE + SBAN N-400 (10 phr) + TPP (5 phr)	543	930

TABLE 4.80 Comparison of FR Formulations in PP: Combinations of Traditional FR's and Nanomers

Components						
Homopolymer PP	73.3	80	77	74	74	68
DBDPO	20	15	15	15	15	15
Sb_2O_3	6.7	5	5	5	5	5
Nanomer I.44PA	0	0	3	6	0	0
Nanomer C.44PA	0	0	0	0	6	12
UL-94 rating @ 1/8″	V-0	Fail	V-2	V-0	V-2	V-0

elongation. Peak heat release is reduced by 50 percent compared to the control sample by including 10 percent of clay. Traditional flame-retarding systems (based on decabromodiphenyl oxide and antimony trioxide) do not affect heat release rate. Most remarkably, the elongation of the nanocomposite clay sample remains close to that of the control (neat polyethylene), not suffering from the significant reduction observed with the filler type additives.

Despite the successes on heat release rate measurements and stability, commercial specifications are not easy to achieve with nanocomposite systems. Especially, the UL-94 test has not been easy to achieve without the inclusion of traditional flame retardants.

Inoue and Hosokawa [76] have used melamine salts and polymer layered silicates to impart flame retardancy to Nylon 6, PBT, and *polyphenylene sulfide* (PPS). Melamine salts of ammonium were pre-intercalated into the synthetic silicate, *fluorinated mica* (FSM), and 10 to 15 percent of the additive was used. The increase in HDT was most noted along with the improved flammability, achieving V-0 ratings.

Combinations of polystyrene nanocomposites with RDP to achieve good flame retardancy while maintaining heat distortion temperature were studied successfully [77]. Heat release rate is reduced but the system ignites faster than polystyrene. The conclusions from the study suggest the need for a more active flame retardant in the gas phase.

Similar work by Lan et al. [79] focused on a deca/antimony system in polypropylene. The study showed that the amount of brominated FR can be reduced if an appropriate PP nanocomposite is used. The new system achieves UL-94 V-0 and significantly lower heat release rate. Table 4.80 shows some results from this system with commercially available organically modified clays [79].

A similar observation of FR-nanocomposite synergy occurs with $Mg(OH)_2$ in an EVA system, as shown in Table 4.81 [78].

The incorporation of the nanomer is claimed to improve mechanical properties and reduce additive migration to the polymer surface (additive bloom), a known detriment of decabromodiphenyl oxide in polypropylene.

Another commercially available nanocomposite filler is Nanofil 15 from Sudchemie. The intercalated ion is distearyldimethylammonium chloride and the additive is recommended for polypropylene and EVA applications.

TABLE 4.81 Formulations of Nanomers in EVA-Combinations with $Mg(OH)_2$

Components						
EVA (wt%)	40	45	42	39	50	47
$Mg(OH)_2$ (wt%)	60	55	55	55	50	50
Nanomer (wt%)	0	0	3	6	0	3
UL94 rating @ 1/8″	V-0	Fail	V-0	V-0	Fail	V-0

TABLE 4.82 Properties of Nanofil 15

Nanofil 15	Active nanofillers for polymer applications	
Composition	Organic modified nanodispersed layered silicate	
Chemical functionality	Long chain hydrocarbon	
Typical technical data	Product form	Powder
	Color	Crème
	Specific gravity	1.8 g/cm^3
	Bulk density	440 g/l
	Average particle size	25 microns
	Moisture content	Less than 2%
	Loss on ignition	Approx. 35%
Intercalation	Distearlydimethylammonium chloride	
Recommended polymer	PP (grafted is preferred)	EVA

The diameter of the fully exfoliated Nanofil varies between 100 and 500 mm at a layer thickness of only 1 mm. This special structure of the layers results in an extraordinarily high aspect ratio of more than 100. Table 4.82 summarizes the properties of Nanofil 15.

Application Example. For wire and cable applications it is often necessary to compound 60 percent of an inorganic hydroxide into the polymer. Cable electrical and insulation properties are severely compromised with such high additive load levels. The incorporation of a nanodispersed layered silicate has a synergistic effect with these fillers, and time to ignition as well as dripping are minimized. Extensive tests during cable production at Kabelwerk Eupen AG (www.eupencable.com) demonstrate the advantages of Nanofil 15 in PE/EVA cable applications: improved flame retardancy with 10 to 15 percent of Nanofil 15, reduction of the level of ATH normally utilized and increased extrusion speed during cable production (Table 4.83) [79].

Figs. 4.40 and 4.41 show the results of a cone calorimeter study based on the formulations described in Table 4.83. The same peak heat release rate can be achieved with a formulation containing 55 percent ATH, by including 5 percent Nanofil 15 to a formulation containing 40 percent ATH. The time to ignition is improved from 128 to 175 s. Experiments were carried out at a heat flux of 35 kW/m^2.

TABLE 4.83 Improving FR of EVA/PE based Cable Compounds with Nanofiller

Polymer	4 parts LDPE (Escorene LLN 1001)		
Flame retardant additive	Aluminum trihydrate (ATH)		
Nanofiller	Nanofil 15		
Processing	Buss Kneader MDK 46 (10 Kg/h at 150°C)		

Sample	Polymer (phr)	Al(OH)$_3$ (phr)	Nanofil 15 (phr)
1	100	—	5
2	45	40	—
3	45	40	5
4	45	55	—
5	45	55	5

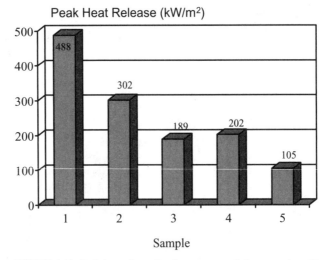

FIGURE 4.40 Peak heat release data from a cone calorimeter study at 35 kW/m² based on the formulations shown in Table 83.

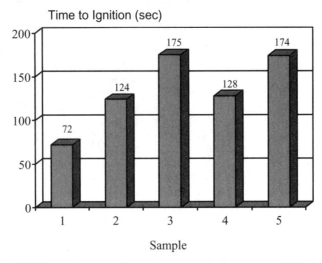

FIGURE 4.41 Time to ignition data from a cone calorimeter study at 35 kW/m² based on the formulations shown in Table 83.

REFERENCES

1. Babrauskas, V., Harris, R.H., Gann, R.G., Levin, B.C., Lee, B.T., Peacock, R.D., Paabo, M., Twilley, W., Yoklavich, M.F., and Clark, H.M., "Fire Hazard Comparison of Fire Retarded and Non-Fire-Retarded Products," NBS Special Publication 749, July 1988.
2. "Additives for Plastics," Marketing Report by Business Communications Company (BCC), May 2001.
3. Davenport, R.E., Fink, U., and Ishikawa, Y., "Flame Retardants," SRI International, November 1999.

4. Troitzsch, J., *International Plastics Flammability Handbook,* 2nd edition, Hanser Publishers, 1990.
5. Price, D., Anthony, G., and Carty, P., "Polymer Combustion, Condensed Phase Pyrolysis and Smoke Formation," in Horrocks, A.R., and Price, D., eds., *Fire Retardant Materials,* CRC Press, 2000.
6. Le Bras, M., Camino, G., Bourbigot, S., and Delobel, R., eds., *Fire Retardancy of Polymers—The Use of Intumescence,* The Royal Society of Chemistry, 1998.
7. Green, J., "Flame Retardants and Smoke Suppressants," in Lutz, J. T., and Grossman, R. F., eds., *Polymer Modifiers and Additives,* Marcel Dekker, 2001.
8. Markezich, R.L., "Chlorine Containing Flame Retardants," in *Flame Retardants 101—Basic Dynamics,* FRCA Spring Conference, 1996, pp. 89–104.
9. Hilado, C.J., *Flammability Handbook for Plastics,* 4th edition, Technomic Publishing Co., 1990.
10. Green, J., *Flame Retardants 92,* Proc. Flame Retard. 92 Conference (5th), 168–175, Elsevier Publishers, London, UK, 1992.
11. Green, J., and Chung, J., "Flame Retarding Poly(butylene) Terephthalate-Properties, Processing Characterization and Rheology," *Journal of Fire Sciences,* Vol. 8(4), 254–265, July/August 1990, and also J. Green, "Flame Retarding PC/ABS Blends with a Brominated Phosphate Ester," *Journal of Fire Sciences,* Vol.9(4), 285–295, July/August 1991.
12. Papazoglou, E., *Flame Retardants 94,* Elsevier Publishers, London, UK, 1994.
13. Papazoglou, E., Seibel, S.R., and Munro, S., "Compounding of Flame Retardant Concentrates for Improved Light Stability," in *Recent Advances in Flame Retardancy of Polymeric Materials,* 9: 144–158, 1998.
14. Atwell, R.W., Hodgen, H.A., Fielding, W.R., Favstritsky, N.A., and Termine, E.J., "Polymers of Brominated Styrene," US Patent 5304618A, 1994.
15. Seibel, S.R., Papazoglou, E., and Beecher, E., "Optimization of Compounding Methods of Flame Retardant Polymers via Co-rotating Twin Screw Extruders" in *Compounding. 2000—Issues, Solutions, Opportunities,* Fire Retard. Chem. Assoc. (FRCA)—Fall Conference, 1997, pp. 183–190B.
16. Pitts, J.J., Scott, P.H., and Powell, D.G., "Thermal Decomposition of Antimony Oxychloride and Mode in Flame Retardancy," *Journal of Cellular Plastics,* 6, 35–37 (1970).
17. Green, J., "Influence of Co-Additives in Phosphorus Based Flame Retardant Systems," *Plast. Compounding,* 10(3), 57, p. 60–64, May/June 1987.
18. Green, J., "Mechanisms of Flame Retardancy and Smoke Suppression—A Review," *Journal of Fire Sciences,* 14(6), 426–442, 1996.
19. (a) Weil, E., "Synergists, adjuvants and antagonists in flame retardant systems," in Grand, A., and Wilkie, C., eds., *Flame Retardancy of Polymeric Materials,* Marcel Dekker, New York, 1999; (b) Weil, E.D., in Lewin, M., Atlas, S.M., and Pearce, E.M., eds., *Phosphorus Based Flame Retardants in Flame Retardant Polymeric Materials,* Vol 2, Plenum Press, New York, 1978, pp. 103–128.
20. Touval, I., "Effect of Antimony Synergists on the Properties of Impact Polystyrene," *Recent Advances in FR Polymeric Matter,* Business Communications, Norwalk, CT, 1990.
21. Touval, I., and Scwartz, R.T, "Effect of Antimony Oxide on the Physical Properties of ABS," FRCA-SPE Conference, San Antonio, Texas, March 13–16, 1994.
22. Carnahan, J., Haaf, W., Nelson, G., Lee, G., Abolins, V., and Shank, P., "Investigations into the Mechanism for Phosphorus Flame Retardancy in Engineering Plastics," *Proceedings of the International Conference on Fire Safety,* 4, 1979, pp. 312–323.
23. Anindya, M., "Melamine Derivatives: An Alternative to Traditional Flame Retardants," *Plastics Engineering,* 57(2): 42–46, 2001.
24. Karpinidis, N., and Zingg, J., "Recent Advances in Flame Retardant Compositions," *The International Conference on Polyolefins, Polyolefins 2002.*
25. Munro, S., and Farner, R., "High Perfromance N-P Flame Retardant for Polypropylene," *Polypropylene 2001,* September 11–13, 2001.
26. Larsen, E., and Ecker, E.L., "Thermal Stability of Fire Retardants: I HBCD," *Journal of Fire Sciences,* Vol. 4, 261–275, July/August, 1986.
27. Kaspersma, J., Doumen, C., Munro, S., and Prins, A., "Flame Retardant Mechanism of Aliphatic Bromine in Polystyrene and Polypropylene," *Polymer Degr. and Stab.,* 77(2), 325–331, 2002.
28. Landry, S.D., "Solving the UV Resistance Problem in FR HIPS," *Plastics Compounding,* 17(6): 48–50, 1994.

29. Iji, M., Serizawa, S., and Yukihiro, K., "New Flame Retarding Plastics Without Halogen and Phosphorus for Electronic Products," *FRCA Conference,* March 1999.
30. Akio, N., "Bromine Free Flame Retardant Polycarbonate," *FRCA Conference,* March 1999.
31. Jackson, W.J. Jr., and Caldwell, J.R., "Antiplasticization II—Characteristics of Antiplasticizers," *Journal of Applied Polymer Science,* 11: 211–226, 1967.
32. Jackson, W.J. Jr., and Caldwell, J.R., "Antiplasticization III—Characteristics and Properties of Antiplasticizable Polymers," *Journal of Applied Polymer Science,* 11: 227–244, 1967.
33. Green, J., "A Phosphorus-Bromine Flame Retardant for Engineering Thermoplastics—A Review," *Journal of Fire Sciences,* 12: 388–408, 1994.
34. Ballisteri, A., Roti, S., Montaudo, G., Paparaldo, S., and Scamporrino, E., "Thermal Decomposition of Flame Retardants, Chlorine-Bromine Antagonism in Mixtures of Halogenated Polymers with Sb_2O_3," *Polymer,* 10: 783–784, 1979.
35. Morgan A.W., "Formulation and Testing of Flame Retardant Systems for Plasticized PVC," in Bhatnagar, V.M., ed., *Advances in Flame Retardants,* Vol. 1, Technomic Publishing Co., Westport, CT, 1972.
36. Moy, P., "FR Characteristics of Phosphate Ester Plasticizers with Antimony Trioxide," *Plastics Engineering,* 61–63, Nov. 1997.
37. Shen, K.K., and Ferm, D.J., "Use of Zinc Borate for Electrical Applications," FRCA, March 1999.
38. Derouet, D., Morvan, F., Brosse, J.C., "Chemical Modification of Epoxy Resins by Dialkyl or Aryl Phosphates: Evaluation of Fire Behaviour and Thermal Stability," *Journal of Applied Polymer Science,* 62(11): 1855–1868, 1996.
39. Liu, Y.L., Hsiue, G.H., and Chiu, Y.S., "Synthesis, Characterization, Thermal and Flame Retardant Properties of Phosphate Based Epoxy Resins," *Journal of Polymer Science, Part A, Polymer Chemistry,* 35(3): 565–574, 1997.
40. Buckingham, M.R., Lindsay, A.J., Stevenson, D.E., Muller, G., Morel, E., Coates, B., and Henry, Y., "Synthesis and Formulation of Novel Phosphorylated Flame Retardant Curatives for Thermoset Resins," *Polymer Degradation and Stability,* 54(2–3):, 311–315, 1996.
41. (a) Levchik, S.V., Camino, G., Luda, M.P., Costa, L., Muller, G., Coates, B., and Henry, Y., "Epoxy Resins Cured with Aminophenylmethylphosphine Oxide. I. Combustion Performance," *Polymers for Advanced Technologies,* 7(11): 823–830, 1996; (b) "II. Mechanism of Thermal Decomposition," *Polymer Degradation and Stability,* 60(1): 169–183, 1998.
42. Costa, L., DiMontelera, L.R., Camino, G., Weil, E.D., and Pearce, E.M., "Flame Retardant Properties of Phenol-Formaldehyde-Type Resins and Triphenyl Phosphate in Styrene Acrylonitrile Copolymers," *Journal of Applied Polymer Science,* 68(7): 1067–1076, 1998.
43. Mandal, H., and Hay, A.S., "Polycyclic Phosphonate Resins: Thermally Crosslinkable Intermediated for Flame-Resistant Materials," *Journal of Polymer Science, Part A, Polymer Chemistry,* 36(11): 1911–1918, 1988.
44. Jang, J., Chung, H., Kim, M., and Kim, Y., "Improvement of Flame Retardancy in Phenolics and Paper/Sludge Phenolic Composites," *Journal of Applied Polymer Science,* 69(10): 2043–2050, 1998.
45. Swett, B., "Unsaturated Polyester Resin Market Overview," *FRCA Conference,* March 1996.
46. Georlette, P., "Applications of Halogen Flame Retardants," in Horrocks, A.R., and Price, D., *Fire Retardant Materials,* Woodhead Publishing, CRC Press, 2000.
47. Burton, J., and Sutker, A., "Thermoplastic Elastomers for Wire and Cable Applications—Flame Retardant Opportunities," *International Wire and Cable Symposium Proceedings,* 1981.
48. Joseph, M., and Lesniewski, A., "Comparison of the Relative Cost Effectiveness of Brominated and Chlorinated Flame Retardants in EPDM," *International Wire and Cable Symposium Proceedings,* 1980.
49. Indyke, D.M., and Pettigrew, F.A., "High Performance Flame Retardants for Wire and Cable Applications," Fall 1991 Conference, Flame Retardant Chemicals Association (FRCA).
50. Gosh, S.N., and Maiti, S., "A Polymeric Flame Retardant Additive for Rubbers," *Journal of Polymer Materials,* 11(1): 49–56, 1994.
51. Derouet, D., Morvan, F., and Brosse, J.-C., "Flame Resistance and Thermal Stability of 1,4-Polydienes Modified by Dialkyl or Diaryl Phosphates," *Journal of Natural Rubber Research,* 11(1): 9–25, 1996.

52. Derouet, D., Radhakrishnan, N., Brosse, J,-C., and Boccaccio, G., "Phosphorus Modification of Epoxidized Liquid Natural Rubber to Improve Flame Resistance of Vulcanized Rubbers," *Journal of Applied Polymer Science,* 52(9): 1309–1316, 1994.

53. Brosse, J.-C., Koh, M.P., and Derouet, D., "Modification Chimique du Polybutadiene-1,2 par le Phosphonate d'Ethyl," *European Polymer Journal,* 19(12): 1159–1165, 1983.

54. Derouet, D., Brosse, J.-C., and Pinazzi, C.P., "Modification Chimique des Polyalkadienes par les Composes du Type CX3POZ2. I. Modification des Polybutadienes-1,2 par le Dichlorure de Trichloromethylphosphonyle," *European Polymer Journal,* 17(7): 763–772, 1981.

55. Camino, G., Guaita, M., and Priola, A., "Study of Flame Retardance in Brominated Liquid Polybutadienes," *Polymer Degradation and Stability,* 12(3): 241–247, 1985.

56. Gilman, J.W., Jackson, C.L., Morgan, A.B., Harris, R., Manias, E., Giannelis, E.P., Wuthenow, M., Hilton, D., and Phillips, S., "Flammability Properties of Polymer Layered-Silicate (Clay) Nanocomposites," *Flame Retardants 2000,* Interscience, London, 2000, p. 49.

57. Kandola, B., "Nanocomposites," in Horrocks, A.R., and Price, D., *Fire Retardant Materials,* CRC Press, 2000.

58. Kawasumi, M., Hasegawa, N., Kato, M., Usuki, A., and Okada, A., "Preparation and Mechanical Properties of Polypropylene Clay Hybrids," *Macromolecules,* 30: 6333, 1997.

59. Giannelis, E.P., "Polymer Layered Silicate Nanocomposites," *Advanced Materials,* 8(1), 29, 1996.

60. Hasegawa, N., Kawasumi, M., Kato, M., Usuki, A., and Okada, A., "Preparation and Properties of Polypropylene Clay Hybrids Using Maleic Anhydride Modified Polypropylene Oligomer," *Journal of Applied Polymer Science,* 67: 87, 1998.

61. Anderson, P., "Optimizing Processing for Nanocomposites," FRCA 2002.

62. Zhu, J., and Wilkie, C., "Thermal and Fire Studies on Polystyrene-Clay Nanocomposites," *Polymer International,* 49(10): 1158, 2000.

63. Moet, A.S., and Alekah, A., "Polymer-Clay Nanocomposites: Polystyrene Grafted onto Montmorillonite Interlayers," *Materials Letters,* 18: 97, 1993.

64. Kojima, Y., Usuki, A., Kawasumi, M., Okada, A., Kuraucji, T., and Kamigaito, O., "Synthesis of Nylon 6-Clay Hybrid by Montmorillonite Intercalated with e-Caprolactam," *Journal of Polymer Science, Part A, Polymer Chemistry,* 31: 983, 1993.

65. Kornmann, X., Berglund, L.A., Sterte, J., and Giannelis, E.P., "Nanocomposites Based on Montmorillonite and Unsaturated Polyester," *Polymer Engineering and Science,* 38(8): 1351, 1998.

66. Blumstein, A., "Polymerization of Adsorbed Monolayers. II. Thermal Degradation of the Inserted Polymer," *Journal of Polymer Science, Part A, Polymer Chemistry,* 3: 2665, 1965.

67. Lee, J., Taketoshi, T., and Giannelis, E.P., "Fire Retardant Polyetherimide Nanocomposites," *Materials Research Society Symposia Proceedings,* 457: 513, 1997.

68. Burnside, S.D., and Giannelis, E.P., "Synthesis and Properties of Two Dimensional Nanostructures by Direct Intercalation of Polymer Melts in Layered Silicates," *Chemistry of Materials,* 5: 1694, 1993.

69. Gilman, J.W., Kashiwagi, T., Nyden, M., Brown, J.E.T., Jackson, C.L., Lomakin, S., Giannelis, E.P., and Manias, E., "Flammability Studies of Polymer Layered Silicate Nanocomposites: Polyolefin, Epoxy and Vinyl Ester Resins," in Al-Malaika, S., Golovoy, A., and Wilkie, C.A., eds., *Chemistry and Technology of Polymer Additives,* Blackwell Science, Oxford, UK, 1999, Chapter 14.

70. Bourbigot, S., Le Bras, M., Dabrowski, F., Gilman, J.W., and Kashiwagi, T., *Fire and Materials,* 24: 201–208, 2000.

71. Wang, J., Du, J., Zhu, J., and Wilkie, C.A., "An XPS Study of the Thermal Degradation and Flame Retardant Mechanism of Polystyrene-Clay Nanocomposites," *Polymer Degradation and Stabilization,* 77(2), 249–252, 2002.

72. Lomakin, S.M., Usachev, S.V., Koverzanova, E.V., Ruban, L.V., Kalinina, I.G., and Zaikov, G.E., "An Investigation of the Thermal Degradation of Polymer Flame Retardant Additives: Triphenylphosphate and Modified Intercalated Triphenyl Phosphate, *10th Annual Conference, Recent Advances in the Fire Retardancy of Polymeric Materials,* Business Communications Co., Norwalk, CT, 1999.

73. Ruban, L., Lomakin, S., and Zaikov, G., "Polymer Nanocomposites with Participation of Layer Aluminium Silicates," in Zaikov, G.E., and Khalturinski, N.A., eds., *Low Flammability Polymeric Materials,* Nova Science Publishers, New York, 1999.

74. Takekoshi, T., Fouad, F., Mercex, F.P.M., and De Moor, J.J.M., US Patent 5773502, issued to General Electric Company, 1998.
75. Okada, K., Japan Patent 11–228748, 1999.
76. Inoue, H., and Hosokawa, T., Japan Patent Application (Showa Denko K.K.), JP 10 81510, 1998.
77. Zhu, J., Uhl, F.M., and Wilkie, C.A., "Fire and Polymers," in Nelson, G.L., and Wilkie, C.A., eds., *ACS Symposium Series 797,* Washington, 2001, pp. 24–33.
78. Lan, T., Qian, G., Liang, Y., and Cho, J.W., FRCA 2002.
79. Beyer, G., "Progress on Flame Retardancy of Polymers with Nanocomposites, presented at Cables 2002, Cologne, March 19–21, 2002.

CHAPTER 5
FIBERS AND FABRICS

Dr. Debbie J. Guckert, Susan L. Lovasic, and Dr. Roger F. Parry
DuPont Personal Protection
Richmond, VA 23261

5.1 THE ROLE OF FABRICS IN FIRE PROTECTION

Fibers and fabrics can be involved in fire protection in several different ways. The most obvious is their use in protective clothing worn by those engaged in fighting fires or those who work in situations where the potential for fire or other thermal hazard is a factor. Fabrics are often a part of the environment where a fire might occur, such as upholstery or floor coverings in a theater or airplane. Fabrics also play a part in garments of individuals who do not anticipate being in a fire situation.

In this chapter we will begin by providing some basic information about fibers and their properties, fibers being the basic precursor materials from which clothing, floor covering, or upholstery fabrics are made. We will then consider the types of fabrics made from these fibers, their characteristics in fire situations, and the tests used to evaluate these fabrics in different applications. Finally, and this will be the major part of this chapter, we will turn to specific applications and example situations in which fire-resistant fabrics might be used. Here we will explore the performance of various fabrics in a variety of circumstances.

5.2 FIBERS AND THEIR PROPERTIES

5.2.1 Materials from Which Fibers Are Formed

Fibers and fabrics can be made from a variety of materials: glass, ceramic, metal, and natural and synthetic organic polymers being the most common. In a fire situation the fabrics from organic polymers have the disadvantage that under severe enough conditions they will burn and contribute to the thermal energy generated in a fire. However, they also have significant advantages over the other options in their potential to offer improved wearer comfort (in protective clothing) and good thermal insulation properties. In recent years, organic polymer-based fabrics have been developed that do not burn in practical fire situations and have become a major factor in fire protection.

5.2.2 Forms of Fibers Available

Fibers are available in two principle forms: continuous filament and staple. Continuous filament fibers are, as the name implies, fibers composed of filaments that are essentially infinite in length. Staple fibers are composed of discrete lengths of filament.

Synthetic organic fibers are usually manufactured in continuous form by an extrusion process through a specially designed multihole plate (or spinneret) producing a multifilament yarn bundle.

Cutting the continuous filaments into appropriate lengths produces the staple form of these fibers. Metal and glass fibers are also produced in continuous form by spinning (extruding) the molten form of the material. Naturally occurring fibers, such as wool and cotton, are produced naturally in staple form. Continuous lengths of staple fibers are produced by a mingling or traditional spinning process, which combines and entangles the short filaments. There are several types of spinning processes used today. Examples include ring, rotor, air-jet, and friction spinning [1, 2].

5.2.3 Fiber Properties

The one key dimension of a fiber is its diameter. Both the diameters of the individual filaments as well as the overall yarn bundle are important. The textile term used to characterize the diameter is *denier,* which is the weight in grams of 9000 m of the filament or fiber yarn bundle. Filament diameter is often termed *denier per filament* (dpf). In addition, because not all fibers are round, their cross-sectional shape can impart other important characteristics, e.g., improved moisture transport, increased fabric bulk (due to lower fiber packing efficiencies), and other specialized functions.

Fiber strength is termed tenacity, with units of *grams per denier* (gpd). A stress-strain curve for a fiber is a plot of tenacity versus elongation. The initial slope of the stress-strain curve is termed modulus, also with units of gpd. For most fibers, another important property is equilibrium moisture content, or the water content at a specified temperature and relative humidity.

Tables 5.1 and 5.2 list several fibers used in textile applications along with their key property ranges [3–7]. A comprehensive treatment of fibers, their structure, and their properties can be found in several books [1, 7].

5.2.4 Flammability Characteristics

The primary characteristics of materials that are considered when dealing with fire are ignition, flame spread, energy or heat release rate (HRR), and the production of smoke, toxic gases, and corrosive products. These characteristics apply to materials that might contribute to a fire. Much of the test work done has been from this perspective. When we discuss fabrics and flammability in subsequent sections, we will provide other perspectives.

One widely used quantitative measure of a material's flammability is its *limiting oxygen index* (LOI). This measurement determines the concentration of oxygen in air that will support combustion once a material has ignited. A specimen for this test is typically mounted vertically in a chimney and the concentration of oxygen in air that will support continued burning is determined. Materials with an LOI greater than 21 percent (the approximate concentration of oxygen in air) are generally considered nonflammable.

Other well-known properties of materials that relate to flammability include the heat capacity (specific heat), the heat of combustion, the char fraction that remains after pyrolysis, and the temperature at which a material would spontaneously combust (ignite without an external ignition source).

A lesser-known characteristic of materials is the *heat release capacity*. Walters and Lyon [8] have developed experimental techniques for measuring this characteristic, which they claim is a basic property of materials, and that it can be estimated from molar contributions. The test involves the programmed heating of small material samples to very high temperatures and the complete combustion of the resulting pyrolysis gases. The primary measurement is the heat release rate.

Table 5.3 shows those material properties related to flammability for some commonly used fiber-forming materials [8–14].

The basic flammability of a material is only one of the items that must be considered. Features such as additives can be incorporated to reduce flammability. The geometry of a product can have a major impact on its tendency to burn. Materials can be combined in a complex system to produce excellent flammability performance.

TABLE 5.1 Properties of Common Fibers Used in Textile Applications [3, 4, 5, 6, 7]

Property	Polyester Regular tenacity	Polyester High tenacity	Polyester Staple	Carbon	Polyamide 6-Nylon	Polyamide 66-Nylon	Viscose rayon	Acrylic	Modacrylic	Poly tetrafluoro-ethylene
Tenacity, dry (g/denier)	2.8–5.7	5.9–9.7	2.7–7.1	24.1	3.3–7.2	2.3–7.5	2.0–3.1	2.0–3.0	1.7–2.6	0.9–2.1
Tenacity, wet (g/denier)	2.8–5.7	5.9–9.7	2.7–7.1	24.1	3.7–6.2	2.0–5.7	1.2–2.2	1.1–2.7	1.5–2.4	0.9–2.1
Tensile strength (kpsi)	50–99	106–168	41–105	550	73–100	40–106	—	30–45	29–45	40–50
Elongation at break (%)	19–42	10–34	18–65	1.6	17–46	25–65	17–20	40–55	45–60	40–62
Elastic Recovery (%)	76@3%	88@3%	81@3%	100	98@10%	88@3%	82@2%	73@3%	99@5%	—
Stiffness (g/denier)	10–30	30	12–17	1500	18–23	5–24	6–17	5–7	3.8	1.0–13.0
Toughness (g-cm)	0.4–1.1	0.5–0.7	0.2–1.1	—	0.67–0.9	0.8–1.25	0.2	0.62	0.5	0.15
Specific Gravity	1.38	1.38	1.38	1.77	1.14	1.14	1.52	1.17	1.34	2.2
Moisture regain (%)										
—@ 70°F, 65% RH	0.4	0.4	0.4	—	3–5	4–4.5	12–14	1–2	0.5–3	0
—@ 70°F, 95% RH	0.6	0.6	0.6	—	3.5–8.5	6.1–8.0	—	2.0–2.5	—	0
Melt temperature (°F)	482	482	482	DNM	419–430	480–500	DNM	DNM	DNM	620
Decomposition t. (°F)	570	570	570	600	600–730	610–750	350	530	—	—
Chemical resistance (acid/alkali)	Good	Good	Good	Good	Fair	Fair	Fair	Good	Good	Excel
Solvent resistance	Excel	Excel	Excel	Excel	Good	Good	Excel	Excel	Excel	Excel
Sunlight resistance	Good	Good	Good	Excel	Good	Good	Good	Good	Excel	Excel
Abrasion resistance	Excel	Excel	Excel	Poor	Excel	Excel	Good	Excel	Good	Good

DNM = Does not melt.

TABLE 5.2 Properties of Common Fibers Used in Textile Applications [3, 4, 5, 6, 7]

	Olefin		Polybenzimidazole	Aramid		Cotton	Silk	Wool	Glass
Property	Polyethylene	Polypropylene		Meta	Para				
Tenacity, dry (g/denier)	3.4–6.8	2.3–7.9	2.6–3.0	4.9–5.1	23	2.1–6.3	2.8–5.2	1.0–1.7	3.1–4.7
Tenacity, wet (g/denier)	3.4–6.8	2.3–7.9	2.1–2.5	3.0–4.1	21.7	2.5–7.6	2.4–4.4	0.9–1.5	3.1–4.7
Tensile strength (kpsi)	261–609	20–50	50	85–110	425	42–125	45–83	17–28	100–150
Elongation at break (%)	2.7	30–100	25–30	25–32	3.4	3–10	13–31	20–50	1.0–2.0
Elastic recovery (%)	—	95@5%	—	—	100@3%	45@5%	33@20%	63@20%	—
Stiffness (g/denier)	1,200	12–25	9–12	80–120	585	42–82	76–117	24–34	—
Toughness (g-cm)	0.7	0.7–3.0	0.4	0.85	—	—	—	—	0.4
Specific gravity	0.97	0.91	1.43	1.38	1.44	1.5	1.25	1.3	2.47
Moisture regain (%)									
—@ 70°F, 65% RH	0	0.01	15	4.3	4.5	7–8	10	14–18	0
—@ 70°F, 95% RH	0	0.01	20	6.5	7.5	—	—	—	—
Melt temperature (°F)	300	320	DNM	DNM	DNM	DNM	DNM	DNM	1,220
Decomposition t. (°F)	650–820	610–750	860	800	900	310	—	230	—
Chemical resistance (acid/alkali)	Excel	Excel	Excel	Good	Good	Fair	Good	Fair	Excel
Solvent resistance	Excel	Excel	Excel	Excel	Excel	Excel	Good	Good	Excel
Sunlight resistance	Good	Good	Fair	Fair	Fair	Good	Fair	Good	Excel
Abrasion resistance	Good	Good	Good	Good	Good	Fair	Good	Fair	Poor

DNM = Does not melt.

TABLE 5.3 Material Properties Related to Flammability [8, 9, 10, 11, 12, 13, 14]

Material	Heat Capacity @ 25°C (cal/g °C)	Thermal Conductivity (W/m K)	Heat of Combustion (cal/g)	Ignition Temperature (°C)	Limiting Oxygen Index (%)	Heat Release Capacity[b] (J/g-K)	% Char in HRR Test
Polyester (polyethylene terephthalate)	0.30	0.141	−5,280	480	20.5	332	5.1
Polyamide (6 nylon)	0.38	0.247	−7,350	450	20.8	487	0
Polyamide (66 nylon)	0.40	0.243	−7,530	532	20.1	615	0
Polyethylene	0.55	0.335–0.523	−11,100	349	17.4	1,676	0
Polypropylene	0.46	0.117	−11,000	550	17.7	1,571	0
Polyvinyl alcohol	0.41	—	−5,950	—	21.6	533	3.3
Polytetrafluoro ethylene	0.25	0.251	−1,200	530	>100	35	0
Acrylic (polyacrylonitrile)	0.36	0.200	−7,670	560	19.9	—	—
Modacrylic (vinyl chloride/acrylonitrile copolymer)	—	—	—	315*	29.5	—	—
Aramid–meta (poly m-phenylene diamine)	0.30	0.250	−6,700	427*	30	52	48.4
Aramid–para (poly(p-phenylene diamine)	0.34	0.043	−8,300	482*	29	302	36.1
PBI (polybenzimidazole)	0.30	0.400	—	~500*	38	36	67.5
Cotton	0.29	0.461	−4,400	400	18–20	—	—
Wool	0.32	0.193	−4,900	600	25	—	—
Rayon	0.30	0.289	−3,900	420	19.7	—	—

* Decomposition temperature.

In the simplest case, a fabric could be made from a single type of fiber with no additives or treatments. The suitability of this material for a given application where a threat of fire exists could then be related to its measured flammability properties (e.g., the LOI or the heat release capacity).

One class of fibers that does not burn is that based on refractory oxides or glass-type materials (although many applied finishes used on these fibers will burn). Fibers and fabrics made from highly aromatic polymers, such as the commercial fibers of PBI® (polybenzimidazole), Nomex® (*meta*-aramid), and Kevlar® (*para*-aramid) are inherently flame resistant. These polymers pyrolize when exposed to high thermal energies leaving a large char residue.

5.2.5 Flame Retardants

Fibers or fabrics that are not inherently flame resistant can be modified by a variety of chemical treatments to impart some flame resistance. Significant research and development efforts have been pursued seeking durable, effective, and low-cost solutions to improved flame-retardant performance. A broad summary of this effort can be found in a text prepared by Atlas and Pearce [15]. It explores the various methods developed for reducing the flammability on a number of fiber- and yarn-forming materials, including cotton and wool.

5.2.5.1 *Cotton*

Various combinations of ammonium salts of the mineral acids have long been used to impart flame resistance to cotton [16]. To this day, phosphoric acid precursors are applied to cotton to provide multiple-use flame resistance. This is done in three ways:

- A reaction with methylol linkages in cotton cellulose. This results in dramatic strength loss to the textile caused by the cross-linking reaction.

- Coating with various binder systems. There is questionable durability from this approach due to the loss of binders to abrasion and laundry chemicals.
- The formulation of a high molecular weight polymer within the cotton by reaction with gaseous ammonia. This has been the most commercially successful procedure.

Although some of these treated fabrics maintain their flame-resistant (FR) qualities after multiple home-laundry cycles, they are not usually as durable when subjected to commercial laundering [18]. In recent years, cotton-based FR fabrics have been developed with improved durability when subjected to the extreme conditions used by commercial laundry processes. The process is known as the ammonia-cure method of applying a flame retardant to cotton material. The cotton fabrics are treated with phosphonium salts then chemically cured in an ammoniator. Two of the better-known commercial products of this type are Proban® and Pyroset® [18]. Another approach to durable flame-retardant cotton is the treatment with phosphorus-based compounds that will react with the cellulose, the main constituent of cotton fiber [19]. Even with these advancements in imparting flame-resistant characteristics to cotton, laundry situations do exist (use of chlorine bleach or pH imbalances) that readily cause the removal of the treatment and its flame-resistance properties.

5.2.5.2 Wool

A similar effort to cotton has been made to develop flame-retardant treatments for wool. Wool, without modification, is generally considered nonflammable with an LOI of ~25, but there are commercial incentives to improve its performance and much research effort has been expended to accomplish this goal.

As with cotton, the simplest technique for improving performance is chemical impregnation. Here a water-soluble salt is added to the wool. Ammonium salts of phosphoric, sulfamic, and boric acids are especially effective [20], but these treatments have not proved durable.

The most effective treatment has involved the salts of zirconium hexafluoride, potassium hexafluoro titanate, and potassium hexafluoro zirconate. This technology, developed by the International Wool Secretariat, is known as the Zirpro® flame-retardant process. The treatment is durable to repeated laundering [21]. A combination of Zirpro® and tetrabromophthalic acid provides additional flame-retardant performance [22].

5.2.5.3 Synthetic Fibers

Several synthetic fibers are inherently nonflammable in air, including polybenzimidazole (LOI near 40) and the aramids (LOI near 30). Most other important commercial synthetic fibers have LOI values at or below 21 percent. Of these, flame-retardant variants have been developed for acrylic and polyester fibers. Attempts to produce commercially viable flame-retardant versions of nylon and polyolefin fibers have been less successful.

Acrylic fibers (polyacrylonitrile) are quite flammable, with an LOI near 20. An inherently nonflammable derivative of this material can be produced by copolymerization with 20 to 60 percent vinyl chloride. This class of copolymers is termed *modacrylic* and is much less flammable than the homopolymer, having LOI values near 30.

Polyester, *polyethylene terephthalate* (PET), accounts for over half of the worldwide production of synthetic fibers. It is often used in blends with cotton, wool, and many other fibers. Unmodified, PET has an LOI of approximately 20 to 21. The reduction of the flammability of PET is accomplished in a similar manner to that of most other noninherent flame-retardant fibers. This involves the application of flame-retardant coatings, or adding components to the polymer chain itself via copolymerization. The compounds used for both coating and copolymerization are usually based on halogens (chlorine and bromine), on phosphorus, or on both [23]. As with other fabrics, topical applications of coatings are less durable, but some treatments can provide durable flame-retardant properties, increasing the LOI to the 28 to 30 range [24]. Flame-retardant fibers based on copolymers of PET with phosphorus-containing compounds have been available for over 20 years [25]. It should be noted, however, that these flame-retardant-treated polyesters are still thermoplastics. Exposure to heat and flame in most fire-threat situations will lead to melting and dripping.

5.3 FABRIC TYPES

There are many fabric types available and diverse methods for producing them. The most common fabric forms include woven, knit, nonwoven, and tufted materials. The connection among these fabric types is the conversion of fibers and yarns into three-dimensional structures of sufficient width and length for use in the manufacture of clothing, upholstery, draperies, carpets, and so forth. Within each general type of fabric, there are many subcategories that offer different characteristics. The exact type and subcategory of fabric selected for a particular application depends on the ultimate end-use needs. Aspects such as durability to use, cost, styling, and design, as well as flammability, must be considered before selecting a particular fabric type. Factors including pattern, weight, and thickness are also of importance in determining the appropriate structure for a particular application. Detailed descriptions of the most commonly available fabric types and their structural features can be found, for example, in Adanur [26].

5.4 FLAMMABILITY OF FABRICS

5.4.1 Characteristics of Burning Fabrics

When we described the burning process earlier, in reference to a material's contribution to a fire, we identified ignition, flame spread, and combustion products as key characteristics. As we move to describe the role of fabrics in fire protection, we will shift our perspective. Of course, some fiber products such as carpeting or draperies should still be considered from the perspective of how much fuel they contribute to a fire. However, here we will consider the heat transfer features of fabrics that are used in clothing or upholstery.

When we consider protective clothing, the first requirement is that its component fabrics not support combustion in the thermal environment of a fire (continue to burn after the fire is removed). The next is that the clothing must provide thermal insulation protection to reduce the potential for convective and radiant heat transfer. This feature is typically achieved in fabrics by utilizing tight weaves, increased thickness, and increased basis weight (weight per unit area). In specialized protective fabrics, this can also be accomplished by surface reflective treatments (e.g., aluminum coatings) to further reduce the radiant component of the thermal threat. In environments where substantial thermal threats exist, such as those found in structural fire fighting, the addition of thick layers of low-density structures or battings is also employed. Here, the low thermal conductivity of air is exploited as a component of the thermal barrier.

Garment fit, which is sometimes overlooked when considering the characteristics of protective clothing, is a key contributing factor in reducing heat transfer in a fire environment. A tight-fitting garment can increase the potential for burn injuries. This is due to the elimination or significant reduction in thickness of the thermal insulating air layer between the skin and exposed garment fabric. Essentially, the direct contact or conductive heat transfer from the exposed heated fabric to the skin becomes more prominent. A protective garment that is significantly oversized can also fail to protect by allowing easy penetration of flames and convective heat, bypassing the primary barrier system and conceivably igniting flammable undergarments.

Other germane adverse effects must also be evaluated in protective fabric systems designs. Factors such as the fabric's potential to store thermal energy must be considered. This can contribute to burn injury well after the external thermal hazard is removed. Additionally, a clothing system's protective characteristics should not be impaired by water, either from a fire hose or by perspiration. Excessive water retention reduces a system's overall thermal insulating performance and can result in scalding and steam burns. Protective clothing systems should be permeable to water vapor to meet wearer comfort standards. Generally, protective clothing should also be as lightweight as possible to reduce physiological thermal stress. This complex list of performance characteristics can require a clothing system that incorporates more than one material layer.

Different materials, in fabric form, respond differently to exposure to a flame or thermal assault.

Some will be totally consumed by burning without melting or dripping. Others will burn readily, but will melt in the process and drip molten polymer (which is usually burning). Still others will not sustain burning, but will form a char residue as the fire raises the material's temperature above its decomposition point. Some will be unaffected by the flame, neither burning nor charring. It is important to consider that the most severe burn injuries occur when clothing has ignited and continues burning. The heat energy transferred to the wearer is the greatest, due primarily to the physical proximity of the fire to the skin.

Fabrics that burn easily and totally provide additional fuel to the fire: No thermal protection is imparted. Additionally, materials that drip as they burn commonly expand the scope of the fire. Usually this is to places below the initial burning area where the fire might not have otherwise spread. In addition, clothing made from materials that melt in a fire present an increased risk for burn injury due to the heat conducted directly to the skin (from the heat released in solidification).

5.4.2 Thermal Performance Tests for Fabrics

5.4.2.1 Protective Clothing

There are a number of performance tests that characterize fabrics used in flame, thermal electric arc, and generalized heat-protective clothing. The key tests for fabrics in these applications are summarized in Table 5.4.

General Flammability Tests. The "Vertical Flammability Test" (ASTM D6413) is the simplest of these assessments and is used as a screening test. It was developed in the 1920s and is widely used today. Its purpose is to answer the question: "Does this fabric burn?" especially in reference to clothing. For this investigation, a section of fabric is mounted vertically and a specified flame is applied to its lower edge for a defined time, then removed. The response of the fabric to the flame exposure is recorded. The length of the fabric that is burned, or charred, is measured. Times for afterflame (the continued burning of the specimen after removal of the test flame) and afterglow (characterized by smoldering after removal of the test flame) are also measured. Additionally, observations regarding melting and dripping of the specimen are recorded. Pass/fail specifications based on this method are established for industrial worker clothing, firefighter turnout gear and flame-retardant station wear, and military clothing. Fig. 5.1 shows a photograph of the vertical flammability test apparatus.

The screening test for flammability used in Europe is the *limited flame spread* test (EN-532). In this test, a piece of fabric is mounted vertically, subjected to a flame directed at the center of the fabric, and the extent to which the flame spreads is noted. It is a pass/fail test. Other European tests and standards for protective clothing are summarized in Table 5.5 [27].

Flash Fire Heat Energy Transfer Tests. The next characterization, typically used to evaluate protective fabrics that pass the vertical flammability criteria, attempts to answer the question: "How much protection does a fabric provide from heat and flame?" The most common test used to respond to this question is the *thermal protective performance test*" or TPP test (NFPA 2112, Section 8.2). This heat transfer assessment, developed in the1970s [28], subjects a test fabric or layered system to a controlled heat flux and attempts to measure its heat transmission response during direct heat energy impingement (stored heat is ignored).

In this test, a flame is directed at a section of fabric mounted in a horizontal position at a specified heat flux (typically 84 kW/m^2). In newer variants of this apparatus, an additional radiant heat flux component is provided through a bank of electrically heated quartz tubes (see Fig. 5.2). The test measures the transmitted heat energy from the source through the specimen using a copper slug calorimeter (located either spaced above or in direct contact with the test specimen). A spacer is used when testing single-layer fabrics but is omitted in tests of multiple-layer systems. The test endpoint is characterized by the time required to attain a predicted second-degree skin burn injury using a simplified model developed by Stoll and Chianta [29]. The value assigned to a specimen in this test, denoted as the TPP value, is the total heat energy required to attain the endpoint, or the direct heat

TABLE 5.4 Test Methods Used to Characterize Fabrics Used for Protective Clothing

Test method	Title	Scope
ASTM D 6413–99	Flame Resistance of Textiles (Vertical Flammability Test)	Test method is used to measure the vertical flame resistance of textiles. As part of the measure of flame resistance, char damage length, afterflame, and afterglow characteristics are evaluated.
ASTM D 4108–87 (discontinued in 1996)	Thermal Protective Performance of Materials for Clothing by Open-Flame Method	Test method rates textile materials for thermal resistance and insulation when exposed to a convective energy of about 2.0 cal/cm^2 sec for a short duration. It is applicable to woven fabrics, knit fabrics, battings, and nonwoven fabrics intended for use as clothing for protection against a chance, short exposure to open flames.
ASTM F 1930–99	Evaluation of Flame Resistant Clothing for Protection Against Flash Fire Simulations Using an Instrumented Manikin	Test method covers quantitative measurements and subjective observations that characterize the performance of single layer garments or protective clothing ensembles in a simulated flash fire environment having controlled heat flux, flame distribution, and duration.
ASTM F1939–99a	Radiant Protective Performance of Flame Resistant Clothing Materials	Test method covers a means of measuring the effect of radiant heat exposure at the standard levels of (a) 0.5 or (b) 2.0 cal/cm^2 sec on a fabric specimen or a fabric assembly specimen. For use with fabrics that are flame resistant and that are used in the manufacture of protective clothing.
ASTM F 1959/ F 1959M–99	Determining the Arc Thermal Performance Value of Materials for Clothing	Test method used to measure the arc thermal performance value of materials intended for use as flame resistant clothing for workers exposed to electric arcs that would generate heat flux rates from 2 to 600 cal/cm^2 sec.
ASTM F 1958/ F 1958M–99	Determining the Ignitability of Non-Flame-Resistant Materials for Clothing by Electric Arc Exposure Method Using Mannequins	Test method used to identify materials that are ignitable and that can continue to burn when exposed to an electric arc, and determines (a) the incident exposure energy that causes ignition, and (b) the probability of ignition. The specimens tested are materials fabricated in the form of shirts. This test shall be used to measure and describe the properties of materials, products or assemblies in response to convective and radiant energy generated by an electric arc under controlled laboratory conditions.
ASTM F 1060–87 (reapproved 1993)	Thermal Protective Performance of Materials for Protective Clothing for Hot Surface Contact	Test method used to rate textile materials for thermal resistance and insulation when exposed for a short period of time to a hot surface with a temperature up to 600°F. It is applicable to woven fabrics, knit fabrics, battings, and sheet structures intended for use as clothing for protection against short exposures to hot surfaces.
ASTM F955–96	Evaluating Heat Transfer Through Materials for Protective Clothing Upon Contact with Molten Substances	Test method covers the evaluation of materials' thermal resistance to molten substance (aluminum, brass, and iron) pour by describing means of measuring heat transfer.

FIGURE 5.1 Vertical flammability test apparatus.

source exposure time to the predicted burn injury multiplied by the incident heat flux. Higher TPP values denote better insulation performance.

Fig. 5.3 shows an overlaid example of the spaced TPP method on inherently flame-resistant *meta*-aramid (203 g/m^2 [6.0 oz/yd^2] Nomex® IIIA fabric) and a chemically treated flame-resistant cotton (305 g/m^2 [9 oz/yd^2] Indura® fabric)-based materials. In this example, we apply an 85.0 W/m^2 (2.03 cal/cm^{2}s) heat energy source for 5.62 s to the Indura® sample (TPP = 2.03 cal/cm^{2}s × 5.62 s or 11.4 cal/cm^2) and 6.05 s to the Nomex® sample (TPP = 2.03 cal/cm^{2}s × 6.05 s or 13.2 cal/cm^2) to reach

TABLE 5.5 Applications: Relevant European Norms for Protective Clothing and Workwear [27]

EN 340	Protective clothing; general requirements
EN 366	Protective clothing; protection from heat and fire. Evaluation of materials and material assemblies exposed to a source of radiant heat.
EN 367	Protective clothing; protection from heat and flames; determination of heat transmission on exposure to flame.
EN 373	Protective clothing; assessment of resistance to molten metal splash.
EN 469	Protective clothing for firefighters.
EN 531	Protective clothing for workers exposed to heat.
EN 532	Protective clothing; protection against heat and flame; test for limited flame spread.
EN 533	Protective clothing; protection against heat and flame; limited flame spread materials and material assemblies.

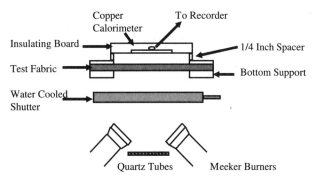

FIGURE 5.2 Thermal protective performance test apparatus.

the predicted second-degree burn injury criteria. The figure also illustrates the stored heat energy transmitted from the sample fabric to the calorimeter at the completion of the TPP measurement (collected after the heat source is removed)—a critical heat contributor to burn injury that is ignored in this test. The difference in the response and shape of the two sample curves is due to difference in mass between them and the result of activation and subsequent depletion of the flame-retardant agent in the treated cotton.

A variation of the TPP test that attempts to address radiant-only heat energy transmission through a specimen is the *radiant protective performance test* or RPP test. This method replaces the combined convective and radiant heat energy source of the TPP unit with one that is essentially only radiant. This test is designed to provide information about the level of protection a fabric would provide in a situation where the hazard is predominantly radiant, such as an aircraft fuel fire or a wildlands fire. For the RPP method, the typical heat flux ranges from 21 kW/m^2 (NFPA 1977) to 84 kW/m^2 (NFPA 1976).

A similar test exists for Europe, e.g., ISO 17492, where the time to reach a specified sensor temperature rise is used in lieu of the Stoll and Chianta predicted second-degree burn injury criteria for determining the endpoint.

Unfortunately, these bench-top fabric tests (TPP, RPP, and the European variants) provide a

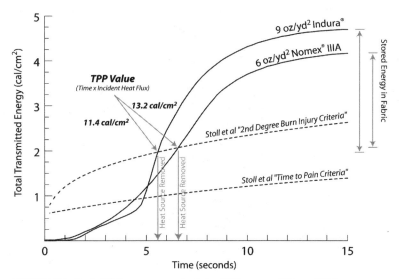

FIGURE 5.3 Typical TPP curves comparing Indura® FR cotton and Nomex® IIIA.

skewed response to the overall protective capability of source materials for garment or clothing systems. This is clearly seen in the stored energy contribution in the example shown above (Fig. 5.3) and in the many unsuccessful attempts to correlate the TPP results with full-scale thermal mannequin testing (described below). These test designs measure only the heat transfer during direct flame plus radiant (TPP), or radiant only (RPP) heat energy impingement. This is a significant omission in determining a protective fabric system's overall response to heat and flame. The stored energy by itself is more than sufficient to lead to predicted skin burn injury. In their current state, the direct heat impingement measurement methods are limited in the ability to address overall protective capabilities. This has been recognized by a U.S. standards organization, ASTM (F23.80 Subcommittee) and activity is currently under way to revise the methods to account for the stored energy effects. Until these new consensus standards are issued, the existing performance standards (e.g., NFPA 2112) will continue to include specifications based on the heat energy impingement-only tests.

A third class of testing has been developed to provide better information about the protective capability of actual clothing ensembles in a flash fire situation rather than only assessing component fabric properties. It is termed the "Instrumented Thermal Mannequin Test" (ASTM F 1930). It was developed in the 1970s, is more complex and costly to execute than its predecessor bench-top methods, and there are only three organizations currently in North America capable of carrying it out (E.I. du Pont de Nemours & Co., Inc., North Carolina State University, and the University of Alberta, Canada). This test method accounts for the stored energy in a fabric system (in a specific garment construction) and provides a better estimate of protective capability.

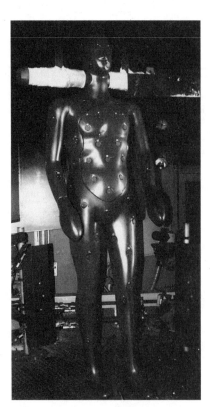

FIGURE 5.4 Instrumented thermal mannequin apparatus.

The core of this full-size test is the instrumented mannequin, the computer-based data acquisition, and the analysis software. The mannequin itself is designed with over 100 thermal sensors, imbedded into the ~1.8 m^2 of surface area, to measure incident heat energy. It is dressed in the clothing system to be evaluated and placed into a propane-based flash fire simulation chamber. The desired exposure heat energy is delivered in the flash fire chamber by providing the appropriate amount of propane fuel gas using a precision gas handling and delivery system. This culminates in a series of high-output *Big Bertha* torches that generate a precisely timed, uniform, ~84 kW/m^2, fuel-rich propane engulfment fireball. Thermal sensor information is acquired during and for a predefined period after the flash fire. An advanced burn injury model is then used to estimate the amount, degree, and location of predicted human skin burn injury from the mannequin. This test has the advantage of assessing actual garments under fairly realistic conditions. Its disadvantages include the absence of body movement, a fairly high cost to implement, and a somewhat limited flash fire exposure environment (convective and radiant heat energy exposure from propane gas combustion). Work is under way to address limited articulation (U.S. Navy Natick Research) and predominantly radiant energy exposures addressing a broader range of firefighter scenarios.

Fig.5.4 shows a photograph of the instrumented thermal mannequin apparatus, and Figs. 5.5, 5.6, and 5.7 show results during and after testing. Fig. 5.8 illustrates a predicted skin burn injury profile using the ASTM F 1930 burn injury model. Note that portions of

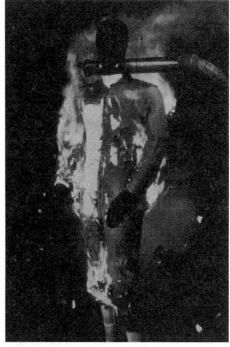

FIGURE 5.5 Flash fire exposure using instrumented thermal mannequin.

FIGURE 5.6 Flammable coverall continues burning after torches extinguish.

the body covered with undergarments in addition to outerwear show limited or no burn injury. Areas not covered (e.g., the head) by clothing received third-degree burns.

Electric Arc Heat Energy Transfer Tests. Although the test methods discussed thus far have chiefly focused on assessment of protective qualities when a fabric is subjected to a typical fire situation, the following two methods deal with a specialized type of fire exposure—electric arcs. Figs. 5.9 and 5.10 show the results of mannequin electric arc testing of an untreated cotton shirt and of a flame-resistant shirt.

The first of these methods, ASTM F-1958, the "Garment Ignition Test," is used to determine the amount of exposure energy (expressed in cal/cm^2) required to reach the 50 percent probability point for ignition of non-flame-resistant clothing materials. Here the material to be assessed is constructed into shirts, placed on a specified mannequin, and exposed to a range of electric arc exposure energies. The mannequin used for this test method is not instrumented since ignition is readily apparent. The test uses independent copper slug calorimeters for determination of incident energy.

The second of these specialized methods, ASTM F 1959, "Panel Electric Arc Test Method," is used to assess the thermal energy protective characteristics of flame-resistant fabrics for electric arc protective clothing. This test apparatus is illustrated in Fig. 5.11. In this test, a series of three single- (or multiple-) layer specimen fabrics is affixed onto instrumented flat panels and placed symmetrically and equally distant from a center *arc electrode*. Each of the three panels incorporates copper slug calorimeters to measure the incident and transported (through the fabric specimen) heat energy. During the testing protocol, electric arc discharges in the center arc electrode of varying heat energies are generated and material responses are recorded. The test outcome is the determination of an *arc thermal performance value* (ATPV) or *break open threshold energy* (E_{bt}). The ATPV is the level

FIGURE 5.7 Nomex® coverall self-extinguishes after torches extinguish.

of incident energy (expressed in cal/cm²) that is the 50 percent predicted probability to result in a second-degree burn injury with the specimen fabric or fabric system. The method uses the Stoll and Chianta criteria detailed in the TPP method above. If an ATPV is not obtainable using the various data selection criteria specified in the method, then an E_{bt} value is substituted. E_{bt} provides an estimate of the highest incident energy (again in cal/cm²) below the second-degree burn injury criteria that did not cause the fabric to fail (generation of a hole or tear of specified dimensions). Note that this method accounts for the stored energy in a fabric or system.

Conductive Energy Transfer Tests—Molten Metals. A separate class of heat energy transfer tests is available to evaluate protective clothing used in the molten metals industry. Here, the primary threat is conductive heat energy transfer from direct molten metal contact. In the United States the primary test procedure used is ASTM F955, "Standard Test Method for Evaluating Heat Transfer Through Materials for Protective Clothing Upon Contact with Molten Substances." It is designed to assess the protective characteristics of fabrics and systems from molten aluminum, brass, and iron. The method

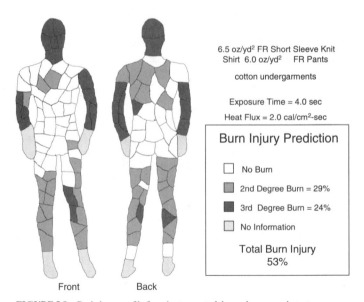

FIGURE 5.8 Body burn profile from instrumented thermal mannequin test.

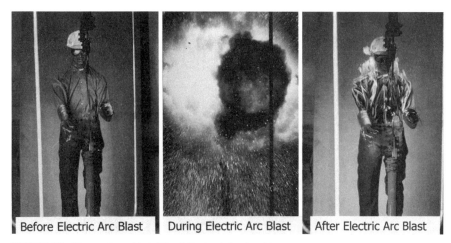

FIGURE 5.9 Mannequin arc blast: 5.7 oz/yd² untreated cotton shirt.

subjects a specimen to a quantity of molten metal, typically 2.2 lb, and evaluates the heat energy transferred as well as subjective information on the fabric's response. The test protocol has the fabric mounted on an instrumented (two copper slug calorimeters) and angled test board (70° from horizontal to allow run-off). The fabric is then subjected to a molten metal exposure, a pour of ~2.2 lb from a height of 12 in. Fabrics are rated primarily according to whether they ignite and if any molten metal adheres to their surface. The output of the copper slug calorimeters is also measured and a time to produce a predicted second-degree burn injury is determined (again using the Stoll and Chianta criteria). Similar test methods are employed in Europe (EN 373).

5.4.2.2 Furnishings

The role of fabrics in the category of "furnishings" can be as upholstery, bedding, carpeting, draperies, wall coverings, and a host of other components used to enhance the living environment. In fire scenarios where these materials are prevalent, the primary considerations are generally how

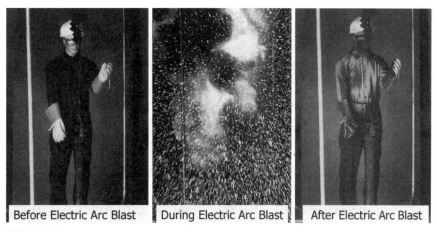

FIGURE 5.10 Mannequin arc blast: 4.5 oz/yd² flame-resistant shirt.

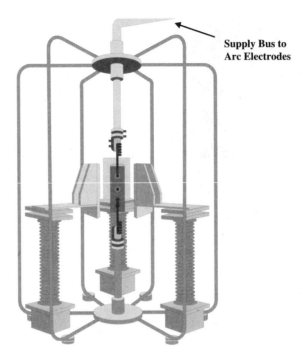

FIGURE 5.11 Panel electric arc test setup.

easily can the item be ignited, how quickly will the fire take hold, and at what rate is heat released once a fire has started. The methods for evaluating objects and materials are thus different from the methods used to evaluate protective clothing. The key test methods used to characterize fabrics in these applications are described in Table 5.6.

Since the mid-1990s, the primary small-scale test method used for furnishing applications is the "Oxygen Consumption Calorimeter," standardized as ASTM E 1354 [30–32]. According to Lyon [33], "heat release rate is considered to be the single most important fire parameter in determining the fire hazard of a material." Lyon has developed a theoretical framework relating HRR, time to ignition, and other characteristics of a burning system to the basic properties of the materials involved. Lyon has also recently added heat release capacity (mentioned in Section 5.2.4 above) to this framework. The HRR test additionally provides quantitative information about ignition and the products of combustion.

An early test to measure heat release is known as the "OSU" (Ohio State University) test. The method is standardized in ASTM E 906. This assessment is similar to the oxygen consumption calorimeter, but uses the temperature rise in the exhaust gases to calculate the heat release rate.

In addition, there are a series of full-scale tests for furniture, mattresses, and other applications. Indeed, there is a large array of tests, methods, and standards for evaluating the broad variety of products of this class that can be involved in fires. A helpful report [34] summarizes over 75 test methods applicable to different types of furniture (see Table 5.7). Hilaldo [35] also provides a useful summary of the more important tests.

A key test for carpets is known as the "Flooring Radiant Panel Test," which is outlined in ASTM E 648 (also NFPA 253-2000). This method measures the tendency of the test material to spread flame. In this method a sample of carpet is positioned horizontally onto the floor of a test chamber and subjected to an inclined radiant panel with an open flame from a pilot burner. The distance burned to "flame out" is noted and reported. The inclined panel provides a gradient of heat flux energy over the length of the test panel from 0.1 to 1.1 W/cm^2. The *critical radiant flux* is the low-

TABLE 5.6 Test Methods Used to Characterize Fabrics Used for Furnishings

Test method	Title	Scope
ASTM D 2863–00 and ISO 4589–2	Measuring the Minimum Oxygen Concentration to Support Candle-Like Combustion of Plastics (Oxygen Index)	Test method describes a procedure for measuring the minimum concentration of oxygen that will just support flaming combustion in a flowing mixture of oxygen and nitrogen. A method is provided for testing flexible sheet or film materials while supported vertically.
ASTM E 1354–99 and ISO 5660 (1993)	Heat and Visible Smoke Release Rates for Materials and Products Using an Oxygen Consumption Calorimeter	Test method is used to determine the ignitability, heat release rates, mass loss rates, effective heat of combustion, and visible smoke development of materials and products. The rate of heat release is determined by measurement of the oxygen consumption as determined by the oxygen concentration and the flow rate in the exhaust product stream.
ASTM E 906–99 (OSU test)	Heat and Visible Smoke Release Rates for Materials and Products	Test method can be used to determine the release rates of heat and visible smoke from materials and products when exposed to different levels of radiant heat using the test apparatus, specimen configurations, and procedures described in this test method.
ASTM E 648–00 and NFPA 253 (2000)	Critical Radiant Flux of Floor-Covering Systems Using a Radiant Heat Energy Source	Procedure for measuring the critical radiant flux of horizontally mounted floor-covering systems exposed to a flaming ignition source in a graded radiant heat energy environment, in a test chamber. A specimen is mounted over underlayment, a simulated concrete structural floor, bonded to a simulated structural floor, or otherwise mounted in a typical and representative way.
ASTM D 2859 and DOC FF 1–70	Ignition Characteristics of Finished Textile Floor Covering Materials	Covers the determination of the flammability of finished textile floor covering materials when exposed to an ignition source under controlled laboratory conditions. The test uses a methenamine pill as the source of ignition.
NFPA 701 (1999)	Fire Tests for Flame Propagation of Textiles and Films	Covers fire safety requirements that apply to flame-resistant materials which are used extensively in the interior furnishing of buildings and transport facilities, in protective clothing for certain occupations and situations, and for protective outdoor coverings such as tarpaulins and tents.

est radiant heat (W/cm^2) that is necessary for a floor covering to sustain flame propagation over its surface. A less stringent test, termed "the pill test" (DOC FF 1–70 found in 16 CFR 1630), provides for the placement and ignition of a methenamine tablet in the center of a subject carpet. The distance that the carpet burns away from the tablet is reported. This test is used to simulate conditions such as a lighted cigarette, match, or fireplace ember falling onto a carpet.

NFPA 701 provides fire tests for flame propagation of textiles and films. Test Method 1 in the standard is applicable for most curtains, drapery, and other window treatment fabrics. It directs that the sample be conditioned, weighed, hung vertically, and exposed to a methane burner flame for 45 sec-

TABLE 5.7 Flammability Test Methods for Mattresses and Upholstered Furniture [34]

Full-scale test	Bench-scale test	Small-scale test	Specifies other test methods	Specification only
ASTM E 1537 Upholstered furniture	ASTM E 84	ASTM D 2863	Arizona Fire Code	ACA 3–4201
ASTM E 1590 Mattress	ASTM E 906	ASTM D 3675	ASTM D 3574	ASTM E 60 Test guide only
Boston BFD IX-10 Upholstered furniture	ASTM E 1352 Furniture—cigarette	ASTM D 4151	ASTM D 4723	ASTM D 3453
Boston BFD IX-11 Mattress	ASTM E 1353 Furniture—cigarette	ASTM D 5238	Connecticut 3748-M-229k—Mattress	BOCA F-307.2
BS 5852: Part 2 Upholstered furniture	ASTM E 1354	ASTM E162	Connecticut 9118-C-342rn—Mattress	BS 7177
BS 6807 Mattress	ASTM E 1474	ASTM E 662	Connecticut 3748-M-34—Mattress	California TB 26
California TB 106 Mattress—cigarette test	BIFMA Cigarette test only	BIFMA Flame tests only	Connecticut 3748-M-353a—Mattress	New Hampshire 3738-01—Mattress
California TB 116 Furniture—cigarette test	BOCA F-307.3	California TB 117 Upholstered furniture	Connecticut 3748-M-360f—Mattress	New Hampshire 3748-03—Mattress
California TB 121 Mattress	BS 5852: Part 1 Furniture	CS 191-53	Delaware, Univ. of Mattress	
California TB 129 Mattress	CAN 2-4.2-M77 Method 27.7—Mattress	FAR 25.853 a & b	District of Columbia Mattress	
California TB 133 Seating furniture	ISO 8191-1 Furniture—cigarette	FAR 25.853 b2 & b3	Maine Mattress	
16 CFR 1632 FF 4-72—mattress	ISO 8191-2 Furniture—match	Federal Test 5903	Maryland Mattress	
FAR 25–59 Aircraft seats	NFPA 260 Furniture—cigarette	Federal Test 5906	Massachusetts 527 CMR 21.00	
Michigan roll-up test Mattress	NFPA 261 Furniture—cigarette	Federal Test 5908	MIL-STD-1623D	
NFPA 266 Upholstered furniture	NFPA 264	MVSS 302 Motor vehicles	Minnesota 7510 Seating furniture	
NFPA 267 Mattress	NFPA 264A	NFPA701 Small-scale	NFPA101	
Nordtest NT 032 Upholstered furniture	NFPA 701 Large-scale	UL 214 Small-scale	NY/NJ Port Authority	
UL 1056 Upholstered furniture	Nordtest NT 037		North Carolina 7210-M-2c, MF-3e, M-ld Mattress	
UL1895 Mattress	UFAC Furniture—cigarette		Rhode Island 4358–06 Mattress	
UL2060 Mattress	UL214 Large-scale		West Virginia Mattress	

Reprinted with permission from Damant, G.H. "Combustabiity Database, Fire Tests Resource Guide," Sleep Products Safety Council, 1993.

onds. When the fabric stops burning, the remaining fabric is weighed and the percent remaining is reported.

5.4.3 Performance Standards

In addition to the material testing methods, there are specific performance standards for materials used in fire hazard situations. We will again consider these specifications in two groups: protective clothing and furnishings. Performance standards tend to change as experience and new materials progress (usually becoming more restrictive). As such, we will provide only a summary of the current specific requirements.

5.4.3.1 Protective Clothing

The *National Fire Protection Association* (NFPA), the *Occupational Safety and Health Administration* (OSHA), *Canadian General Standards Board* (CSGB), and *European Community on Standardization* (EN) have developed standards for garments worn by firefighters, electrical workers, and industrial workers. These standards establish performance requirements for materials, components, and garment design to protect workers from fire and electrical arc hazards. A summary of standards for thermal protective apparel is presented in Table 5.8.

North American Standards for General Flash Fire Garments

The U.S NFPA has issued two standards for garments to protect industrial workers against injury from a flash fire. The first of these, NFPA 2112, specifies minimum standards for the performance of the materials used to construct the garments. The second, NFPA 2113, provides standards for selection, care, use, and maintenance of these garments. These standards do not apply to protective clothing used for fire fighting (e.g., wildlands, structural, proximity, or technical rescue). They also do not apply to protective clothing used for electric arc thermal flash, radiological, or biological hazards. These hazards are covered in separate NFPA standards.

NFPA 2112 establishes standards based on the following material tests described above:

Vertical flame test: maximum char length and afterflame time are specified with no melting or dripping permitted.

Thermal protective performance test: minimum TPP values are specified for testing with and without an air space between fabric and sensor.

Instrumented manikin test: the FR fabric to be evaluated is constructed into a standard coverall pattern. The predicted body burn injury results for exposure to a specific heat energy level must be less than a specified percentage.

Heat resistance and thermal shrinkage resistance tests: the materials of construction must not ignite, melt, or drip, or, for fabrics, exceed a specified shrinkage when exposed to a specified temperature for a specific length of time.

Joining requirements: sewing thread used in the garment must be made of an inherently flame-resistant fiber and have a melting point ≥ 500°F.

NFPA 2113 provides the following:

- A hazard assessment is required of the industrial workplace.
- Protection level, physical characteristics of FR fabrics, static buildup, garment construction and design, fabric comfort, and conditions under which garment will be used are factors that must be taken into account for clothing selection.
- Establishes requirements on how FR garments are worn (e.g., sleeves down and buttoned), the need for eye, head, neck, and foot protection, and the use of FR overgarments.
- Cleaning and maintenance requirements for protective garments.

TABLE 5.8 North American Performance Standards for Thermal Protective Clothing

Test method	Title	Scope
NFPA 2112 (2001)	Flame-Resistant Garments for Protection of Industrial Personnel Against Flash Fire	Specifies the minimum design, performance, and certification requirements, and test methods for new flash fire protective garments. (Does not apply to firefighter clothing.)
NFPA 2113 (2001)	Selection, Care, Use, & Maintenance of Flame-Resistant Garments for Protection of Industrial Personnel Against Flash Fire	Specifies the minimum requirements for the selection, care, use, and maintenance of flash fire protective garments meeting the requirements of NFPA 2112.
NFPA 1971 (2000)	Protective Ensemble for Structural Fire Fighting	Specifies minimum requirements for the design, performance, testing, and certification of the elements of the protective ensemble including coats, trousers, helmets, gloves, footwear, and interface items for protection from the hazards of structural fire.
NFPA 70E (2000)	Electrical Safety Requirements for Employee Workplaces	Standard consists of safety related installation requirements, work practices, maintenance requirements, and requirements for special equipment.
NFPA 1976 (2000)	Protective Ensemble for Proximity Fire Fighting	Specifies minimum requirements for the design, performance, testing, and certification of the elements of the protective ensemble for protection from the hazards of proximity fire fighting.
NFPA 1951 (2001)	Protective Ensemble for Urban Technical Rescue	Specifies minimum requirements for the design, performance, testing, and certification of the elements of the protective ensemble for protection for urban technical rescue.
NFPA 1851 (2001)	Selection, Care and Maintenance of Structural Fire Fighting Ensembles	Specifies minimum requirements for the selection, care, use, and maintenance of structural firefighting ensembles meeting NFPA 1971.
NFPA 1977 (1998)	Protective Ensemble for Wildland Fire Fighting	Specifies minimum requirements for the design, performance, testing, and certification of the elements of the protective ensemble for protection from the hazards of wildland fire fighting.
CAN/CGSB-155.20-2000	Workwear for Protection Against Hydrocarbon Flash Fire	Specifies the minimum design, performance, and certification requirements, and test methods for new flash fire protective garments.
CAN/CGSB-155.21-2000	Practices for the Provision and Use of Workwear for Protection Against Hydrocarbon Flash Fire	Specifies the minimum requirements for the selection, care, use, and maintenance of flash fire protective garments meeting the requirements of CAN/CGSB-155.20-2000.
CAN/CGSB-155.1-98	Structural Fire Fighting Protective Clothing	Specifies the minimum design, performance, certification requirements, and test methods for structural fire fighting clothing.
CAN/CGSB-155.22-97	Forest Fire Fighting Protective Clothing	Specifies the minimum design, performance, certification requirements, and test methods for forest fire fighting clothing.
OSHA 29 CFR 1910.269	Electric Power Generation, Transmission, and Distribution—Special Industries	Prohibits clothing that, when exposed to electric arcs, could increase the extent of injury that would be sustained by the employee.
OSHA 29 CFR 1910.132	Personal Protective Equipment for General Industry	Specifies employer responsibility in selection, care, and maintenance of protective clothing based on risk assessment of hazards to employees.
OSHA 29 CFR 1910.156	Fire Brigades	Specifies garment requirements based on the 1975 edition of NFPA 1971. Requirements less stringent than corresponding NFPA standards.

The precursors to the NFPA 2112 and 2113 performance standards are also active in Canada, and are outlined in Canadian General Standards Board requirements CGSB-155.20 and CGSB-155.21. These were used as models for development. The European Community for Standardization (EN) has issued performance standards of similar scope (EN 340, EN 367, EN 532, and EN 533).

U.S. Standards for Fire Fighting Garments

NFPA 1971, "Standard on Protective Ensemble for Structural Fire Fighting 2000 Edition," provides performance standards for garments, helmets, gloves, hoods, and footwear used in structural fire fighting. In addition to the fabric and thread requirements specified in NFPA 2112, the standard also specifies performance requirements for water repellency and blood-borne pathogen resistance for firefighter protection. The TPP requirement is also much higher (>35 cal/cm^2) for the multicomponent turnout gear. A new test to measure thermal insulation under compression (Conductive Compressive Heat Resistance [CCHR]) has also been added to this newest edition.

Stull [36] has provided a critique of the 1997 version of the standard that suggests testing be included that incorporates both conductive (contact) and pure radiation sources of heat flux (RPP). In the 2000 edition, conductive testing has been added, along with more stringent preconditioning requirements, and a *total heat loss* requirement. The total heat loss test quantifies the ability of a clothing system to allow heat to transfer to the outside environment. This requirement precludes the use of nonbreathable moisture barriers. All standards based on the current TPP test will likely be revised as this test method is updated to account for the stored energy in the materials (see Sec. 5.4.2.1).

U.S. Standards for Electrical Safety

Another standard, NFPA 70E, "Standard for Electrical Safety Requirements for Employee Workplaces 2000 Edition," provides a comprehensive standard of electrical safety requirements for the majority of workplaces. A key feature of this standard related to protective clothing is the requirement that a job hazard assessment be conducted before work is started near electrical equipment. This establishes the physical flash protection boundary where protection from a thermal electric arc hazard is required. It also sets up the minimum arc thermal protective value required of a clothing ensemble to be worn by workers when inside the defined boundary. The standard also provides critical safety guidance for installation, work practice, maintenance, and special equipment. The personal protective equipment requirements are found in the safety-related work practices section of the standard. The scope of NFPA 70E excludes several specific workplaces—one of which is a workplace under the exclusive control of electric utilities where electrical workers are involved in generation, transmission, and/or distribution of electric power.

ASTM F 1506, "Standard Performance Specification for Flame Resistant Textile Materials for Wearing Apparel for Use by Electrical Workers Exposed to Momentary Electric Arc and Related Thermal Hazards," applies to all electrical workers and sets minimum requirements on woven and knit fabrics used to make arc flash protective clothing. Performance requirements include breaking load, tear resistance, colorfastness, laundry shrinkage, and flammability (initially and after multiple launderings). Additionally, performance values in electric arc thermal testing are required to be listed for each specific garment.

In the United States, the Federal Occupational Safety and Health Administration (OSHA) provides additional directives for electric power generation, transmission, and distribution workers' apparel. Regulation 29 CFR 1910.269, Section 1, part 6 sets the apparel requirements for workers who may be exposed to the hazards of flames or electric arcs. In essence, it states under subpart (iii) "The employer shall ensure that each employee who is exposed to the hazards of flames or electric arcs does not wear clothing that, when exposed to flames or electric arcs, could increase the extent of injury that would be sustained by the employee." This subpart also specifically prohibits clothing made from several common fibers, including nylon, acetate, polyester, and rayon, either alone or in blends unless the employer can demonstrate that the fabric has been treated to withstand the conditions that may be encountered. As with the other performance standards described, this

OSHA standard is expected to be revised since new consensus standards have been developed since the 1994 issuance of 29 CFR 1910.269.

5.4.3.2 Furnishings

In 1972, the U.S. Department of Commerce published the fire performance standard, DOC FF 4–72, which applies to all mattresses sold for household use. This standard is based on a test in which nine cigarettes are placed both on the bare surface of a mattress and on the mattress between two cotton bed sheets. The mattress fails the standard if smoldering from any cigarette spreads more than 2 in from the nearest point of the cigarette. This standard remains the current primary benchmark for home mattresses.

California Technical Bulletin 129 is a full-scale fire test of mattresses and mattress systems. It is being developed with the intention of providing improved standards for bedding used in public institutions, such as prisons, college dormitories, nursing homes, hotels, and hospitals. It is anticipated to become a standard by 2004. The test is performed in a fire test room capable of measuring heat release rate. A mattress, with or without support and bedding, is subjected to a propane gas T-burner for 180 s. The main test criterion is that the peak heat release rate not exceed 100 kW. A number of mattresses are now available that will pass the CAL 129 test criteria.

California Technical Bulletins 116 and 117 provide a series of tests and standards for materials used in upholstered furniture. They are aimed primarily at residential furniture. Both flame and smoldering tests are included, but neither requires the use of flame-resistant fabric for home use. These standards are widely used throughout the United States. A revised draft of CAL 117 (February 2002) has added a specification for an open-flame ignition test of a fabric-covered cushion assembly, where the weight loss of the assembly is measured.

California Technical Bulletin 133, ASTM E 1357, and NFPA 266 are full-scale fire tests for seating furniture for use in public spaces. These tests are performed in a fire test room. In the case of CAL 133, the tested furniture item is ignited by means of a square propane gas burner for 80 s. The test measures heat release, mass loss, carbon monoxide, and smoke release. The key performance measure is a peak heat release rate of less than 80 kW.

All carpet sold in the United States must meet, at a minimum, Federal Government flammability standards. Per the Carpet and Rug Institute's web site, carpets and large rugs (24 ft^2 in area or larger) sold in the United States must pass the "Pill Test" (DOC FF 1-70). This test requires that no more than one out of the eight carpet specimens assessed shall burn more than 3 in from the burning pill. Many localities dictate that carpeting used in public places, such as health care facilities, meet additional specifications based on the Flooring Radiant Panel Test (ASTM E 648 and NFPA 253).

Small rugs (less than 24 ft^2 in area) are subject to DOC FF 2-70. The small rugs that fail this test must be specially labeled with a warning indicating that they are "flammable."

Curtain, drapery, and other window treatment fabrics are tested according to NFPA 701. This test method includes pass/fail criteria that have been adopted as standards in various localities. The primary specification requires that less than 40 percent of the fabric structure burn when ignited at the bottom in a free hanging configuration.

In 1971 the U.S. Federal *Department of Transportation* (DOT) adopted a mandatory Motor Vehicle Safety Standard 302 that sets a maximum burning rate for textile and plastic fabrics used for seating upholstery. It is primarily aimed at preventing fires caused by cigarettes.

The U.S. *Federal Railroad Administration* (FRA), Amtrak, and the NFPA have issued the current standards relating to materials used in passenger trains. These standards and guidelines all specify test requirements and criteria for flammability and smoke emission. For the most part, they are based on ASTM laboratory tests. Table 5.9 provides a summary of these tests and criteria for the different categories of materials used [37]. Most of these tests, however, do not include the measurement of a material's characteristic HRR. The exception is NFPA 130. It specifies that seat cushioning materials and mattresses used in passenger rail transport meet requirements for flame spread and HRR, as measured by ASTM D-3675.

TABLE 5.9 U.S. Flammability and Smoke Emission Requirements for Passenger Rail [37] (Materials with Fiber or Fabric Components)

Category	Function of material	Flammability		Smoke emission	
		Test procedure	Performance criteria	Test procedure	Performance criteria
Passenger seats, sleeping and dining car components	Cushions, mattresses Seat upholstery, curtains, mattress ticking and covers	ASTM D-3675 FAR 25.843 (Vertical)	$I_s \leq 25$ Flame time ≤ 10 s Burn length ≤ 6 in	ASTM E-662 ASTM E-662	$D_s (1.5) \leq 100$ $D_s (4.5) \leq 175$ $D_s (4.0) \leq 250$
Panels	Wall, ceiling, partition, tables, and shelves	ASTM E-162	$I_s \leq 35$	ASTM E-662	$D_s (1.5) \leq 100$ $D_s (4.0) \leq 200$
Flooring covering		ASTM E-648 ASTM E-162	$CRF \geq 5$ kW/m^2 $I_s = 25$	ASTM E-662	
Insulation	Thermal, acoustic	ASTM E-162	$I_s \leq 25$	ASTM E-662	$D_s (4.0) \leq 100$

Notes: I_s = Flammability Index, CRF = Critical Radiant Flux, D_s = Optical Density.

The U.S. *Federal Aviation Administration* (FAA) specifies all tests and requirements for the fire-related properties of materials used in aircraft operating within the United States. The specification details are found in the FAA document, *Aircraft Materials Fire Test Handbook* [38]. Included are tests for vertical flammability and heat release rate (OSU test). The FAA also requires upholstered seats in commercial aircraft to pass an "Oil Burner Test," FAR 25 Amendment 25–59. This severe test essentially requires that a fire blocking fabric be used with polyurethane seat cushions.

5.5 APPLICATIONS—PROTECTIVE CLOTHING

In this section, we will examine in more detail where protective clothing is used and how it functions. We will include coverage of clothing designs that provide thermal protection against flash fires, fire fighting hazards, electric arc flash hazards, and molten metal contact.

The selection of the appropriate protective clothing for any job or task is usually dictated by an analysis or assessment of the hazards presented. The expected activities of the wearer as well as the frequency, magnitude, and types of exposure, are typical variables that input into this determination. For example, a firefighter, expected to extinguish a fire, is exposed to a variety of burning materials, varying durations of high levels of radiant heat, direct conductive hazards, and potential flashover events. Specialized multilayer fabric systems are thus used to meet the thermal challenges presented. This results in protective gear that is usually fairly heavy and bulky, designed for relatively short wearing periods, and essentially provides the highest levels of protection against any fire situation. In contrast, an industrial worker who has to work frequently in areas where the possibility of a flash fire exists would have a very different set of hazards and requirements. In many cases, a flame-resistant coverall worn over cotton work clothes adequately addresses the hazard; a firefighter turnout system would be inappropriate. One system is chosen to permit escape from a fire situation; the other is chosen to permit entry into a fire situation.

It can sometimes be difficult to accurately assess the thermal hazard a job or task presents. In this case, the energy level of a typical fire hazard can at times be estimated by examining clothing worn in actual related fires [39]. Table 5.10 shows the typical energy loads from a variety of thermal hazards estimated from returned garments of Nomex® fabric worn by exposed workers.

TABLE 5.10 Estimates of Heat Flux in Typical Fire Situations [39]

Exposure Description	Estimated* total exposure (cal/cm^2)	Time** (sec)	Estimated heat flux (cal/cm^2 sec)
Brush fire	12	8	1.5
Oil well flash fire	4	2	2
Solvent vapor-ashover	16	5	3.2
Apartment building fire	19	12	1.6
Back draft from gas line leak	7	6	1.2
Auto gas tank explosion	14	<5	>2.8

* Total exposure estimated from condition of Nomex® garments exposed to actual fire.
** Time estimated by eyewitnesses.

5.5.1 Burn Injuries

Every year, thousands of workers are injured or killed from fire exposure. Additionally, many firefighters suffer serious burn injuries each year. In order to understand how protective clothing functions, it is important to understand how burn injuries occur.

The severity of a burn injury depends on the depth of skin that is damaged. Burns are generally categorized according to three levels of severity. First-degree burns provide damage to 20 to 50 μm of the outer layer of skin (epidermis), causing pain, redness, and swelling. Second-degree burns provide damage to both the outer skin and underlying tissue layers (100 to 250 μm encompassing the epidermis and dermis), causing pain, redness, swelling, and blistering. First- and second-degree burns are fully recoverable and leave little or no scarring. Third-degree burns produce damage that extends deeper into the tissues (250 to 2000 μm, which includes the hypodermis), causing naturally irreversible tissue destruction. The skin is dry, gray, charred, and may feel leathery. Third-degree burns destroy all layers of the skin, which is replaced by scar tissue. Research done by Stoll and Chianta gives an indication of typical skin temperatures resulting in the different degrees of skin burn injury [29]. Burn injuries usually occur when skin temperature exceeds 44°C.

The extent and severity of a burn depends on the skin temperature that is reached and the length of time the skin is exposed to that temperature. Fig. 5.12 shows the combinations of time and temperature required to produce a second-degree burn [40]. The *Society of Fire Protection Engineers*

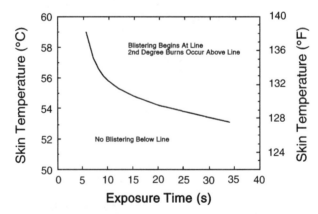

FIGURE 5.12 Time and temperature to produce second-degree burn. (Reprinted with permission from Lawson, J. R., "Fire Fighters' Protective Clothing: and Thermal Environments of Structural Fire Fighting" in *Performance of Protective Clothing: Sixth Volume, ASTM STP 1273*, 1997.)

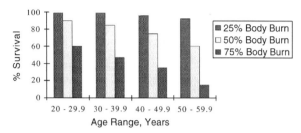

FIGURE 5.13 Burn injury—chance of survival.

(SFPE) provides a guide that presents a survey of methods for predicting injury to humans from thermal energy exposure [41]. The SFPE study provides data and equations for predicting time/irradiance combinations that will produce pain. More complicated models are also evaluated. Of these, the *damage burn integral* model is the one most widely accepted for predicting first- and second-degree burn injury. There are no widely accepted models for predicting the onset of third-degree burn injury.

Another factor to be considered is the age and health of the burn victim. A study by the American Burn Association in the early 1990s estimated the effect of age on the survival rate of burn victims subjected to burn injury over different amounts of the body. Results from that study for the four age groups covering the traditional working ages over three burn injury levels are shown in Fig. 5.13 [42].

5.5.2 Flash Fires

A flash fire is an unexpected, extreme fire situation. Fire conditions can be characterized in terms of air temperature and the level of thermal radiation. A flash fire would correspond to an *emergency* situation. Because they are unexpected, it is difficult to be prepared for a flash fire.

Examples of flash fire situations would include that where a structural firefighter was caught in a room that reached the flashover point, an industrial worker trapped in the middle of a chemical spill that ignites, or a wildlands firefighter combating a forest fire who gets caught in a major wind shift.

5.5.2.1 Design Criteria for Flash Fire Protective Clothing

As identified earlier, there are numerous workplace injuries and fatalities yearly in the United States due to fire and ignition of clothing. Of these, the most severe injuries are usually associated with clothing that has ignited and continues to burn well after the original flash fire exposure. The critical point is that everyday clothing can ignite and/or melt causing the clothed areas to be more severely burned than exposed skin. For situations where there is a flash fire exposure risk, it is crucial that workers wear the appropriate protective equipment. A key part of this equipment is flame-resistant clothing: materials that will not ignite and continue to burn from exposure to flames. This clothing is designed to maintain a barrier to isolate the wearer from the thermal exposure, provide valuable escape time, and minimize potential burn injury. In most cases, a worker's survival, extent of injury, recovery time, and quality of life after the exposure depend on the protection provided by the flame-resistant clothing.

Those in charge of selecting and providing flash fire protective clothing for the industrial workplace have multiple criteria to consider. The first is compliance to relevant government safe workplace standards, a given. However, many of these regulations only require the employer to assure that "flame-resistant" clothing is worn. Basically, the clothing worn by the employee should not ignite and continue to burn if exposed to a flash fire. Adherence to this requirement still permits selection from a wide range of FR garments.

Other key areas must be considered when specifying flame-resistant clothing. These include protection level, durability, wearer comfort, heat stress potential, garment styling, ease of movement, and life cycle cost. All must be considered, even if not given the same weighting. Most flame-resistant clothing used in industrial workplaces is meant to be a last line of defense. It is meant to permit escape from the unlikely occurrence of a system failure that results in the release and/or ignition of a flammable material. As such, the clothing system selected for daily wear must allow the worker to perform their ordinary job tasks in a satisfactory manner while still affording protection. There are select tasks where the thermal hazard level and probability of occurrence is known to be high and/or where the ability of the wearer to escape the fire is hindered. In these cases, specially designed flame-resistant garment systems are usually employed.

There typically are trade-offs among the various selection criteria. For example, increasing fabric weight will afford higher levels of thermal protection and increase the wear durability of the garment. However, increasing fabric weight will also decrease wearer comfort in hot and/or humid work environments. In addition, heavier weight fabric garments are more expensive. Increasingly, worker acceptance of the protective clothing (for comfort) is becoming a critical element to maximize adherence to employer's protective clothing programs.

Several elements are important when considering the protection levels afforded by flame-resistant garments. One of these is the durability of the flame-resistant property of the fabric used in the garment. The nature of the materials' flame resistance must be understood in order to maintain the protective capabilities of a garment for its expected service life. Materials that are topically treated for flame resistance must be more carefully monitored, as many of these treatments can be washed or worn away. Maintenance can be simplified somewhat with the selection of inherently flame-resistant fibers/fabrics in the protective garments. The wearer can be assured that the flame-resistant characteristics are not affected by washing or wearing, as long as the fabric used in the garment is intact (and not contaminated with flammable materials).

Another factor to consider when designing a garment for protection from flash fire exposures is fabric construction. Using a flame-resistant fabric with a tightly woven or knit construction will generally improve its protection level. Increasing the thickness of the fabric will also afford additional protection due to the increased mass (where more fiber is used) or lowered density (where air is utilized for improved thermal insulation). Additionally, wearing several layers of flame-resistant garments or nonmelting undergarments have shown to increase the protection level afforded. In this case, the additional layers add both mass and air between the thermal threat and the skin.

As mentioned earlier (see Sec. 5.4.3.1), there are several performance standards for flame-resistant garments to be used to protect industrial personnel from flash fire exposures. These are NFPA 2112 and 2113 in the United States and CGSB 155.20 and 155.21 in Canada.

5.5.2.2 Performance of Fabrics as Flash Fire Protective Clothing

Table 5.11 lists common commercial fabrics used for thermal protective clothing in industrial applications. As was noted in Sec. 5.4.2.1, there are several tests that have been used to characterize these fabrics, including the vertical flame, TPP, and thermal mannequin tests. Of these, the thermal mannequin test provides the most useful data.

Table 5.12 provides typical vertical flame and TPP data for fabric constructions used for protective clothing, and Table 5.13 provides instrumented thermal mannequin data. Fig. 5.14 shows a plot of data indicating how different fabrics respond to increased flame exposure.

5.5.3 Structural and Wildlands Fires

The role of protective clothing in structural fire scenarios relates primarily to clothing for firefighters. Structural fires are distinguished from flash fires in the sense that, from a firefighter's standpoint, a structural fire is quite evident, planned for, and approached systematically. A flash fire is unexpected. The overall strategies and objectives are also different. In a structural fire, the objective is to control and ultimately extinguish the fire. In a flash fire, the objective is to escape without injury.

TABLE 5.11 Common Fabrics Used for Fire Protective Clothing in Industrial Applications

Fabric name/source	Description
Banwear®/ Itex, Colorado	88% cotton/12% nylon blend. Base fabric treated with flame retardant (FR) chemical
Dale Antiflame® Dale A/S, Norway	100% cotton treated with Pyrovatex® FR chemical
Firewear™ Springeld LLC, NY	55% FR modacylic/45% cotton
Indura® Westex, Inc., Illinois	100% cotton treated with Proban® FR chemical
Kermel®/FR Viscose Rhodia, France and Lensing AG, Germany	50% polyamide-imide/50% FR viscose
Nomex® IIIA DuPont, Delaware	Blend of 93% meta-aramid/5% para-aramid/2% antistatic fiber
Nomex® Comfortwear DuPont, Delaware, and Lenzing AG, Germany	65% Nomex® IIIA/35% FR rayon
Kevlar®/PBI® DuPont, Delaware, and Celanese, NC	60% para-aramid/40% polybenzimidizole
Tuffweld® Southern Mills, Georgia	60% FR rayon/40% para-aramid
Indura® Ultrasoft® Westex, Illinois	88% cotton/12% nylon. Base fabric treated with Proban® FR chemical

Firefighter protective clothing assumes a dual role in that it is expected to provide protection in both situations.

5.5.3.1 Typical Scenario for Firefighters

In choosing or designing appropriate protective apparel for fire fighting it is important to consider all the threats that are faced. A firefighter's clothing ensemble is designed to give a degree of protection from the thermal environment produced by fire. It affords a certain degree of protection from radiative, convective, and conductive thermal assaults. Burn injuries here are directly related to the firefighter's thermal exposure, actions within the thermal environment, physiological aspects that regulate heat buildup in the body, available moisture at the skin surface and its temperature, and the performance of the total protective clothing ensemble.

Lawson describes a typical burn injury scenario for a firefighter [40, 43]. The alarm comes in to the fire station. Firefighters respond by donning their turnout gear (including helmet, hood, coat and pants, and boots). On arrival at the fire scene they don a *self-contained breathing apparatus* (SCBA) and gloves, and proceed to pull hose lines into the involved structure, often up a flight of stairs. In most cases, they will have already exerted a large amount of energy and will be sweating profusely. As they reach the fire, they are subjected to intense radiative and convective heat energy. As their body temperature rises, their ability to function decreases. The sweat from their bodies also affects their clothing. As the moisture load increases, the thermal performance of their garment degrades due to changes in heat capacity and thermal conductivity. The potential for scalding and steam burns increases.

TABLE 5.12 Vertical Flammability and TPP Test Results

Fabrics	Actual weight (oz/yd²)	Vertical Flammability*			TPP rating (cal/cm²)
		Char length (inches)	Afterglow (seconds)	Afterflame (seconds)	
100% untreated cotton	—	>12	—	—	—
Untreated polyester/cotton	—	>12	—	—	—
Nomex® IIIA aramid	4.7	2.85	0.7	0	11.6
Nomex® IIIA aramid	5.9	2.95	0.7	0	14.1
Nomex® IIIA aramid	8.1	1.50	3.5	0	16.7
Nomex® Comfortwear aramid/FR rayon	4.5	2.00	0.7	0	9.6
Nomex® IIIA aramid turnout system	19.5	—	—	—	42
Nomex Omega® turnout system 40	16.6	—	—	—	44
Nomex Omega® turnout system 50	18.1	—	—	—	52
Nomex®/Kevlar® turnout system	18.5	—	—	—	44
Indura® FRT cotton	7.9	3.75	0.6	0	9.8
Indura® FRT cotton	10.1	3.50	0.5	0	13.2
Banwear® FRT cotton/nylon	7.0	5.20	0.6	0	7.3
Banwear® FRT cotton/nylon	10.5	4.25	0.5	0	10.3
Firewear® FR modacrylic/cotton	6.1	4.90	25.6	0	9.1
Firewear® FR modacrylic/cotton	9.7	4.25	49.2	0	9.7
Kevlar®/PBI®	4.3	0.70	16.5	0	10.3
Kevlar®/PBI®	5.7	0.65	29.0	0	12.3

* Average of warp and fill direction results.

TABLE 5.13 Instrumented Thermal Mannequin Flash Fire Exposure Test Results

Fabrics	Nominal weight (oz/yd²)	Total percent predicted body burn at various exposures:		
		3 Seconds	4 Seconds	5 Seconds
Untreated cotton	5.5	96	96	96
Indura® FRT cotton	9.0	8.3	80	80
Ultrasoft® FRT cotton/nylon blend (88/12)	9.5	8.7	81.7	90.7
Kevlar®/PBI® blend (60/40)	4.5	31.0	—	—
Nomex® IIIA aramid	4.5	38.0	51.7	—
Nomex® IIIA aramid	6.0	29.0	44.3	58
Nomex® IIIA aramid	7.5	19.3	36.7	53.3
Nomex® IIIA/FR rayon blend (65/35)	4.5	—	48.3	—
Nomex® IIIA/FR rayon blend (65/35)	5.5	—	46.7	—
Nomex® IIIA/ FR rayon blend (65/35)	6.5	—	38.3	—

Conditions: One home laundering; 100% cotton underwear (t-shirt and briefs) worn under test garment; heat flux of 2 cal/cm² sec; test garment is standard pattern coverall; conducted per ASTM F-1930; results are average of 3 replicates.

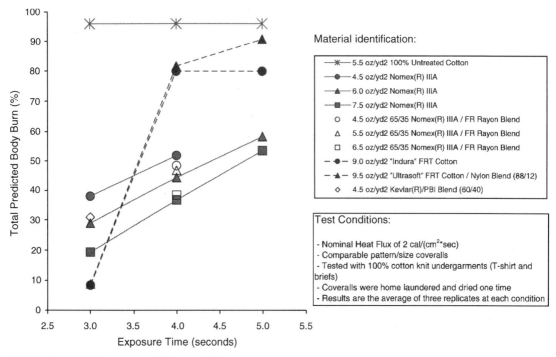

FIGURE 5.14 Body burn response with increased flame exposure.

The flow of hot gases from a doorway or through a window may be well above 500°C (932°F) and may extend tens of meters down a corridor or across an adjoining room ceiling. The thermal radiation from a room's open doorway or window may reach levels that will cause instantaneous burn injuries to exposed skin, charring or ignition of protective clothing fabrics, and potential burn injuries to protected skin. Surface temperatures of solids within this staging zone may easily exceed 200°C (392°F), and touching these surfaces without adequate protection could also result in an immediate burn injury.

The threat of burn injury becomes obvious as the firefighter reaches the fire.

Many burn injuries can be accelerated in the fire-fighting scenario above. The degree that a firefighter is preheated and perspiring before entering the fire scene can shorten the protection offered by their gear. Moisture saturation of the clothing under the turnout will affect the performance of the moisture barrier in the ensemble and increase the heat transfer rate due to direct conduction with water (scalding). In addition, the straps from the SCBA can compress and reduce the thermal protective performance of the underlying material, creating localized burns from increased direct heat conduction. This is especially evident in the shoulder area where the compressive loads are the greatest.

5.5.3.2 Range of Thermal Conditions Encountered

It is helpful to describe the conditions a firefighter must face in terms of the thermal radiation that is encountered and the resulting rise in air temperature. One such classification [44, 45] is illustrated in Table 5.14.

The term "routine" describes a condition where one or two objects, such as a wastebasket or a bed are burning in a room. Both the thermal radiation and the resulting air temperature are little more than would be encountered on a hot summer day. Firefighter protective clothing is more than capable of providing the necessary thermal protection required for this situation for an extended period of time.

TABLE 5.14 Overview of Thermal Conditions Encountered by Firefighters [44, 45]

	Thermal radiation range in cal/cm² sec (kW/m²)	Air temperature °F (°C)	Tolerance time
Routine	<0.04 (<1.7)	68–158 (20–70)	10–20 minutes
Ordinary	0.04–0.3 (1.7–12.6)	158–572 (70–300)	1–10 minutes
Emergency	0.3–5.0 (12.6–209)	572–2192 (300–1200)	15–20 seconds

The "ordinary" range includes temperatures that would be encountered in a more serious fire or being next to a "flashed-over" room—similar to that described in Sec. 5.5.3.1. Generally, turnout gear should afford the firefighter from 1 to 10 min of reliable thermal protection under these conditions. The higher end of this range (300°C) is quite hot, but the firefighter would not stay in such conditions for long.

The term "emergency" describes the severe and unusual conditions of being caught in a "flashed-over" room or a flash fire. In emergency situations, the incident heat flux can exceed 2.0 cal/cm²s (84 kW/m²) and temperatures can exceed the limitations of the individual textiles in the protective clothing. The function of the turnout gear here is to provide time to escape with minimal injuries. In most cases, the gear is damaged beyond repair and would be removed from service.

5.5.3.3 Design Criteria for Firefighter Protective Clothing

Our primary focus will be on the fabrics used in the components of firefighter turnout gear, but proper protection must include the entire ensemble. This includes boots, gloves, helmets, hoods, belts, straps, and requisite hardware. The typical multicomponent firefighter garment is designed to provide durable thermal protection in a wide range of environments. Typical turnout gear is composed of several layers that include an outer shell, a thermal liner, a moisture barrier, and an inner liner component. The overall performance of this hybrid garment is dependent on the composite performance of these components and their assemblage. Figs. 5.15 and 5.16 show typical firefighter turnout gear and the layers that make up the hybrid fabric system.

The outer shell is usually composed of a blend of *meta*-aramid and *para*-aramid fibers, or a blend of *para*-aramid fibers with other thermally resistant fibers. The primary function of the outer shell is to protect inner components from direct flame exposure. It is also designed to protect against mechanical and physical hazards that may result in rips, tears, or abrasive wear. The functionality of this component layer must be maintained under the wide range of conditions that the firefighter might encounter.

The thermal liner component is typically constructed of a lightweight face cloth quilted to a single layer or multiple layers of a nonwoven fabric or batting structure. This imparts most of the thermal protection in a turnout garment by providing a thick, low-density structure. Superior thermal insulation results from the increased thickness of what is essentially an engineered air layer. The level of insulation is directly related to the thickness of the thermal liner. The downside of this insulating layer is the accompanying reduction of heat loss capability by the wearer. This can lead to a heat stress hazard if the body is unable to regulate its internal temperature. The insulation layer prevents the body from removing the excess heat generated by the intense physical activity required for fighting fires.

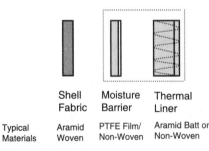

FIGURE 5.15 Firefighters turnout coat components.

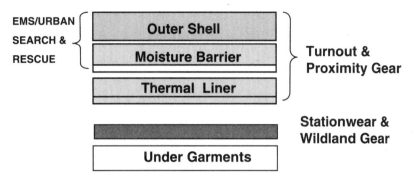

FIGURE 5.16 Firefighter turnout systems.

The moisture barrier is composed of a vapor-permeable film laminated to a lightweight woven or nonwoven fabric. Initially, the moisture barriers were designed to keep external water from saturating the garment; however, the increasing role of firefighters as first responders has increased the risk for chemical and blood-borne pathogen exposure. New moisture barriers now provide resistance to these hazards as well as offering "breathability" to reduce the risk of heat stress.

Torvi and Hadjisophocleous, with the National Research Council of Canada [46], have begun a research project to evaluate the useful lifetimes of protective clothing for firefighters. Their initial report provides a comprehensive literature review.

5.5.3.4 Heat Transfer Model for Firefighter's Protective Clothing

Mell and Lawson [47] are working to develop a computer model that will predict heat transfer through protective clothing. The current state of the model reasonably predicts temperatures within the fabrics, but overestimates actual fabric surface temperatures. It is currently based on fabrics that are assumed to be dry and at temperatures well below their point of degradation. A planar model is assumed with one-dimensional heat transfer.

An experimental apparatus to measure temperatures within the fabric layers has also been developed. It is designed to allow bench-top testing of firefighter turnout composite fabric systems' efficiency. This is one of several active fire-modeling programs at the Building and Fire Research Laboratory at the *National Institute of Standards and Technology* (NIST).

5.5.3.5 Performance of Fabrics in Firefighter Protective Clothing

The performance criteria for firefighter structural protective clothing are currently defined in the United States by NFPA 1971. This standard defines the thermal, physical, and exposure tests that each component and their base materials must pass. The thermal tests include vertical flame, dimensional stability, and TPP testing.

Stull [48] suggests that the fire environment encountered by firefighters is quite complicated and that additional tests should be carried out to characterize protective clothing. In response, methods to characterize both contact thermal protection and radiant heat protection have been developed for a series of fabrics that are typically used for firefighter protective clothing. Table 5.15 provides an illustrative sampling of the performance data.

As with testing of fabrics used for protection in industrial flash fire situations, the use of instrumented thermal mannequin testing provides the most reliable data for firefighter protective clothing. Currently this testing is limited to static direct flash fire exposure. However, work is under way at several research facilities (Worcester Polytechnic Institute, NIST, U.S. Navy) aimed at assessing burn injury predictions from radiant-only exposures and the effect of articulation.

Table 5.16 provides generalized direct flash fire mannequin exposure test data for three levels of firefighter protective clothing: clothing for wildlands fires (aramid outer shell); search and rescue

TABLE 5.15 Comparison of Contact, Radiant, and Thermal Protective Performance of Composite Fabrics for Firefighter Protective Clothing [48]

Composite fabrics:						
Outer shell	Nomex® III	Nomex® III	Nomex® III	PBI®/Kevlar®	PBI®/Kevlar®	PBI®/ Kevlar®
Weight (oz/yd^2)	6.0	6.0	6.0	7.5	7.5	7.5
Moisture barrier	E89 Crosstech®	E89 Crosstech®	Neoprene polycotton	E89 Crosstech®	E89 Crosstech®	Neoprene polycotton
Thermal liner	Caldura® SL	Q9	Q9	Caldura® SL	Caldura® SL	Aralite®
Total weight (oz/yd^2)	17.5	19.4	27.4	19.2	21.2	26.6
Thickness (mm)	3.18	5.44	5.36	2.92	5.11	4.17
Time to second degree burn (sec):						
Contact (CPP)	17.6	25.9	30.0	16.2	25.4	23.3
Radiant (RPP)	16.0	18.8	19.6	17.2	20.3	17.9
Thermal (TPP)	18.1	22.8	26.4	20.1	25.6	21.0

Reprinted with permission from Stull, J.O., "Comparison of Conductive Heat Resistance and Radiant Heat Resistance with TPP of Firefighter Protective Clothing," in *Performance of Protective Clothing, Sixth Volume, ASTM STP 1273*, 1977.

gear (outer shell plus a moisture barrier); and full firefighter turnout gear (complete system of outershell, thermal liner, and moisture barrier).

5.5.4 Electric Arcs

The brief but intense radiative thermal hazard delivered by an electric arc flash presents a different set of problems for workers. In 1982, Ralph Lee raised the issue of injuries caused by electric arc

TABLE 5.16 Instrumented Thermal Mannequin Test Results for Firefighter Protective Clothing

Type of protective garment:	Wildlands fire fighting	Search and rescue	Firefighter turnout gear
Typical construction of garment:			
Outer shell	7–8 oz/yd^2 aramid blend	7–8 oz/yd^2 aramid blend	7–8 oz/yd^2 aramid blend
Moisture barrier	None	4–5 oz/yd^2 PTFE	4–5 oz/yd^2 PTFE
Thermal liner	None	None	5–7 oz/yd^2 aramid batt
Thermal Mannequin Predicted Burn Injury (%)*			
6 sec exposure to 2 cal/cm^2 sec:			
2nd degree burn:	38	7	0
3rd degree burn:	16	0	0
Total	54	7	0
10 sec exposure to 2 cal/cm^2 sec:			
2nd degree burn:	18	23	6
3rd degree burn:	58	20	0
Total	76	43	6

* All testing was done with the head of the Thermal Mannequin covered.

TABLE 5.17 Comparison of Electric Arc and Flash Fire Exposures

Exposure elements	Electric arc	Flash fire
Incident energy (cal/cm^2)	1 to >100	1 to 30
Radiant heat energy (%)	90	30–50
Convective heat energy (%)	10	50–70
Exposure time (sec)	0.01 to >1	1 to 15
Concussive forces	High	Variable
Ionized air generation	High	Moderate
Smoke/fumes	Yes	Yes
Molten metal spatter	Yes	No
Potential for reoccurrence	Reclosing	Reignition
Intensity limiting factors	Electrical system	Fuel, air
Exposure level estimation	System parameters permit estimates	Unpredictable, difficult to estimate

blast burns [49]. A 1995 study [50] suggested that as many as 80 percent of electrical injuries are burns that result from exposure to the thermal energy of electric arcs.

An electric arc usually results from an electrical system fault that allows for the striking and sustained passage of substantial electric current through ionized air. It typically lasts for a very short time (<1 s) and generates high-temperature plasma that provides a mostly radiant energy exposure. The resulting thermal energy is capable of melting or igniting everyday clothing. It has the characteristics of an explosion. Injury can be the result of electrocution, physical assault from the concussive forces of the blast, skin burn injury from the intense radiant thermal energy or direct contact with the hot plasma, inhalation of toxic gases, molten metal splattering, exposure to fires from secondary items like hot or ignited transformer oils, and the melting or ignition of clothing. Table 5.17 compares the characteristics of electric arcs to flash fires.

5.5.4.1 Design Criteria for Electric Arc Protective Clothing

As is the case for flash fires, numerous injuries and fatalities are caused yearly by exposure to the thermal energy from electric arcs. Here, too, the most severe injuries are usually associated with clothing that ignites and continues to burn well after the original thermal exposure. Flame-resistant clothing is clearly needed for workers exposed to potential electric arc flash hazards.

Unlike flash fire exposure hazards where the exact thermal threat is difficult to estimate, the thermal hazard potential of an electric arc flash can be estimated beforehand for most job tasks. The available fault current, voltage, and clearing times for the subject electrical systems are generally known in advance, either by the worker or a responsible engineer. This information, along with other parameters such as number of phases, area geometry where the arc may occur, and the proximity of the worker to the arc can be applied to available calculation models to generate an estimate of the thermal hazard present. The appropriate protective clothing systems can then be selected based on this estimated thermal threat.

Neal et al. [50] and Doughty et al. [51] pioneered the arc flash test methods that have resulted in the development of the current ASTM standard procedures for assessing the thermal arc protection level of materials for clothing, ASTM F-1958 and F-1959. In addition, their work proposes a hazard analysis procedure to estimate the level of incident energy. They also present a list of general guidelines for thermal electric arc protective clothing. These include:

- Protective clothing selection must be based on probable worst-case exposure for a task.
- Flammable fabrics should be avoided as the outer layer of protective clothing.
- Outer flame-resistant layers must not have openings that expose flammable underlayers.
- Tight-fitting clothing should be avoided.

- If the outer flame-resistant layer of a multilayer thermal electric arc protection clothing system is susceptible to breaking open due to the high energy of the arc blast, then the inner layer(s) must be flame resistant.
- Fabrics made of thermoplastic (meltable) synthetic fibers should be avoided in the layer next to the skin.
- Increased fabric thickness will normally improve thermal protection.
- Multiple layers of clothing are generally more effective than single layers.

The NFPA 70E standard is also a useful guide to assess electric arc thermal hazard levels. It incorporates risk category classifications and the appropriate flame-resistant clothing options.

Several flame-resistant garments are available that can provide protection from electric arcs. The key considerations are similar to those for flash fire protective clothing (see Sec. 5.5.2.1), i.e., protection level, durability, wearer comfort, heat stress potential, garment styling, ease of movement, and life cycle cost. Again, the clothing system selected must permit the worker to perform job tasks in a satisfactory manner while still affording appropriate protection.

5.5.4.2 Performance of Fabrics in Arc Protective Clothing

Most flame-resistant fabrics commercially offered for protection against flash fires are also generally used in garments designed for electric arc flash hazards. Their performance is measured with the ASTM F1959 "Standard Test Method for Determining the Arc Thermal Performance Value of Materials for Clothing." As noted earlier, the method generates an arc thermal performance value (ATPV) or energy to break open (E_{bt}) for the fabric or fabric system tested. Features such as fabric weight and layering have a direct effect on these values. Table 5.18 provides ATPV values and some E_{bt} data for common flame-resistant fabrics that were generated during the initial development of the ASTM method [51].

Typically, garments and ensembles are selected for a job or task based on the thermal hazard potential as determined by the risk analysis process. Use of these calculation methods for determining the hazard potential will yield the total heat energy value, expressed as calories per square cen-

TABLE 5.18 Arc Testing Results for Typical Flame Resistant Fabrics [51]

Single layer fabrics	Weight (oz/yd^2)	APTV (cal/cm^2)	APTV or E_{bt} per unit weight (cal/cm^2)/(oz/yd^2)	E_{bt} (cal/cm^2)
Nomex® IIIA	4.5	5	1.1	—
	6	6	1.0	—
	7.5	7	0.9	—
	9	9	1.0	—
Indura FRT® cotton	7.5	6	0.8	—
	9.3	7.9	0.8	—
TufStuf® cotton/nylon	7.2	6.6	0.9	—
	9.2	11.3	1.1	—
Firewear® FR modacrylic/cotton	5.5	6.4	1.2	—
	7	—	1.0	6.8
	9.5	—	1.2	11
Kevlar®/PBI®	4.6	6	1.3	—
	6.1	6.8	1.1	—
Two layer fabrics				
Cotton + Nomex®	4.2/4.8	—	1.2	10.6
	4.2/6.2	—	1.3	13.3
Cotton + 60% Nomex®/40% Kevlar®	4.2/7.8	18.4	1.5	—

timeter for a potential thermal electric arc flash exposure. To provide the necessary level of performance, this value is usually matched against the ATPV or E_{bt} value of the protective garment ensemble. It should be noted that the ATPV represents the 50 percent probability of the prediction of a second-degree burn injury with a selected fabric or system. Therefore, the selection garments with ATPV higher than the estimated threat may be prudent to add an additional margin to the protective performance a specific garment or system delivers.

The NFPA 70E standard offers an alternative to direct calculation of the arc hazard potential. A matrix of hazard risk category classifications is available that lists generic tasks versus system voltage classes. The task lists assume that the equipment is energized and work is to be done inside the calculated *flash protection boundary*. The hazard category is then applied to a separate protective clothing and personal protective equipment matrix to determine the appropriate protective ensembles. Although much simpler to apply, the use of this method can result in overly protective garments that can interfere with the execution of the required task, may reduce the wearer's comfort, may increase the potential for heat stress, and can result in higher cost clothing systems.

5.5.5 Molten Metals

Another specialized situation requiring thermal protective clothing is that of workers in the metals industries. These personnel can be exposed to high material temperatures, radiant heat energy, and splashes of molten metals. The direct conductive heat energy threat poses the biggest challenge. For example, in the aluminum industry, the working temperatures of molten metal reach 1400°F (760°C), whereas in the steel industry they reach 2550°F (1400°C). In addition, welders will encounter a greater range of temperature hazards, although on a somewhat scaled down level of threat.

5.5.5.1 Design Criteria for Molten Metal Protective Clothing

The primary hazard in this industry is the threat of a molten metal splash onto clothing (assuming that the worker's body is totally protected). The first requirement for protective clothing is that it must not ignite and continue to burn should molten metal make contact. The second is that the molten metal should not stick to the clothing. Most protective clothing ensembles are incapable of mitigating the conductive heat energy transfer from sources at this temperature. Molten metals sticking to clothing generally results in serious burn injuries. For similar reasons, the garment fabric components should not split open or shrink excessively. ASTM F955 (see Sec. 5.4.2.1) attempts to quantify these required characteristics.

The remaining considerations are similar to those for flash fire protective clothing (see Sec. 5.5.2.1), i.e., durability, wearer comfort, heat stress potential, garment styling (appropriate for the task), ease of movement, and cost. Again, the clothing system selected must permit a worker to perform their job tasks in a satisfactory manner while still affording appropriate protection.

5.5.5.2 Performance of Fabrics in Molten Metal Protective Clothing

Several recent studies have developed data that characterize the performance of fabrics used for protective clothing in the molten metals industry. Barker [52] has examined over 30 fabrics in two versions of ASTM F955, one that poured molten aluminum at 1400°F (760°C) and the other at 1760°F (960°C). The test fabrics in this investigation included five *primary protective fabrics* and 28 *secondary protective fabrics*.

In the molten metal industry, primary protective fabric is intended for protective apparel used for limited times. This would typically be for especially hazardous tasks where a significant exposure risk to molten substance splash, radiant heat, and flame is likely. Barker tested flame-retardant (Zirpro®)-treated wool, untreated wool, aluminized PBI® (polybenzimidazole), Kevlar® (*para*-aramid), and Kevlar®/carbon fiber blends. The testing found that all of the samples shed the molten aluminum, did not ignite, and maintained their integrity in the test. However, only the thickest wool products (850 g/m² [25 oz/yd²]) "provided sufficient thermal insulation to ensure long-term protection from second-degree burns."

Clothing from secondary protective fabrics is designed for continuous wear in environments where intermittent exposure to molten metal splash, radiant heat, and flame sources is possible. Barker found that none of the secondary protective fabrics provided burn injury protection in these tests. This is attributed to their thickness and lighter weights (200 to 575 g/m^2 [6 to 17 oz/yd^2]). In general, the Zirpro®-treated wool fabrics were least affected by the molten metal, followed by the polyvinyl-alcohol (Vinal) based products, flame-retardant-treated cotton, and finally untreated cotton. Additionally, the heavier fabrics seemed to perform better than the lighter fabrics. The Aluminum Association [53] recommends protective clothing made from either flame-retardant treated wool, flame retardant-treated cotton, or Vinal-based fabrics.

The European counterpart to ASTM F955 is EN 373. This test method focuses mainly on the iron industry where testing is specified for molten iron at 2550°F (1400°C). EN-373 does, however, incorporate provisions for testing with molten aluminum. Mäkinen [54] reports on a variety of fabrics subjected to molten iron and provides a summary of other work using this method. Mäkinen has tested combinations of outerwear and underwear with the test fabrics. This work indicates that the best performing combinations involved either wool or flame-retardant-treated (Pyrovatex®) cotton outerwear in combination with a flame-retardant-treated underwear. As in the Barker study, there appears to be a general tendency of better performance from the heavier fabrics.

5.5.6 Soiling and Cleaning of Protective Clothing

There are two important factors that are often overlooked which can significantly affect the performance of fabrics designed to provide protection from fire. These are garment soiling through use and the process used for cleaning. Many soils and contaminants encountered in use are flammable and can impair the performance properties of a flame-resistant fabric. Nonvolatile organic compounds, such as petroleum products, constitute one of the most common contaminants found on these garments. When cleaning is considered, care must be given to selecting appropriate cleaning agents. The residue from many detergents and fabric softeners can impair protection. Furthermore, aggressive home and most industrial laundering procedures can, over a short period of time, remove a variety of topically applied flame-retardant treatments.

Several studies have been carried out to measure the impact that use and laundering have on properties of flame-resistant fabrics. Mäkinen [55] has measured the char lengths, afterflame times, and afterglow times of fabrics that included firefighters' suits aged from 4 to over 10 years old. Permeability and physical properties were also measured. Mäkinen concludes "that suits which are over ten years old and, particularly ones soiled with combustible substances, may be in risky condition. No linear degradation was found. . . . Mere laundering does not give sufficient information about the effect of wear."

Stull [56] has studied the effect of different types of laundering and dry cleaning techniques on the removal of contaminants and the performance of the protective apparel. He found that storing the garments in a ventilated closet at slightly (+5°C) elevated room temperature is quite effective in eliminating volatile organic contaminants. This procedure, combined with regular conventional laundering, could provide adequate cleaning while having little negative effect on the garments' performance.

These studies indicate that soil, especially from petroleum-based oils and greases, can be detrimental to the fire protection characteristics of fabrics; and proper laundering can maintain high levels of performance.

5.6 APPLICATIONS—FURNISHINGS

As discussed earlier, fibers and fabrics are important components in assessing the risk of fire. This is especially true when they are used in upholstery, carpets, draperies, and other assorted interior furnishings. The 1998 Federal Emergency Management Agency report, "Fire in the United States,

1989–1998 [57] identifies and ranks the most common scenarios for fire death in the U.S. This survey found that "The leading cause of (residential) death in 1998 is smoking, as in all *National Incident Reporting System* (NFIRS) years, at 21 percent. Most of the smoking deaths come from cigarettes dropped on upholstered furniture or bedding, often by someone who has been drinking." When all ignition sources are considered, furnishings and mattresses account for ~40 percent of U.S. fire deaths. The 1995 CBUF report [58] from the Commission of the European Communities found that "fire statistics show that the majority of European casualties are due to fires in upholstered furniture."

In this section, we explore the role of fibers and fabrics in fire prevention and propagation. We will consider their application in public facilities, residential homes, and general transportation.

5.6.1 Structures

The development of fire-resistant furnishings and fabrics that enable fire resistance is primarily driven by regulatory activity. Historically, the resulting laws and requirements have been applied more strictly to public institutions than to furnishings for residential homes.

Public institutions have different sets of complicating characteristics that influence the impact of a fire event. For example, in prisons, there will be restraints to egress in a fire; in hospitals, some patients will need assistance to leave a fire; in theaters, large crowds of people can be involved and normally adequate exits may be blocked. All of these factors impact a design of safe facilities.

It is especially clear that in these applications the entire environment must be considered. Features include:

- The size, shape, and ventilation of the structure or room
- Who might be involved
- The likely sources of ignition
- The fuel load present, e.g., the types and amounts of materials and furnishings
- The location and types of fire protection systems in place (e.g., sprinklers)
- Routes and methods for escape, and
- The types of clothing that are likely to be involved

5.6.1.1 Fire Response Characteristics

Hilaldo [59] identifies 10 *fire response characteristics* that describe the response of a material when exposed to fire. They include:

1. Smolder susceptibility—smoldering combustion is the slow propagation of a combustion wave through porous fuel, characterized by relatively low temperatures and incomplete oxidation.
2. Ignitability—the ease of ignition, especially by a small flame or spark.
3. Flash-fire propensity—a fire that spreads with extreme rapidity, as might occur with the ignition of a pool of flammable liquid.
4. Flame spread—the progress of flame over a surface.
5. Heat release—the heat produced by the combustion of a given quantity of material.
6. Fire endurance—the time during which a material maintains its design integrity under specified burning conditions.
7. Ease of extinguishment—the ease with which burning can be extinguished for a specific material.
8. Smoke evolution—the generation of visible, nonluminous, airborne suspension of particles, usually expressed in terms of light obscuration.
9. Toxic gas evolution—gases that are poisonous or destructive to body tissues.
10. Corrosive gas evolution—gases that corrode otherwise stable materials, especially metals.

These characteristics are important considerations for fabrics and other materials that make up the furnishings of any institutional space.

5.6.1.2 Room Fire Scenarios

In the early 1990s, the Commission of European Communities initiated a study within the European Fire Research Programme, "CBUF—Combustion Behaviour of Upholstered Furniture" [58]. The study involved 11 laboratories and over 50 scientists. As part of that report, the authors distinguish smoldering from flaming fires.

Nonflaming or Smoldering Fires. When furniture is subject to a small radiant heat source (being placed too close to a space heater, for example) or having a cigarette dropped into a crevice, a self-sustaining smoldering fire can occur. This results in the slow release of toxic thermal decomposition products that can produce hazardous conditions in the room over a period of an hour or more.

Flaming Fires. When an object, such as an item of upholstered furniture, has caught fire and is burning, the gases in the room will separate early into distinct hot and cold layers. The upper layer contains the hot, toxic combustion gases and opaque smoke. The lower layer contains, at least initially, relatively fresh air. The interface between the two is termed the "thermal discontinuity." A first-approach hazard analysis would note that a person could escape from this fire by staying below the thermal discontinuity.

The thermal discontinuity descends as the fire grows and, in most situations, a mixing occurs between the layers such that the layer of toxic gases descends to the level of the furniture in the room (2 to 3 ft). In small, poorly ventilated rooms, as little as 1 lb of fire-consumed furniture composite material could incapacitate an occupant.

At a certain stage in the development of a room (or contained) fire, all of the exposed surfaces will reach their ignition temperature more or less simultaneously and the fire spreads quickly to encompass the entire space. This stage is termed *flashover*. Once reached, the chance of survival for an occupant is minimal.

Clearly, resistance to ignition is important, but the burning rate (or more accurately, the HRR) is also important, assuming that ignition will happen in some cases. The CBUF report expresses the size of a fire in terms of HRR. In part, HRR is important because it allows time for an alert person to escape, and in part, because the fire can spread to other rooms, slower burning will again provide time for occupants of those rooms to escape. HRR will also determine how quickly flashover is reached. CBUF estimates that flashover will occur when a fire reaches an HRR of about 1000 kW in a small room.

5.6.1.3 Models for Structural Fires

Many attempts have been made to model room fires based on both theoretical and empirical data. As data have been acquired in recent years from full-scale tests of furnishings, computer models of those fires have developed. It is estimated that over 50 such models have been actively supported [60].

CFAST, for example, is a multiroom model developed at NIST. [61] Application of this model permits the prediction of various fire characteristics, including flashover. Indeed, in spite of the range in complexity and results, many models are able to predict flashover within the precision of available experimental data [62].

CBUF has developed three models that predict room fire behavior where upholstered furniture or mattresses are involved [58]. The models use data for components of the furniture obtained from the cone calorimeter (oxygen consumption calorimeter) test. As part of that extensive research program, CBUF has also evaluated the CFAST model along with a field modeling program, JASMINE, and found both useful.

5.6.1.4 The Performance of Fabrics in Fire Situations

Numerous tests have been run in a variety of laboratories to determine the fire characteristics of rooms containing upholstered furniture. Among the most comprehensive is the CBUF study cited

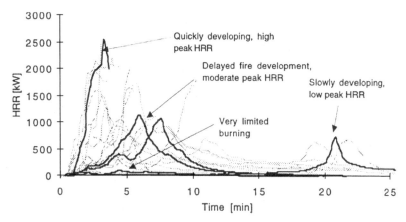

FIGURE 5.17 Heat release rate as a function of time. (Reprinted with permission from CBUF Consortium, *Fire Safety of Upholstered Furniture—the Final Report of the CBUF Research Programme*, European Commission Measuring and Testing Report EUR 16477 EN, 1995.)

above. In that study, data were gathered in a full-size instrumented room (ISO 9705) utilizing the furniture calorimeter (NT FIRE 032) and the cone calorimeter (ISO 5660). The CBUF study [58] explored a number of variables that might affect the nature of a fire. In this section, we summarize their findings as they relate to the effect of upholstery fabrics on the fire.

Fig. 5.17 provides the type of HRR versus time data obtained in room fire tests from upholstered chairs. Note that the peak HRR equals or exceeds the 1000-kW level estimated to produce flashover in three of the four cases, but the time to reach that peak differs significantly. Table 5.19 summarizes the series of upholstered chair fire data from the CBUF test program [58]. All of the chairs were the same style and size. They each used foam cushions made from one of two types that had similar burning characteristics, making the primary variable the fabric used for upholstering. Fig. 5.18 shows the effects of the fabric materials most clearly. The best performing single fabric was the plain wool cover, for which the peak HRR was delayed by almost 10 min over all the others. The addition of flame retardant to both the cotton and the polyester both depressed the peak HRR and delayed the time to reach the peak value. These studies also show that the use of an aramid interliner with a cot-

TABLE 5.19 Room Fire Tests of Upholstered Furniture [58]

Fabric	Interliner	Peak HRR [kW]	Time to peak HRR [sec]	EHC [MJ/kg]	Peak SPR [m²/s]	HCN peak [ppm]	CO peak [ppm]
Polyester	—	1054	130	17.05	7.38	61	735
Polyester	—	1181	209	15.28	7.0	—	—
FR polyester	—	868	170	14.48	6.95	68	601
Cotton	—	872	190	17.82	3.64	—	587
FR cotton	—	832	210	16.47	5.27	33	655
FR cotton	—	850	—	17.15	2.93	—	621
FR cotton	—	664	656	16.35	3.0	15	557
Cotton	Kevlar®	887	900	18.01	4.16	30	1105
Polyacrylic–FR backing	—	1176	230	18.23	10.96	35	—
Wool	—	867	830	16.98	4.14	95	589

Notes: HRR = Heat Release Rate; EHC = Effective Heat of Combustion; SPR = Smoke Production Rate. Chairs had timber frames, were fully upholstered to the ground, and had loose seat and back cushions.

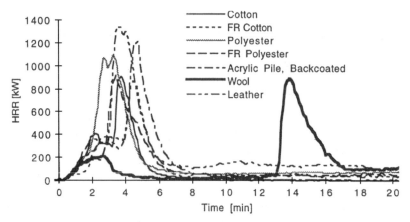

FIGURE 5.18 Heat release rate for various cover fabrics. (Reprinted with permission from CBUF Consortium, *Fire Safety of Upholstered Furniture—the Final Report of the CBUF Research Programme,* European Commission Measuring and Testing Report EUR 16477 EN, 1995.)

ton cover provides a substantial delay in the time to reach peak HRR, although the burning is just as intense as without the interliner (discussed in the next section).

The CBUF study also collected data on smoke production and toxic gas emissions for a wide range of fabrics. The studies showed that a wool-covered chair produced the lowest peak values, but the addition of a fire retardant to the cover fabric had little effect on smoke production. Cyanide gas peak levels for chairs covered with the different fabrics were also discussed in this study. While the concentration for the wool-covered chair was higher than the others, it was not enough to affect the escape time.

The CBUF study also collected data from furniture calorimeter tests (NT FIRE 032). However, this series of tests did not include the specially constructed upholstered chairs used in the room scale tests to study the effect of cover fabrics. Bench-scale cone calorimeter test data were also presented in this study.

5.6.1.5 The Role of Fire-Blocking Layers

A fire-blocking fabric (or interliner) is a fabric that is used to provide a barrier between a flammable object or material, such as a polyurethane seat cushion, and the flame from a fire. In the case of a chair, the usual construction would include a cushion, the fire-block, and an outer upholstery fabric. If an ignition source manages to ignite the outer fabric, the fire-block is designed to keep the flame from reaching the cushion, which contains the greatest part of flammable material, and prevent or delay the development of a large fire.

In 1992, the California Bureau of Home Furnishings and Thermal Insulation published a fire test protocol for upholstered furniture used in public occupancies (hotels, hospitals, airport lounges, etc.). The test, developed in conjunction with NIST, is designated California Technical Bulletin 133 (CAL 133). It involves subjecting the full-scale item to an arsonlike ignition source (an 18-kW gas burner) in an instrumented room. The standard specifies limits to heat release rate, total heat evolved, smoke obscuration, and carbon monoxide level. The key specification requires the peak heat release rate not to exceed 80 kW. As can be seen from the data in the previous section, upholstered chairs made from standard materials do not meet this requirement. The addition of fire-blocking interliners, however, can provide one approach to meeting the CAL 133 test requirements. Fig. 5.19 shows identical chairs, with and without a fire-blocking interliner, burned according to the CAL 133 protocol.

Examples of furniture with and without a fire-block are also found in the CBUF study [54]. The test results of two similar chairs (cotton fabric, high-resilience polyurethane foam), one having an

FIGURE 5.19 Chairs tested using CAL 133 protocol, showing the impact of a fire blocking interliner.

aramid interliner (Kevlar®) and the other without, showed a peak HRR 24 min later for the interliner-equipped chair. Replacement of the aramid interliner with a glass fiber material (Unigard®) resulted in no ignition. When the Unigard® interliner was used with a polyester cover fabric, the peak HRR was delayed more than 18 min. In other cases, the use of an interliner actually stopped the fire from growing.

Other studies have provided more extensive data comparing various fabrics coupled with different fire-blocking layers [63, 64]. Table 5.20 provides data from a NIST study that evaluates 27 combinations of cover fabric, fire-block, and cushion in full-scale furniture calorimeter tests. In these tests, two HRR peaks are observed, one within the first 3 min of the test while the burner is active, and a second peak after the burner is turned off. These tests indicate that while most of the constructions would not pass the CAL 133 requirements for peak HRR, some combinations do pass, including a cotton cover fabric. The modacrylic (acrylonitrile-vinyl chloride copolymer)/nylon cover fabric performed well with all three fire-block interlayers. The results from a similar study, found in Table 5.21, compare eight cover fabrics over a nonwoven aramid fire-block. These tests show no second HRR peak (burner off). The best performing cover fabric in this study is also the modacrylic/nylon blend [64].

These investigations indicate that with proper design and a fire-block interlayer, furniture can be constructed that will meet the stringent CAL 133 standards. This is critical to the furnishings industry as standards similar to CAL 133 are being adopted in other sectors. The transportation industry has already adopted comparable requirements (see Sec. 5.6.2).

5.6.1.6 Home Furnishings

In 1999, there were ~383,000 fires in residential structures in the United States with nearly 3000 associated fatalities and 16,000 injuries [65] (see Tables 5.22 and 5.23). As noted above, furnishings and mattresses accounted for ~40 percent of U.S. fire deaths, many of these related to smoking. The

TABLE 5.20 Furniture Calorimeter Tests of Upholstered Furniture with Fire-block Barriers [63]

			During burner exposure		After burner exposure	
Fabric	Interliner	Cushion	Peak HRR [kW]	Time to peak [sec]	Peak HRR [kW]	Time to peak [sec]
100% polyester	Nonwoven aramid	Cal 117	60	128	83	938
(9.0 oz/yd^2)	Knitted glass—A	Cal 117	84	138	20	640
	Woven glass	Cal 117	100	128	12	750
100% nylon	Nonwoven aramid	Cal 117	ca. 100	—	ca. 300	—
(11.5 oz/yd^2)	Knitted glass—A	Cal 117	141	138	476	1098
	Woven glass	Cal 117	135	139	32	900
	Woven glass	Cal 117	147	138	<20	860
75% modacrylic/	Nonwoven aramid	Cal 117	ca. 35	—	None	—
25% nylon	Knitted glass—A	Cal 117	35	128	None	—
(11.0 oz/yd^2)	Woven glass	Cal 117	35	138	None	—
38% polyester/	Nonwoven aramid	Cal 117	ca. 60	—	ca. 40	—
62% cotton	Knitted glass—A	Cal 117	177	108	326	1808
(9.0 oz/yd^2)	Woven glass	Cal 117	117	118	22	370
100% cotton	Nonwoven aramid	Cal 117	66	138	106	958
(9.0 oz/yd^2)	Knitted glass—A	Cal 117	90	128	32	1138
	Woven glass	Cal 117	70	125	<10	640
100% polypropylene	Nonwoven aramid	Cal 117	109	199	378	670
(8.0 oz/yd^2)	Nonwoven aramid	Cal 117	151	148	242	478
	Knitted glass—A	Cal 117	130	138	188	668
	Woven glass	Cal 117	150	128	744	648
	Nonwoven aramid	IFR	235	118	272	538
	Knitted glass—A	IFR	96	138	229	598
	Woven glass	IFR	136	128	326	1048
100% polypropylene	Nonwoven aramid	Cal 117	175	128	290	598
w. FR backcoat	Knitted Glass—B	Cal 117	82	138	57	1438
(12.0 oz/yd^2)	Woven glass	Cal 117	128	138	473	1038
	Nonwoven aramid	IFR	112	108	419	828
	Nonwoven aramid	IFR	165	138	108	1169
	Knitted Glass—B	IFR	98	138	35	1088
	Knitted Glass—B	IFR	92	120	23	4900
	Woven glass	IFR	144	138	72	1318

Notes: HRR = Heat Release Rate.
Chairs were CAL 133 steel mockup frames with seat and back cushions plus full thickness arm cushions.
Interliners: Nonwoven aramid = 2.0 oz/sq yd Kevlar® Z-11; Knitted Glass—A = 7.3 oz/sq yd knitted glass/charring fiber blend; Knitted Glass—B = 13.8 oz/sq yd knitted glass/charring fiber blend embedded w. halogen FR resin; Woven glass = 3.5 oz/sq yd woven glass fiber.
Cushions: Cal 117 = polyurethane foam w. low level halogen FR; IFR = polyurethane w. medium level melamine resin.

trend since 1977 is quite promising, showing a steady decrease from a high of 6015 civilian fire deaths in the home in 1978 to the 1999 level of 2885.

A typical home fire would be similar to the room scenarios described earlier, with some bias toward smoldering fires. At this time the standards for home use (primarily CAL 116 and 117) are only designed to avoid the use of highly flammable materials in furniture. The use of materials less likely to contribute to fires in the home will only increase as the value of more stringent regulation for institutional furniture becomes evident and relatively inexpensive products are available to meet those regulations.

TABLE 5.21 Furniture Calorimeter Tests of Upholstered Furniture with Fire-block Barriers [64]

Fabric	Interliner	Cushion	Peak HRR [kW]	time to peak during burner exposure [sec]
Thermoplastic:				
100% FR polyester	Nonwoven aramid	Cal 117	107	82
100% nylon	Nonwoven aramid	Cal 117	147	90
100% nylon	Nonwoven aramid	Cal 117	118	90
63% nylon/37% polyester	Nonwoven aramid	Cal 117	150	76
Charring:				
52% wool/48% nylon	Nonwoven aramid	Cal 117	110	73
100% cotton	Nonwoven aramid	Cal 117	133	70
60% cotton/32% wool/8% nylon	Nonwoven aramid	Cal 117	122	75
60% cotton/30% wool/6% rayon/4% nylon	Nonwoven aramid	Cal 117	106	70
75% modacrylic/25% nylon	Nonwoven aramid	Cal 117	75	82

Notes: HRR = Heat Release Rate.
Chairs were CAL 133 steel mockup frames with seat and back cushions plus full thickness arm cushions.
Interliners: Nonwoven aramid = 2.0 oz/sq yd Kevlar® Z-11.
Cushions: Cal 117 = polyurethane foam with low level halogen FR.

5.6.1.7 Draperies and Carpets

Although mattresses and upholstered furniture are the main textile sources of fire in buildings, curtains and carpets can also contribute. Curtains are typically lightweight and easily ignited, acting almost as a fuse.

As noted in Sec. 5.4.2.2, NFPA 701 provides test methods and standards for materials used for draperies and other window treatments. The standard is used in many public localities to preclude the use of highly flammable materials in draperies. Until recently, the standard only applied to single-layer fabrics, but evaluations of the flammability in full-scale tests showed that single-layer fabrics that passed NFPA 701 could provide significant hazards when used in multilayer combinations [66]. This is especially true when a charring fabric, such as flame-retardant-treated cotton, is combined with a melting/dripping fabric such as flame-retardant-treated polyester. A sample of these results is illustrated in Table 5.24.

TABLE 5.22 Estimates of 1999 Fires and Property Loss by Property Use [65]

Type of fire	Number of fires	Property Loss (millions of dollars)
Fires in structures	523,000	8,490
Fires in highway vehicles	345,000	1,149
Fires in other vehicles	23,500	175
Fires outside of structures	64,000	123
Fires in brush, grass wildlands	498,000	—
Fires in rubbish	226,500	—
All other fires	143,000	87
Total	1,823,000	10,024

Reprinted with permission from NFPA, Fire Loss in the United States During 1999. Copyright ©2000, National Fire Protection Association, Quincy, MA 02269.

TABLE 5.23 Estimates of 1999 Civilian Fire Deaths and Injuries by Property Use [65]

Property Use	Civilian deaths	Civilian injuries
Residential (total)	2,920	16,425
One- and two-family dwellings	2,375	11,550
Apartments	520	4,500
Other residential	25	375
Non-residential structures	120	2,100
Highway vehicles	450	1,600
Other vehicles	20	250
All other	60	1,500
Total	3,570	21,875

Reprinted with permission from NFPA, Fire Loss in the United States During 1999. Copyright ©2000, National Fire Protection Association, Quincy, MA 02269.

Benisek and others [67, 68] have evaluated a number of carpets using the Tablet (DOC FF 1-70) and the Flooring Radiant Panel Tests (NFPA 253, ASTM E 648). Their findings, summarized in Table 5.25, show the effect of the different materials used to construct the carpets. Zirpro®-treated wool and the 80/20-wool/nylon blend show the best performance and pass the NFPA 253 standards, as would the untreated versions of these two materials. Viscose rayon, polyester, and polypropylene all fail the NFPA 253 standards. Wool is the only material tested in these evaluations with a corresponding flame-retardant version.

TABLE 5.24 Flammability of Curtains and Drapery Assemblies [66]

Drapery materials	Unit weight (oz/yd^2)	Flames to ceiling?	Time* (sec)
Single panels:			
Heavy FR cotton	8.4	No	—
Light FR cotton	4.1	No	—
Heavy FR polyester	7.1	No	—
Double Panels:			
Heavy FR polyester + light FR cotton	7.1/4.1	Yes**	110
Heavy FR polyester + light FR polyester	7.1/3.1	No	—
Heavy FR cotton + light FR polyester	8.4/3.1	Yes	260
Heavy FR cotton + light FR cotton	8.4/4.1	Yes	230
Heavy FR cotton + heavy FR cotton	8.4/8.4	Yes	290
Light FR cotton + light FR cotton	4.1/4.1	Yes**	260

Notes:
* Time for flames to reach the ceiling.
** Flames did not reach ceiling if flame was applied in center of panel, but did if applied to edge.
Test Method: 8 ft long by 16 ft wide panels were pleated to cover about 5 ft width and hung next to a gypsum wall in a large (40 ft x 90 ft x 25 ft high) room. A bunsen burner was placed just below the fabric at either the center of the panel or at an edge. The burner produced a diffusion flame 8 to 12 inches high. It was applied for five minutes.
Portions of this work are reproduced from the July–August edition of the *Building Standards* magazine, copyright ©1988, with the permission of the publisher, the International Conference of Building Officials.

TABLE 5.25 Effect of Fiber Material on the Flammability of Carpet Samples [67, 68]

Fiber material	Table test (pass/fail)	Critical radiant flux (W/cm^2)	Smoke emission (% trans–min)
Wool, Zirpro® treated	—	>1.0	0
80/20 wool/nylon (Zirpro® treated)	—	>1.0	0
80/20 wool/nylon	Pass	0.70	77
Wool	Pass	0.67	46
Acrylic	Fail	0.31	1959
Nylon, continuous filament	—	0.24	490
Viscose rayon	Pass	0.15	330
Polyester	—	<0.125	3005
Polypropylene	Fail	<0.125	4845

Reprinted with permission from Benisek, L., and Phillips, W.A., "Evaluation of Carpets in the National Bureau of Standards Flooring Radiant Panel," *Textile Research Journal*, Vol. 53, #1, January 1983, p. 36–43.

5.6.2 Transportation

Another industry segment where fabrics are involved in fires and fire protection is transportation. Fires are a significant danger in most areas of transportation. Here we examine fabrics in use for the three major transportation segments: rail, aircraft, and automobile.

5.6.2.1 Rail

Two recent articles [37, 69] provide a comprehensive review of the development of methods of fire protection in passenger rail transportation, citing 102 references. They find that "Fire safety in any application, including transportation systems, requires a multifaceted systems approach. The effects of vehicle design, material selection, detection and suppression systems, and emergency egress and their interaction, on the overall fire safety of the passenger trains must be considered."

Passenger trains today contain increasing quantities of combustible materials. Although the basic shell is stainless steel, the interiors are a vast assortment of potentially flammable materials and ignition sources. Passenger cars provide a good example. They will typically contain seat cushioning and upholstery, coverings for floors, walls and ceilings, window glazing (generally polycarbonate), curtains and gaskets, and insulation for pipes and wiring. Further, passengers bring aboard quantities of flammable belongings adding to the potential fuel load. Specialized cars and first-class sections also provide additional flammable material potential in a variety of geometries (additional cushioning, compartmentalized sections, bedding materials, additional storage, etc.).

The Federal Railroad Administration, Amtrak, and the National Fire Protection Association have issued standards relating to materials used in passenger trains. These specify test requirements and criteria for flammability and smoke emission based on ASTM laboratory tests (see Sec. 5.4.3.2). None of these tests, as the authors point out, includes measurement of the most important material characteristic, the heat release rate. Nevertheless, current standards have contributed to a historical fire record that has been very good, with few serious passenger train fires.

Research at NIST is under way to develop a systems approach to fire safety analysis for passenger trains. The approach involves four steps [69]:

1. Defining the application
2. Calculating the fire performance of the application
3. Defining specific fire scenarios for the application, and
4. Evaluating the suitability of the proposed system design

TABLE 5.26 Cone Calorimeter Test Data for Selected Passenger Train Materials [69]

Material	Time to ignition (sec)	Peak heat release rate (kW/m^2)	Smoke emission (m^2/kg)
Seat cushion			
Foam	14	80	30
Interliner	5	30	300
Fabric cover	11	420	225
PVC cover	7	360	770
Seat cushion			
Foam	14	80	30
Interliner	5	30	300
Fabric cover	8	265	400
Wall covering			
Wool carpet	30	655	510
Wool fabric	21	745	260
Fabrics			
Wool curtain/nylon window drape	13	310	380
Polyester window drape	20	175	810
Wool blanket	11	170	560
Modacrylic blanket	17	18	n.a.
Cotton pillow fabric	24	340	570
Floor coverings			
Nylon carpet	10	245	350

Cone calorimeter tests carried out as prescribed in ASTM E 1354. The heat flux level was 50 kW/m^2.

The heart of this process is a computer-based fire model of the passenger car itself. The key input to the model is the HRR data for the contents of a specific car. NIST has determined HRR data for a broad selection of furnishing materials from typical passenger trains (see Table 5.26). They note the wide range of HRR measured for the different materials and point out the need to evaluate the total context [69].

The next step in this program is to perform full-scale tests to confirm the model results.

5.6.2.2 *Aircraft*

Fire in any aircraft can have deadly consequences. Where proper design of emergency exits can provide easy egress from a theater or passenger train, aircraft fires present a more complex situation. For many years, fire had not been a prevailing safety concern of the aircraft industry, primarily due to the materials and design concerns of the day. The lightweight structure needed to allow the aircraft to perform its mission also permitted impact energy to be transmitted to the aircraft occupants. Death was much more likely to occur due to impact trauma than to fire, but today's turbine-powered aircraft structures are strengthened by heavier gauge metals and advanced composites so that occupants can potentially survive a crash. Post-crash fire has now become the predominant cause of injuries or death [70].

Since 1988, the *Federal Aviation Agency* (FAA) and most other aviation authorities worldwide have implemented numerous modifications to aircraft fire safety standards that significantly improve fire safety. In a recent report [71], 24 civil transport aircraft accidents and another 9 incidents, which occurred during the period 1987 to 1996, were analyzed for their fire safety implications. Of the more than 3000 occupants of the 24 crashes, there were 1451 fatalities. The FAA accredits the lack of a greater number of fatalities to these modified fire safety standards.

The typical scenario involving a fire following a crash on land starts with the catastrophic failure of the fuel system and the release of large quantities of fuel onto the ground. The fuel is subsequently ignited by a variety of available ignition sources (exposed electrical circuits, sparks and friction-generated heat, etc.). The resulting intense fuel fire quickly penetrates the cabin where interior materials become involved. Passengers have a very limited time available to evacuate as the heat, hot toxic combustion gases, and opaque smoke overwhelm the cabin. The times required for evacuation typically range from 2 to 5 min.

Fibers and fabrics are involved in aircraft fires in much the same way as they are in passenger trains; seat cushion upholstery, floor and wall carpeting, blankets and pillows, and occupant clothing and carry-on luggage provide the basic fuel load. Additionally, the fibrous materials used for thermal and acoustical insulation are often involved.

This scenario provides the ideal situation for the application of fire-blocking fabrics. These materials have demonstrated the ability to extend the time that it takes for the seat cushions, a large component of the cabin, to begin to burn. This single feature can provide additional critical time to escape. This feature was recognized by the FAA and became the basis for the Seat Cushion "Fire Blocking" Rule of 1984, requiring that all cabin seat cushions in transport aircraft meet a large oil burner test. The result of this rule is that most seat cushions are fire-blocked.

The FAA is continuing its quest for increasing the survival of passengers from fires on the ground. Among their projects for the future are the development of almost noncombustible materials and increasing the ability of the shell of the aircraft to protect the interior cabin from the ground fire.

5.6.2.3 Automobiles

Automotive fires are often overlooked as a fire problem area requiring protection solutions. Statistically, from 1989 to 1998, motor vehicle fires accounted for ~17 percent of the overall deaths in the United States that are attributed to fire. They also represent nearly 24 percent of all reported fires [57]. Although the total number of vehicle fires and related deaths and injuries are trending down, the property value loss has trended up ~14 percent.

The causes of these fires are fairly diverse: accidents, mechanical failure, human error, and arson. Most are caused by mechanical and design failures. This includes items such as broken fuel lines, engine overheating, defective exhaust components (especially catalytic converters), and other assorted mechanical failures that lead to accidents (e.g., defective or blown tires).

The majority of fire deaths occur with accidents. However, a large percentage (estimated at 14 to 30 percent of fire deaths and injuries) comes from preventable human error. Cigarettes dropped on upholstery, inattentive driving, stopping or parking over combustible materials with a hot catalytic converter, and the misuse of gasoline are at the top of the list.

Few standards address automotive fire safety when compared to other industry segments. For vehicle interiors, the 1972 Federal Motor Vehicle Safety Standards and Regulations Standard 302 is the industry guide. It specifies a horizontal flame test for all materials used in the interiors of automobiles and provides details of the test methods. It applies to seat cushions, seat backs, headlining, trim panels, floor carpeting, belts, and other fabrics. It is aimed primarily at fires caused by smoking and specifies limits on the rate of flame spread. Several other standards have been issued that relate to fuel tank integrity and similar structural characteristics [72].

This limited regulatory response to the automotive fire challenge appears matched by the absence of programs aimed at further reducing the impact of vehicle fires.

5.7 CHALLENGES FOR THE FUTURE

In the United States, fire still claims thousands of lives each year. Unfortunately, inappropriate fabrics and materials are frequently involved. In recent years, the death toll has been falling. This is due in large part to scientists and engineers who work to greatly increase our understanding of the nature of fire and the materials that are involved.

This improved understanding has led to an approach to fire safety that is based on an overall systematic environmental analysis, be it the dynamics and interactions of an aircraft and associated materials in a fire, the response of a room and its contents to a fire, or the evaluation of situational requirements of protective clothing ensembles worn by firefighters. This analysis technique has led to testing regimes that better evaluate and predict a material's system performance in a fire. This test method development must continue, as many unknowns remain in the determination of material characteristics that are key to providing safer environments.

We expect the systems approach and improved understandings to result in a continuing process of improved standards development. This is the required groundwork that provides the critical guidance for fire safety development to manufacturers and designers.

As we better understand total systems and develop tests to fit that understanding, we will be better able to design materials to meet those needs. This will sometimes involve totally new materials, improved versions of familiar materials, and better ways of combining and structuring these materials. In every case, we evolve to an improved understanding of both the system and the materials involved.

ACKNOWLEDGMENTS

The authors would like to acknowledge the assistance of the following individuals in the development of this chapter: Dr. John Gallini, Dr. Hamid Ghorashi, Dr. Richard Young, and Herman H. Forsten.

REFERENCES

1. Adanur, S., *Wellington Sears Handbook of Industrial Textiles,* Wellington Sears Co., Woodhead Publishing Ltd, Cambridge, UK, 1995, pp. 75–82.
2. Fourné, F., *Synthetic Fibers: Handbook for Plant Engineering, Machine Design and Operation,* Hanser Publishers, Munich, 1998, Chap. 4.
3. "Manmade Fiber Chart for 2000," *Textile World,* 150(8): August 2000.
4. "Manmade Fiber Chart for 1990," *Textile World,* 140(8): August 1990.
5. Billmeyer Jr., F.W., *Textbook of Polymer Science,* John Wiley & Sons, New York, NY, 1984.
6. Grayson, M., *Encyclopedia of Textiles, Fibers, and Nonwoven Fabrics,* Interscience Publishers, New York, 1984.
7. Morton, W.E., and Hearle, J.W.S., *Physical Properties of Textile Fibers,* 3rd edition, The Textile Institute, Manchester, UK, 1993.
8. Walters, R., and Lyon, R.E., "Calculating Polymer Flammability From Molar Group Contributions," National Technical Information Service, DOT/FAA/AR-01/31, Springfield, VA, Sept. 2001.
9. Hilaldo, C.J., *Flammability Handbook for Plastics,* 5th edition, Technomic Publishing, Lancaster, PA, 1998, Chap. 2.
10. Mark, H.F., et al., "Combustion of Polymers and Its Retardation," in Atlas, S.M., and Pearce, E.M., editors, *Flame-Retardant Materials,* Plenum Press, New York, NY, 1975.
11. Cote, A.E., and Linville, J.L., editors, *Fire Protection Handbook, 17th Edition,* National Fire Protection Association, Quincy, MA, 1991, Appendix A.
12. Fourné, F., *Synthetic Fibers—Machines and Equipment, Manufacture, Properties,* Hanser Publishers, Munich, Germany, 1998.
13. *Technical Guide—Kevlar® Aramid Fiber,* E.I. DuPont de Nemours, Inc., Wilmington, DE, 2000.
14. *Technical Guide for Nomex® Brand Fiber,* E.I. DuPont de Nemours, Inc., Wilmington, DE, 1999.
15. Atlas, S.M., and Pearce, E.M., editors, *Flame-Retardant Materials,* Plenum Press, New York, NY, 1975.
16. Lyons, J., *The Chemistry and Uses of Fire Retardants,* John Wiley & Sons, New York, 1970.

17. Loftin, D.H., "The Durability of Flame Resistant Fabrics in an Industrial Laundry Environment," in *Performance of Protective Clothing: Fourth Volume, STP 1133,* ASTM, Philadelphia, PA, 1992, pp. 775–784.
18. Jackson, D., "FR Cotton Is Still on Top," *Safety & Protective Fabrics,* 18–21, March 1993.
19. Wakelyn, P.J., Rearick, W., and Turner, J., "Cotton and Flammability—Overview of New Developments," *American Dyestuff Reporter,* 87(2): 13–21, Feb. 1998.
20. Benisek, L., "Flame Retardance of Protein Fibers," in Atlas, S.M., and Pearce, E.M., editors, *Flame-Retardant Materials,* Plenum Press, 1975.
21. Benisek, L., Meyers, P., Palin M., and Woollin, R., "Latest Developments in Wool Aircraft Textile Furnishings," *Fire Safety Journal,* 17(3): 187–203, 1991.
22. Benisek, L., and Craven, P.C., "Evaluation of Metal Complexes and Tetrabromophthalic Acid as Flame Retardants for Wool," *Textile Research Journal,* 53: 438, 1983.
23. Lawton, E.L., and Setzer, C.J., "Flame-Retardant Polyethylene Terephthalate Fibers," in Atlas, S.M., and Pearce, E.M., editors, *Flame-Retardant Materials,* Plenum Press, 1975.
24. Kim, Y.-H., "Durable Flame–Retardant Treatment of Polyethylene Terephthalate (PET) and PET/Cotton Blend Using Dichlorotribromophenyl Phosphate as New Flame Retardant for Polyester," *Journal of Applied Polymer Science,* 81: 793–799, 2001.
25. Lewis, P., "The Advantages of Inherently Flame Retardant Polyesters," *Technical Textiles International,* 6(5): 20–21, 1997.
26. Adanur, S., *Handbook of Weaving,* Technomic Publishing, Lancaster, PA, 2001.
27. *IBENA® Textilwerke Beckmann GmbH,* Peterskamp 20, D-46414 Rhede, Germany, Website: http://www.ibena.de/protect/english/Protective_Clothing/protective_clothing.html
28. Behnke, W.P., "Thermal Protective Performance Test for Clothing," *Fire Technology,* 13: Feb. 1977, pp. 6–12.
29. Stoll, A.M., and Chianta, M.A., "Method and Rating Systems for Evaluation of Thermal Protection," *Aerospace Medicine Journal,* 40: 1232–1238, 1969.
30. Babrauskas, V., "Modern Test Methods for Flammability," NISTIR 4326, National Institute of Standards and Technology, Gaithersburg, MD, June 1990.
31. Babrauskas, V., "Ten Years of Heat Release Research with the Cone Calorimeter," *Japan Symposium on Heat Release and Fire Hazard, First Proceedings,* III, Tsubuka, Japan, May 1993, pp. 1–8.
32. Babrauskas, V., and Grayson, S.J., *Heat Release in Fires,* Elsevier Applied Science, New York, NY, 1992.
33. Lyon, R.E., "Advanced Fire Safe Materials for Aircraft Interiors," in Beall, K.A., editor, *U.S./Japan Government Cooperative Program on Natural Resources (UJNR), Fire Research and Safety, 13th Joint Panel Meeting, Volume 2,* March 13–20, 1996, Gaithersburg, MD, 1997, pp. 249–259.
34. Damant, G.H., "Combustibility Database Fire Tests Resource Guide," Sleep Products Safety Council, Burtonsville, MD, 1993.
35. Hilaldo, C.J., *Flammability Handbook for Plastics,* 5th edition, Technomic Publishing, Lancaster, PA, 1998, Chap. 4.
36. Stull, J.O., "Comparison of Conductive Heat Resistance and Radiant Heat Resistance with Thermal Protective Performance of Fire Fighter Protective Clothing," *Performance of Protective Clothing: Sixth Volume, STP 1273,* ASTM, Philadelphia, PA, 1997, pp. 248–268.
37. Peacock, R.D., Reneke, P., Jones, W., Bukowski R., and Babrauskas, V., "Concepts for Fire Protection of Passenger Rail Transportation Vehicles: Past, Present, Future," *Fire and Materials,* 19: 71–87, 1995.
38. Horner, A., ed., "Aircraft Materials Fire Test Handbook," DOT/FAA/AR-00-12, Federal Aviation Administration, Fire Safety Section, Washington, DC, April 2000.
39. Behnke, W.P., "Predicting Flash Fire Protection of Clothing from Laboratory Tests Using Second-Degree Burn to Rate Performance," *Fire and Materials,* 8(2): 57–63, 1984.
40. Lawson, J.R., "Fire Fighters' Protective Clothing and Thermal Environments of Structural Fire Fighting," *Performance of Protective Clothing: Sixth Volume, STP 1273,* ASTM, Philadelphia, PA, 1997.
41. *Engineering Guide: Predicting 1st and 2nd degree Skin Burns from Thermal Radiation,* Society of Fire Protection Engineers, Bethesda, MD, 2000.

42. Saffle, J.R., Davis, B., and Williams, P., "Recent Outcomes in the Treatment of Burn Injury in the United States," *Journal of Burn Care & Rehabilitation,* 14(10): Part 1, May/June 1995, pp. 219–232.
43. Lawson, J.R., "Thermal Environments of Structural Fire Fighting," NIST SP 911, National Institute of Standards and Technology, Gaithersburg, MD, Feb. 1997.
44. Abbott, N.J., and Schulman, S., "Protection from Fire: Nonflammable Fabrics and Coatings," *Journal of Coated Fabrics,* 6: 48–62, July 1976.
45. Barker, R.L., Guerth, C., and Behnke, W., "Measuring the Thermal Energy Stored in Firefighter Protective Clothing," *Performance of Protective Clothing: Seventh Volume: STP 1386,* ASTM, Philadelphia, PA, 2000, pp. 33–44.
46. Torvi, D.A., and Hadjisophocleous, G.V., "Development of Methods to Evaluate the Useful Lifetimes of Firefighters' Protective Clothing," *Performance of Protective Clothing: Seventh Volume: STP 1386,* ASTM, Philadelphia, PA, 2000, pp. 117–129.
47. Mell, W.E., and Lawson, J.R., "Heat Transfer Model for Fire Fighter's Protective Clothing," *Fire Technology,* 36(1): 39–68, Feb. 2000.
48. Stull, J.O., "Comparison of Conductive Heat Resistance and Radiant Heat Resistance with TPP of Fire Fighter Protective Clothing," *Performance of Protective Clothing: Sixth Volume: STP 1273,* ASTM, Philadelphia, PA, 1997, pp. 248–268.
49. Lee, R., "The Other Electrical Hazard: Electric Arc Blast Burns," *IEEE Transactions on Industry Applications,* 1A-18(3): 246, 1982.
50. Neal, T.F., Bingham, A.H., and Doughty, R.L., "Protective Clothing Guidelines for Electric Arc Exposure," *IEEE Paper No. PCIC-96-34,* New York, NY, 1996.
51. Doughty, R.L., Neal, T.F., Bingham, A.H., and Dear, T.A., "Protective Clothing & Equipment for Electric Arc Exposure," *IEEE Paper No. PCIC-97-35,* New York, NY, 1997.
52. Barker, R.L., "Resistance of Protective Fabrics to Molten Aluminum and Bath Slash and Their Comfort Properties," Report submitted to The Aluminum Association, Inc., Washington, DC, Oct. 2000.
53. Peterson, W.S., ed., *Guidelines for Handling Molten Aluminum,* 2nd edition, The Aluminum Association, Washington, DC, Dec. 1990.
54. Mäkinen, H., Laiho, H., and Pajunen, P., "Evaluation of the Protective Performance of Fabrics and Fabric Combinations Against Molten Iron," in *Performance of Protective Clothing: Sixth Volume: STP 1273,* ASTM, Philadelphia, PA, 1997, pp. 225–237.
55. Mäkinen, H., "The Effect of Wear and Laundering on Flame-Resistant Fabrics," in *Performance of Protective Clothing, Fourth* Volume: *STP 1133,* ASTM, Philadelphia, PA, 1992, pp. 754–765.
56. Stull, J.O. et al., "Evaluating the Effectiveness of Different Laundering Approaches for Decontaminating Structural Fire Fighting Protective Clothing," in *Performance of Protective Clothing: Fifth Volume: STP 1237,* ASTM, Philadelphia, PA, 1996, pp. 447–468.
57. *Fire in the United States 1989–1998, Twelfth Edition,* Federal Emergency Management Agency, United States Fire Administration, National Fire Data Center, Emmitsburg, MD, FA-216, 2001.
58. Sundstrom, B., ed., *Fire Safety of Upholstered Furniture—The Final Report of the CBUF Research Programme,* European Commission Measuring and Testing Report EUR 16477 EN, Boras, 1995.
59. Hilaldo, C.J., *Flammability Handbook for Plastics,* 5th edition, Technomic Publishing, Lancaster, PA, 1998, Chap. 3.
60. Friedman, R., *Survey of Computer Models for Fire and Smoke,"* 2nd edition, Factory Mutual Research Corp., Norwood, MA, 1991.
61. Peacock, R.D., Forney, G., Reneke, P., Portier R., and Jones, W., "CFAST, The Consolidated Model of Fire and Smoke Transport," NIST Tech Note 1299, National Institute of Standards and Technology, Gaithersburg, MD, 1993.
62. Peacock, R.D., Reneke, P., Bukowski, R., and Babrauskas, V., "Defining Flashover for Fire Hazard Calculations," *Fire Safety Journal,* 32: 331–345, 1999.
63. Ohlemiller, T.J., and Shields, J.R., "Behavior of Mock-ups in the California Technical Bulletin 133 Test Protocol: Fabric and Barrier Effects," NISTIR 5653, National Institute of Standards and Technology, Gaithersburg, MD, May 1995.
64. Forsten, H.R., "Use of Small-Scale Testing to Predict Cal 133 Performance, *Fire and Materials,* 21: 153–160, 1997.

65. Karter, Jr., M.J., "Fire Loss in the United States During 1999," NFPA, Quincy, MA, Sept. 2000.
66. Belles, D.W., and Beitel, J.J., "Fire Performance of Curtains and Drapery Assemblies," *Building Standards,* 57(4): 4–9, Jul.-Aug. 1988.
67. Benisek, L., and Phillips, W.A., "Evaluation of Carpets in the National Bureau of Standards Flooring Radiant Panel," *Textile Research Journal,* 53(1): 36–43, Jan. 1983.
68. Benisek, L., Meyers, P., Palin, M., and Woollin, R., "Latest Developments in Wool Aircraft Textile Furnishings," *Fire Safety Journal,* 17: 187–203, 1991.
69. Bukowski, R.W., Peacock, R., Reneke, P., Averill, J., and Marcos, S., "Development of a Hazard-Based Method for Evaluating the Fire Safety of Passenger Trains," *International Interflam Conference, 8th Proceedings, Vol. 2,* Interscience Communications Ltd., London, UK, 1999, pp. 853–864.
70. Brenneman, J.J, "Aviation," in Cote, A.E., and Linville, J.L., eds., *Fire Protection Handbook, 17th Edition,* National Fire Protection Association, Quincy, MA, 1991, Chap. 33: Sec. 8.
71. Hill, R.G., and Blake, D.R., "A Review of Recent Civil Air Transport Accidents/Incidents and Their Fire Safety Implications," *Proceedings, 1998 International Fire & Cabin Safety Research Conference.* Atlantic City, NJ, November, 1998.
72. Darmstadter, N., "Motor Vehicles," in Cote, A.E., and Linville, J.L., eds., *Fire Protection Handbook, 17th Edition,* National Fire Protection Association, Quincy, MA, 1991.

APPENDIX

Banwear® is a registered trademark of Itex, Inc., Colorado.

Dale Antiflame® is a registered trademark of Dale A/S, Norway.

Firewear® is a registered trademark of Springfield LLC, New York.

Indura® is a registered trademark of Westex, Inc., Illinois.

Indura® Ultrasoft® is a registered trademark of Westex, Inc., Illinois.

Kermel® is a registered trademark of Rhodia, France.

Kevlar® is a registered trademark of the E.I. DuPont Co., Delaware.

Lenzing FR® is a registered trademark of Lenzing Fibers, Austria.

Nomex® IIIA is a registered trademark of the E.I. DuPont Co., Delaware.

Nomex® Comfortwear is a registered trademark of the E.I. DuPont Co., Delaware.

PBI® is a registered trademark of Celanese, North Carolina.

Proban® is a registered trademark of Rhodia, France.

Pyroset® is a registered trademark of Cytec Australia Holdings Pty. Limited, Australia.

Pyrovatex® is a registered trademark of Ciba, Switzerland.

Tuffweld® is a registered trademark of Southern Mills, Georgia.

Zirpro® is a registered trademark of International Wool Secretariat, Australia.

CHAPTER 6
STRUCTURAL MATERIALS

Nestor R. Iwankiw, Jesse J. Beitel, and Richard G. Gewain
Hughes Associates, Inc.
3610 Commerce Drive, Suite 817
Baltimore, Maryland 21227-1652

6.1 STRUCTURAL MATERIALS USED IN CONSTRUCTION

6.1.1 INTRODUCTION

Fire protection and performance of structural elements used in construction have long been a concern. The great fire of London in 1666 showed how a fire can spread throughout a city. Over time, this has been demonstrated in the United States by fires such as the Great Chicago Fire of 1871, Baltimore City fire of 1904, and the fires following the San Francisco earthquakes in 1906. These fires showed how a small fire could grow to encompass a building, then a group of buildings, and finally a portion or a complete section of a city.

A concept known today as fire resistance was devised to prevent these types of fires from occurring. Fire resistance is the property of a material or assembly to withstand fire and continue to perform its structural function and/or provide containment of a fire. Thus, when this concept is applied to structural elements such as walls, floors, etc., then these building elements must resist the fire itself and help keep the fire from spreading. In concept, the use of fire resistance was to keep sections of cities from burning down by way of using fire-resistant walls and roofs, etc., on buildings that would be exposed to a fire. In addition, it became a goal to use fire-resistant construction to confine a fire within a single building and thus the structural elements had to be protected or they could fail and cause collapse of the building and potentially expose surrounding buildings. Later, the goal became the confinement of the fire to a single compartment within the building. To accomplish this goal, structural and nonstructural elements must be protected from fire and provide barriers to the spread of fire.

6.1.2 Construction Materials

In general, the primary structural elements of a building are the structural frame system to include the exterior walls, roof, interior floors/ceilings, interior walls or partitions, columns, roof trusses/beams, etc. These building elements are the primary building elements for structural support of the building and as such must exhibit some fire resistance depending on the use and type of building under consideration. Many of these same elements perform a second function, when required, of providing barriers to the spread of fire. Over time, each of these building elements have been designed and tested such that specific designs or construction techniques will provide, for that particular member, fire resistance for some duration of time.

Fire resistance of nonstructural members used on or in a building is also a means of controlling the potential spread of fire. In many cases, certain interior walls, while not load bearing, are used to provide compartmentation, thus limiting the spread of a fire within a building. Other building elements, such as doors, penetration seals, joints, windows, etc., may have to exhibit some degree of fire resistance.

The various types of structural and nonstructural building elements encompass many different types of construction materials. The materials can typically involve concrete of various types such as normal, lightweight, etc.; steel in a multitude of forms, shapes, and sizes; wood, again in a multitude of forms and sizes; or masonry such as clay brick and concrete masonry units. While this list is not all-inclusive, these materials, either alone or in combination with each other or other materials, provide building elements that will withstand fire and continue to perform their function.

6.2 DEVELOPMENT OF FIRE RESISTANCE TESTING

6.2.1 Historical Fire Events

Perhaps, in the United States, the beginning of the realization of the severe impact of fires in urban areas started with the Great Chicago Fire of October 8–10, 1871. At that time, the United States was becoming urbanized and industrialized, with the accompanying growth of larger and more densely populated urban centers. This conflagration resulted in the deaths of about 300 Chicagoans; 100,000 were left homeless, 18,000 buildings burned, and property losses amounted to $200 million. A depiction of this Great Chicago Fire is shown in Fig. 6.1. Even much earlier, in Europe, there are records of the great fire in London, England, which occurred in 1666.

Another significant fire event in the United States was the fire following the Great San Francisco Earthquake of 1906, which not only alerted everyone to the huge dangers of strong earthquake ground motions, but also again reaffirmed the potential fire hazards in our growing cities. The 1906 post-earthquake fires caused tremendous damage and deaths, in addition to those directly resulting from the strong earthquake. Fig. 6.2 shows the fire spreading in central San Francisco. Since that date, the danger of post-earthquake fires in cities has been recognized as a substantial hazard that may, under some circumstances, cause as much destruction as the seismic event itself.

During the modern times of the last 50 years, there have been numerous major fires that will be long remembered for their enormous destruction. In terms of simple frequency of occurrence, fires in low-rise buildings, particularly family residences, are by far the most common. According to the fire statistics from the National Fire Protection Association [1] during the 1990s, these most frequent home residence fires account for over 90 percent of all the fires in the United States. Fires in taller buildings of seven stories or more number over 10,000 cases per year and have caused annual civilian fatalities of 30 to 110, and annual property damage of $25 to $150 million.

Some of the significant fires of the last half of the 20th century caused destruction to the following multistory buildings:

FIGURE 6.1 1871 Great Chicago Fire.

FIGURE 6.2 Fire spread in 1906 San Francisco after the Great Earthquake.

- Andraus Building in Sao Paulo, Brazil, 31 stories, reinforced concrete framing, Feb. 24, 1972
- Joelma Building in Sao Paulo, Brazil, 25 stories, reinforced concrete framing, Feb. 1, 1974
- MGM Grand Hotel in Las Vegas, 26 stories, mixed steel and concrete construction, Nov. 21, 1980
- CESP 2 in Sao Paulo, 21 stories, reinforced concrete framing, May 21, 1987, core collapse
- First Interstate Bank in Los Angeles, 62 stories, structural steel framing, May 4, 1988
- One Meridian Plaza in Philadelphia, 38 stories, structural steel framing, Feb. 23, 1991

More recently, the destruction of the *World Trade Center* (WTC) Towers and complex in New York and the damage to the Pentagon in Washington, DC, on Sept. 11, 2001, have raised the issue of fire-induced structural collapses. The May 2002 FEMA 403 Report documents the WTC events of that day, particularly the collapse of the 110-story WTC 1 and 2 Towers and the 47-story WTC 7 steel buildings [2]. Further technical research and studies performed by federal government agencies, mostly by the *National Institute for Standards and Technology* (NIST), on the implications of the WTC disaster are ongoing, and new information and developments on this are expected in the near future [3]. Further attention to potential advances in fire engineering and safety will be addressed, among other subjects.

One recent NIST study on the assessment of national needs and capabilities for structural fire resistance has identified 22 cases of multistory buildings (defined as having four or more stories) where fire was the direct cause of partial or total structural collapse [4]. Five of these cases occurred on Sept. 11, 2001 (WTC 1, 2, 5, 7, and the Pentagon), which were counted as separate incidents. All of the 22 fire-induced failures had occurred since 1970, with 15 being in the United States. One of the most important findings of the study was that partial or total fire-induced collapse can occur in all types of construction with all types of materials exhibiting failures. This information provides further documentation as to the importance of fire protection for structural elements used in buildings.

6.2.2 Early Fire Resistance Test Procedures

Some of the earliest recorded fire resistance tests were in the 1790s in London, England [5]. These were carried out to determine the relative performance of fireproofing systems consisting of iron plates and stucco. The tests were conducted in response to the need to provide better fire protection and prevent collapse of buildings as occurred during the London fire. The tests were conducted for 1 to 2 h, using furnaces fueled by wood shavings.

The earliest recorded fire tests in the United States were in Denver in 1890. These tests were performed to select from among three proposed floor systems for use in the multistory Denver Equitable building. Tests were conducted using coal-fired furnaces that produced temperatures of approximately 1500°F for a 24-h duration.

Systematic fire tests of structural elements continued to occur; floors were evaluated in New York City in 1896 and wall systems were evaluated in 1901. The first permanent fire resistance test station was founded by Professor Ira Woolson at Columbia University in New York City.

The ASTM E 119 fire test standard used in North America to evaluate the fire resistance of structural elements was initially promulgated as ASTM C 119 in 1918 [6]. This test standard allowed for the systematic testing and comparison of results for the various fire-resistance-rated materials and assemblies. This test has changed very little over time and remains the basis for fire resistance testing in North America.

6.2.3 Standard Fire Resistance Tests

ASTM E 119, "Standard Test Method for Fire Tests of Building Construction and Materials" (or its equivalent ANSI/UL 263 and NFPA 251), has been the traditional standard fire resistance test in the United States for building materials, members, and assemblies since 1918. The comparable international and European standard is ISO 834. ASTM E 119 is applicable to individual beams, columns, floors, roofs, walls, and other building elements of any material. Each test assembly, or member, is subjected to a standard fire exposure in a furnace compartment of a certain duration and severity, the so-called standard time-temperature curve. The ASTM E 119 standard time-temperature fire curve is shown in Fig. 6.3, and its discrete temperatures in 5-min increments are tabulated in Table 6.1. A more severe, faster rising, standard time-temperature relationship is described in ASTM E 1529 [7], which is intended to be representative of hydrocarbon pool fires whose temperatures plateau at about 1100°C (2000°F).

Usually, the fire test duration will not exceed 3 to 4 h, given that these are the normal maximum code requirements for building fire ratings. A standard fire, such as that from E 119, prescribes ever-increasing temperatures with time duration that approaches or maintains a high constant temperature. This standard fire exposure implies an endless fuel supply and adequate ventilation for intense fire continuation in a given chamber/location, all simulating a hot postflashover stage. In contrast, a natural or real fire reaches, at some finite time, a decay period and eventual burnout within a particular area that is highly dependent on the fuel load and ventilation present at that time. Approximate time equivalencies have been established between the standard and natural fire exposures, which serve as an underlying basis for today's prescribed code requirements for fire ratings that are derived from E 119. If adequate suppression measures (such as automatic sprinklers) or fire barriers are not present to contain the fire within its room of origin, the fire can normally spread to other locations or floors in a building.

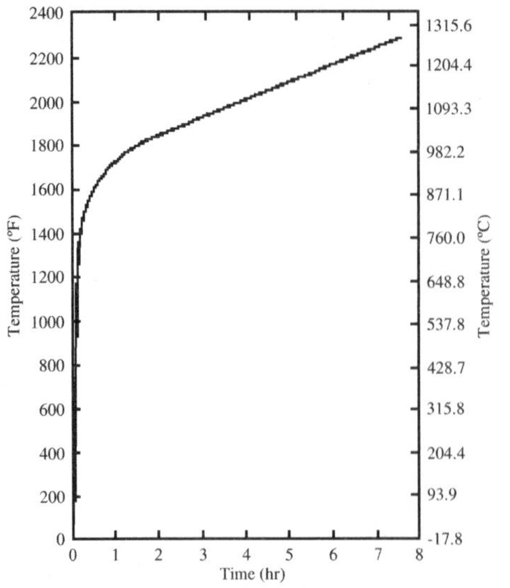

FIGURE 6.3 Standard ASTM E 119 time-temperature fire curve. (From Ref. 8. With permission, Society of Fire Protection Engineers [SFPE], *Handbook of Fire Protection Engineering*, 3rd edition, 2002.)

The E119 standard fire curve may be mathematically approximated by the following Eq. (6.1) for any needed computational applications:

$$T = 750[1 - e^{-3.79553\sqrt{t_h}}] + 170.41\sqrt{t_h} + T_o \qquad (6.1)$$

where T = temperature, °C
T_o = ambient temperature, °C
t_h = time, h

The maximum size of the fire test frame and its capabilities will vary by the individual laboratory facilities. Consequently, the fire specimen dimensions are quite constrained relative to full-scale actual construction. For example, *Underwriters Laboratories, Inc.* (UL), whose fire ratings per E 119 (or equivalently UL 263) are commonly recognized by the U.S. building code authorities, has approximately 14 × 17-ft furnace plan dimensions for beams, floor, and roof assemblies, while the wall and column furnace accommodates, and requires, specimens 10 × 10 ft and 9 ft high, respectively.

The clear implication is that while ASTM E 119 addresses large-scale testing, its use is limited to numerous compartment-type fire tests of individual and smaller members and assemblies, which may not be real, full-scale replicates of actual construction or of real fires. Continuous or larger beams, entire floor or roof systems, deep trusses or plate girders, or long columns and walls have not been fire tested in accordance with the E 119 standard due to practical size considerations.

The limitations of such a standard fire test are many, as with the use of any limited experiments for more general purposes. These issues include the specimen boundary conditions in the furnace (degree of end restraint), furnace size, realism of the standard fire time-temperature curve, specimen size and span effects, extrapolation and interpolation of test results, furnace air pressures, and the magnitude of the applied test loads. The degree of actual restraint depends on the particular structural member, its end connection, and the adjacent and overall building framing system. Because these actual structural continuity conditions cannot be duplicated in a size-limited test furnace with specimen boundary conditions that are ill-defined, restrained and unrestrained fire rating classifications have been developed in E 119 to approximate these effects. These restrained and unrestrained fire ratings are peculiar only to ASTM E 119 and to the U.S. code practice, and they are not used in other countries.

ASTM E 119 is strictly a comparative standardized test of selected structural and product features within a limited furnace space; it is not an accurate overall measure or predictor of actual member, assembly, or building performance under an uncontrolled real fire and other expected concur-

TABLE 6.1 Standard E 119 Time-Temperature Values [7]

Time, hr.:min.	Temperature, deg. Fahr.	Area above 68°F base		Temperature, deg. Cent.	Area above 20°C base	
		deg. Fahr.-min	deg. Fahr.-hr.		deg. Cent.-min.	deg. Cent.-hr.
0:00	68	00	0	20	00	0
0:05	1000	2330	39	538	1290	22
0:10	1300	7740	129	704	4300	72
0:15	1399	14,150	236	760	7860	131
0:20	1462	20,970	350	795	11,650	194
0:25	1510	28,050	468	821	15,590	260
0:30	1550	35,360	589	843	19,650	328
0:35	1584	42,860	714	862	23,810	397
0:40	1613	50,510	842	878	28,060	468
0:45	1638	58,300	971	892	32,390	540
0:50	1661	66,200	1103	905	36,780	613
0:55	1681	74,220	1237	916	41,230	687
1:00	1700	82,330	1372	927	45,740	762

TABLE 6.1 *Continued*

Time, hr.:min.	Temperature, deg. Fahr.	Area above 68°F base		Temperature, deg. Cent.	Area above 20°C base	
		deg. Fahr.-min	deg. Fahr.-hr.		deg. Cent.-min	deg. Cent.-hr.
1:05	1718	90,540	1509	937	50,300	838
1:10	1735	98,830	1647	946	54,910	915
1:15	1750	107,200	1787	955	59,560	993
1:20	1765	115,650	1928	963	64,250	1071
1:25	1779	124,180	2070	971	68,990	1150
1:30	1792	132,760	2213	978	73,760	1229
1:35	1804	141,420	2357	985	78,560	1309
1:40	1815	150,120	2502	991	83,400	1390
1:45	1826	158,890	2648	996	88,280	1471
1:50	1835	167,700	2795	1001	93,170	1553
1:55	1843	176,550	2942	1006	98,080	1635
2:00	1850	185,400	3091	1010	103,020	1717
2:10	1862	203,330	3389	1107	112,960	1882
2:20	1875	221,330	3689	1024	122,960	2049
2:30	1888	239,470	3991	1031	133,040	2217
2:40	1900	257,720	4295	1038	143,180	2386
2:50	1912	276,110	4602	1045	153,390	2556
3:00	1925	294,610	4910	1052	163,670	2728
3:10	1938	313,250	5221	1059	174,030	2900
3:20	1950	332,000	5533	1066	184,450	3074
3:30	1962	350,890	5848	1072	194,950	3249
3:40	1975	369,890	6165	1079	205,500	3425
3:50	1988	389,030	6484	1086	216,130	3602
4:00	2000	408,280	6805	1093	226,820	3780
4:10	2012	427,670	7128	1100	237,590	3960
4:20	2025	447,180	7453	1107	248,430	4140
4:30	2038	466,810	7780	1114	259,340	4322
4:40	2050	486,560	8110	1121	270,310	4505
4:50	2062	506,450	8441	1128	281,360	4689
5:00	2075	526,450	8774	1135	292,470	4874
5:10	2088	546,580	9110	1142	303,660	5061
5:20	2100	566,840	9447	1149	314,910	5248
5:30	2112	587,220	9787	1156	326,240	5437
5:40	2125	607,730	10,129	1163	337,630	5627
5:50	2138	628,360	10,473	1170	349,090	5818
6:00	2150	649,120	10,819	1177	360,620	6010
6:10	2162	670,000	11,167	1184	372,230	6204
6:20	2175	691,010	11,517	1191	383,900	6398
6:30	2188	712,140	11,869	1198	395,640	6594
6:40	2200	733,400	12,223	1204	407,450	6791
6:50	2212	754,780	12,580	1211	419,330	6989
7:00	2225	776,290	12,938	1218	431,270	7188
7:10	2238	797,920	13,299	1225	443,290	7388
7:20	2250	819,680	13,661	1232	455,380	7590
7:30	2262	841,560	14,026	1239	467,540	7792
7:40	2275	863,570	14,393	1246	479,760	7996
7:50	2288	885,700	14,762	1253	492,060	8201
8:00	2300	907,960	15,133	1260	504,420	8407

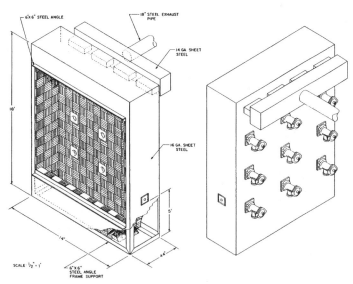

FIGURE 6.4 Sketch of E 119 furnace.

rent loading and related variables. The resultant fire-resistance rating for a material or assembly is expressed in the number of hours that the assembly or component was able to withstand exposure to the standard fire before a limiting E 119 criterion was reached.

6.2.4 Test Equipment

The needed laboratory equipment to conduct an E 119 type of fire resistance test is a controlled furnace chamber of adequate dimensions to accommodate the minimum required specimen sizes, instrumentation to measure temperatures (thermocouples) in the furnace and in the specimen and to measure specimen deflections, and the means to apply superimposed loading to the specimen, as needed. All these individual testing requirements are given in ASTM E 119 and the other standards that establish the particular fire-resistance ratings for the type of assembly being evaluated. Fig. 6.4 provides a sketch of a test furnace.

6.2.5 Failure Criteria

The ASTM E 119 ratings have been cited in the U.S. building codes [9, 10] to limit the potential spread of fire. One critical test limit of E 119 is reached at the time when the specimen can no longer support its maximum applied design load. Another limit point, or the only one if tested without loading, is the limiting temperature of the specimen under fire exposure. For floor, wall, and roof construction, an additional E 119 acceptance criterion exists for the maximum temperature rise on the unexposed surface of the specimen, or ignition of a cotton wool pad. The standard E 119 test thereby evaluates the relative heat transmission characteristics and structural integrity of specimens under a common and well-controlled fire exposure. Table 6.2 summarizes the pertinent temperature endpoint criteria of E 119 for the various types of structural members and assemblies. These critical temperatures have been selected as conservative estimates of the maximum allowed reduction in strength of the structural members under elevated temperatures. However, in reality, they do not directly account for the loading conditions that actually may be present or expected, wherein members may not be supporting their maximum applied design load.

The E 119 fire test is intended to explicitly demonstrate the adequacy of the fire protection

TABLE 6.2 ASTM E 119 Temperature Endpoint Criteria

Structural assembly or member	Temperature location	Maximum temperature °F(°C)
Walls and partitions (bearing and non-load-bearing)	Average*	250 (139)
	Single point*	325 (181)
Steel columns	Average maximum	1000 (538)
	Single point maximum	1200 (649)
	Average*	250 (139)
	Single point*	325 (181)
Steel Beam		
Floor/roof assemblies and loaded beams	Average maximum	1100 (593)
	Single point maximum	1300 (704)
	Pre-stressing steel max.	800 (426)
	Reinforcing steel max.	1100 (593)
	Open-web steel joists max.	1100 (593)
Steel beams and girders, not loaded	Average maximum	1000 (538)
	Single point maximum	1200 (649)

* Maximum temperature increase on the unexposed surface of the assembly.

material or system under fire conditions as well as large deflections or distortions, deterioration, delamination, and/or detachment from the base specimen material. The E 119 standard fire test was originally developed to be a method for evaluating the fire protection material in a restrained condition, which was considered to be the most severe exposure for the fire protection material. This is true for columns, which have a cast-in-place concrete cap at each end, and the fire protection material is applied tight against these end caps. For beams, the fire protection is applied tight against the beam hangar in the furnace. In the case of load-bearing walls, the wall and its fire protection is not restrained along its vertical edges. For non-load-bearing walls, all four edges are restrained.

By E 119 definition, the limiting beam temperature for an unrestrained condition must not be exceeded throughout the entire rating classification time. For a restrained rating, the beam and its assembly must support the load for the full rating time, and the beam temperature criteria must not be exceeded at one-half of the classification time, or 1 h, whichever is greater.

Table 6.3 (from ASTM E 119, Table X3.1) relates actual construction conditions encountered in steel, concrete, and wood buildings to these restrained and unrestrained fire classifications.

Regardless of its several limitations and uncertainties, the E 119 standard has historically served as a safe and conservative, albeit somewhat crude, benchmark measure of the relative fire resistance of different building elements and materials. UL and other laboratories have conducted many E 119 tests, and their various fire resistance directories contain published fire ratings and listings of many proprietary products that are annually updated.

However, it is recognized that the ASTM E 119 fire test standard and its ratings are not intended to be predictive of actual fire endurance times and performance in buildings under real, uncontrolled, natural fire conditions, or other unusual conditions that may be encountered.

While ASTM E 119 covers the standard fire testing of individual structural elements and members, none of it is directly applicable to end connections or interior splices. The U.S. building codes also have no specific language that covers connections of any type or material. Thus, in contrast to the extensive E 119 test database on such members and assemblies, there are relatively little other fire test results available on any structural connections. This is now a recognized knowledge gap that needs future research attention.

6.2.5.1 Superimposed Loading

ASTM E 119 requires maximum superimposed loading to be applied during the duration of the fire test for bearing walls and partitions, columns, floor and roof assemblies, and beams to match the

TABLE 6.3 Restrained and Unrestrained Construction Systems

I. **Wall bearing:**
Single span and simply supported end spans of multiple bays[a]
(1) Open-web steel joists or steel beams, supporting concrete slab precast units, or metal decking .... unrestrained
(2) Concrete slabs, precast units, or metal decking .... unrestrained
Interior spans of multiple bays:
(1) Open-web steel joists, steel beams, or metal decking, supporting continuous concrete slab .... restrained
(2) Open-web steel joists or steel beams, supporting precast units or metal decking .... unrestrained
(3) Cast-in-place concrete slab systems .... restrained
(4) Precast concrete where the potential thermal expansion is resisted by adjacent construction[b] .... restrained

II. **Steel framing:**
(1) Steel beams welded, riveted or bolted to the framing members .... restrained
(2) All types of cast-in-place floor and roof systems (such as beams and slabs, flat slabs, pan joists, and waffle slabs) where the floor or roof system is secured to the training members .... restrained
(3) All types of prefabricated floor or roof systems where the structural members are secured to the training members and the potential thermal expansion of the floor or roof system is resisted by the framing system or the adjoining floor or roof construction[b] .... restrained

III. **Concrete framing:**
(1) Beams securely fastened to the framing members .... restrained
(2) All types of cast-in-place floor or roof systems (such as beam-and-slabs, flat slabs, pan joists, and waffle slabs) where the floor system is cast with the training members .... restrained
(3) Interior and exterior spans of precast systems with cast-in-place joists resulting in restraint equivalent to that which would exist in condition III(1) .... restrained
(4) All types of prefabricated floor or roof systems where the structural members are secured to such systems and the potential thermal expansion of the floor or roof systems is resisted by the framing system or the adjoining floor or roof construction[b] .... restrained

IV. **Wood construction:**
All types .... unrestrained

Source: SFPE Table 4-9-1 and ASTM, p. 1-176.
[a] Floor and roof systems can be considered restrained when they are led into walls with or without tie beams, the walls being designed to resist thermal thrust from the floor or roof system.
[b] For example, resistance to potential thermal expansion is considered to be achieved when
  1. Continuous structural concrete topping is used.
  2. The space between the ends of precast units or between the ends of units and the vertical face of supports is filled with concrete or mortar, or
  3. The space between the ends of precast units and the vertical faces of supports, or between the ends of solid or hollow cone stub units, does not exceed 0.25% of the length for normal weight concrete members or 0.1% of the length for structural lightweight concrete members.

specimen's maximum design strength. Nonbearing walls and partitions and an alternate method for rating unloaded columns are also available to be used without superimposed loads. An extensive database of fire resistance results has been generated on this basis.

While the full, superimposed loading has long been considered conservative, newer probability assessments of extreme loading combinations with fire demonstrate that it is indeed true. A more statistically probable load combination with a significant fire exposure would be the "arbitrary point in time" gravity load, which is about 0.24 to 0.40 of the maximum lifetime live load. For design, this companion action coefficient is taken as 0.5. Thus, use of such a reduced load combination would dictate superimposed loading on the specimen during a fire of one-half of what has traditionally been used under E 119.

6.2.5.2 Restrained and Unrestrained Assemblies

Restrained and unrestrained classifications pertain to ASTM E 119 tests on beams, floors, and roofs, and depend on whether the test arrangements allowed for the free thermal expansion of the tested specimen (unrestrained test) or not (restrained test). In practical terms, the unrestrained ratings will always require as much, or more, fire protection than their restrained counterparts, often up to a twofold difference. The following summarizes how these E 119 ratings are developed.

- The ASTM E 119 test protocol for unloaded structural steel (and composite steel/concrete) columns requires longitudinal expansion of the applied fire protection material to be restrained (this conservative requirement can result in earlier fall-off of the fire protection and faster heating of the tested steel beam). This type of test, when applied to an unloaded beam member, results in a single unrestrained beam rating based on the period of fire exposure where the average measured temperature at any section of the steel beam remains under 1000°F, and the measured temperature at any single location of the steel beam remains under 1200°F. While this type of test is commonly conducted to develop column fire ratings, it is rarely conducted for beams.

- ASTM E 119 tests on loaded structural steel (and composite steel/concrete) beams are always restrained and result in two ratings: 1) restrained beam rating based on the period of fire exposure where the beam sustains the applied design load, but not more than twice the corresponding unrestrained beam rating, and provided the latter is 1 h or more and 2) unrestrained beam rating based on the period of fire exposure where the average measured temperature at any section of the steel beam remains under 1100°F, and the measured temperature at any single location of the steel beam remains under 1300°F.

ASTM E 119 tests on floor and roof assemblies are generally loaded to full design levels. While the assemblies could be tested either in the unrestrained or restrained condition around the floor/roof perimeter of the furnace, almost all assemblies are tested in the restrained condition. Whenever the tested floor/roof assembly contains a structural steel beam, both restrained and unrestrained assembly tests will result in an unrestrained beam rating (based on the same temperature criteria specified for loaded restrained beam tests) in addition to the assembly ratings. For any assembly rating period, the unexposed surface of the tested floor/roof should neither develop conditions that will permit ignition of cotton waste, nor the average temperature rise on the unexposed surface of the tested floor/roof be allowed to exceed 250°F. A restrained assembly test will result in two assembly ratings: 1) restrained assembly rating based on the period of fire exposure where the assembly sustains the applied design load, but not more than twice the corresponding unrestrained assembly rating, and provided the later is 1 h or more and 2) unrestrained assembly rating based on the same temperature criteria specified for unrestrained beam rating, except for steel structural members spaced 4 ft or less on center, where the criterion for the average measured temperature of all such members remaining under 1100°F applies.

These restrained and unrestrained fire ratings are peculiar only to ASTM E 119 and to the U.S. code practices and are not used in other countries. Given the ambiguity of these definitions, these classifications have continued to raise related design and interpretation questions in the past.

Appendix X3, Table X3.1 of ASTM E 119 provides guidance on the classification of beams and floor and roof systems in construction as restrained or unrestrained. In most practical cases, structural steel beams and floor systems within steel-framed buildings are classified as being restrained. Further, and more recent, information on the recommended use of restrained ratings for steel construction has been published [11].

6.2.6 Special Hazard Resistance Tests ("High-Rise" Curves)

In the late 1980s, due to the new and unique hazards that were recognized for petrochemical spill fires by the petroleum industry, a new standard of fire exposure was developed to more closely represent this type of fire. ASTM E 1529, "Standard Test Methods for Determining Effects of Large Hydrocarbon Pool Fires on Structural Members and Assemblies" (UL 1709), provides this special fast-rising fire curve, wherein the furnace temperature reaches a plateau of approximately 1100°C (2000°F) after about 5 min, thereby applying a sudden and intense thermal shock to the structural assembly that is being evaluated. This empirical fire was developed on the basis of engineering judgment and experience to simulate a typical large hydrocarbon pool fire that is burning in the open, or in some other situation with access to full ventilation.

This quick-starting fire results in more severe thermal exposure to both the structural fire protection material and the structure itself, compared to E 119. In particular, this might cause premature fire protection material fall-off, and concrete members might be more prone to explosive spalling under such conditions. There are also some thermocouple instrumentation differences between these two fire standards. Of the standard fire tests that have been conducted on similar assemblies or members, those exposed to an ASTM E 1529 (UL 1709) fire have produced much shorter fire resistance times than the ASTM E 119 (UL 263) tests.

However, as with E 119, E 1529 is not intended to reproduce any specific fire, but rather is a standard fire test that is to be used for general comparative and data-acquisition purposes. As such, E 1529 is also not considered a completely predictive test of fire resistance.

6.3 STRUCTURAL MATERIALS AND FIRE

6.3.1 Reaction of Structural Materials to Fire

The degree of combustibility is one broad and important fire classification of building materials. Combustible materials will not only degrade at higher temperatures, but also ignite and burn, thereby adding to the fuel contents during a fire. Noncombustible materials will degrade under the higher temperatures of a fire, but will not typically burn in the context of building fires.

Building materials in the ensuing section will be considered within the context of their use as primary, load-bearing elements that are necessary to preserve the structural safety of the building in preventing partial or total collapse. The traditional building materials are steel, concrete, and wood. Wood is the only combustible material of these three, with steel and concrete being noncombustible. In either case, visible damage/distortions and degradation of the thermal and mechanical properties of all building materials occur under prolonged elevated temperatures. Deflections of structural members during the longer duration and hot fires can reach many inches or several feet. This is an order of magnitude greater than the small elastic deflections, usually no more than 1 in., which are normally contemplated for design service. Hence, this property deterioration and the effects of large thermal deformations on the load-carrying capabilities of the materials during a fire are the two important effects that need to be included in the fire resistance analysis or engineering of a building. The material ductility at ultimate strength and high temperatures, or its maximum mechanical strain, may govern the structural integrity of those frames and connections that offer significant restraint to thermal expansion, wherein the thermal strains will be approximately equal to the mechanical strain, but opposite in sense.

The structural fire response can be evaluated in one of three ways: first, empirically, through standardized or special-purpose fire tests and ratings; second, through elementary calculation methods that encompass heat transfer principles and the residual strength of individual structural elements in a simplified manner; or third, through more comprehensive and sophisticated modeling of fires and the resulting framing behavior. The most common standard fire tests indirectly include the material and fire protection characteristics in the derived individual member or assembly ratings, but often without any instrumented data that would provide clear correlations to their underlying material property changes with temperature. Such additional experimental data, which would be of more general interest, but not specifically required by the test standard, is not usually recorded during a standard fire test due to the extra expenses and irrelevance to the assembly's derived fire rating. Often, the standard fire endurance rating times are based exclusively on material temperatures not exceeding a given critical temperature. More realistic, full-scale fire tests of multistory frames, such as those in Cardington, UK [12], or of parking garages are conducted relatively infrequently due to their high costs.

Application of either the simple or more advanced fire resistance calculations will require a more explicit representation of the material properties at elevated temperatures. The following sections present the basic material response and property variations at high temperatures of fire of steel and concrete materials to enable fire resistance calculations and evaluations. Empirical data, as well as convenient engineering representations of these material properties, are given. These are usually presented only in SI/metric dimensional units, given the fact that most of the referenced international work in this field has been reported in those terms.

6.3.1.1 Steel

Steel is a noncombustible material that is available in various product types: structural (hot-rolled), reinforcing, prestressing, or cold-formed. As with any other material, exposure to elevated temperatures leads to a temporary decrease in the strength and stiffness of steel. Such prolonged degradation adversely affects the resulting deformations and load-carrying capabilities of steel during the fire exposure, i.e., deformations are increased and strength and stiffness are reduced. Steel thermal properties are also affected, such as the coefficient of thermal expansion, specific heat, and conductivity, which affect the heat transfer and temperature profile calculations.

After exposures to temperatures in excess of about 600°C for more than 15 min or so, unprotected steel will quite visibly deform, twist, and buckle. At and above such high temperatures, the crystalline/metallurgical structure of typical carbon-based steels for buildings also undergoes a transformation. These metallurgical changes are inconsequential during the fire itself relative to the significant decrease in strength at these elevated temperatures, but the significant heating effects of the fire could permanently alter the nature of the steel material. Usually, steel that has experienced a hot and prolonged fire will have such extensive damage that replacement, rather than its straightening and repair, will be prudent on cost considerations alone, so the long-term metallurgical changes will be irrelevant.

Steel that has not been exposed to very high temperatures, as just described, for a prolonged time will not be significantly deformed, its metallurgy will not be affected after cooling, and hence, it will regain its original thermomechanical properties. Hence, such members can normally be effectively straightened and repaired, as necessary, and put back into normal service. A maximum steel temperature of 600°C can serve as a convenient boundary in this regard. If the steel has not exceeded this temperature, it will return to its original metallurgical and mechanical properties upon cooling. Hence, for such fire exposures, the temperature-dependent changes in material properties are only transient.

Because of steel's higher strengths and resulting lighter-weight construction relative to concrete, unprotected steel frames may be vulnerable to fire distress under some conditions, unless this risk is properly evaluated and managed. Modern U.S. and other building codes contain prescriptive criteria for the determination of when and what steel fire protection is required for the various possible types of construction, heights, areas, and occupancies. When necessary, steel members can be insulated from the damaging fire heat effects through various means of fire protection, such as spray-on materials, intumescent paints, membrane/gypsum boards, or concrete encasement or filling, all with the purpose of delaying the temperature rise in the steel and its corresponding material property degradation.

Stress-strain curves for a typical hot-rolled steel at different elevated temperatures are given in Fig. 6.5. Cold-formed and prestressing steel will have slightly different strength-reduction patterns, and there is some natural variability in the material testing results (see Figs. 6.6 and 6.7). However, the basic trend is that at higher temperatures, both the yield and ultimate strength decrease, until almost total strength depletion occurs beyond 700°C. The same steel strength curves and values may be used for both tension and compression loadings. For design use, the scatter in these yield and ultimate strength reductions of steel for the various kinds of steel can be conveniently represented as a strength fraction relative to its ambient temperature value (sometimes called a strength retention factor) by the following linear reduction equations [Eq. (6.2)] [13] for structural steel, reinforcing steel, and prestressing steel.

$$k_{y,T} = (905 - T)/690 \quad \text{structural steel}$$
$$k_{y,T} = (720 - T)/470 \quad \text{reinforcing steel} \quad (6.2)$$
$$k_{y,T} = (700 - T)/550 \quad \text{prestressing steel}$$

Similar convenient formulations for analysis or design are available from various other sources (codes, textbooks, and references). Engineering judgment should be used to decide whether the relatively small differences in the reported properties among the different steel types, or their mathemat-

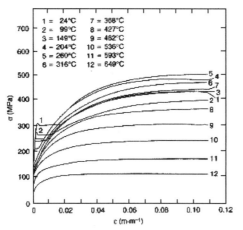

FIGURE 6.5 Typical stress-strain curves for hot-rolled steel at elevated temperatures (From Ref. 8, Fig 1.10.8). (With permission, Society of Fire Protection Engineers [SFPE], *Handbook of Fire Protection Engineering,* 3rd edition, 2002.)

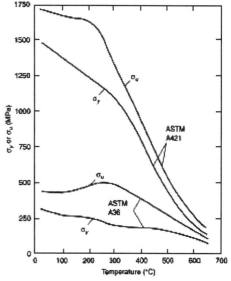

FIGURE 6.6 Ultimate and yield strengths for structural steel (ASTM A 36) and prestressing steel (ASTM A 421) at elevated temperatures (Ref. 8, Fig. 1.10.10). (With permission, Society of Fire Protection Engineers [SFPE], *Handbook of Fire Protection Engineering,* 3rd edition, 2002.)

ical representations, warrant separate design considerations, in view of the much larger uncertainties inherent to the fire-exposure problem itself.

Young's modulus of elasticity E, which affects elastic structural stiffness (deflections and distortions) and elastic buckling, similarly declines with temperature. Because the significant structural fire effects occur in the large deflection and nonlinear plastic range, the small elastic deflection considerations are usually negligible. However, the effect of a lower elastic modulus on stability is important. Fig. 6.8 shows the temperature variation of Young's modulus for structural steels and reinforcing bars.

Again, despite the data scatter for particular types of steels and tests, an empirical design equa-

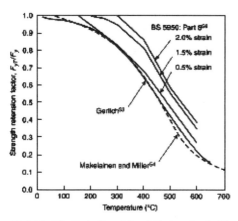

FIGURE 6.7 Reduction of the yield strength of cold-formed steel at elevated temperatures (Ref. 8, Fig. 1.10.13). (With permission, Society of Fire Protection Engineers [SFPE], *Handbook of Fire Protection Engineering,* 3rd edition, 2002.)

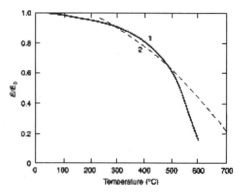

FIGURE 6.8 Effect of temperature on the modulus of elasticity of structural steel and reinforcing bars. (With permission, Society of Fire Protection Engineers [SFPE], *Handbook of Fire Protection Engineering,* 3rd edition, 2002.)

tion [Eq. (6.3)] can be formulated for the residual modulus of elasticity fraction of steel relative to its value at room temperature (20°C):

$$k_{E,T} = 1.0 + T/[2000 \ln(T/1100)] \qquad 0 < T \leq 600°C$$
$$k_{E,T} = 690(1 - T/1000)/(T - 53.5) \qquad 600 < T \leq 1000°C \qquad (6.3)$$

Strength and modulus of elasticity are the principal mechanical properties of relevance to fire engineering. As indicated by this information, a steel member will lose about one-half of its strength and stiffness when its average overall temperature reaches 550 to 600°C. Three key thermal properties are also involved in representing steel's behavior under fire exposures, namely, the coefficient of thermal expansion, specific heat, and thermal conductivity.

The coefficient of thermal expansion, or equivalently, the thermal strain $\Delta L/L$, governs the amount of thermally induced expansion ΔL in a member. This property increases with temperature. The thermal expansion should be included in any advanced modeling work to more accurately determine restraint levels in the surrounding structure, and the resulting reactions and displacements, while for simplified fire resistance calculations, this expansion is usually ignored. For steel, the coefficient of thermal expansion value is usually taken as 11.5×10^{-6}/°C at ambient temperatures. The Eurocode 3 recommends a slightly higher value of 14.0×10^{-6}/°C. At elevated temperatures, the thermal strain can then be approximated [14] for design by the linear function:

$$\Delta L/L = 14 \times 10^{-6}(T - 20) \qquad (6.4)$$

The density of steel essentially remains constant with temperature at 7850 kg/m³. The steel specific heat c_p is often taken as approximately a constant value of 600 J/kg·°C, but it actually varies between 400 and 700, with the highest value plateau being reached at about 700°C (see Fig. 6.9). A possible numerical complication is that the specific heat has an abrupt spike to 2000 to 5000 J/kg at about 700°C, when the steel undergoes a metallurgical phase change. However, this very limited discontinuity is often ignored for the sake of simplicity. Also depicted in Fig. 6.9 is steel's thermal conductivity in W/m·°C, which reduces linearly from 54 at 0°C to its minimum value of 27.3 at 800°C.

The previous coverage was limited to the common steel products used for building construction, such as those given in the AISC [15], ACI, PCI, and AISI design standards and manuals. Caution should be exercised in extrapolating these properties of the commonly used steels to any special, heat-treated, high-strength alloy steels or cables, which may require a separate evaluation of their high-temperature effects. Such high-strength steels would have yield strengths in excess of 500 MPa, or about 70 ksi, and are seldom used in ordinary building construction.

6.3.1.2 Concrete

Concrete is a noncombustible material with a relatively low thermal conductivity. Its structural design often results in heavier and more massive members compared to steel framing, which pro-

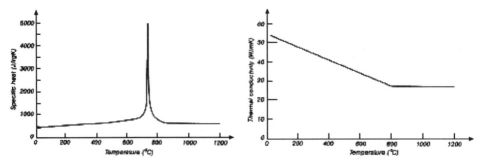

FIGURE 6.9 Specific heat and thermal conductivity of steel at elevated temperatures. (From Ref. 13. With permission, John Wiley & Sons, Limited, from Buchanan, Andrew H., *Structural Design for Fire Safety*, Copyright 2001.)

TABLE 6.4 Minimum Width and Cover for Reinforced Concrete Members

		Beams	Columns	Slabs	Walls
0.5 hours	Width	80	150	75	75
	Cover	20	20	15	15
1.0 hours	Width	120	200	95	75
	Cover	30	35	20	15
1.5 hours	Width	150	250	110	100
	Cover	40	30	25	25
2 hours	Width	200	300	125	100
	Cover	50	35	35	25
3 hours	Width	240	400	150	150
	Cover	70	35	45	25
4 hours	Width	280	450	170	180
	Cover	80	35	55	25

vides a desirable heat sink for absorption of the heat from the fire. Because of this thermal mass effect as well as for its primary load-bearing capabilities, concrete has been, and continues to be, used for thermal insulation and/or fire barriers. Nevertheless, similar to other construction materials, concrete also experiences property degradation as well as visible cracking or spalling damage with increasing temperatures. The property variations are dependent on the weight density (lightweight or normal weight), its compressive strength level (high or normal), water content (water-cement ratio), and the type of aggregate and reinforcing (bars or fibers) in the concrete. Higher internal moisture content will beneficially delay the temperature rise in the material, but it may also cause explosive spalling in some cases through pore pressure buildup, thereby not only reducing the member cross section, but also directly exposing any interior steel reinforcing.

Concrete structures can be unreinforced (plain), reinforced, prestressed, and composite. When concrete is commonly used with reinforcing or prestressing steel, or in composite designs, the temperatures developed in this companion steel material govern the fire resistance ratings of the concrete member, because the concrete is usually so much larger and massive compared to the steel reinforcing. Typically, the so-called concrete cover, or distance from the interior steel to the outside concrete surface, provides the fire protection. Table 6.4 shows the usual, empirically derived cover value requirements for a reinforced concrete column. Fiber-reinforced concrete will be covered later.

The density of *normal-weight concrete* (NWC) is about 2200 kg/m^3 (about 150 lb/ft^3), while that for *lightweight concrete* (LWC) is approximately 0.67 of this value, or about 100 lb/ft^3. These densities effectively remain constant up to a temperature of about 800°C, when normal weight density begins to rapidly deteriorate by approximately 25 to 50 percent.

For normal-strength concretes, the compressive strength retention ratio for LWC is given in Fig. 6.10, and a comparable plot for NWC is shown in Figs. 6.11 and 6.12. Similar to steel, the strength reductions are minimal up to about 300°C (about 600°F), and the strength losses greatly increase at temperatures hotter than this.

At room temperature, this baseline concrete compressive strength f_c' can vary between about 20 to 50 MPa (or 3000 to 7000 psi). For LWC, this upper bound strength is only 40 MPa, dependent upon the concrete mix characteristics, such as water-cement ratio, age of the concrete, and the amount and type of aggregate. The tensile strength of concrete is negligible and usually ignored at higher temperatures, much as is the practice for normal temperature design. It should also be noted that the maximum compressive strain at ultimate strength for concrete is much lower than that for steel, roughly by a factor of 10. This means that concrete's inherent ductility at ambient and higher temperatures is limited.

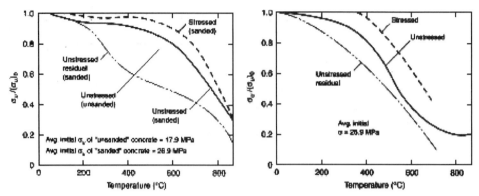

FIGURE 6.10 Compressive strength retention factors for lightweight concrete (Ref. 8, Fig. 1.10.19). (With permission, Society of Fire Protection Engineers [SFPE], *Handbook of Fire Protection Engineering*, 3rd edition, 2002.)

FIGURE 6.11 Compressive strength retention factors for normal weight concrete (NWC) with carbonate aggregate (Ref. 8, Fig. 1.10.18). (With permission, Society of Fire Protection Engineers [SFPE], *Handbook of Fire Protection Engineering*, 3rd edition, 2002.)

The modulus of elasticity ratio at higher temperatures relative to ambient conditions is shown in Fig. 6.13 for both normal and lightweight aggregates. The room temperature E_o baseline value can vary from 5000 to 35,000 MPa, depending upon the same previously mentioned concrete mix factors as for strength.

The coefficient of thermal expansion variation with temperature is given in Fig. 6.14. NWC exhibits greater and increasing propensity for thermal expansion than LWC.

Thermal conductivity also mainly depends on the nature of the aggregate in the concrete mix. NWC has a higher conductivity than LWC (see Fig. 6.14).

The specific heat of concrete is about 840 J/kg·°C for LWC, and varies between 1000 and 1200 J/kg·°C for NWC through the temperature range up to 800°C. A useful combination parameter is the volume specific heat, which is the product of specific heat and material density. This volume specific heat for concrete is illustrated in Fig. 6.15, with LWC having the lower values.

In general, LWC has a lower thermal conductivity, lower specific heat, and lower thermal expansion at higher temperatures than NWC. For these reasons, LWC is preferred for fire resistance purposes.

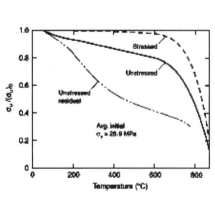

FIGURE 6.12 Compressive strength reduction for NWC with silicate aggregate. (From Ref. 8, Fig. 1.10.17. With permission, Society of Fire Protection Engineers [SFPE], *Handbook of Fire Protection Engineering*, 3rd edition, 2002.)

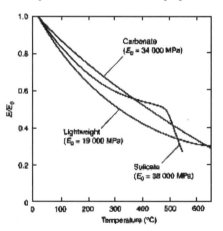

FIGURE 6.13 Effect of temperature on the concrete modulus of elasticity. (With permission, Society of Fire Protection Engineers [SFPE], *Handbook of Fire Protection Engineering*, 3rd edition, 2002.)

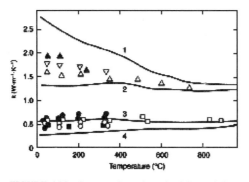

FIGURE 6.14 Concrete thermal conductivity variation with temperature (Ref. 8, Fig. 1.10.24). (With permission, Society of Fire Protection Engineers [SFPE], *Handbook of Fire Protection Engineering*, 3rd edition, 2002.)

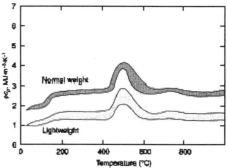

FIGURE 6.15 Volume specific heat of concrete (Ref. 8, Fig. 1.10.23). (With permission, Society of Fire Protection Engineers [SFPE], *Handbook of Fire Protection Engineering*, 3rd edition, 2002.)

All these empirical data and their scatter have been reduced and simplified to convenient mathematical expressions for engineering applications, as with steel. Buchanan [9] provides the following sets of design equations [(6.5) to (6.8)], based on BS 8110 and/or EC2 (1993), for concrete compressive strength, modulus of elasticity, and thermal strain variations, respectively, with temperature. As with steel, other simplified design formulations for these concrete properties are available from other sources.

$$k_{c,T} = 1.0 \quad \text{for } T < 350°C$$
$$k_{c,T} = (910 - T)/560 \quad \text{for } T > 350°C \tag{6.5}$$

$$k_{c,T} = 1.0 \quad \text{for } T < 500°C$$
$$k_{c,T} = (1000 - T)/500 \quad \text{for } T > 500°C \tag{6.6}$$

$$k_{E,T} = 1.0 \quad \text{for } T < 150°C$$
$$k_{E,T} = (700 - T)/550 \quad \text{for } T > 150°C \tag{6.7}$$

$$\Delta L/L = 18 \times 10^{-6} T \quad \text{for siliceous aggregate concrete}$$
$$\Delta L/L = 12 \times 10^{-6} T \quad \text{for calcareous aggregate concrete} \tag{6.8}$$
$$\Delta L/L = 8 \times 10^{-6} T \quad \text{for lightweight concrete}$$

An even simpler and constant (temperature invariant) set of concrete material properties is given in ASCE 29–99 (see Table 6.5) [16].

TABLE 6.5 Concrete Material Properties

	Normal weight[a]	Structural lightweight[b]
Thermal conductivity, k_c	0.95 Btu/h-ft-°F (1.64 W/m·K)	0.35 Btu/h-ft-°F (0.61 W/m·K)
Specific heat, c_c	0.20 Btu/h-ft-°F (0.84 kJ/kg·K)	0.20 Btu/h-ft-°F (0.84 kJ/kg·K)
Density, ρ_c	145 lb/ft³ (2323 kg/m³)	110 lb/ft³ (1762 kg/gm³)
Moisture content, m (percent by volume)	4	5

[a] Normal weight concrete is carbonate or siliceous aggregate concrete, as defined in Chapter 2.
[b] Structural lightweight concrete is lightweight or sand-lightweight concrete as defined in Chapter 2, with a minimum density (unit weight) of 110 pounds per cubic foot (1762 kilograms per cubic meter).

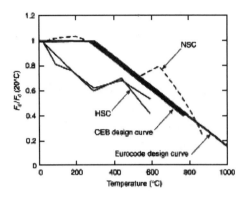 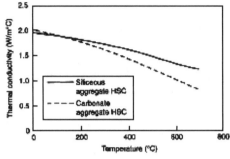

FIGURE 6.16 Compressive strength retention factors for HSC (Ref. 8, Fig. 1.10.27). (With permission, Society of Fire Protection Engineers [SFPE], *Handbook of Fire Protection Engineering,* 3rd edition, 2002.)

FIGURE 6.17 Thermal conductivity of HSC (Ref. 8, Fig. 1.10.28a). (With permission, Society of Fire Protection Engineers [SFPE], *Handbook of Fire Protection Engineering,* 3rd edition, 2002.)

High-strength concrete (HSC), often defined as having a compressive strength f_c' greater than about 55 MPa (8000 psi), has unique temperature dependence, different than that for either regular NWC or LWC. HSC is more susceptible to explosive spalling, which may occur when it is exposed to severe fire conditions. The compressive strength retention characteristics of HSC are given in Fig. 6.16, thermal conductivity in Fig. 6.17, and specific heat in Fig. 6.18.

Fiber-reinforced concrete represents another class of concrete material. Discontinuous steel and polypropylene fibers are added to the concrete mix to enhance its strength and ductility. In general, there is relatively less high-temperature information available on fiber-reinforced concrete. However, it is recognized that the addition of polypropylene fibers can be effective in minimizing explosive spalling during a fire. There are some available data on *steel-fiber-reinforced concrete* (SFRC) at elevated temperatures; see Fig. 6.19 for its compressive strength at elevated temperatures.

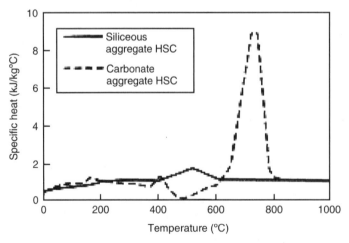

FIGURE 6.18 Specific heat of HSC (Ref. 8, Fig. 1.10–28b). (With permission, Society of Fire Protection Engineers [SFPE], *Handbook of Fire Protection Engineering,* 3rd edition, 2002.)

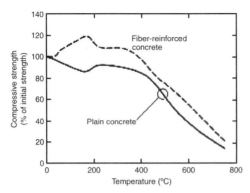

FIGURE 6.19 Effect of temperature on compressive strength of SFRC (Ref. 8, Fig. 1.10.25, pp. 1–171). (With permission, Society of Fire Protection Engineers [SFPE], *Handbook of Fire Protection Engineering*, 3rd edition, 2002.)

6.3.1.3 Wood

Timber construction may be of two distinct categories: heavy and light. Heavy timber uses glue-laminated timber or large-dimension sawn timber for the principal structural beams, columns, decks, or trusses, whereas light timber consists of the smaller sizes of wood framing, such as wall studs and floor joists. Wood is primarily used in residential and low-rise construction. It is a combustible and orthotropic material, which has different properties in its transverse and longitudinal directions dependent on the wood grain orientations.

Strength along the wood grain is much higher than perpendicular to it, the maximum strength is usually different for compression and tension, and the properties vary with the different species and grades of wood products. In addition, it is well established that wood strength declines with time under long duration loads. Chapter 7 has much more detailed information on wood and wood products, and the following contains just some basic facts.

An illustration of wood framing damage after a severe fire is shown in Figs. 6.20 and 6.21. Moisture content, rate of charring, and the grain orientation are the principal parameters that affect wood's high-temperature properties. The dry density of clear wood at room temperature ranges from 300 to 700 kg/m^3, which decreases by 10 percent at 200°C, and then abruptly declines by 80 percent at about 350°C. Its modulus of elasticity parallel to the grain varies from 5.5 to 15.0 $\times$ 10^3 MPa, while the compressive strength varies from 13 to 70 MPa.

Fig. 6.22 shows the thermal reductions in modulus of elasticity and compressive strength of dry, clear wood. Tensile strength reduction with temperature is similar to compressive strength, but is slightly less rapid. Effectively, at and beyond temperatures of 300°C, much of the strength and stiffness of wood is lost.

FIGURE 6.20 Wood fire damage 1. (From Ref. 13. With permission, John Wiley & Sons, Limited, from Buchanan, Andrew H., *Structural Design for Fire Safety*, Copyright 2001.)

FIGURE 6.21 Wood fire damage 2. (From Ref. 13, p. 275. With permission, John Wiley & Sons, Limited, from Buchanan, Andrew H., *Structural Design for Fire Safety,* Copyright 2001.)

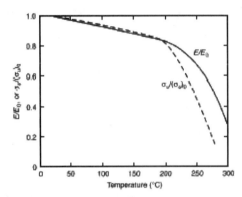

FIGURE 6.22 Effect of temperature on the modulus of elasticity and compressive strength of wood (Ref. 8, Fig. 1.10.33). (With permission, Society of Fire Protection Engineers [SFPE], *Handbook of Fire Protection Engineering,* 3rd edition, 2002.)

The relationship of thermal conductivity with temperature is given in Fig. 6.23, while specific heat is given in Fig. 6.24. The sudden spike in the latter at about 100°C represents the heat required to evaporate the interior moisture in the wood.

The coefficient of thermal expansion varies from (3.2 to 4.6) $\times$ 10^{-6}/°C along the wood grain, and (21.6 to 39.4) $\times$ 10^{-6}/°C transverse to the grain.

When wood burns, the wood surface ignites and forms a layer of char that effectively insulates the solid, and combustible, wood inside. This char layer thickness results in a reduction of the effective wood cross section available for structural load resistance. The rate of charring is affected by the density of wood and its moisture content, as shown in Fig. 6.25. Various design

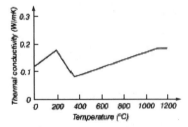

FIGURE 6.23 Variation of thermal conductivity of wood with temperature. (From Ref. 13, Fig. 10.4; Ref. 17, Knudson, 1975. With permission, John Wiley & Sons, Limited, from Buchanan, Andrew H., *Structural Design for Fire Safety,* Copyright 2001.)

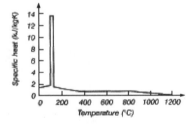

FIGURE 6.24 Variation of specific heat of wood with temperature. (From Ref. 13, Fig. 10.5; Ref. 18, Konig, 1999. With permission, John Wiley & Sons, Limited, from Buchanan, Andrew H., *Structural Design for Fire Safety,* Copyright 2001.)

recommendations for charring rate during a standard fire resistance test may be found in the building codes of the United States, Europe, New Zealand, and Australia.

Connections for wood members are made with either metal fasteners or adhesives. These timber connections behave very differently to fire exposures, but relatively little focused research has been conducted on them. Metal fasteners (nails, screws, and bolts) have been observed to work well at elevated temperature exposures if they are adequately protected from the fire either by shielding within the wood itself, or by application of additional fire protection materials to the connection. Timber members with adhesively bonded joints, which are combustible, generally behave similarly to the wood member itself. However, some adhesives, such as elastomerics and epoxies, are very vulnerable to deterioration at higher temperatures and should not be relied upon under fire conditions. Further fire research on timber connections is warranted.

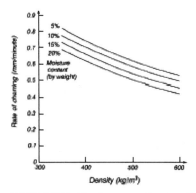

FIGURE 6.25 Charring rate as a function of density and moisture content. (From Ref. 13, Fig. 10.24; Ref. 19, Lie, 1972. With permission, John Wiley & Sons, Limited, from Buchanan, Andrew H., *Structural Design for Fire Safety*, Copyright 2001.)

6.3.1.4 Connections

Structural end connections or splices between members have not been specifically or extensively studied or tested under fire conditions, even though connections are commonly recognized as the critical link to the safety of all structures under all loading exposures. The standard ASTM E 119 column, floor, and beam tests for fire ratings all assess only the members themselves, and not their end connections or internal splices. Large-scale fire tests, such as that in Cardington [12], have yielded some global experimental observations on framing system and connection performance, as has anecdotal evidence from past fires and some limited past studies. However, this type of real fire data is relatively sparse, and much more information is needed.

Indications are that below approximately 600°C, weld and high-strength steel bolt strengths (ASTM A 325 and A 490) are not substantially affected. However, a comprehensive set of limit states and detailed constitutive properties on the behavior of high-strength steel bolts, weldments, miscellaneous connection attachments, or reinforcing details are not yet well known throughout the full-range of elevated temperatures. This paucity of data is compounded by the large variety of possible structural connections, of moment resisting (rigid), and simple shear or axial type, and the different available geometrical and connection size configurations for each.

Equally important are the thermal forces in the framing system that can be induced in these connections and members (compressive restraint or catenary tension) during a fire, and the connections' ability to transfer these axial forces and sustain the accompanying deformations. The design forces and bending moments assumed for normal building loadings may change dramatically during the course of a fire event. For example, a restrained beam member that is heated will initially try to expand, and compressive axial forces (thrust) will be induced from the constraint present in the surrounding structure, acting in combination with the applied bending moments and shears from the existing loads. As the fire continues and the member properties degrade, the beam will vertically sag and locally deform or buckle, giving rise to the so-called catenary action, or tensile membrane forces, that accompany such large deformations. Therefore, the initial compressive axial thrust in the beam from the heating later transitions to an axial tension, both of which must be adequately transferred by the adjacent connections to the other members of the structure in order to avoid collapse. All these structural and fire-engineering issues are not explicitly considered in conventional structural design. It is not well established under what conditions connectors will fracture at elevated temperatures, and the associated thermally induced forces and deformations that should be considered.

One of the strong recommendations in the FEMA 403 Report on the WTC disaster of Sept. 11, 2001, was to obtain a better understanding of connection framing system response to fire conditions.

The fire resistance and level of fire protection warranted for structural connections and framing systems, and its design implementation, has yet to be rigorously determined, both for steel and all other building materials.

6.3.1.5 Trusses

Trusses can be of any building material, but usually are constructed of either steel or wood. Most of the primary steel building framing consists of manufactured members with solid webs having various standard cross-sectional profiles: I-shape (wide flange), channels, hollow structural sections or tubes, angles, or tees. In order to reduce weight or to economically accommodate designs for longer spans or heavier loads, built-up members consisting of combinations of standard steel shapes and/or plates, plate girders, trusses or joists (so-called open-web members) are sometimes used in place of rolled beams in floor systems. Steel joists are standard manufactured products, while plate girders and trusses are built-up members, fabricated from individual steel plates or shapes, and are usually of more substantial depth. In concept, the fire resistance of such open-web trusses and joists is ultimately dependent on the behavior of their individual components, so the same generic steel material properties that were previously discussed would be applicable for any needed analyses of these members. Additional considerations of the component interactions and interconnections within the truss/joist member may be necessary to verify that any localized failures of the chords or diagonals do not compromise the overall truss/joist member stability.

A reality of the open-web truss/joist construction is that typically their weight is less than that of comparable solid-web members. In addition, their fire-exposed perimeter would be the same, or more, due to the increased member depth. Hence, for an equivalent member load-carrying capacity, trusses and joists would have a lower relative *weight-to-heated-perimeter* (W/D) ratio, thereby indicating their greater sensitivity to higher temperatures than comparable solid-web steel members. UL had conducted many ASTM E 119 fire tests on steel joist floor systems, and the resultant fire ratings are published in the UL *Fire Resistance Directory* [12]. These floor joist ratings can be, and have been, conservatively applied to similarly protected built-up trusses, since trusses have a higher W/D ratio compared to joists.

6.4 PROTECTION OF STRUCTURAL MATERIALS FROM FIRE

In order to control and moderate the damaging temperature rise that occurs during a fire exposure in the primary load-bearing elements of the structural framing, additional insulating materials are usually added as a coating, encasement, envelope, or protective membrane. In addition, other special methods and systems have been developed for this same purpose, such as flame shielding and water-filled tubular columns.

This fire protection provides the structural member or assembly with a certain fire resistance, as typically measured by a fire endurance rating time in a standard fire test, such as ASTM E 119. These protective materials and methods for fire resistance come in many different forms and products that are readily commercially available. These major types of fire protection materials and methods are subsequently described: gypsum, spray-applied materials, intumescents, concrete, and other special fire-resistive systems.

The insulating materials most widely used are mineral fiber and expanded aggregate coatings, such as vermiculite and perlite, which are called spray-applied fire resistive-materials as a group. UL has recently started referring to all the spray-applied fire protection coatings generically as *spray-applied fire resistive materials* (SFRMs), unless they are of the mastic or intumescent type.

The best energy-absorbing materials are gypsum and concrete due to their bound water. Intumescent types of coating materials, applied as paint, expand upon exposure to high temperatures to form an insulating layer.

The selection of the insulation type and thickness for fire protection should not be determined in isolation, but is recommended to be performed in conjunction with the architectural, structural, economical, and common construction needs and practices, product availabilities, and rational size increments. Often, these practical considerations dictate reasonable uniformity and consistency of the structural fire protection designs throughout the building, with as little variation as possible for the different members.

6.4.1 Fire-Resistive Materials

6.4.1.1 Gypsum

Gypsum is a noncombustible material, produced in the form of flat boards or plaster that consists of approximately 21 percent by weight of chemically combined water. This water content greatly contributes to the gypsum products' effectiveness as a fire-resistive barrier. Much of the background information in this section, and much more, on gypsum can be found in the *Fire Resistance Design Manual,* Gypsum Association, Washington, DC (available online at: <http://gypsum.org/securepubs/download.asp>) [21].

The thermodynamic reaction of gypsum board or gypsum plaster during a fire exposure produces a slow release of its internal water content as steam, thereby effectively retarding the heat transmission from the source to the protected structural member (Fig. 6.26). When gypsum-protected wood or steel structural members are exposed to a fire, this slow process of water release as steam, known as calcination, acts as a thermal barrier until all the internal water has evaporated. The temperature directly behind the plane of calcination is only slightly higher than that of boiling water (212°F), which is significantly lower than the temperature at which steel begins losing strength or wood ignites. Once calcination is completed, the in-place calcined gypsum continues to act as a physical shield to protect the underlying structural members from direct exposure to flames.

The ASTM C 36 standard describes two types of gypsum board, regular and Type X or improved Type X, each providing a different degree of fire resistance. A gypsum board without any special

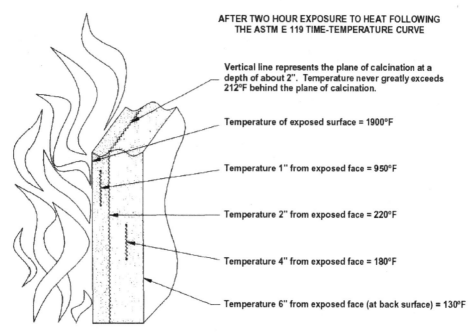

FIGURE 6.26 Fire protection characteristics of Type X gypsum board. (With permission, Gypsum Association Document GA-235, Gypsum Board Typical Mechanical and Physical Properties, *Fire Resistance Design Manual,* Document GA-600.)

formulation additives for fire resistance is called regular gypsum. Although regular gypsum has some degree of natural fire resistance, when fire-resistance-rated systems are specified, Type X or improved Type X gypsum board is typically required to achieve the rating. The Type X board core contains special additives to further enhance the natural fire resistance of regular gypsum. Type X gypsum board is defined in ASTM C 36 as gypsum board that provides not less than 1-h fire resistance for boards $5/8$ in. thick, or not less than $3/4$-h fire-resistance rating for boards $1/2$ in. thick, applied parallel with and on each side of load-bearing 2 × 4 wood studs spaced 16 in. on center with 6d coated nails, 1 in. long, 0.095-in.-diameter shank, $1/4$-in.-diameter heads, spaced 7 in. on center with gypsum board joints staggered 16 in. on each side of the partition, and tested in accordance with the requirements of ASTM E 119.

Where $3/4$- or 1-in. gypsum board is described as Type X in proprietary systems, the board manufacturer should be consulted to determine what specific products are required for such an application, since the composition of the product board core varies from one manufacturer to another.

Some of the main properties of gypsum board, such as potential heat, specific heat, and conductance, are given in Table 6.6.

The gypsum board is usually configured to form a protective membrane or envelope surrounding the structural member or assembly. In the case of a column, a box enclosure is formed with the board, of appropriate thickness and installation, to achieve the necessary fire resistance required by the applicable building code. For a floor or roof assembly, gypsum board can be used as the lay-in parts of a suspended ceiling to provide a thermal barrier to the floor and roof system above the ceiling, again in accordance with fire-rated design and listing requirements, such as those contained in the UL *Fire Resistance Directory* or the aforementioned *Fire Resistance Design Manual* of the Gypsum Association. It is important that not only the proper thickness and type of gypsum board is used in a fire-resistive assembly, but that all of its installation details fully comply with the rating and code requirements in order to maintain the integrity of the protective envelope prior to and during the fire.

6.4.1.2 Spray-Applied Materials

Spray-applied fire-resistive materials are coatings that are applied directly to the surface of the steel member as insulation from heat exposures. UL has recently started referring to all the spray-applied fire protection coatings generically as SFRMs, unless they are of the mastic or intumescent type.

As with most other construction materials, proper product and field quality control, inspection, and applications are very important factors in the successful fire performance of the product and assembly. Many of the current issues and questions faced by the fire protection material industry are directly related to their product application, quality control procedures, and implementation.

Standard fire-testing procedures require the rated assembly to be built and protected in accordance with these published recommendations. Upon successful completion of the test, the configuration of the assembly, as well as the manufacturer's recommendations for the product application procedure, become a part of the "listing" for the particular design. The fire listing information (such as that found in the UL *Fire Resistance Directory*) depicts exactly how the member or assembly is to be designed and mandates compliance with all the manufacturer's instructions for the SFRM application. Essentially, all the fire protection manufacturers list field application procedures such as the following:

1. Application shall be in accordance with the manufacturer's listing and recommendations.
2. All surfaces receiving SFRM shall be thoroughly cleaned of oil, grease, dirt, loose paint, loose mill scale, or any other matter that will impair bond.
3. All clips, hangers, supports, or sleeves are to be installed prior to application of fire protection material.
4. All ducts, pipes, conduits, etc., shall be installed after the application of SFRM.
5. Generally, the SFRM is applied to galvanized steel deck. If bond seal is required, or if unclassified painted steel deck is used, a metal lath may be required to assure proper bond strength to the steel.

TABLE 6.6 Selected Gypsum Board Properties

Surface Burning Characteristics (Independent of thickness) (ASTM E 84—CAN/ULC-S102)		
Board type	Flame spread	Smoke developed
Gypsum wallboard	10–15	0
Gypsum lath	10	0
Gypsum sheathing	10–20	0

Fire Resistance (ASTM E 119—CAN/ULC-S101-M)
See the Gypsum Association *Fire Resistance Design Manual*.
Noncombustibility (core) (ASTM E 136—CAN/ULC-S114-M)
Pass

Potential Heat (From NFPA 220, Appendix C)			
Thickness in. (mm)	Board type	Potential heat, weight basis	
		(Btu/lb)	(kJ/kg)
3/8 (9.5)	gypsum lath	310	721
3/8 (9.5)	gypsum wallboard	760	1770
3/8 (9.5)	gypsum wallboard, paper removed	−270	−628
1/2 (12.7)	gypsum wallboard	650	1512

MISCELLANEOUS

Thermal Properties (typical) (From ASHRAE Handbook of Fundamentals. 75°F (24°C) mean temperature)						
	Resistance (R)		Conductance (R)		Specific Heat	
Thickness in. (mm)	°F-ft²-hr/Btu	K·m²/W	Btu/hr-ft²-°F	W/m²·K	Btu/lb-°F	J/kg·K
3/8 (9.5)	0.32	0.06	3.10	16.7	0.26	1090
4/10 (10.2)	0.36	0.063	2.78	15.8	0.26	1090
1/2 (12.7)	0.45	0.079	2.22	12.6	0.26	1090
5/8 (15.9)	0.56	0.099	1.78	10.1	0.26	1090

Weight per Unit Area (for use in calculating dead loads)		
Thickness in. (mm)	Weight	
	psf	kg/m5
1/4 (6.4 mm)	1.2	5.86
5/16 (7.9 mm)	1.3	6.35
3/8 (9.5 mm)	1.4	6.84
1/2 (12.7 mm)	2.0	9.77
5/8 (15.9 mm)	2.5	12.21
3/4 (19.0 mm)	3.0	14.65
1 (25.4 mm)	4.0	19.53

From Fire Resistance Design Manual, Gypsum Association.

Many ASTM standards also address the quality control and expected in situ performance of the relevant fire protection products. For example, the SFRM commonly used to protect structural steel would also undergo the following standard tests, the results of which would be reported in the individual SFRM product literature:

- ASTM E 84, "Standard Test Method for Surface Burning Characteristics of Building Materials"
- ASTM E 605, "Standard Test Methods for Thickness and Density of Sprayed Fire-Resistive Material Applied to Structural Members"
- ASTM E 736, "Standard Test Method for Cohesion/Adhesion of Sprayed Fire-Resistive Materials Applied to Structural Members"
- ASTM E 759, "Standard Test Method for Effect of Deflection on Sprayed Fire-Resistive Material Applied to Structural Members"
- ASTM E 760, "Standard Test Method for Effect of Impact on Bonding of Sprayed Fire-Resistive Material Applied to Structural Members"
- ASTM E 761, "Standard Test Method for Compressive Strength of Sprayed Fire-Resistive Material Applied to Structural Members"
- ASTM E 859, "Standard Test Method for Air Erosion of Sprayed Fire-Resistive Materials Applied to Structural Members"
- ASTM E 937, "Standard Test Method for Corrosion of Steel by Sprayed Fire-Resistive Material Applied to Structural Members"

None of these standards, however, address impact or dynamic loadings, such as those that occurred on the Sept. 11, 2001, disasters from the terrorist attacks.

Present inspection procedures are not always adequate to verify that the fire protection material actually applied to a building is of the proper chemical composition, density, and thickness as the material listed in the fire-rated design. Cohesion and adhesion of direct application SFRM to the steel substrate has become a periodic problem in the field because of the rapid increase in the number and type of lightweight insulating materials, especially of the SFRM category.

Unfortunately, there are no current standards or building code requirements wherein the various fire protection products can be assessed in terms of long-term durability, aging, corrosion, and resistance to impact, vibrations, abrasion, air erosion, etc. An evaluation of their performance under standard fire exposures and fire ratings is usually the primary code consideration.

6.4.1.2.1 Cementitious and Mineral Fiber Materials. SFRMs are intended to insulate steel from heat. They generally fall into two broad categories: mineral fiber and cementitious materials. These popular commercial products have proprietary formulations and, therefore, it is imperative to closely follow the manufacturer's recommendations for mixing and application.

The most attractive fire protection systems are those coatings that both insulate and absorb energy. The insulating materials most widely used are mineral fiber and expanded aggregate coatings, such as vermiculite and perlite, which are called SFRMs as a group. UL has recently started referring to all the spray-applied fire protection coatings generically as SFRMs, unless they are of the mastic or intumescent type. Accordingly, it may now be difficult to immediately distinguish between the very light density, fibrous material mixed with water at the nozzle and the cementitious materials mixed in a hopper and transported wet to the nozzle for application. More uncertainties and variables are present when the fire protection material is delivered in a dry form and is mixed with water at the nozzle on the project job site.

Mineral fiber and vermiculite acoustical plaster on metal lath are two of the SFRMs that are frequently used on steel columns, beams, and floor joists. The lower density mineral fiber is a highly efficient and lightweight fire protection material. The mineral fiber mixture combines the fibers, mineral binders, air, and water. The mineral fiber fire protection material is spray-applied with specifically designed equipment that feeds the dry mixture of mineral fibers and various binding agents to a spray nozzle, where water is added to the mixture as it is sprayed on the surface to be protected.

In the final cured form, the mineral fiber coating is usually lightweight, noncombustible, chemically inert, and a poor conductor of heat (low thermal conductivity insulator).

The most commonly used sprayed-fiber fire protection materials are efficient and inexpensive, but they are also relatively weak and fragile. These lower density fiber materials are soft and can be easily removed from the steel by accident, as well as intentionally to provide for hangers, clamps, mounting of electrical boxes, steel conduit, ductwork, etc. These and many other mineral fiber materials are not suitable for exterior use, where they would be exposed to the weather. The more durable sprayed-fiber coatings have a dry density greater than 20 lb/ft^3. These are Portland-cement-based, medium-density products that are suitable for use under limited (indirect) weather and higher humidity exposure conditions (such as open parking garages), and that can endure some limited physical abuse during construction.

The vermiculite acoustical plaster—or cementitious—fire protection material is composed of gypsum and perlite or vermiculite lightweight aggregates. Some manufacturers have substituted polystyrene beads for the vermiculite aggregate. The more desirable cementitious products contain Portland cement as the basic binder in the product. Gypsum is calcined to obtain the base material for gypsum plasters. Water is added to the gypsum during mixing of the fire-resistive coating. Some formulations use magnesium oxychloride or oxysulfate, calcium aluminate, phosphate, or ammonium sulfate. Various additives and foaming agents could be added into the mixture.

When cement is exposed to fire, heat is absorbed in removing the water of hydration and the absorbed water. Spray-applied mineral cementitious systems typically have weight densities from 15 to 50 lb/ft^3.

In addition to low-density SFRMs, there are several other fire-resistive spray-on materials that fit into two major coating categories: high-density inorganic systems and medium-density inorganic systems.

The high-density inorganic systems include products that contain magnesium oxychloride, or a vermiculite-cement system. These materials have densities ranging from 40 to 80 lbs/ft^3. For the same amount of material, magnesium oxychloride will have over 2^1/$_2$ times the water content of gypsum. This chemically bound water is released when the cement is heated.

Several medium-density inorganic systems that have been successfully tested contain magnesium oxychloride and have densities of 20 to 27 lbs/ft^3. Several cementitious coating products in the 20 to 27 lb/ft^3 density range have also demonstrated not only excellent fire resistance, but also excellent serviceability.

All these SFRM products are required to be, and have been, free of the carcinogenic asbestos since the early 1970s.

6.4.1.2.2 Intumescents and Mastic Coatings. Intumescent and mastic coatings can be categorized as light organic materials. Whereas the more traditional fire protection materials typically experience a slight shrinkage during a fire exposure, an intumescent coating chars, foams, and expands upon heating to about 15 to 30 times its original volume. The intumescent mechanism involves the interaction of four types of compounds: a carbon source, an intumescent, a blowing agent, and a resin. When the coating becomes sufficiently heated, the carbon source reacts with a dehydrating agent to form a char, which is simultaneously expanded by gases released from the blowing agent. The resin binder prevents the gases from escaping. As a result, an insulating layer is produced that can be more than 100 times thicker than the original coating. To retain this insulating layer or char in many tested samples, reinforcing is required for the flange tips of a steel section.

Extensive research and development over the last decade has led to greatly improved formulations in new products that do not use traditional compounds. The final total thickness of intumescent coatings typically ranges from 0.03 to 0.4 in. for rated construction, which is much less than the typical ^{3}/$_8$- to ^{1}/$_2$-in. minimum thickness of SFRM or gypsum board protection.

The compounds of intumescent systems can generally be placed into four categories:

1. Inorganic acid, or material yielding an acid at temperatures of 212 to 570°F
2. Polyhydric material rich in carbon
3. Organic amine or amide, as a flowing agent

4. Halogenated material. In addition, various binders and additives are mixed in to provide specific physical properties of the total system. In many instances, the system would be made of several coats with different properties and functions; for example, a topcoat will provide a durable and aesthetic finished surface, and the base coat will provide a strong bond to the steel substrate.

Again, intumescent types of coating materials, applied as paint, expand upon exposure to high temperatures to form an insulating layer. Thus, these products must have adequate spacing from other materials so as to reach their required expansion when exposed to the fire.

Organic (mastic) systems function by means of a complex series of reactions: intumescence, sublimation, ablation, and heat-absorbing chemical and physical reactions. Some of these systems require reinforcement at the flange tips to maintain the char in-place.

Although all of the intumescent and mastic products are relatively expensive, they provide many benefits, including lighter weight, durable surfaces, and good adhesion.

6.4.1.3 Concrete

Ordinary concrete is the heaviest inorganic material, with a density ranging from about 100 to 150 lb/ft^3 for lightweight and normal-weight aggregates, respectively. In addition to its many structural load-bearing applications, concrete often also serves a separate, or dual, function as a fire-protective material.

The best energy-absorbing materials are gypsum and concrete. Each of these energy-absorbing materials also release water of crystallization when exposed to high temperatures. According to Lie [22], the fire endurance time of concrete is increased by approximately 3 percent for each percent of entrapped water. Although gypsum is superior in this regard, concrete provides a much tougher and more durable protection. Magnesium oxychloride cements also act as an energy-absorber; they contain $3^1/_2$ times the amount of water as gypsum.

Concrete and the other higher density spray-on materials are more durable as long-term fire protection, but there is the expected tradeoff in associated weight and costs that is often the main consideration in this selection.

6.4.1.4 Masonry

Masonry can be used in much the same way as concrete for both load-bearing and fire protection applications. Similarly, its relatively heavy weight and higher costs compared to the lighter SFRMs or gypsum products often make masonry a less attractive choice in terms of economy for simple building fire protection needs.

6.4.2 Fire-Resistive Systems

Besides the particular fire protection material products described in Sec. 6.4.1 that are commonly used for prescriptive code-based designs, special systems or configurations of materials can be used to control and slow the temperature rise in structural framing materials when exposed to heating. These are called fire-resistive systems, in contrast to just fire protection materials. Flame shielding and water-filled columns are two of the innovative systems that have been successfully used for the fire protection of steel-framed buildings. These system protection methods normally are considered to be in the domain of performance-based design for fire safety and require additional calculations and/or documentation to justify their acceptance to the authority having jurisdiction (AHJ). Hence, flame shielding and water-filled columns are usually not the first, simplest, or quickest choice for fire protection, and are typically only used for unique projects or conditions in terms of architecture, occupancy, risk, etc. These more advanced methods have been used more internationally than in the United States.

In the November, 1996, issue of ASCE's *Civil Engineering Magazine,* the article titled "Bare Bones Buildings" by William Baker, Hay Iyengar, Ronald Johnson, and Robert Sinn of Skidmore, Owings, and Merrill described several innovative fire-engineering solutions in major European buildings that permitted the use of exposed exterior structural steel. These projects were the Broadgate Exchange House and One Ludgate Place, both in London, and the Hotel Arts in Barcelona, Spain. Even though the use of performance-based fire design in the United States is an exception rather than the rule, a number of notable projects have also been designed and built using some of these more advanced fire-engineering concepts. A few of the unique domestic projects in this regard include:

- One Liberty Plaza, New York—flamed shielding of exterior steel
- Great Platte River Bridge, Nebraska—flame shielding of exposed steel
- John Deere Building, Moline, Illinois—flame shielding
- U.S. Steel Building, Pittsburgh, Pennsylvania—water-filled HSS columns

6.4.2.1 Water-filled Columns

One of the special systems that can be used to offer fire protection is water-cooled columns, often used in combination with steel hollow structural sections (HSS) that can serve as "piping" for the liquid. As shown in Figs. 6.27 and 6.28, this concept depends on hydraulics and thermodynamics in the circulation of cold water within the steel columns when exposed to a fire. The mechanism provides a continuing heat sink to the fire exposures in the form of the circulating water, or another liquid, within the columns, which keeps the structural steel itself cool enough to avoid strength and stiffness degradation. Such liquid-filled columns theoretically and under idealized circumstances can

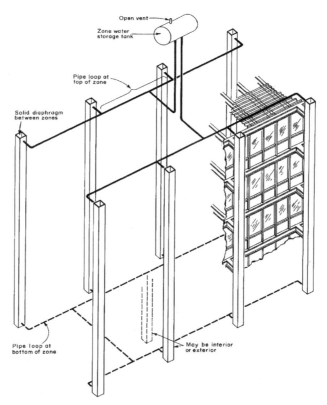

FIGURE 6.27 Schematic layout of typical piping in a liquid-filled column fire-protective system.

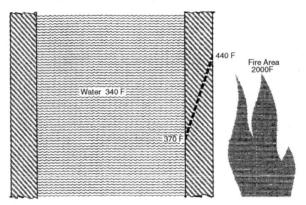

FIGURE 6.28 Representative temperature variations in cross-section of liquid-filled steel HSS column.

maintain unlimited fire resistance, as long as the liquid supply and circulation are maintained. This system also requires that the liquid in the columns contain a rust inhibitor and in the cold-weather regions an antifreeze agent, and perhaps some other additives to avoid biological stagnation and growth of microorganisms in the water supply.

The 64-story USX (U.S. Steel) building in Pittsburgh was the first building in the United States to employ water-filled HSS columns in 1970. Several other buildings in the United States and worldwide have also successfully used this concept. The initial design cost and necessity to maintain a functional mechanical piping network in this way, without leakage, freezing, and steel corrosion, are the potential difficulties and risks with this fire protection method, which would usually not be considered a passive protection system, such as SFRM, gypsum, or concrete. The nature of liquid-filled columns more resembles the active features provided by automatic sprinklers, though the latter is a fire-suppression measure. Its past actual building usage has been good, without reported problems.

6.4.2.2 Flame Shielding

The principle of flame shielding relies on providing a physical barrier to the direct flame impingement on the structural member that is to be considered fire resistive. Fig. 6.29 schematically illustrates how the exterior radiation heat temperature contours are lower the farther they are from the direct fire source inside the building. Thus, the flame shield through and around the window opening serves as a sacrificial element that not only protects the member from exposure to the hotter fire temperatures, but also provides some air/distance separation between the fire and the member, which is well known to be beneficial in moderating the temperature rise in the structural member itself. The flame-shielding concept was successfully used to protect exposed spandrel steel beams in the 54-story One Liberty Plaza building in New York City, among others.

The popularity of architecturally exposed steel has been increasing, given the inclination of many notable architects to aesthetically express the structural form of the building in this way. An analytical method is available and can be effectively used to determine when fire-unprotected steel is acceptable for building exteriors. The justification for such use of exposed steel without any fire protection is based on an evaluation of the potential temperature increase in the exterior steel due to a fire inside the building, including any shielding effects, with the flames impinging on the exterior exposed steel through the window openings and transferring heat by radiation (see Fig. 6.29).

This methodology, originally developed by Margaret Law ("Fire-Safe Structural Steel, A Design Guide," AISI, March 1979), involves the calculations of burning combustibles in rooms adjacent to the exterior walls, intensity and rate of burning in a room, the flame exposure outside of the

STRUCTURAL MATERIALS **6.31**

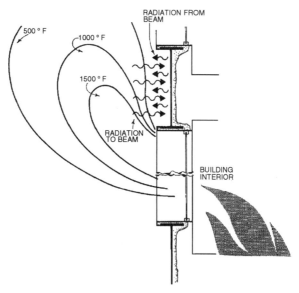

FIGURE 6.29 Typical exterior heat radiation temperatures developed from interior building fire.

windows, flame impingement on the exposed structural steel members, and, finally, the resulting temperatures in the outside steel. The typical flame shapes and fire radiation configurations for the basic conditions of forced draft and without forced draft are illustrated in Fig. 6.30. Three different possible column locations are shown that each require correspondingly different analyses, with column C having the least severe fire exposure from the window opening. If the resulting computed steel temperature is less than a critical temperature of 1000°F, the design is considered to be adequate for fire safety. Positioning of appropriate supplemental flame shielding may be helpful to

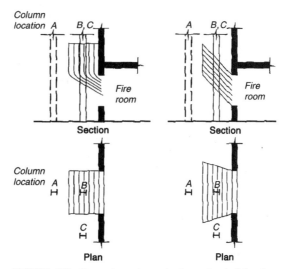

FIGURE 6.30 Exposed exterior steel columns in a building fire, without and with through draft.

prevent more direct flame impingement on exterior load-carrying members. A more current and readily available reference for this analysis procedure may be found in the ECCS *Model Code on Fire Engineering* [23]. A number of prominent projects that utilized flame shielding were given previously.

Sometimes, a building owner or code jurisdiction will require additional fire test verification of these analyses, particularly for larger landmark structures.

6.5 DETERMINATION OF FIRE RESISTANCE BY TESTING

Standard fire tests in conformance with ASTM E 119 and several possible ancillary standards provide the basis for many of the prescriptive structural fire resistance requirements in the building codes. In the majority of projects, these ratings alone, or in combination with available semiempirical design aids, are adequate to satisfy the code requirements. In some situations, wherein a unique structural assembly, occupancy, or fire hazard is encountered, or a more accurate fire safety assessment is desired, special-purpose fire tests and/or analyses can be conducted to determine the necessary fire resistance characteristics as a performance-based design.

6.5.1 Steel Construction

6.5.1.1 Columns

The E 119 standard subjects the test column to the standard fire on all sides of its profile for the full length of the member. Most columns have been fire tested without superimposed column loads, and, consequently, the only E 119 acceptance criterion is the limiting steel temperature of 1000°F (538°C) for arithmetic average within the section, or 1200°F (649°C) at any single section location. For convenient design applications, columns that have been tested and assigned a fire endurance rating per ASTM E 119 are usually arranged in summary listings showing the steel section and type of fire protection. The fire-rated steel columns are of various shapes (wide-flange, tubular, tee, pipe), protected by spray-on materials, intumescent or mastic coatings, gypsum board, or concrete encasement. One rated column design, UL Design No. X737, is given in Fig. 6.31 as a sample listing, while Table 6.7 lists several typical UL steel column designs. The details of this tested assembly and design are all provided therein, including the fire protection material product, construction, application or installation procedures, insulation thickness, and minimum column size.

The laboratory constrains the representative steel column size to be compatible with its experimental facility capabilities, and to be 9 ft high. The most frequently tested column shapes have been the W 10 × 49 for the lighter weight steel members and the W 14 × 228 or W 14 × 233 for the heavier sections. The listing gives the minimum column size necessary for the applicable fire rating, i.e., for the member that was tested. Larger members than the minimum steel size may be conservatively used with the given design fire protection. However, if a lighter steel section is to be actually used for the column, more fire protection will be required, as will be discussed later in Sec. 6.6.1.1. The reason for this tradeoff is the reduced heat sink capabilities of lighter members with smaller W/D ratio, which require more insulation than heavier members for the same fire exposure conditions. This adjustment of fire protection thickness is mandatory only when the fire-rated column design is to be extrapolated to a column size that is smaller than minimum size tested, hence for an increased thickness requirement. Otherwise, simplified column protection thickness formulas given in the UL *Fire Resistance Directory,* or in ASCE/SFPE 29–99, allow for calculated interpolation of insulation thickness, provided that the column shape of interest is within the W/D, insulation product and thickness, and fire rating period of columns that have been fire tested. This caveat reasonably prescribes that the correlation formula for column protection

Design No. X737
Rating — 2 and 3 Hr.

Typical W 10 x 49

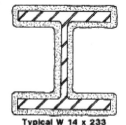
Typical W 14 x 233

1. **Spray-Applied Fire Resistive Materials*** — See table below for appropriate thickness. Applied in several coats to steel surfaces which must be clean and free of dirt, loose scale and oil. Min avg density is 22 pcf with min ind density of 19 pcf. For method of density determination refer to General Information Section.

Rating Hr	Min Thkns In.
2	1-1/2
3	2

SOUTHWEST VERMICULITE CO —Type 7TB.
W R GRACE & CO - CONN
CONSTRUCTION PRODUCTS DIV —Types 3300, 3306.

2. **Metal Lath** — 3.4 lb per sq yd galv or painted steel lath.
   a. **Contour Spray Application** — Secured to column by bending tight around flanges a min of 3-1/2 in. toward web of column. Optional for 2 Hr rating, required for 3 Hr rating.
   b. **Boxed Column Application** — Lath lapped 1 in. and tied together with 18 SWG galvanized steel wire spaced vertically 6 in. OC.
3. **Steel Column** — Min size of column, W10x49 with outside dimensions of 10x10 in., a flange thickness of 9/16 in., a web thickness of 5/16 in. and a cross-sectional area of 14.4 sq in.

*Bearing the UL Classification Mark

FIGURE 6.31 UL column design X737. (With permission, Underwriters Laboratories, Inc., *Fire Resistance Directory*, 2002, Northbrook, IL.)

TABLE 6.7 Selected UL Steel Column Designs

Assembly rating (hr)	Type of protection	Column types	UL design number
1,2,3	Gypsum wallboard	W, HSS	X528
2	W		X516, X518, X520
3	X		509, X510, X513
3/4, 1, 1 1/2, 2, 3, 4	Spray-applied fire resistive material	HSS, Pipe	X771, Y707
1, 1 1/2, 2, 3, 4		W	X772, X829, Y708, Y725
		W, HSS, Pipe	X790, X795
		HSS, Pipe	X827

Source: AISC Manual, p. 2-44.

[a] The referenced assemblies are some commonly used Underwriters Laboratories (UL) assemblies used for conventional steel framed structures. For additional assemblies the reader should reference the UL Fire Resistance Directory.

[b] For additional design requirements such as beam spacing, concrete strength, density, reinforcing, and clear cover, minimum metal deck gauge, maximum deck span, shear connector requirements, design stress limitations, etc., see the specific referenced assembly in the Underwriters Laboratories (UL) directory.

thickness be applied only within the fire-tested range, without extrapolation beyond the bounds of the test data.

Column fire ratings are used not only for actual building columns, but also for other members that are designed for compression loads, such as truss members and bracing.

6.5.1.2 Beams and Girders

Beams are primary bending members, usually consisting of rolled shapes or built-up sections, such as trusses or plate girders. Beams that are deeper and frame into the columns are referred to as girders. Generically, all these bending members will subsequently be referred to as beams. Beams commonly support floor slabs, and, as such, the standard fire test includes a portion of the accompanying floor system. Thus, in contrast to the E 119 column test wherein only the column is exposed to fire on all of its sides and the member is unloaded, a beam will have its top side (flange) shielded from direct flame exposures by the floor segment, while it is usually subjected to superimposed gravity loads during the fire.

Beam fire ratings can be derived in one of two ways: 1) from a floor or roof assembly rating, or 2) from an individual beam test. The latter type of beam test includes a representative section of the floor or roof. In both of these alternatives, maximum design gravity loads are superimposed

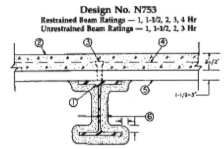

Design No. N753
Restrained Beam Ratings — 1, 1-1/2, 2, 3, 4 Hr
Unrestrained Beam Ratings — 1, 1-1/2, 2, 3 Hr

1. **Steel Beam** — W8x28 min size.
2. **Normal Weight or Lightweight Concrete** — Compressive strength, 2500 psi. For normal weight concrete either carbonate or siliceous aggregate may be used. Unit weight, 148 pcf. For lightweight concrete, unit weight 109 pcf.
3. **Shear Connector** — (Optional) — Studs, 3/4 in. diam headed type or equivalent per AISC specifications. Welded to the top flange of beam through the steel floor units.
4. **Welded Wire Fabric** — 6x6, W1.4 x W1.4.
5. **Steel Floor and Form Units** — 1-1/2 to 3 in. deep corrugated, fluted or cellular units, welded to beam. Max usage of cellular units shall consist of a 1:1 blend with fluted units.
6. **Spray-Applied Fire Resistive Materials*** — Prepared by mixing with water and liquid concentrate according to instructions on each bag of material and spray-applied to beam surfaces which are free of dirt, loose scale, and oil. The crests of corrugated and fluted units above the beam shall be filled to capacity with Spray-Applied Fire Resistive Materials. Min avg density of 27 pcf with min ind density of 25 pcf. For method of density determination, refer to Design Information Section.

Rating Hr	Min Thkns In.	
	Unrestrained Beam Rating Hr	Restrained Beam Rating Hr
1	7/16	7/16
1-1/2	11/16	1/2
2	15/16	3/4
3	1-7/16	1-1/4
4	—	1-11/16

ALBI MFG, DIV OF
STANCHEM INC — Type DS 30.
*Bearing the UL Classification Mark

FIGURE 6.32 UL beam design N753. (With permission, Underwriters Laboratories, Inc., *Fire Resistance Directory*, 2002, Northbrook, IL.)

TABLE 6.8 Common Steel Beam Fire Rated Designs

Beam-only designs—roof[a,b]

Assembly rating					
Restrained (hr)	Unrestrained (hr)	Type of protection system	Roof insulation type	Metal deck depth[c] (in)	UL design number
1, 1½, 2, 3	1, 1½, 2, 3	Spray-applied fire resistive material	Rigid	1½	S715, S733
1, 1½, 2, 3, 4	1, 1½, 2, 3, 4		Rigid	1½	S701, S721, S724, S729, S734, S805
			Rigid or insulating fill	1½	S735

Source: From Ref. 10, pp. 2-43.

[a] The referenced assemblies are some commonly used Underwriters Laboratories (UL) assemblies used for conventional steel framed structures. For additional assemblies the reader should reference the UL Fire Resistance Directory.

[b] For additional design requirements such as beam spacing, concrete strength, density, reinforcing and clear cover, minimum metal deck gauge, maximum deck span, shear connector requirements, design stress limitations, etc., see the specific referenced assembly in the Underwriters Laboratories (UL) directory.

[c] Metal deck depth for some assemblies is shown as a minimum and deeper decks may be substituted. Refer to the specific UL assembly for additional information.

Beam-only designs—floor[a,b]

Assembly rating			Concrete			
Restrained (hr)	Unrestrained (hr)	Type of protection system	Min. thickness above deck flutes (in)	Type	Metal deck depth (in)	UL design number
2	2	Gypsum	2½	NW	1½	N501, N502
3	2	Wallboard	2½	NW	1½	N505
1, 1½, 2, 3, 4	1, 1½, 2, 3	Spray-applied fire resistive material	2½	NW or LW	1½, 2, 3	N706, N734, N739, N823
					1 5/16, 1½, 2, 3	N708, N772, N782

[a] The referenced assemblies are some commonly used Underwriters Laboratories (UL) assemblies used for conventional steel framed structures. For additional assemblies the reader should reference the UL Fire Resistance Directory.

[b] For additional design requirements such as beam spacing, concrete strength, density, reinforcing and clear cover, minimum metal deck gauge, maximum deck span, shear connector requirements, design stress limitations, etc., see the specific referenced assembly in the Underwriters Laboratories (UL) directory.

on the assembly during the fire test. As with columns, a minimum beam size is given in the fire-rated design, and substitution of a different size is allowed only for shapes that have a greater W/D value than for the beam specified. A sample beam listing, UL Design N753, is shown in Fig. 6.32. Other popular steel beam-only designs for both floor and roof construction fire ratings are given in Table 6.8.

Due to practical laboratory considerations, the minimum beam size in most UL fire-rated designs is nominally an 8 or 10-in. deep wide flange, in the range of a W8 × 28 and similar shapes. This is relatively small and lightweight section compared to the beam size requirements typically necessary for most steel floor systems in buildings, so beam section substitutions are usually necessary. In Sec.

6.6.1.2, an empirically derived interpolation equation is given that can be used to adjust spray-applied thickness requirements for unrestrained beams as a function of W/D.

Substitution of noncomposite beams is allowed unequivocally for a listed composite beam. However, composite beams may only be substituted into rated designs for other composite beams. The justification for this limitation is that the fire exposure places more severe demands on a fully loaded composite beam than its noncomposite counterpart.

Substitution of beam-only designs into floor or roof assembly designs is permitted only for assemblies that have an equivalent or greater heat dissipation capacity of the floor or roof construction specified, as compared to the heat dissipation of the floor or roof assembly in the referenced beam-only design.

As discussed previously and shown in Table 6.3, two kinds of fire ratings are provided: restrained and unrestrained. The unrestrained rating is based on temperature endpoints only, while the restrained rating is governed principally by load-carrying capability. By the E 119 definition:

> A restrained condition in fire tests, as used in this Standard, is one in which expansion at the supports of a load-carrying element resulting from the effects of fire is resisted by forces external to the element. An unrestrained condition is one in which the load-carrying element is free to expand and rotate at its supports.

Further discussion of these concepts and their acceptance criteria may be found in the E 119 standard and in the UL Directory. The important distinction for design application purposes is that unrestrained ratings are more conservative and require usually greater or at least the same insulation thickness than what is required for the equivalent restrained ratings. The meaning and interpretation of restrained and unrestrained ratings has been somewhat confusing and uncertain in practice, despite the guidance offered by Table 6.3 and by other research and references, which indicate that most common steel framing can be considered to be restrained for these purposes [11].

Beam-only tests are considered restrained specimens, and both restrained and unrestrained ratings are derived from these. Floor and roof assembly designs contain only unrestrained beam ratings because the beam is considered to be only a part of the restrained assembly. Often, beam-only ratings are used to comply with code requirements that specify a longer fire rating time for the beam than the floor assembly, or that require restrained beam ratings that a particular assembly listing does not include.

6.5.1.3 Walls

The fire resistance ratings of steel stud and panel walls can be evaluated under E 119. The listed wall designs are contained in the UL *Fire Resistance Directory* under the U, V, or W categories. The type of fire protection that can be employed for these steel wall systems includes gypsum board, lath and plaster, SFRM, and concrete or masonry.

6.5.1.4 Floor/Ceiling and Roof/Ceiling Assemblies

Steel floor/ceiling and roof/ceiling assemblies include various systems with rolled shapes or joists. Both restrained and unrestrained assembly ratings are usually provided in each UL design, along with an unrestrained beam rating that is no greater than the assembly rating. A sample UL listing of a floor/ceiling assembly, UL Design D923, is given in Fig. 6.33. Table 6.9 lists some popular UL fire rated steel floor and roof assembly designs. The selected information in Table 6.9, and that in Tables 6.7 and 6.8, should be independently verified and used as indicated in the UL *Fire Resistance Directory,* and is not meant to preclude the selection of other rated assemblies and fire protection methods.

ASTM E 119 requires the fire testing of these assemblies to occur with full design loading. In addition to the resulting restrained and unrestrained assembly ratings, an unrestrained beam rating is provided from an assembly test. Various designs with protected and unprotected steel deck, LWC and NWC floor topping, deck and insulation products are catalogued and available for use. As discussed

Design No. D923
Restrained Assembly Ratings — 3/4, 1, 1-1/2, 2 or 3 Hr
(See Items 2, 7, 8, 9 and 12)
Unrestrained Assembly Rating — 0 Hr (See Item 4)
Unrestrained Beam Ratings — 1, 1-1/2, 2 and 3 Hr
(See Items 5, 8 and 12)

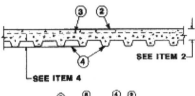

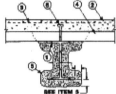

1. **Supports** — W8x28 min size steel beams.
2. **Normal Weight or Lightweight Concrete** — Normal weight concrete, carbonate or siliceous aggregate, 147 to 153 pcf unit weight, 3500 psi compressive strength, vibrated. Lightweight concrete, expanded shale, or slate aggregate by rotary-kiln method, or expanded clay aggregate by rotary-kiln or sintered-grate method; 3000 psi compressive strength, vibrated, 4 to 7 percent entrained air.

Restrained Assembly Rating Hr	Concrete (Type)	Concrete Unit Weight pcf	Concrete Thkns In.
1	Normal Weight	—	3-1/2
1-1/2	Normal Weight	—	4
2	Normal Weight	—	4-1/2
3	Normal Weight	—	5-1/4
3/4 or 1 (See Item 6)	Lightweight	107-113	2-1/2
1	Lightweight	107-113	2-5/8
1-1/2	Lightweight	107-113	3
2**	Lightweight	107-113	3-1/4
2**/4**	Lightweight	107-116	3-1
2	Lightweight	107-120	3-1/2
3	Lightweight	107-113	4-3/16
3	Lightweight	107-120	4-7/16

*For use with 2 or 3 in. steel floor and form units only.
**When optional Items 7 and 11 are used, the Restrained Assembly Ratings of the 3-1/4 in. lightweight concrete thicknesses are reduced to 1-1/2 h.

3. **Welded Wire Fabric** — 6x6, 10x10 SWG.
4. **Steel Floor and Form Units*** — Composite 1-1/2, 1-5/8, 2 or 3 in. deep galv units of 4-1/2 in. deep noncomposite galv units. Fluted units may be uncoated or phosphatized/painted. Min gauges are 22 MSG for fluted and 20/20 MSG for cellular units. The following combinations of units may be used:
(1) All 18, 24, 26, 28 or 36 in. wide cellular.
(2) All fluted.
(3) One or two 3 in. deep, 12 in. wide, 18/18 MSG min cellular, alternating with 3 in. deep fluted or other cellular.
(4) Any blend of fluted and 18, 24, 26, 28 or 36 in. wide cellular.
(5) 3 in. deep, 30 in. wide cellular with 8-1/8 in. wide valley along side joints may be used when 3/8 in. diam reinforcing bars are placed 1-1/2 in. to each side of side joints and 1 in. above bottom of unit.
(6) Corrugated, 1-5/16 in. deep, 30 in. wide, 24 MSG min galv units with shear wires factory welded to deck corrugations. Welded to supports 12 in. OC through welding washers. For shear wire spacing of 8 in. or less, the steel deck stress shall not exceed 20 KSI. For shear wire spacing greater than 8 in. OC but less than or equal to 12 in. OC, steel deck stress shall not exceed 12 KSI.
Components for field-assembled cellular metal raceway units:
Raceway bottom — 24 or 36 in. wide Types 212 VS, 312VS.
Raceway cover plate — Types CP-12, CP-16.
Raceway divider — Types DC-20, DC-25.
Raceway isolation trough — Types T-20, T-25, T-30.
CHENG HO STEEL CO LTD —24 or 36 in. wide Types LF2, LF3.
CONSOLIDATED SYSTEMS INC —24 in. wide Types CFD-2.

CFD-3; 24, 30 or 36 in. wide Type CFD-1.5; 24 or 36 in. wide Types Mac-Lok 2, Mac-Lok 3; 24 in. wide, Types B2C, B2FC, NC, NFC; 30 in. wide Type B3C; 12 in. wide Mac-Way cellular 45MOW, 2-633MTWA, 2-633MTWV, 3-633MTWA, 3-633MTWV+.
EPIC METALS CORP —24 in. wide Types EC150, ECP150, EC300, ECP300, EC366, ECP366, EC150, BC300 inverted; 30 in. wide Types ECB150, ECBR150; 36 in. wide Type EC266.
H H ROBERTSON —QL Types, 24 in. wide 3 or 3 inverted, UKX, UKX-3 2 in. 99, AKX, 21 or 21 inverted, 121, NKX, TKX; 24 or 30 in. wide GKX, GKX-A; 36 in. wide 99, AKX, WKX; 24, 26 or 36 in. wide NKX; 1.5NKC, NKC, AKX, 2 or 3 in TKC; 12 in. wide noncomposite Sec. 12; 17 in. wide 21, 26 or 28 in. wide UKX, 87.5 cm wide. Side joints of QL, 99, 121, WKX, TKX, TKC, and Metric units–QT-77-900; QLC-78-900 may be welded together 60 in. O.C. Side joints of 99, AKX, WKX, GKX, GKX-A, TKX and Metric units–QL-77-900 and QLC-78-900 may be fastened together with min 1 in. long No. 12x14 self-drilling, self-tapping steel screws 36 in. O.C.
HAMBRO STRUCTURAL SYSTEMS, DIV OF CANAM STEEL CORP —36 in. wide, 1-1/2 in. Type P3615HB. The max superimposed load for Type P3615HB units shall not exceed 250 PSF. For single spans, the use of the units shall be limited to 5 ft, 6 in., 6 ft, 0 in. and 6 ft, 6 in. max spans for the Nos. 22, 20 and 18 gauge units, respectively. For multiple spans, No. 18 gauge units may be used on max 7 ft, 6 in. spans with a max total superimposed load of 240 pcs.
MARLYN STEEL DECKS INC — Type 1.5 CF, 2.0 CF or 3.0 CF.
ROOF DECK INC —36 in. wide Types LOK 1-1/2, LOK 1-1/2 R; 24 in. wide Types LOK-2, LOK-3.
UNITED STEEL DECK INC —24 in. wide, Types 1-1/2, 2 or 3 in. LOK-Floor, LOK-Floor Cell; 36 in. wide Types 2 or 3 in. LOK-Floor, LOK-Floor Cell; 24 in. wide Types N-LOK, N-LOK Cell; 24, 30 or 36 in. wide Types 1-1/2 in. B-LOK, B-LOK Cell.
VALLEY JOIST — 24 or 36 in. wide Types WVC 1-1/2 or WVC 2.
VERCO MFG CO —Formlok Types 24 or 36 in. wide W2, W3; 24 in. wide B, BR, BN; 30 in. wide B, BR; 12 in. wide W2 or W3 units may be blended with 24 or 36 in. wide W2 or W3, respectively.
VULCRAFT, DIV OF NUCOR CORP —24, 30 or 36 in. wide Types 1.5VL, 1.5VLI, 1.5VLP; 24 or 36 in. wide Types 2VLI, 2VLP, 3VLI, 3VLP.
WALKER SYSTEMS INC —24 in. wide, Type 2 or 3 in. WDR.
WHEELING CORRUGATING CO, DIV OF WHEELING-PITTSBURGH STEEL CORP —30 in. wide Types SB-150, -150N, -150NR, -150R; 30 in. wide Types SB-B16LF, -B16LFR; 24 or 36 in. wide Types P20LF, SB-P21LF, -P31LF; 24 in. wide Types P20LF, SB-200, -300; 24 in. wide Types 1.5 SB, 1.5 SBR; 24 or 36 in. wide Types 2.0 SB, 3.0 SB; Type SB-B16LFR may be phos/ptd; 30 in. wide Types 1-1/2 in. V-Grip, 1-1/2 in. RV-Grip; 24 or 36 in. wide Types 212V-Grip, 312V-Grip; 36 in. wide Types 212VW3-Wireway, 312VW3-Wireway.

Spacing of weld attaching units to supports shall be 12 in. OC for 12, 24 and 36 in. wide units, four welds per sheet for 30 in. wide units, 6 in. OC for 18 in. wide and Sec. 12 units. Unless noted otherwise adjacent units button-punched or welded together 36 in. OC along side joints. Adjacent 18 in. wide units welded together 30 in. OC along side joints. For 3 Hr Rating, units with overlapping type side joints welded together 24 in. OC max.

+12 in. wide, 1-1/2 in. deep Mac-Way units may be blended with 24 in. wide B2C or 30 in. wide B3C units in a blend of one cell to one or more fluted units. 12 in. wide, 2 in. deep Mac-Way units may be blended with 36 in. wide Mac-Lok 2 units in a blend of one cell to one or more fluted units. 12 in. wide, 3 in. deep Mac-Way units may be blended with 36 in. wide Mac-Lok 3 units in a blend of one cell to one or more fluted units. The side edge of the fluted unit is placed on the top of the side edge of the Mac-Way unit and the two are welded together with welding washers spaced a max of 32 in. O.C. for Mac-Lok 2 or 3 units and a max of 24 in. O.C. for the B2C or B3C units. The Unrestrained Assembly Rating is 1-1/2 h for the following units with the stated limitations:
(a) 1-1/2 in. deep, 24 in. wide, 22 MSG or thicker fluted with clear spans not more than 7 ft, 8 in.
(b) 1-1/2 in. deep, 24 in. wide, 20 MSG or thicker fluted with clear spans not more than 8 ft, 8 in.
(c) 1-1/2 in. deep, 24 in. wide, 16 MSG or thicker fluted and 18/18 MSG or thicker cellular with clear spans not more than 9 ft, 11 in.
(d) 3 in. deep, 36 in. wide, 18 MSG or thicker fluted and 24 in. wide, 20/18 MSG or thicker cellular with clear spans not more than 13 ft, 2 in.

FIGURE 6.33 (Part 1) UL floor/ceiling design D923.

5. **Spray-Applied Fire Resistive Materials*** — Applied by mixing with water and spraying in more than one coat to the beam to the final thicknesses shown below. When fluted or corrugated steel floor units are used, crest areas above the beam shall be filled with Spray-Applied Fire Resistive Materials. Beam surfaces shall be clean and free of dirt, loose scale and oil. Min avg and min ind density of 23 and 21 pcf, respectively.

Unrestrained Beam Rating Hr	Restrained Assembly Rating Hr	Spray Applied Fire Resistive Mtl Thkns In
1	1, 1-1/2, 2	1/2
1-1/2	1, 1-1/2, 2, 3	3/4

    SOUTHWEST VERMICULITE CO —Type 7MP or FP-2

6. **Shear-Connector Studs** — (Optional) — Studs 3/4 in. diam by 3 in. long, for 1-1/2 in. deep form units to 5-1/4 in. long for 3 in. deep form units, headed type or equivalent per AISC specifications. Welded to the top flange of the beam through the steel form units.

7. **Electrical Inserts*** — (Not shown) Classified as "Outlet Boxes and Fittings Classified for Fire Resistance".
    H H ROBERTSON —Present Inserts. For use with 2-1/2 in. lightweight concrete topping over Q1-WKX steel floor units. Installed over factory-punched holes in floor units per accompanying installation instructions. Spacing shall not be more than one insert in each 14 sq ft of floor area with spacing along floor units not less than 48 in. O.C. The holes cut in insert cover for passage of wires shall be no more than 1/8 in. larger diam than wire. Restrained Assembly Rating is 3/4 h with Tapmate II-FS-1 and 1 h with Tapmate II-FS-2 inserts.
    H H ROBERTSON —Tapmate II-FS-t, II-FS-2; Series KEB.
    WALKER SYSTEMS INC —After set inserts. Single-service after set inserts installed per accompanying installation instructions in 2-1/2 in. diam hole core-drilled through min 3-1/4 in. thick concrete topping to top of cell of any min 3 in. deep cellular steel floor unit specified under Item 3. Spacing shall be no more than one insert in each 10 sq ft of floor area in each span with a min center to center spacing of 16 in. If the high potential and low potential raceways of the cellular steel floor unit are separated by a valley filled with concrete, the center to center spacing of the high potential and low potential single-service after set inserts may be reduced to a min of 7-1/2 in. Restrained Assembly Rating is 2 h or less with internally protected Type 436 after set insert with Type M4-, M6- or M8- Series single-service activation fitting.
    WALKER SYSTEMS INC —Internally protected Type 436 after set insert with Type M4-, M6- or M8- Series single -service activation fitting.

8. **Mineral and Fiber Boards*** — (Optional, not shown). Applied over concrete floor with no restriction on board thickness. When mineral and fiber boards are used, the unrestrained beam rating shall be increased by a minimum of 1/2 hr. See Mineral and Fiber Board (CERZ) category for names of manufacturers.

9. **Roof Covering Materials*** — (Optional, not shown) — Consisting of materials compatible with insulations described herein which provide Class A, B or C coverings. See Built-Up Roof Covering Materials in Building Materials Directory.

10. **Insulating Concrete** — (Optional, not shown) — Various types of insulating concrete prepared and applied in the thickness indicated:
    A. **Vermiculate Concrete** — (Optional, not shown).
        1. Blend 6 cu ft of Vermiculate Aggregate* to 94 lb Portland Cement and air entraining agent. Min thickness of 2 in. as measured to the top surface of the structural concrete or foamed plastic (Item 11) when it is used.
            SIPLAST INC
            VERMICULITE PRODUCTS INC
        2. Blend 3.5 cu ft. or Type NVC Concrete Aggregate* or Type NVS Vermiculate Aggregate* to 94 lb Portland Cement. Slurry coat 1/8 in. thickness beneath foamed plastic (Item 11) when used, 1 in. min topping thickness.
            ELASTIZELL CORP OF AMERICA
            SIPLAST INC
            VERMICULITE PRODUCTS INC
        Vermiculite concrete may be covered with Roof Covering Materials (Item 9).
    B. **Cellular Concrete** — Roof Topping Mixture* — Concentrate mixed with water and Portland Cement per manufacturer's specifications. Cast dry density 28-day min compressive strength of 190 psi as determined with ASTM C495-66.
        CELCORE INC —Cast dry density of 31 (+ or -) 3.0 pcf.
        CELLULAR CONCRETE L L C —Cast dry density of 37 (+ or -) 3.0 pcf.
        ELASTIZELL CORP OF AMERICA —Type II. Mix #1 of cast dry density 39 (+ or -) 3.0 pcf, Mix #2 of cast dry density 40 (+ or -) 3.0 pcf, Mix #3 of cast dry density 47 (+ or -) 3.0 pcf.
        LITE-CRETE INC —Cast dry density of 29 (+ or -) 3.0 pcf.
    C. **Cellular Concrete** — Roof Topping Mixture* — Concentrate mixed with water and Portland Cement per manufacturer's specifications. 28-day min compressive strength of 190 psi as determined with ASTM C495-66.
        SIPLAST INC —Mix No. 1 or 2. Cast dry density of 32 to 35 pcf for Mix No. 1 or 36 to 39 pcf for Mix No. 2.
    D. **Perlite Concrete** — 6 cu ft of Perlite Aggregate* to 94 lb of Portland Cement and 1-1/2 pt air entraining agent. Min thickness 2 in. as measured to the top surface of structural concrete or foamed plastic (Item 11A) when it is used. See Perlite Aggregate (CFFX) in Fire Resistance Directory for names of manufacturers.
    E. **Cellular Concrete** — Roof Topping Mixture* — Foam concentrate mixed with water, Portland Cement and UL Classified Vermiculite Aggregate per manufacturer's application instructions. Cast dry density of 30 to 36 pcf and 28-day compressive strength of min 250 psi as determined in accordance with ASTM C495-86.
        CELLULAR CONCRETE L L C — Mix No. 3.
        SIPLAST INC —Mix No. 3.
    F. **Floor Topping Mixture*** — (Optional, not shown) — Approx 4.5 gal of water to 41 lbs of NVS Premix floor topping mixture. Slurry coat 1/8 in. thickness beneath foamed plastic (Item 11) when used, 1 in. min topping thickness.
        SIPLAST INC
        Floor Topping Mixture may be covered with Built-Up or Single Membrane Roof Covering.

11. **Foamed Plastic*** — (Optional, not shown) — For use only with vermiculite (Item 10A) or cellular (Item 10C) concretes or Floor-Topping Mixture (Item 10F) -rigid polystyrene foamed plastic insulation having slots and/or holes sandwiched between vermiculite concrete slurry which is applied to the normal or lightweight concrete surface and vermiculite concrete topping (Item 10A). Max thickness to be 8 in.
    SIPLAST INC
    VERMICULITE PRODUCTS INC

11A. **Foamed Plastic*** — (Optional) — Nom 24 by 48 in., 48 by 48 in. or 30 by 60 in. by max 8 in. thick polystyrene foamed plastic insulation boards with holes symmetrically placed having a max density of 2.0 pcf. For use only with cellular concrete roof topping mixture (Item 10B).
    STARRFOAM MFG INC

11B. **Foamed Plastic*** — (Optional, not shown) — For use only with cellular concrete. Nom 24 by 48 by max 8 in. thick polystyrene foamed plastic insulation boards having a density of 1.0 + 0.1 pcf encapsulated within cellular concrete topping (Item 10B). Each insulation board shall contain six nom 3 in. diam holes oriented in two rows of three holes each with the holes spaced 12 in. OC, transversely and 16 in. OC longitudinally. See Foamed Plastic* (BRYX) category in the Building Materials Directory or Foamed Plastic* (CCVW) category in the Fire Resistance Directory for list of manufacturers.

12. **Foamed Plastic*** — (Optional, not shown). Polyisocyanurate roof insulation applied over concrete floor with no restriction on insulation thickness. When polyisocyanurate insulation is used, the unrestrained beam rating shall be increased by a minimum of 1/2 hr. See Foamed Plastic (CCVW), category in the Fire Resistance Directory for list of manufacturers.

*Bearing the UL Classification Mark

FIGURE 6.33 (Part 2) UL floor/ceiling design D923. (With permission, Underwriters Laboratories, Inc., *Fire Resistance Directory*, 2002, Northbrook, IL.)

in Sec. 6.5.1.2 for beams, beam substitutions are usually necessary and may be made on the basis of a simple equation to be given later. The UL *Fire Resistance Directory* had limited the application of this adjustment equation for beam insulation thickness to unrestrained ratings, but now has approved its use for both restrained and unrestrained ratings. The previous section discussion on the differences and meaning of restrained and unrestrained classifications, as defined in ASTM E 119, again applies.

6.5.1.5 Trusses

Because of the aforementioned size constraints in fire test laboratories, there are no, or very limited, direct fire test data on full-scale truss assemblies. UL does not have any published fire ratings specifically for large trusses. Nevertheless, in order to enable a rational fire resistance assessment of trusses, acceptable methods have been developed to overcome this limitation by applying other existing information from ASTM E 119 tests.

There are three fundamental approaches to fire protect a steel truss:

1. Using a membrane fire-resistant ceiling system
2. Providing individual protection for each truss element, usually with spray-on material, considering each as a column member
3. Enclosing the entire truss assembly for its entire depth and span with fire-resistant materials

TABLE 6.9 Common Steel Floor and Assembly Designs

Floor-ceiling assemblies[a,b]

Assembly rating		Type of protection system	Concrete		Metal deck depth (in)	UL design number
Restrained (hr)	Unrestrained (hr)		Min. thickness above deck flutes (in)	Type		
1, 1½, 2, 3	1, 1½, 2, 3	Acoustical ceiling membrane	based upon required rating	NW or LW NW	1½, 2, 3	D216
2, 3	2, 3				1½	D218
1½, 2	1½, 2	Gypsum wallboard ceiling membrane	2½	NW	1½, 2, 3	D502
2	1½		2	LW	2, 4½, 6, 7½	D501
1, 1½, 2, 3	1, 1½, 2, 3	Spray-applied fire resistive material	2	NW or LW	2, 3	D743
			2½	NW or LW	9/16, 15/16, 1 15/16	D780
					1½, 2, 3	D759, D832
1, 1½, 2, 3, 4	1, 1½, 2, 3, 4		2½	NW or LW	1½, 2, 3	D739, D767, D779, D858
			3¼	LW	1, 1½, 2, 3	D782
2	1, 1½		2½	LW	1, 1½, 2, 3	D752
2, 3, 4	1, 1½, 2, 3		2½	NW	1½, 1 5/18	D744
3, 4	1½, 2		3¼	LW	1½, 2, 3	D754
1, 1½, 2, 3	1, 1½, 2, 3	Spray-applied fire resistive material w/unprotected deck	based upon required rating	NW or LW	1½, 1 5/8, 2, 3	D902
					1½, 2, 3	D916, D925

[a] The referenced assemblies are some commonly used Underwriters Laboratories (UL) assemblies used for conventional steel framed structures. For additional assemblies the reader should reference the UL Fire Resistance Directory.

[b] For additional design requirements such as beam spacing, concrete strength, density, reinforcing and clear cover, minimum metal deck gauge, maximum deck span, shear connector requirements, design stress limitations, etc., see the specific referenced assembly in the Underwriters Laboratories (UL) directory.

TABLE 6.9 (Continued)
Roof-ceiling assemblies[a,b,c]

Assembly rating		Type of protection system	Roof insulation type	Metal deck depth[d] (in)	UL design number
Restrained (hr)	Unrestrained (hr)				
1	3/4	Acoustical ceiling membrane	Rigid	1 1/2	P254
			Rigid	1 1/2	P214
1	1		Insulating fill	9/16, 5/16, 1 15/16	P246, P255
				9/16	P261
1, 1 1/2	1, 1 1/2		Rigid	1, 1 1/2	P250
				1 1/2	P230
				1, 1 1/2, 2, 3	P225
			Insulating fill	15/16, 1 5/16, 1 1/2	P231
1, 1 1/2, 2	1, 1 1/2, 2		Insulating fill	9/16, 3/4, 1 1/4	P251
2	2		Rigid	1, 1 1/2	P237
1 1/2, 2	1 1/2, 2	Plaster w/metal lath membrane	Rigid	1 1/2	P404
1	1	Gypsum wallboard ceiling membrane	Insulating fill	1 5/16	P509
2	2		Rigid	1 1/2	P514
3/4, 1, 1 1/2, 2	3/4, 1, 1 1/2, 2	Spray-applied fire resistive material	Rigid	1 1/2, 3	P701
			Rigid	1 1/2	P711, P740, P741
				1 1/2, 3	P714, P717, P725, P739, P819
1, 1 1/2, 2	1, 1 1/2, 2		Insulating fill	9/16, 15/16, 1 5/16, 1 1/2	P921
			Insulating fill	9/16, 15/16, 1 5/16, 1 1/2	P927
1, 1 1/2, 2, 3	1, 1 1/2, 2		Rigid	1 1/2, 3	P719
				1 1/2	P723, P733
1, 1 1/2, 2, 3	1, 1 1/2, 2, 3		Rigid	1 1/2, 3	P732
2	1 1/2		Rigid	1 1/2	P718

[a] The referenced assemblies are some commonly used Underwriters Laboratories (UL) assemblies used for conventional steel framed structures. For additional assemblies the reader should reference the UL Fire Resistance Directory.
[b] For additional design requirements such as beam spacing, concrete strength, density, reinforcing and clear cover, minimum metal deck gauge, maximum deck span, shear connector requirements, design stress limitations, etc., see the specific referenced assembly in the Underwriters Laboratories (UL) directory.
[c] For roof designs that incorporate structural concrete slabs, D-series assemblies can be used provided that the roof insulation type, density and the appropriate D-series assembly modifications are in accordance with the UL directory.
[d] Metal deck depth for some assemblies is shown as a minimum and deeper decks may be substituted. Refer to the specific UL assembly for additional information.

Membrane protection is accomplished by specifying a fire-rated ceiling assembly contained in published listings. The envelope enclosure method of protection is illustrated in Fig. 6.34 for a staggered truss. Layers of rated gypsum wallboard (Type X) fully enclose the truss, with the board thickness determined by the required fire rating. Table 6.10 gives guidance based on past test data for such envelope protection. The individual truss element protection is a conservative approach that treats each member as a column, using the published column listings.

The nature of the truss fire protection to be used will be influenced or code mandated by the type of truss and its structural function. A transfer truss is a critical structural member that carries loads from multiple floor levels above and/or below the truss. A staggered truss system is primarily used in residential occupancy buildings to provide column-free interior spaces. These are story-high

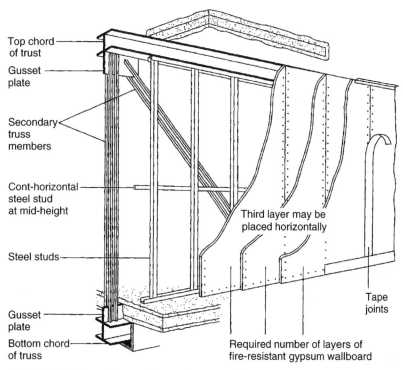

FIGURE 6.34 Staggered truss envelope enclosure protection.

trusses that span the full building width at alternating column lines on each floor. Thus, this type of truss supports loads from two floors at both its top and bottom chords. Interstitial trusses are used to create deep floor/ceiling concealed spaces, often for placement of mechanical or electrical equipment in health-care facilities. The interstitial trusses support the floor above and the loads from the equipment in the concealed space, and may be considered as being analogous to deep, open-web, joist floor systems for application of membrane protection.

Once the truss structural function is defined, the appropriate fire protection system (ceiling membrane, individual element, or enclosure) can be appropriately selected. The fire protection method

TABLE 6.10 Guidelines for Gypsum Wallboard Protection for Envelope Protection of Steel Trusses

Fire endurance	Gypsum ×	Wallboard type
60	5/8″ (16 mm)	5/8″ (16 mm)
120	1 1/4″ (32 mm)	—
180	—	1 1/2″ (35 mm)

Source: From Ref. 8, Table 4-9.6. With permission, Society of Fire Protection Engineers [SFPE], *Handbook of Fire Protection Engineering,* 3rd edition, 2002.

TABLE 6.11 Typical Fire Protection Methods for Steel Trusses

Truss type	Fire protection method		
	Membrane	Envelope	Individual element
Transfer		X	X
Staggered		X	X
Individual	X	X	X

Source: From Ref. 7, Table 4-9.7, p. 4-186. With permission, Society of Fire Protection Engineers [SFPE], *Handbook of Fire Protection Engineering,* 3rd edition, 2002.

typically used for each truss type is given in Table 6.11. The ceiling membrane protection is really only applicable to interstitial trusses, whereas the envelope and individual member protection methods typically can be used for any steel truss.

6.5.2 Concrete Construction

ASTM E 119 is applicable to all building materials, and hence includes fire-tested assemblies of concrete floor and wall systems. Similar to their steel counterparts, these concrete assemblies are listed in UL's *Fire Resistance Directory,* and may be easily used to comply with the prescriptive building code requirements for fire and life safety.

6.5.2.1 Floor-Ceiling Assemblies

The J and K series of UL Designs cover the listed concrete floor-ceiling systems with membrane protection, spray-on materials, or unprotected. The tested assembly details and fire endurance times are explicitly given, as for the other materials.

6.5.2.2 Walls

Concrete walls and partitions are included in the U, V, or W series of UL Designs.

6.5.3 Penetrations and Joints

In order to complement the E 119 fire ratings for isolated members or assemblies, which were usually tested without any penetrations, the ASTM E 814 (ANSI/UL 1479), "Standard Test Method for Fire Tests of Through-Penetration Firestops," was first published in 1981 [24]. This ASTM standard addresses the fire safety implications of the many architectural, electrical, and/or mechanical system openings that commonly become necessary in actual construction through fire barriers (walls and floors) with a required fire-resistive rating, and determines corresponding penetration firestop ratings. The F rating measures the time during which flame passage through the firestop system is prevented, at which time it must be accompanied by a successful hose stream test. The T rating requires the temperature rise on the unexposed surface of the wall or floor, on the penetrating item, and on the penetration fill material flame to not exceed 325°F above ambient, as well as all the F criteria. A third criterion, L ratings, determines the amount of air leakage. The IBC 2000 Code, Sections 711.3.1.2 and 711.4.1.2 [9], cover the required firestopping of through-penetrations in vertical and horizontal assemblies by reference to ASTM E 814 and its F and T ratings to preserve the original fire rating times of the underlying members and assemblies.

In a similar manner and purpose to through-penetrations of fire-rated structural systems, ASTM E 1966 (ANSI/UL 2079) "Standard for Tests for Fire Resistance of Building Joint Systems" was subsequently first issued in 1994 to address construction joint systems, such as floor-to-floor, wall-to-wall, floor-to-wall, and head of wall [25, 26]. The important presumption and prerequisite of this test standard is that both of the construction elements meeting at the joint are fire rated, and do not contain any unprotected openings, such as windows. Fire-resistive joints are commonly required by the codes to have fire resistance ratings of no less than the fire rating of the adjacent wall and/or floor assemblies.

There is new standards development work in progress at ASTM to address the fire resistance ratings for perimeter joints, which occur along the outside of building floor systems at their junction with the curtain walls. However, most curtain wall products are not rated for fire resistance because of their use as exterior, non-load-bearing building facades. Because of this fact, and the presence of window openings, the previously referenced ASTM and UL standards are not directly applicable to these products. Consequently, to address this particular need to evaluate the fire rating of a perimeter joint (gap) at the intersection of a nonrated curtain wall containing windows with a rated floor system, UL developed in the late 1990s a separate fire rating classification for perimeter fire containment systems. UL defines this type of system as the following:

> A perimeter fire containment system is a specific construction consisting of a floor with an hourly fire endurance rating, an exterior curtain wall with no hourly fire endurance rating, and the fill material installed between the floor and curtain wall to prevent the vertical spread of fire in a building.

This type of fire test is a combination (or as UL states, an "assimilation") of ANSI/UL 2079 and NFPA 285 "Standard Method of Test for the Evaluation of Flammability Characteristics of Exterior Non-Load-Bearing Wall Assemblies Containing Combustible Contents Using the Intermediate-Scale, Multistory Test Apparatus" [27]. The UL Perimeter Fire Containment System listings show two kinds of fire ratings: an integrity rating, which is analogous to the F rating of ASTM E 814 (ANSI/UL 1479), and an insulation rating similar to the temperature-based T ratings discussed previously. A maximum linear opening width is given for each rated system assembly, along with an indication on the cyclic movement capabilities of the joint, if any, in accordance with its intended service functions. These anticipated movements are dependent on the type of curtain wall and its attachment details to the primary framing, the expected deflections of the building due to wind, and other applied loads.

The ASTM draft document on perimeter joints that is still under committee ballot is tentatively titled "Standard Test Method for Determining the Fire-Endurance of Perimeter Joint Protection Using the Intermediate-Scale, Multi-Story Test Apparatus" [28]. Its original basis is in NFPA 285, but with particular adaptation to evaluate perimeter joints, using F and T ratings similar to those in ASTM E 814, and with provisions for joint cyclic movement. The *intermediate-scale, multistory test apparatus* (ISMA) customized for this usage is similar to that in NFPA 285.

6.6 DETERMINATION OF FIRE RESISTANCE BY CALCULATION

Use of the previously described and temperature-dependent thermophysical material properties, shape geometry, and fundamental heat transfer and structural principles, in combination with previous fire test data, has provided adequate justification for several relatively easy calculation methods of fire resistance. The simpler computational methods, such as those in ASCE/SFPE-29, are usually at least semiempirically based on standard fire test results, and they enable an efficient and generally conservative way to provide fire resistance ratings for members and assemblies that do not directly match historical tests. In this manner, the prescriptive code requirements can be conveniently and safely met. Higher order computer analyses and natural fire simulations can also be used as performance-based design alternatives to achieve a more accurate and/or innovative solution to overall fire safety.

In the following sections, various computational approaches to the determination of the fire resistance of steel and concrete construction are summarized.

6.6.1 Steel Construction

Design for structural safety and for the possible strength limit states, in general, requires that the structural resistance be no less than the applied load effects. During a fire duration time, this strength limit can be symbolically expressed in Eq. (6.9) as:

$$R_{\text{fire}} \geq L_{\text{fire}} \qquad (6.9)$$

where L_{fire} represents all the load effects (structural forces and bending moments resulting from these applied demands) expected to be simultaneously acting during the fire event, and R_{fire} is the available structural resistance under the particular high temperature conditions, including the degraded material properties. The applied load effects should be developed from those load combinations that are probabilistically expected to be present during the relatively rare fire exposure, and is likely to be less than the full-design live load that is normally specified. ASCE 7-99 recommends using 50 percent of the normal design live load, in combination with dead, in such a fire analysis and design, considering that fire is a rare, extreme event. Other international codes and standards use comparable load reductions in combination with fire exposures. It should be recognized that the uncertainties of the real fire ignition, intensity, duration, spread, and heating distribution effects to the affected structural members (L_{fire}) are qualitatively large relative to the variability of the structural fire resistance (R_{fire}).

The fire and other concurrent load effects will change with time during a fire, as will the resistance due to its degradation in material properties with increasing temperatures. Thus, a fire analysis is time dependent, similar to the time-history response of a structure to an earthquake. A beam may change from a pure bending member at ambient temperatures to one with combined bending and axial compression, and, finally, during the large deflection and high temperature stages, it may experience combined bending and axial tension (catenary action). The adjacent member connections must also accordingly accommodate these transmitted forces, moments, and distortions, with the latter implying adequate ductility to avoid failures.

The total strain is composed primarily of a mechanical and a thermal part. If the member is fully restrained so that no total strain occurs, all the thermal expansion effects are converted into an equal and opposite mechanical strain. These mechanical strains induce internal reaction forces and moments in the structure, which lead to either ultimate strength or strain becoming the governing limit state. Similar to seismic design, the ultimate strain, or deformation, may become the failure limit for the more brittle construction materials that do not have sufficient ductility, comparable to earthquake design considerations. For the opposite extreme, in the case of a simply connected but effectively unrestrained member and with the assumption of relatively light applied loads, all the thermal expansion will be free to occur as total strain; there will be little mechanical strain, and its accompanying internal forces and moments, present. In this latter case, even though the thermal distortions will be large, because they are simply due to unrestrained thermal elongations, the structure is less likely to experience a catastrophic failure. Unfortunately, the reality in most buildings lies in between these extremes and requires a more in-depth analysis for an accurate determination.

The relatively simple fire resistance calculations for individual structural members presented in this section do not directly invoke Eq. (6.9), but are developed from best-fit regression equations of the E 119 fire test data. These models and equations also do not directly include the thermal elongation effects of the fire on the member and its surrounding structure. In addition, the critical high temperature limits cited by ASTM E 119 may not, by themselves, signify a real structural limit state of the member, or assembly, in the absence of additional evaluation of the real fire exposure, of the concurrent building loading effects and continuity of the exposed area(s) with the surrounding structural floor and/or lateral framing system. For such cases, Eq. (6.9) provides the essential underlying criterion to assess strength adequacy under fire conditions, both for structural members and entire framing systems. More sophisticated analyses for members and frames need to rely on this basic limit state more explicitly.

When loaded specimens are permitted and used, the ASTM E 119 standard conservatively assumes that the full design load is present during the standard fire. All of its derived ratings and calculations have that same presumption, as well as all the other stated limitations of the E 119 standard and test ranges. The following member-based fire resistance calculations are likewise limited to the single element exposure within a compartment fire domain, and do not include any interactions, structural or thermal, with any adjacent members, elements, or framing.

In contrast to the material properties and their variations with temperatures that were predominantly given in SI/metric units, most of these design equations are presented in U.S. customary units due to their U.S. application origins.

The fundamental parameter in both the simpler and some of the more advanced calculations of fire resistance in steel members is the W/D ratio, where W is the weight per unit length of steel shape and D is the heated perimeter of the inside of the fire protection material (or, equivalently, the heated perimeter of the outside of the steel).

This W/D ratio characterizes the thermal mass resistance of the member under fire, with high ratios indicating better fire-resistance capability, and vice versa. Members with larger W/D values will experience a slower temperature rise under equivalent heat exposures and other conditions than ones with a lower W/D. The explanation for this well-known and verified trend is that a heavier steel shape will provide a greater heat sink than a lighter member. In addition, a smaller value of its heated perimeter means that there is less surface area available for heat transfer, which is a more favorable situation to limiting temperature rise in the steel.

The heated perimeter will depend on the nature of the fire exposure (flames on all sides or one-side unexposed), protection profile (contour or box) that is used, and the geometry of the steel shape itself. Another similar variable representing the member mass and fire exposure dimensions that is commonly used for steel hollow structural section (HSS or tubular) and pipe products is A/P, where A is the cross-sectional area of shape and P is the heated perimeter (= D).

Internationally, this steel member property ratio is often expressed for all steel shapes in terms of P/A, or "section factor = F/V." The W/D and A/P (or its reciprocal P/A, or F/V) are all equivalent, with the inclusion of the steel density constant and conversions of the appropriate units.

The W/D and/or A/P properties have been compiled for many of the standard steel shapes and are widely used in evaluating both column and beam substitutions for fire ratings. Fig. 6.35 shows the heated perimeter determination for steel columns, while Fig. 6.36 shows this for steel wide-

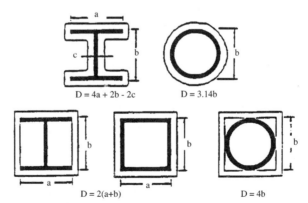

FIGURE 6.35 Heated perimeter for steel columns. (With permission, BOCA Guidelines for Determining Fire Resistance Ratings of Building Elements, Country Club Hills, IL.)

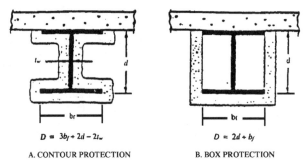

FIGURE 6.36 Heated perimeter calculations for wide-flange beams and guides. (With permission, BOCA Guidelines for Determining Fire Resistance Ratings of Building Elements, Country Club Hills, IL.)

flange beams and girders. A major difference between the beam and column heated perimeter, D, is whether the entire outside perimeter is used, as for columns, or the three-sided, fire-exposed perimeter for beams.

The W/D or A/P ratios have been tabulated for the common steel shapes, and are given in Tables 6.12 to 6.14 for a group of steel wide-flange columns, HSS (tubes), and wide-flange beams, respectively. The U.S. customary dimensional units for these are pounds/(inch·foot) for W/D, and inches for A/P.

There may be some slight differences in these numerical values between references depending upon if and how the minor effects of the shape fillet radii are included. Other similar or more current W/D tabulations may be found in the available design manuals and handbooks, such as those from AISC, AISI, Gypsum Association, etc. For nonstandard, built-up, or new shapes, the straightforward dimensional calculations, as shown in Figs. 6.35 and 6.36, can be independently made to determine the appropriate member D, and W/D.

Several numerical methods from the literature are readily available for calculating the temperature rise and, consequently, the fire endurance rating in individual steel members. The equations, graphs, or charts are derived from simplified applications of heat transfer theory. Two relatively easy and popular formulations are presented here to determine the temperature time-history for unprotected and protected steel members.

As a practical consideration, usually a minimum spray-on material, or other protection, thickness of $3/8$ to $1/2$ in. is provided once it has been determined that the steel is required to be protected. Likewise, the increments of steel fire protection thickness for spray-on (SFRM) and similar materials are commonly no less than $1/8$ in., and the tested or calculated requirements are conservatively rounded-up to the nearest $1/8$-in. thickness. Smaller minimum, or total, steel fire protection material thickness is not feasible due to the relatively coarse tolerances of the protection material and of its application to the steel. However, intumescent or mastic coatings are applied to a much smaller final thickness and finer increments.

Note that on a relative percentage basis, $1/8$ in. or so, any deviations in SFRM thickness on the unconservative, or less than required side, are more significant for members and assemblies that specify a lower thickness (say $1/2$ in.), than for those that need over 1 in.

6.6.1.1 Unprotected Steel

One step-by-step calculation method adaptable to spreadsheet programming is based on the heat energy equilibrium over a small time step between the exposed surface area and the temperature rise in the steel, under the assumption that the steel is a lumped mass at uniform temperature. Convection and radiation are the assumed heat transfer mechanisms. This quasi-steady state and

TABLE 6.12 W/D for Wide-Flange Columns

Weight-to-heated-perimeter ratios (W/D) for typical structural steel wide-flange columns

Structural shape	Contour profile (W/D)	Box profile (W/D)	Structural shape	Contour profile (W/D)	Box profile (W/D)
W14 × 730	6.62	9.05	W12 × 87	1.20	1.76
W14 × 665	6.14	8.46	W12 × 79	1.10	1.61
W14 × 605	5.69	7.89	W12 × 72	1.00	1.48
W14 × 550	5.26	7.35	W12 × 65	0.91	1.35
W14 × 500	4.86	6.83	W12 × 58	0.91	1.31
W14 × 455	4.49	6.35	W12 × 53	0.84	1.20
W14 × 426	4.24	6.02	W12 × 50	0.89	1.23
W14 × 398	4.00	5.71	W12 × 45	0.81	1.12
W14 × 370	3.76	5.38	W12 × 40	0.72	1.00
W14 × 342	3.51	5.04			
W14 × 311	3.23	4.66	W10 × 112	1.78	2.57
W14 × 283	2.97	4.31	W10 × 100	1.61	2.33
W14 × 257	2.72	3.97	W10 × 88	1.43	2.08
W14 × 233	2.49	3.65	W10 × 77	1.26	1.85
W14 × 211	2.28	3.35	W10 × 68	1.13	1.66
W14 × 193	2.10	3.09	W10 × 60	1.00	1.48
W14 × 176	1.93	2.85	W10 × 54	0.91	1.34
W14 × 159	1.75	2.60	W10 × 49	0.83	1.23
W14 × 145	1.61	2.39	W10 × 45	0.87	1.24
W14 × 132	1.52	2.25	W10 × 39	0.76	1.09
W14 × 120	1.39	2.06	W10 × 33	0.65	0.93
W14 × 109	1.27	1.88			
W14 × 99	1.16	1.72	W8 × 67	1.34	1.94
W14 × 90	1.06	1.58	W8 × 58	1.18	1.71
W14 × 82	1.20	1.68	W8 × 48	0.99	1.44
W14 × 74	1.09	1.53	W8 × 40	0.83	1.23
W14 × 68	1.01	1.41	W8 × 35	0.73	1.08
W14 × 61	0.91	1.28	W8 × 31	0.65	0.97
W14 × 53	0.89	1.21	W8 × 28	0.67	0.96
W14 × 48	0.81	1.10	W8 × 24	0.58	0.83
W14 × 43	0.73	0.99	W8 × 21	0.57	0.77
			W8 × 18	0.49	0.67
W12 × 336	4.02	5.56			
W12 × 305	3.70	5.16	W6 × 25	0.69	1.00
W12 × 279	3.44	4.81	W6 × 20	0.56	0.82
W12 × 252	3.15	4.43	W6 × 16	0.57	0.78
W12 × 230	2.91	4.12	W6 × 15	0.42	0.63
W12 × 210	2.68	3.82	W6 × 12	0.43	0.60
W12 × 190	2.46	3.51	W6 × 9	0.33	0.46
W12 × 170	2.22	3.20			
W12 × 152	2.01	2.90	W5 × 19	0.64	0.93
W12 × 136	1.82	2.63	W5 × 16	0.54	0.80
W12 × 120	1.62	2.36			
W12 × 106	1.44	2.11	W4 × 13	0.54	0.79
W12 × 96	1.32	1.93			

With permission, BOCA Guidelines for Determining Fire Resistance Ratings of Building Elements, Country Club Hills, IL.

TABLE 6.13 A/P for Structural Tubing (HSS)

Area-to-heated-perimeter ratios (A/P) for typical round, square, and rectangular structural tubing

Round pipe columns standard steel pipe			Extra-strong steel pipe columns			Double extra-strong steel pipe columns		
Nominal diameter (inches)	Thickness (inches)	A/P ratio	Nominal diameter (inches)	Thickness (inches)	A/P ratio	Nominal diameter (inches)	Thickness (inches)	A/P ratio
12	0.375	0.36	12	0.500	0.48	8	0.875	0.79
10	0.365	0.35	10	0.500	0.48	8	0.864	0.75
8	0.322	0.31	8	0.500	0.47	5	0.750	0.65
6	0.280	0.27	6	0.432	0.40	4	0.674	0.57
5	0.258	0.25	5	0.375	0.35	3	0.600	0.50
4	0.237	0.22	4	0.337	0.31			
3.5	0.226	0.21	3.5	0.318	0.29			
3	0.216	0.20	3	0.300	0.27			

Area-to-heated-perimeter ratios (A/P) for typical round, square, and rectangular structural tubing

Square structural tubing					
Nominal size each side (inches)	Thickness (inches)	A/P ratio	Nominal size each side (inches)	Thickness (inches)	A/P ratio
16	⁵⁄₈	0.58	7	⁹⁄₁₆	0.49
16	½	0.48	7	½	0.44
14	⁵⁄₈	0.58	7	³⁄₈	0.34
14	½	0.47	7	⁵⁄₁₆	0.29
14	³⁄₈	0.36	7	¼	0.24
12	⁵⁄₈	0.57	6	⁹⁄₁₆	0.48
12	½	0.47	6	½	0.43
12	³⁄₈	0.36	6	³⁄₈	0.34
10	⁵⁄₈	0.56	6	⁵⁄₁₆	0.29
10	⁹⁄₁₆	0.51	6	¼	0.23
10	½	0.46	6	³⁄₁₆	0.18
10	³⁄₈	0.35	5	½	0.42
10	⁵⁄₁₆	0.30	5	³⁄₈	0.33
9	⁵⁄₈	0.55	5	⁵⁄₁₆	0.28
9	⁹⁄₁₆	0.51	5	¼	0.23
9	½	0.46	5	³⁄₁₆	0.18
9	³⁄₈	0.35	4	½	0.40
9	⁵⁄₁₆	0.29	4	³⁄₈	0.32
8	⁵⁄₈	0.54	4	⁵⁄₁₆	0.27
8	⁹⁄₁₆	0.50	4	¼	0.22
8	½	0.45	4	³⁄₁₆	0.17
8	³⁄₈	0.35	3	⁵⁄₁₆	0.26
8	⁵⁄₁₆	0.29	3	¼	0.22
8	¼	0.24	3	³⁄₁₆	0.17

Rectangular structural tubing					
Nominal size (inches)	Thickness (inches)	A/P ratio	Nominal size (inches)	Thickness (inches)	A/P ratio
16 × 12	⁵⁄₈	0.58	14 × 10	³⁄₈	0.36
16 × 12	½	0.47	12 × 8	⁵⁄₈	0.56
16 × 8	½	0.47	12 × 8	⁹⁄₁₆	0.51
14 × 10	⁵⁄₈	0.57	12 × 8	½	0.46
14 × 10	½	0.47			

TABLE 6.13 (*Continued*)
Area-to-heated-perimeter ratios (A/P) for typical round, square, and rectangular structural tubing

Rectangular structural tubing					
Nominal size (inches)	Thickness (inches)	A/P ratio	Nominal size (inches)	Thickness (inches)	A/P ratio
12 × 8	3/8	0.35	8 × 6	1/4	0.24
12 × 6	5/8	0.55	8 × 4	9/16	0.48
12 × 6	9/16	0.51	8 × 4	1/2	0.43
12 × 6	1/2	0.46	8 × 4	3/8	0.34
12 × 6	3/8	0.35	8 × 4	5/16	0.29
10 × 8	5/8	0.55	8 × 4	1/4	0.23
10 × 8	9/16	0.51	7 × 5	1/2	0.43
10 × 8	1/2	0.46	7 × 5	3/8	0.34
10 × 8	3/8	0.35	7 × 5	5/16	0.29
10 × 8	5/16	0.29	7 × 5	1/4	0.23
10 × 8	1/4	0.24	6 × 5	1/2	0.43
10 × 6	5/8	0.54	6 × 5	3/8	0.33
10 × 6	9/16	0.50	6 × 5	5/16	0.28
10 × 6	1/2	0.45	6 × 5	1/4	0.23
10 × 6	3/8	0.35	6 × 5	3/16	0.18
10 × 6	5/16	0.29	6 × 4	1/2	0.42
10 × 5	5/8	0.54	6 × 4	3/8	0.33
10 × 5	9/16	0.49	6 × 4	5/16	0.28
10 × 5	1/2	0.45	6 × 4	1/4	0.23
10 × 5	3/8	0.34	6 × 4	3/16	0.18
10 × 5	5/16	0.29	6 × 3	3/8	0.32
9 × 7	5/8	0.50	6 × 3	5/16	0.28
9 × 7	9/16	0.45	6 × 3	1/4	0.23
9 × 7	1/2	0.35	6 × 3	3/16	0.17
9 × 7	3/8	0.29	5 × 3	1/2	0.40
9 × 7	5/16	0.24	5 × 3	3/8	0.32
9 × 6	5/8	0.54	5 × 3	5/16	0.27
9 × 6	9/16	0.49	5 × 3	1/4	0.22
9 × 6	1/2	0.45	5 × 3	3/16	0.17
9 × 6	3/8	0.34	4 × 3	5/16	0.27
9 × 6	5/16	0.29	4 × 3	1/4	0.22
9 × 5	9/16	0.49	4 × 3	3/16	0.17
9 × 5	1/2	0.44	4 × 2	5/16	0.26
9 × 5	3/8	0.34	4 × 2	1/4	0.22
9 × 5	5/16	0.29	4 × 2	3/16	0.17
8 × 6	9/16	0.49	3.5 × 2.5	1/4	0.22
8 × 6	1/2	0.44	3.5 × 2.5	3/16	0.17
8 × 6	3/8	0.34	3 × 2	1/4	0.21
8 × 6	5/16	0.29	3 × 2	3/16	0.16

With permission, BOCA Guidelines for Determining Fire Resistance Ratings of Building Elements, Country Club Hills, IL.

one-dimensional heat flow solution for an unprotected steel member results in the following change of steel temperature equation [Eq. (6.10)]:

$$\Delta T_s = \frac{F}{V}\frac{1}{\rho_s c_s}\{h_c(T_f - T_s) + \sigma\varepsilon(T_f^4 - T_s^4)\}\Delta t \tag{6.10}$$

where ΔT_s = change in steel temperature in the time step, °C or K
F = surface area of unit length of the member, m²
V = volume of steel in unit length of the member, m³
ρ_s = density of steel, kg/m³

TABLE 6.14 W/D for Wide-Flange Beams

Weight-to-heated-perimeter ratios (W/D) for typical wide-flange beam and girder shapes

Structural shape	Contour profile	Box profile	Structural shape	Contour profile	Box profile
W36 × 300	2.47	3.33	W24 × 162	1.85	2.57
× 280	2.31	3.12	× 146	1.68	2.34
× 260	2.16	2.92	× 131	1.52	2.12
× 245	2.04	2.76	× 117	1.36	1.91
× 230	1.92	2.61	× 104	1.22	1.71
× 210	1.94	2.45	× 94	1.26	1.63
× 194	1.80	2.28	× 84	1.13	1.47
× 182	1.69	2.15	× 76	1.03	1.34
× 170	1.59	2.01	× 68	0.92	1.21
× 160	1.50	1.90	× 62	0.92	1.14
× 150	1.41	1.79	× 55	0.82	1.02
× 135	1.28	1.63	W21 × 147	1.83	2.60
W33 × 241	2.11	2.86	× 132	1.66	2.35
× 221	1.94	2.64	× 122	1.54	2.19
× 201	1.78	2.42	× 111	1.41	2.01
× 152	1.51	1.94	× 101	1.29	1.84
× 141	1.41	1.80	× 93	1.38	1.80
× 130	1.31	1.67	× 83	1.24	1.62
× 118	1.19	1.53	× 73	1.10	1.44
			× 68	1.03	1.35
W30 × 211	2.00	2.74	× 62	0.94	1.23
× 191	1.82	2.50	× 57	0.93	1.17
× 173	1.66	2.28	× 50	0.83	1.04
× 132	1.45	1.85	× 44	0.73	0.92
× 124	1.37	1.75			
× 116	1.28	1.65	W18 × 119	1.69	2.42
× 108	1.20	1.54	× 106	1.52	2.18
× 99	1.10	1.42	× 97	1.39	2.01
			× 86	1.24	1.80
W27 × 178	1.85	2.55	× 76	1.11	1.60
× 161	1.68	2.33	× 71	1.21	1.59
× 146	1.53	2.12	× 65	1.11	1.47
× 114	1.36	1.76	× 60	1.03	1.36
× 102	1.23	1.59	× 55	0.95	1.26
× 94	1.13	1.47	× 50	0.87	1.15
× 84	1.02	1.33	× 46	0.86	1.09
			× 40	0.75	0.96
			× 35	0.66	0.85

TABLE 6.14 (*Continued*)
Weight-to-heated-perimeter ratios (W/D) for typical wide-flange beam and girder shapes

Structural shape	Contour profile	Box profile	Structural shape	Contour profile	Box profile
W16 × 100	1.56	2.25	W10 × 112	2.14	3.38
× 89	1.40	2.03	× 100	1.93	3.07
× 77	1.22	1.78	× 88	1.72	2.75
× 67	1.07	1.56	× 77	1.52	2.45
× 57	1.07	1.43	× 68	1.35	2.20
× 50	0.94	1.26	× 60	1.20	1.97
× 45	0.85	1.15	× 54	1.09	1.79
× 40	0.76	1.03	× 49	0.99	1.64
× 36	0.69	0.93	× 45	1.03	1.59
× 31	0.65	0.83	× 39	0.90	1.40
× 26	0.55	0.70	× 33	0.77	1.20
			× 30	0.79	1.12
W14 × 132	1.83	3.00	× 26	0.69	0.98
× 120	1.67	2.75	× 22	0.59	0.84
× 109	1.53	2.52	× 19	0.59	0.78
× 99	1.39	2.31	× 17	0.54	0.70
× 90	1.27	2.11	× 15	0.48	0.63
× 82	1.41	2.12	× 12	0.38	0.51
× 74	1.28	1.93			
× 68	1.19	1.78	W8 × 67	1.61	2.55
× 61	1.07	1.61	× 58	1.41	2.26
× 53	1.03	1.48	× 48	1.18	1.91
× 48	0.94	1.35	× 40	1.00	1.63
× 43	0.85	1.22	× 35	0.88	1.44
× 38	0.79	1.09	× 31	0.79	1.29
× 34	0.71	0.98	× 28	0.80	1.24
× 30	0.63	0.87	× 24	0.69	1.07
× 26	0.61	0.79	× 21	0.66	0.96
× 22	0.52	0.68	× 18	0.57	0.84
			× 15	0.54	0.74
W12 × 87	1.44	2.34	× 13	0.47	0.65
× 79	1.32	2.14	× 10	0.37	0.51
× 72	1.20	1.97			
× 65	1.09	1.79	W6 × 25	0.82	1.33
× 58	1.08	1.69	× 20	0.67	1.09
× 53	0.99	1.55	× 16	0.66	0.96
× 50	1.04	1.54	× 15	0.51	0.83
× 45	0.95	1.40	× 12	0.51	0.75
× 40	0.85	1.25	× 9	0.39	0.57
× 35	0.79	1.11			
× 30	0.69	0.96	W5 × 19	0.76	1.24
× 26	0.60	0.84	× 16	0.65	1.07
× 22	0.61	0.77			
× 19	0.53	0.67	W4 × 13	0.65	1.05
× 16	0.45	0.57			
× 14	0.40	0.50			

With permission, BOCA Guidelines for Determining Fire Resistance Ratings of Building Elements, Country Club Hills, IL.

c_s = specific heat of steel, J/(kg·K)
h_c = convective heat transfer coefficient, W/(m²·K)
σ = Stefan-Boltzmann constant, 56.7×10^{-12} kW/(m²·K⁴)
ε = resultant emissivity
T_f = temperature in the fire environment, K
T_s = temperature of the steel, K
Δt = small time step, s

Usually, a time step on the order of 30 seconds or less will be adequate, and a time-temperature response curve can be obtained from the summation of the temperature changes during the time steps. Care must be taken not to use a larger time step, because this method will then become mathematically divergent, or inaccurate. The fire temperature for use in this equation at any given time can be calculated from the standard E 119 curve or from any other parametric or natural fire. A mathematical representation of the fire time-temperature will be convenient herein, as given previously for the E 119 standard fire. Equally important to the accuracy of this solution is to obtain realistic thermal material properties for the required input variables. Because most of the insulating materials are proprietary, these published values are often estimated, and there are inherent uncertainties in their temperature variations. However, the fire effects variability and real-temperature exposures are admittedly the much greater unknowns.

6.6.1.2 Protected Steel

A similar one-dimensional heat balance equation and iterative solution may be derived for the case of a steel member subjected to a fire that is protected by a contour insulation material along the profile of the steel shape. The same idealizations of steady state fire conditions, lumped mass and uniform steel temperature apply, so these greatly simplified heat transfer analyses do not include the effects of any in-plane, or out-of-plane, spatial temperature variations in the steel. In addition, it is assumed that the external surface of the insulation is at the same temperature as the fire gases, and that the insulation interior is at the same temperature as the steel. Equation 6.11 gives this general equation for one-dimensional heat transfer by conduction to protected steel:

$$\Delta T_s = (F/V)(k_i/d_i\rho_s c_s)\{\rho_s c_s/(\rho_s c_s + (F/V)d_i\rho_i c_i/2)\}(T_f - T_s)\Delta t \qquad (6.11)$$

where ΔT_s = change in steel temperature in the time step, °C or K
F = surface area of unit length of the member, m²
V = volume of steel in unit length of the member, m³
k_i = thermal conductivity of the insulation, W/(m·K)
d_i = thickness of the insulation, m
ρ_s = density of steel, kg/m³
ρ_i = density of the insulation, kg/m³
c_s = specific heat of steel, J/kg·K
c_i = specific heat of the insulation, J/kg·K
h_c = convective heat transfer coefficient, W/(m²·K)
σ = Stefan-Boltzmann constant, 56.7×10^{-12} kW/(m²·K⁴)
ε = resultant emissivity
T_f = temperature in the fire environment, K
T_s = temperature of the steel, K
Δt = small time step, s

If the steel insulation is of low mass and specific heat, then the more general Eq. (6.11) for temperature change can be simplified by eliminating the entire bracketed factor, thereby ignoring the heat capacity of the insulation. A criterion for this simplification has been established, but is herein omitted because Eq. (6.11) is easy enough to check whether, and how, this possible simplification affects the results. Again, a small enough time step is necessary, and 30 s or less remains a reasonable increment. If moisture is present in the insulation material that will delay the steel temperature increase, this effect may be included by appropriate modification of the thermal insulation property input.

These two equations have generally been shown to produce good results and agreement with experimental data. However, the complete member (all sides) fire exposure and lumped mass simplification for both the unprotected and protected cases is more suitable for interior steel columns than beams, because the latter normally supports a concrete floor that acts as both a heat shield and sink at the top of the beam. Thus, a beam with a floor will have a thermal gradient ranging from the highest beam temperature on the bottom exposed surface to the lowest at its interface with the floor, and not the constant steel temperature assumed in these heat balance equations. Likewise, an exterior, or perimeter, column would typically have its one side shielded from the interior fire or open to the outside, thereby violating the assumption of uniform heat exposure of the member.

Further adjustments (such as "effective" thermal property input from other calibrations) of these equations or more refined heat transfer analyses, employing the finite element or finite difference methods, may be warranted for such members (floor beams and perimeter/external columns) to improve the accuracy of the results. Likewise, the steady-state assumption will be violated for fires that have high rates of temperature changes, and further reductions in the time step interval will not correct the solution. Otherwise, the uniform exposure and lumped mass simplification could serve as a conservative approximation for the member temperature predictions or at least a reasonable estimate for the temperature of the exposed surface of the member that is closest to the fire.

If structural loading effects are to be included in a straightforward manner, these should be accounted for in establishing the correct temperature limit for the steel member under consideration through use of its strength reduction versus temperature relationship. A lightly loaded steel member is able to sustain a higher temperature rise before failure than one that is heavily loaded because the former can safely experience a much larger loss of strength than the latter. Thus, a load or a demand-to-capacity ratio, which is the applied load divided by the member's ultimate strength at ambient temperatures, is often used to modify the critical temperatures at which failure due to heating effects will occur. The lower is this load ratio, or the higher is its reciprocal, the member reserve strength, the higher will be the critical temperature endpoint.

In the following sections, additional simple member design criteria for fire resistance of steel columns, beams, and trusses are outlined. These provide easy calculation methods for determining steel members' fire protection requirements for various fire endurance times, and vice versa, primarily for the purpose of rationally substituting steel member sizes and extending the applications of the published fire test results. However, since these expressions were all empirically derived from past E 119 and comparable tests, the equations have the same inherent limitations as those that were described previously for the E 119 fire test itself, i.e., standard fire time-temperature curve exposure, maximum applied floor loading, specimen size and boundary conditions, etc.

More advanced analytical methods that extend beyond the single member and the other simplifying idealizations of this section are starting to emerge as part of the new performance-based design alternatives. The EuroCode and codes of some other countries have expanded coverage of computational fire design criteria in their building codes, including consideration of natural fires, applied loadings, and actual framing system boundary conditions. Computer software is available that can perform higher-level analyses of real fires, the structural resistance, and interaction with this fire in three dimensions. These model simulations can produce time-histories of temperature distributions, their structural effects, and predictions of any structural vulnerability. All these calculations and analyses are very dependent on the accuracy of the thermomechanical properties of the steel and its insulation material that are used as input. Such advanced techniques are beyond the scope of this section, and will only be briefly outlined later.

6.6.1.3 Columns

Table 6.15 provides the simple equations in U.S. customary units to estimate the fire endurance times of various steel columns and their fire protection. These expressions were curve-fit from E 119 fire test data, and include unprotected members; those with gypsum wallboard, spray-on (wide flange and hollow sections), concrete-covered and encased columns, and concrete-filled hollow steel columns, both reinforced and unreinforced. Limitations exist on their application relative to a range of member shapes (W/D), protection thickness, rating times, and other physical bounds of the underlying tests upon which the correlations were established. These equations are used to supplement the existing database of E 119 standard column fire test results and allow for efficient steel member substitutions and protection thickness adjustments.

The UL *Fire Resistance Directory* allows calculation of the increase of the required insulation thickness for wide-flange columns that are smaller than the members that are listed as the minimum size. Unless otherwise provided for in the UL design itself, Eq. (6.12) provides a linear scaling expression that relates the SFRM thickness to W/D, with an apparent factor of safety of 1.25 included, which is otherwise similar to the comparable formula in Table 6.15:

$$x_2 = 1.25(x_1)\left(\frac{W_1}{D_1}\right)\left(\frac{D_2}{W_2}\right) \tag{6.12}$$

where x_2 = thickness of coating for smaller wide-flange section
x_1 = thickness of coating used on rated steel section
W_2 = weight per foot of smaller wide-flange section
W_1 = weight per foot of rated steel section
D_2 = perimeter of smaller steel section at interface with coating
D_1 = perimeter of rated steel section at interface with coating

W/D and A/P are the main common variables in the Table 6.15 column equations that relate the member fire endurance time to the column protection thickness and properties. The C_1 and C_2 insulation material constants must be established for each different SFRM. Representative values of these two constants for the cementitious material identified in UL Designs X701, X704, X722, and X723 are 69 and 31, respectively. For the mineral fiber SFRM, based on UL Designs X801, X807, X818, X821, and X822, $C_1 = 63$ and $C_2 = 42$. The UL *Fire Resistance Directory,* which is annually updated, contains much more information and properties for these column protection calculations in specific designs. All this design listing information has now also been made readily available for search and downloading through the UL Web site, as well as the Web sites of many of the commercial product suppliers.

The details of the Type X gypsum wallboard attachment to the steel column are critical to preserve their fire protection function. Figs. 6.37 and 6.38 illustrate such recommended wallboard attachment, both with sheet steel column covers and with steel stud/screw attachment, respectively.

The ASCE/SFPE 29–99 standard covers these and the other simpler calculation methods and also provides their explicit limitations and equivalent formulations in SI units. The only subtle difference between ASCE/SFPE 29–99 and Table 6.15 is that the fire endurance times R are expressed in minutes in the latter, and in hours in the former. Before application of any of these equations, the user should independently confirm their meaning and range of applicability to the particular construction from the original standard or source reference.

The concrete-protected steel columns and associated variables are illustrated in Fig. 6.39. If all the reentrant spaces of the wide-flange steel column are filled with concrete, this is considered to be an encased condition, and the H parameter includes the additional term shown in Table 6.15. This formulation assumes noncomposite column action, wherein the concrete encasement serves only as a thermal barrier to the steel, and is not relied upon for structural load-bearing capabilities. The typical equilibrium moisture content of concrete by volume (m) is 4 percent for NWC and 5 percent for LWC.

A concrete-filled steel HSS column of square or round shape is shown in Fig. 6.40. It may be filled only with plain concrete (unreinforced), or additionally with steel fiber or bar reinforcement. Table 6.16 provides the fire resistance formulas for these based on Canadian research [29]. The key computational difference among these cases may be found in the *a* parameter, which varies for the different reinforcement cases, shape geometry, and type of concrete aggregate (LWC or NWC). As

TABLE 6.15 Equations for Estimating the Fire Endurance of Structural Steel Columns

Member/protection	Solution	Symbols
Column/unprotected	$R = 10.3 \, (W/D)^{0.7}$, for $W/D < 10$ $R = 8.3 \, (W/D)^{0.8}$, for $W/D \leq 10$ (for critical temperature of 1000°F)	R = fire endurance time (min) W = weight of steel section per linear foot (lb/ft) D = heated perimeter (in.)
Column/gypsum wallboard	$R = 130 \left(\dfrac{hW'/D}{2} \right)^{0.75}$ where $W' = W + \left(\dfrac{50hD}{144} \right)$	h = thickness of protection (in.) W' = weight of steel section and gypsum wallboard (lb/ft)
Column/spray-applied materials and board products—wide flange shapes	$R = [C_1(W/D) + C_2]h$	C_1 & C_2 = constants for specific protection material
Column/Spray-applied materials and board products—hollow sections	$R = C_1 \left(\dfrac{A}{P} \right) h + C_2$	C_1 & C_2 = constants for specific protection material The A/P ratio of a circular pipe is determined by $A/P \text{ pipe} = \dfrac{t(d - t)}{d}$ where d = outer diameter of the pipe (in.) t = wall thickness of the pipe (in.) The A/P ratio of a rectangular or square tube is determined by $A/P \text{ tube} = \dfrac{t(a + b - 2t)}{a + b}$ where a = outer width of the tube (in.) b = outer length of the tube (in.) t = wall thickness of the tube (in.)
Column/concrete cover	$R = R_0(1 + 0.03m)$ where $R_0 = 10(W/D)^{0.7} + 17 \left(\dfrac{h^{1.6}}{k_c^{0.2}} \right)$ $\cdot \left\{ 1 + 26 \left[\dfrac{H}{\rho_c c_c h(L + h)} \right]^{0.8} \right\}$ $D = 2(b_f + d)$	R_0 = fire endurance at zero moisture content of concrete (min.) m = equilibrium moisture content of concrete (% by volume) b_f = width of flange (in.) d = depth of section (in.) k_c = thermal conductivity of concrete at ambient temperature (Btu/hr·ft·°F)
Column/Concrete encased	for concrete-encased columns use $H = 0.11W + \dfrac{\rho_c c_c}{144}(b_f d - A_s)$ $D = 2(b_f + d)$ $L = (b_f + d)/2$	H = thermal capacity of steel section at ambient temperature (= 0.11 W Btu/ft·°F) c_c = specific heat of concrete at ambient temperature (Btu/lb·°F) L = inside dimension of one side of square concrete box protection (in.) A_s = cross-sectional area of steel column (in.²)

Source: From Ref. 7. With permission, Society of Fire Protection Engineers [SFPE], *Handbook of Fire Protection Engineering*, 3rd edition, 2002.

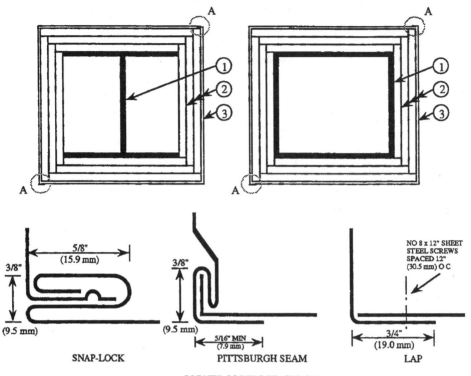

CORNER JOINT DETAILS (A)

Notes:
1. Structural steel column, either wide-flange or tubular shapes
2. Type X gypsum wallboard. For single-layer applications, the wallboard shall be applied vertically with no horizontal joints. For multiple-layer applications, horizontal joints shall be permitted at a minimum spacing of 8 feet (2.4 m), provided that the joints in successive layers are staggered at least 12 inches (304.8 mm). The total required thickness of wallboard shall be determined on the basis of the specified fire-resistance rating and the weight and heated perimeter of the column. Alternatively, for fire resistance ratings of two hours or less, one of the required layers of gypsum wallboard shall be applied to the exterior of the sheet steel column covers with 1 inch (25.4 mm) long Type S screws spaced 1 inch (25.4 mm) from the wallboard edge and 8 inches on center. For such installations, 0.016 inch (0.4 mm) minimum thickness galvanized steel corner beads with 1-1/2 inch (203.2 mm) legs shall be attached to the wallboard with Type S screws spaced 12 inches (304.8 mm) on center.
3. For fire resistance ratings of three hours or less, the column covers shall be fabricated from 0.024 inch (0.6 mm) minimum thickness galvanized or stainless steel. For four hour fire resistance ratings, the column covers shall be fabricated from 0.024 inch (0.6 mm) minimum thickness stainless steel. The column covers shall be erected with the snap lock or Pittsburgh seam details. Alternatively, for fire resistance ratings of two hours or less, column covers fabricated from 0.027 inch (0.7 mm) minimum thickness galvanized or stainless steel shall be erected with lap joints located anywhere around the perimeter of the column cover. The lap joint shall be secured with 1/2 inch (12.7 mm) long No. 8 sheet metal screws spaced 12 inches (304.8 mm) on center. The column covers shall be provided with a minimum expansion clearance of 1/8 inch per linear foot (10.4 mm/m) between the ends of the cover and any restraining construction.

FIGURE 6.37 Gypsum wallboard protected structural steel columns with sheet steel column covers (4 h or less). (With permission, ASCE/SFPE 29–99, Standard Calculation Methods for Structural Fire Protection, Structural Engineering Institute of the American Society of Civil Engineers, Reston, VA, 1999.)

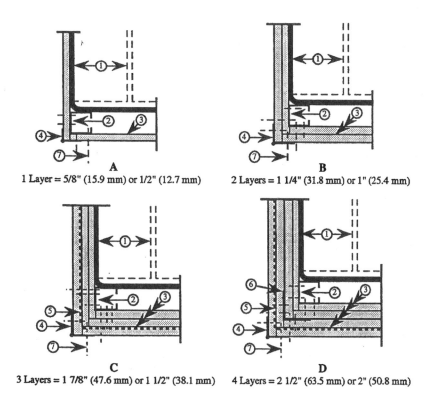

A 1 Layer = 5/8" (15.9 mm) or 1/2" (12.7 mm)

B 2 Layers = 1 1/4" (31.8 mm) or 1" (25.4 mm)

C 3 Layers = 1 7/8" (47.6 mm) or 1 1/2" (38.1 mm)

D 4 Layers = 2 1/2" (63.5 mm) or 2" (50.8 mm)

Notes:
1. Structural steel column, either wide-flange, pipe, or tubular shapes.
2. One and five-eighths inch (15.9 mm) deep studs fabricated from 0.021 inch (0.5 mm) minimum thickness galvanized steel with 1-5/16 (33.3 mm) or 1-7/16 (36.5 mm) inch legs and 1/4 inch (6.4 mm) stiffening flanges. The length of the steels studs shall be 1/2 inch (12.7 mm) less than the height of the assembly.
3. Type X Gypsum wallboard. For single-layer applications, the wallboard shall be applied vertically with no horizontal joints. For multiple-layer applications, horizontal joints shall be permitted at a minimum spacing of 8 feet (2.4 m), provided that the joints in successive layers are staggered at least 12 inches (304.8 mm). The total required thickness of wallboard shall be determined on the basis of the specified fire resistance rating and the weight and heated perimeter of the column.
4. Galvanized steel corner beads (0.016 inch (0.4 mm) minimum thickness) with 1-1/2 inch (38.1 mm) legs attached to the wallboard with 1 inch (25.4 mm) long Type S screws spaced 12 inches (304.8 mm) on center.
5. No. 18 SWG steel tie wire spaced 24 inches (610 mm) on center.
6. Sheet metal angles with 2 inch (50.8 mm) legs fabricated from 0.021 inch (0.5 mm) minimum thickness galvanized steel.
7. Type S screws 1 inch (25.4 mm) long shall be used for attaching the first layer of wallboard to the steel studs and the third layer to the sheet metal angles at 24 inches (609.6 mm) on center. Type S screws 1-3/4 inches (44.5 mm) long shall be used for attaching the second layer of wallboard to the steel studs and the fourth layer to the sheet metal angles at 12 inches (304.8 mm) on center. Type S screws 2-1/4 inches (57.2 mm) long shall be used for attaching the third layer of wallboard to the steels studs at 12 inches (304.8 mm) on center.

FIGURE 6.38 Gypsum wallboard protected structural steel columns with steel stud/screw attachment system (3 h or less). (With permission, ASCE/SFPE 29–99, Standard Calculation Methods for Structural Fire Protection, Structural Engineering Institute of the American Society of Civil Engineers, Reston, VA, 1999.)

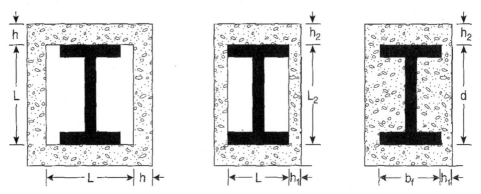

FIGURE 6.39 Concrete-encased wide-flange steel columns.

would be expected, the reinforced concrete-filled HSS provide longer fire endurance times and load-carrying capacity than the unreinforced counterparts for either the square or circular column shapes.

As previously mentioned, the fire resistive criteria for steel columns may be applied to other steel compression members, such as individually protected truss elements and bracing.

6.6.1.4 Beams

Similar to columns, W/D is the key section property that governs the fire endurance time of the beam member. A heavier beam, or one with a greater W/D ratio, may be directly substituted for the lighter members shown in fire-rated designs. However, doing so without compensating for the more favor-

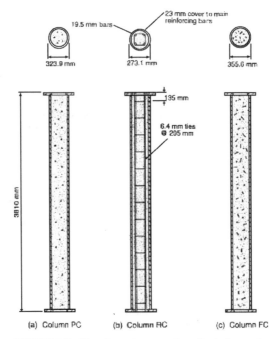

FIGURE 6.40 Elevation and cross section of typical concrete-filled HSS columns used in fire test.

TABLE 6.16 Canadian Fire Resistance Formulas for Concrete-Filled Steel HSS Columns [29]

$$R = 0.58a \frac{(f_c + 2.90)}{(KL - 3.28)} D^2 (D/C)^{0.5}$$

In SI units:

$$R = a \frac{(f_c + 20)}{60(KL - 1000)} D^2 (D/C)^{0.5}$$

Where:
- R = fire resistance rating in hours,
- a = 0.07 for circular columns filled with siliceous aggregate concrete,
- = 0.08 for circular columns filled with carbonate aggregate concrete,
- = 0.06 for square or rectangular columns filled with siliceous aggregate concrete,
- = 0.07 for square or rectangular columns filled with carbonate aggregate concrete,
- f'_c = specified 28-day compressive strength of concrete in kips per square inch (megapascals),
- KL = column effective length in feet (millimeters),
- D = outside diameter for circular columns in inches (millimeters),
- = outside dimension for square columns in inches (millimeters),
- = least outside dimension for rectangular columns in inches (millimeters), and
- C = compressive force due to unfactored dead load and live load in kips (kilonewtons).

Values of constant "a"

Aggregate type*	Filling type	Steel reinforcement	Circular columns	Square columns
S	PC	N/A	0.070	0.060
S	FC	≈2%	0.075	0.065
S	RC	1.5%–3%	0.080	0.070
		3%–5%	0.085	0.070
N	PC	N/A	0.080	0.070
N	FC	≈2%	0.085	0.075
N	RC	1.5%–3%	0.090	0.080
		3%–5%	0.095	0.085

* Type S concrete is made with siliceous coarse aggregate; Type N concrete is made with carbonate coarse aggregate.
PC: plain concrete
RC: steel-bar-reinforced concrete
FC: steel-fiber-reinforced concrete

able thermal resistance characteristics of the higher W/D beam is inefficient. Thus, an empirical relationship was developed from fire test results to determine the tradeoffs in SFRM protection requirements for beams as a function of their W/D.

This simple linear equation effectively proportions the spray-on material thickness relative to that in a listed UL design, as in Eq. (6.13):

$$h_1 = \left(\frac{W_2/D_2 + 0.6}{W_1/D_1 + 0.6} \right) h_2 \tag{6.13}$$

where h = thickness of spray-applied fire protection, in
W = weight of steel beam, lb/ft
D = heated perimeter of steel beam

and where the subscripts 1 = substitute beam and required thickness protection, and 2 = beam and protection thickness specified in the referenced tested design or tested assembly

Limitations of this equation are noted as follows:

1. $W/D \geq 0.37$.
2. $h \geq 3/8$ in. (9.5 mm).
3. The unrestrained beam rating in the referenced tested design or tested assembly is at least 1 h.

In order to realize cost savings, this equation may be used to determine the permissible reduction in beam spray-on thickness for a larger W/D beam relative to the shape listed in a given UL design. Both UL and ASCE/SFPE 29–99 recognize Eq. (6.13), and UL has very recently approved the use of this equation for both unrestrained and restrained beams. Previously, UL had limited its application to unrestrained beam cases only.

6.6.1.5 Trusses

As discussed under steel truss fire testing, three generic methods of fire protecting them are available, depending on the truss type: fire-rated ceilings, individual element protection, and truss enclosures.

The individual truss member protection criteria, member substitutions, and associated calculations would be identical to those discussed for columns if the truss members will be simultaneously exposed to fire on all sides. If the truss chord member supports a floor or roof, thereby partially reducing its heated perimeter D, it is rational to include this modification in determining the chord's effective W/D as a fire-resistant column. For truss ceiling and enclosure protection, strict compliance with the applicable fire-rated assembly design is necessary, and there are few, if any, easy computational adjustments that can be made.

6.6.1.6 Connections

Because there are no standard fire tests or well-documented fire performance data for connections, there are also no related simple calculations. It has been implicitly assumed by most in practice that the underlying intent is to fire protect connections to at least the same level, and in the same manner, as its adjacent structural member with the highest fire rating.

More sophisticated tests and analyses of high-temperature effects on welded and bolted steel connections and those of other materials and types are warranted in order to develop this needed knowledge base.

6.6.2 Concrete Construction

As with other construction materials, the various types of concrete structures can be analytically assessed for fire resistance through utilization of heat transfer and structural response models. These analyses can be simplified, or can be more elaborate, as needed and warranted for the problem under consideration. The thermal analysis provides the temperature distributions in the concrete members, which subsequently are used to determine the temperature-dependent mechanical properties and their effects on the computed structural resistance of the member or subassembly during a fire event. Both simple and continuous floor beams and slabs, columns, and walls of reinforced concrete may be evaluated, as necessary, using these rational methods. A higher level of complexity is introduced when a framing subassemblage, or the entire building frame, is to be analyzed for its response under fire. General or special-purpose computer software can be used for this purpose, often in a performance-based design setting for a project.

Often, just simplified design tables and charts based on past fire tests and analyses are used to quickly determine fire endurance times for members and assemblies of reinforced concrete construction, such as those given in Chap. 2 of ASCE-29. These fire ratings essentially depend on the overall size and type of member, type of concrete, and the cover distance provided from the external concrete surface to the outermost layer of steel reinforcing. Similar fire protection design aids are also available in Chap. 9 of the PCI design handbook for precast and prestressed concrete [30].

6.7 APPLICATION OF FIRE RESISTANCE RATINGS

The various building codes that are enforced throughout the United States and other countries throughout the world use fire resistance as one part of the fire protection requirements for buildings. Its specific application is dependent on many factors, especially the philosophical approach used by the code officials. In some countries, other forms of fire protection such as sprinklers are required and can replace or supplant fire-resistive construction. The degree to which this occurs varies widely throughout the world. In the United States, fire resistance continues to be required in many instances, but not all.

The construction of the structural systems and their protection must, however, be seriously considered when designing a new building or retrofitting an existing facility. As stated earlier, several instances of fire-induced structural collapse have been identified, and that, coupled with the requirements for protecting occupants in a building either as they evacuate or as they wait for assistance, requires the use of fire-resistive construction for the structural elements as a minimum.

Thus, a thorough understanding of materials used in this application, as well as their performance in a fire exposure, must be attained so as to maximize their use while providing life and property protection from fire.

REFERENCES

1. Hall, J.R., Jr., "High-Rise Building Fires," NFPA Report, Quincy, MA, September 2001.
2. FEMA 403, "World Trade Center Building Performance Study: Data Collection, Preliminary Observations, and Recommendations," FEMA, Washington, DC, May 2002.
3. ASCE-SEI (2003), *The Pentagon Building Performance Report,* American Society of Civil Engineers–Structural Engineering Institute, Reston, VA, January, 2003.
4. Iwankiw, N. and Beitel, J. (2002), "Analysis of Needs and Existing Capabilities for Full-Scale Fire Resistance Testing," Hughes Associates, Report NIST GCR 02-843, December, 2002.
5. Babrauskas, V., and Williamson, R.B., "The Historical Basis of Fire Resistance Testing—Part II," *Fire Technology,* 304–316, Nov. 1978.
6. ASTM E 119, "Standard Fire Test Method for Building Construction and Materials," ASTM, West Conshohocken, PA.
7. ASTM E 1529, "Standard Test Methods for Determining Effects of Large Hydrocarbon Pool Fires on Structural Members and Assemblies," ASTM, West Conshohocken, PA.
8. Society of Fire Protection Engineers (SFPE), *SFPE Handbook of Fire Protection Engineering,* 3rd edition, SFPE, Bethesda, MD, 2002.
9. International Code Council (ICC), *International Building Code 2000.*
10. NFPA 5000, *Building Construction and Safety Code,* 2003 edition, NFPA, Quincy, MA, 2003.
11. Gewain, R.G., and Troup, E.W.J., "Restrained Fire Resistance Ratings in Structural Steel Buildings," *AISC Engineering Journal,* AISC, Chicago, IL, Second Quarter, 2001.
12. Newman, G.M., Robinson, J.F., and Bailey, C.G., *Fire Safe Design: A New Approach to Multi-Storey Steel-Framed Buildings,* Steel Construction Institute, Berkshire, UK, 2000.
13. Buchanan, A.H., *Structural Design for Fire Safety,* John Wiley & Sons, Ltd., 2001.
14. EC3 (1995), *Eurocode 3: Design of Steel Structures, ENV 1992-1-2, General Rules, Structural Fire Design,* European Committee for Standardization, Brussels, Belgium.

15. *AISC Manual of Steel Construction, Load and Resistance Factor Design,* 3rd edition, AISC, Chicago, IL, 2001.
16. ASCE/SFPE 29–99, "Standard Calculation Methods for Structural Fire Protection," Structural Engineering Institute of the American Society of Civil Engineers, Reston, VA, 1998.
17. Knudson, R.M., and Scneiwind, A.P., "Performance of Wood Structural Members Exposed to Fire," *Forest Products Journal,* 25(2): 23–32 1975.
18. Konig, J. and Walleij, L., "One-dimensional Charring of Timber Exposed to Standard and Parametric Fires in Initially Unprotected Situations," Report No. 19908029, Tratek, Swedish Institute for Wood Technology Research, Stockholm, Sweden, 1999.
19. Lie, T.T., *Fire and Buildings,* Applied Science Publishers Ltd., London, UK, 1972.
20. Underwriters Laboratories, Inc., *Fire Resistance Directory,* Volumes 1 and 2, Northbrook, IL, 2002.
21. Gypsum Association, *Fire Resistance Design Manual,* Gypsum Association, Washington, DC.
22. ASCE Manual and Reports on Engineering Practice No. 78 (1992), Lie, T.T., ed., *Structural Fire Protection,* American Society of Civil Engineers, New York, NY, 1992.
23. European Convention for Constructional Steelwork (ECCS), "Model Code on Fire Engineering," 1st edition, No. 111, Technical Committee 3, Brussels, Belgium, May 2001.
24. ASTM E 814, "Standard Test Method for Fire Tests of Through-Penetration Fire Stops," ASTM, West Conshohocken, PA.
25. ASTM E 1966, "Standard for Tests for Fire-Resistance of Building Joint Systems," ASTM, West Conshohocken, PA.
26. Underwriters Laboratories, Inc., UL 2079, "Standard for Test for Fire Resistance of Building Joint Systems," Northbrook, IL.
27. NFPA 285, "Standard Method of Test for the Evaluation of Flammability Characteristics of Exterior Non-Load-Bearing Wall Assemblies Containing Combustible Components Using the Intermediate-Scale, Multi-Story Test Apparatus," NFPA, Quincy, MA.
28. ASTM Draft, "Standard Test Method for Determining the Fire-Endurance of Perimeter Joint Protection Using the Intermediate-Scale, Multi-Story Test Apparatus," ASTM, West Conshohocken, PA.
29. Kodur, V.K.R., and MacKinnon, D.H. "Design of Concrete-Filled Hollow Structural Steel Columns for Fire Endurance," *AISC Engineering Journal,* First Quarter, 2000, pp. 13–24, AISC, Chicago, IL, 2000.
30. PCI, *PCI Design Handbook, Precast and Prestressed Concrete,* 5th edition, Precast/Prestressed Concrete Institute, 1999.

CHAPTER 7
WOOD AND WOOD PRODUCTS

Marc Janssens, Ph.D.
Director, Department of Fire Technology
Southwest Reasearch Institute
San Antonio, TX 78238-5166

Bradford Douglas, P.E.
Director, Engineering
American Forest and Paper Association
Washington, DC 20036

7.1 UNITS

Fire science and engineering are relatively new disciplines, and the S.I. system of units is used throughout the world. However, the U.S. customary system is still the system of choice for structural engineering calculations and design in the United States. Because Secs. 7.3, 7.6, and 7.8 deal with structural fire protection, U.S. customary units are used in these sections. S.I. units are used in the remaining sections of the chapter.

7.2 INTRODUCTION

7.2.1 Wood

Wood is one of the oldest and most widely used materials in the world, with many advantages as a building material. Wood is natural, obtained from renewable resources, clean, easy to work with, and strong and lightweight. Most wood is easy to fasten with nails, glue, and other connectors, and easy to paint or treat with preservatives. Production of wood products requires very little process energy compared with other materials. The disposal of untreated wood provides no serious environmental difficulties. Recycling is possible.

In addition to structural components, wood and wood-based materials are widely used for nonstructural applications such as flooring, paneling, doors, cladding, and furniture in all sorts of buildings. The fire design of these products is also addressed in this chapter.

7.2.2 Forestry

Forestry is a global industry. Much of the world's economy depends on the use of wood as a raw material, including the building industry. Total annual production of logs is over 3 billion cubic meters, with a global average of approximately 0.7 cubic meters per person per year.

Wood is a renewable resource. Commercial forests in the U.S. are managed to provide sustained yield production on a long-term basis. A comprehensive report entitled U.S. Forests Facts & Figures reported that in 1996 U.S. timberlands achieved a net annual growth of about 0.7 billion cubic meters compared with an annual harvest of approximately 0.5 billion cubic meters [1].

Virtually every country with significant forest resources has policies that directly affect forest practices. The United States is highly regulated through various environmental statutes and state authorities. It is the only country with a far-reaching endangered species act that has had large-scale impacts on forest management.

Certification and other programs designed to ensure sustainable forest practices have emerged as a new factor in global fiber supply. Programs designed to address concerns over forest sustainability take several forms. The largest sustainable forestry program in the United States is the *Sustainable Forestry Initiative®* (SFI) of the American Forest & Paper Association [2]. Adopted in 1994, SFI is based on the premise that responsible environmental policy and sound business practices can be integrated to the benefit of companies, their shareholders, customers, and the people they serve. The SFI program is a comprehensive system of principles, objectives, and performance measures that integrates the perpetual growing and harvesting of trees with the protection of wildlife, plants, soil, and water quality.

7.2.3 Wood and Carbon

The carbon in wood is an important part of the global carbon cycle. Carbon is absorbed during the process of photosynthesis, which uses energy from solar radiation to convert carbon dioxide from the atmosphere into cellulose and other chemicals in wood. No wood lasts forever. The carbon in wood is eventually released to the atmosphere when wood burns or decays, or is eaten by bugs. In a natural forest the cycle may take 40 to 100 years from capture of carbon to eventual release depending on the age of trees and many other factors. Using wood in construction products can increase the storage life of the carbon for another 50 years or more [2A].

Fossil fuels also contain carbon, obtained in the same way from the atmosphere using solar energy, many millions of years ago. The level of carbon dioxide in the atmosphere has been increasing steadily in recent years because of two human activities: burning of fossil fuel and destruction of forests. The only long-term method of stabilizing atmospheric carbon dioxide levels is to reduce the use of fossil fuels and move to a sustainable solar energy economy, which can include the use of wood both as fuel and as a raw material for many uses.

7.2.4 Wood and Fire

Everyone knows that wood burns. Wood is a valuable fuel, with more than half of the annual global wood harvest used as fuel. Wood stores solar energy captured by the leaves of trees during the process of photosynthesis. Every kilogram of dry wood contains approximately 20 MJ of energy, which is released when wood burns.

This chapter summarizes many studies on the ignition and combustion of wood and wood-based products, considering heat release rate, combustion products, and the performance of fire-retardant-treated wood. The chapter also describes the structural performance of timber members exposed to fire.

7.3 WOOD AS A CONSTRUCTION MATERIAL (U.S. CUSTOMARY UNITS)

A large number of wood-based construction products can be manufactured from wood. This section describes the most common of these products.

7.3.1 Sawn Timber

Cylindrical logs are converted to rectangular-shaped sawn lumber in sawmills. Sawmilling usually produces rough surfaces, so sawn timber is often planed smooth to make a more attractive and dimensionally accurate product. Structural sawn lumber consists of four broad groups.

7.3.1.1 Dimension Lumber

Dimension lumber refers to members of rectangular cross section from 2 to 4 in. (nominal) in thickness and 2 in. or more (nominal) in width. Visually graded dimension lumber is primarily intended for conventional and engineered applications and is further separated into four categories:

Structural light framing	2 to 4 in. (nominal) thick, 2 to 4 in. (nominal) wide
Light framing	2 to 4 in. (nominal) thick, 2 to 4 in. (nominal) wide
Studs	2 to 4 in. (nominal) thick, 2 in. (nominal) or wider
Structural joists and planks	2 to 4 in. (nominal) thick, 5 in. (nominal) or wider

Mechanically graded dimension lumber, primarily intended for engineered applications, is divided into two categories:

Machine-stress-rated lumber (MSR)	2 in. (nominal) or less thick, 2 in. (nominal) or wider
Machine-evaluated lumber (MEL)	2 in. (nominal) or less thick, 2 in. (nominal) or wider

7.3.1.2 Beams and Stringers

Beams and stringers are products of rectangular cross section that are 5 in. (nominal) or more in thickness with width more than 2 in. greater than thickness. These members, such as $6 \times 10, 6 \times 12, 8 \times 12, 8 \times 16$, and 10×14, are intended primarily to resist bending loads applied to the narrow face.

7.3.1.3 Posts and Timbers

Posts and timbers are products of square or rectangular cross section that are 5 in. (nominal) or more in thickness, but with width not more than 2 in. greater than thickness. These members, such as $6 \times 6, 6 \times 8, 8 \times 10$, and 12×12, are intended primarily to resist longitudinal loads.

7.3.1.4 Decking

Decking refers to lumber from 2 to 4 in. (nominal) thick, intended for use as floor, roof, or wall sheathing. Decking is primarily intended for application in the flatwise direction with the wide face of the decking in contact with supporting members. The narrow face of decking may be flat, tongue-and-grooved, or spline-and-grooved for interconnection of the decking members.

7.3.2 Panel Products

A large number of wood-based panel products are manufactured for various applications. Panel products consist of at least three main groups.

7.3.2.1 Wood Structural Panels

Wood-based panel products are composed of thin wood layers cross-aligned to give strength and dimensional stability in both directions. Wood structural panels are bonded with waterproof adhesives. Classified under this designation are composite panels, *oriented strand board* (OSB), and plywood. For structural-use purposes, these panels are considered interchangeable.

Composite panels are wood structural panels composed of wood veneers and wood-based materials to optimize engineering properties for specific applications. These panels typically include five wood veneer layers located on the outermost surfaces and at the panel center. Reconstituted wood fiber is sandwiched between the veneer layers.

OSB panels are mat-formed wood structural panels composed of thin rectangular wood strands arranged in cross-aligned layers with surface layers normally arranged in the long panel direction. OSB can be recognized by the distinctly bonded chips that appear on the sheet faces.

Plywood panels are wood structural panels composed of thin wood veneers arranged in cross-aligned layers. There are usually three, five, or seven veneers (sometimes called plies) glued together.

7.3.2.2 Particle Board Panels

A number of wood-based panel products manufactured from particles of various sizes and shapes, glued and pressed together to make rigid boards, are known collectively as *particle boards*. Fibers are used to make high-density fiberboard (hardboard), *medium-density fiberboard* (MDF), and low-density fiberboard. The product commonly referred to as particleboard is made from particles up to ³/₈ in. long.

Fiberboard is a structural and decorative panel made of wood or cane cellulosic fibers produced by interfelting fibers followed by consolidation under heat and pressure. Fiberboard may be used in roof and wall applications to reduce noise, to enhance energy efficiency, and to provide corner bracing. Other materials may be added to the fiber to improve the fiberboard performance, such as increased strength and water resistance.

Hardboard is a sheet panel product that is most commonly used for interior prefinished paneling and exterior siding. Hardboard sheets are composed of interfelted lignocellulosic wood fibers that are consolidated under heat and pressure to a minimum density of 31 lb/ft^3. Other materials may be added to enhance physical properties such as stiffness, hardness, finish, abrasion resistance, moisture protection, strength, and durability. Common applications include siding, prefinished paneling, and cabinets.

Particleboard is a sheet panel product that is comprised of small wood particles usually bonded together with a formaldehyde resin. Typical uses include, but are not limited to, exterior construction and manufactured home decking.

7.3.3 Engineered Wood Products

7.3.3.1 Structural Glued Laminated Timber (Glulam)

Structural glued laminated timber, or glulam, is an engineered, stress-rated product, composed of specially selected and prepared wood laminations bonded together with adhesives. The grain of all laminations is approximately parallel longitudinally. The individual laminations do not exceed 2 inches in net thickness. Glued laminated timbers are used as large beams, columns, and arches as well as other curved shapes. The development of resorcinol and other synthetic resin glues with high moisture resistance has expanded the uses of glued laminated timber to bridges, marine construction, and other applications involving direct exposure to the weather.

7.3.3.2 Structural Composite Lumber (SCL)

Structural composite lumber is a group of products primarily used as structural members. Classified under this designation are *laminated veneer lumber* (LVL) and *parallel strand lumber* (PSL).

LVL is a composite of wood veneer sheet elements with the wood fiber primarily oriented along the length of the member. Veneers are under ¹/₄ in. thick.

PSL is a composite of wood strand elements with wood fibers primarily oriented along the length of the member. The smallest dimension of the strands is ¹/₄ in. or less and the lengths of the strands are at least 150 times the least dimension.

7.3.3.3 Prefabricated Wood I-Joists

A prefabricated wood I-joist is a structural member manufactured using sawn lumber or SCL flanges and wood structural panel webs bonded together with exterior exposure adhesives, forming an "I" cross-sectional shape.

7.3.3.4 Wood Trusses

A wood truss is a framework of wood elements joined together at the ends, creating a series of triangles that form a structural member. Timber trusses can be made in several styles. Long-span roof and bridge trusses can be made with large timber members connected with steel bolts, rods, or plates. Small-span trusses, manufactured from dimension lumber connected with steel plates, are very economical for residential construction.

7.3.4 Other Materials

7.3.4.1 Wood Logs

Harvested from trees, unprocessed logs may be used to make some simple structures such as bridges. Smaller diameter logs or poles are often used as building materials. They require a minimum of processing. Usually they have the bark removed. If used in close proximity to the ground, they are often preservative treated.

Timber pole is a round, tapered wood log with its larger (butt) end embedded in the ground. Timber poles are often used for the framework of industrial and agricultural buildings.

Timber pile is a round, tapered wood log with its small (tip) end embedded in the ground. Timber piles are generally used as part of foundation systems.

7.3.4.2 Gypsum Board

Gypsum board is a group of products widely used as lining material. Gypsum board products include wallboard, plasterboard, and sheathing. The thickness of board is usually between $3/8$ and $3/4$ in. The most common boards consist of a gypsum plaster core with paper facings on both sides. The paper provides most of the flexural and tensile strength of these boards. The density of the gypsum core can be reduced with air entrainment. Boards required to have good fire resistance are reinforced with glass fibers in the gypsum core and, sometimes, with additives such as vermiculite to reduce shrinkage under fire exposure.

Gypsum plaster is calcium sulfate dihydrate, with two water molecules for each calcium sulfate molecule, being 20.9 percent water by mass. When gypsum is heated in a fire, the hydration process is reversed in an endothermic decomposition reaction that occurs between 100 and 120°C, as gypsum is converted back to calcium sulfate hemihydrate (plaster of Paris) accompanied by a major loss of strength:

$$CaSO_4 \cdot 2H_2O \rightarrow CaSO_4 \cdot 1/2 H_2O + 1 1/2 H_2O \tag{7.1}$$

If calcium sulfate hemihydrate is heated to higher temperatures, complete dehydration occurs in a second reaction:

$$CaSO_4 \cdot 1/2 H_2O \rightarrow CaSO_4 + 1/2 H_2O \tag{7.2}$$

In addition to the water of crystallization described above, at room temperatures there is free moisture in voids between the gypsum crystals.

When gypsum is heated in a fire, both the free water and the water of crystallization will be driven off, absorbing energy, which contributes to the excellent fire performance of gypsum board.

7.3.4.3 Insulating Materials

Many different products are used for insulation in timber buildings, especially in the cavities of light wood-frame construction. The most common insulating materials are glass fiber, rock fiber, cellulosic fiber, and plastic foam.

7.3.4.4 Adhesives

Many of the products previously described (glulam, SCL, and panel products) are manufactured with adhesives. A variety of adhesives can be used. The most common and most durable adhesives are based on phenol resorcinol and melamine urea formaldehyde. Glued laminated timber with these types of adhesives has excellent behavior in fires, so that charring occurs at the same rate as for solid sawn timber of the same size. Casein adhesive has been shown to give slightly faster charring at the gluelines in fires. Other adhesives such as *polyvinyl acetate* (PVA) and epoxies do not perform as well at elevated temperatures.

7.4 PHYSICAL AND CHEMICAL CHARACTERISTICS (S.I. UNITS)

Some understanding of the physical structure and the chemical composition of wood facilitates the understanding of its fire properties, in particular its burning rate under various exposure conditions. Therefore, a brief description of the physical and chemical nature of wood is given below. It is partly based on one of the many excellent publications on this subject [3]. Finally, this section is concluded with a discussion on the thermal degradation of wood and its main components and flame-retardant treatments of wood.

7.4.1 Botanical Categories

Woods, and the trees that produce them, are subdivided into two botanical categories: softwoods and hardwoods. Softwood trees are characterized by needlelike leaves. They are commonly referred to as evergreens (most remain green the year around) or conifers (most bear scaly cones). Hardwood trees have broad leaves that change color and drop in the fall. The physical structure as well as the chemical composition is considerably different between softwoods and hardwoods. Consequently, the botanical category of a wood product affects its fire performance.

Worldwide, the volume of hardwoods is about twice that of softwoods. Softwood trees dominate the forests in the Northern Hemisphere. Hardwoods are found all over the globe. The densest hardwood forests are located in tropical regions. Softwoods are used predominantly for structural applications, whereas most of the hardwoods go into interior finish such as paneling, furniture, etc.

7.4.2 Physical Structure

The macroscopic structure of softwood and hardwood stems is very similar. The stems consist of a core of wood (xylem) covered by a protective layer of bark. The xylem consists of long fiberlike cells that are oriented preferentially in one direction (vertical) referred to as the grain. Water, minerals, and nutrients are taken up by the roots and are transported in the form of sap through the outer part of the xylem to the leaves. Photosynthesis in the leaves uses the water and CO_2 (from the atmosphere) to form various sugars. A solution of the sugars into water (also referred to as sap) is moved to various parts of the tree through the inner layer of the bark (phloem).

The growth process is limited to a thin layer between the xylem and the phloem, called the cambium. New cells are created on either side of the cambium through division of existing cells. Thus, both the phloem and the xylem are growing. The cambium forms a ring with a continuously increasing diameter. The growth rate is fastest in the spring, slows down in the summer, and (usually) stops in the fall. Thus, the wood that is produced at the end of the growth season is denser than that produced earlier in the season. This leads to the typical pattern of annual growth rings that can be observed on a cross section of a stem.

After a number of years, the cells in the inner part of the xylem die. The wood in this part of the xylem is referred to as heartwood. The remaining (outer) part of the xylem is called sapwood. From the moment the first heartwood is formed, it expands together with the cambial layer. After many years, all cells in the xylem may eventually die. Trees are harvested long before this happens. Therefore, commercial lumber usually consists of a mixture of primarily heartwood and some sapwood.

The mechanical properties of heartwood and sapwood of the same stem are very similar. However, heartwood is often denser than sapwood. In addition, heartwood contains a variety of products of sugar decomposition originating in the cambium. These products can be removed by physical or chemical extraction processes, and are therefore commonly referred to as extractives. The extractives clog tiny passages in the cell walls. Consequently, they significantly reduce the permeability of wood. The chemical composition of some extractives is very different from that of the main components of wood: cellulose, hemicellulose, and lignin. Thus, because of the aforementioned physical and chemical phenomena associated with the presence of extractives, the fire behavior of heartwood may be somewhat different from that of sapwood. For example, it was shown that most of the scatter could be eliminated in a correlation of gross heat of combustion with lignin content if extractives were removed from the oxygen bomb samples [4].

7.4.3 Moisture

Wood, under practical conditions, always contains a certain amount of moisture. Because of the large amount of energy required to evaporate water (~2250 kJ/kg for free water at atmospheric pressure), moisture dramatically improves the fire performance of wood. Moisture content is related to the oven dry mass and is defined as

$$u = \frac{m_u - m_0}{m_0} \qquad (7.3)$$

where u = moisture content, kg/kg
m_u = mass at moisture content u, kg
m_0 = oven dry mass, kg

The equilibrium moisture content of wood is a function of the temperature and relative humidity of the environment in which it is exposed, and can be calculated or obtained from tables [5]. It is very common to condition fire test specimens to equilibrium at a temperature of 23°C and a relative humidity of 50 percent. The corresponding equilibrium moisture content for wood is approximately 9.2 percent. When used inside buildings, the moisture content of wood is usually between 5 and 15 percent.

Cells of wood of a living tree (green wood) are partially filled with liquid water. The moisture content for green sapwood can be up to 250 percent! When kiln-drying green lumber, the liquid water in the cells gradually evaporates. The point at which all liquid water in the cells has evaporated but the cell walls are still saturated with water is termed the *fiber saturation point* (FSP). The moisture content at this point is typically around 30 percent, irrespective of species. When dried further, water desorbs from the cell walls and remains present in the cells as a vapor in equilibrium with the moisture in the cell walls. Ultimately, when all moisture has left the cell walls, the wood is oven dry. Note that water in the cell walls is adsorbed and not absorbed. Thus, extra heat in addition to the heat of evaporation is required to drive off the moisture. This extra energy is referred to as the heat of wetting.

7.4.4 Chemical Composition

The elemental composition of dry wood consists of about 50 percent carbon, 6 percent hydrogen, 44 percent oxygen, and small amounts of nitrogen and some inorganic compounds. The principal elemental constituents are combined into a number of natural polymers: cellulose, hemicellulose, and lignin.

Cellulose consists of a large number of glucose molecules joined in a chainlike polymer. Glucose ($C_6H_{12}O_6$), is the principal sugar generated by photosynthesis. The chemical formula of cellulose is $(C_6H_{10}O_5)_n$, i.e., two molecules of water are generated for every pair of glucose molecules linked together. The degree of polymerization n may be as high as 30,000. Various other sugars produced in the leaves are combined to branched-chain polymers called hemicelluloses. The degree of polymerization is generally only a few hundred. Lignin is a stable high molecular weight polymer that is phenolic in nature. It acts as a binding agent within and between cell walls. The lignin content, in general, is significantly higher in softwoods as illustrated in Table 7.1 [6].

The average composition of a large number of wood species has been documented in the literature [7]. The literature values are in percent by mass of hollocellulose and Klason lignin for dry wood after removal of the extractives. Holocellulose is the total polysaccharide, i.e., all cellulose and hemicellulose. Klason lignin is the residue after solubilizing the carbohydrate with a strong mineral acid. Klason lignin does not include the acid-soluble lignin. Softwood lignins are insoluble.

TABLE 7.1 Chemical Composition of Dry Wood

Type of wood	Cellulose (%)	Hemicelluloses (%)	Lignin (%)
Hardwood	40–44	23–40	18–25
Softwood	40–44	20–32	25–35

However, small fractions of hardwood lignins (10 to 20 percent of the total lignin content) are soluble so that the literature values are slightly too low.

7.4.5 Thermal Decomposition and Pyrolysis

The three main components of wood have quite different thermal degradation characteristics. This is illustrated by *thermogravimetric analysis* (TGA), showing that the constituents decompose to release volatiles over different temperature ranges [8], typically:

Cellulose	240–350°C
Hemicellulose	200–260°C
Lignin	280–500°C

Consequently, the thermal degradation characteristics of wood shift toward higher temperatures with increasing lignin content. This explains why the surface temperature at ignition is significantly higher for softwoods than for hardwoods (see below). In addition, lignin decomposes to volatiles for about 50 percent of its mass and is therefore responsible for most of the char. The charring rate of wood is very sensitive to the presence of inorganic impurities, such as fire retardants.

Pyrolysis of porous char-forming solids, such as wood, exposed to fire is a very complex process. Figure 7.1 identifies the major physical and chemical phenomena involved in the pyrolysis of a burning slab of wood.

Under practical conditions of use, wood products always contain a certain percentage of moisture. When exposed to fire, the temperature of the wood will rise to a point when the moisture starts

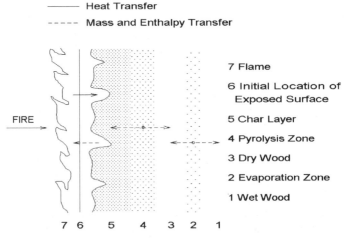

FIGURE 7.1 Heat and mass transfer in a burning slab of wood.

to evaporate. Because the water is adsorbed to the cell walls (at least if the moisture content is below the fiber saturation point), evaporation requires more energy than needed to boil free water and may occur at temperatures exceeding 100°C. The water vapor partly migrates toward and escapes through the exposed surface. A fraction also migrates in the opposite direction and recondenses at a location where the temperature is below 100°C.

The dry wood (zone 3 in Fig. 7.1) further increases in temperature until the fibers begin to degrade. The thermal degradation starts around 200 to 250°C. The volatiles that are generated again travel primarily toward the exposed side, but also partly in the opposite direction. They consist of a combustible mixture of gases, vapors, and tars. Roberts has shown that the composition of the volatiles has an average molecular formula of $(CH_2O)_n$ [9]. A solid carbon char matrix remains. The volume of the char is smaller than the original volume of the wood. This results in the formation of cracks and fissures that greatly affect the heat and mass transfer between the flame and the solid. The combustible volatiles that emerge from the exposed surface mix with ambient air and burn in a luminous flame.

Under certain conditions, oxygen may diffuse to the surface and lead to char oxidation. The exposed surface recedes as combustion progresses due to the char contraction and possible char oxidation.

7.4.6 Fire-Retardant Treatments

Wood products are often treated with fire-retardant chemicals to accomplish an improved level of fire performance, e.g., to meet specific building code requirements for interior finish materials and combustible structural members. Fire-retardant chemicals can be applied through pressure treatment or as part of a surface coating.

Pressure treatment involves impregnation of wood with a solution of fire-retardant chemicals in an autoclave. Most fire-retardants increase the thermal dehydration reactions that occur during thermal decomposition, so that more char and fewer combustible volatiles are produced [10]. The most common fire retardants for pressure treatment of wood are phosphate salts and boron compounds (borax and boric acid). These chemicals are water soluble and not resistant to leaching in wet and humid environments. Formulations that are resistant to leaching have been developed for outdoor use.

Depending on the chemical that is used, fire-retardant impregnations either slightly reduce the charring rate of wood exposed under standard fire conditions or do not affect it at all [11]. Pressure treatment with fire retardants is much more effective in improving the surface flammability characteristics of wood products.

Pressure treatment reduces the strength of wood [12]. Problems in the 1980s with degradation of fire-retardant-treated plywood and lumber used in attic spaces resulted in the development of a series of standards to evaluate the mechanical properties of softwood plywood (ASTM D 5516, "Standard Method for Evaluating the Mechanical Properties of Fire-Retardant Treated Softwood Plywood Exposed to Elevated Temperatures") and lumber (ASTM D 5664, "Standard Method for Evaluating the Effects of Fire-Retardant Treatments and Elevated Temperatures on Strength Properties of Fire-Retardant Treated Lumber") under elevated temperature and humidity conditions. Because every formulation is unique and proprietary, information about the strength reduction must be obtained from the manufacturer of the fire-retardant chemical.

Pressure treatments cannot be applied to some wood species with low permeability or to existing structures. In those cases, applying a fire-retardant coating to the surface is the only option. Two types of fire-retardant coatings are used for wood products. The first type of coating is based on the same chemicals that are used for impregnation. These coatings result in improved surface flammability characteristics, but do not have a significant effect on the charring rate and fire resistance. The second type consists of coatings that intumesce and form an insulating low-density char layer when exposed to fire. Intumescent coatings improve both surface flammability and fire resistance of wood structures.

7.5 THERMAL PROPERTIES (S.I. UNITS)

7.5.1 Wood and Char

The response of a slab of wood exposed to a fire in terms of when it will ignite and how fast it will subsequently burn is controlled primarily by the rate of heat transfer into the solid. A significant part of the incident heat flux from the fire usually consists of radiation. The surface absorptivity and emissivity determine which fraction of the incident radiation is actually absorbed. The net heat flux at the surface that is conducted into the solid consists of the absorbed radiative and convective heat fluxes. The primary material properties that affect conduction heat transfer are density, specific heat, and thermal conductivity for wood, partially charred wood, and char; the heat of pyrolysis of wood; and the heat of vaporization of water. These properties are discussed in detail in the sections that follow. This material, except for the section on surface absorptivity and emissivity, is largely based on a paper by Janssens [13].

7.5.1.1 Surface Absorptivity and Emissivity

Melinek studied the darkening of irradiated wood [14]. Measurements of reflectivity r as a function of time for wood specimens exposed at various irradiance levels were correlated assuming a first-order equation. Melinek's experiments were for irradiance from high-temperature tungsten filament lamps, and no spectral effects were taken into account.

Wesson made an extensive number of measurements of absorptivity (α) as a function of wavelength of the incident radiation [15]. From these measurements, he derived average values for the absorptivity according to the spectral distribution of the source. For flame sources, almost no difference was found between different wood species, and the overall average was 0.76. For radiation from tungsten lamps or the sun, absorptivity was much lower and varied for different species. Fortunately, the latter is not representative of the range of thermal exposure in fires. Wesson made all his absorptivity measurements at ambient temperature and did not consider any changes in surface appearance (darkening) due to heating.

Vovelle et al. related absorptivity of wood surfaces directly to pyrolysis [16]. The Arrhenius constants were obtained via TGA and were found to be fairly independent of species.

In this section, absorptivity α, reflectivity r, and emissivity ε are used interchangeably. Indeed, most investigators assumed Kirchoff's law is valid for global values of α, ε, and r ($\alpha = \varepsilon = 1 - r$). This is not entirely correct, as wood surfaces do not behave as gray bodies. In addition, spectral distribution of irradiance from heat sources used in piloted ignition tests is only approximately gray at best. Nevertheless, the assumption that Kirchoff's law is valid is reasonable, at least for an engineering analysis.

Wood is slightly diathermanous. However, the importance of the effect of diathermancy on the surface temperature rise is unclear. It certainly seems to be much less critical for wood products than for translucent plastics.

Based on Melinek's data, Janssens calculated that the absorptivity changes from 0.8 to 0.95 after 150 s of exposure to a radiant heat flux of 12.6 kW/m^2 [17]. Using the model proposed by Vovelle et al., Janssens determined that the absorptivity of pine increases from 0.8 to 0.84 after 210 s of exposure to a radiant heat flux of 15 kW/m^2. In conclusion, based on these calculations and on Wesson's data, Janssens recommended using an average value of $\alpha = \varepsilon = 0.88$ (average between Wesson's value of 0.76 at ambient temperature and 1.0 for a black surface of charred wood) for the preheat period before ignition, and a value of $\alpha = \varepsilon = 1.0$ during flaming combustion.

7.5.1.2 Density

Density is the ratio of mass over volume. The mass changes as wood dries and loses moisture, and as wood is converted to char. The volume changes as wood shrinks due to moisture loss and charring.

7.5.1.2.1 Mass Loss. Many investigators have used first-order Arrhenius-type reaction kinetics to model the thermal degradation of wood. The corresponding equation describing mass loss as a function of temperature is given by

$$\frac{dm}{dt} = A \cdot m \cdot \exp\left(-\frac{E}{RT}\right) \tag{7.4}$$

where m = mass, g
t = time, s
A = frequency factor, s^{-1}
E = activation energy, kJ/kmol
R = universal gas constant, 8.314 kJ/kmol·K
T = absolute temperature, K

The frequency factor and activation energy can be determined by fitting Eq. (7.4) to experimental mass loss data for wood exposed at elevated temperatures. Atreya conducted an extensive survey of studies to determine kinetic parameters for wood in this manner [18], and recommended the following values for large samples of wood: $A = 10^8$ s^{-1} and $E = 125$ kJ/mol. These values are consistent with a low-temperature reaction scheme proposed for cellulose by Shafizadeh [19].

More complex kinetic models have been developed. For example, Alves and Figueiredo proposed a scheme of six independent reactions [20]. Although such a complex scheme has the potential of being more accurate, in practice this is offset by the uncertainties of the larger number of kinetic parameters that need to be determined. Perhaps the only "complex" scheme that offers an improvement over a single reaction is that proposed by Parker [21]. This scheme involves three reactions, one for each of the main components of wood. Parker's scheme has the advantage over a single reaction that it accounts for composition effects on wood ignition and combustion (for example, the effect of the higher lignin content in softwoods vs. hardwoods). It does not have the drawback of other complex schemes, because Parker made very detailed measurements of the kinetic parameters of cellulose, hemicelluloses, and lignin.

Another approach to describe the thermal degradation of wood is by expressing the mass (instead of the mass loss rate) as a function of temperature. The simplest version of such a model assumes that wood is converted to char at a particular temperature, T_p. Although this approach has proven to be useful in calculating the fire resistance of structural wood members (see below), Nurbakhsh demonstrated that different temperatures are required to conserve mass and energy depending on the heating conditions at the surface [22]. A more accurate model involves specifying mass as function of temperature over the range that is pertinent to fires. Such a model will be discussed below.

Mass can be expressed in the form of the ratio of mass to oven dry mass of the wood. This ratio is denoted Z. Initially, Z is equal to $1 + u$, where u is the moisture content. When the temperature of the wood rises, Z gradually decreases to 1 as the moisture in the wood evaporates. White measured the charring rate of and temperature profile in slabs of eight different wood species exposed to ASTM E 119 standard fire conditions [23]. All species were tested at various moisture contents, ranging from 5 to 18 percent. White's data show that all moisture evaporates over a narrow temperature range (10 to 20°C) and initiates at a species-dependent temperature, which varies from 100°C to approximately 160°C. Numerous measurements indicated that pyrolysis is insignificant below 200°C [24, 25]. Therefore, Z is equal to 1 for dry wood below 200°C.

Above 200°C, Z decreases continuously as a function of temperature due to pyrolysis. The proper temperature function has to be established based on some accurate and reliable experimental data. The measurements reported by Beall [24] and Slocum et al. [25] were found suitable for this purpose.

Beall measured the mass loss and dimensional changes of 10 × 10 × 10-mm oven dry cubes of six wood species as a function of temperature up to 600°C for three heating rates (1, 10, and 50°C/min) [24]. The mass loss curves obtained by Beall for softwoods were significantly different

TABLE 7.2 Z_{600} for Beall's Data

Species	Y_l (kg/kg)	Predicted Z_{600} (kg/kg)	Measured Z_{600} (kg/kg)
White oak	0.27	0.29	0.31
Hard maple	0.23	0.25	0.25
Southern pine	0.29	0.24	0.25
Douglas fir	0.27	0.23	0.24
Basswood	0.21	0.23	0.22
Redwood	0.37	0.27	0.29

from those for hardwoods. The endpoint Z at 600°C, denoted as Z_{600}, was also significantly different between the two botanical groups. Based on Beall's data, Z_{600} can be predicted reasonably well based on the Klason lignin content Y_l (in kg/kg):

$$\text{Softwoods: } Z_{600} = 0.12 + 0.41 Y_l \tag{7.5}$$

$$\text{Hardwoods: } Z_{600} = 0.02 + Y_l \tag{7.6}$$

Values of Y_l for a large number of wood species can be found in the literature [7]. Table 7.2 lists these literature values for the species tested by Beall, and compares predicted and measured Z_{600}.

White reported final char mass for his experiments [23]. His thermocouple measurements indicated that average char temperature at the end of the tests was 615 ± 18°C. This was close enough to 600°C so that White's measurements can be used to check Eqs. (7.5) and (7.6). The results are shown in Table 7.3. Agreement is even better than for Beall's data and within 1 percent.

Slocum et al. conducted the same type of experiments as Beall for two hardwoods up to 800°C [25]. Their measurements indicated that Z decreases by 0.02 between 600 and 800°C. This is approximately 25 percent of the decrease between 400 and 600°C. Therefore, further mass loss at temperatures exceeding 800°C is expected to be negligible. Finally, based on the data reported by Beall and by Slocum et al., generic mass loss curves as a function of temperature can be developed for softwoods and hardwoods, respectively. The breakpoints of these piecewise linear curves are presented in Table 7.4, where $\Delta Z \equiv 0.3 - Z_{600}$.

Table 7.4 shows that softwoods lose more of their mass at higher temperatures than hardwoods. This results from the higher lignin content in softwoods, which decomposes at higher temperatures than cellulose and hemicellusoses.

7.5.1.2.2 Dimensional Changes due to Drying. The swollen volume due to moisture content u is given by [6]

$$V_u = V_0(1 + 0.00084 \rho_0 u) \tag{7.7}$$

TABLE 7.3 Z_{600} for White's Data

Species	Y_l (kg/kg)	Predicted Z_{600} (kg/kg)	Measured Z_{600} (kg/kg)
Engelmann spruce	0.29	0.24	0.24
Western red cedar	0.37	0.27	0.26
Southern pine	0.29	0.24	0.24
Redwood	0.37	0.27	0.28
Hard maple	0.23	0.25	0.25
Yellow poplar	0.20	0.22	0.23
Red oak	0.25	0.27	0.26
Basswood	0.21	0.23	0.22

TABLE 7.4 Generic Mass Loss Curves

T (°C)	Z softwoods (kg/kg)	Z hardwoods (kg/kg)
200	1.00	1.00
250	0.95–0.5ΔZ	0.79–0.5ΔZ
300	0.78–ΔZ	0.48–ΔZ
350	0.54–ΔZ	0.42–ΔZ
400	0.40–ΔZ	0.38–ΔZ
600	0.30–ΔZ	0.30–ΔZ
≥800	0.28–ΔZ	0.28–ΔZ

where V_u = volume at moisture content u, m³
V_0 = oven dry volume, m³
ρ_0 = oven dry density, kg/m³

Swelling is the greatest in the tangential direction, and can be twice that of the radial direction. Swelling in the longitudinal direction is negligible. The ring orientation in wood members exposed to fire is usually not known, although the exposure is typically perpendicular to the grain. For this reason, rather than considering separate expansion or contraction factors for the three dimensions, only two directions are considered; parallel to the grain (or longitudinal) and perpendicular to the grain (average of radial and tangential). The corresponding expansion or contraction factors are expressed in m/m and denoted f_l and f_p, respectively. The factor for expansion due to moisture perpendicular to the grain can be estimated from

$$f_p = \sqrt{1 + 0.00084\rho_0 u} \tag{7.8}$$

This factor will always be greater than one. The longitudinal factor for expansion due to moisture is equal to one.

7.5.1.2.3 Thermal Expansion of Wood. According to Kollman and Côté [6], the thermal expansion factor in the longitudinal direction f_l is estimated from

$$f_l = 1 + 3.75 \cdot 10^{-6}(T - T_r) \tag{7.9}$$

where T is the temperature (°C) and T_r is the reference temperature (20°C).

Thermal expansion factors for softwoods and low-density hardwoods ($\rho_0 < 600$ kg/m³) in the radial direction is calculated from

$$f_r = 1 + \rho_0 \cdot 5.5 \cdot 10^{-8}(T - T_r) \tag{7.10}$$

and in the tangential direction from

$$f_t = 1 + \rho_0 \cdot 8.2 \cdot 10^{-8}(T - T_r) \tag{7.11}$$

Thermal expansion factors for high-density hardwoods ($\rho_0 \geq 600$ kg/m³) in the radial direction is given by

$$f_r = 1 + \rho_0 \cdot 4.5 \cdot 10^{-8}(T - T_r) \tag{7.12}$$

and in the tangential direction by

$$f_t = 1 + \rho_0 \cdot 5.8 \cdot 10^{-8}(T - T_r) \tag{7.13}$$

In all cases, an average thermal expansion factor perpendicular to the grain is obtained from

$$f_p = \sqrt{f_r \cdot f_t} \tag{7.14}$$

The combined effect from swelling due to moisture and thermal expansion is estimated by multiplying the expansion factors in Eqs. (7.8) and (7.14).

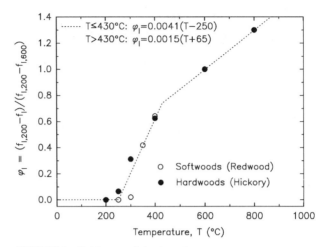

FIGURE 7.2 Shrinkage parallel to the grain.

7.5.1.2.4 Char Contraction. At temperatures exceeding 200°C, thermal degradation results in a decrease of the volume. A char contraction factor as a function of temperature can be estimated based on data by Beall and Slocum et al. Because heating rates under standard fire exposure are generally between 1 and 10°C/min, averages of the data at these two rates are considered. The contraction measurements are plotted against temperature in the form of new variables φ_l and φ_p, which are defined below. The resulting temperature functions are shown in Figs. 7.2 and 7.3 for φ_l and φ_p, respectively.

$$\varphi_l \equiv \frac{f_{l,200} - f_l}{f_{l,200} - f_{l,600}} \qquad (7.15)$$

$$\varphi_p \equiv \frac{f_{p,200} - f_p}{f_{p,200} - f_{p,600}} \qquad (7.16)$$

The values for f_l and f_p at 200°C can be calculated from the equations in the previous section at $T = 200$. Beall found that $f_{l,600}$ is nearly constant and equal to 0.82 [24]. Beall also correlated char density at 600°C as a function of ρ_0:

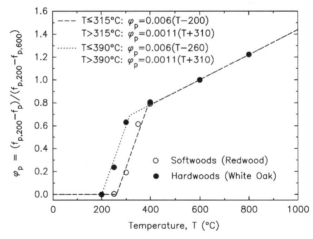

FIGURE 7.3 Shrinkage perpendicular to the grain.

$$\rho_{c,600} = 0.75\rho_0 - 63 \tag{7.17}$$

This correlation is consistent with char density measurements by Evans [26]. Finally, $f_{p,600}$ can then be obtained from

$$f_{p,600} = \sqrt{\frac{Z_{600}\rho_0}{f_{l,600}\rho_{c,600}}} \tag{7.18}$$

7.5.1.2.5 Summary. Density is the ratio of mass to volume. Because Z, f_l, and f_p are all normalized to oven dry mass and volume at reference temperature $T_r = 20°C$, density is given by

$$\rho(T) = \frac{Z(T)}{f_l(T)f_p^2(T)}\rho_0 \tag{7.19}$$

7.5.1.2.6 Examples

Example 7.1 Calculate the density at 20°C of Douglas fir with a moisture content of 10 percent and oven dry density of 460 kg/m³.

Z at 20°C is equal to $1 + u = 1.1$ and $f_l = 1.0$.

Equation 7.8 ⇨ $f_p = \sqrt{1 + 0.00084 \times 460 \times 0.1} = 1.01914$ m/m

Equation 7.19 ⇨ $\rho(20) = \dfrac{1.1}{1.0 \times 1.01914^2}\, 460 = 487$ kg/m³

Example 7.2 Calculate the density at 200°C of Douglas fir with oven dry density of 460 kg/m³.

Z at 200°C is equal to 1.0, because all moisture has evaporated and pyrolysis has not started.

Equation 7.9 ⇨ $f_l(200) = 1 + 3.75 \cdot 10^{-6}(200 - 20) = 1.00068$ m/m

Equation 7.10 ⇨ $f_r(200) = 1 + 460 \times 5.5 \cdot 10^{-8}(200 - 20) = 1.00455$ m/m

Equation 7.11 ⇨ $f_t(200) = 1 + 460 \times 8.2 \cdot 10^{-8}(200 - 20) = 1.00679$ m/m

Equation 7.14 ⇨ $f_p(200) = \sqrt{1.00455 \times 1.00679} = 1.00567$ m/m

Equation 7.19 ⇨ $\rho(200) = \dfrac{1}{1.00068 \times 1.00567^2}\, 460 = 455$ kg/m³

Example 7.3 What is the density at 400°C of partially charred Douglas fir with oven dry density of 460 kg/m³?

The Klason lignin content, $Y_l = 0.27$, follows from Table 7.2.

Equation 7.5 ⇨ $Z_{600} = 0.12 + 0.41 \times 0.27 = 0.23$ kg/kg

Table 7.4 ⇨ $Z(400) = 0.4 - (0.30 - 0.23) = 0.33$ kg/kg

Figure 7.2 ⇨ $\varphi_l(400) = 0.0041(400 - 250) = 0.615$

Equation 7.15 ⇨ $f_l(400) = 1.00068 - 0.615(1.00068 - 0.82) = 0.8896$ m/m

Equation 7.17 ⇨ $\rho_{c,600} = 0.75 \times 460 - 63 = 282$ kg/m³

Equation 7.18 ⇨ $f_{p,600} = \sqrt{\dfrac{0.23 \times 460}{0.82 \times 282}} = 0.6764$ m/m

Figure 7.3 ⇨ $\varphi_p(400) = 0.0011(400 + 310) = 0.781$

Equation 7.16 ⇨ $f_p(400) = 1.00567 - 0.781(1.00567 - 0.6764) = 0.7485$ m/m

Equation 7.19 ⇨ $\rho(400) = \dfrac{0.33}{0.8923 \times 0.7485^2}\, 460 = 305$ kg/m³

7.5.1.3 Specific Heat

7.5.1.3.1 Wood. Literature data for specific heat of wood as a function of temperature are accurate and consistent. This is because it is much easier to measure specific heat, even at higher temperatures, than, for example, thermal conductivity (see below). The following equations are taken from one of the most extensive surveys of wood thermal properties available [27]

$$c_u(T) = \frac{c_0(T) + 4187u}{1 + u} + \Delta c(T, u) \quad (7.20)$$

with

$$c_0(T) = 1159 + 3.86T \quad (7.21)$$

and

$$\Delta c(T, u) = (23.55T - 1326u + 2417)u \quad (7.22)$$

where $c_u(T)$ = specific heat of wood with moisture content u at temperature T, J/kg·°C
$\Delta c(T,u)$ = correction, J/kg·°C
$c_0(T)$ = specific heat of oven dry wood at temperature T, J/kg·°C

Δc accounts for the fact that below the fiber saturation point, water is bound to the cell walls and is not free.

7.5.1.3.2 Char. A temperature function for specific heat of wood char has been proposed by Widell [28]. As shown in Fig. 7.4, Widell's curve agrees quite well with data for amorphous carbon and graphite used by others [21, 29].

7.5.1.3.3 Partially Charred Wood. The equations for wood are used at temperatures below 200°C. The equation for char is used at temperatures exceeding 800°C. In between these temperatures, it is suggested that specific heat be determined by interpolation between the wood and char value via

$$c(T) = \zeta c_0(T) + (1 - \zeta)c_c(T) \quad (7.23)$$

with

$$\zeta = \frac{Z - Z_{800}}{1 - Z_{800}} \quad (7.24)$$

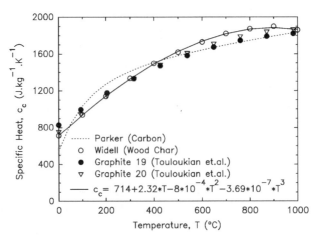

FIGURE 7.4 Specific heat of char.

where $c_c(T)$ is the specific heat of char at temperature T (J/kg·°C), and ζ is the degree of conversion from wood to char.

7.5.1.3.4 Examples

Example 7.4 Calculate the specific heat at 20°C of Douglas fir with a moisture content of 10 percent and oven dry density of 460 kg/m³.

Equation 7.21 ⇨ $c_0(20) = 1159 + 3.86 \times 20 = 1236$ J/kg · °C

Equation 7.22 ⇨ $\Delta c(20,0.1) = (23.55 \times 20 - 1326 \times 0.1 + 2417) \times 0.1 = 276$ J/kg · °C

Equation 7.20 ⇨ $c_{0.1}(20) = \dfrac{1236 + 4187 \times 0.1}{1.1} + 276 = 1780$ J/kg · °C

Example 7.5 Calculate the specific heat at 200°C of Douglas fir with oven dry density of 460 kg/m³.

Equation 7.21 ⇨ $c_0(200) = 1159 + 3.86 \times 200 = 1931$ J/kg · °C

Example 7.6 What is the specific heat at 400°C of partially charred Douglas fir with oven dry density of 460 kg/m³?

Table 7.4 ⇨ $Z(800) = 0.28 - (0.30 - 0.23) = 0.21$ kg/kg

Equation 7.24 ⇨ $\zeta = \dfrac{0.33 - 0.21}{1.0 - 0.21} = 0.1519$

Figure 7.4 ⇨ $c_c(400) = 714 + 2.32 \times 400 - 8 \cdot 10^{-4} \times 400^2 - 3.69 \cdot 10^{-7} \times 400^3 = 1490$ J/kg · °C

Equation 7.21 ⇨ $c_0(400) = 1159 + 3.86 \times 400 = 2703$ J/kg · °C

Equation 7.23 ⇨ $c(400) = 0.1519 \times 2703 + (1 - 0.1519) \times 1490 = 1674$ J/kg · °C

7.5.1.4 Thermal Conductivity

7.5.1.4.1 Dry Wood at Ambient Temperature.
Available data for thermal conductivity at ambient temperature are consistent. Therefore, one of the many correlations of oven dry density reported in the literature can be used to determine this property. Instead of an existing correlation, a model is proposed. The model has the advantage that it can be extended to areas where there is major scatter in literature values, i.e., for wet wood at high temperature and for char. The model was originally developed by Kollman and Malmquist [30] and refined by Fredlund [31] through the addition of a radiation term.

Dry wood consists of solid fiber in the cell walls, and air in the cell cavities. Porosity π_g is the fraction of the total volume occupied by air and is given by

$$\pi_g = 1 - \frac{\rho_0}{\rho_s} \qquad (7.25)$$

where the bulk density ρ_s is equal to 1460 kg/m³ [32].

An upper limit for thermal conductivity of dry wood is obtained for a system with the same porosity, composed of alternating air and solid layers arranged in parallel.

$$k_{max} = \pi_g k_g + \pi_s k_s \qquad (7.26)$$

where k_{max} = upper limit of thermal conductivity of dry wood, W/m·°C
π_g = fraction of total volume of wood occupied by gas, m³/m³
k_g = thermal conductivity of gas filling the void spaces, W/m·°C
π_s = fraction of total volume of wood occupied by wood fibers, m³/m³
k_s = thermal conductivity of wood fiber, W/m·°C

A lower limit is obtained for a system of layers in series.

$$k_{min} = \frac{k_g k_s}{\pi_g k_s + \pi_s k_g} \qquad (7.27)$$

where k_{min} is the lower limit of the thermal conductivity of dry wood (W/m·°C).

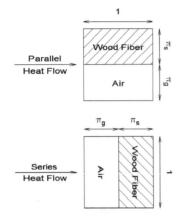

FIGURE 7.5 Upper and lower limits of thermal conductivity of dry wood.

These two extremes are illustrated in Fig. 7.5. The real value falls between these limits and can be obtained as a weighted average with weighing factor ξ. To arrive at the total apparent conductivity, Fredlund added a term to account for radiative heat transfer in the cell cavities [31].

$$k_r = \frac{4\pi_g \sigma (T + 273)^3 d_p}{1 - \pi_g} \quad (7.28)$$

where k_r = apparent thermal conductivity accounting for radiation, W/m·°C

σ = Boltzmann constant, 5.67×10^{-8} W/m²·K⁴

d_p = average diameter of the cell cavity, m

With an average softwood tracheid diameter of 35 μm [33], d_p is estimated as

$$d_p = 3.5 \cdot 10^{-5} \sqrt{\pi_g} \quad (7.29)$$

Finally, the thermal conductivity of dry wood is estimated from

$$k_0 = \zeta k_{max} + (1 - \xi) k_{min} + k_r \quad (7.30)$$

Figure 7.6 compares thermal conductivity as a function of oven dry density calculated according to Eq. (7.30), with correlations from the literature [27, 34, 35]. With $k_g = 0.026$ W/m·°C, $k_s = 0.42$ W/m·°C, and $\xi = 0.58 + 10^{-4} \cdot \rho_0$ (all approximating the values suggested by Kollman and Malmquist [30], and Fredlund [31]), the model agrees very well with the correlation recommended by TenWolde et al. [27].

7.5.1.4.2 Wet Wood at Ambient Temperature. The same model can be used as for dry wood, but with the addition of bound water as a third constituent. Figure 7.7 illustrates the concept. The equations are modified as follows

$$\pi_s = \frac{\rho_0}{(1 + 0.00084 \rho_0 u) \rho_s} \quad (7.31)$$

$$\pi_w = \frac{\rho_0 u}{(1 + 0.00084 \rho_0 u) \rho_w} \quad (7.32)$$

$$\pi_g = 1 - \pi_s - \pi_w \quad (7.33)$$

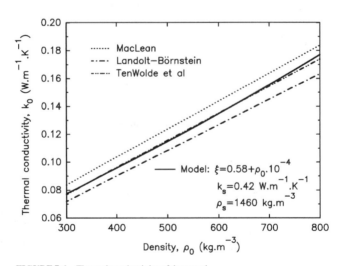

FIGURE 7.6 Thermal conductivity of dry wood.

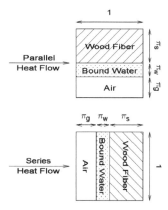

FIGURE 7.7 Upper and lower limits of thermal conductivity of wet wood.

$$k_{max} = \pi_g k_g + \pi_s k_s + \pi_w k_w \tag{7.34}$$

$$k_{min} = \frac{k_g k_s k_w}{\pi_g k_s k_w + \pi_s k_g k_w + \pi_w k_g k_s} \tag{7.35}$$

where ρ_w = density of bound water, kg/m^3

k_w = thermal conductivity of bound water, W/m·°C

π_w = fraction of the total volume of wood occupied by bound water, m^3/m^3

Note that the density of bound water ρ_w is higher than that of free water. It is a function of u according to [36]

$$\rho_w = 1298 - 1132u + 1766u^2 \tag{7.36}$$

k_r and k_u are still calculated from Eqs. (7.28) and (7.30), respectively. Model predictions are compared to correlations by TenWolde et al. [27] in Fig. 7.8. The calculations were performed with the same values for k_g and k_s as in the previous section, and with $k_w = 0.8$ W/m·°C and $\xi = 0.58 + 10^{-4} \rho_0 + 0.5u$.

7.5.1.4.3 Wood at Elevated Temperature. Many investigators have suggested a linear relationship between the thermal conductivity of dry wood and temperature. However, there is major disagreement on the slope of the temperature function. Recommended values vary from nearly 0 to as much as 1 percent increase per °C. There is some evidence that the thermal diffusivity of dry wood is independent of temperature [21, 37]. With this assumption, the thermal conductivity of dry wood can be calculated as a function of temperature from

$$k_0(T) = \frac{k_0(T_r)\rho_0(T)c_0(T)}{\rho_0(T_r)c_0(T_r)} \tag{7.37}$$

where $k_0(T)$ is the thermal conductivity of oven dry wood at temperature T (W/m·°C).

k_0, ρ_0, and c_0 at the reference temperature $T_r = 20$°C are determined from Eqs. (7.30), (7.19), and (7.21), respectively. Because ρ decreases only slightly with temperature, and c_0 is a linear function of temperature, Eq. (7.37) indicates that k_0 is indeed a linear function of temperature. The equation agrees reasonably well with some experimental data [38, 39], and its slope falls within the range of values reported in the literature.

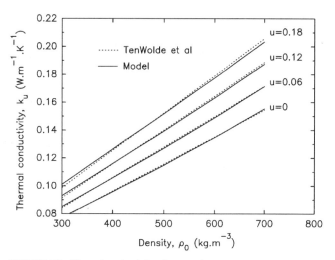

FIGURE 7.8 Thermal conductivity of wet wood.

Using Eq. (7.37) and the model for thermal conductivity of dry wood outlined above, it is possible to determine how k_s varies with temperature. The following linear function resulted in good agreement between the conductivity model and Eq. (7.37)

$$k_0(T) = 0.42 + 1.3 \cdot 10^{-3} (T - T_r) \tag{7.38}$$

This function was obtained with the following equation for thermal conductivity of air based on literature data [40].

$$k_g(T) = 0.024 + 7.05 \cdot 10^{-5} T - 1.59 \cdot 10^{-8} T^2 \tag{7.39}$$

Finally, the model for wet wood at ambient temperature can be extended to higher temperature by using Eqs. (7.38) and (7.39) for k_s and k_g, respectively, and by using the following expression for thermal conductivity of bound water

$$k_w(T) = (0.66 + 2 \cdot 10^{-3} T - 8 \cdot 10^{-4} T^2) \frac{\rho_w}{1000} \tag{7.40}$$

Equation (7.40) is based on literature data [40], with an empirical adjustment for increased density of bound water over free water.

7.5.1.4.4 Char and Partially Charred Wood. As wood is converted to char its structure remains largely intact, until at high temperatures large cracks occur due to thermal stresses [41]. Therefore, thermal conductivity of wood char can be modeled in the same way as for dry wood. However, both the bulk density and thermal conductivity have to be adjusted.

Cutter and McGinnes [42] measured char bulk density for the same species used earlier by Beall. Figure 7.9 shows the results. Breakpoints of piecewise linear curves are listed in Table 7.5. These data indicate that bulk density decreases slightly as wood is converted to char. A distinct difference between softwoods and hardwoods can be observed.

Elemental analysis of wood char indicates that more than 90 percent consists of carbon [43]. Therefore, literature values for thermal conductivity of amorphous carbon are a good indication of the value to use for k_s of wood char. Unfortunately, these literature values vary over more than two orders of magnitude [44]. This is probably because the thermal conductivity of carbon depends strongly on its degree of graphitization, which is a function of the maximum temperature to which it has been heated.

Very few data are available for thermal conductivity of wood char at elevated temperatures. The few measurements that were reported [45, 46] were used in combination with data for carbon fiber felts [47] and the model for thermal conductivity of dry wood, to arrive at the following expression

$$k_s = 0.33 + 1.6 \cdot 10^{-4} T + 1.08 \cdot 10^{-7} T^2 \tag{7.41}$$

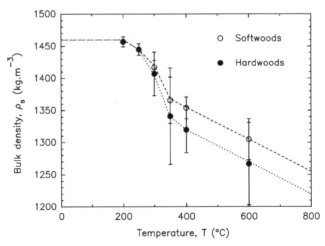

FIGURE 7.9 Bulk density of charred wood.

TABLE 7.5 Bulk Density of Charred Wood

T (°C)	ρ_s Softwoods (kg/m³)	ρ_s Hardwoods (kg/m³)
200	1460	1460
250	1445	1445
300	1420	1405
350	1368	1340
400	1355	1320
600	1305	1270

The resulting predictions for thermal conductivity of wood char at ambient temperature as a function of char density are compared to literature data in Fig. 7.10.

To determine the thermal conductivity of partially charred wood, the same model is used with ρ_s from Fig. 7.9, and k_s as a weighted average of the bulk values for wood and char using ξ as the weighing factor.

7.5.1.4.5 Examples

Example 7.7 Calculate the thermal conductivity at 20°C of Douglas fir with a moisture content of 10 percent and oven dry density of 460 kg/m³.

Equation 7.31 ⇨ $\pi_s = \dfrac{460}{(1 + 0.00084 \times 460 \times 0.1) \times 1460} = 0.3033 \text{ m}^3/\text{m}^3$

Equation 7.36 ⇨ $\rho_w = 1298 - 1132 \times 0.1 + 1766 \times 0.1^2 = 1202 \text{ kg/m}^3$

Equation 7.32 ⇨ $\pi_w = \dfrac{460 \times 0.1}{(1 + 0.00084 \times 460 \times 0.1) \times 1202} = 0.0368 \text{ m}^3/\text{m}^3$

Equation 7.33 ⇨ $\pi_g = 1 - 0.3033 - 0.0368 = 0.6599 \text{ m}^3/\text{m}^3$

Equation 7.34 ⇨ $k_{max} = 0.6599 \times 0.026 + 0.3033 \times 0.42 + 0.0368 \times 0.8 = 0.1740 \text{ W/m} \cdot \text{°C}$

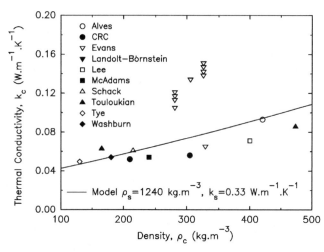

FIGURE 7.10 Thermal conductivity of wood char at T_r.

Equation 7.35 ⇨ $k_{min} = \dfrac{0.026 \times 0.42 \times 0.8}{0.6599 \times 0.42 \times 0.8 + 0.3033 \times 0.026 \times 0.8 + 0.0368 \times 0.026 \times 0.42}$

$= 0.0382$ W/m · °C

Equation 7.28 ⇨ $k_r = \dfrac{4 \times 0.6599 \times 5.67 \cdot 10^{-8} \times (293)^3 \times 3.5 \cdot 10^{-5} \times \sqrt{0.6599}}{1 - 0.6599}$

$= 0.0003$ W/m · °C

The weighing factor is given by $\xi = 0.58 + 10^{-4} \times 460 + 0.5 \times 0.1 = 0.676$

Equation 7.30 ⇨ $k = 0.676 \times 0.1740 + (1 - 0.676) \times 0.0382 + 0.0003 = 0.1303$ W/m · °C

Example 7.8 Calculate the thermal conductivity at 200°C of Douglas fir with oven dry density of 460 kg/m³.

Equation 7.25 ⇨ $\pi_g = 1 - \dfrac{460}{1460} = 0.6849$ m³/m³ ∴ $\pi_s = 1 - 0.6849 = 0.3151$ m³/m³

Equation 7.38 ⇨ $k_s = 0.42 + 1.3 \cdot 10^{-3} \times (200 - 20) = 0.6540$ W/m · °C

Equation 7.39 ⇨ $k_g = 0.024 + 7.05 \cdot 10^{-5} \times 200 - 1.59 \cdot 10^{-8} \times 200^2 = 0.0376$ W/m · °C

Equation 7.26 ⇨ $k_{max} = 0.6849 \times 0.0376 + 0.3151 \times 0.6540 = 0.2318$ W/m · °C

Equation 7.35 ⇨ $k_{min} = \dfrac{0.0376 \times 0.6540}{0.6849 \times 0.6540 + 0.3151 \times 0.0376} = 0.0535$ W/m · °C

Equation 7.28 ⇨

$k_r = \dfrac{4 \times 0.6849 \times 5.67 \cdot 10^{-8} \times (473)^3 \times 3.5 \cdot 10^{-5} \times \sqrt{0.6849}}{1 - 0.6849} = 0.0015$ W/m · °C

The weighing factor is given by $\xi = 0.58 + 10^{-4} \times 460 = 0.626$

Equation 7.30 ⇨ $k = 0.626 \times 0.2318 + (1 - 0.626) \times 0.0535 + 0.0015 = 0.1666$ W/m · °C

Example 7.9 What is the thermal conductivity at 400°C of partially charred Douglas fir with oven dry density of 460 kg/m³?

Table 7.5 ⇨ $\rho_s = 1355$ kg/m³

Equation 7.25 ⇨ $\pi_g = 1 - \dfrac{305}{1355} = 0.7749$ m³/m³ ∴ $\pi_s = 1 - 0.7749 = 0.2251$ m³/m³

Equation 7.38 ⇨ $k_s = 0.42 + 1.3 \cdot 10^{-3} \times (400 - 20) = 0.4940$ W/m · °C for wood

Equation 7.41 ⇨ $k_s = 0.33 + 1.6 \cdot 10^{-4} \times 400 + 1.08 \cdot 10^{-7} \times 400^2 = 0.4113$ W/m · °C for char

Average ⇨ $k_s = 0.1519 \times 0.4940 + (1 - 0.1519) \times 0.4113 = 0.4238$ W/m · °C

Equation 7.39 ⇨ $k_g = 0.024 + 7.05 \cdot 10^{-5} \times 400 - 1.59 \cdot 10^{-8} \times 400^2 = 0.0497$ W/m · °C

Equation 7.26 ⇨ $k_{max} = 0.7749 \times 0.0497 + 0.2251 \times 0.4238 = 0.1339$ W/m · °C

Equation 7.35 ⇨ $k_{min} = \dfrac{0.0497 \times 0.4238}{0.7749 \times 0.4238 + 0.2251 \times 0.0497} = 0.0620$ W/m · °C

Equation 7.28 ⇨

$k_r = \dfrac{4 \times 0.7749 \times 5.67 \cdot 10^{-8} \times (673)^3 \times 3.5 \cdot 10^{-5} \times \sqrt{0.7749}}{1 - 0.7749} = 0.0073$ W/m · °C

The weighing factor is given by $\xi = 0.58 + 10^{-4} \times 305 = 0.611$

Equation 7.30 ⇨ $k = 0.611 \times 0.1339 + (1 - 0.611) \times 0.0620 + 0.0073 = 0.1132$ W/m · °C

7.5.1.5 Heat of Wetting and Vaporization

The integral heat of wetting in kilojoules per kilogram of water is given by [6]

$$\Delta h_w = \frac{92.1}{0.07 + u} \qquad (7.42)$$

The heat of vaporization in kilojoules per kilogram of water can be expressed as a function of temperature based on the steam tables.

$$\Delta h_v = 2552 - 2.93T \qquad (7.43)$$

The corresponding pressure in pascals can be calculated via the Antoine correlation [48].

$$P = \exp\left(16.23 - \frac{3774}{T + 225}\right) \qquad (7.44)$$

where P is pressure (Pa).

7.5.1.6 Heat of Pyrolysis

Atreya conducted an extensive review of the literature to find experimental data for the heat of pyrolysis of wood [18]. Reported values range from 750 kJ/kg (endothermic) to −18,840 kJ/kg (exothermic). Large exothermic values would lead to thermal runaway, a phenomenon that has not been observed. Kung and Kalelkar pointed out that the specific heats of active wood and char are considerably higher than the specific heat of volatiles [49]. Hence, although the decomposition reactions may be endothermic at low temperatures, they can be exothermic at high temperatures.

Atreya compared the predictions of a numerical pyrolysis model with small-scale burning rate data for slabs of eight different species of wood exposed to a constant radiant heat flux [18]. He used different values of the heat of pyrolysis, ranging from +125 kJ/kg to −125 kJ/kg. For some tests, best agreement was obtained for an endothermic heat of pyrolysis, but for other tests, exothermic values gave better results. Based on this analysis, Atreya recommended a value of $\Delta h_p = 0$ kJ/kg.

7.5.2 Other Materials

7.5.2.1 Wood Panel Products and Engineered Wood Products

Thermal properties of glulam are comparable to those of the solid wood of the laminates. TenWolde et al. recommend using the same equations as for solid wood of the same density to determine the specific heat of plywood and particleboard, but notes that the specific heat of fiberboard is different [27]. The same investigators suggest reducing the thermal conductivity of wood by 14 percent for plywood, 25 percent for particleboard, and 35 percent for fiberboard.

7.5.2.2 Gypsum Board

Thomas conducted an extensive review of the thermal properties of gypsum board. Recommended values for mass loss, specific heat, and thermal conductivity of gypsum board are presented in Table 7.6.

7.5.2.3 Insulation

The thermal property data discussed in this section are based on data reported by Bénichou and Sultan for glass and rock fiber insulation [50].

The mass of glass fiber insulation decreases approximately linearly to 94 percent at 300°C, and remains constant at higher temperatures. The mass of rock fiber insulation decreases more gradually to 94 percent at 1100°C.

For both glass and rock fiber insulation, there is a slight increase of the specific heat up to 300 to 350°C. The specific heat of glass fiber insulation is approximately 900 J/kg.

The thermal conductivity of glass fiber insulation increases similarly to rock fiber insulation from 0.022 W/m·°C at 24°C to 0.204 W/m·°C at 515°C. Beyond 515°C the thermal conductivity of glass

TABLE 7.6 Recommended Thermal Properties for Gypsum Board

T (°C)	Mass (%)	T (°C)	Specific heat (J/kg · °C)	T (°C)	Thermal conductivity (W/m · °C)
0	100	0	950	0	0.25
80	100	100	950	70	0.25
125	95	110	52,450	130	0.13
540	93.5	110	18,120	300	0.13
650	91	140	950	800	0.18
1000	90	140	3390	1000	0.35
—	—	220	3390	1000	0.78
—	—	≥600	950	4000	10

fiber insulation increases very rapidly to 1.587 W/m·°C at 632°C. The thermal conductivity of rock fiber insulation continues to increase at approximately the same rate as at low temperatures up to 1100°C.

7.6 MECHANICAL PROPERTIES (U.S. CUSTOMARY UNITS)

Knowledge of mechanical properties of wood is essential for designing wood structures. Wood is different from many materials in that the mechanical properties are different in different grain directions (parallel and perpendicular to the grain), in tension and compression, and in small and large-size specimens. Figure 7.11 shows several important ways in which wood can be loaded, each producing a different failure mode.

7.6.1 Properties at Normal Temperatures

7.6.1.1 Design Values

Structural design with timber requires information on the design strength of the material. In the United States there are two recognized formats for engineering design of wood—the traditional *allowable stress design* (ASD) method and the new *load and resistance factored design* (LRFD) method. Design values for ASD are contained in the *Allowable Stress Design (ASD) Manual for Engineered Wood Construction*, 2001 Version [51]. Design values for LRFD are contained in the *Load and Resistance Factor Design (LRFD) Manual for Engineered Wood Construction*, 1996 Version [52].

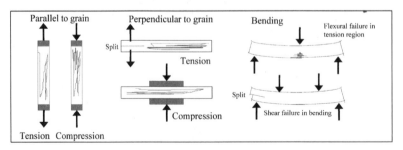

FIGURE 7.11 Loading of wood members in different directions.

7.6.2 Properties at Elevated Temperatures

7.6.2.1 Sources

The most comprehensive review on the effects of moisture content and temperature on the mechanical properties of wood was conducted by Gerhards [53], who reported the results of many previous studies. The most significant of this information has been summarized in the *Wood Handbook* [5] and forms the basis of this section.

7.6.2.2 Temperature Range

In most conditions, pyrolysis of wood begins at approximately 200°C and the remaining wood fiber converts to char by 300°C. The range of interest for fire engineering is therefore from room temperature to approximately 300°C.

7.6.2.3 Effect of Moisture Content

The strength of wood at elevated temperatures is not well understood. In addition to temperature, the effect of moisture content is very important in timber members exposed to fires. For this reason, the influence of moisture content on elevated temperature response will be included in this section, wherever possible.

Moisture content in wood is sensitive to the test method and size of the test specimen. Most often, test specimens are at a certain moisture content before the test and are allowed to dry out when heated during the test. As a specimen dries, some moisture may migrate into the interior of the specimen; moisture gradients depend on the size of the specimen. If wood is heated to temperatures above 100°C, all free moisture will evaporate after some time, depending on the permeability of the particular species.

7.6.2.4 Steam Softening

It is well known from the furniture and boat-building industries that hot moist wood can be bent into curved shapes using steam bending. Steam bending occurs because wood becomes plastic in compression under certain combinations of temperature and moisture content.

When wood is heated in a fire, conditions that produce softening of the wood may occur for a short period of time. If the moisture content decreases, the wood will "set" even though temperatures may continue to increase. The effect of wood softening will be very different for large-cross-section members versus small-cross-section members. In a large-cross-section member, conditions to produce softening occur in a thin layer. This layer progresses into the wood at about the same rate as char formation, thus having little effect on the overall strength and stiffness of the member. However, a small-cross-section member may experience these conditions over a large proportion of the cross section, in which case the member may deform plastically in compression or bending. In small-cross-section members, these conditions only occur for a short period of time, so if the assembly can resist applied loads, the wood will recover some of its strength and stiffness.

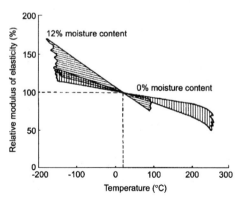

FIGURE 7.12 Modulus of elasticity of wood parallel to grain vs. temperature (*Wood Handbook*).

7.6.2.5 Parallel to Grain Properties

7.6.2.5.1 Modulus of Elasticity. Figure 7.12 represents the effect of elevated temperature on the modulus of elasticity of wood as reported in the *Wood Handbook*. The effect of temperature on the modulus of elasticity parallel to the grain is roughly linear up to 200°C.

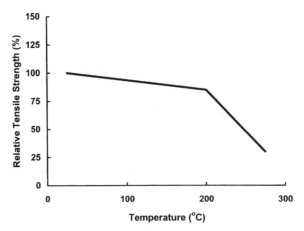

FIGURE 7.13 Tension strength of wood parallel to grain vs. temperature [54].

7.6.2.5.2 Tension Strength. Figure 7.13 shows the effects of elevated temperature on the tensile strength of dry wood as reported by Schaffer [54]. Tensile strength reduces linearly to approximately 85 percent of the initial room temperature strength at about 200°C. At temperatures above 200°C, tensile strength reduces linearly at a more rapid rate.

7.6.2.5.3 Compression Strength. Figure 7.14 represents the effect of elevated temperatures on compressive strength parallel to grain as reported in the *Wood Handbook*. These results are for both dry wood and wood at approximately 12 percent moisture content. There are no reported compression test results for moist wood over 60°C where plastic behavior might be expected. As discussed earlier in this chapter, plastic behavior of wood in compression can become important in some timber structures exposed to fire.

7.6.2.5.4 Bending Strength. Bending behavior in wood can be best described from an understanding of tension and compression behavior, as described in an earlier section. The effect of elevated temperatures on bending behavior is, in theory, predictable from the information presented in Figs. 7.12, 7.13, and 7.14. Dry wood will lose strength and stiffness at the rate given in those figures. However, moist wood is much more complicated. If conditions become suitable for plastic behavior, large strains will occur in the compression zone, resulting in a relocation of the neutral axis, leading to large deformations. The shaded areas on Fig. 7.15 represent the results reported in the *Wood Handbook*.

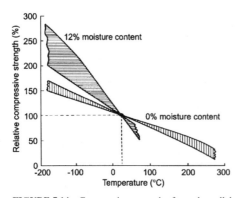

FIGURE 7.14 Compression strength of wood parallel to grain vs. temperature (*Wood Handbook*).

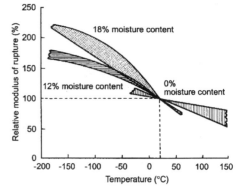

FIGURE 7.15 Bending strength of wood parallel to grain vs. temperature (*Wood Handbook*).

7.6.2.5.5 Other Properties. Gerhards reports the results of an analysis of other strength properties of wood at elevated temperatures, generally below 100°C [53]. Although this report provides useful information, it should be limited to in-use applications at or below 100°C.

7.7 REACTION TO FIRE (S.I. UNITS)

The subject of this section is the reaction to fire of wood, i.e., ignition, heat release, production of smoke and toxic products of combustion, and surface flame spread. In North America, reaction to fire is also referred to as material flammability.

7.7.1 Ignition

7.7.1.1 Piloted Ignition

When a combustible material is exposed to a constant external heat flux (radiative, convective, or a combination), its surface temperature starts to rise. The temperature inside the solid also increases with time, but at a slower rate. Provided the net flux into the material is sufficiently high, the surface temperature eventually reaches a level at which pyrolysis begins. The fuel vapors generated emerge through the exposed surface and mix with air in the boundary layer. Under certain conditions this mixture exceeds the lower flammability limit and ignites. The initiation of flaming combustion as described above is termed flaming ignition. For some materials or under certain conditions, combustion is not in the gas phase but in the solid phase. In such cases, no flame can be observed and the surface is glowing. This quite different phenomenon is termed glowing ignition.

This section deals with piloted ignition of wood, i.e., flaming combustion of wood exposed to an external (radiant) heat source initiated near a small pilot located in the wake. This pilot may be a small gas flame (premixed or diffusion), an electric spark, or a glowing wire. It is small enough so that locally it does not provide any significant heat flux to the sample in addition to the external heat flux.

Piloted ignition of wood has been studied extensively over the past 40 to 50 years. These studies usually involved laboratory-scale experiments to measure the time to ignition at different levels of incident heat flux from a radiant panel.

7.7.1.1.1 Heating Regimes. A distinction has to be made between two heating regimes:

1. Short duration (minutes) exposure to high heat fluxes (≥ 10 kW/m²)

2. Prolonged (hours) exposure to low heat fluxes (<10 kW/m²)

Piloted ignition experiments usually involve heating regime 1, and lead to flaming ignition, which at low heat fluxes may be preceded by some glowing. Experiments are intended to simulate the response of the material to the heat transfer from a fire that has already been initiated, for example, an interior finish material exposed to the heat flux from a burning object in a compartment or an exterior cladding of a structure adjacent to a burning building.

Although the experimental data are scarce, there is an indication that heating regime 2 generally leads to glowing ignition with a possible transition to flaming combustion after additional prolonged thermal exposure. In this case, experiments are intended to assess the ignition propensity of a material exposed to a low-intensity heat source, i.e., ignition of the material will initiate a fire.

7.7.1.1.2 Piloted Ignition Properties. The data resulting from a series of piloted ignition experiments consist of ignition times over a range of heat flux levels. How can this information be used to predict if and when a material will ignite when exposed to the dynamic conditions of a real fire? The most common approach involves analysis of the data based on a simplified model to extract material properties. The properties in conjunction with a variant of the model can then be used to predict ignition behavior in real fires. Three properties are commonly used to describe the ignition behavior of a material.

1. The first property quantifies a critical condition for ignition. Piloted ignition occurs when the lower flammability limit is reached in the fuel-air mixture around the pilot. Consequently, a critical mass flux criterion seems to be logical. This has been proposed [55], but, unfortunately, it is not very practical. The most common criterion is based on the assumption that ignition occurs when a critical temperature at the surface, T_{ig}, is reached.

2. The second property is the minimum heat flux for ignition $\dot{q}''_{min}$. This heat flux is just sufficient to heat the material surface to T_{ig} for very long exposure times (theoretically ∞). The minimum heat flux is not a true material property, because it depends on the rate of convective cooling from the surface. This, in turn, depends primarily on the orientation, size, and flow field around the exposed surface. Because these are different in a small-scale test versus a real fire, the minimum heat flux determined based on test data is an approximate value. To make the distinction, it is referred to as the critical heat flux for ignition, $\dot{q}''_{cr}$. Since the rate of convective cooling in a small-scale test is generally smaller than in a real scale, $\dot{q}''_{cr}$ is a conservative estimate of $\dot{q}''_{min}$.

3. The third property is the thermal inertia $k\rho c$. This property is a measure of how fast the surface temperature of a material rises when exposed to heat. A material with lower $k\rho c$ will ignite faster than a material with higher $k\rho c$ and the same T_{ig} exposed to the same heat flux.

Ignition properties can be determined by direct measurements. T_{ig} can be measured with fine thermocouples attached to the exposed surface of ignition test specimens, or by using an optical pyrometer. $\dot{q}''_{cr}$ can be determined by bracketing, i.e., by conducting experiments at incrementally decreasing heat flux levels until ignition does not occur within a specified period (usually 10 or 20 min). $k\rho c$ can be determined by measuring thermal conductivity, density, and specific heat separately. However, because k and c are temperature dependent, measurements at elevated temperature are needed.

Because it is very tedious to measure T_{ig} and $k\rho c$ directly, it is much more common to determine ignition properties on the basis of an analysis of time-to-ignition data obtained over a range of heat fluxes. The analysis is based on a simple heat-conduction model, which assumes that the solid is inert (negligible pyrolysis before ignition) and thermally thick (heat wave does not reach the back surface before ignition). Janssens conducted an extensive survey of procedures to analyze piloted ignition data of wood [55]. A few additional procedures have been published after this survey was completed [56–58]. Ngu applied seven of these procedures to analyze ignition data for seven New Zealand wood products [59]. Several methods have been developed to analyze ignition data of materials that do not show thermally thick behavior [56, 60, 61].

7.7.1.1.3 Piloted Ignition Data. A number of investigators measured T_{ig} for a range of wood products [17, 18, 57, 62–66]. Reasonably constant values were found for each material at heat fluxes at and above 25 kW/m². All studies reported a significant increase of T_{ig} at lower heat fluxes (50 to 150°C at 15 kW/m²). This is because pyrolysis and char formation at the surface are no longer negligible for ignition times exceeding 3 min. Janssens recommended the following values for oven dry wood at high heat fluxes [17]:

Softwoods: $T_{ig} = 350–365°C$

Hardwoods: $T_{ig} = 300–310°C$

These values are consistent with the measurements of other investigators. Janssens found that T_{ig} increases by approximately 2°C per percent increase in moisture content [17]. This is also consistent with other studies [57, 62]. Babrauskas proposed a value of $T_{ig} = 250°C$ for low heat fluxes [68]. This value is independent of moisture content because the wood dries out for prolonged exposure to a low heat flux.

The commonly accepted minimum heat flux for piloted ignition of wood is 12.5 kW/m². Law first used this value based on earlier work by Simms [69] in a procedure to calculate safe separation distances between buildings [70]. $\dot{q}''_{cr}$ for untreated wood products measured in small-scale tests over the past 30 years ranges from 10 to 18 kW/m². The value depends primarily on wood species (softwood vs. hardwood) and moisture content. Because flow conditions and convective cooling vary slightly, different critical heat fluxes have been obtained with different ignition test apparatuses. Dietenberger, for example, obtained critical heat flux values of 14.3 and 18.8 kW/m² for conditioned redwood in the cone calorimeter and lateral ignition and flame spread test (LIFT) apparatus, respectively [56].

Finally, Janssens found that the apparent $k\rho c$ obtained from analysis of piloted ignition data of wood is approximately equal to the literature values at a temperature halfway between ambient and T_{ig} [17]. Janssens' $k\rho c$ values ranged approximately from 0.09 to 0.4 kJ/m⁴·K²·s for dry wood with densities between 330 and 810 kg/m³, and increased by 30 to 40 percent for specimens conditioned to equilibrium at 23°C and 50 percent relative humidity before testing.

7.7.1.2 Autoignition

Autoignition of wood exposed to radiant heat is similar to piloted ignition, except that the hot surface triggers ignition of the flammable mixture of volatiles and air in the boundary layer. Hence, T_{ig} for autoignition is much higher than for piloted ignition. Abu-Zaid measured 510 and 550°C for Douglas fir with 0 and 17 percent moisture content, respectively [62]. His data indicate that the critical heat flux for autoignition is between 30 and 40 kW/m². This is consistent with the value of 33 kW/m² proposed by Law [70] based on earlier work by Simms [69].

7.7.1.3 Spontaneous Ignition

Spontaneous ignition or self-heating occurs when the heat generated by slow oxidation in a fuel exposed to air exceeds the heat losses to the surroundings. This leads to an increase in temperature, which in turn accelerates the chemical reactions and eventually leads to thermal runaway and glowing or flaming ignition. This process usually takes a long time (hours, days, or even longer). Whether spontaneous ignition occurs depends on the type, size, and porosity of the fuel array and the temperature of the surrounding air. Porous wood products such as piles of sawdust, fiberboard, and cellulosic fiber insulation are more susceptible to spontaneous ignition than solid wood and have been reported to ignite at temperatures as low as 80°C [71]. Ignition for piles of sawdust contaminated with oil has been observed at even lower temperatures [72].

7.7.2 Heat Release and Charring Rate

Heat release rate is the single most important variable in fire hazard [73]. Heat release rate at different heat fluxes can be measured in a bench-scale calorimeter. The most common devices used for this purpose are the OSU calorimeter (ASTM E 906, "Standard Test Method for Heat and Visible Smoke Release Rates from Materials and Products"), the cone calorimeter (ASTM E 1354, "Standard Test Method for Heat and Visible Smoke Release Rates for Materials and Products Using an Oxygen Consumption Calorimeter") and the fire propagation apparatus (ASTM E 2058, "Standard Test Methods for Measurement of Synthetic Polymer Material Flammability Using a Fire Propagation Apparatus").

Charring rate is closely related to heat release rate. The charring rate determines how fast the load-bearing section of structural wood members shrinks to a critical level. Charring rate under standard fire conditions can be measured in a furnace that is used for fire resistance testing. Fire resistance tests in North America are conducted according to ASTM E 119, "Standard Test Methods for Fire Tests of Building Construction and Materials." Fire resistance test standards in other parts of the world are based on ISO 834, "Fire Resistance Tests—Elements of Building Construction."

7.7.2.1 Heat Release Rate Data

A heat release rate curve measured in a bench-scale calorimeter is usually bimodal. Figure 7.16 shows a typical example of heat release rates measured in the cone calorimeter at different heat flux

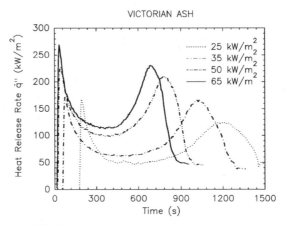

FIGURE 7.16 Typical heat release rate curves for wood measured in the cone calorimeter.

levels. Shortly after ignition, the heat release rate rises rapidly to the first peak. A protective char layer builds up as the pyrolysis front moves inward. The char layer forms an increasing thermal resistance between the exposed surface and the pyrolysis front, resulting in a continuous decrease of the heat release rate after the first peak. At some point, the surface recedes at approximately the same rate as the pyrolysis front and the heat release rate becomes steady. Bench-scale calorimeter specimens are usually backed by high-temperature insulation. The heat release rate will start to rise again when the pyrolysis front approaches the back surface. After the second peak, the heat release rate drops rapidly as flaming ceases and the char residue continues to smolder.

Janssens and Tran published extensive surveys of heat release rate data of wood [74, 75]. Janssens' measurements of the first peak heat release rate for a number of conditioned wood products are given in Table 7.7 [17]. Peak values at 50 kW/m² for untreated wood products range approximately from 180 to 230 kW/m². Wood can be treated with fire retardants to reduce the value well below 100 kW/m². The plywood in Table 7.7 was treated to obtain a Class I or A rating in the Steiner tunnel test (see below).

TABLE 7.7 Typical Peak Heat Release Rates of Wood Products

Material	Irradiance			
	25 kW/m²	35 kW/m²	50 kW/m²	65 kW/m²
Western red cedar	150	193	214	236
Redwood	129	169	200	230
Radiata pine	140	158	179	205
Douglas fir	155	192	207	223
Victorian ash	158	187	228	269
Blackbutt	176	195	231	274
Douglas fir plywood	170	180	197	215
Oriented strand board	177	168	198	224
Southern pine plywood	156	156	203	246
Particleboard	—	191	222	264
Fire-retardant treated plywood	—	62	81	101

7.7.2.2 Heat Release Properties

The two properties that are related to heat release rate are the effective heat of combustion $\Delta h_{c,\text{eff}}$ (MJ/kg) and the heat of gasification Δh_g (MJ/kg). These properties are described below.

7.7.2.2.1 Heat of Combustion. The effective heat of combustion is the ratio of heat release rate to mass loss rate measured in a small-scale calorimeter:

$$\Delta h_{c,\text{eff}} \equiv \frac{\dot{Q}''}{\dot{m}''} \tag{7.45}$$

where $\dot{Q}''$ = heat release rate per unit exposed area, kW/m²
$\dot{m}''$ = mass loss rate per unit exposed area, g/m²·s

$\Delta h_{c,\text{eff}}$ of wood is reasonably constant during flaming combustion, and increases rapidly to approximately 32 MJ/kg due to smoldering of the char. Typical values for the flaming phase based on Janssens' data are given in Table 7.8 [76]. Tran presented the following correlation between $\Delta h_{c,\text{eff}}$ and incident heat flux based on measurements in the OSU calorimeter [75]:

$$\Delta h_{c,\text{eff}} = 9.95 + 0.068 \dot{q}_e'' \tag{7.46}$$

with $\Delta h_{c,\text{eff}}$ expressed in megajoules per kilogram and $\dot{q}_e''$ in kilowatts per square meter.

The symbol $\Delta h_{c,\text{eff}}$ is used to make a distinction between this property and the lower calorific value measured in an oxygen bomb calorimeter, $\Delta h_{c,\text{net}}$. The latter is measured in a small container under high pressure and in pure oxygen, conditions that are not representative of real fires. The conditions in the new types of bench-scale calorimeters resemble those in real fires much more closely. For some fuels, in particular gases, $\Delta h_{c,\text{eff}}$ is nearly identical to $\Delta h_{c,\text{net}}$. However, for charring solids such as wood, $\Delta h_{c,\text{eff}}$ is significantly lower and equal to the heat of combustion of the volatiles during flaming combustion. The net heat of combustion of the volatiles is slightly higher for lignin than for the other constituents. Parker measured 14.7 MJ/kg for lignin, 13.8 MJ/kg for cellulose, and even lower values for some hemicelluloses of Douglas fir [21]. This is offset by the fact that more energy is required to generate volatiles from lignin, so that $\Delta h_{c,\text{eff}}$ is independent of lignin content [77]. The heat of combustion of wood measured in an oxygen bomb, which includes the heat released by the char, increases with lignin content as follows [4]

$$\Delta h_{c,\text{gross}} = 17.6 + 8.5 Y_l \tag{7.47}$$

where $\Delta h_{c,\text{gross}}$ is the gross heat of combustion of extractive-free wood (MJ/kg) and Y_l is the Klason lignin content (kg/kg).

The gross heat of combustion is higher than the net heat of combustion by the heat required to evaporate the water in the products of combustion. The difference is approximately 1.3 MJ/kg.

7.7.2.2.2 Heat of Gasification. The second material property is heat of gasification Δh_g, defined as the net heat flow into the material required to convert one mass unit of solid material to volatiles. The net heat flux into the material can be obtained from an energy balance at the surface of the

TABLE 7.8 Typical Heat of Combustion and Heat of Gasification Values of Wood Products

Material	$\Delta h_{c,\text{eff}}$ (MJ/kg)	Δh_g (MJ/kg)	$\Delta h_g / \Delta h_{c,\text{eff}}$
Western red cedar	13.1	3.27	0.25
Redwood	12.6	3.14	0.25
Radiata pine	11.9	3.22	0.27
Douglas fir	12.0	2.64	0.22
Victorian ash	11.7	2.57	0.22
Blackbutt	10.6	2.54	0.24

specimen. Typically, a sample exposed in a bench-scale calorimeter is heated by external heaters and by its own flame. Heat is lost from the surface in the form of radiation. Because of the small sample size, the flame flux is primarily convective, and flame absorption of external heater and specimen surface radiation can be neglected. Hence, Δh_g can be defined as

$$\Delta h_g \equiv \frac{\dot{q}''_{net}}{\dot{m}''} = \frac{\dot{q}''_e + \dot{q}''_f - \dot{q}''_l}{\dot{m}''} \qquad (7.47)$$

where Δh_g = heat of gasification, MJ/kg
$\dot{q}''_e$ = heat flux from external sources, kW/m²
$\dot{q}''_f$ = heat flux from flame, kW/m²
$\dot{q}''_l$ = heat losses from exposed surface, kW/m²

Using peak heat fluxes and mass loss rates in Eq. (7.47) leads to Δh_g values that range approximately from 5 to 8 MJ/kg [76]. Using averages leads to equally unrealistic values that range from 7 to 12 MJ/kg. The heat of gasification of wood actually varies with time. Typical average values measured by Janssens are given in Table 7.8. Peak values are approximately 70 percent higher than the averages. These results are consistent with theoretical calculations [78] and dynamic measurements by other investigators [79].

7.7.2.3 Charring Rate

Numerous studies have shown that charring rates of exposed wood members under standard fire exposure conditions are comparable, regardless of whether the test is conducted according to ASTM E 119 or ISO 834. An extensive review of charring rate data for wood was published by Hall et al. in 1976 [80]. The review indicated that an average charring rate of 0.6 mm/min is appropriate for design purposes. Since publication of the review, four major studies on the charring rate of wood have been reported. These studies are summarized next.

As part of his doctoral research, Robert White tested slabs of eight different wood species in a furnace according to ASTM E 119 [23]. The species that were tested included four softwoods and four hardwoods. They were selected on the basis of a factorial design, maximizing the relative effects of density, permeability, and chemical composition. Thermocouples were embedded into the specimens to measure temperature at various depths from the exposed surface. Tests were conducted after conditioning of specimens to equilibrium at 30, 50, 65, and 80 percent relative humidity and a temperature of 23 to 27°C. Based on his data, White concluded that the location of the char front x_c is a slightly nonlinear function of time given by

$$t = m x_c^{1.23} \qquad (7.48)$$

where x_c = distance of the 288°C isotherm from the initially exposed surface, mm
m = coefficient, min/mm$^{1.23}$
t = time, min

White correlated m against dry density, moisture content, char contraction factor, and other characteristics of the test specimens. Later, White and Nordheim reanalyzed the data and indicated that the effect of moisture is species dependent [81]. The fraction of water vapor generated that migrates toward the cold side increases with permeability. Because a flow of enthalpy is associated with this mass transfer, charring rate also increases slightly with permeability. White and Nordheim presented a linear regression equation for m as a function of dry density ρ_0, char contraction factor f_c, and moisture content u. All coefficients in this equation are constant, except for the moisture coefficient Z_i, which is species dependent. The equation was developed using White's measured values for f_c. This makes it difficult to apply the equation to species for which f_c values are not available. When using the equations in Sec. 7.5.2.1.4 to determine f_c values for the eight species tested by White, the following regression equation is obtained

$$m = 0.091 + 0.53 \cdot 10^{-3} \rho_0 + 0.233 f_c + Z_i u \qquad (7.49)$$

TABLE 7.9 Values of f_c and Z_i for the Wood Species Tested by White

Species	f_c	Z_i
Engelmann spruce	0.70	0.0224
Western red cedar	0.78	0.0111
Southern pine	0.68	0.0042
Redwood	0.76	0.0165
Hard maple	0.68	0.0059
Yellow poplar	0.66	0.0105
Red Oak	0.71	0.0162
Basswood	0.69	0.0060

where ρ_0 = oven dry density, kg/m³

f_c = char contraction factor

Z_i = empirical parameter

u = moisture content, %

The values for f_c and Z_i for the species tested by White are given in Table 7.9. Figure 7.17 compares measured and predicted values of m. Application of Eq. (7.49) for an arbitrary species is still difficult because values for Z_i are generally not available for species other than those listed in Table 7.9. However, it is usually sufficient to estimate Z_i based on how the permeability of the arbitrary species compares to those listed in Table 7.9.

Lache exposed specimens of two softwoods and three hardwoods to the German standard fire endurance test DIN 4102 [82]. The main objective of this study was to evaluate the effect on charring rate of species, moisture content, and density. Lache measured charring rates that were in the range of 0.55 to 0.80 mm/min. The results were consistent with previous measurements, with one exception: Lache's charring rates for Beech were much higher than previously reported values for high-density hardwoods. A unique feature of this research was that charring rates were measured in various ways. Consistency was found between the different methods. A novel technique consisted of the continuous measurement of displacement of steel needles that were pneumatically pushed through the char layer. The needles protruded through the furnace wall opposite the specimen.

Gardner and Syme exposed eight species of glulam beams to the standard heating conditions in accordance with the Australian fire resistance test standard AS 1530 Part 4 [83]. Density of the test

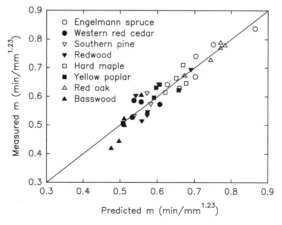

FIGURE 7.17 Comparison of measured and predicted values of m.

specimens ranged from 440 to 1050 kg/m³. Moisture content was between 8 and 15 percent. Thermocouples were inserted in the beams at various depths. Assuming a char base temperature of 288°C, char depth was calculated as a function of time based on the thermocouple data. The results were correlated against actual density of the test specimens, with the following result

$$x_c = 1.61 + \frac{413t}{\rho} \tag{7.50}$$

where x_c = char depth, mm

t = time, min

ρ = density, kg/m³

Finally, Collier made extensive measurements of the charring rate of Radiata pine at a moisture content of 12 percent, using AS 1530 Part 4 exposure conditions [84]. The measured charring rates appeared to be a linear function of density: $\beta = 0.75$ mm/min at $\rho_0 = 370$ kg/m³, $\beta = 0.7$ mm/min at $\rho_0 = 470$ kg/m³, and $\beta = 0.65$ mm/min at $\rho_0 = 570$ kg/m³. Based on this work, the design charring rate for Radiata pine in New Zealand was changed from 0.6 to 0.65 mm/min.

Parts 1 and 2 of Eurocode 5 gives the following equation to adjust the charring rate obtained under standard fire conditions to a room fire [85]:

$$\beta_{par} = 1.5\beta_0 \frac{5F - 0.4}{4F - 0.08} \tag{7.51}$$

with

$$F = \frac{A_v}{A_t}\sqrt{H_v} \tag{7.52}$$

where β_{par} = parametric charring rate, mm/min

β_0 = charring rate under standard fire exposure conditions, mm/min

F = ventilation factor, m$^{1/2}$

A_v = total area of vertical openings, m²

A_t = total area of floor, walls, and ceiling, m²

H_v = average height of vertical openings, m

β_{par} is valid between $t = 0$ min and $t = t_0 = 0.0006q_{tot}/F$, where q_{tot} (MJ/m²) is the design fire load density related to the total area of floor, walls, and ceiling of the compartment. Beyond t_0 the parametric charring rate decreases linearly to 0 mm/min at $t = 3t_0$.

To avoid the need for a detailed heat transfer calculation, Annex E of the Eurocode recommends generic temperature profiles. These profiles specify the temperature as a function of distance from the char front x. It is assumed the temperature at the char front T_p is equal to 300°C. This temperature is close to the value of 288°C commonly accepted in North America. The following generic temperature profile is given for when the member behaves as a semi-infinite solid:

$$T = T_i + (T_p - T_i)\left(1 - \frac{x}{a}\right)^2 \tag{7.53}$$

where T = temperature, °C

T_i = initial temperature, °C

T_p = char front temperature, 300°C

x = distance from the char front, mm

a = thermal penetration depth, 40 mm

A similar equation with an exponential term in place of a power term was reported by Schaffer [86]. Equation (1) is based primarily on German measurements of temperature profiles of wood slabs and beams exposed in a furnace according to the ISO 834 standard temperature-time curve. Janssens and White demonstrated that Equation (7.53) fits White's temperature data reasonably well, but recommended 35 mm for the value of a [87].

7.7.2.4 Pyrolysis Models

More than 50 different mathematical models for the pyrolysis of wood have been developed since World War II [18, 45, 76, 88–136]. These models range from simple approximate analytical equations to very complex numerical solutions of the conservation equations. They vary widely in complexity depending on the physical and chemical phenomena that are included and the simplifying assumptions that are made. Some address both heat and mass transfer, whereas others completely ignore migration of water and/or fuel vapors. There are two main application areas for such models: (1) use of wood fuel for energy generation, and (2) fire performance of wood.

Ten of the models in the second category were specifically developed for structural applications [96, 100, 103, 105, 112, 121, 128, 130, 136]. The remaining models in the second category were developed to predict the flammability of wood in building fires or the burning behavior of forest fuels.

It is relatively easy to write down a comprehensive set of model equations [137]. The main equation expresses the conservation of energy as follows:

$$\rho c_p \frac{\partial T}{\partial t} + \nabla \cdot (\rho_g \bar{v}_g c_g T) = \nabla \cdot (k \nabla T) - \dot{r}_v (\Delta h_v + \Delta h_w) - \dot{r}_p \Delta h_p \tag{7.54}$$

where ρ = density of wood, partially charred wood, or char, kg/m^3

c_p = specific heat of wood, partially charred wood, or char, J/kg·K

T = temperature, K

t = time, s

ρ_g = density of volatiles, kg/m^3

$\bar{v}_g$ = velocity vector of the volatiles, m/s

c_g = specific heat of volatiles, J/kg·K

k = thermal conductivity of wood, partially charred wood, or char, W/m·K

$\dot{r}_v$ = vaporization rate of water, kg/s

Δh_v = heat of vaporization of water, J/kg

Δh_w = heat of wetting, J/kg

$\dot{r}_p$ = generation rate of pyrolyzates, kg/s

Δh_p = heat of pyrolysis, J/kg

Solving the equations is not so easy. Moreover, obtaining material properties can be a monumental task.

7.7.2.5 Smoldering Combustion

Smoldering is a slow exothermic surface reaction between a solid fuel and oxygen in the air. Oxygen is needed to support smoldering combustion, but it is consumed at a much smaller rate than in flaming fires. Smoldering fires involve a low rate of mass loss per unit time, but a larger share of lost mass is released as products of incomplete combustion, in particular *carbon monoxide* (CO), than in flaming fire conditions. Smoldering can occur in porous cellulosic materials such as fiberboard, and progress at velocities between 1.3×10^{-4} and 1.3×10^{-3} mm/s depending on the rate of air supply [118].

7.7.3 Products of Combustion

7.7.3.1 Smoke

The production rate of smoke by wood products in a fire is relatively low compared to plastics. Janssens measured average specific extinction areas in the cone calorimeter between 6 and 76 m^2/kg for different wood products and heat fluxes ranging from 25 to 65 kW/m^2. This is below the value of 110 m^2/kg for black PMMA, which is a low-smoke-producing plastic that is commonly used to check the heat release rate measurements in the cone calorimeter.

7.7.3.2 Toxicity

The primary toxic gas that is generated in the combustion of wood is carbon monoxide. Concentrations are very low under overventilated conditions (of the order of 100 ppm) [67]. They increase dramatically (to several percent) when air supply is at or below stoichiometric, although the concentration under those conditions seem to be independent of the fuel [67A].

The University of Pittsburgh, or UPitt, method is used to demonstrate compliance with the requirement in the New York City building code that no product shall be more toxic than wood. The UPitt method is the only toxicity test relying on animal response (bioassay) that is specified by a code or regulation in the United States. A small sample of the product is heated in a muffle furnace, and four mice are exposed to the products of combustion diluted with air. The furnace temperature is ramped at a rate of 5°C/min. The test is terminated after 30 min. The objective is to find the quantity of the product in grams that results in 50 percent mortality of the test animals. A product meets the requirements if this quantity, referred to as the LC_{50}, is equal to or greater than 19.5 g (the value generically assigned to wood).

7.7.4 Surface Flame Spread

Flames can spread over a solid surface in two modes. The first mode is referred to as wind-aided flame spread. In this mode, flames spread in the same direction as the surrounding airflow. Typical examples are upward flame spread over a vertical wall and flame spread over a ceiling. The region ahead of the pyrolysis front is heated primarily by the flame of the burning fuel section. Due to this heat feedback mechanism, wind-aided flame spread may be extremely rapid (typically 0.1 to 1 m/s). Therefore, the propensity for wind-aided flame spread is a key factor in the overall fire hazard assessment of a lining material.

The second mode is referred to as opposed-flow flame spread, which occurs when flames spread in the opposite direction of the surrounding airflow or against gravity. Heat transfer from the flame to the virgin fuel is significant only over a small zone close to the flame foot. Consequently, opposed-flow flame spread is much slower than wind-aided spread (typically 0.1 to 1 mm/s).

7.7.4.1 Wind-Aided Flame Spread

The Steiner tunnel test is the most common reaction-to-fire test method prescribed by U.S. model building codes. The primary intent of the test is to assess the wind-aided flame spread propensity of interior finish materials. The Steiner tunnel test is described in ASTM E 84, "Standard Test Method for Surface Burning Characteristics of Building Materials." The apparatus consists of a long tunnel-like enclosure measuring 8.7 × 0.45 × 0.31 m (25 × 1½ × 1 ft). The test specimen is 7.6 m (24 ft) long and 0.51 m (1.67 ft) wide, and is mounted in the ceiling position. It is exposed at one end, designated as the burner end, to a 79-kW (5000 Btu/min) gas burner. There is a forced draft through the tunnel from the burner end with an average initial air velocity of 1.2 m/s (240 ft/min). The measurements consist of flame spread over the surface and smoke obscuration in the exhaust duct of the tunnel. Test duration is 10 min. An extended 30-min test is used to qualify fire-retardant-treated wood products. A *flame-spread index* (FSI) is calculated on the basis of the area under the curve of flame tip location versus time. The FSI is 0 for an inert board, and is normalized to approximately 100 for red oak flooring.

The classification of linings in the model building codes is based on the FSI. There are three classifications: Class A, or I, for products with FSI ≤ 25; Class B, or II, for products with 25 < FSI ≤ 75; and Class C, or III, for products with 75 < FSI ≤ 200. Class A, or I, products are generally permitted in enclosed vertical exits. Class B, or II, products can be used in exit access corridors, and Class C, or III, products are allowed in other rooms and areas. Table 7.10 gives FSI values for a wide range of wood products taken from "DCA 1: Flame Spread Performance of Wood Products" [138].

TABLE 7.10 FSI Values for Wood Products

Material[1]	ASTM E-84 flame spread[2]	Source[3]	Material[1]	ASTM E-84 flame spread[2]	Source[3]
Lumber			*Softwood plywood (exterior glue[6])*		
Birch, Yellow	105–110	UL	Cedar 3/8″	70–95	APA
Cedar, Pacific Coast Yellow	78	CWC	Douglas Fir 1/4″	150	APA
Cedar, Western Red	70	HPVA	Douglas Fir 5/16″	115–155	APA
Cedar, Western Red	73	CWC	Douglas Fir 3/8″	110–150	APA
Cherry 3/4″	76	HPVA	Douglas Fir 1/2″	130–150	APA
Cottonwood	115	UL	Douglas Fir 5/8″	95–130	APA
Cypress	145–150	UL	Hemlock 3/8″	75–160	APA
Elm 3/4″	76	HPVA	Southern Pine 1/4″	95–110	APA
Fir, Douglas	70–100	UL	Southern Pine 3/8″	100–105	APA
Fir, Douglas 3/4″ flooring	83–98	WEY	Southern Pine 5/8″	90	APA
Fir, Amabilis (Pacific Silver)	69	CWC	Redwood 3/8″	95	UL
Fir, White	65	HPVA2	Redwood 5/8″	75	UL
Gum, Red	140–155	UL			
Hem-Fir Species Group[5]	60	HPVA2	*Hardwood plywood[7]*		
Hemlock, West Coast	60–75	WEY, UL	Ash 3/4″–Particleboard Core	134	HPVA
Larch, Western	45	HPVA2	Birch 1/4″–Douglas Fir Veneer Core	135–173	HPVA
Maple (flooring)	104	CWC	Birch 1/4″–Fuma Veneer Core	127	HPVA
Oak, Red or White	100	UL	Birch 3/4″–Douglas Fir Veneer Core	114	HPVA
Oak, Red 3/4″	84	HPVA	Birch 3/4″–High Density Veneer Core	114	HPVA
Oak, White 3/4″	77	HPVA	Birch 3/4″–Particleboard Core	124	HPVA
Pecan 3/4″	84	HPVA	Birch 3/4″–MDF Core	134	HPVA
Pine, Eastern White	85	CWC	Honduras Mahogany 3/4″–Particleboard Core	105	HPVA
Pine, Idaho White	72	HPVA	Lauan 11/64″	167	NIST
Pine, Idaho White	82	WEY	Lauan 1/4″	150	HPVA
Pine, Lodgepole	98	WEY	Oak 1/4″–Douglas Fir Veneer Core	153	HPVA
Pine, Northern White	120–215	UL	Oak 3/4″–MDF Core	123	HPVA
Pine, Ponderosa[4]	105–230	UL			
Pine, Ponderosa	115	HPVA2	*Particleboard*		
Pine, Red	142	CWC	3/16″ (Aromatic Cedar Flakeboard)	156	HPVA
Pine, Southern Yellow	130–195	UL	3/8″	200	UL
Pine, Sugar	95	HPVA2	1/2″	135	HPVA
Pine, Western White	75	UL	1/2″	156	NIST
Poplar, Yellow	170–185	UL	5/8″	153	NIST
Redwood	70	UL	11/16″	168	UL
Redwood 3/8″	102	UL	3/4″	145	UL
Spruce, Engelmann	55	HPVA2	3/4″ (Exterior Glue[5])	88–98	APA2
Spruce, Northern	65	UL			
Spruce, Sitka	74	CWC	*Medium density fiberboard–MDF*		
Spruce, Western	100	UL	3/8″	140	UL
Walnut	130–140	UL	7/16″	125	HPVA
Walnut 3/4″	101	HPVA	5/8″	120	HPVA
			11/16″	140	UL
Oriented strand board, waferboard (exterior glue[6])			3/4″	140	HPVA
5/16″	127–138	APA2	23/4″	140	HPVA
7/16″	86–150	APA	3/4″	130	HPVA
1/2″	74–172	APA2	1″	90	UL
3/4″	147–158	APA2			
			Shakes and shingles		
			Western Red Cedar Shakes 1/2″	69	HPVA
			Western Red Cedar Shingles 1/2″	49	HPVA

TABLE 7.10 *Continued*

Reprinted courtesy, American Forest & Paper Association, Washington, D.C.

[1] Thickness of material tested is one-inch nominal except where indicated.

[2] The ASTM E-84 test method has been revised a number of times during the years referenced by the source reports. However, the E-84 test apparatus has changed little over this period. Slightly different flame spread indices, usually lower, result when recent E-84 flame spread calculation techniques are applied to older wood product data. These changes in flame spread indices are not sufficient to change the flame spread class for the wood products described in this report.

[3] Sources:
APA-APA-The Engineered Wood Association, Research Reports 128, Revised, August 1979.
APA2-APA-The Engineered Wood Association Test Results
CWC-*Wood and Fire Safety*, Canadian Wood Council, 1991.
HPVA-Hardwood Plywood and Veneer Association, Test Reports, 202, 203, 335, 336, 337, 592, and 596; Special flame spread performance tests, Aug. 1974; T9234, T9237, T9317, T9344, T9354, May 1995; T9422, T9430, T9431, T9453, T9665, Feb/July 1997.
HPVA2-Hardwood Plywood and Veneer Association, March/April 1995; October/December 2000.
NIST-National Institute of Standards and Technology (formerly National Bureau of Standards), Technical Notes 879 and 945.
UL-Underwriter's Laboratory, UL 527, May 1971; Subject 723, Assignment 71SC509, Mar 15 & 16, 1971; Assignment 84NK1898, File R10917, Mar 9, 1984.
WEY-Weyerhaueser Fire Laboratory, 1973, 1987, January & February 1988.

[4] Average of 18 tests was 154 with three values over 200.

[5] The Hem-Fir Species Group represents six species: Californian Red Fir, Grand Fir, Nobel Fir, Pacific Silver Fir, Western Hemlock, and White Fir. The reported flame spread index represents a product containing a mixture of these species. When lumber is from a single species refer to the specific species flame spread index.

[6] Exposure 1 or exterior.

[7] Flame spread of plywood is affected by the species of the face veneer but can also be influenced by the species of the underlying core veneer. Various panel constructions involving certain core species show a relatively high degree of variability and potential to yield flame spread values above 200. Panel constructions involving cores of aspen, sumauma, yellow poplar and white fir have exhibited this behavior with average flame spread indices ranging from 78 to 259. Other factors, in addition to species, including material and process variables related to specific manufacturers can also affect flame spread. Thus, for plywood panels with certain core species, test data from the actual manufacturer is particularly important in establishing the flame spread classification of the product.

Copyright © 1997, 1998, 2001, 2002
American Forest & Paper Association, Inc.

7.7.4.2 Opposed-Flow Flame Spread

An analytic solution for opposed-flow spread of a laminar diffusion flame over a thin fuel sheet was first obtained by deRis [139]. Although the functional form of the solution was correct, approximations made by deRis resulted in some errors in the proportionality constants. More recently, Delichatsios pointed this out and obtained the exact solution [140]. Nevertheless, the solution is still commonly referred to as the "deRis equation." Quintiere found that the deRis equation is also valid for turbulent flame spread over thick fuel sheets [141], and simplified it to the following form:

$$V = \frac{\phi}{k\rho c (T_{ig} - T_s)^2} \quad (7.55)$$

where V = opposed-flow flame spread velocity, mm/s

ϕ = flame heating parameter, kW²/m³

$k\rho c$ = thermal inertia, kJ²/m⁴·K²·s

T_{ig} = ignition temperature, °C

T_s = surface temperature ahead of the flame front, °C

Quintiere and Harkleroad used the LIFT apparatus for ignition tests to determine $k\rho c$ and T_{ig} for a wide range of materials [142]. Analysis of LIFT flame spread data following Eq. (7.52) resulted in values of ϕ for the same set of materials. These data together with Eq. (7.52) can be used to predict opposed-flow flame spread rates over these materials in real fire conditions.

Of considerable interest is the question under what conditions opposed-flow flame spread ceases. A convenient criterion has been used by Quintiere and Harkleroad, namely the minimum surface temperature for spread $T_{s,\min}$ [142]. It is a characteristic of the material and can be measured in the LIFT apparatus. If the surface temperature just ahead of the pyrolysis front is lower than $T_{s,\min}$, the gas phase heat conduction from the flame is insufficient to raise the fuel temperature to T_{ig}. $T_{s,\min}$ for

PMMA is equal to ambient temperature. Consequently no externally imposed heat flux is needed to sustain opposed-flow spread over a PMMA surface.

Janssens obtained values of ϕ and $T_{s,min}$ for 10 wood species from an analysis of LIFT data [17]. Rather than using the method of Quintiere and Harkleroad, a new and improved procedure was developed for this purpose. The flame heating parameter of specimens conditioned at 23°C and 50 percent ranged from 1.7 to 8.8 kW²/m³. $T_{s,min}$ for the same specimens ranged from 73 to 183°C, and was in all cases higher than $T_{s,min}$ for dry specimens. The general conclusion from these measurements is that flames spread very slowly over wood surfaces in the direction against that of the surrounding air flow or against gravity.

7.7.4.3 Room/Corner Fire Growth

Room/corner tests have been used for more than two decades to assess the fire performance of wall and ceiling linings in a realistic configuration and scale. Three standard room/corner test protocols are now available (ISO 9705, "Fire Tests—Full-Scale Room Test for Surface Products"; NFPA 265, "Standard Methods of Fire Tests for Evaluating Room Fire Growth Contribution of Textile Wall Coverings"; and NFPA 286, "Standard Methods of Fire Tests for Evaluating Contribution of Wall and Ceiling Interior Finish to Room Fire Growth"), and are specified in codes and regulations for qualifying interior finishes. The three room/corner test standards are very similar and are briefly described in the following three paragraphs.

The apparatus consists of a room measuring 3.6 m (12 ft) deep by 2.4 m (8 ft) wide by 2.4 m (8 ft) high, with a single ventilation opening (doorway) measuring approximately 0.8 m (30 in.) wide by 2 m (80 in.) high in the front wall. Walls and ceiling are lined for standard tests according to ISO 9705. For tests according to the NFPA standards, the interior surfaces of all walls (except the front wall) are covered with the test material. NFPA 286 is also suitable for evaluating ceiling finishes (see below).

The test material is exposed to a propane burner ignition source, located on the floor in one of the rear corners of the room opposite the doorway. The burner is placed directly against (ISO 9705 and NFPA 286) or at a distance of 50 mm (2 in.) (NFPA 265) from the specimen mounted on the back and sidewalls. The ISO burner consists of a steel sandbox measuring $0.17 \times 0.17 \times 0.17$ m ($6.7 \times 6.7 \times 6.7$ in.) Propane is supplied to the burner at a specified rate such that a net heat release rate of 100 kW is achieved for the first 10 min of the test, followed by 300 kW for the remaining 10 min (20 min test duration, unless terminated when flashover occurs). The NFPA burner consists of a steel sandbox measuring $0.305 \times 0.305 \times 0.152$ m ($12 \times 12 \times 6$ in.), with the top surface positioned 0.305 m (12 in.) above the floor of the room. Propane is supplied at a specified rate such that a net heat release rate of 40 kW is achieved for the first 5 min of the test, followed by 150 kW (NFPA 265) or 160 kW (NFPA 286) for the remaining 10 min (15 min test duration unless terminated when flashover occurs). A fundamental difference between NFPA 265 and NFPA 286 is the fact that the flame from the burner alone just touches the ceiling in NFPA 286, making it suitable for assessing the fire performance of interior ceiling finish, an application for which NFPA 265 is unsuitable. This effect is somewhat due to the higher energy release rate of the NFPA 286 burner, but primarily due to the burner being in direct contact with the walls, thereby reducing the area over which the flames can entrain air and increasing the overall flame height.

Heat release rate is measured on the basis of oxygen consumption. Instrumentation for measuring rate of heat release and smoke production is installed in the exhaust duct connected to a fume collection hood located outside the room immediately adjacent to the doorway. The duct instrumentation consists of thermocouples for measuring exhaust gas temperature, a bidirectional probe for measuring exhaust gas velocity, a collimated white light or monochromatic laser light system for measuring smoke obscuration, and probes for sampling oxygen and carbon oxides. The room contains a single heat flux gauge in the center of the floor. The NFPA standards also specify that seven thermocouples be installed in the upper part of the room and doorway to measure the temperature of hot gases that accumulate beneath the ceiling and flow through the doorway. In addition to quantitative heat release and smoke production rate measurements, time to flashover (if it occurs) is one of the main results of a room/corner test. Different criteria are commonly used to define flashover, e.g., upper layer temperature of 600°C (1100°F), flames emerging through the doorway, heat flux to the floor of 20 kW/m², etc.

Two programs were conducted in the late 1980s and early 1990s to evaluate the performance of wood products when tested according to NFPA 286 [143, 144]. With one exception, flashover occurred for all untreated wood products within a few minutes after the burner increase from 40 to 160 kW. Flashover occurred 34 s before the increase for oriented strand board. Flashover did not occur for ASTM E 84 Class A or I fire-retardant-treated plywood. Very few tests have been conducted on wood products according to NFPA 265, but one study indicates that the results are not too different from those for NFPA 286 [145].

A compilation of room test data by White et al. indicates that flashover occurs for untreated wood products after 2 to 3 min of exposure in an ISO 9705 room test with walls and ceiling lined [146]. However, wood products can be treated with fire retardants so that flashover does not occur within the 20-min test period. The same study concludes that the time to flashover in the ISO 9705 test with material on walls only is most consistent with the ASTM E 84 flame spread classification.

7.8 FIRE RESISTANCE (U.S. CUSTOMARY UNITS)

This section deals with the fire resistance of wood structures and consists of two parts. The first part describes two calculation methods to determine the time to failure of unprotected large wood members exposed under standard fire conditions. A simple method to determine the fire resistance of protected light wood-frame assemblies is discussed in the second part.

7.8.1 Exposed Wood Members

Large wood members have long been recognized for their ability to maintain structural integrity while exposed to fire. Early mill construction from the 19th century utilized massive timbers to carry large loads and to resist structural failure from fire. The superior fire performance of heavy timbers can be attributed to the charring effect of wood. As wood members are exposed to fire, an insulating char layer is formed that protects the core. Thus, beams and columns can be designed so that a sufficient cross section of wood remains to sustain the design loads for the required duration of fire exposure. A standard fire exposure is used for testing for building code compliance.

7.8.1.1 Concepts of Heavy Timber Fire Design

As illustrated in Fig. 7.18 for a section of a beam exposed on three sides, at fire exposure time t, the initial breadth B and depth D of a member are reduced to b and d, respectively. The original section is rectangular. However, because the corners are subject to heat transfer from two directions, charring is faster at these corners. This has the effect of rounding the corners; therefore, shortly after ignition, the remaining cross section is no longer rectangular. The boundary between the char layer and the remaining wood section is quite distinct, and corresponds to a temperature of approximately 550°F. The remaining wood section is heated over a narrow region that extends to a maximum of approximately 1.5 in. from the char front. The inner core of the remaining wood section is at ambient (or initial) temperature. A section smaller than the original section is capable of supporting the design load because of the safety margin provided in cold design. The original section is stressed only to a fraction of maximum capacity. Failure occurs when the remaining cross section is stressed beyond maximum capacity.

For members stressed in bending during fire exposure, failure occurs when bending capacity is exceeded due to reduction in section modulus S. For members stressed in tension parallel-to-grain during fire exposure, failure occurs when tension capacity is exceeded due to reduction in cross-sectional area A.

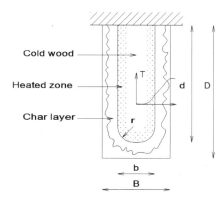

FIGURE 7.18 Heavy timber beam section exposed on three sides to fire.

For members stressed in compression parallel-to-grain during fire exposure, the failure mode is a function of the column slenderness ratio (L_e/D). The column slenderness ratio changes with exposure time. For short-column members ($L_e/D \approx 0$) stressed in compression during fire exposure, failure occurs when compressive capacity is exceeded due to reduction in cross-sectional area A. For long-column members ($L_e/D \approx \infty$) stressed in compression during fire exposure, failure occurs when critical buckling capacity is exceeded due to reduction in the moment of inertia I. Current code-accepted design procedures in the *2001 National Design Specification® (NDS®) for Wood Construction* [147] and the *1996 Standard for Load and Resistance Factor Design (LRFD) for Engineered Wood Construction* [148] contain a single column equation, which is used to calculate a stability factor C_p, approximating the column capacity for all slenderness ratios based on the calculated interaction of theoretical short- and long-column capacities.

7.8.1.2 Empirical Design Method

One code-accepted design method for fire-resistive exposed wood members used in North America is based on analysis conducted by T.T. Lie at the National Research Council of Canada in the 1970s [149]. This empirical method was first recognized by U.S. model building codes in 1984 through a *Council of American Building Official* (CABO) National Evaluation Board Report [150]. In subsequent years, the method was adopted directly into the building codes by the then-three model code organizations, allowing engineers and architects to include fire-rated heavy timber members in their projects without conducting expensive standard fire resistance tests [151].

The method assumes a charring rate of 1.42 in/h, and accounts for a reduction in strength and stiffness caused by heating of a small zone progressing approximately 1.5 in. ahead of the char front. Lie reported that studies showed that the ultimate strength and stiffness of various woods, at temperatures that uncharred wood normally reaches in fires, reduces to about 0.85 to 0.90 of the original strength and stiffness. To account for this effect, reductions to strength and stiffness properties were implemented by uniformly reducing strength and stiffness values over the remaining cross section by a factor α. Furthermore, a factor k was introduced to account for the ratio of design strength to ultimate strength. To obtain conservative estimates, Lie recommended a k factor of 0.33 based on a safety factor of 3, and an α factor of 0.8 to account for strength and stiffness reductions.

For simplification, the method ignores the increased rate of charring at the corners, and assumes that the remaining section is rectangular. With this assumption, initial breadth B and depth D of a member after t minutes of fire exposure are reduced to b and d, respectively, as shown in Fig. 7.19. Both b and d are a function of exposure time t and charring rate β.

7.8.1.2.1 Beams.
The method assumes that a beam fails when the reduction in cross section results in a critical value for the section modulus S being reached. Assuming a safety factor reduction of k, a load factor of Z, and a uniform reduction in strength properties of α, the critical section is determined from:

$$kZ\left(\frac{BD^2}{6}\right) = \alpha\left(\frac{bd^2}{6}\right) \tag{7.56}$$

Given the initial dimensions B (width) and D (depth) and the charring rate β, an equation can be derived to solve for the fire endurance time t. The roots to the resulting equations must be solved iteratively. To avoid these cumbersome iterative procedures, Lie approximated his solutions with a

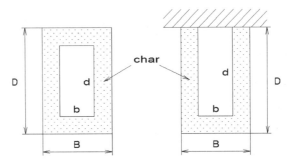

FIGURE 7.19 Reduction of load-bearing section due to charring.

set of simple equations that allow for a straightforward calculation of fire endurance time as a function of member size for a realistic range of member dimensions. The method approximates the solutions for $\alpha = 0.8$ and $k = 0.33$ to:

$$t = 2.54ZB\left(4 - \frac{2B}{D}\right) \quad \text{for 4-sided beam exposure} \quad (7.57)$$

$$t = 2.54ZB\left(4 - \frac{B}{D}\right) \quad \text{for 3-sided beam exposure} \quad (7.58)$$

with

$$Z = 1.3 \quad \text{for } R < 0.5 \quad (7.59)$$

$$Z = 0.7 + \frac{0.3}{R} \quad \text{for } R \geq 0.5 \quad (7.60)$$

where R is the ratio of applied to allowable load, t is in minutes, and all dimensions are in inches.

7.8.1.2.2 Columns. As noted in the previous section, column failure mode depends on the slenderness ratio. Short columns fail when the reduction in cross section results in a critical value for the cross-sectional area A being reached. Assuming a safety factor reduction of k, a load factor of Z, and a uniform reduction in strength properties of α, the critical section is determined from:

$$kZ(BD) = \alpha(bd) \quad (7.61)$$

Long columns fail when the reduction in cross section results in a critical value for the moment of inertia I being reached. Assuming a safety factor reduction of k, a load factor of Z, and a uniform reduction in strength properties of α, the critical section is determined from:

$$kZ\left(\frac{BD^3}{12}\right) = \alpha\left(\frac{bd^3}{12}\right) \quad (7.62)$$

where D denotes the narrowest dimension of a column section and buckling is assumed to occur in the weakest direction.

Again, given the initial dimensions B (widest dimension), D (narrowest dimension), and the charring rate β, an equation can be derived to solve for the fire endurance time t for short columns and long columns. Again, to avoid the cumbersome iterative solution of these equations, Lie approximated his solutions with a set of simple equations. The method approximates the solutions for $\alpha = 0.8$ and $k = 0.33$ to:

$$t = 2.54ZD\left(3 - \frac{D}{B}\right) \quad \text{for 4-sided column exposure} \quad (7.63)$$

$$t = 2.54ZD\left(3 - \frac{D}{2B}\right) \quad \text{for 3-sided column exposure} \quad (7.64)$$

where Z for short columns ($K_e l/D \leq 11$) follows from

$$Z = 1.5 \quad \text{for } R < 0.5 \quad (7.65)$$

$$Z = 0.9 + \frac{0.3}{R} \quad \text{for } R \geq 0.5 \quad (7.66)$$

and Z for long columns ($K_e l/D > 11$) is given by

$$Z = 1.3 \quad \text{for } R < 0.5 \quad (7.67)$$

$$Z = 0.7 + \frac{0.3}{R} \quad \text{for } R \geq 0.5 \quad (7.68)$$

where R is the ratio of applied to allowable load, t is in minutes, and all dimensions are in inches.

7.8.1.2.3 Examples

Example 7.10 Douglas-fir glued laminated timber beams span $L = 18$ ft, and are spaced at $s = 6$ ft. The design loads are $q_{live} = 100$ lb/ft² and $q_{dead} = 25$ lb/ft². Timber decking nailed to the compression edge of the beams provides lateral bracing. Calculate the fire resistance time using Eq. (7.58).

For the structural design of the wood beam, calculate the maximum induced moment.
Calculate beam load:

$$w_{total} = s(q_{dead} + q_{live}) = (6)(25 + 100) = 750 \text{ lb/ft}$$

Calculate maximum induced moment:

$$M_{max} = w_{total} L^2/8 = (750)(18^2)/8 = 30{,}375 \text{ ft·lb}$$

Select a 6³/₄ × 13¹/₂-in. 24F visually graded Douglas-fir glulam beam with a tabulated bending stress F_b equal to 2400 lb/in².

Calculate beam section modulus:

$$S = bd^2/6 = (6.75)(13.5^2)/6 = 205.0 \text{ in}^3$$

Calculate the adjusted allowable bending stress (assuming $C_D = 1.0$: $C_M = 1.0$: $C_t = 1.0$: $C_L = 1.0$: $C_V = 0.98$)

$$F_b' = F_b C_D C_M C_t \text{ (lesser of } C_L, C_V) = 2400(1.0)(1.0)(1.0)(0.98) = 2343 \text{ lb/in}^2 \quad \text{(NDS 5.3.1)}$$

Calculate design resisting moment:

$$M' = F_b' S = (2343)(205.0)/12 = 40{,}032 \text{ ft·lb}$$

Structural check: $M' \geq M_{max}$ 40,032 ft·lb 30,375 ft·lb √

For the fire design of the wood beam, the loading is unchanged. Therefore, the maximum induced moment is unchanged.
Calculate the ratio of applied load to allowable load:

$$R = 30{,}375/40{,}032 = 0.76$$

Calculate the fire endurance time:

$$Z = 0.7 + 0.3/0.76 = 1.10 \text{ [Eq. (7.60)]}$$

$$t = 2.54ZB\,(4\,B/D) = 2.54\,(1.10)(6.75)(4 - 6.75/13.5) = 66 \text{ min [Eq. (7.58)]}$$

Example 7.11 A Southern pine glued laminated timber column with an effective column length, $L_e = 168$ in. The design loads are $P_{snow} = 16{,}000$ lb and $P_{dead} = 6000$ lb. Calculate the failure time using Eq. (7.63).

For the structural design of the wood column, calculate the maximum induced compression stress f_c.

Calculate column load:

$$P_{load} = P_{dead} + P_{snow} = 8000 + 16,000 = 22,000 \text{ lb}$$

Select an $8^{1}/_{2} \times 9^{5}/_{8}$-in. combination #48 Southern pine glulam column with a tabulated compression parallel-to-grain stress F_c, equal to 2200 lb/in², and a tabulated modulus of elasticity E, equal to 1,700,000 lb/in².

Calculate column area:

$$A = bd = (9.625)(8.5) = 81.81 \text{ in}^2$$

$$I = bd^3/12 = (9.625)(8.5)^3/12 = 492.6 \text{ in}^4$$

Calculate the adjusted allowable compression stress (assuming $C_D = 1.15$: $C_M = 1.0$: $C_t = 1.0$):

$$E' = E(C_M)(C_t) = 1,700,000(1.0)(1.0) = 1,700,000 \text{ psi (NDS 5.3.1)}$$

$$F_{cE} = 0.418 E'/(L_e/d)^2 = 0.418(1,700,000)/(168/8.5)^2 = 1819 \text{ psi (NDS 3.7.1.5)}$$

$$F_c^* = F_c(C_D)(C_M)(C_t) = 2200(1.15)(1.0)(1.0) = 2530 \text{ psi (NDS 3.7.1.5)}$$

$c = 0.9$ for glued laminated timbers (NDS 3.7.1.5)

$$\alpha_c = F_{cE}/F_c^* = 1819/2530 = 0.7190$$

$$C_p = \frac{1 + \alpha_c}{2c} - \sqrt{\left(\frac{1 + \alpha_c}{2c}\right)^2 - \frac{\alpha_c}{c}} = 0.6186 \text{ (NDS 3.7.1.5)}$$

$$F_c' = F_c^*(C_p) = 2530(0.6186) = 1565 \text{ psi (NDS 5.3.1)}$$

Calculate the resisting column compression capacity:

$$P' = F_c' A = (1565)(81.81) = 128,043 \text{ lb}$$

Structural check: $P' \geq P_{load}$ 128,043 lb $\geq$ 22,000 lb ✓

For the fire design of the wood column, the loading is unchanged. Therefore, the maximum induced column load is unchanged.

Calculate the ratio of applied load to allowable load:

$$R = 22,000/128,043 = 0.17$$

Calculate the fire endurance time, given $K_e L_e/d = 19.8$:

$Z = 1.3$ [Eq. (7.67)]

$t = 2.54 ZD(3 - D/B) = 2.54(1.3)(8.5)(3 - 8.5/9.625) = 59$ min [Eq. (7.63)]

7.8.1.3 Mechanics-Based Design Method

The method for calculating fire endurance of exposed, large wood members, developed by Lie, is based on actual fire test results and sound engineering. However, the final prediction equations are based on empirical solutions fit to beam and column test data; therefore, application of the Lie method is limited. A new mechanics-based design method was deemed necessary to permit the calculation of fire endurance for exposed, large wood members for other loading conditions and fire exposures not considered by Lie. As a result, a new mechanics-based design method has been developed and verified in a publication entitled "Technical Report 10: Calculating the Fire Resistance of Exposed Wood Members" [152]. This new method has been adopted into the fire design provisions of the 2001 *NDS*.

The new, mechanics-based design procedure calculates the capacity of exposed wood members using basic wood-engineering mechanics. Actual mechanical and physical properties of the wood are used and member capacity is directly calculated for a given period of time. Section properties are computed assuming an effective char rate β_{eff} at a given time t. Reductions in strength and stiffness of wood directly adjacent to the char layer are addressed by accelerating the char rate 20 percent. Average member strength properties are approximated from existing accepted procedures used to

TABLE 7.11 Effective Char Rates and Char Layer Thicknesses (for $\beta_n = 1.5$ inches/hour)

Required fire endurance (h)	Effective char rate, β_{eff} (in/h)	Effective char thickness, α_{char} (in)
1-Hour	1.80	1.8
1½-Hour	1.67	2.5
2-Hour	1.58	3.2

calculate design properties. Finally, wood members are designed using accepted engineering procedures found in the *NDS* for allowable stress design.

7.8.1.3.1 Char Rate. The effective char rate to be used in this procedure can be estimated from published nominal 1-h char rate data using the following equation:

$$\beta_{\text{eff}} = \frac{1.2\beta_n}{t^{0.187}} \qquad (7.69)$$

where β_{eff} = effective char rate, adjusted for exposure time t, in/h

β_n = nominal char rate, linear char rate based on 1-h exposure, in/h

t = exposure time, h

A nominal char rate β_n of 1.5 in/h is commonly assumed for solid-sawn and glued-laminated softwood members. For $\beta_n = 1.5$ in/h, the effective char rates β_{eff} and effective char layer thicknesses a_{char} for each exposed surface are given in Table 7.11.

Section properties can be calculated using standard equations for area, section modulus, and moment of inertia using reduced cross-sectional dimensions. Dimensions are reduced by the effective char layer thickness a_{char} for each surface exposed to fire.

7.8.1.3.2 Member Strength. For fire design, the estimated member capacity is evaluated against the loss of cross-sectional and mechanical properties as a result of fire exposure. Although the loss of cross-sectional and mechanical properties are addressed by reducing the section properties using the effective char layer thickness, the average member strength properties must be determined from published allowable design stresses. The average member capacity of a wood member exposed to fire for a given time t can be estimated using the average member strength and reduced cross-sectional properties. For solid-sawn and glued-laminated wood members, the average member capacity can be approximated by multiplying the allowable design capacity R by the factors K, given in Table 7.12.

Axial/bending stress interactions can be calculated using this procedure. All member strength and cross-sectional properties should be adjusted before solving the interaction calculations. The interaction calculations should then be conducted in accordance with appropriate *NDS* provisions.

TABLE 7.12 Allowable Design Stress to Average Ultimate Strength Adjustment Factor

Member Capacity	K
Bending moment capacity, in-lb	2.85
Tensile capacity, lb	2.85
Compression capacity, lb	2.58
Beam buckling capacity, lb	2.03
Column buckling capacity, lb	2.03

7.8.1.3.3 Design of Members. Once member capacity has been determined using effective section properties and member strength approximations, the wood member can be designed using accepted *NDS* design procedures for the following loading condition:

$$D + L \leq KR_{ASD} \tag{7.70}$$

where D = design dead load, lb
L = design live load, lb
R_{ASD} = nominal allowable design capacity, lb
K = factor to adjust from nominal design capacity to average ultimate capacity

7.8.1.3.4 Design Procedures for Timber Decks. Under U.S. building codes, timber decks must consist of planks that are at least 2 in. thick. The planks span the distance between supporting beams, and can be arranged in different ways depending on available lengths. Usually, a single or double tongue-and-groove (T&G) joint is used to connect adjoining planks, but splines or butted joints are also common.

In order to meet building code requirements for a given fire resistance rating, a timber deck needs to maintain its thermal separation function and load-carrying capacity for the specified duration of exposure to standard fire conditions. The thermal separation requirement limits the temperature rise on the unexposed side of the deck to 250°F above ambient temperature over the entire surface area, or 325°F above ambient temperature at a single location. When these limits cannot be met by decking alone, additional floor coverings can be used to increase the thermal separation time.

To meet the load-carrying capacity requirement, a deck must carry the specified load for the required endurance time. 2001 *NDS* fire design provisions also apply to design of timber decks. Single and double T&G decking should be designed as an assembly of wood beams fully exposed on one face. Butt-jointed decking should be designed as an assembly of wood beams partially exposed on the sides and fully exposed on one face. To compute the effects of partial exposure on the sides of decking, the char rate for this limited exposure should be reduced to 33 percent of the effective char rate.

7.8.1.3.5 Special Provisions for Glued Laminated Beams. For glued laminated timber bending members rated for 1-h fire endurance, an outer tension lamination shall be substituted for a core lamination on the tension side for unbalanced beams and on both sides for balanced beams. For glued laminated timber bending members rated for $1^{1}/_{2}$- or 2-h fire endurance, two outer tension laminations are substituted for two core laminations on the tension side for unbalanced beams and on both sides for balanced beams.

7.8.1.3.6 Examples

Example 7.12 Check the glued laminated timber beam sized in Example 7.10 for a 1-h fire resistance time using the mechanics design method.

From Example 7.10,

$$M_{max} = 30{,}375 \text{ ft·lb}$$

For fire design, determine the loss in cross section for a 1-h exposure:

$$a = 1.8 \text{ in. (NDS 16.2.1)}$$

Calculate beam section modulus exposed on three sides:

$$S = (b - 2a)(d - a)^2/6 = (6.75 - 3.6)(13.5 - 1.8)^2/6 = 71.9 \text{ in}^3$$

Calculate the adjusted allowable bending stress (assuming C_D = N/A: C_M = N/A: C_t = N/A: C_L = 1.0: C_V = 0.98)

$$F_b' = F_b \text{ (lesser of } C_L \text{ or } C_V) = 2400(0.98) = 2343 \text{ lb/in}^2 \text{ (NDS 5.3.1)}$$

Calculate strength-resisting moment:

$$M' = (2.85)F_b' S = (2.85)(2343)(71.9)/12 = 40{,}010 \text{ ft·lb (NDS 16.2.2)}$$

Fire check: $M' \geq M_{max}$ 40,010 ft·lb $\geq$ 30,375 ft·lb $\checkmark$

Example 7.13 Check the glued laminated timber column sized in Example 7.11 for a 1-h fire resistance time using the mechanics design method.

From Example 7.11,

$$P_{load} = 22{,}000 \text{ lb}$$

For fire design, determine the loss in cross section for a 1-h exposure

$$a = 1.8 \text{ in. (NDS 16.2.1)}$$

Calculate column area A and moment of inertia I for column exposed on four sides:

$$A = (b - 2a)(d - 2a) = (9.625 - 3.6)(8.5 - 3.6) = 29.52 \text{ in}^2$$
$$I = (b - 2a)(d - 2a)^3/12 = (9.625 - 3.6)(8.5 - 3.6)^3/12 = 59.07 \text{ in}^4$$

Calculate the adjusted allowable compression stress (assuming C_D = N/A: C_M = N/A: C_t = N/A):

$$F_{cE} = (2.03)0.418E'/(L_e/d)^2 = (2.03)(0.418)(1{,}700{,}000)/(168/(8.5 - 3.6))^2 = 1227 \text{ psi (NDS 16.2.2)}$$
$$F_c^* = (2.58)F_c = (2.58)(2200) = 5676 \text{ psi (NDS 16.2.2)}$$
$$\alpha_c = F_{cE}/F_c^* = 1227/5676 = 0.2162$$
$$C_p = \frac{1 + \alpha_c}{2c} - \sqrt{\left(\frac{1 + \alpha_c}{2c}\right)^2 - \frac{\alpha_c}{c}} = 0.2106 \text{ (NDS 3.7.1.5)}$$
$$F_c' = 5676(0.2106) = 1195 \text{ psi (NDS 5.3.1)}$$

Calculate the resisting column compression capacity:

$$P' = F_c' A = (1195)(29.52) = 35{,}280 \text{ lb}$$

Fire check: $P' \geq P_{load}$ 35,280 lb $\geq$ 22,000 lb √

Example 7.14 Solid sawn hem-fir timbers used as heavy timber truss webs. The total design tension loads from a roof live and dead load are P_{total} = 3500 lb. Calculate the required section dimensions for a 1-h fire resistance time.

For the structural design of the wood timber, calculate the maximum induced tension stress f_t.
Calculate tension load:

$$P_{load} = 3500 \text{ lb}$$

Select a nominal 6 × 6 (5½ × 5½ in.) hem-fir #2 grade timber with a tabulated tension stress F_t equal to 375 lb/in².

Calculate timber area:

$$A_s = bd = (5.5)(5.5) = 30.25 \text{ in}^2$$

Calculate the adjusted allowable tension stress (assuming C_D = 1.25: C_M = 1.0: C_t = 1.0):

$$F_t' = F_t C_D C_M C_t = 375(1.25)(1.0)(1.0) = 469 \text{ lb/in}^2 \text{ (NDS 4.3.1)}$$

Calculate the resisting tension capacity:

$$P' = F_t' A_s = (469)(30.25) = 13{,}038 \text{ lb}$$

Structural check: $P' \geq P_{load}$ 14,180 lb $\geq$ 3500 lb √

For the fire design of the timber tension member, the loading is unchanged. Therefore, the total load is unchanged. The fire resistance must be calculated.

Calculate column area A for a tension member exposed on four sides:

$$A_f = (b - 2a)(d - 2a) = (5.5 - 3.6)(5.5 - 3.6) = 3.61 \text{ in}^2$$

Calculate the adjusted allowable tension stress (assuming C_D = N/A: C_M = N/A: C_t = N/A):

$$F_t' = (2.85) F_t = (2.85)(375) = 1069 \text{ lb/in}^2 \text{ (NDS 16.2.2)}$$

Calculate the resisting tension capacity:
$$P' = F_t' A_f = (1069)(3.61) = 3858 \text{ lb}$$
Fire check: $P \geq P_{\text{load}}$ $3858 \text{ lb} \geq 3500 \text{ lb}$ ✓

Example 7.15 Hem-fir tongue-and-groove timber decking spans $L = 6$ ft. A single layer of $^3/_4$-in. sheathing is installed over the decking. The design loads are $q_{\text{live}} = 40$ lb/ft² and $q_{\text{dead}} = 10$ lb/ft².

Calculate deck load:
$$w_{\text{total}} = B(q_{\text{dead}} + q_{\text{live}}) = (5.5 \text{ in}/12 \text{ in/ft})(50 \text{ lb/ft}^2) = 22.9 \text{ lb/ft}$$

Calculate maximum induced moment:
$$M_{\text{max}} = w_{\text{total}} L^2/8 = (22.9)(6^2)/8 = 103 \text{ ft·lb}$$

Select nominal 3×6 ($2^1/_2 \times 5^1/_2$ in.) hem-fir commercial decking with a tabulated bending stress F_b equal to 1350 lb/in².

Calculate beam section modulus:
$$S_s = bd^2/6 = (5.5)(2.5)^2/6 = 5.73 \text{ in}^3$$

Calculate the adjusted allowable bending stress (assuming $C_D = 1.0$: $C_M = 1.0$: $C_t = 1.0$: $C_F = 1.04$):
$$F_b' = F_b C_D C_M C_t C_F = 1350(1.0)(1.0)(1.0)(1.04) = 1404 \text{ lb/in}^2 \text{ (NDS 4.3.1)}$$

Calculate resisting moment:
$$M' = F_b' S_s = (1404)(5.73)/12 = 670 \text{ ft·lb}$$

Structural check: $M' \geq M_{\text{max}}$ $670 \text{ ft·lb} \geq 103 \text{ ft·lb}$ ✓

For the fire design of the timber deck, the loading is unchanged. Therefore, the maximum induced moment is unchanged. The fire resistance must be calculated.

Calculate beam section modulus exposed on one side:
$$S_f = (b)(d - a)^2/6 = (5.5)(2.5\ 1.8)^2/6 = 0.45 \text{ in}^3$$

Calculate the adjusted allowable bending stress (assuming $C_D = $ N/A: $C_M = $ N/A: $C_t = $ N/A: $C_F = 1.04$):
$$F_b' = F_b C_F = 1350(1.04) = 1404 \text{ lb/in}^2$$

Calculate resisting moment:
$$M' = (2.85) F_b' S_f = (2.85)(1404)(0.45)/12 = 150 \text{ ft·lb (NDS 16.2.2)}$$

Fire check: $M' \geq M_{\text{max}}$ $150 \text{ ft·lb} \geq 103 \text{ ft·lb}$ ✓

7.8.1.4 Connections

Where fire endurance is required, connectors and fasteners must be protected from fire exposure by wood, fire-rated gypsum board, or any coating approved for the required endurance time.

7.8.2 Protected Wood-Frame Construction

Lightweight wood-frame construction is commonly used in residential and commercial structures. For both new and existing construction, building codes often require structural elements such as exterior walls, load-bearing partitions, floor/ceiling assemblies, and roofs to achieve a minimum fire endurance rating. Membrane protection and protective insulation barriers are typically used to enhance the fire resistance of the assembly.

7.8.2.1 Listed Fire-Rated Assemblies

Historically, protected assemblies have been tested in accordance with standardized fire test procedures and assigned a fire endurance rating based on time to failure. ASTM Standard E 119 is normally used. Many sources are available for obtaining information on the fire endurance of assemblies. Generally, publications from recognized testing laboratories provide a good source for fire endurance ratings of assemblies that have been tested. Building codes and regulators regularly accept assemblies included in these publications as having the identified fire endurance rating.

7.8.2.2 Nonlisted Fire-Rated Assemblies

To permit use of "nonlisted" assemblies, a methodology for calculating the fire endurance of load-bearing and non-load-bearing floor, wall, ceiling, and roof assemblies has been adapted for use in, and is recognized by, U.S. building codes. As a result, regulators routinely accept the fire endurance ratings developed under this *component additive method* (CAM) calculation methodology described in a publication entitled "DCA 4: Component Additive Method for Calculating and Demonstrating Assembly Fire Endurance" [153].

The original methodology for calculating fire endurance ratings of assemblies by CAM was developed in the early sixties by the *National Research Council of Canada* (NRCC). The methodology resulted from detailed review of 135 standard fire test reports on wood stud walls, 73 test reports on wood-joist floor assemblies, and the "Ten Rules of Fire Resistance Rating" by Dr. Tibor Harmathy [154], an eminent NRCC fire researcher. Review of the fire tests provided assigned time values for contribution to fire endurance ratings for each separate component of an assembly. The "Ten Rules" provided a method for combining the individual contributions to obtain the overall fire endurance rating of an assembly.

In developing the methodology, the fire endurance of the assembly was separated into the fire endurance contribution of the exposed membrane and the time to destruction of the framing members. As a result, the calculated fire endurance would equal the sum of (1) the contribution of the fire-exposed membrane, (2) the time to failure of the framing members, and, if applicable, (3) any additional protection due to use of cavity insulation or reinforcement of the membrane.

The times assigned to protective wall and ceiling coverings are given in Table 7.13. These times are based on the ability of the membrane to remain in place during fire tests. This "assigned time"

TABLE 7.13 Time Assigned to Protective Membrane

Description of finish	Time (min)
3/8 inch Douglas fir plywood, phenolic bonded	5
1/2 inch Douglas fir plywood, phenolic bonded	10
5/8 inch Douglas fir plywood, phenolic bonded	15
3/8 inch gypsum board	10
1/2 inch gypsum board	15
5/8 inch gypsum board	20
1/2 inch Type X gypsum board	25
5/8 inch Type X gypsum board	40
Double 3/8 inch gypsum board	25
1/2 + 3/8 inch gypsum board	35
Double 1/2 inch gypsum board	40

Reprinted courtesy of American Forest & Paper Association, Washington, D.C.

Notes:
1. On walls, gypsum board shall be installed with the long dimension parallel to framing members with all joints finished. However, 5/8 inch Type X gypsum wallboard may be installed horizontally with the horizontal joints unsupported.
2. On floor/ceiling or roof/ceiling assemblies, gypsum board shall be installed with the long dimension perpendicular to framing members and shall have all joints finished.

TABLE 7.14 Time Assigned to Wood-Frame Components

Description of frames	Time (min)
Wood studs, 16 inches on center	20
Wood joists, 16 inches on center	10
Wood roof and floor truss assemblies, 24 inches on center	5

Reprinted courtesy of American Forest & Paper Association, Washington, D.C.

should not be confused with the "finish rating" of a membrane. By standard definition, the finish rating is the time it takes for the temperature to rise 250°F on the unexposed surface of a material when the material is exposed to a heat source following the ASTM E 119 time-temperature curve. As shown in Table 7.13, some pairs of membranes have been tested resulting in greater fire endurance times than the sum of the ratings of the individual membranes, in accordance with Harmathy's first rule.

The times assigned to wood studs and joists were determined based upon the time it takes for the framing members to fail after failure of the protective membrane. The fire endurance time assigned to framing members is given in Table 7.14. These times are based on the ability of framing members to provide structural support when subjected to the ASTM E 119 fire endurance test without the benefit of a protective membrane. These time values were derived, in part, from results of full-scale tests of unprotected wood studs and floor joists where the structural elements were loaded to design capacity. They apply to all framing members and do not increase if, for example, 2 × 6 studs are used rather than 2 × 4 studs, as implied by Harmathy's fifth rule.

7.8.2.2.1 Wall Assemblies. Additional fire endurance can be provided to wall assemblies by the use of mineral fiber, paper, or foil-faced glass fiber insulation batts. The time assigned to each type of insulation in contributing additional fire endurance to the assembly is presented in Table 7.15. For a wall or partition where only plywood is used as the membrane on the side assumed to be exposed to the fire, insulation must be used within the assembly.

In developing this methodology, it was also determined that the primary function of the membrane on the unexposed side of an exterior wall is to keep the insulation in place and prevent the transmission of heat. Fire endurance of wall assemblies is consistently dependent upon the fire-exposed side membrane. As a result, it is considered very reasonable to substitute various exterior cladding materials as the membrane on the unexposed side or exterior wall assemblies. Therefore, where a fire endurance rating for an exterior wall is to be determined using CAM, any combination of sheathing, paper, and exterior finish listed in Table 7.16 may be used, or the outer membrane may

TABLE 7.15 Time Assigned for Additional Protection

Description of additional protection	Time (min)
Add to the fire endurance rating of wood stud walls if the spaces between the studs are filled with rockwool or slag mineral wool batts weighing not less than 1/4 lb./sq. ft. of wall surface.	15
Add to the fire endurance rating of non-load-bearing wood stud walls if the spaces between the studs are filled with fiberglass batts weighing not less than 1/4 lb./sq. ft. of wall surface.	5

Reprinted courtesy of American Forest & Paper Association, Washington, D.C.

TABLE 7.16 Membrane on Exterior Face of Wall

Any combination of sheathing, paper (if required) and exterior finish listed below may be used

Sheathing	Paper	Exterior finish
5/8 inch T & G lumber 5/16 inch exterior grade plywood 1/2 inch gypsum board	Sheathing paper	Lumber siding Wood shingles and shakes 1/4 inch ext. grade plywood 1/4 inch hardboard Metal siding Stucco on metal lath Masonry veneer
None		3/8 inch ext. grade plywood

Reprinted courtesy of American Forest & Paper Association, Washington, D.C.

consist of any membrane combination that is assigned a time for contribution to fire endurance of at least 15 min in Table 7.13.

7.8.2.2.2 Roof and Floor/Ceiling Assemblies. In the case of a roof or floor/ceiling assembly, fire testing is normally done with exposure from below the assembly. To comply with this calculation methodology, floor and roof assemblies must have a protective membrane in conformance with Table 7.13. The upper membrane must consist of a subfloor or roof deck and finish in conformance with Table 7.17. Alternatively, any combination of membranes listed in Table 7.13, with a fire endurance rating of at least 15 min, may be used on the unexposed (upper) side. If the proposed assembly is a ceiling with an attic above, most building codes allow elimination of the upper membrane.

7.8.2.2.3 Examples

Example 7.16 Determine the fire endurance rating of a wall assembly having one layer of $^5/_8$-in. Type X gypsum wallboard attached to wood studs on the fire-exposed side.

From Table 7.13:

$^5/_8$-in. Type X gypsum wallboard has an assigned fire endurance time of 40 min.

From Table 7.14:

Wood studs spaced 16 in. on center have a fire endurance time of 20 min.

TABLE 7.17 Flooring or Roofing Membrane

Assembly	Structural members	Subfloor or roof deck	Finish flooring or roofing
Floor	Wood	1/2 inch plywood or 11/16 inch T&G softwood	Hardwood or softwood flooring on building paper. Resilient flooring, parquet floor, felted-synthetic-fiber floor coverings, carpeting, or ceramic tile on 3/8 in. thick panel-type underlay. Ceramic tile on 1-1/4 in. mortar bed.
Roof	Wood	1/2 inch plywood or 11/16 inch T&G softwood	Finish roofing material with or without insulation.

Reprinted courtesy of American Forest & Paper Association, Washington, D.C.

Adding the fire endurance times of the components:

$5/8$-in. Type X gypsum wallboard = 40 min
Wood studs at 16 in. on center = 20 min
―――――――――――――――――――――――――――――
Combined fire endurance time = 60 min

If the wall is exposed to fire from both sides (e.g., for interior fire-rated partitions), each surface of the framing member would be required to be fire protected with at least 40 min of membrane coverings in Table 7.13. If the proposed wall is assumed to be exposed to fire from one side only, as in some exterior wall applications, the fire exposure is assumed to be from the interior, which would require a total contribution of 40 min from the membrane coatings from Table 7.13.

Example 7.17 Determine the fire endurance rating of a floor/ceiling assembly having wood joists spaced 16 in. on center and protected on the bottom side (ceiling side) with two layers of $1/2$-in. Type X gypsum wallboard and having a $1/2$-in. plywood subfloor on the upper side (floor side).

From Table 7.13:

$1/2$-in. Type X gypsum wallboard has an assigned time of 25 min per layer.

From Table 7.14:

Wood joists spaced 16 in. on center have a fire endurance time of 10 min.

Adding the fire endurance times of the components:

$1/2$-in. Type X gypsum wallboard (face layer) = 25 min
$1/2$-in. Type X gypsum wallboard (base layer) = 25 min
Wood joists at 16 in. on center = 10 min
―――――――――――――――――――――――――――――――――――――――
Combined fire endurance time = 60 min

REFERENCES

1. AF&PA and Clemson University, "U.S. Forests Facts & Figures," American Forest & Paper Association, Washington, DC, 2001.
2. AF&PA, "Sustainable Forestry Initiative (SFI) Program," American Forest & Paper Association, Washington, DC, 2002.
2A. Buchanan, A.H., and Levine, S.B., "Building Materials and Global Carbon Emissions," Proceedings of the International Wood Engineering Conference, New Orleans, Louisiana, vl, October 1996: 349–356.
3. Haygreen, J., and Bowyer, J., *Forest Products and Wood Science,* 2nd edition, Iowa State University Press, Ames, IA, 1989.
4. White, R., "Effect of Lignin Content and Extractives on the Higher Heating Value of Wood," *Wood and Fiber Science,* 19: 446–452, 1987.
5. Forest Products Laboratory, *Wood Handbook—Wood as an Engineering Material,* U.S. Department of Agriculture, Forest Service, Forest Products Laboratory, Madison, WI, 1999.
6. Kollman, F., and Côté, W., *Principles of Wood Science and Technology,* vol. 1: *Solid Wood,* Springer Verlag, Berlin, Germany, 1984.
7. Petterson, R., "The Chemical Composition of Wood," *Advances in Chemistry Series 207,* R. Rowell, ed., American Chemical Society, 1984, Washington, DC, pp. 57–126.
8. Roberts, A., "A Review of Kinetic Data for the Pyrolysis of Wood and Related Substances," *Combustion and Flame,* 14: 261–272, 1970.
9. Roberts, A., "Problems Associated with the Theoretical Analysis of Burning Wood," presented at Thirteenth Symposium (International) on Combustion, Combustion Institute, Pittsburgh, PA 1970.

10. LeVan, S., "Chemistry of Fire Retardancy," *Advances in Chemistry Series 207*, R. Rowell, ed., American Chemical Society, Washington, DC, 1984, pp. 531–574.
11. Schaffer, E., "Effect of Fire-Retardant Impregnations on Wood Charring Rate," *Journal of Fire and Flammability/Fire Retardant Chemistry*, 1: 96–109, 1974.
12. LeVan, S., and Winandy, J., "Effect of Fire-Retardant Treatments on Wood Strength: A Review," *Wood and Fiber Science*, 22: 113–131, 1990.
13. Janssens, M., "Thermo-Physical Properties for Wood Pyrolysis Models," presented at Pacific Timber Engineering Conference, Gold Coast, Australia, CSIRO, Sydney, Australia 1994.
14. Melinek, S., "The Darkening of Irradiated Wood," Fire Research Station, Borehamwood, UK, 1968.
15. Wesson, H., "The Piloted Ignition of Wood under Radiant Heat," University of Oklahoma, Norman, OK, 1971.
16. Vovelle, C., Akrich, R., and Delbourgo, R., "Temperature Evolution Prior to Ignition of Cellulosic Materials (in French)," presented at First Specialists Meeting (International) of the Combustion Institute, Combustion Institute, Pittsburgh, PA. 1981.
17. Janssens, M., "Thermophysical Properties of Wood and their Role in Enclosure Fire Growth," The University of Ghent, Ghent, Belgium, 1991.
18. Atreya, A., "Pyrolysis, Ignition and Fire Spread on Horizontal Surfaces of Wood," *Division of Applied Sciences*, Harvard University, 1983, Cambridge, MA, p. 407.
19. Shafizadeh, F., "The Chemistry of Wood Pyrolysis and Combustion," *Advances in Chemistry Series 207*, R. Rowell, ed., American Chemical Society, 1984, Washington, DC, pp. 489–529.
20. Alves, S., and Figueiredo, J., "Pyrolysis Kinetics of Lignocellulosic Materials by Multistage Isothermal Thermogravimetry," *Journal of Analytical Applied Pyrolysis*, 13: 123, 1988.
21. Parker, W., "Prediction of the Heat Release Rate of Wood," George Washington University, Washington, DC, 1988.
22. Nurbakhsh, S., "Thermal Decomposition of Charring Materials," Michigan State University, East Lansing, MI, 1989.
23. White, R., "Charring Rates of Different Wood Species," University of Wisconsin, Madison, WI, 1988.
24. Beall, F., "Properties of Wood during Carbonization under Fire Conditions," presented at Wood Technology: Chemical Aspects (Symposium Series 43), 1977.
25. Slocum, D., McGinnes, E., and Beall, F., "Charcoal Yield, Shrinkage, and Density Changes during Carbonization of Oak and Hickory Woods," *Wood Science*, 11: 42–47, 1978.
26. Evans, D., "Density of Wood Charcoal," Technical Report #14, Harvard University, Cambridge, MA, 1976.
27. TenWolde, A., McNatt, J.D., and Krahn, L., "Thermal Properties of Wood and Wood Panel Products for Use in Buildings," ORNL/USDA-21697/1, Oak Ridge National Laboratory, Oak Ridge, TN, 1988.
28. Widell, T., "Thermal Investigations into Carbonization of Wood," Report No. 199, Academy of Engineering Sciences, Stockholm, Sweden, 1948.
29. Touloukian, Y., Powell, R., Ho, C., and Klemens, P., *Thermal Conductivity, Nonmetallic Solids*, Plenum Press, New York, NY, 1970.
30. Kollman, F., and Malmquist, L., "Über die Wärmeleitzahl von Holz und Holzwerkstoffen," *Holz als Roh- und Werkstoff*, 14: 201–204, 1956.
31. Fredlund, B., "A Model for Heat and Mass Transfer in Timber Structures during Fire," Lund University, Lund, Sweden, 1988.
32. Stamm, A., and Hansen, L., "The Bonding Force of Cellulosic Materials for Water," *Journal of Physical Chemistry*, 41: 1007–1016, 1937.
33. Siau, J., *Transport Processes in Wood*, Springer Verlag, Berlin, Germany, 1984.
34. MacLean, J., "Thermal Conductivity of Wood," *ASHRAE Journal*, 47: 323–354, 1941.
35. Hausen, R., *Landolt-Börnstein: Zahlenwerte und Funktionen aus Physik, Chemie, Astronomie, Geophysik, Technik* (in German), Part 4b. Springer Verlag, Berlin, Germany, 1972.
36. Stamm, A., and Seborg, R., "Absorption Compression on Cellulose and Wood," *Journal of Physical Chemistry*, 39: 133–142, 1934.
37. Martin, S., "Simple Radiant Heating Method for Determining Thermal Diffusivity of Cellulosic Materials," *Journal of Applied Physics*, 31: 1101–1104, 1960.

38. Kanter, K., "The Thermal Properties of Wood (Translated from Russian)." *Nauka i Tekhnika*, 6: 17–18, 1957.
39. Zoufel, R., "Determination of the Thermal Conductivity and the Specific Thermal Capacity of Wood at High Temperatures," presented at 3rd International Symposium on Fire Protection of Buildings, Eger, Hungary, 1990.
40. Touloukian, Y., Liley, P., and Saxena, S., *Thermal Conductivity, Nonmetallic Gases and Liquids*, Plenum Press, New York, NY, 1970.
41. Zicherman, J., "A Study of Wood Morphology and Microstructure in Relation to its Behavior in Fire Exposure," The University of California at Berkeley, Berkeley, CA, 1978.
42. Cutter, B., and McGinnes, A., "A Note on Density Change Patterns in Charred Wood," *Wood and Fiber*, 13: 39–44, 1981.
43. Roberts, A., "Ultimate Analysis of Partially Decomposed Wood Samples," *Combustion and Flame*, 4: 345–346, 1964.
44. Ho, C., Powell, R., and Liley, P., "Thermal Conductivity of the Elements," *Journal of Physical and Chemical Reference Data*, 3: 111–117, 1974.
45. Alves, S., and Figueiredo, J., "A Model for Pyrolysis of Wet Wood," *Chemical Engineering Science*, 44: 2861–2869, 1989.
46. Schack, A., *Der Industrielle Wärmeübergang*, Stahleisen Verlag, Düsseldorf, Germany, 1940.
47. Tye, R., Lander, L., and Meiler, K., *The Thermal Transport Properties of Fibrous Carbon and Graphite Products*, vol. 19, Plenum Press, New York, NY, 1985.
48. Reid, R., Prausnitz, J., and Sherwood, T., *The Properties of Gases and Liquids*, McGraw-Hill, New York, NY, 1973.
49. Kung, H., and Kalelkar, A., "On the Heat of Reaction in Wood Pyrolysis," *Combustion and Flame*, 20: 91–103, 1973.
50. Bénichou, N., and Sultan, M., "Thermal Properties of Components of Lightweight Wood-Framed Assemblies at Elevated Temperatures," presented at Fire and Materials 2001, Seventh International Conference, San Francisco, CA, Interscience Communications, London, England, 2001.
51. AF&PA, *Allowable Stress Design (ASD) Manual for Engineered Wood Construction—ANSI/AF&PA NDS-2001*, American Forest & Paper Association, American Wood Council, Washington, DC, 2001.
52. AF&PA, *Load and Resistance Factor Design (LRFD) Manual for Engineered Wood Construction*, American Forest & Paper Association, American Wood Council, Washington, DC, 1996.
53. Gerhards, C., "Effect of Moisture Content and Temperature on the Mechanical Properties of Wood: An Analysis of Immediate Effects," *Wood and Fiber*, 14: 4–36, 1982.
54. Schaffer, E., "Effect of Pyrolytic Temperatures on the Longitudinal Strength of Dry Douglas Fir," *Journal of Testing and Evaluation*, 1: 319–329, 1973.
55. Janssens, M., "Piloted Ignition of Wood: A Review," *Fire and Materials*, 15: 151–167, 1991.
56. Dietenberger, M., "Ignitability Analysis using the Cone Calorimeter and LIFT Apparatus," presented at Twenty-Second International Conference on Fire Safety, Columbus, OH, Product Safety Corporation, 1996.
57. Moghtaderi, B., Novozhilov, V., Fletcher, D., and Kent, J., "A New Correlation for Bench-Scale Piloted Ignition Data of Wood," *Fire Safety Journal*, 29: 41–59, 1997.
58. Spearpoint, M., and Quintiere, J., "Predicting the Piloted Ignition of Wood in the Cone Calorimeter Using an Integral Model: Effect of Species, Grain Orientation, and Heat Flux," *Fire Safety Journal*, 36: 391–415, 2001.
59. Ngu, C., "Ignition Properties of New Zealand Timber," Fire Engineering Research Report 01/5, University of Canterbury, Christchurch, NZ, 2001.
60. Grenier, A., and Janssens, M., "Improved Method for Analyzing Ignition Data of Composites," presented at Twenty-Third International Conference on Fire Safety, Millbrae, CA, Product Safety Corporation, 1997.
61. Mikkola, E., and Wichman, I., "On the Thermal Ignition of Combustible Materials," *Fire and Materials*, 14: 87–96, 1989.
62. Abu-Zaid, M., "Effect of Water on Ignition of Cellulosic Materials," Michigan State University, East Lansing, MI, 1988.
63. Atreya, A., Carpentier, C., and Harkleroad, M., "Effect of Sample Orientation on Piloted Ignition of Wood," presented at First International Symposium on Fire Safety Science, Gaithersburg, MD, Hemisphere Publishing Co., New York, NY, 1985.

64. Fangrat, J., Hasemi, Y., Yoshida, M., and Hirata, T., "Surface Temperature at Ignition of Wooden Based Slabs," *Fire Safety Science,* 27: 249–259, 1996.
65. Urbas, J., and Parker, W., "Surface Temperature Measurements on Burning Wood Specimens in the Cone Calorimeter and Effect of Grain Orientation," *Fire and Materials,* 17: 205–208, 1993.
66. Yudong, L., and Drysdale, D., "Measurement of the Ignition Temperature of Wood," *Fire Safety Science,* 1: 25–30, 1992.
67. Hirschler, M., and Smith, G., "Investigation of a Smoke Toxicity Fire Model for Use on Wood," presented at Third International Symposium on Fire Safety Science, Edinburgh, Scotland, Elsevier Science Publishers, New York, NY, 1991.
67A. Garciek, D.T., and Lattimer, B.Y., "Effect of Combustion Conditions on Species Production," *The SFPE Handbook of Fire Protection Engineering,* 3rd edition, SFPE, Bethesda, MD, pp. 2-54–2-82, 2002.
68. Babrauskas, V., "Ignition of Wood: A Review of the State of the Art," presented at Interflam 2001, Edinburgh, Scotland, September 17–19, Interscience Communications, London, England, 2001.
69. Simms, D., "Ignition of Cellulosic Materials by Radiation," *Combustion & Flame,* 4: 293–300, 1960.
70. Law, M., "Heat Radiation from Fires and Building Separation," Fire Research Technical Paper No. 5, HMSO, London, UK, 1963.
71. Kubler, H., Wang, Y., and Barkalow, D., "Generation of Heat in Wood between 80 and 130°C," *Holzforschung,* 39: 85–89, 1985.
72. Bowes, P., "Self-Heating: Evaluating and Controlling the Hazards," HMSO, London, UK, 1984.
73. Babrauskas, V., and Peacock, R., "Heat Release Rate: The Single Most Important Variable in Fire Hazard," *Fire Safety Science,* 18: 255–272, 1992.
74. Janssens, M., "Rate of Heat Release of Wood Products," *Fire Safety Journal,* 17: 217–238, 1991.
75. Tran, H., "Chapter 11: Wood Materials. Part B. Experimental Data on Wood Materials," *Heat Release in Fires,* V. Babrauskas and S. Grayson, eds., Elsevier Applied Science, New York, NY, 1992, pp. 357–372.
76. Janssens, M., "Cone Calorimeter Measurements of the Heat of Gasification of Wood," presented at Interflam '93, Oxford, England, Interscience Communications, London, England, 1993.
77. Fangrat, J., Hasemi, Y., Yoshida, M., and Kikuchi, S., "Relationship between Heat of Combustion, Lignin Content, and Burning Weight Loss," *Fire and Materials,* 22: 1–6, 1998.
78. Sibulkin, M., "Heat of Gasification for Pyrolysis of Charring Materials," presented at First International Symposium on Fire Safety Science, Gaithersburg, MD, Hemisphere Publishing Co., New York, NY, 1985.
79. Urbas, J., "Non-Dimensional Heat of Gasification Measurements in the Intermediate Scale Rate of Heat Release Apparatus," *Fire and Materials,* 17: 119–123, 1993.
80. Hall, G., Saunders, R., Allcorn, R., Jackman, P., Hickey, M., and Fitt, R., "Fire Performance of Timber—A Literature Survey," Timber Research and Development Association, High Wycombe, UK, 1976.
81. White, R., and Nordheim, E., "Charring Rate of Wood for ASTM E119 Exposure," *Fire Technology,* 28: 5–30, 1992.
82. Lache, M., "Abbrandverhalten von Holz (in German)," *Holzbau Technik,* vols. 4 and 5, Informationsdienst, Munich, Germany, 1991.
83. Gardner, W., and Syme, D., "Charring of Glued-Laminated Beams of Eight Australian-Grown Timber Species and the Effect of 13 mm Gypsum Plasterboard Protection on their Charring," Technical Report No. 5, ODC 843.4-015, N.S.W. Timber Advisory Council, Sydney, Australia, 1991.
84. Collier, P., "Charring Rates of Timber," Study Report No. 42, Building Research Association of New Zealand, Judgeford, New Zealand, 1992.
85. CEN, "Eurocode 5: Design of Timber Structures," ENV 1995-1-2, CEN, Brussels, Belgium, 1994.
86. Schaffer, E., "Structural Fire Design: Wood," FPL Research Paper 450, U.S. Department of Agriculture, Forest Service, Forest Products Laboratory, Madison, WI, 1984.
87. Janssens, M., and White, R., "Short Communication: Temperature Profiles in Wood Members Exposed to Fire," *Fire and Materials,* 18: 263–265, 1994.
88. Bamford, C., Crank, J., and Malan, D., "The Combustion of Wood. Part I," *Proceedings of the Cambridge Philosophical Society,* 42: 166–182, 1946.

89. Squire, W., and Foster, C., "A Mathematical Study of the Mechanism of Wood Burning," Technical Progress Report No. 1, NBS Contract No. CST-362, Southwest Research Institute, San Antonio, TX, February 3, 1961.
90. Tinney, E., "The Combustion of Wooden Dowels in Heated Air," presented at Tenth Symposium (International) on Combustion, Combustion Institute, Pittsburgh, PA, 1965.
91. Kanury, A.M., "Burning of Wood—A Pure Transient Conduction Model," *Journal of Fire & Flammability,* 2: 191–205, 1971.
92. Panton, R., and Rittman, J., "Pyrolysis of a Slab of Porous Material," presented at Thirteenth Symposium (International) on Combustion, Combustion Institute, Pittsburgh, PA, 1971.
93. Havens, J., Hashemi, H., Brown, L., and Welker, R., "A Mathematical Model of the Thermal Decomposition of Wood," *Combustion Science and Technology,* 5: 91–98, 1972.
94. Kanury, A.M., "Rate of Burning of Wood (A Simple Thermal Model)," *Combustion Science and Technology,* 5: 135–146, 1972.
95. Kung, H., "A Mathematical Model of Wood Pyrolysis," *Combustion and Flame,* 18: 185–195, 1972.
96. Knudsen, R., and Schniewind, A., "Performance of Structural Wood Members Exposed to Fire," *Forest Products Journal,* 5: 23–32, 1973.
97. Maa, P., and Bailie, R., "Influence of Particle Sizes and Environmental Conditions on High Temperature Pyrolysis of Cellulosic Material—I (Theoretical)," *Combustion Science and Technology,* 7: 257–269, 1973.
98. Thomas, P., and Nilsson, L., "Fully Developed Compartment Fires: New Correlations of Burning Rates," Fire Research Note No. 979, Fire Research Station, Borehamwood, Herts, England, August 1973.
99. Tamanini, F., "A Study of the Extinguishment of Wood Fires," *Division of Engineering and Applied Physics,* Harvard University, Cambridge, MA, 1974, p. 223.
100. Hadvig, S., and Paulsen, O., "One-Dimensional Charring Rates in Wood," *Journal of Fire & Flammability,* 7: 433–449, 1976.
101. Fan, L., Fan, L., Miyanami, K., Chen, T., and Walawender, W., "A Mathematical Model for Pyrolysis of a Single Particle," *The Canadian Journal of Chemical Engineering,* 55: 47–53, 1977.
102. Kansa, E., Perlee, H., and Chaiken, R., "Mathematical Model of Wood Pyrolysis," *Combustion and Flame,* 29: 311–324, 1977.
103. White, R., and Schaffer, E., "Application of the CMA Program to Wood Charring," *Fire Technology,* 14: 279–290, 1978.
104. Hoffman, F., "Study of the Thermal Behavior of Wood as Affected by Hygroscopic Moisture (in German)," Carolo-Wilhelmina University, Braunschweig, Germany, 1979, p. 112.
105. Hadvig, S., "Charring of Wood in Building Fires," Technical University of Denmark, Lyngby, Denmark, 1981.
106. Handa, A., Morita, M., Sugawa, O., Ishii, T., and Hayashi, K., "Computer Simulation of the Oxidative Pyrolysis of Wood," *Fire Science and Technology,* 2: 109–116, 1982.
107. Delichatsios, M., and deRis, J., "An Analytical Model for the Pyrolysis of Charring Materials," FMRC Report No. RC83-BP-5, Factory Mutual Research Corporation, Norwood, MA, May 1983.
108. Pyle, D., and Zaror, C., "Heat Transfer and Kinetics in the Low Temperature Pyrolysis of Solids," *Chemical Engineering Science,* 79: 147–158, 1984.
109. Chan, W.R., Kelbon, M., and Krieger, B., "Modeling and Experimental Verification of Physical and Chemical Processes during Pyrolysis of a Large Biomass Particle," *Fuel,* 64: 1505–1513, 1985.
110. Miller, C., and Ramohalli, K., "A Theoretical Heterogeneous Model of Wood Pyrolysis," *Combustion Science and Technology,* 46: 249–265, 1986.
111. Villermaux, J., Antoine, B., and Soulignac, F., "A New Model for Thermal Volatilization of Solid Particles Undergoing Fast Pyrolysis," *Chemical Engineering Science,* 41: 151–157, 1986.
112. Gammon, B., "Reliability Analysis of Wood-Frame Wall Assemblies Exposed to Fire," *Graduate Division,* University of California, Berkeley, CA, 1987, p. 377.
113. Wichman, I., and Atreya, A., "A Simplified Model For the Pyrolysis of Charring Materials," *Combustion and Flame,* 68: 231–247, 1987.
114. Capart, R., Falk, L., and Gelus, M., "Pyrolysis of Wood Macrocylinders under Pressure Application of a Simple Mathematical Model," *Applied Energy,* 30: 1–13, 1988.

115. Ragland, K., Boerger, J., and Baker, A., "A Model of Chunkwood Combustion," *Forest Products Journal,* 38: 27–32, 1988.
116. Purnomo, D., Aerts, J., and Ragland, K., "Pressurized Downdraft Combustion of Woodchips," presented at Twenty-Third Symposium (International) of Combustion, Combustion Institute, Pittsburgh, PA, 1990.
117. Koufopanos, C., and Papayannakos, N., "Modeling the Pyrolysis of Biomass Particles: Studies on Kinetics, Thermal and Heat Transfer Effects," *The Canadian Journal of Chemical Engineering,* 69: 907–915, 1991.
118. Ohlemiller, T., "Chapters 2–11: Smoldering Combustion," *SFPE Handbook of Fire Protection Engineering,* P. DiNenno, ed., NFPA, Quincy, MA, 1995.
119. Delichatsios, M., and Chen, Y., "Asymptotic, Approximate, and Numerical Solutions for the Heatup and Pyrolysis of Materials Including Reradiation Losses," *Combustion and Flame,* 92: 292–307, 1993.
120. DiBlasi, C., "Analysis of Convection and Secondary Reaction Effects within Porous Solid Fuels Undergoing Pyrolysis," *Combustion Science and Technology,* 90: 315–340, 1993.
121. Fredlund, B., "Modelling of Heat and Mass Transfer in Wood Structures During Fire," *Fire Safety Journal,* 20: 39–69, 1993.
122. Mardini, I., Lavine, A., and Dhir, V., "Experimental and Analytical Study of Heat and Mass Transfer in Simulated Fuel Element during Fires," presented at Heat Transfer in Fire and Combustion Systems, 1993.
123. Mehaffey, J., Cuerrier, P., and Carrisse, G., "Model for Predicting Heat Transfer through Gypsum-Board/Wood Stud Walls Exposed to Fire," *Fire and Materials,* 18: 297–305, 1994.
124. Shrestha, D., Cramer, S., and White, R., "Time-Temperature Profile across a Lumber Section Exposed to Pyrolytic Temperatures," *Fire and Materials,* 18: 211–220, 1994.
125. Suuberg, E., Milosavlevic, I., and Lilly, W., "Behavior of Charring Materials in Simulated Fire Environments," NIST-GCR-94-645, National Institute of Standards and Technology, Gaithersburg, MD, June 1994.
126. Albini, F., and Reinhardt, E., "Modeling Ignition and Burning Rate of Large Woody Natural Fuels," *International Journal of Wildland Fire,* 5: 81–91, 1995.
127. Caballero, J., Font, R., Marcilla, A., and Conesa, J., "New Kinetic Model for Thermal Decomposition of Heterogeneous Materials," *Industrial and Engineering Chemistry Research,* 34: 806–812, 1995.
128. Clancy, P., Beck, V., and Leicester, R. "Time-Dependent Probability of Failure of Wood Frames in Real Fire," presented at Fourth Fire and Materials Conference, Crystal City, VA, 1995.
129. Moghtaderi, B., Novozhilov, V., Fletcher, D., and Kent, J., "An Integral Model for the Transient Pyrolysis of Solid Materials," *Fire and Materials,* 21: 7–16, 1997.
130. Tavakkol-Khah, M., and Klingsch, W., "Calculation Model for Predicting Fire Resistance Time of Timber Members," presented at Fire Safety Science, Fifth International Symposium, Melbourne, Australia, Elsevier Science Publishers, New York, NY, 1997.
131. Ritchie, S., Steckler, K., Hamins, A., Cleary, T., Yang, J., and Kashiwagi, T., "Effect of Sample Size on the Heat Release Rate of Charring Materials," presented at Fire Safety Science, Fifth International Symposium, Melbourne, Australia, Elsevier Science Publishers, New York, NY 1997.
132. Yuen, R., Casey, R., DeVahl-Davis, G., Leonardi, E., Yeoh, G., Chandrasekaran, V., and Grubits, S., "Three-Dimensional Mathematical Model for the Pyrolysis of Wet Wood," presented at Fire Safety Science, Fifth International Symposium, Melbourne, Australia, Elsevier Science Publishers, New York, NY, 1997.
133. deRis, J., and Yan, Z., "Modeling Ignition and Pyrolysis of Charring Fuels," presented at Fifth Fire and Materials Conference, San Antonio, TX, 1998.
134. Jia, F., Galea, E., and Patel, M., "Numerical Simulation of the Mass Loss Process in Pyrolyzing Char Materials," *Fire and Materials,* 23: 71–78, 1999.
135. Spearpoint, M., and Quintiere, J., "Predicting the Burning of Wood Using an Integral Model," *Combustion and Flame,* 123: 308–325, 2000.
136. Janssens, M., "Modeling of the Thermal Degradation of Structural Wood Members Exposed to Fire," presented at Structures in Fire SiF '02, University of Canterbury, Christchurch, NZ, 2002.
137. Wichman, I., "A Continuum-Mechanical Derivation of the Conservation Equations for the Pyrolysis and Combustion of Wood," Research Report 591, Technical Research Centre of Finland, Espoo, Finland, March 1989.

138. AF&PA, "DCA No. 1: Flame Spread Performance of Wood Products," American Forest & Paper Association, American Wood Council, Washington, DC, 2002.
139. deRis, J., "Spread of a Laminar Diffusion Flame," presented at 12th Symposium (International) on Combustion, Combustion Institute, Pittsburgh, PA, 1969.
140. Delichatsios, M., "Exact Solution for the Rate of Creeping Flame Spread over Thermally Thin Materials," *Combustion Science & Technology*, 44: 257–267, 1986.
141. Quintiere, J., "An Approach for Modeling Wall Fire Spread in a Room," *Fire Safety Journal*, 3: 201, 1981.
142. Quintiere, J., and Harkleroad, M., "New Concepts for Measuring Flame Spread Properties," NBSIR 84-2943, National Bureau of Standards, Gaithersburg, MD, 1984.
143. Gardner, W., and Thomson, C., "Flame Spread Properties of Forest Products," *Fire and Materials*, 12: 71–85, 1988.
144. Tran, H. and Janssens M., "Wall and Corner Fire Tests on Selected Wood Products," *Journal of Fire Sciences*, 9: 106–124, 1991.
145. Hirschler, M., and Janssens, M., "Heat and Smoke Measurements of Construction Materials Tested in a Room-Corner Configuration According to NFPA 265," presented at Twenty-Seventh International Conference on Fire Safety, Millbrae, CA, 1999.
146. White, R., Dietenberger, M., Tran, H., Grexa, O., Richardson, L., Sumathipala, K., and Janssens, M., "Comparison of Test Protocols for Standard Room/Corner Test," *Fire and Materials*, 23: 139–146, 1999.
147. AF&PA, "National Design Specification (NDS®) for Wood Construction," American Forest & Paper Association, American Wood Council, Washington, DC, 2001.
148. AF&PA, "Standard for Load and Resistance Factor Design (LRFD) for Engineered Wood Construction—AF&PA/ASCE 16-95," American Forest & Paper Association, American Wood Council, Washington, DC, 1995.
149. Lie, T., "A Method for Assessing the Fire Resistance of Laminated Timber Beams and Columns," *Canadian Journal of Civil Engineering*, 4: 161–169, 1977.
150. CABO, "Design of One-Hour Fire-Resistive Exposed Wood Members," Report No. NRB-250, Council of American Building Officials National Evaluation Board, 1984.
151. AF&PA, "DCA No. 2: Design of Fire-Resistive Exposed Wood Members," American Forest & Paper Association, American Wood Council, Washington, DC, 2001.
152. AF&PA, "Calculating the Fire Resistance of Exposed Wood Members: Technical Report No. 10," American Forest & Paper Association, American Wood Council, Washington, DC, 1999.
153. AF&PA, "DCA No. 4: CAM for Calculating and Demonstrating Assembly Fire Endurance," American Forest & Paper Association, American Wood Council, Washington, DC, 2001.
154. Harmathy, T., "Ten Rules of Fire Endurance Rating," *Fire Technology*, 1: 93–102, 1965.

CHAPTER 8
LIQUIDS AND CHEMICALS

A. Tewarson
FM Global Research, Norwood, MA

G. Marlair
INERIS, Verneuil-en-Halatte, France

8.1 INTRODUCTION

Chemicals present as liquids in their original form or on heating or dissolving with solvents are generally identified as fluids. Use of fluids to enhance and improve life is a widespread practice worldwide, for example, through manufacturing of:

- Pharmaceuticals
- Agrochemicals such as pesticides and fertilizers
- Cosmetics and perfumes
- Cleaning agents
- Solvents
- Process chemicals
- Polymers
- Fire retardants
- Fuels for internal combustion engines, boilers, and furnaces
- Heat transfer agents such as transformer oils
- Hydraulic fluids

There are large numbers of chemicals that are available as fluids for use by various industries, records of which can be found in the Chemical Abstract Service (CAS) (U.S.A.), Chemtrec (Canada), and European Chemicals Bureau (EINECS) (Europe). These fluids consist of both inorganic and organic compounds. Most of the inorganic fluids are nonflammable; however, when heated, some may develop self-sustained exothermic decomposition similar to that occurring in fires (e.g., NPK-type fertilizers), while others can be hazardous to life and the environment as poisons and pollutants. Most of the organic fluids are flammable as they consist of carbon, hydrogen, oxygen, nitrogen, sulfur, and halogen atoms, attached to each other via saturated, unsaturated, linear, branched, and ring types of chemical bonds. Consequently, organic fluids burn in fires, releasing heat, smoke, toxic, and corrosive compounds that are hazardous to life, property, and the environment. In addition, some of the liquids and chemicals are poisonous and hazardous to the environment even without burning.

Inorganic and organic fluids are used as single or multicomponent mixtures with a variety of additives, including fire retardants. They are stored and carried in small and large containers and tanks on the ground (trains, trucks, buses, and automobiles), over water (rivers, lakes, and oceans by boats and tankers), and in air (airplanes), and pumped through pipes under pressure. Because of the varieties of ways in which fluids are stored, transported, and utilized, numerous accidents have occurred due to their release from the storage tanks, containers, and pipes. These accidents have been responsible for the contamination of land, water, and air, release of toxic and flammable vapor clouds, explosions, and pool and jet fires.* Studies on these accidents and protection from them and fundamental understanding of pool and jet fires have been reviewed in Refs. 1 through 5. Examples of some of the accidents involving release of fluids are discussed in the following sections.

8.1.1 Accidents Involving Fluids Stored in Warehouses

Several accidents have occurred in chemical warehouses with severe consequences for the safety of life, property, and the environment. Some examples are the following:

- A fire at Sandoz Schweizerhalle works near Basle, Switzerland, in November 1986 contaminated the atmosphere, the surrounding soil, and the Rhine River (more than 500 km and a long-term ecological effect of at least 10 years) [6–10]. Ignition of Prussian blue pigment started the fire that propagated to the stored agrochemicals (there were 20 pesticides). Heavy, black smoke containing phosphoric esters and mercaptans was released, exposing residents of Basle, who experienced headaches, nausea, and respiratory irritation; however, no long-term serious illnesses were found. Water used to fight the fire caused severe ecological damage to the Rhine River.

    Detailed descriptions of the warehouse content, course of events, observations, and modeling on the dispersion of the fire plume and inhalation toxicity issues associated with the fire at Sandoz Schweizerhalle are described in Ref. 8. The chemicals involved in this fire are listed in Table 8.1, which is taken from Ref. 8.

- A fire at Woodkirk, Yorkshire, UK, polluted water, land, and air as a result of the release of agrochemicals [11]. The fire started with *octyl phenol* (OP) and propagated to other chemicals stored in the warehouse [Agroxone, Bronocot, and the herbicides Reglone (diquat) and Gramoxone (paraquat)] in plastic liners within steel drums.

- A fire following an explosion at an Alabama industrial distribution warehouse contaminated local watercourses [12]. The warehouse contained 18,000 gallons of a termite killer containing 44 percent pure chlorpyrifos and 25 tons of pressurized cans containing 1 percent Orthene, which all burned. The fire resulted in a massive plume of dense smoke.

- Three violent explosions, followed by a fire, occurred on November 5, 1997, at Hoechst's chemical warehouse in Antananarivo, Madagascar. Two hundred people were exposed to heat and fire products. The unburned chemicals and fire products contaminated the soil and water surrounding the site [13]; 155 tons of chemicals comprising 41 products were burned, including 10 tons of pesticides.

- A fire occurred at a fertilizer warehouse in Nantes, France, in 1987 [10]. The warehouse contained *ammonium nitrate* (AN) and NPK-type (also called ternary type) fertilizers. A stock of 20 tons of NPK, an inorganic product, was set on fire and resulted in the evacuation of some 15,000 people from the town center. Fortunately, the toxic cloud moved towards the ocean and did not affect the people.

    A similar type of accident on January 26, 2002, occurred in Murcia, Spain, which is close to the Mediterranean coast [14].

- A fire occurred on December 17, 1995, at Somerset West, near Cape Town in the Republic of South Africa [15–18]. The fire affected an abnormally huge sulfur storage area (15,000

---

* Fires are classified into three categories based on the orientation of burning surfaces: (1) pool fires: combustible surfaces with horizontal orientation; (2) wall fires: combustible surfaces with vertical orientation, and (3) jet fires: noncombustible surfaces with small openings, such as ruptures in a pipe, through which combustible fluid under pressure is released. A combustible fluid spilled accidentally can be soaked into a porous solid such as sand, soil, mat, or carpet. If a fire occurs, a flame will spread over the porous solid soaked with the combustible liquid both in a horizontal direction (pool fire) or vertical direction (wall fire).

TABLE 8.1 Chemicals Involved in the Fire at Sandoz Schweizerhalle Works Near Basle, Switzerland on November 1986 [8]

Types of chemicals	Active moiety	Concentration (%)	Amount (metric to)
Agrochemicals:			
Organophosphates	Disulfoton	92	304
	Disulfoton	50	29
	Thiometon	50	206
	Thiometon	25	107
	Etrimfos	65	89.3
	Etrimfos	10	16
	Propetamphos	92	69
	Fenitrothion	96	10.8
	Ethylparathion	5	14.7
	Ethylparathion		9
	Quinalphos		0.6
Chlorinated organic compounds	Metoxuron	97	11.9
	Tetradifon		2.3
	Endosulfan		2.0
	Endosulfan		2.0
	Captafol		0.16
	Dichlorvos		0.1
Organic mercury compounds	Ethoxyethyl-Hg-OH	16	8.6
	Phenyl-Hg-acetate		1.5
Other agrochemicals	Dinitro-o-kresol	90	73.2
	Oxadixyl	97	26
	Others		1.5
Total agrochemicals			982.66
Others: (dyestuffs, solvents, emulgators, stabilizers, raw materials)			354.09

tons distributed on half the size of a soccer playground). The sulfur was being used in the manufacture of fertilizers and ammunition.

The initial fire development was wrongly identified as a bush fire, a recurrent fire cause in this rural region. It was estimated that 14,000 tons of SO_2 were released to the environment for some 20 hours. Consequently, the toxic plume affected a whole town downwind of the fire. The people suffered severe injuries due to inhalation of irritant combustion products (SO_2), and 2 to 12 people died in the vicinity immediately or several days after the exposure.

Because of the fire hazard and environmental contamination from fires involving chemicals stored in warehouses, several research projects have been funded by the ad hoc National and Community Authorities of the European Union (Commission of the European Communities), such as the R&D programs named COMBUSTION, MISTRAL, and TOXFIRE. As an example, the TOXFIRE program, entitled more completely Guidelines for Management of Fires in Chemical Warehouses, was carried out during the period 1993–1996 [19]. Examples of findings obtained in the field in the MISTRAL program are given in Refs. 20 through 22. Databases for the chemical warehouse fires have also been developed up to 1993 [23].

8.1.2 Accidents Involving Release of Fluids Contained in Vessels, Tanks, and Pipes

Many fires have occurred involving fluids in ground storage tanks, in oil-carrying tankers on ocean surfaces, in transport by trucks and trains, in fuel tanks in automobiles, in petroleum oil fields and refineries, in hydraulic systems, and elsewhere. Some examples are:

- Fire involving containers carrying malathion (a pesticide) near Port Rashid, Dubai, probably started from spontaneous combustion [24]. The container was eventually incinerated at sea.

- The 1989 oil spill from the Exxon Valdez tanker onto waters of Prince William Sound in Alaska [25]. This is an example of one of the risks of oil drilling and transportation. These types of accidents have occurred on many ocean surfaces, releasing natural crude oil or its refined products in oil spills. Oil contamination of land and water is an environmental hazard to life. In situ burning of spilled oil, as very large liquid pool fires, has been used as one of the techniques to reduce the impact on the environmental pollution, although it is regarded as a response method of last resort [25, 26].

- Uncontrolled gas and oil blowout fires of 610 wells in Kuwait set by Iraqis in February 1991 [27, 28]. The last burning oil well was extinguished and capped in November 1991. The continuous release of smoke and heat from these fires during the 9-month period caused severe environmental and health problems.

- Creation of hazardous conditions due to boilover of water film at the bottom of large storage tanks containing multicomponent fuels (particularly crude oils or other heavy oils) [29]. Boilover is defined as the explosive vaporization of water. Although the fuel burning by itself is similar to that of a single fuel, the presence of water introduces effects that are caused by the transfer of heat from the fuel to the water underneath. This heat transfer in depth induces boiling and splashing of water.

- Intense, radiative-heat-emitting fireballs resulting from the release of hydrocarbon-based fuels into the atmosphere [4, 30]. Fireballs are considered as one of the major hazards of the modern chemical industry and have been responsible for loss of life, damage to many industrial buildings, chemical plants, trains, and trucks, and emergency evacuations of cities and towns. Fireballs have been examined in various studies over the past two decades and quantitative information has been developed on the maximum diameter, elevation, lifetime, surface temperature, and emissive power [4, 30]. This information has been used successfully in the quantitative risk assessment, providing a fast screening tool for the analysis of possible accident scenarios and implementation of adequate fire protection.

- Jet fires such as those due to release of hydraulic fluid sprays, streams, or mists because of leaks or breaks in the hydraulic systems or tanks and pipes [4, 5, 31, 32]. Jet fires are created by the encounter of fluids with ignition sources such as an electrical spark, flame, or hot surface. The most common source of leakage, for example, in hydraulic systems, is from fittings, valves, steel-reinforced rubber hoses, and steel and copper pipes [31, 32]. Engineering modifications of the systems and utilization of fire-resistant hydraulic fluids have minimized the risk of hydraulic fluid fires [31].

- Large, turbulent, diffusion flames caused by the accidental release of hydrocarbon vapors in processing environments [4]. Intentional disposal of large quantities of unwanted liquid vapors and gases by burning them in a flare has been used traditionally in the petroleum industry. There are three types of flaring of liquid vapors and gases in the petroleum industry [4]: (1) production flaring in the production oil fields; (2) process flaring in petrochemical plants, oil refineries, and gas processing plants; and (3) emergency flaring for the safe disposal of large volumes of combustible vapors of liquids and gases, such as from a fire, power failure, or overpressure in a process vessel.

8.2 PROPERTIES ASSOCIATED WITH THE IGNITION, COMBUSTION, AND FLAME SPREAD BEHAVIORS OF FLUIDS

The ease with which fluids ignite, burn, spread the flame, and release heat and fire products are characterized by the following fluid properties [3–36]:

- Density ρ, heat capacity c, and molecular weight M
- Distillation and vapor pressure
- Boiling point* T_b
- Flash point† T_f
- Fire point‡ T_{fr}
- Autoignition temperature§ T_a
- Heat of vaporization ΔH_v
- Heat of gasification ΔH_g
- Upper and lower flammability limits (UFL and LFL)
- Flame height X_f
- Net heat of complete combustion ΔH_{ncc}
- Chemical, convective, and radiative heat of combustion ΔH_{ch}, ΔH_{con}, and ΔH_{rad}, respectively
- Yields of products y_j
- Smoke point¶ L_s

The fluid property data are available in the literature [34], and are included in Table 8.2 as examples.

8.3 VAPORIZATION AND BOILING CHARACTERISTICS OF FLUIDS

Vaporization of a fluid is a surface mass transfer phenomenon, although intense heating may produce bubbles within the fluids [34]. Upon heating a fluid body, internal convective currents develop with intensity dependent on the heating rate, viscosity, surface tension, gravity, and geometry of the body [34].

The ease with which vapors can be produced by heating a fluid is known as its volatility. A fluid is considered to be highly volatile if its vapor pressure at a given temperature is high (i.e., the boiling point at a given pressure is low) and its heat of gasification is low. The process of fluid vaporization and boiling is characterized by [34]:

- Boiling point T_b, K
- Latent heat of vaporization, $\int_{T_0}^{T_b} c\, dT$, where T_0 is the ambient temperature (K) and c is the heat capacity (kJ/kg·K);
- Heat of vaporization ΔH_v (kJ/kg).

---

 * Boiling point for a single-component fluid is generally defined as the temperature at which the vapor pressure equals one standard atmosphere.
 † Flash point is the minimum temperature at which a fluid gives off sufficient vapors to form an ignitable mixture with air near the surface of the liquid or within the test vessel used. Flash points are reported as open- or closed-cup flash points [33, 34].
 ‡ Fire point is the lowest temperature at which a fluid in an open container will give off enough vapors to continue to burn once ignited. It is generally slightly above the open-cup flash point [33, 34].
 § Autoignition temperature is a rapid, self-sustaining, sometimes audible gas-phase reaction of the fluid or its decomposition products with an oxidant. A readily visible yellow or blue flame usually accompanies the reaction [33, 34].
 ¶ Smoke point L_s is the minimum laminar axisymmetric diffusion flame height at which smoke just escapes from the flame tip. It has been used for decades to express the smoke formation characteristics of gases, liquids, and solids [35].

The sum of the heat of vaporization and latent heat of vaporization is defined as the heat of gasification ΔH_g (kJ/kg):

$$\Delta H_g = \int_{T_0}^{T_b} c\, dT + \Delta H_v \tag{8.1}$$

The values of T_b, c, and ΔH_v are measured by several standard test methods, for example:

- ASTM E 1269-95 and E 793-95: standard test methods for the measurement of c and ΔH_v values [37–39];
- ASTM D 1120-94, ASTM D 2887-97, and ASTM D 86-96: standard test methods for the measurement of T_b values [40–42]. The ASTM D 1120-94 test method determines the equilibrium boiling points of fluids [40]. The equilibrium boiling point indicates the temperature at which the sample will start to boil in a cooling system under equilibrium conditions at atmospheric conditions. In ASTM D 2887-97 and ASTM D 86-96 [41, 42], a fluid sample is injected into a *gas chromatograph* (GC). The temperature for 99.5 percent of the total integrated GC detector response is used as the T_b value of the fluid. The temperature for 0.5 percent of the total integrated GC detector response is used as the *initial boiling point* (IBP) value of the fluid.

The data measured for the same fluid property by different ASTM standard test methods, however, do not agree with each other [43]. Examples of the ΔH_v and T_b values for fluids, obtained from some of these methods and reported in Ref. 34, are listed in Table 8.2 along with the molecular weights (M) of the fluids.

TABLE 8.2 Flammability Properties of Gases and Liquids [34]

Gas/liquid	Composition	M (kg/kmole)	T_b (°C)	T_a (°C)	ΔH_v (kJ/kg)	ΔH_{gcc} (MJ/kg)	T_f(°C) Closed	T_f(°C) Open	Flamm limits (%) Lower	Flamm limits (%) Upper	$(\Delta H_v \cdot M/T_b)/R$
Alkanes											
Methane	CH_4	16	−162	637	509	50.2	—	—	5.3	15.0	8.81
Ethane	C_2H_6	30	−89	472	489	47.6	—	−135	3.0	12.5	9.57
Propane	C_3H_8	44	−42	450	426	46.4	—	−104	2.2	9.5	9.76
n-Butane	C_4H_{10}	58	0	288	386	45.9	—	−60	1.9	8.4	9.88
i-Butane	C_4H_{10}	58	−10	462	366	45.9	−117	—	1.8	8.4	9.71
n-Pentane	C_5H_{12}	72	36	243	365	45.5	—	−49	1.4	7.8	10.22
i-Pentane	C_5H_{12}	72	13	420	371	45.5	—	−51	1.4	7.6	11.23
n-Hexane	C_6H_{14}	86	69	225	365	45.2	−22	—	1.2	7.4	11.04
i-Hexane	C_6H_{14}	86	69	—	365	45.2	−29	—	1.0	7.0	—
n-Heptane	C_7H_{16}	100	98	204	365	45.0	−4	—	1.2	6.7	11.83
i-Heptane	C_7H_{16}	100	98	—	365	45.0	−18	—	1.0	6.0	—
n-Octane	C_8H_{18}	114	125	206	298	44.9	13	—	0.8	3.2	10.26
i-Octane	C_8H_{18}	114	125	—	298	44.9	−12	—	1.0	6.0	—
n-Nonane	C_8H_{20}	128	151	205	288	44.8	31	—	0.7	2.9	10.46
n-Decane	$C_{10}H_{22}$	142	174	201	360	44.7	44	—	0.6	5.4	13.76
n-Undecane	$C_{11}H_{24}$	156	196	—	308	44.6	—	65	0.7	12.3	12.32
n-Dodecane	$C_{12}H_{26}$	170	216	203	293	44.6	72	—	0.6	12.3	12.25
Kerosene	$C_{14}H_{30}$	198	232	260	291	44.0	49	—	0.6	5.6	13.72
Alkenes											
Ethylene	C_2H_4	29	−104	490	516	47.3	−121	—	2.7	28.6	10.28
Propene	C_3H_6	42	−48	455	437	45.9	−108	—	2.1	11.1	9.81
1-Butene	C_4H_8	56	−6	385	398	45.4	−80	—	1.6	9.9	10.04
1-Pentene	C_5H_{10}	70	30	275	314	46.9	—	−18	1.4	9.7	8.72
Hexene	C_6H_{12}	84	67	245	388	47.5	—	—	—	—	11.53

TABLE 8.2 (*Continued*)

Gas/liquid	Composition	M (kg/kmole)	T_b (°C)	T_a (°C)	ΔH_v (kJ/kg)	ΔH_{gcc} (MJ/kg)	T_f(°C) Closed	T_f(°C) Open	Flamm limits (%) Lower	Flamm limits (%) Upper	$(\Delta H_v \cdot M/T_b)/R$
Cycloparaffins											
Cyclopropane	C_3H_6	42	−34	498	588	46.3	−95	—	2.4	10.4	12.43
Cyclobutane	C_4H_8	56	13	210	483	44.8	−65	—	1.1	—	11.38
Cyclopentane	C_5H_{10}	70	49	361	443	44.3	−37	—	2.0	—	11.58
Cyclohexane	C_6H_{12}	84	81	245	358	43.9	−20	—	1.3	7.8	10.22
Cycloheptane	C_7H_{14}	99	119	—	376	43.7	9	—	1.2	—	11.31
Dimethylcyclohexane	C_8H_{16}	112	119	232	300	46.3	11	—	—	—	10.31
Aromatics											
Benzene	C_6H_6	78	80	498	432	40.7	−11	—	1.2	7.1	11.48
Toluene	C_7H_8	92	110	480	362	410.0	4	7	1.3	6.8	10.46
m-Xylene	C_8H_{10}	106	139	528	343	41.3	25	—	1.1	7.0	10.61
o-Xylene	C_8H_{10}	106	141	464	347	41.3	17	24	1.0	6.0	10.69
p-Xylene	C_8H_{10}	106	137	529	339	41.3	25	—	1.1	7.0	10.54
bi-Phenyl	$C_{12}H_{10}$	154	254	540	—	40.6	113	124	—	—	—
Naphthalene	$C_{10}H_8$	128	218	526	316	40.3	79	88	0.9	5.9	9.91
Anthracene	$C_{13}H_{10}$	166	340	540	310	40.0	121	196	0.6	—	10.10
Ethylbenzene	C_8H_{10}	106	136	432	320	43.1	15	24	1.0	—	9.98
Butylbenzene	$C_{10}H_{14}$	134	173	410	277	43.7	49	63	0.8	5.9	10.01
Alcohols											
Methanol	CH_3OH	32	64	385	1101	20.8	12	16	6.7	36.5	12.57
Ethanol	C_2H_5OH	46	78	363	837	27.8	13	22	3.3	19.0	13.19
n-Propanol	C_3H_7OH	60	97	432	686	31.3	15	29	2.2	13.5	13.38
i-Propanol	C_3H_7OH	60	82	399	667	33.1	12	—	2.0	11.8	13.56
n-Butanol	C_4H_9OH	74	117	343	621	36.1	29	43	1.4	11.3	14.17
i-Butanol	C_4H_9OH	74	107	405	578	36.1	28	—	1.7	—	13.54
2-Pentanol	$C_5H_{11}OH$	88	119	343	575	—	—	41	1.2	—	15.52
i-Amyl alcohol	$C_5H_{11}OH$	88	130	300	501	35.3	43	46	1.2	10.0	13.16
3-Pentanol	$C_5H_{11}OH$	88	118	435	575	—	34	39	1.2	—	15.56
n-Hexanol	$C_6H_{13}OH$	102	157	—	458	36.4	45	74	1.2	—	13.07
Cyclohexanol	$C_6H_{13}OH$	102	161	300	460	36.6	68	—	1.2	—	13.00
n-Heptanol	$C_7H_{15}OH$	116	176	—	439	39.8	—	71	—	—	13.64
1n-Octanol	$C_8H_{17}OH$	130	196	—	408	40.6	81	—	—	—	13.60
2n-Octanol	$C_8H_{17}OH$	130	180	—	419	40.6	74	82	—	—	14.46
Nonanol	$C_9H_{19}OH$	144	214	—	403	40.3	—	—	—	—	14.33
i-Decanol	$C_{10}H_{21}OH$	158	235	288	373	—	—	—	—	—	14.12
Carbonyls											
Formaldehyde	CH_2O	30	97	430	826	18.7	93	—	7.0	73.0	8.05
Formaldehyde in 37% in	H_2O	—	97	424	826	18.7	54	93	7.0	—	8.05
Acetaldehyde	C_2H_4O	44	21	204	570	25.1	−38	—	4.0	57.0	10.26
Allyl alcohol	C_3H_6O	58	95	378	684	31.9	21	24	2.5	18.0	12.93
i-Butyraldehyde	C_4H_8O	72	61	230	444	33.8	−40	−24	2.5	—	11.51
Crotonaldehyde	C_4H_6O	70	102	232	490	34.8	13	—	2.1	15.5	11.00
Diethyl Acetaldehyde	$C_6H_{12}O$	76	118	—	500	—	21	—	—	—	11.70
Ethyl Hexaldehyde	$C_8H_{16}O$	128	163	—	325	39.4	—	52	—	—	11.48
Paraldehyde	$C_6H_{12}O_3$	132	124	238	328	—	17	36	1.3	—	13.11
Salicylaldehyde	$C_7H_6O_2$	122	196	—	396	—	78	—	—	—	12.39
Benzaldehyde	C_7H_6O	106	179	192	362	—	64	74	—	—	10.21

TABLE 8.2 (Continued)

Gas/liquid	Composition	M (kg/kmole)	T_b (°C)	T_a (°C)	ΔH_v (kJ/kg)	ΔH_{gcc} (MJ/kg)	T_f(°C) Closed	T_f(°C) Open	Flamm limits (%) Lower	Flamm limits (%) Upper	$(\Delta H_v \cdot M/T_b)/R$
Ketones											
Acetone	C_3H_6O	58	56	465	521	29.1	−18	−9	2.6	12.8	11.05
2-Butanone	C_4H_8O	72	80	404	443	33.8	−2	1	1.8	9.5	10.87
Diethyl ketone	$C_5H_{10}O$	86	101	450	380	33.7	—	13	—	—	10.51
Methyl i-butyl ketone	$C_6H_{12}O$	100	116	533	345	35.2	23	24	1.4	7.6	10.66
Dipropyl ketone	$C_7H_{14}O$	114	144	533	317	38.6	—	—	—	—	10.42
Methyl n-Propyl ketone	$C_5H_{10}O$	86	102	340	376	33.7	7	16	1.5	8.2	10.37
Methyl vinyl ketone	C_4H_6O	70	81	—	440	—	−7	—	—	—	10.46
Acids											
Formic acid	CH_2O_2	46	101	540	502	5.7	69	—	—	—	7.42
Acetic acid	$C_2H_4O_2$	60	118	464	405	14.6	40	57	5.4	—	7.48
Benzoic acid	$C_7H_6O_2$	122	250	570	270	24.4	121	—	—	—	7.58
Miscellaneous											
Camphor	$C_{10}H_{16}O$	152	204	466	265	38.8	66	93	0.6	3.5	10.16
Carbon disulfide	CS_2	76	47	125	—	13.6	30	—	1.3	50.0	—
m-Creosol	C_7H_8O	108	203	559	—	34.6	86	—	1.1	—	—
O-Creosol	C_7H_8O	108	191	599	—	34.1	81	—	1.3	—	—
P-Creosol	C_7H_8O	108	202	559	—	34.1	86	—	1.0	—	—
Furan	C_4H_4O	68	31	—	399	—	−35	—	2.3	14.3	10.73
Pyridine	C_5H_5N	70	114	482	449	35.0	20	—	1.8	12.4	11.02
Aniline	C_6H_7N	93	183	617	434	36.5	76	91	1.3	—	10.64
Acetal	$C_6H_{14}O_2$	118	103	230	277	31.8	−21	—	1.6	10.4	10.46
P-Cymene	$C_{10}H_{14}$	134	176	436	283	43.9	47	63	0.7	5.6	10.16
O-Dichloro benzene	$C_6H_4Cl_2$	146	180	648	—	19.3	66	74	2.2	9.2	—
1,1-Dichloroethylene	$C_2H_2Cl_2$	96	37	460	—	19.3	—	−10	5.6	11.4	—
1,2-Dichloroethylene	$C_2H_2Cl_2$	96	61	—	—	—	6	—	9.7	12.8	—
Monochlorobenzene	C_6H_5Cl	112	132	674	—	—	32	38	1.8	—	—
Resorcinol	$C_6H_6O_2$	110	276	608	—	26.0	127	—	1.4	—	—
Ethylformate	$C_3H_6O_2$	74	54	455	—	22.5	−20	−12	2.7	16.4	—
Ethylacetate	$C_4H_8O_2$	88	77	427	—	25.9	−4	−1	2.2	11.4	—
Methylpropionate	$C_4H_8O_3$	104	80	469	—	22.2	−2	—	2.4	13.0	—
Acrolein	C_3H_4O	56	53	235	—	29.1	—	−26	2.8	31.0	—
Acrylonitrile	C_3H_3N	53	77	481	—	24.5	—	0	3.0	17.0	—
n-Amyl acetate	$C_7H_{14}O_2$	130	149	360	—	33.5	24	27	1.1	7.5	—
1-Amyl acetate	$C_7H_{14}O_2$	130	143	379	—	—	25	38	1.0	7.5	—
1,3-Butadiene	C_4H_6	54	−4	420	—	—	−76	—	2.0	11.5	—
n-Butyl acetate	$C_6H_{12}O_2$	116	127	421	—	30.0	22	32	1.4	7.6	—
n-Butyl ether	$C_8H_{18}O$	130	141	194	—	39.7	25	38	1.5	—	—
Dimethylether	C_2H_6O	46	−24	350	—	31.6	−41	—	3.4	26.7	—
Divinyl ether	C_4H_4O	70	39	360	—	—	−30	—	1.7	27.0	—
Diethyl ether	$C_4H_{10}O$	74	35	160	—	37.4	−45	—	1.9	36.5	—
Gasoline	—	—	33	371	—	44.1	−45	—	1.4	6.8	—
Naphtha	—	—	177	246	—	—	41	—	0.8	5.0	—
Petroleum ether	—	—	78	288	—	—	−18	—	1.4	5.9	—

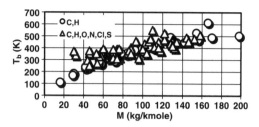

FIGURE 8.1 Relationship between the boiling points and molecular weights of fluids. (Data are taken from Ref. 34, Table 8.2.)

The values of both ΔH_g and ΔH_v for fluids with carbon, hydrogen, oxygen, and other atoms are higher than the values for fluids with only carbon and hydrogen atoms [43]. The difference is probably due to the latent heat of vaporization. The latent heat of vaporization is the major contributor toward the ΔH_g value for fluids consisting of carbon and hydrogen atoms and seems to be approximately constant for the majority of these fluids (average value is 131 kJ/kg) [43]. For fluids consisting of carbon, hydrogen, oxygen, and other atoms, on the other hand, ΔH_v is the major contributor toward the ΔH_g value [43].

The boiling process requires overcoming the intermolecular forces. Thus, as the molecules get larger (increase in molecular weight) boiling points increase because of increase in the intermolecular forces [44], such as shown in Fig. 8.1. With some exceptions, for fluids containing carbon and hydrogen atoms, the boiling point increases by 20 to 30°C for each carbon that is added to the chain [44]. A comparison of the data for fluids with carbon-hydrogen atoms with those containing carbon-hydrogen-other atoms in Fig. 8.1 with similar molecular weight show that the boiling points are different. These differences are accounted for by differences in [44]:

- Intermolecular forces: dipole-dipole interactions and van der Waals forces
- Ionic bonding
- Hydrogen bonding

According to Trouton's rule [45], the ratios of molar heats of vaporization to the boiling points of nonpolar fluids are approximately constant:

$$\frac{\Delta H_v M}{T_b} \approx 92 \ (\text{kJ/mol} \cdot \text{K}) \tag{8.2}$$

Figure 8.2 shows the plot of $(\Delta H_v M)$ versus T_b for the fluids where data from Ref. 34 (Table 8.2) have been used. For nonpolar fluids, the data satisfy the Trouton's rule. The general idea underlying Trouton's rule is that nonpolar fluids have essentially similar random molecular configurations [45]. Therefore, on vaporization to the gaseous state, they occupy roughly similar molal volumes [45].

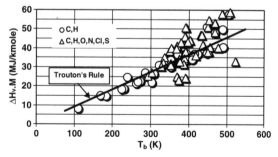

FIGURE 8.2 Relationship between the molal heats of vaporization and boiling points of fluids. (Data are taken from Ref. 34, Table 8.2.)

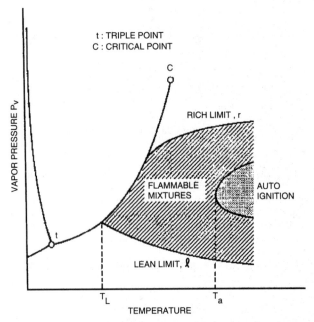

FIGURE 8.3 Phase change diagram for combustible fluid vapor-air mixture taken from Ref. 34 by permission of SFPE.

8.4 IGNITION CHARACTERISTICS OF FLUIDS

When a fluid is heated, vapors emanating from the surface mix with ambient air and form a combustible mixture, which can autoignite or can be ignited by a hot surface, spark, or pilot flame. The formation of a flammable mixture depends on many factors, such as air movement, fluid temperature, closed or open fluid reservoir, reservoir geometry, molecular weight of the vapors relative to that of air, and the location of the igniter [34].

The fluid ignition process is shown in Fig. 8.3, which is taken from Ref. 34. As the temperature of the fluid is increased, the fluid vapor pressure increases, and at a fluid temperature of T_L, a lean limit mixture is formed, defined as the *lean limit of flammability* or *lower flammability limit* (LFL)* [34]. T_L is related to the flash point T_f [34]. With further increase in the temperature, the fluid reaches the fire point, T_{fr}, where fluid vapor-air mixture ignites if a small heat source, such as a pilot flame, is present. In the absence of the small heat source (pilot flame), the fluid temperature continues to increase, reaching the autoignition temperature T_a, and the fluid vapor-air mixture ignites by itself.

For a fluid temperature greater than T_L, the fluid vapor-air mixture remains flammable until the *rich limit of flammability* or the *upper flammability limit* (UFL)† is reached, beyond which the mixture becomes nonflammable because it becomes too fuel rich.

The fluid ignition process is characterized by [33, 34] the following:

- Flash point T_b, °C
- Autoignition temperature T_a, °C

---

* LFL is defined as the lowest volume percent of the fluid vapor in the mixture with air that will barely support flame spread away from the pilot flame [33, 34].
† UFL is defined as the highest volume percent of the fluid vapor in the mixture with air that will barely support flame spread away from the pilot flame [33, 34].

8.4.1 Flash Points of Fluids

Flash point is the minimum temperature at which a fluid gives off sufficient vapors to form an ignitable mixture with air near the surface of the liquid or within the test vessel used. Flash points are reported as open- or closed-cup flash points [33, 34].

For the measurement of flash points of fluids, open or closed cups are used in several standard test methods such as those specified in the "Globally Harmonized System (GHS) for Classification and Labeling of Chemicals"* as [46–48]

1. *Association Francaise de Normalisation* (AFNOR)
    a. French Standard NF M 07-019
    b. French Standards NF M 07-011/NF T 30-050/NF T 66-009
    c. French Standard NF M 07-036
2. Deutscher Normenausschuss
    a. Standard DIN 51755 (flash points below 65°C)
    b. Standard DIN 51758 (flash points 65 to 165°C)
    c. Standard DIN 53213 (for varnishes, lacquers, and similar viscous liquids with flash points below 65°C)
3. International Standards
    a. ISO 1516
    b. ISO 1523
    c. ISO 3679
    d. ISO 3680
4. State Committee of the Council of Ministers for Standardization, Moscow GOST 12.1.044-84
5. British Standards Institution
    a. British Standard BS EN 22719
    b. British Standard BS 2000 Part 170
6. American Society for Testing and Materials
    a. ASTM D 3829-93, "Standard Test Method for Flash Point by Small Scale Closed Tester"
    b. ASTM D 56-93, "Standard Test Method for Flash Point by Tag Closed Tester"
    c. ASTM D 3278-96, "Standard Test Method for Flash Point of Liquids by Setaflash Closed-Cup Tester"
    d. ASTM D 93-96, "Standard Test Method for Flash Point by Pensky-Martens Closed-Cup Tester"

The flash points of liquids are measured typically in an apparatus shown in Fig. 8.4 [49] (used in ASTM D 93-96). The test is performed in a closed 54-mm-wide and 56-mm-deep brass cup heated electrically. The cover of the cup has provisions for introducing a thermocouple, a stirrer, and a shutter with a pilot flame. The liquid is stirred in the cup as it is heated. The shutter has a control mech-

---

* GHS has been developed for the safe use, transport, and disposal of liquids and chemicals [46]. The international mandate that provided the impetus for developing GHS was the 1992 *United Nations Conference on Environment and Development* (UNCED). GHS has been developed to (1) enhance the protection of mankind, property, and the environment by providing an internally comprehensive system for hazard communication; (2) provide a recognized framework for those countries without an existing system; (3) reduce the need for testing and evaluation; and (4) facilitate international trade in liquids and chemicals whose hazards have been properly assessed and identified on an international basis.

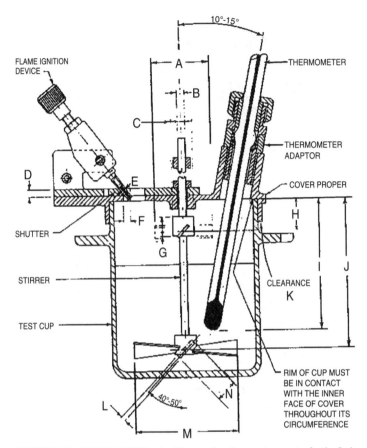

FIGURE 8.4 ASTM D 93-97 Pensky-Martens closed cup test apparatus for the flash point of fluids. (Figure taken from Ref. 49.)

anism to lower the flame into the vapor space of the test cup in 0.5 s, keep it in the lowered position for 1 s, and quickly raise it to its upward position.

The cup has a marker enabling it to be filled with a fixed volume of the liquid. The liquid is heated to a temperature below the flash point and a pilot flame is applied at a temperature that is a multiple of 5°C. The test has a repeatability of 5°C and a reproducibility of 10°C.

Examples of the open-cup and closed-cup flash points for fluids reported in Ref. 34 are listed in Table 8.2. The open-cup flash point is higher than the closed-cup flash point, as shown in Fig. 8.5 (data taken from Ref. 34; Table 8.2). The flash points and boiling points are interrelated, as shown in Fig. 8.6 for the relationship between the closed-cup flash points and boiling points (data taken from Ref. 34; Table 1). The data in Fig. 8.6 suggest that $T_f = 0.75 T_b$.

8.4.2 Autoignition Temperature of Fluids

Autoignition temperature is a rapid, self-sustaining, sometimes audible gas-phase reaction of the fluid or its decomposition products with an oxidant. A readily visible yellow or blue flame usually accompanies the reaction [33, 34]. The autoignition temperature is measured typically in apparatuses such as shown in Fig. 8.7 and specified in ASTM E 659-78 [50].

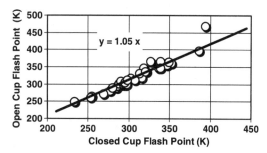

FIGURE 8.5 Relationship between the open cup and closed cup flash points of fluids. (Data taken from Ref. 34, Table 8.2.)

In the test, 10 ml of the fluid is injected into a uniformly heated 500-ml glass flask containing air at a predetermined temperature, measured by a thermocouple located at the center of the flask, as shown in Fig. 8.7. The contents of the flask are observed in a dark room for 10 min following the insertion of the sample or until autoignition occurs. Autoignition is evidenced by the sudden appearance of a flame inside the flask and by a sharp rise in the temperature of the gas mixture. The lowest internal flask temperature at which hot-flame ignition occurs for a series of prescribed sample volumes is taken to be the hot-flame autoignition temperature (T_a) of the fluid in air at atmospheric pressure. The repeatability of the test is 2 percent and the reproducibility is 5 percent.

The T_a values of fluids reported in the literature [34] are included in Table 8.2. There is no consistent relationship between T_a and T_b or M, except that at higher T_b and M values, T_a values become approximately constant.

8.4.3 Hazard Classification of Fluids Based on Ignition Resistance

The following systems are commonly used for the hazard classification of fluids based on their T_f and T_b values for transportation, waste disposal, storage, handling, and emergency response:

- The *National Fire Protection Association* (NFPA) system (NFPA **30**) [48]
- The U.S. *Department of Transportation* (DOT) system [49 *Code of Federal Regulations* (CFR), Part 173.120 (c)] [48]
- The *American National Standards Institute* (ANSI) system (ANSI Z129.1) [48]
- EC Directives (67/548/EEC and parent directive 1999/45/EC)
- The GHS system [46–48]

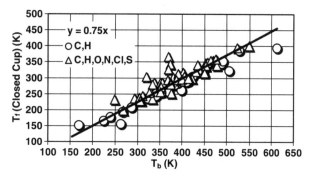

FIGURE 8.6 Relationship between the closed cup flash points and boiling points of fluids. (Data taken from Ref. 34, Table 8.2.)

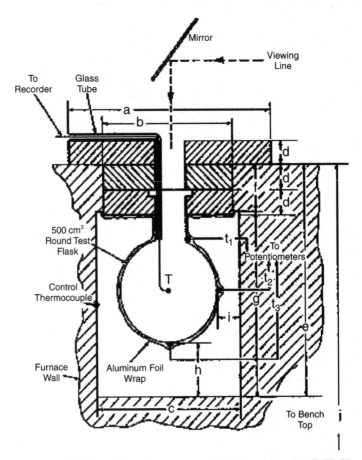

FIGURE 8.7 Test apparatus specified in the ASTM standard test method E 659–78 for the measurement of autoignition temperatures of fluids. (Figure taken from Ref. 50.)

The hazard classification criteria used in these systems are listed in Table 8.3. The criteria are selected because of the following reasons [48]:

1. **Transportation:** It involves extraneous risk factors not present in fixed storage facilities, including increased risk of mechanical damage, variable environmental impacts in transit, lack of continuous fire protection, and increased emergency response. Owing to the potential for elevated temperatures in shipping containers and other places, it is believed that fluids may attain a temperature of 60°C without deliberate heating. Thus, flammable fluids are identified as fluids having $T_f < 60°C$.

   The DOT definitions and regulations are generally (but not completely) aligned with the GHS system. The U.S. *Occupational Safety and Health Administration's* (OSHA) *Hazard Communications Standards* (HCS) requires all liquids and chemicals in the workplace to be labeled in a manner that warns of hazards presented by them. A fluid with $T_f > 93°C$ or that meets certain exceptions is not regulated by DOT and is classified as neither flammable nor combustible.

2. **Waste disposal:** Characteristics of hazardous waste are used in the *Resource Conservation and Recovery Act* (RCRA) regulations. Fluids having $T_f < 60°C$ are considered as capable of generat-

TABLE 8.3 Hazard and Labeling Criteria for Fluids[a]

Hazard category	UNCED "Harmonization" GHS [46, 47, 48]			NFPA 30/704 [48]			DOT Class 3 [48]		ANSI Z129.1 Labeling [48]	
	Criteria (°C)	Labeling	Class	Hazard rating	Criteria (°C)[b]	Packing group	Criteria (°C)	Hazard level	Criteria[c] (°C)
1	IBP ≤ 35	Extremely flammable liquid and vapors	1A	4	T_b <38; T_f < 23	I	IBP ≤ 35	Extremely flammable	T_f ≤ −7 or T_b ≤ 35; T_f ≤ 61
			1B	3	T_b ≥ 38; T_f < 23				
			1C	3	23 ≤ T_f < 38				
2	IBP > 35; T_f < 23	Highly flammable liquid and vapors	II	2	38 ≤ T_f < 60	II	IBP > 35; T_f < 23	Flammable	T_b > 35; T_f ≤ 61
3	IBP > 35; 23 ≤ T_f ≤ 60	Flammable liquid and vapors	IIIA	2	60 ≤ T_f < 93	III	IBP > 35; 23 ≤ T_f ≤ 61	Combustible	61 < T_f < 93
			IIIB	1	T_f ≥ 93				
4	60 < T_f ≤ 93	Combustible liquid	0	0	5 minute T_{ig} > 816				

[a] IBP: initial boiling point; T_b: boiling point; T_f: closed cup flash point.
[b] For single component fluids it is defined as the temperature at which the vapor pressure is equal to one atmosphere. For mixtures that do not have a constant boiling point, the 20% evaporation point of a distillation performed in accordance with ASTM E 86 is considered to be the boiling point.
[c] Boiling point assumed to be IBP.

ing ignitable vapors in situations of elevated temperatures because of biological activities in landfills.

3. Storage: Regulatory agencies such as OSHA and consensus standards-setting organizations such as NFPA and *American Petroleum Institute* (API) cover the storage and handling of hazardous liquids and chemicals. There is latitude for variation in the fire hazard classification systems depending on the codes of practice imposed by the regulatory agencies of individual countries. In the United States, the most comprehensive classification system is published by NFPA ("Flammable and Combustible Liquids Code," NFPA 30). The NFPA hazard ratings of liquids and chemicals are compiled in NFPA 325 ("Fire Hazard Properties of Flammable Liquids, Gases and Volatile Solids") and NFPA 49 ("Hazardous Chemicals Data"). They are frequently reported on *materials safety data sheets* (MSDS).

4. *Emergency response* (ER): The response hazards are very different from those from transportation, waste disposal, and storage. For example, during a building fire, even liquids with very high T_f values can present serious fire hazards. It is, therefore, necessary for ER fire hazard classification systems to identify fluids that are able to burn as opposed to those simply having high T_f values. Water solubility is another factor that needs to be considered. Information useful for the ER fire hazard classification is discussed later in the sections dealing with the pool fires and release rates of vapors, heat, and fire products.

In practice, the T_f and T_b values of single-component fluids are subject to variability depending on the test method, purity of the fluids, and ambient pressure at which the "standard" data are applied [48]. Mixtures of fluids introduce further variability depending on the precise composition and the definition of boiling point used [48]. There are different flash point methods recognized by different authorities. Also, flash point is generally not an accurate measure of the lowest temperature at which a liquid forms flammable mixture in a closed container and typically overestimates the lowest temperature at which flammable mixtures are produced [48]. Thus, fire hazard classification is subject to errors, especially for those fluids with T_f and T_b values close to the values between the two groups or classes of fluids.

8.5 FLAMMABILITY CHARACTERISTICS OF FLUIDS

The flammability of fluids is characterized by [33, 34]:

- Lower flammability limit (LFL), defined as the lowest volume percent of the fluid vapor in the mixture with air that will barely support flame spread away from the pilot flame
- Upper flammability limit (UFL), defined as the highest volume percent of the fluid vapor in the mixture with air that will barely support the flame

The LFL and UFL values are measured in apparatuses such as shown in Fig. 8.8 and specified in ASTM E 681-98 [51]. The apparatus consists of a glass test vessel about 5 liters in volume, an insulated chamber equipped with a source of controlled air temperature, an ignition device with an appropriate power supply, a magnetic stirrer, and a cover equipped with the necessary operating connections and components.

The vessel is heated to the desired temperature; after a period of equilibration, the vessel is evacuated, and pressure inside the vessel is measured. A measured volume of the fluid is introduced into the vessel by a hypodermic syringe. The stirring mechanism is activated to agitate the fluid and produce a large surface area for evaporation. After all the fluid has evaporated, the pressure of the fluid vapor is measured. Air is then introduced until the pressure in the vessel is atmospheric, which is also measured. The fuel concentration is calculated from the ratios of the pressures of the fluid vapor and its mixture with air.

The high-energy source is activated for 1 s and flame propagation is observed in the test vessel.

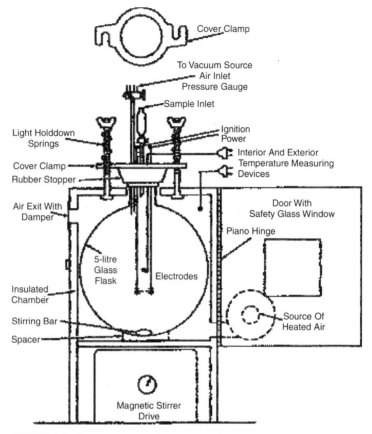

FIGURE 8.8 Test apparatus specified in the ASTM standard test method E 681-98 for the measurement of lower and upper flammability limits of fluids. (Figure taken from Ref. 51.)

Fluid sample volume is varied to find the minimum sample volume L_1 that gives flame propagation* and the maximum sample volume L_2 below L_1 that does not give flame propagation. The difference between L_1 and L_2 is a measure of the variability of the procedure for the sample being studied. In a similar manner, the highest fluid sample volume U_1 is determined for flame propagation, and the least volume U_2 above U_1 that will not propagate a flame. The LFL and UFL are then expressed as $(L_1 + L_2)/2$ and $(U_1 + U_2)/2$, respectively.

The test method is limited to an initial pressure of 101 kPa (1 atm) or less with a practical lower pressure limit of approximately 13.3 kPa (100 mmHg). The maximum operating temperature of this equipment is approximately 150°C, although tests can be performed up to 280°C. LFL and UFL values of gases and liquids are available in the literature [34], examples of which are included in Table 8.2. The LFL and UFL values of fluids are interrelated as shown in Fig. 8.9, where data are taken from Ref. 34 (Table 1). The expression for their relationship is included in the figure.

---

* Propagation of flame is defined in the test as the upward and outward movement of the flame front from the ignition source to the vessel walls or at least to within 13 mm of the wall, which is determined by visual observations [51]. By outward is meant a flame front that has a horizontal component to the movement away from the ignition source [51].

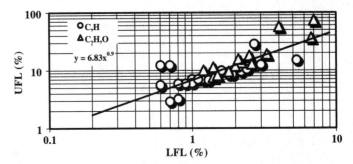

FIGURE 8.9 Relationship between the lower and upper flammability limits of fluids. (Data taken from Ref. 34, Table 8.2.)

The LFL value of a fluid is related to ΔH_v, T_f, and T_b (Clausius-Clapeyron relationship) [34]:

$$\ln\left(\frac{1}{\text{LFL}}\right) \geq \frac{1}{R}\left(\frac{\Delta H_v \cdot M}{T_b}\right)\left(\frac{T_b - T_f}{T_f}\right) \tag{8.3}$$

where R is the universal gas constant (8.314 kJ/kmol·K). Values of $(\Delta H_v M/T_b)/R$ [Eq. (8.3)] are listed in Table 8.2. Thus, from Eq. (8.3):

$$\ln\left(\frac{1}{\text{LFL}}\right) \geq 11\left(\frac{T_b - T_f}{T_f}\right) \tag{8.4}$$

The relationship in Eq. (8.4) is strongly dependent on the transient convective-diffusion process that plays a crucial role in determining the T_f values [34]. In addition, LFL is an empirical extrinsic parameter whose dependence on the fundamental properties of the system is not known quantitatively [34]. As a result, the experimental ln (1/LFL) values are higher than the calculated values from Eq. (8.4), as shown in Fig. 8.10, where data are taken from Ref. 34 (Table 1).

8.6 COMBUSTION CHARACTERISTICS OF FLUIDS

After ignition and flame spread over the surface, the fluid burns as a pool fire or a wall fire (fluid-soaked solid materials). The heat flux from the flame transferred to the surface is the source of

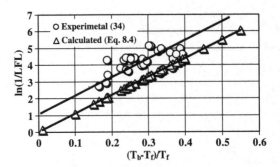

FIGURE 8.10 Relationship between the lower flammability limit, boiling point, and flash point (closed cup) for fluids. (Data taken from Ref. 34, Table 8.2.)

energy for continued combustion. The flame heat flux increases with pool diameter or wall height, with dominant mode of heat transfer changing from conductive to convective to radiative. Flame heat flux reaches its limiting value for large pool or wall fires and decreases with further increase in the pool diameter or wall height. With increase in the flame heat flux, release rates of fluid vapors, heat, and products increase, resulting in the increase in fire intensity, and thus the fire hazard.

As noted before, the emergency response hazards for liquid fires are very different from the hazard classification of the fluids based on T_f and T_b values (utilized for the transportation, waste disposal, and storage of fluids). For example, during a building fire, even liquids with very high T_f values can present serious fire hazards depending on the combustion characteristics of the fluids. Therefore, it is necessary to develop hazard classifications of fluids for emergency response based on their combustion characteristics rather than on their T_f and T_b values.

The combustion of fluids has been characterized in small- and large-scale pool, wall, and spray fires [1–4, 25, 26, 28–32, 35, 36, 43, 52–64]. The following characterize the combustion process for the fluids:

- Release rate of fluid vapors $\dot{m}''_f$ (kg/m^2·s), ΔH_g (kJ/kg), net flame heat flux received by the fluid surface $\dot{q}''_n$ (kW/m^2), flame heat flux received by the fluid surface $\dot{q}''_f$ (kW/m^2), and surface reradiation loss $\dot{q}''_{rr}$ (kW/m^2)
- Chemical heat release rate and its convective and radiative components, $\dot{Q}''_{ch}$, $\dot{Q}''_{con}$, and $\dot{Q}''_{rad}$, respectively (all in kW/m^2), chemical heat of combustion and its convective and radiative components ΔH_{ch}, ΔH_{con}, and ΔH_{rad}, respectively (all in MJ/kg), and the heat release parameter HRP ($\Delta H_{ch}/(\Delta H_g$, MJ/MJ)
- Release rates of products $\dot{G}''_f$ (kg/m^2·s)
- Yields of products y_j (kg/kg)
- Smoke point L_s (mm)
- Flame height X_f (m)

The combustion characteristics of fluids in terms of their fire properties are generally characterized in small-scale tests. Several small-scale standard test methods are available for such characterization. The most commonly used standard test methods are ASTM E 1354/ISO 5660 (cone calorimeter) [65] and ASTM E 2058 (fire propagation apparatus) [66]. The fluid samples* are placed typically in a 100-mm-diameter and 50-mm-deep Pyrex or aluminum dish, for example, at the location marked "sample" in Fig. 8.11 (ASTM E 2058).

For evaluating the ignition characteristics of the fluids, the samples contained in the Pyrex dish or soaked onto the cloth are exposed to external heat flux in the range of 10 to 60 kW/m^2 in the presence of a pilot flame, and time to ignition is measured. The time to ignition versus external heat flux relationship is used to derive the ignition properties of fluids consisting of [35, 36]: (1) *critical heat flux for ignition*† (CHF) and (2) *thermal response parameter*‡ (TRP).

For evaluating the combustion characteristics of the fluids, samples are exposed to external heat flux in the range of 0 to 50 kW/m^2 with air containing oxygen, concentration in the range of 0 to 50 percent by volume and flowing around the sample inside the quartz tube. Measurements are made for the release rates of fluid vapors, heat, and products, and flame heights. Fire properties of fluids are then derived from the heat and mass balances.

---

* Fluid sample is poured into the empty dish or soaked onto cloth contained inside the Pyrex dish. For long steady-state combustion, fluid-soaked cloth is found to be ideal.

† CHF is the external heat flux value at which there is no ignition for 600 s under quiescent airflow condition.

‡ TRP is a combination of fluid properties and is expressed as $\Delta T_{ig} \sqrt{k\rho c}$, where ΔT_{ig} is the ignition temperature above ambient temperature (K), k is the thermal conductivity of the fluid (kW/m·K), ρ is the density of the fluid (kg/m^3), and c is the heat capacity of the fluid (kJ/kg·K). It relates to the ignition delay at a specified heat exposure.

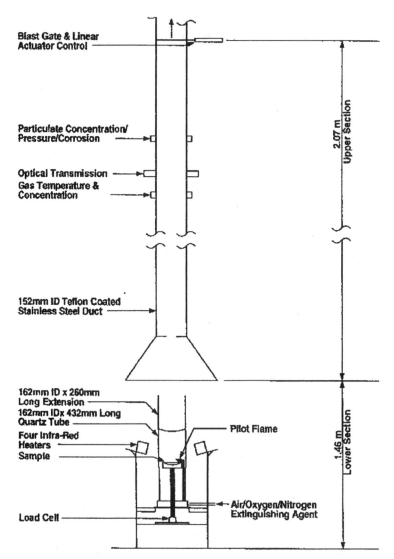

FIGURE 8.11 Fire Propagation Apparatus specified in the ASTM E 2058 Standard Test Method. (From Ref. 66.)

8.6.1 Release Rate of Fluid Vapors in Pool Fires

The release rate of fluid vapors measured in the pool fires is expressed as the ratio of the net heat flux minus the surface reradiation loss to the heat of gasification [35, 36]:

$$\dot{m}''_f = \dot{q}''_n / \Delta H_g \tag{8.5}$$

$$\dot{q}''_n = \dot{q}''_e + \dot{q}''_f - \dot{q}''_{rr} \tag{8.6}$$

where $\dot{q}''_n$ is the net heat flux (kW/m²), $\dot{q}''_e$ is the external heat flux (kW/m²), $\dot{q}''_f$ is the flame heat flux (kW/m²), and $\dot{q}''_{rr}$ is the surface reradiation loss (kW/m²). $\dot{q}''_{rr}$ is negligibly small for fluids with low molecular weights and low boiling points.

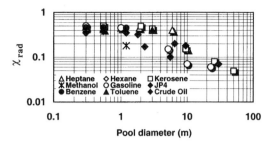

FIGURE 8.12 Radiative component of the combustion efficiency versus the pool diameter for the combustion of fluids. (Data taken from Ref. 1.)

In pool fires, there are three different modes of heat transfer from the flame: conduction, convection, and radiation. Heat transfer by conduction is through the pool rim (the edge effect) and is associated with the condensed-phase transformation [1, 2, 4, 53]. Conduction is the major mode of heat transfer for the combustion of fluids in very small pool diameters (about 0.004 to 0.030 m) [2]. In this pool diameter range, mass loss rate decreases with increase in the pool diameter.

Heat transfer by convection, driven by the flow movements induced in the surroundings, occurs at all stages, but is of particular importance at the early stages of fire growth, when flame is small and the radiative contribution is low [2]. It is the major mode of heat transfer for pool fires with moderate pool diameters in the range of about 0.030 to 0.20 m [2]. In this range, release rate of fluid vapors is almost independent of the pool diameter.

For pool diameters >0.25 m, radiative heat transfer contribution increases with pool diameter [1, 2, 4, 53]. For example, the convective component of the flame flux to the pool surface $\dot{q}''_{con}$ decreases from about 54 to 5 percent as pool diameter increases from about 0.15 to 0.50 m. Release rate of fluid vapors increases rapidly with increase in the pool diameter for these pool sizes.

For pool diameters >1 m, radiative component of the flame heat flux, $\dot{q}''_{fr}$, becomes the dominant factor in the control of release rate of fluid vapors [1, 2, 4, 53]. In this range, release rate of fluid vapors is also affected by the presence of a cool, fuel-rich region near the pool surface that attenuates the $\dot{q}''_{fr}$ values.* For the pool diameter in the range of about 0.5 to 3 m, radiation-dominating release rate of fluid vapors reaches its limit.

Beyond about 3 m, the release rate of fluid vapors decreases with further increase in the diameter due to decrease in the $\dot{q}''_{fr}$ values. The overall flame radiation also decreases as a thick layer of soot surrounds the flame that blocks radiation from the flame and diffusion of air into the combustion zone. The decrease is indicated by the decrease in the radiative component of the combustion efficiency χ_{rad}, as shown in Fig. 8.12.

8.6.1.1 Release Rates of Fluid Vapors in Small-Scale Pool Fires

The relationship in Eq. (8.5) under steady-state condition has been utilized to derive the following combustion characteristics of the fluids from the data measured in the ASTM E 2058 apparatus [35, 36]:

- *Determination of heat of gasification of fluids:* Tests are performed in the heat flux range of 10 to 60 kW/m² in air with less than 10 percent of oxygen concentration flowing around the sample. Under this condition, $\dot{q}''_f = 0$ and ΔH_g values are determined from the measured $\dot{m}''_f$ values and Eq. (8.5). The ΔH_g values determined from this technique have been reported in the literature [35, 36]:

---

* This phenomenon is called radiative energy blockage [2].

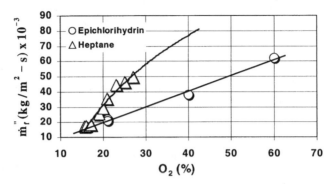

FIGURE 8.13 Release rate of fluid vapors versus the oxygen concentration in air flowing around the sample in the ASTM E 2058 apparatus at INERIS and FM Global Research. No external heat flux was used in the tests. Increase in the release rate of fluid vapors is due to increase in the flame heat flux transferred to the fluid surface. Epichlorihydrin data are courtesy of Eka Chimie SA (Akzo Noble Group)

- *Determination of flame heat flux:* Flame radiation scaling technique is used for the determination of flame heat flux expected in large-scale fires. Tests are performed in air with 20 to 60 percent oxygen concentration flowing around the sample, without the external heat flux. Under this condition, $\dot{q}''_e = 0$ and $\dot{q}''_f$ values are determined from the measured $\dot{m}''_f$ values and Eq. (8.5) with known ΔH_g values. Typical data obtained from this technique are shown in Fig. 8.13 for the combustion of epichlorihydrin, C_3H_5ClO, and heptane, C_7H_{16}. (These data were measured in the ASTM E 2058 apparatus at INERIS and FM Global Research, respectively.) The increase in the release rate of fluid vapors with oxygen concentration is due to increase in the $\dot{q}''_f$ values. For heptane, $\dot{m}''_f$ values measured in the large pool (1.2 to 10 m in diameter) fires in normal air are in the range of $(75 \text{ to } 81) \times 10^{-3}$ kg/m²·s (Table 8.4) [35, 36, 54, 59].

8.6.1.2 Release Rates of Fluid Vapors in Large-Scale Pool Fires

Release rates of fluid vapors $\dot{m}''_f$ have been measured for a variety of fluids with known ΔH_g values in several large-scale pool fires [35, 36, 54, 59–61]. Table 8.4 lists the measured values of $\dot{m}''_f$ for fluids with known ΔH_g values that are also included in the table. The table also includes the estimated $\dot{q}''_n$ values from $\dot{m}''_f$, ΔH_g, and Eq. (8.5). The $\dot{q}''_n$ values of the fluids do not show many variations, suggesting that in large-scale pool fires flame heat flux is independent of the chemical nature of the fluids.

In Table 8.4, the average value of $\dot{q}''_n = 33$ kW/m², and since $\dot{q}''_{rr}$ for the fluids is very small, $\dot{q}''_n \approx \dot{q}''_f$. It thus appears that in large-scale pool fires, flame heat flux for fluids is significantly lower than for the solids ($\dot{q}''_f$ values are in the range of 61 to 71 kW/m² [35]).

The combustion characteristics of fluids in large-scale fires depend on the thermophysical properties of the fluids, as indicated by the following relationships derived from Eqs. (8.1), (8.2), and (8.5), and $\dot{q}''_n = 33$ kW/m²:

$$\dot{m}''_f \approx 33/\Delta H_g \approx 33/\left(\Delta H_v + \int_{T_0}^{T_b} c\,dT\right) \tag{8.7}$$

$$1/\dot{m}''_f \approx 2.78 \left(\frac{T_b}{M}\right) + 0.030 \int_{T_0}^{T_b} c\,dT \tag{8.8}$$

Thus, in large-scale pool fires, $\dot{m}''_f$ values of fluids are governed by M (molecular weight), T_b (boiling point), c (heat capacity), and T_0 (ambient temperature).

TABLE 8.4 Heat of Gasification, Steady State Release Rate of Fluid Vapors, and Estimated Net Flame Heat Flux for the Combustion of Fluids in Large Pool Fires

Fluid and heat of gasification	Pool size (m)	Mass loss rate × 10³ (kg/m²·s) [35, 36]	[59]	[54]	[60, 61]	$\dot{q}_n''$(kW/m²) (estimated)
Heptane; ΔH_g = 493 kJ/kg	1.6	—	81	—	—	40
	2.4	—	79	—	—	39
	1.2–10	75	—	—	—	37
Hexane; ΔH_g = 481 kJ/kg	3	—	79	—	—	38
	0.75–10	77	—	—	—	37
Octane; ΔH_g = 550 kJ/kg	1.0	—	—	—	69	38
Dodecane; ΔH_g = 770 kJ/kg	0.94	36	—	—	—	28
Benzene; ΔH_g = 543 kJ/kg	0.75–6.0	81	—	88	—	44
Toluene; ΔH_g = 513 kJ/kg	1.0	—	—	—	68	35
	1.6	—	64	—	—	33
Xylene; ΔH_g = 503 kJ/kg	1.22	67	—	—	—	34
	5.4	—	60	86	—	38
	22.3	—	62	—	—	31
Kerosine; ΔH_g = 446 kJ/kg	30–50	—	65	—	—	29
Gasoline; ΔH_g = 500 kJ/kg	3	—	60	—	—	30
	5.4	—	70	—	—	35
	22.3	—	62	—	—	31
JP-4; ΔH_g = 500 kJ/kg	1.0–5.3	67	—	—	—	34
JP-5; ΔH_g = 500 kJ/kg	0.60–17	75	—	—	—	38
Transformer fluids; ΔH_g = 871 kJ/kg	2.37	27	—	—	—	24
Methanol; ΔH_g = 9600 kJ/kg	1.2–2.4	25	—	—	24	27
Ethanol; ΔH_g = 1000 kJ/kg	5.0	—	—	—	30	30
Acetone; ΔH_g = 632 kJ/kg	1.52	38	—	—	—	24
Toluene diisocyanate; ΔH_g = 870 kJ/kg	0.3	—	—	—	23	20
	1.0	—	—	—	34	30
	1.5	—	—	—	39	34
	2.0	—	—	—	33	29
Adiponitrile ΔH_g = 1000 kJ/kg	1.0	—	—	—	36	36
	1.5	—	—	—	35	35
	2.0	—	—	—	30	30
Acetonitrile; ΔH_g = 571 kJ/kg	0.7	—	—	—	63	36
	1.0	—	—	—	58	33
					Average	33
					Standard deviation	5

8.6.1.3 Release Rates of Fluid Vapors in Mesoscale Pool Fires of Oils

Study of the burning of oil spills is relatively easier compared to solids, as it requires minimum equipment and because oil is gasified during combustion and the need for physical collection, storage, and transport of recovered product is reduced [25]. Burning of oil spills in place normally produces a visible smoke plume containing soot and other combustion and pyrolysis products released during burning of the spilled oil. Various types of oils have been burned over salt water in mesoscale pool fires simulating oil spills [25, 26, 67, 68].

Table 8.5 lists the data measured in the mesoscale pool fire tests for Louisiana crude oil. These data show that the average values for the mesoscale pool fire tests of crude oil are: surface regression rate = 0.056 ± 0.009 mm/s, $\dot{m}_f''$ = 0.048 ± 0.009 kg/m²·s, and ΔH_g = 0.69 MJ/kg [using the $\dot{m}_f''$ value in Eq. (8.5) with $\dot{q}_n''$ = 33 kW/m²].

TABLE 8.5 Release Rates of Fluid Vapors and Heat and Smoke Yields from Burning of Louisiana Crude Oil[a] [25, 26, 67, 68]

Burn diameter (m)	Burn time (s)	Oil consumed (kg)	Surface regression rate (mm/s)	$\dot{m}_f''$ (kg/m²·s)	$\dot{Q}_{ch}''$ (kW/m²)	y_s (kg/kg)
Small-scale pool fire tests							
0.6	—	—	0.016	0.020	697	0.080
	—	—	0.016	0.020	697	0.084
	—	—	0.017	0.021	736	0.078
2.0	—	—	0.037	0.044	1560	0.141
	—	—	0.038	0.045	1590	0.138
	—	—	0.039	0.047	1640	—
Mesoscale pool fire tests							
6.9	1548	2645	0.054	0.046	1925	0.137
	651	1270	0.062	0.052	2195	—
	1156	820	0.023	0.019	799	—
	1122	1760	0.050	0.042	1765	0.079
	1012	—	0.062	0.053	2210	0.146
	1045	1775	0.054	0.046	1910	0.137
9.6	448	1765	0.064	0.054	2270	—
12.0	404	2745	0.071	0.060	2500	0.103
	993	4840	0.051	0.043	1790	—
	1188	6015	0.053	0.044	1860	0.121
	1020	5645	0.057	0.049	2030	—
14.7	645	4600	0.049	0.042	1755	—
15.2	319	2520	0.052	0.044	1825	—
17.2	935	9800	0.054	0.045	1900	—
	1000	11,840	0.061	0.051	2145	0.127
	820	—	0.059	0.050	2095	—
	900	—	0.063	0.054	2240	0.103
	848	—	0.063	0.053	2235	0.111
	755	—	0.060	0.051	2120	0.102
	885	—	0.061	0.052	2180	0.101
	820	—	0.059	0.050	2095	0.118

[a] Louisiana crude oil properties: specific gravity: 0.8451; carbon mass fraction: 0.862; hydrogen mass fraction: 0.134; sulfur mass fraction: 0; ΔH_{ch} (Cone Calorimeter): 41.9 MJ/kg; ΔH_g (Cone Calorimeter): 1.68 MJ/kg; y_s (Cone Calorimeter): 0.062 kg/kg.

8.6.2 Heat Release Rate

Heat release rate is expressed as the heat of combustion times the release rate of fluid vapors [35, 36]:

$$\dot{Q}_i'' = \Delta H_i \dot{m}_f'' \tag{8.9}$$

where subscript i is chemical, convective, and radiative. From Eqs. (8.5) and (8.9):

$$\dot{Q}_{ch}'' = (\Delta H_{ch}/\Delta H_g)\dot{q}_n'' = \chi(\Delta H_{net}/\Delta H_g)\dot{q}_n'' \tag{8.10}$$

where ΔH_{net} is the net heat of complete combustion (MJ/kg) and χ is the combustion efficiency. The ratio $\Delta H_{ch}/\Delta H_g$ is defined as the *heat release parameter* (HRP).

8.6.2.1 Heat Release Rates in Small-Scale Pool Fires

Measurements for the heat release rate in the ASTM E 1354/ISO 5660 (cone calorimeter) [65] and ASTM E 2058 apparatus [66] and relationships in Eqs. (8.9) and (8.10) have been utilized to derive the combustion properties of the fluids. The combustion properties of fluids associated with the heat release rate, derived from the measurements in the ASTM E 2058 apparatus [35, 36, 43, 60, 61, 63, 69, 70], are listed in Tables 8.6 to 8.12.

The following correlation between heat of combustion and molecular weight of chemicals and fluids has been reported in the literature [35, 36, 69, 70]:

$$\Delta H_i = h_i \pm m_i/M \tag{8.11}$$

where h_i is the mass coefficient for the heat of combustion (MJ/kg) and m_i is the molar coefficient for the heat of combustion (MJ/kmol). Both h_i and m_i depend on the chemical structure of the fluids; m_i values become negative as a result of the introduction of oxygen, nitrogen, and sulfur atoms into the structures [35, 36, 69, 70]. Values of h_i and m_i for generic fluids and chemicals have been

TABLE 8.6 Combustion Properties of Saturated Aliphatic Hydrocarbon Gases and Fluids [35, 36, 43, 69, 70]

Hydrocarbons (gases and fluids)	Composition C	Composition H	M (kg/kmole)	s (kg/kg)	L_s (m)	ΔH_i(MJ/kg) ncc	ΔH_i(MJ/kg) chem	ΔH_i(MJ/kg) con	ΔH_i(MJ/kg) rad	Yield (kg/kg) CO	Yield (kg/kg) Smoke
Methane	1	4	16	17.1	NA	50.0	49.5	42.0	7.5	0.001	0.001
Ethane	2	6	30	16.0	0.243	47.1	45.7	34.1	11.6	0.001	0.013
Propane	3	8	44	15.6	0.162	46.0	43.7	31.2	12.5	0.005	0.024
Butane	4	10	58	15.4	0.160	45.4	42.6	29.6	13.0	0.007	0.029
Pentane	5	12	72	15.3	0.139	45.0	42.0	28.7	13.3	0.008	0.033
Hexane	6	14	86	15.2	0.118	44.8	41.5	28.1	13.4	0.009	0.035
Heptane	7	16	100	15.1	0.123	44.6	41.2	27.6	13.6	0.010	0.037
Octane	8	18	114	15.1	0.118	44.5	41.0	27.3	13.7	0.010	0.038
Nonane	9	20	128	15.0	0.110	44.4	40.8	27.0	13.8	0.011	0.039
Decane	10	22	142	15.0	0.110	44.3	40.7	26.8	13.9	0.011	0.040
Undecane	11	24	156	15.0	0.110	44.3	40.5	26.6	13.9	0.011	0.040
Dodecane	12	26	170	14.9	0.108	44.2	40.4	26.4	14.0	0.011	0.041
Tridecane	13	28	184	14.9	0.106	44.2	40.3	26.3	14.0	0.012	0.041
Tetradecane	14	30	198	14.9	0.109	44.1	40.3	26.2	14.1	0.012	0.042
Hexadecane	16	34	226	14.9	0.118	44.1	40.1	26.0	14.1	0.012	0.042
Methylbutane	5	12	72	15.3	0.113	45.0	40.9	27.2	13.7	0.012	0.042
Dimethylbutane	6	14	86	15.2	0.089	44.8	40.3	26.3	14.0	0.014	0.046
Methylpentane	6	14	86	15.2	0.094	44.8	40.3	26.3	14.0	0.014	0.046
Dimethylpentane	7	16	100	15.1	0.096	44.6	39.9	25.7	14.2	0.015	0.049
Methylhexane	7	16	100	15.1	0.109	44.6	39.9	25.7	14.2	0.015	0.049
Isooctane	8	18	114	15.1	0.070	44.5	39.6	25.3	14.3	0.016	0.052
Methylethyl-pentane	8	18	114	15.1	0.082	44.5	39.6	25.3	14.3	0.016	0.052
Ethylhexane	8	18	114	15.1	0.093	44.5	39.6	25.3	14.3	0.016	0.052
Dimethylhexane	8	18	114	15.1	0.089	44.5	39.6	25.3	14.3	0.016	0.052
Methylheptane	8	18	114	15.1	0.101	44.5	39.6	25.3	14.3	0.016	0.052
Cyclopentane	5	10	70	14.7	0.067	44.3	39.2	24.1	15.1	0.018	0.055
Methylcyclo-pentane	6	12	84	14.7	0.052	43.8	38.2	23.0	15.2	0.019	0.061
Cyclohexane	6	12	84	14.7	0.087	43.8	38.2	23.0	15.2	0.019	0.061
Methylcyclo-hexane	7	14	98	14.7	0.075	43.4	37.5	22.3	15.2	0.021	0.066
Ethylcyclohexane	8	16	112	14.7	0.082	43.2	36.9	21.7	15.2	0.021	0.069
Dimethylcyclo-hexane	8	16	112	14.7	0.057	43.2	36.9	21.7	15.2	0.021	0.069
Cyclooctane	8	16	112	14.7	0.085	43.2	36.9	21.7	15.2	0.021	0.069

TABLE 8.7 Combustion Properties of Unsaturated Aliphatic Hydrocarbon Gases and Fluids [35, 36, 43, 69, 70]

Hydrocarbons (gases and fluids)	Composition C	H	M (kg/kmole)	s (kg/kg)	L_s (m)	ΔH_i(MJ/kg) ncc	chem	con	rad	Yield (kg/kg) CO	Smoke
Ethylene	2	4	28	14.7	0.106	48.0	41.5	27.3	14.2	0.013	0.076
Propylene	3	6	42	14.7	0.029	46.4	40.5	25.6	14.9	0.017	0.070
Butylene	4	8	56	14.7	0.019	45.6	40.0	24.8	15.2	0.019	0.067
Pentene	5	10	70	14.7	0.053	45.2	39.7	24.2	15.5	0.021	0.065
Hexene	6	12	84	14.7	0.063	44.9	39.4	23.9	15.5	0.021	0.064
Heptene	7	14	98	14.7	0.073	44.6	39.3	23.7	15.6	0.022	0.063
Octene	8	16	112	14.7	0.080	44.5	39.2	23.5	15.7	0.022	0.062
Nonene	9	18	126	14.7	0.084	44.3	39.1	23.3	15.8	0.022	0.062
Decene	10	20	140	14.7	0.079	44.2	39	23.2	15.8	0.022	0.061
Dodecene	12	24	168	14.7	0.080	44.1	38.9	23.1	15.8	0.023	0.061
Tridecene	13	26	182	14.7	0.084	44	38.9	23	15.9	0.023	0.061
Tetradecene	14	28	196	14.7	0.079	44	38.8	22.9	15.9	0.023	0.060
Hexadecene	16	32	224	14.7	0.080	43.9	38.8	22.8	16	0.023	0.060
Octadecene	18	36	252	14.7	0.075	43.8	38.7	22.8	15.9	0.023	0.060
Cyclohexene	6	10	82	14.2	0.044	43	35.7	20.2	15.5	0.029	0.085
Methylcyclohexene	7	12	96	14.3	0.043	43.1	35.8	19.8	16	0.029	0.085
Pinene	10	16	136	14.1	0.024	36	33.5	18.9	14.6	0.039	0.114
Acetylene	2	2	26	13.2	0.019	47.8	36.7	18.7	18	0.042	0.096
Heptyne	7	12	96	14.3	0.035	44.8	36	18.8	17.2	0.036	0.094
Octyne	8	14	110	14.4	0.030	44.7	35.9	18.9	17	0.036	0.094
Decyne	10	18	138	14.4	0.043	44.5	35.9	18.9	17	0.035	0.094
Dodecyne	12	22	166	14.5	0.030	44.3	35.9	18.9	17	0.035	0.094
1,3 Butadiene	4	6	54	14	0.015	44.6	33.6	15.4	18.2	0.048	0.125

TABLE 8.8 Combustion Properties of Aromatic Hydrocarbon Fluids [35, 36, 43, 69, 70]

Hydrocarbons (fluids)	Composition C	H	M (kg/kmole)	s (kg/kg)	L_s (m)	ΔH_i(MJ/kg) ncc	chem	con	rad	Yield (kg/kg) CO	Smoke
Benzene	6	6	78	13.2	0.007	40.1	27.6	11.2	16.5	0.067	0.181
Toluene	7	8	92	13.4	0.006	39.7	27.7	11.2	16.5	0.066	0.178
Styrene	8	8	104	13.2	0.006	39.4	27.8	11.2	16.6	0.065	0.177
Ethylbenzene	8	10	106	13.6	0.005	39.4	27.8	11.2	16.6	0.065	0.177
Xylene	8	10	106	13.6	0.006	39.4	27.8	11.2	16.6	0.065	0.177
Indene	9	8	116	13	0.008	39.2	27.9	11.3	16.6	0.065	0.176
Propylbenzene	9	12	120	13.7	0.009	39.2	27.9	11.3	16.6	0.065	0.175
Trimethylbenzene	9	12	120	13.7	0.006	39.2	27.9	11.3	16.6	0.065	0.175
Cumene	9	12	120	13.7	0.006	39.2	27.9	11.3	16.6	0.065	0.175
Naphthalene	10	8	128	12.9	0.005	39	27.9	11.3	16.6	0.065	0.175
Tetralin	10	12	132	13.5	0.006	39	27.9	11.4	16.5	0.064	0.174
Butylbenzene	10	14	134	13.8	0.007	39	27.9	11.4	16.5	0.064	0.174
Diethylbenzene	10	14	134	13.8	0.007	39	27.9	11.4	16.5	0.064	0.174
p-Cymene	10	14	134	13.8	0.007	39	27.9	11.4	16.5	0.064	0.174
Methylnaphthalene	11	10	142	13.1	0.006	38.9	28.0	11.4	16.6	0.064	0.174
Pentylbenzene	11	16	148	13.9	0.009	38.8	28.0	11.4	16.6	0.064	0.173
Dimethylnaphthalene	12	12	156	13.2	0.006	38.8	28.0	11.4	16.6	0.064	0.173
Cyclohexylbenzene	12	16	160	13.7	0.007	38.7	28.0	11.4	15.5	0.064	0.173
Diisopropylbenzene	12	18	162	14	0.007	38.7	28.0	11.4	16.6	0.064	0.173
Triethylbenzene	12	18	162	14	0.006	38.7	28.0	11.4	16.6	0.064	0.173
Triamylbenzene	21	36	288	14.3	0.007	38.1	28.2	11.6	16.6	0.063	0.169

TABLE 8.9 Combustion Properties of Fluids Containing Carbon, Hydrogen, and Oxygen Atoms [35, 36, 43, 69, 70]

Fluids	Composition C	H	O	M (kg/kmole)	s (kg/kg)	L_s (m)	ΔH_i(MJ/kg) ncc	chem	con	rad	Yield (kg/kg) CO	Smoke
Methyl alcohol	1	4	1	32	6.4	0.305	20.0	19.1	16.1	3.0	0.001	0.001
Ethyl alcohol	2	6	1	46	9.0	0.190	27.7	25.6	19.0	6.5	0.001	0.008
n-Propyl alcohol	3	8	1	60	10.3	0.155	31.8	29.0	20.6	8.5	0.003	0.015
Isopropyl alcohol	3	8	1	60	10.3	0.155	31.8	29.0	20.6	8.5	0.003	0.015
n-Butyl alcohol	4	10	1	74	11.1	0.141	34.4	31.2	21.6	9.6	0.004	0.019
Isobutyl alcohol	4	10	1	74	11.1	0.141	34.4	31.2	21.6	9.6	0.004	0.019
sec-Butyl alcohol	4	10	1	74	11.1	0.141	34.4	31.2	21.6	9.6	0.004	0.019
ter-Butyl alcohol	4	10	1	74	11.1	0.141	34.4	31.2	21.6	9.6	0.004	0.019
n-Amyl alcohol	5	12	1	88	11.7	0.131	36.2	32.7	22.2	10.4	0.005	0.022
Isobutyl carbinol	5	12	1	88	11.7	0.131	36.2	32.7	22.2	10.4	0.005	0.022
sec Butyl carbinol	5	12	1	88	11.7	0.131	36.2	32.7	22.2	10.4	0.005	0.022
Methylpropyl carbinol	5	12	1	88	11.7	0.131	36.2	32.7	22.2	10.4	0.005	0.022
Dimethylethyl carbinol	5	12	1	88	11.7	0.131	36.2	32.7	22.2	10.4	0.005	0.022
n-Hexyl alcohol	6	14	1	102	12.1	0.125	37.4	33.7	22.7	11.0	0.006	0.024
Dimethylbutyl alcohol	6	14	1	102	12.1	0.125	37.4	33.7	22.7	11.0	0.006	0.024
Ethylbutyl alcohol	6	14	1	102	12.1	0.125	37.4	33.7	22.7	11.0	0.006	0.024
Allyl alcohol	3	6	1	58	9.5	0.159	31.4	28.6	20.4	8.2	0.003	0.014
Cyclohexanol	6	12	1	100	11.7	0.124	37.3	33.6	22.6	11.0	0.005	0.024
Acetone	3	6	1	58	9.5	0.176	29.7	27.9	20.3	7.6	0.003	0.014
Methylethyl ketone	4	8	1	72	10.5	0.169	32.7	30.6	22.1	8.6	0.004	0.018
Cyclohexanone	6	10	1	98	11.2	0.164	35.9	33.7	24.1	9.6	0.005	0.023
di-Acetone alcohol	6	12	2	116	9.5	0.161	37.3	35.0	24.9	10.1	0.006	0.026
Ethyl formate	3	6	2	74	6.5	0.137	20.2	19.9	13.5	6.3	0.003	0.011
n-Propyl formate	4	8	2	88	7.8	0.114	23.9	23.4	15.4	8.0	0.005	0.019
n-Butyl formate	5	10	2	102	8.8	0.099	26.6	26.0	16.7	9.3	0.007	0.025
Methyl acetate	3	6	2	74	6.5	0.137	20.2	19.9	13.5	6.3	0.003	0.011
Ethyl acetate	4	8	2	88	7.8	0.114	23.9	23.4	15.4	8.0	0.005	0.019
n-Propyl acetate	5	10	2	102	8.8	0.099	26.6	26.0	16.7	9.3	0.007	0.025
n-Butyl acetate	6	12	2	116	9.5	0.093	28.7	28.0	17.8	10.2	0.008	0.029
Isobutyl acetate	6	12	2	116	9.5	0.093	28.7	28.0	17.8	10.2	0.008	0.029
Amyl acetate	7	14	2	130	10	0.086	30.3	29.5	18.6	11.0	0.009	0.033
Cyclohexyl acetate	8	14	2	142	10.2	0.083	31.5	30.6	19.1	11.5	0.01	0.035
Octyl acetate	10	20	1	172	11.2	0.077	33.6	32.6	20.2	12.5	0.012	0.039
Ethyl acetoacetate	6	10	3	130	7.4	0.086	30.3	29.5	18.6	11.0	0.009	0.033
Methyl propionate	4	8	2	88	7.8	0.114	23.9	23.4	15.4	8.0	0.005	0.019
Ethyl propionate	5	10	2	102	8.8	0.099	26.6	26.0	16.7	9.3	0.007	0.025
n-Butyl propionate	7	14	2	130	10.0	0.086	30.3	29.5	18.6	11.0	0.009	0.033
Isobutyl propionate	7	14	2	130	10.0	0.086	30.3	29.5	18.6	11.0	0.009	0.033
Amyl propionate	8	16	2	144	10.5	0.082	31.6	30.8	19.2	11.6	0.01	0.035
Methyl butyrate	5	10	2	102	8.8	0.099	26.6	26.0	16.7	9.3	0.007	0.025
Ethyl butyrate	6	12	2	116	9.5	0.093	28.7	28.0	17.8	10.2	0.008	0.029
Propyl butyrate	7	14	2	130	10.0	0.086	30.3	29.5	18.6	11.0	0.009	0.033
n-Butyl butyrate	8	16	2	144	10.5	0.082	31.6	30.8	19.2	11.6	0.01	0.035
Isobutyl butyrate	8	16	2	144	10.5	0.082	31.6	30.8	19.2	11.6	0.01	0.035
Ethyl laurate	14	28	1	228	12.0	0.196	37.2	35.6	26.5	9.1	0.008	0.031
Ethyl oxalate	4	6	4	102	6.1	0.224	28.7	27.7	21.3	6.4	0.001	0.003
Ethyl malonate	5	8	4	132	7.7	0.210	32.2	31.0	23.4	7.5	0.003	0.015
Ethyl lactate	5	10	3	118	7.0	0.214	30.8	29.6	22.5	7.1	0.001	0.01
Butyl lactate	7	14	3	146	8.5	0.206	33.3	32.0	24.1	7.9	0.004	0.018
Amyl lactate	8	16	3	160	9.0	0.203	34.3	32.9	24.7	8.2	0.005	0.021
Ethyl carbonate	5	10	3	118	7.0	0.214	30.8	29.6	22.5	7.1	0.001	0.01
Monoethyl ether	4	10	2	90	8.4	0.232	26.7	25.8	20.0	5.8	0.001	0.007
Monoethyl ether acetate	6	12	3	132	7.8	0.204	32.2	31.0	23.2	7.7	0.001	0.011
Monoethyl ether diacetate	6	10	4	146	6.1	0.208	33.3	32.0	24.2	7.9	0.001	0.009
Glycerol triacetate	9	14	6	218	6.0	0.195	36.9	35.4	26.3	9.1	0.002	0.011
Benzaldehyde	7	6	1	106	10.4	0.010	32.4	21.2	8.1	13.2	0.062	0.166
Benzyl alcohol	7	8	1	108	10.8	0.010	32.6	22.9	9.8	13.1	0.050	0.137
Cresylic acid	8	8	1	136	9.1	0.015	34.0	25.1	11.6	13.5	0.039	0.107
Ethyl benzoate	9	10	2	150	9.6	0.029	34.5	27.4	14.1	13.3	0.030	0.084

TABLE 8.10 Combustion Properties of Fluids Containing Carbon, Hydrogen, and Nitrogen Atoms [35, 36, 43, 69, 70]

	Composition			M	s	L_s	ΔH_i(MJ/kg)				Yield (kg/kg)	
Fluids	C	H	N	(kg/kmole)	(kg/kg)	(m)	ncc	chem	con	rad	CO	Smoke
Diethylamine	4	11	1	73	14.6	0.089	38.0	34.0	21.3	12.6	0.012	0.039
n-butylamine	4	11	1	73	14.6	0.089	38.0	34.0	21.3	12.6	0.012	0.039
Sec-butylamine	4	11	1	73	14.6	0.089	38.0	34.0	21.3	12.6	0.012	0.039
Triethylamine	6	15	1	101	14.6	0.085	39.6	35.3	22.0	13.3	0.014	0.044
Di-n-butylamine	8	19	1	129	14.6	0.083	40.6	36.1	22.4	13.7	0.014	0.047
Tri-n-butylamine	12	27	1	185	14.7	0.082	41.6	37.0	22.9	14.1	0.015	0.049
Pyridine	5	5	1	79	12.6	0.022	32.2	24.0	11.5	12.5	0.037	0.104
Aniline	6	7	1	93	12.9	0.018	33.8	25.0	11.7	13.3	0.043	0.119
Picoline	6	7	1	93	12.9	0.018	33.8	25.0	11.7	13.3	0.043	0.119
Toluidine	7	9	1	107	13.2	0.014	34.9	25.8	11.9	13.9	0.048	0.13
Dimethylaniline	8	11	1	121	13.3	0.013	35.7	26.4	12.1	14.3	0.051	0.139
Quinoline	9	7	1	129	12.5	0.012	36.1	26.7	12.1	14.5	0.052	0.143
Quinaldine	10	9	1	143	12.7	0.011	36.7	27.1	12.2	14.8	0.055	0.149
Butylaniline	10	15	1	149	13.6	0.009	37.0	27.2	12.2	15.0	0.056	0.151

reported in the literature [35, 36, 69, 70]. With increase in the M values, $\Delta H_i \approx h_i$, and thus heat of combustion becomes approximately constant with a weak dependency on M. For fluids and chemicals for which ΔH_i values are not available, they have been estimated from Eq. (8.11) and included in Table 8.12.

8.6.2.2 Heat Release Rates in Large-Scale Pool Fires

Although heat release rate is measured only in a limited number of large-scale pool fires, release rate of fluid vapors is one of the most common measurements. Because ΔH_i (i = chemical, convective, and radiative) values are independent of large pool sizes, Eq. (8.9) is routinely used to calculate the heat release rate in large-scale pool fires.

The heat release rates for large-scale pool fires of fluids can also be estimated from the HRP values of the fluids and Eq. (8.10), since the average value of $\dot{q}''_n = 33$ kW/m². Examples of such esti-

TABLE 8.11 Combustion Properties of Fluids Containing Carbon, Hydrogen, and Sulfur Atoms [35, 36, 43, 69, 70]

	Composition			M	s	L_s	ΔH_i(MJ/kg)				Yield (kg/kg)	
Fluids	C	H	S	(kg/kmole)	(kg/kg)	(m)	ncc	chem	con	rad	CO	Smoke
Hexyl mercaptan	6	14	1	118	12.2	0.062	33.0	30.1	17.9	12.2	0.012	0.040
Heptyl mercaptan	7	16	1	132	12.5	0.063	33.7	30.4	18.1	12.3	0.013	0.044
Decyl mercaptan	10	22	1	174	13.0	0.062	34.9	31.1	18.4	12.7	0.016	0.051
Dodecyl mercaptan	12	26	1	202	13.3	0.063	35.5	31.4	18.6	12.8	0.017	0.054
Hexyl sulfide	12	26	1	202	13.3	0.063	35.5	31.4	18.6	12.8	0.017	0.054
Heptyl sulfide	14	30	1	230	13.4	0.061	35.9	31.6	18.7	13.0	0.018	0.057
Octyl sulfide	16	34	1	258	13.6	0.061	36.3	31.8	18.8	13.1	0.019	0.059
Decyl sulfide	20	42	1	314	13.8	0.062	36.8	32.1	18.9	13.2	0.020	0.061
Thiophene	4	4	1	84	9.8	0.016	31.9	23.4	10.8	12.6	0.031	0.086
Methylthiophene	5	6	1	98	10.5	0.014	33.2	24.1	10.9	13.2	0.039	0.107
Thiophenol	6	6	1	110	10.6	0.013	34.1	24.6	11.0	13.6	0.045	0.122
Thiocresol	7	8	1	124	11.1	0.011	34.9	25.0	11.0	14.0	0.050	0.135
Cresolmethylsulfide	8	11	1	155	11.6	0.011	36.2	25.7	11.1	14.5	0.058	0.155

TABLE 8.12 Combustion Properties of Fluids and Chemicals (and Pesticides) [60, 61, 63]

Fluids	Composition	M (kg/kmole)	ΔH_t(MJ/kg) ncc	chem	con	rad	CO	Yield (kg/kg) Smoke	HCl	HCN	NO_x
			Fluids and chemicals with carbon, hydrogen, and nitrogen atoms								
Acetonitrile	C_2H_3N	41	29.6	29.0	16.0	13.0	0.025	0.004	—	0.002	0.002
Acrylonitrile	C_2H_3N	41	29.9	29.0	23.0	6.6	0.025	0.026[a]	—	0.001	0.019
Unsymmetrical dimethylhydrazine[a]	$C_2H_8N_2$	60	29.1	21.9	11.0	11.3	0.024	0.036[a]	—	—	—
Ethylenediamine[a]	$C_2H_8N_2$	60	28.6	29.7	14.2	15.3	0.024	0.036	—	—	—
Diethyltriamine[a]	$C_4H_{13}N_3$	103	34.8	32.2	15.5	16.5	—	0.044[a]	—	—	—
Adiponitrile	$C_6H_8N_2$	108	33.1	31.1	22.1	9.1	0.045	0.045[a]	—	0.001	0.017
Hexamethylenediamine	$C_6H_{16}N_2$	116	35.3	32.6	15.7	19.3	0.029	0.045[a]	—	0.010	0.018
			Fluids and chemicals with carbon, hydrogen, and oxygen atoms								
Tetrahydrofuran	C_4H_8O	72	32.2	30.2	16.0	14.2	0.021	—	—	—	—
			Fluids and chemicals with carbon, hydrogen, oxygen, and nitrogen atoms								
Ethylisonicotate	$C_8H_9O_2N$	151	26.2	24.3	12.8	9.2	0.029	0.142[a]	—	—	0.002
Toluene diisocyanate	$C_9H_6O_2N_2$	174	23.6	19.3	11.1	8.3	0.052	0.141[a]	—	0.001	0.013
Isoproturon	$C_{12}H_{18}ON_2$	206	32.8	23.9	7.8	16.1	0.056	0.115	—	0.007	0.009
Diphenylmethane-diisocyanate	$C_{15}H_{10}O_2N_2$	250	27.1[a]	19.6	13.7[a]	9.4[a]	0.042	0.154[a]	—	0.006	0.004
			Fluids and chemicals with carbon, hydrogen, and halogen atoms								
Chloropropane	C_3H_5Cl	76	23.0	10.8	6.9	3.9	0.076	0.196[a]	0.349	—	—
3-Chloropropene	C_3H_5Cl	76	23.0	10.8	4.3	6.5	0.076	0.179	0.650	—	—
Dichloromethane	CH_2Cl_2	85	6.0	2.0	1.6	0.4	0.088	0.081	0.220	—	—
1,3-Dichloropropene	$C_3H_4Cl_2$	111	14.2	5.6	2.1	3.5	0.090	0.169	0.191	—	—
Monochlorobenzene	C_6H_5Cl	113	26.4	11.2	3.6	7.6	0.083	0.232	—	—	—
Trifluoromethylbenzene	$C_6H_5CF_3$	146	18.7	10.1	8.1	5.7	0.069	0.185[a]	—	—	—

TABLE 8.12 (Continued)

Fluids	Composition	M (kg/kmole)	ΔH_f (MJ/kg) ncc	chem	con	rad	Yield (kg/kg) CO	Smoke	HCl	HCN	NO_x
Fluids and chemicals with carbon, hydrogen, oxygen, and nitrogen atoms											
Epichlorhydrin	C_3H_5OCl	92	—	—	—	—	0.067	0.032	0.351	—	—
Ethylmonochloroacetate	$C_4H_7O_2Cl$	122	15.7	14.1	10.1	3.9	0.019	0.138[a]	—	—	—
2,4 D Herbicide	$C_8H_6O_3Cl_2$	221	11.5	4.5	2.8	1.7	0.074	0.163	0.121	—	—
Fluids and chemicals with carbon, hydrogen, oxygen, and sulfur atoms											
Methylthiopropionyl-aldehyde	C_4H_8OS	104	25.0	23.8	18.8[a]	5.9[a]	0.001	0.005[a]	—	—	—
Fluids and chemicals with carbon, hydrogen, nitrogen, and halogen atoms											
Metatrifluoromethyl-phenylacetonitrile	$C_9H_6NF_3$	185	16.0	12.1	8.8	3.3	0.058	0.168[a]	—	0.006	0.011
Fluids and chemicals with carbon, hydrogen, oxygen, nitrogen, sulfur, halogen, and phosphorous atoms											
Dimethoate	$C_5H_{12}NO_3PS_2$	229	27.9[a]	23.1[a]	11.3[a]	9.2[a]	0.068[a]	0.185[a]	—	0.008	0.002
Diuron	$C_9H_{10}ON_2Cl_2$	233	20.3	10.2	3.9	6.3	0.080	0.159	0.144	0.012	0.025
Chlormephos	$C_5H_{12}O_2S_2ClP$	235	19.1	13.9	6.1	7.8	0.075	0.055	0.151	—	—
Aclonifen	$C_{12}H_9N_2O_3Cl$	264	19.7	14.3	9.9	4.4	0.063	0.186[a]	—	0.007	0.015
Mancozebe	$(C_4H_6N_2S_4$-Mn$)_x$Zn$_{0.i}$	269	14.0	9.5	4.2	5.3	—	—	—	0.008	0.008
Folpel	$C_9H_4O_2NSCl_3$	296	9.1	3.6	1.8	1.8	0.062	0.205	0.138	0.001	0.005
Chlorfenvinphos	$C_{12}H_{24}O_4Cl_3P$	360	18.0	7.7	2.7	5.0	0.110	0.288	0.072	—	—

[a] Data estimated from Eqs. 8.11 and 8.15.

TABLE 8.13 Composition, Molecular Weight, Heat Release Parameter and Estimated Heat Release Rate for Large-Scale Pool Fires of Selected Fluids [43]

Fluid	Composition	M (kg/kmole)	HRP (MJ/MJ)	$\dot{Q}''_{ch}$(kW/m²) Estimated[b]
Gasoline	a	a	85	2805
Hexane	C_6H_{14}	86	83	2739
Heptane	C_7H_{16}	100	75	2475
Octane	C_8H_{18}	114	68	2244
Nonane	C_9H_{20}	128	64	2112
Decane	$C_{10}H_{22}$	142	59	1947
Undecane	$C_{11}H_{24}$	156	55	1815
Dodecane	$C_{12}H_{26}$	170	52	1716
Tridecane	$C_{13}H_{28}$	184	50	1650
Kerosene	$C_{14}H_{30}$	198	47	1551
Hexadecane	$C_{16}H_{34}$	226	44	1452
Mineral oil	a	466	72	2376
Motor oil	a	a	62	2046
Corn oil	a	a	54	1782
Benzene	C_6H_6	78	75	2475
Toluene	C_7H_8	92	82	2706
Xylene	C_8H_{10}	106	67	2211
Methanol	CH_4O	32	19	627
Ethanol	C_2H_6O	46	33	1089
Propanol	C_3H_8O	60	46	1518
Butanol	$C_4H_{10}O$	74	58	1914

[a] These fluids are complex mixtures with variable chemical compositions, manufacturer, origin, and others.
[b] Heat release rate estimated from Eq. 8.10 with average $\dot{q}''_n$ = 33 kW/m² (Table 8.4).

mates are included in Table 8.13. From Eqs. (8.8) and (8.9), the chemical heat release rate for large-scale pool fires of fluids can also be expressed as:

$$1/\dot{Q}''_{ch} \approx \frac{\dfrac{2.78 T_b}{M} + 0.030 \displaystyle\int_{T_0}^{T_b} c\, dT}{\Delta H_{ch}} \tag{8.12}$$

Thus, the heat release rate in large-scale pool fires is governed by ΔH_{ch}, T_b, M, and c values of the fluids.

8.6.2.3 Heat Release Rate in Mesoscale Burning of Oils

The $\dot{Q}''_{ch}$ values are not measured directly in these large tests for oils. They are, however, derived from the measured surface regression rates and densities of the oils or the $\dot{m}''_f$ values, measured in the tests, using Eq. (8.5) and the ΔH_{ch} values (obtained from the small-scale pool fire tests for the oils). The $\dot{Q}''_{ch}$ values derived in this fashion for the mesoscale pool fire tests for Louisiana crude oil are listed in Table 8.5. The data in the table show that the average value of $\dot{Q}''_{ch} = 1993 \pm 338$ kW/m² for the mesoscale pool fire tests of the crude oil, which is very similar to the rates predicted for the higher molecular weight hydrocarbons in Table 8.13. The agreement is expected because for large-scale pool fires, $\dot{q}''_n$ = 33 kW/m², irrespective of the chemical natures of the fluids.

8.6.2.4 Heat Release Rates in Large-Scale Spray Fires

Heat release rates have also been measured in large-scale spray fires created in a variety of ways. For example, in the tests using the ISO 15029-3 test method, fluids under high pressure were sprayed at an angle from a nozzle into the flame of a propane gas burner that was located inside a chamber [32, 71]. The chamber was attached to a sampling duct where heat release rate was measured. In the research study for the development of the FM Global Approval Standard Class Number 6930, the fluids under high pressure were sprayed from a nozzle in a vertical direction in normal air under the fire products collector [72, 73]. The nozzle was located in the middle of a 15-kW propane burner with a 0.14-m diameter. The products were captured along with air in the sampling duct of the fire products collector where measurements were made for the heat release rate.

Heat release rate data measured for the hydraulic and other fluids following ISO 15029-3 test method [71] and under the fire products collector [72] are listed in Tables 8.14 and 8.15. Other data included in the tables also taken from these references are for the nozzle pressure, fluid density ρ, nozzle exit velocity U_0, fluid mass flow $\dot{m}_f$, net heat of complete combustion ΔH_{ncc}, chemical heat of combustion ΔH_{ch}, and combustion efficiency χ.

The data in Tables 8.14 and 8.15 indicate that the combustion efficiency χ of fluids in high-

TABLE 8.14 Spray Combustion Data for Hydraulic Fluids from the ISO 15029 Tests [71]

Burner diameter (mm)	Nozzle pressure (MPa)	$\dot{m}_f$ (g/s)	ΔH_{ch} (MJ/kg)	$\dot{Q}_{ch}$ (kW)	χ
Water (34%)-Glycol, ΔH_{ncc} = 13.0 MJ/kg					
10	5	NI	NI	NI	NI
	10	NI	NI	NI	NI
25	5	44.4	5.9	260	0.45
	10	44.7	8.1	360	0.62
100	5	58.5	6.5	380	0.50
	10	50.7	11.1	560	0.85
Water (48%)-Glycol, ΔH_{ncc} = 9.4 MJ/kg					
10	5	NI	NI	NI	NI
	10	NI	NI	NI	NI
25	5	51.7	3.5	180	0.37
	10	144.9	1.0	150	0.11
100	5	66.7	4.8	320	0.51
	10	81.7	4.9	400	0.52
Glycol (5% Water), ΔH_{ncc} = 26.7 MJ/kg					
10	5	20.1	24.8	500	0.93
	10	27.7	23.5	650	0.88
	15	34.8	23.0	800	0.86
25	5	24.4	25.4	620	0.95
	10	33.5	25.4	850	0.95
	15	44.1	22.7	1000	0.85
100	5	29.8	24.8	740	0.93
	10	40.1	22.4	900	0.84
	15	42.1	23.8	1000	0.89

TABLE 8.14 (*Continued*)

Burner diameter (mm)	Nozzle pressure (MPa)	$\dot{m}_f$ (g/s)	ΔH_{ch} (MJ/kg)	$\dot{Q}_{ch}$ (kW)	χ
\multicolumn{6}{c}{Phosphate Ester, ΔH_{ncc} = 30.0 MJ/kg}
10	5	NI	NI	NI	NI
	10	NI	NI	NI	NI
	15	NI	NI	NI	NI
25	5	25.2	22.2	560	0.74
	10	33.7	26.7	900	0.89
	15	43.7	25.2	1100	0.84
100	5	28.5	24.6	700	0.82
	10	36.6	27.3	1000	0.91
	15	46.5	25.8	1200	0.86
\multicolumn{6}{c}{Polyol Ester, ΔH_{ncc} = 36.9 MJ/kg}
10	5	20.4	22.5	460	0.61
	10	34.1	10.0	340	0.27
	15	NI	NI	NI	NI
25	5	22.6	24.4	550	0.66
	10	33.1	33.2	1100	0.90
	15	39.3	34.3	1350	0.93
100	5	26.2	22.9	600	0.62
	10	33.9	32.5	1100	0.88
	15	40.8	34.3	1400	0.93
\multicolumn{6}{c}{Polyalkylene Glycol, ΔH_{ncc} = 41.3 MJ/kg}
10	5	19.0	36.8	700	0.89
	10	25.4	40.5	1030	0.98
	15	33.2	40.1	1330	0.97
25	5	22.9	39.3	900	0.95
	10	29.3	40.9	1200	0.99
	15	34.6	40.5	1400	0.98
100	5	26.2	34.3	900	0.83
	10	31.6	40.5	1280	0.98
	15	37.0	40.5	1500	0.98
\multicolumn{6}{c}{Mineral Oil, ΔH_{ncc} = 42.9 MJ/kg}
10	5	18.9	36.0	680	0.84
	10	28.1	41.6	1170	0.97
	15	33.9	40.7	1380	0.95
25	5	22.2	34.7	770	0.81
	10	29.8	40.3	1200	0.94
	15	36.8	40.7	1500	0.95
100	5	26.7	35.6	950	0.83
	10	31.9	40.7	1300	0.95
	15	36.5	41.1	1500	0.96

NI: no ignition.

TABLE 8.15 Spray Combustion Data for Fluids [72]

Fluid	Nozzle pressure (Mpa)	ρ (kg/m³)	$\dot{m}_f$ (g/s)	U_0 (m/s)	$\dot{Q}_{ch}$ (kW)	ΔH_{ncc} (MJ/kg)	ΔH_{ch} (MJ/kg)	χ
Polyol esters								
#1	6.9	924	5.85	55.5	220	36.6	35.8	0.98
	5.2		5.05	47.9	188		35.4	0.97
	3.5		4.07	38.6	151		35.3	0.97
	1.7		2.70	25.6	90		29.5	0.81
#2	6.9	922	5.87	55.8	216	35.7	35.0	0.98
	5.2		4.97	47.3	183		35.1	0.98
	3.5		4.11	39.1	151		35.0	0.98
#3	6.9	911	5.84	56.2	232	40.3	37.8	0.94
	5.2		4.92	47.4	194		37.5	0.93
	3.5		4.15	40.0	158		36.3	0.90
	1.7		1.80	17.3	62		25.0	0.62
#4	6.9	924	6.11	58.0	228	37.0	35.5	0.96
	5.2		5.00	47.5	191		36.3	0.98
	3.5		4.26	40.4	158		35.3	0.96
	1.7		2.70	25.4	76		23.9	0.65
Phosphate esters								
#5	6.9	1130	5.13	39.8	158	ND	29.3	ND
#6	6.9	1110	5.20	41.1	166	31.8	29.7	0.93
	5.2		4.54	35.9	129		29.2	0.92
	3.5		3.72	29.4	102		28.8	0.91
	1.7		2.72	21.5	70		22.3	0.70
#7	6.9	1110	4.86	38.4	157	32.0	30.3	0.95
	5.2		4.51	35.6	130		29.5	0.92
	3.5		3.80	30.0	108		29.0	0.91
	1.7		2.80	22.0	72		21.8	0.68
Water-in-oil emulsion								
#8	6.9	920	5.60	53.4	164	27.6	27.1	0.98
	3.5		4.58	43.7	115		23.9	0.87
	1.7		3.68	35.1	94		ND	ND
Poly glycol-in-water								
#9	6.9	1078	5.91	48.1	65	11.0	10.5	0.95
	3.5		4.21	34.2	45		10.2	0.93
	1.7		3.21	26.1	26		7.7	0.70
#10	6.9	1078	5.34	43.4	63	11.9	11.2	0.94
	3.5		4.10	33.4	47		11.0	0.92
	1.7		3.18	25.9	27		8.1	0.68
#11	6.9	1079	6.05	49.2	78	14.7	12.3	0.84
	3.5		4.44	36.1	56		12.0	0.82
	1.7		3.23	26.3	32		9.4	0.64
#12	6.9	1073	6.14	50.2	70	12.1	10.9	0.90
	3.5		4.67	38.2	50		10.2	0.84

TABLE 8.15 (*Continued*)

Fluid	Nozzle pressure (Mpa)	ρ (kg/m³)	$\dot{m}_f$ (g/s)	U_0 (m/s)	$\dot{Q}_{ch}$ (kW)	ΔH_{ncc} (MJ/kg)	ΔH_{ch} (MJ/kg)	χ
			Mineral oil					
#13	6.9	874	5.71	57.2	255	46.0	44.6	0.97
	5.2		5.18	52.0	230		44.3	0.96
	3.5		4.45	44.7	202		44.1	0.96
	1.7		3.00	30.1	131		41.4	0.90
			Methanol					
#14	6.9	791	4.00	44.4	81	20.0	19.6	0.99
	5.2		3.20	35.5	64		19.7	0.99
	3.5		2.60	28.8	52		19.4	0.97
	1.7		2.24	24.9	44		19.6	0.98
			Ethanol					
#15	6.9	789	3.70	41.0	100	26.8	26.4	0.99
	5.2		3.20	35.6	86		26.2	0.98
	3.5		2.71	30.0	72		26.0	0.97
	1.7		2.20	24.5	58		26.3	0.98
			Heptane					
#16	6.9	684	3.52	45.1	151	44.6	40.2	0.90
	5.2		3.10	39.8	136		41.0	0.92
	3.5		2.47	31.7	108		40.7	0.91
	1.7		1.90	24.4	79		39.2	0.88

pressure spray fires (high nozzle-exit velocities) is significantly higher than in pool fires. The high pressure increases the fineness of fluid atomization [5]. For highly volatile fluids (methanol, ethanol, and heptane), χ is independent of pressure or U_0, as these fluids atomize easily. The combustion of fluids injected into a flame is not as efficient as the combustion of fluids injected at the center of a ring burner.

8.6.3 Release Rates of Products

The release rate of a product is equal to the release rate of fluid vapors multiplied by the yield of the product [35, 36]:

$$\dot{G}''_j = y_j \dot{m}''_f \quad (8.13)$$

where $\dot{G}''_j$ is the release rate of product j (kg/m²·s) and y_j is the yield of product j (kg of product/kg of fluid vapors). From Eqs. (8.5) and (8.13):

$$\dot{G}''_j = (y_j/\Delta H_g)\dot{q}''_n = f_j(\Psi_j/\Delta H_g)\dot{q}''_n \quad (8.14)$$

where f_j is the generation efficiency of product j and ψ_j is the maximum possible stoichiometric yield of product j (kg/kg).

8.6.3.1 Release Rates of Products in Small-Scale Pool Fires

Measurements for the release rates of products in the ASTM E 1354/ISO 5660 (cone calorimeter) [65] and ASTM E 2058 apparatus [66] and relationships in Eqs. (8.13) and (8.14) have been utilized to derive the combustion properties of the fluids associated with the release of products. The combustion properties of fluids associated with the release of major products of incomplete combustion (CO and smoke), derived from the measurements in the ASTM E 2058 apparatus [35, 36, 43, 60, 61, 63, 69, 70], are listed in Tables 8.6 to 8.12.

A correlation between y_s and M has also been established, similar to the correlation for the heat of combustion [35, 36, 69, 70]:

$$y_j = a_j \pm b_j/M \quad (8.15)$$

where a_j is the mass coefficient for the product yield (kg/kg) and b_j is the molar coefficient for the product yield (kg/kmol). Both a_j and b_j values are available in the literature for generic fluids and chemicals [35, 36, 69, 70]. The a_j and b_j values depend on the chemical structures of the fluids; b_j values become negative because of the introduction of oxygen, nitrogen, and sulfur atoms into the structures [35, 36, 69, 70]. With increase in the M value, the value of $y_j \approx a_j$ becomes approximately constant with a weak dependency on M.

8.6.3.2 Release Rates of Products in Large-Scale Pool Fires of Fluids

The release rates of products in large-scale pool fires of fluids with known y_j values can be estimated from Eq. (8.13) and measured $\dot{m}_j''$ values or from the following relationship [Eqs. (8.8) and (8.13), with $\dot{q}_n'' = 33$ kW/m²], without performing the large-scale pool fire tests:

$$1/\dot{G}_j'' \approx (1/y_j)\left[2.78(T_b/M) + 0.030 \int_{T_0}^{T_b} c\,dT\right] \quad (8.16)$$

For the estimations, however, values of y_j, M, c, and T_b are needed.

8.6.3.3 Release Rate of Smoke in Mesoscale Burning of Oils

The y_s values measured in the mesoscale pool burning of Louisiana crude oil taken from Refs. 25, 26, 67, and 68 are listed in Table 8.5 The data show that the average value of $y_s = 0.115 \pm 0.019$ kg/kg. The dependency of y_s values on the pool diameter or fire size for crude oil is similar to the dependency of the radiative fraction of the combustion efficiency (χ_{rad}). This similarity is shown in Fig. 8.14, where data for y_s values are taken from Ref. 25 and data for χ_{rad} values are taken from Ref. 1 (Fig. 8.12). The similarity is expected as the dependency of χ_{rad} values on smoke concentration in flames is well documented [2, 35, 36, 53, 55, 69, 70].

The particle size distribution of smoke aerosols released from in situ burning of oil spills is also important as it governs smoke plume dispersion and health effects. Because of the irregular shape of smoke particles released from the combustion of fuels, which consist of agglomerates of small spherules, the particles are classified by many different methods such as [74]:

- Dispersed material in terms of an equivalent geometric, projected area
- Property of particles, such as settling rate, optical scattering cross section, or ratio of electric charge to mass

For the characterization of the smoke particles from the mesoscale pool fire tests of oils, *equivalent aerodynamic diameter* has been used [25]. The equivalent aerodynamic diameter of an irregu-

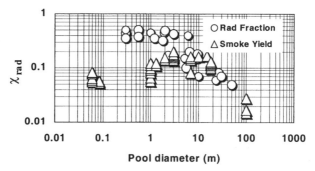

FIGURE 8.14 Yield of smoke for crude oil and radiative fraction of the combustion efficiency for hydrocarbons versus the pool diameter for large-scale fluid fires. (Data for the y_s values are taken from Ref. 25 and data for χ_{rad} values are taken from Ref. 1.)

larly shaped particle is the diameter of a smooth spherical particle having a unit density of 1 g/cm³, with the same terminal velocity as the smoke particle falling in air under the influence of gravity [25]. The data measured in the crude oil burning tests [25, 26, 67, 68] are shown in Fig. 8.15. The data show that about 65 percent of smoke aerosols collected in the tests have aerodynamic diameter <1 μm and about 80 to 90 percent have aerodynamic diameter ≤ 10 μm. The U.S. Environmental Protection Agency uses particulate matter below 10 μm as a parameter to gauge particulate pollution in ambient air [25].

8.7 SMOKE POINT

Smoke emission characteristics of materials have been expressed for decades by smoke point, defined as a minimum laminar axisymmetric diffusion flame height (or fuel volumetric or mass flow rate) at which smoke just escapes from the flame tip [35, 36]. Smoke point (L_s) values have been measured or derived for numerous gases, liquids, and solids and reported in the literature, some of which are taken from Refs. 35, 36, 43, 69, and 70, and listed in Tables 8.6 to 8.11.

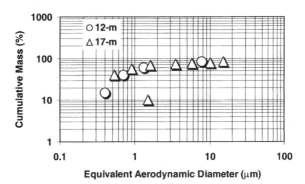

FIGURE 8.15 Smoke aerosol size distribution from the mesoscale pool fire tests of crude oils. (Data taken from Refs. 25, 26, 67, and 68.)

The smoke formation tendency of materials is inversely proportional to L_s. The following expressions have been reported [35, 36, 69, 70] for the relationship between the combustion efficiency χ and its convective (χ_{con}) and radiative (χ_{rad}) components, generation efficiencies of CO and smoke (f_j) and L_s (for $0 > L_s \leq 0.240$ m):

$$\chi = 1.51 L_s^{0.10} \quad (8.17)$$

$$\chi_{rad} = 0.41 - 0.85 L_s \quad (8.18)$$

$$\chi_{con} = \chi - \chi_{rad} \quad (8.19)$$

$$f_{CO} = -[0.0086 \ln(L_s) + 0.0131] \quad (8.20)$$

$$f_s = -[0.0515 \ln(L_s) + 0.0700] \quad (8.21)$$

Equations (8.17) to (8.21) provide reasonable estimates of heats of combustion and yields of CO and smoke; it is, however, necessary to know the values of net heat of complete combustion ΔH_{net} and maximum possible stoichiometric yields ψ_j of CO and smoke.*

NOMENCLATURE

a_j	Mass coefficient for product yield, kg/kg
ANSI	American National Standards Institute
API	American Petroleum Institute
b_j	Molar coefficient for the product yield, kg/kmol
c	Heat capacity, MJ/kg·K
CHF	Critical heat flux, kW/m²
DOT	Department of Transportation
EC	European Commission
f_j	Generation efficiency of product j (−)
$\dot{G}''_j$	Release rate of product j, kg/m²·s
GHS	Globally Harmonized System for Classification and Labeling of Chemicals
h_i	Mass coefficient for heat of combustion, MJ/kg
ΔH_{ch}	Chemical heat of combustion, MJ/kg
ΔH_{con}	Convective heat of combustion, MJ/kg
ΔH_g	Heat of gasification, MJ/kg
ΔH_{gcc}	Gross heat of complete combustion, MJ/kg
ΔH_{rad}	Radiative heat of combustion, MJ/kg
ΔH_{net}	Net heat of complete combustion, MJ/kg
ΔH_v	Heat of vaporization, kJ/kg
HCS	Hazard Communications Standards
HRP	Heat release parameter $\Delta H_{ch}/\Delta H_g$, MJ/MJ
IBP	Initial boiling point, °C
ILO	International Labor Organization
IOMC	Interorganization Program for the Sound Management of Chemicals
MSDS	Material safety data sheet
L_s	Smoke point, mm

---

* $\Delta H_{ch} = \chi \Delta H_{ncc}$; $\Delta H_{con} = \chi_{con} \Delta H_{ncc}$; $\Delta H_{rad} = \chi_{rad} \Delta H_{ncc}$; $y_{CO} = \psi_{CO} f_{CO}$; $y_s = \psi_s f_s$.

LFL	Lower flammability limit, %
$\dot{m}''_f$	Release rate of fluid vapors, kg/m²·s
m_i	Molar coefficient for heat of combustion, MJ/kmol
M	Molecular weight, kg/kmol
OECD	Organization for Economic Cooperation and Development
OSHA	Occupational Safety and Health Administration
$\dot{q}''_f$	Flame heat flux per unit surface area, kW/m²)
$\dot{q}''_{fc}$	Convective component of the flame heat flux, kW/m²
$\dot{q}''_{fr}$	Radiative component of the flame heat flux, kW/m²
$\dot{q}''_n$	Net flame heat flux, kW/m²
$\dot{q}''_{rr}$	Surface reradiation loss, kW/m²
$\dot{Q}''$	Heat release rate per unit surface area, kW/m²
R	Universal gas constant, 8.314 kJ/kmol·K
RCRA	Resource Conservation and Recovery Act
s	Stoichiometric mass air-to-fuel ratio, kg/kg
SDS	Safety data sheet
T_a	Autoignition temperature, °C
T_b	Boiling point, °C
T_f	Flash point, °C
T_{fr}	Fire point, °C
T_0	Ambient temperature, °C
TRP	Thermal response parameter, kW·s$^{1/2}$/m²
U_0	Nozzle exit velocity of the fluid, m/s
UFL	Upper flammability limit, %
UNCED	United Nations Conference on Environment and Development
UNCETDG	United Nations Committee of Experts for the Transport of Dangerous Goods
UNSETDG	United Nations Subcommittee of Experts for the Transport of Dangerous Goods
X_f	Flame height, m
y_j	Yield of product j, kg/kg
ρ	Density, kg/m³
χ	Combustion efficiency
ψ_J	Maximum possible stoichiometric yield of product j, kg/kg

Subscripts

ch	Chemical
con	Convective
f	Fuel or combustible fluid vapors
g	Gasification
i	Component of heat
j	Product
rad	Radiative
s	Smoke

REFERENCES

1. Koseki, H., "Large-Scale Pool Fires: Results of Recent Experiments," *Fire Safety Science—Proceedings of the Sixth International Symposium,* M. Curtat, ed., International Association for Fire Safety Science, France, 2000, pp. 115–132.
2. Joulain, P., "The Behavior of Pool Fires: State of the Art and New Insights," *Twenty-Seventh Symposium (International) on Combustion,* The Combustion Institute, Pittsburgh, PA, 1998, pp. 2691–2706.
3. Chatris, J.M., Quintela, J., Folch, J., Planas, E., Arnaldos, J., and Casal, J., "Experimental Study of Burning Rate in Hydrocarbon Pool Fires," *Combustion and Flame,* 126: 1373–1383, 2001.
4. Beyler, C.L., "Fire Hazard Calculations for Large, Open Hydrocarbon Fires," *SFPE Handbook of Fire Protection Engineering,* 3rd edition, The National Fire Protection Association Press, Quincy, MA, 2002, Sec. 3, Chap. 11, pp. 3-268-3-314.
5. Chigier, N.A., "The Atomization and Burning of Liquid Fuel Sprays," *Progress in Energy and Combustion Science,* 2: 97–114, 1976.
6. "Sandoz Counts Rhine Costs," *Chemical Industry (London),* Issue 23, 805, December 7, 1987.
7. Capel, P.D., Giger, W., Reichert, P., and Wanner, O., "Accident Input of Pesticides into the Rhine River," *Environmental Science and Technology,* 22: 992–997, 1988.
8. Suter, K.E., Gruntz, U., and Schlatter, C., "Analytical and Toxicological Investigations of Respiratory Filters and Building Ventilation Filters Exposed to Combustion Gases of the Chemical Warehouse Fire in Schweizerhalle," *Chemosphere,* 19(7), 1019–1109, ISSN 0045-6535, 1989.
9. Aresu de Sui, H., "Instructive Fires—Sandoz at Schweizerhalle (Switzerland)" *Revue Belge du Feu,* Number 84, February 1987.
10. Marlair, G., "Prevention at the Design Stage of Storage Premises," *Storing and Warehousing Toxic Material Seminar,* EFE Paris, December 16–17, 1997.
11. Stansfield, R.A., "The Diquat Incident at Woodkirk, Yorkshire," I Chem E, *Loss Prevention Bulletin,* Issue 132: 17–20, 1996.
12. Parks, D., "Chlorpyrifos Explosion in US Alabama Warehouse," *Pesticide News,* Issue 38: 3, December 1997.
13. Raveloharifera, F., and Buffin, D., "Accident will Happen," *Pesticide News,* Issue 38: 3, December 1997.
14. Accident at Fertiberia SA, Ascombreras Valley, Murcia, "Self Sustained Decomposition of NPK 15-15-15, 26 to 30 January 2002," Protection Civil Region de Murcia, Communication to C.C.A. Seveso, 11 April 2002.
15. Heinrich Munnik, S.D.O., "South African Sulfur Fire," *Fire Report, Fire International,* September 1996, pp. 13–16.
16. Commission of Inquiry into the Sulfur Fire at Somerset West: Official Inquiry Report (DESAI Report), South African Ministry of the Environment, 1997, www.environment.gov.za/reports/desai.
17. de Ronde, N., "Bushfire Causes Chemical Fire Inferno," *IFFN,* Number 14, January 1996.
18. Batterman, S.A., Cairncross, E., and Huang, H.L., "Estimation and Evaluation of Exposures from a Large Sulfur Fire in South Africa," *Environmental Research Section A,* 81: 316–333, Academy Press, New York, NY, 1999.
19. Markert, F., "Assessment and Mitigation of the Consequences of Fires in Chemical Warehouses," *Safety Science,* 30, No. 1–2: 33–44, Oct.–Nov. 1998.
20. Cole, T., and Wicks, P.J., eds., Industrial Fires: Workshop Proceedings, European Commission Report EUR 15340 EN, PIN-46 (1993), Apeldoorn, The Netherlands, March 11–12, 1993.
21. Cole, T., and Wicks, P.J., eds., Industrial Fires II: Workshop Proceedings, European Commission Report EUR 15967 EN, PIN-46 (1994), Cadarache, France, May 17–18, 1994.
22. Industrial Fires III: Workshop Proceedings, Major Industrial Fires, European Commission Report EUR 17477 EN, Riso, Denmark, September 17–18, 1996.
23. Koivisto, R., and Nielsen, D., "Fire-Database on Warehouse Fire Accidents," *Warehouse Fire Seminar,* TNO, Apeldoorn, the Netherlands, March 9–10, 1993.
24. "Toxic Cargo Fire Threat," *Fire Prevention,* Issue 203: 16, 1987.
25. Evans, D.D., Mulholland, G.W., Baum, H.R., Walton, W.D., and McGrattan, K.B., "In Situ Burning of Oil Spills," *Journal of Research of the National Institute of Standards and Technology,* 106: 231–278, 2001.

26. Jason, N.H., ed., "In Situ Burning Oil Spill," *Workshop Proceedings,* NIST SP 867, National Institute of Standards and Technology, Gaithersburg, MD, August 1994.
27. U.S. Environmental Protection Agency, Report to Congress, United States Gulf Environmental Technical Assistance from January 27 to July 31, 1991 Under Public Law 102-27, Section 309, October 1991.
28. Evans, D.D., Madrzykowski, D., and Haynes, G.A., "Flame Heights and Heat Release Rates of 1991 Kuwait Oil Field Fires," *Fire Safety Science, Proceedings of the Fourth International Symposium,* T. Kashiwagi, ed., International Association for Fire Safety Science, Canada, 1994, pp. 1279–1289.
29. Garo, J.P., Gillard, P., Vantelon, J.P., and Fernandez-Pello, A.C., "On the Thin Layer Boilover," *Fire Safety Science, Proceedings of the Sixth International Symposium,* M. Curtat, ed., International Association for Fire Safety Science, France, 2000, pp. 579–590.
30. Makhvilasze, G.M., Roberts, J.P., and Yakush, S.F., "Modeling and Scaling of Fireballs from Single and Two-Phase Hydrocarbon Releases," *Fire Safety Science, Proceedings of the Sixth International Symposium,* M. Curtat, ed., International Association for Fire Safety Science, France, 2000, pp. 1125–1136.
31. Totten, G.E., and Webster, G.M., "Fire Resistance Testing Procedures: A Review and Analysis," *Fire Resistance of Industrial Fluids,* STP 1284, G.E. Totten and J. Reichel, eds., American Society for Materials and Testing, West Conshohocken, PA, 1996, pp. 42–60.
32. Holmstedt, G., and Persson, H., "Spray Fire Tests with Hydraulic Fluids," *Fire Safety Science, Proceedings of the First International Symposium,* C.E. Grant and P.J. Pagni, eds., International Association for Fire Safety Science, Hemisphere Publishing Corporation, New York, NY, 1986, pp. 869–879.
33. "Fire Hazard Properties of Flammable Liquids, Gases, and Volatile Solids," National Fire Protection Association (NFPA) Standard No. 325-1991.
34. Kanury, A.M., "Ignition of Liquid Fuels," *SFPE Handbook of Fire Protection Engineering,* 3rd edition, The National Fire Protection Association Press, Quincy, MA, 2002, Sec. 2, Chap. 8, pp. 2-188–2-199.
35. Tewarson, A., "Generation of Heat and Chemical Compounds in Fires," *SFPE Handbook of Fire Protection Engineering,* 3rd edition, The National Fire Protection Association Press, Quincy, MA, 2002, Sec. 3, Chap. 4, pp. 3-82–3-161.
36. Tewarson, A., "Flammability," *Physical Properties of Polymers Handbook,* J.E. Mark, ed., The American Institute of Physics, Woodbury, N.Y. 1996, Chap. 42, pp. 577–604.
37. ASTM E 1269-95, "Standard Test Method for Determining Specific Heat Capacity by Differential Scanning Calorimetry," American Society for Testing and Materials, West Conshohocken, PA, August 1995.
38. ASTM E 793-95, "Standard Test Method for Enthalpies of Fusion and Crystallization by Differential Scanning Calorimetry," American Society for Testing and Materials, West Conshohocken, PA, December 1995.
39. Jodeh, S., and. Abu-Isa, I.A., " Determination of Boiling Point, Heat Capacity, and Heat of Vaporization for Engine Compartment Fluids Using Differential Scanning Calorimetry (DSC)," *Eighth International Polymer Characterization,* Denton, TX, January 10–14, 2000.
40. ASTM D1120-94, "Standard Test Method for Boiling Point of Engine Coolants," American Society for Testing and Materials, West Conshohocken, PA, September 1994.
41. ASTM D 2887-97, "Standard Test Method for Boiling Range Distribution of Petroleum Fractions by Gas Chromatography," American Society for Testing and Materials, West Conshohocken, PA, October 1997.
42. ASTM D 86-96, "Standard Test Method for Distillation of Petroleum Products," American Society for Testing and Materials, West Conshohocken, PA, June 1996.
43. *SFPE Hanbook of Fire Protection Engineering,* 3rd edition, The National Fire Protection Association Press, Quincy, MA, 2002.
44. Morrison, R.T., and Boyd, R.N., *Organic Chemistry,* Allyn and Bacon, Inc., Boston, MA, 1962.
45. Paul, M.A., *Physical Chemistry,* D.C. Heath and Company, Boston, MA, 1962.
46. "The Globally Harmonized System for Hazard Classification and Communication," Final Report, Interorganization Program for the Sound Management of Chemicals (IOMC) Coordinating Group for the Harmonization of Chemical Classifications (CG/HCCS), October 2001 (*http://www.ilo.org/public/english/protection/safework/ghs/*).
47. "Globally Harmonized System for the Classification and Labeling of Chemicals (GHS)," report ST/SG/AC·10·30, United Nations, New York & Geneva, United Nations Publications ISBN 92-1-116840-6, 1993.

48. Britton, L.G., "Survey of Fire Hazard Classification Systems for Liquids," *Process Safety Progress,* 18(4): 225–234, Winter 1999.
49. ASTM D 93-97, "Standard Test Method for Flash Point by Pensky-Martens Closed Cup Tester," American Society for Testing and Materials, West Conshohocken, PA, October 1997.
50. ASTM E 659-78, "Standard Test Method for Autoignition Temperature of Liquid Chemicals," American Society for Testing and Materials, West Conshohocken, PA, 1994.
51. ASTM E 681-98, "Standard Test Method for Concentration Limits of Flammability of Chemicals (Vapors and Gases)," American Society for Testing and Materials, West Conshohocken, PA, December 1998.
52. Blinov, V.I., and Khudiakov, G.N., "Certain Laws Governing Diffusive Burning of Liquids," *Doklady Academii Nauk SSSR,* 113: 1094–1098, 1957 (English translation by U.S. Army, NTIS number AD296762, 1961).
53. Hottel, H.C., "Certain Laws Governing Diffusive Burning of Liquids," *Fire Research Abstract and Reviews,* 1: 41–43, 1959.
54. Burgess, D.S., Strasser, A., and Grumer, J., "Diffusive Burning of Liquid Fuels in Open Trays," *Fire Research Abstracts and Reviews,* 3: 177, 1961.
55. Drysdale, D.D., *An Introduction to Fire Dynamics,* John Wiley and Sons, New York, N.Y. 1985.
56. Tanaka, T., Soutome, Y., Kabasawa, Y, and Fujizuka, M., "Preliminary Test for Full Scale Compartment Fire Test (Lubricant Oil Fire Test: Part 1)," *Fire Safety Science, Proceedings of the First International Symposium,* C.E. Grant and P.J. Pagni, eds., International Association for Fire Safety Science, Hemisphere Publishing Corporation, New York, NY, 1986, pp. 799–808.
57. Fujizuka, M., Soutome, Y., Kabasawa, Y., and Morita, J., "Full Scale Compartment Fire Test with Lubricant Oil (Lubricant Oil Fire Test: Part 2)," *Fire Safety Science, Proceedings of the First International Symposium,* C.E. Grant and P.J. Pagni, eds., International Association for Fire Safety Science, Hemisphere Publishing Corporation, New York, NY, 1986, pp. 809–818.
58. Yamaguchi, T., and Wakasa, K., "Oil Pool Fire Experiment," *Fire Safety Science, Proceedings of the First International Symposium,* C.E. Grant and P.J. Pagni, eds., International Association for Fire Safety Science, Hemisphere Publishing Corporation, New York, NY, 1986, pp. 911–918.
59. Koseki, H., "Combustion Properties of Large Liquid Pool Fires," *Fire Technology,* 25(3): 241, 1989.
60. Marlair, G., Cwiklinski, C., Marliere, F., Breulet, H., and Costa, C. "A Review of Large-Scale Fire Testing Focusing on the Fire Behavior of Chemicals," Inteflam, 1996, Cambridge, UK, 1996.
61. Marlair, G., "Internal Research," INERIS, Verneuil-en-Halatte, France, 1997.
62. Kuwana, K., Suzuki, M., Dobashi, R., and Hirano, T., "Effects of Liquid Distribution on Flame Spread Over Porous Solid Soaked with Combustible Liquid," *Fire Safety Science, Proceedings of the Sixth International Symposium,* M. Curtat, ed., International Association for Fire Safety Science, France, 2000, pp. 637–648.
63. Costa, C., Treand, F., Moineault, F., and Gustin, J.L., "Assessment of Thermal and Toxic Effects of Chemical and Pesticide Pool Fires Based on Experimental Data Obtained Using the Tewarson Apparatus," *Process Safety and Environmental Protection, Transactions of the Institute of Chemical Engineers, Part B,* 77(33):154–164, May 1999.
64. Ahmad, T., and Faeth, G.M., "Fire Induced Plumes Along A Vertical Wall: Part III. The Turbulent Combusting Plume," Technical Report for Grant No. 5–9020, National Institute of Science and Technology, Gaithersburg, MD, March 1978.
65. ASTM E 1354-99, "Standard Test Method for Heat and Visible Smoke Release Rates for Materials and Products Using Oxygen Consumption Calorimeter," American Society for Testing and Materials, West Conshohocken, PA, 1999.
66. ASTM E 2058, "Standard Test Methods for Measurement of Synthetic Polymer Flammability Using a Fire Propagation Apparatus (FPA)," American Society for Testing and Materials, West Conshohocken, PA, 2000.
67. Walton, W.D., "In Situ Burning of Oil Spills: Mesoscale Experiments," Technical Report NISTIR 5192 and 5266, National Institute of Standards and Technology, Gaithersburg, MD, September and November 1993.
68. Evans, D.D., Walton, W.D., Baum, H.R., Notarianni, K.A., Lawson, J.R., Tang, H.C., Keydel, K.R., Rehm, R.G., Madrzykowski, D., Zile, R.H., Koseki, H., and Tennyson, E.J., "In-Situ Burning of Oil Spills: Mesoscale Experiments," *Proceedings of the Fifteenth Arctic and Marine Oil Spill Program,* Technical Seminar, Edmonton, Alberta, Canada, June 10–12, 1992, NIST 51, National Institute of Standards and Technology, Gaithersburg, MD.

69. Tewarson, A., "Prediction of Fire Properties of Materials Part 1: Aliphatic and Aromatic Hydrocarbons and Related Polymers," Technical Report, J.I.OK3R3.RC, Factory Mutual Research Corporation, Norwood, MA, July 1986.
70. Tewarson, A., "Smoke Point and Fire Properties of Materials," Technical Report, J.I.OK3R3.RC, Factory Mutual Research Corporation, Norwood, MA, May 1988.
71. Simonson, M., Milovancevic, M., and Persson, H., "Hydraulic Fluids in Hot Industry: Fire Characteristics and Fluid Choice," Technical Report SP RAPP 1998:37, RALF Project 96-1612, SP Swedish National Testing and Research Institute, Fire Technology, Boras, Sweden, 1998.
72. Khan, M.M., "Spray Flammability of Hydraulic Fluids," *Fire Resistance of Industrial Fluids,* STP 1284, G.E. Totten and J. Reichel, eds., American Society for Testing and Materials, West Conshohocken, PA, 1996, pp. 133–147.
73. FM Global Approval Standard "Flammability Classification of Industrial Fluids," Class Number 6930, FM Global, Norwood, MA, January 2002.
74. "Particulate Polycyclic Organic Matter," Committee on Biological Effects of Atmospheric Pollutants, National Academy of Sciences, Washington, DC, 1972.

CHAPTER 9
MATERIALS IN MILITARY APPLICATIONS

Usman Sorathia
Naval Surface Warfare Center, Carderock Division
9500 MacArthur Boulevard
West Bethesda, Maryland 20817-5700

The technical views expressed in this paper are the opinions of the contributing authors, and do not represent any official position of the U.S. Navy.

9.1 INTRODUCTION

Fiber-reinforced polymer (FRP) composites offer inherent advantages over traditional materials with regard to high strength-to-weight ratio, design flexibility, corrosion resistance, low maintenance, and extended service life. FRP materials can be used for both new construction and repair/rehabilitation of existing structures.

Military applications in the U.S. Department of Defense employ materials ranging from commodity plastics, such as polyethylene and polyvinyl chloride in tubes, pipes, and electrical cables, to polyester-based coastal mine hunters (MHC-51), to high-temperature ceramics in jet blast deflectors aboard aircraft carriers, to titanium in deep submersibles. Materials technology, in some limited sense, is driving the next generation of our lightweight, fast-moving, and maneuverable fighters, tanks, ships, weapons, and unmanned vehicles.

In general, materials can be classified as metals, polymers, and ceramics. There is significant research and development being conducted in high-temperature ceramics and high-strength metals at several of our national laboratories. Studies have shown that demands to reduce weight and improve specific structural characteristics of advanced fighters, naval ships and submarines, tanks, and weapons can often be met through the use of organic-matrix-based composite structures. During the past 10 years, there has been "a resurgence of interest" in the development and application of composites to both primary and secondary load-bearing structures as well as machinery components in naval ships and submarines. This new interest in composite materials is due to increased need for a corrosion-free, lightweight, and affordable low-cost alternative to metallic components. A significant technical issue that has limited composite use in mass transit, the aircraft industry, and on board naval ships and submarines is the combustible nature, and hence the fire, smoke, and toxicity of organic-matrix-based composite materials.

This chapter addresses several technical issues related to fire performance of organic-matrix-based composite materials and their use in military applications. The inherent combustible nature of such materials requires careful considerations of flammability characteristics and mitigating such effects in case of fire. This chapter also presents important fire properties of several conventional and advanced composite materials, such as flame spread, smoke generation, combustion gas generation, ignitability, heat release rates (HRRs), fire resistance, and structural integrity during fire. Passive fire protection, such as fire insulation and intumescent coatings, to contain fire and limit fire spread in composite structures is also discussed. The use of composites is widespread across all agencies such as the Army, Navy, and USAF. This chapter also presents selected regulations and associated fire test methods.

9.2 COMPOSITES IN MILITARY APPLICATIONS

Composite materials have been used in the marine industry for over 50 years, and their use is increasing as their burning behavior is better understood and regulations evolve to reflect current technology. The current applications include widespread use in the hulls of yachts, pleasure craft, and racing boats, and certain specialized applications such as lifeboats, pipe, deck grating, and various other components. Composites are also common in small commercial fishing vessels and passenger vessels. Interest in the use of composite materials for larger vessels has been increasing in recent years, primarily for high-speed craft. Their corrosion resistance, low maintenance, and ease of repair make them attractive alternatives to the traditional shipbuilding materials, such as steel and aluminum.

Until recently, the use of composite materials for military applications was limited to aerospace and USAF high-performance applications. Current sea-borne applications of composite materials in the U.S. Navy include sonar bow domes and windows, and coastal mine-hunter MHC-51 hulls [1]. There is a resurgence of interest for the use of composites in military applications including naval vessels, army combat vehicles, and unmanned vehicles. Current and potential composite applications in surface ships and submarines are shown in Figs. 9.1 and 9.2 [2]. These include lightweight foundations, deckhouses, masts, machinery components, composite piping, gratings, stanchions, vent screens, ventilation ducts, etc. A recent notable large composite application is the *Advanced Enclosed Mast/Sensor* (AEM/S) system, which has been installed on USS *Radford* as shown in Fig. 9.3 [3]. The same concept was utilized in the AEM/S system for LPD-17, which is shown in Fig. 9.4 [3].

For military aircraft and space applications, composites are typically manufactured with graphite reinforcement and a variety of matrices whose choice is dependent upon environmental and temperature exposure of the component or the weapon system platform. Some of the thermosetting matrices employed include standard epoxies and higher use temperature polyimides in the forms of the PMR resins and AFR-700B. Some of these applications include empennages in the F-15, F-16, and F-22; secondary wing structures of the BI-B; portions of the fuselage of the F-22 and B-2; wing components of the F-22; and various engine components. Boeing projections for the structural

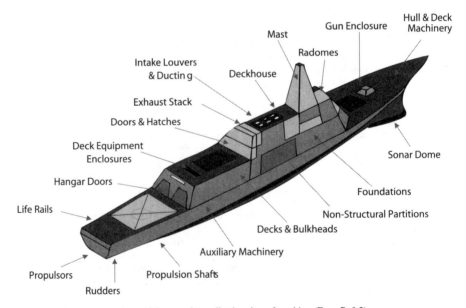

FIGURE 9.1 Current and potential composite applications in surface ships. (From Ref. 2)

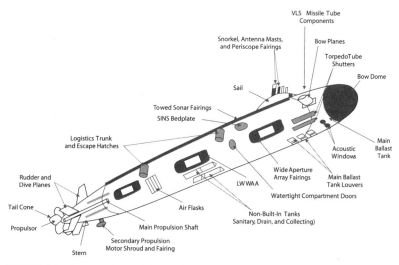

FIGURE 9.2 Current and potential composite applications in submarines. (From Ref. 2)

weight fraction of polymer composites in subsonic commercial airplanes show increases in use from about 7 percent currently to about 20 percent over the next 15 years. Usage of advanced materials in Boeing 767 is shown in Fig. 9.5. Composite usage in military aircraft C 17A is shown in Fig. 9.6.

Figure 9.7 shows the army's composite armor vehicle (tank with wheels). The composite block shown in this figure is a multifunctional integral armor with a modular design for rapid and easy repair. This armor consists of functional layers that include ceramic, metal matrix, and polymer composites.

FIGURE 9.3 Composite mast installed aboard USS *Radford* (DD968). (From Ref. 3)

9.4 CHAPTER NINE

FIGURE 9.4 LPD landing platform dock with AEM/S system. (From Ref. 3)

FIGURE 9.5 Use of advanced composite materials in Boeing 767. (D. Baker, AFRL/MLS-OL, Advanced Composites Office, Hill AFB, UT)

MATERIALS IN MILITARY APPLICATIONS   9.5

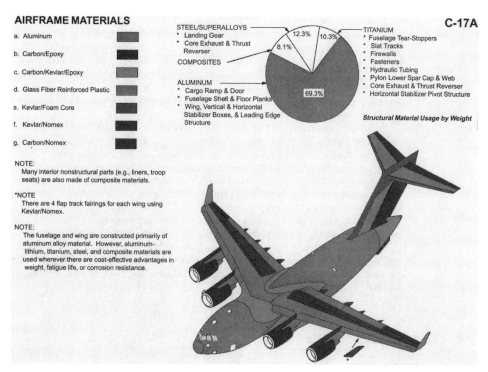

FIGURE 9.6 Use of advanced composite materials in C-17A. (D. Baker, AFRL/MLS-OL, Advanced Composites Office, Hill AFB, UT)

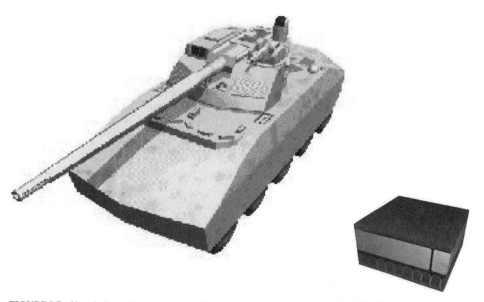

FIGURE 9.7 Use of advanced composite materials in Army's composite armor vehicle. (W. Chin, Army Research Laboratory, Aberdeen Proving Ground, MD)

In order to take advantage of improved high tensile strength and modulus of elasticity, some space components of rocket motor cases and rigid tubular structures of space station have been fabricated from liquid crystalline aromatic heterocyclic rigid-rod materials. These polymers represent an improvement in materials processing technology, since the polymers are biaxially oriented in the melt during the extrusion process and exhibit twice the crush resistance when compared to uniaxially oriented extruded tubes. Weight savings and resulting fuel efficiency are driving the use of advanced lightweight materials by airframe manufacturers and other civil transportation industries.

In infrastructure applications, fiber-reinforced, organic-matrix-based composites are very attractive materials of construction due to their strength, relatively light weight that facilitates on-site handling, and anticipated long-term weather resistance. The Federal government has budgeted $78 billion over the next 20 years for major infrastructure rehabilitation, since nearly 200,000 bridges and highways in the United States are deficient or obsolete. Composites are being considered for various infrastructure uses such as building reinforcement to enhance earthquake resistance, highway overpass reinforcement and repair, as well as foot and highway bridge construction. In such applications, the composites may take a variety of forms. In reinforcement and structural repair applications, for example, the composite might be a thin flat sheet, composed of carbon fibers and epoxy resin, held to the repaired surface (typically concrete) by an adhesive. In bridge and pier construction, the physical form of the composite structural elements is highly variable with the specific design, encompassing pultruded beams, honeycomb deck structures, and filament wound tubes. The resins are limited by cost considerations to high-volume, low-cost polymer types.

9.3 POLYMER COMPOSITES

Polymer composites are engineered materials in which the major component is a high-strength fibrous reinforcement and the minor component is an organic resin binder, often referred to as matrix resin. Thermoset polymers comprise the majority of composite resins and consist primarily of the chemical families that include *polyester* (PE), *vinyl ester* (VE), *epoxy* (EP), *bismaleimide* (BMI), *phenolic* (PH), *cyanate ester* (CE), *silicones* (SI), *polyimides* (PI), *phthalonitriles* (PN), etc. Thermoset polymers are cross-linked and do not melt or drip. High-temperature engineering thermoplastic resins used in military applications include *polyetherimide* (PEI), *polyphenylene sulfide* (PPS), *polyether sulfone* (PES), *polyaryl sulfone* (PAS), *polyether ether ketone* (PEEK), *polyether ketone ketone* (PEKK), etc.

Reinforcing fibers include E and S glass, aramid (Kevlar), carbon, quartz, polyethylene (Spectra), phenylene benzobisoxazole (PBO, Zylon), and boron. These fibers are used either alone or as hybrids in the form of woven rovings, fabrics, unidirectional tapes, bundles (tows), or chopped to various lengths. Selected properties for some of these fibers are given in Table 9.1 [4, 5]. Selected composite properties are presented in Table 9.2 [6–9]. Mechanical properties of various glass-reinforced vinyl esters are also presented in Table 9.3 [10]. These included nonbrominated epoxy vinyl ester resin (1167), brominated bisphenol A epoxy vinyl ester resin (1168), epoxy novolac vinyl ester resin (1169), and bisphenol A epoxy vinyl ester resin (1170). The vinyl ester composites were fabricated

TABLE 9.1 Comparison of Fiber Properties [4, 5]

Property	PBO AS	Spectra 1000	Kevlar 29	Kevlar 49	Kevlar 129	S-glass	E-glass	Carbon-PAN
Tensile strength (GPa)	5.8	2.9	2.9	3.0	3.3	4.5	3.5	2.4–4.8
Tensile modulus (GPa)	180	113	70	112	96	85	72	227–393
Tensile elongation (%)	3.5	2.9	3.6	2.4	3.3	5.7	4.8	0.38–2.0
Density (g/cm^3)	1.54	0.97	1.44	1.44	1.44	2.48	2.6	1.7–1.8

TABLE 9.2 Representative Composite Mechanical Properties [6, 7, 8, 9]

Properties	Polyester	Vinyl ester	Epoxy (250°F)	Epoxy (350°F)	Phenolic	Bismaleimide	PPS
Reinforcement	Glass cloth	Glass roving	Glass cloth	Glass cloth	Glass cloth	Glass fabric style 7781	Glass
Tensile strength, 10^3 psi	50	55	63	70	44	60.7	54
Tensile modulus, 10^6 psi	4.5	6.3	3.4	4.5	3.1	4.28	3.1
Flexural strength, 10^3 psi	80	111	78	88	66	92.5	65
Flexural modulus, 10^6 psi	3.0	3.8	3.2	3.5	3.5	4.51	2.8
Compressive strength, 10^3 psi	50	—	61	71	45	69.7	41
Max. service temp., °F	150–200	200–250	200	350–375	350	500	430

to a thickness of 0.25 in., with a fiber content of about 60 percent by weight using 24 oz/yd² glass woven roving. The resins were cured at ambient temperatures by the use of cobalt naphthenate, dimethyl aniline, and methyl ethyl ketone peroxide as promoters and catalyst. The composites were further postcured at 160°F for 6 h.

Fiber-reinforced polymer composites can be engineered to provide strength, stiffness, weight, and assembly advantages over conventional monolithic materials, but they also pose fire safety concerns due to the combustibility of the organic polymer constituents. A brief description of conventional and advanced matrix resins used in military applications is given below.

9.3.1 Conventional and Advanced Matrix Resins

Polyester (PE). Polyesters are produced by the reaction of dihydric alcohols (glycols) and dicarboxylic acids. Polyesters are classified as saturated or unsaturated, depending on the presence or absence of reactive double bonds in the linear polymer. Saturated polyesters, such as ethylene glycol terephthalate, find their greatest use in fibers and films. Unsaturated polyesters are used principally with fibrous reinforcements, glass fiber being the reinforcement of choice. Isophthalic polyesters are generally used in marine applications due to better corrosion resistance and mechanical properties. Some trade names include Hetron (Ashland Chemical Co.), Stypol (Freeman Chemical Corp.), and Silmar (Silmar Div.). The unsaturated polyester industry produces almost 1.4 billion pounds of resin [6]. Of the total volume, over 75 percent is utilized in the reinforced plastics market, 30 percent of which goes in marine applications. In use for over 50 years, polyester resin manufacturers are continually producing high-performance grades with improved fire retardancy.

Vinyl ester (VE). Vinyl ester resins are the reaction products of epoxy resins with ethylenically unsaturated carboxylic acids. The most common vinyl esters are made by esterifying a diepoxide

TABLE 9.3 Mechanical Characterization of Different Glass/Vinyl Ester (VARTM) Composites [10]

Test	Tensile (ASTM D 638)		Compression (ASTM D 695)			Isoipescu Shear (ASTM D 5379M)	
	Modulus Gpa	Strength Mpa	Modulus Gpa	Strength Mpa	Poisson Ratio	Modulus Gpa	Strength MPa
Epoxy VE (1167)	25.9	441	30.7	368	0.159	4.1	95
Brominated VE (1168)	27.5	432	27.4	324	0.157	3.7	89
Ep-Novolac VE (1169)	21.7	330	29.0	309	0.171	4.4	101
Ep-VE (1170)	20.6	385	27.4	218	0.173	4.6	82

resin, such as the diglycidyl ether of bisphenol A, with a monocarboxylic unsaturated acid such as methacrylic acid or acrylic acid. Generally, the vinyl ester is diluted, like polyester resins, with styrene to about 45 to 50 percent, and the resin mixture is cured by the use of peroxides, cobalt naphthenate, and dimethyl aniline. Methacrylic acid is most commonly used for vinyl ester resins intended for composite applications. Vinyl ester resins cure by addition polymerization with no loss of volatiles during curing. Some trade names include Hetron (Ashland Chemical), Derakane, and Momentum (Dow Chemical).

Epoxy (EP). Epoxy resins are characterized by the epoxide group (oxirane rings). The most widely used epoxy resins are based on diglycidyl ethers of bisphenol A such as DER 332 (Dow Chemical) and Epon 828 (Shell Chemical Co.). Multifunctional epoxies are a broad class of materials that contain two or more epoxy groups on the same molecule, such as MY 720 (Ciba-Geigy), and are used for prepregs and composite applications. Novolac epoxies are usually prepared by reacting phenol or substituted phenol with formaldehyde and then reacting that product with epichlorohydrin. Epoxy Novolacs can be cured to a high cross-link density and are used where high-temperature resistance and improved chemical resistance are required. The curing agents for epoxy resins include amines, anhydrides, polyamides, and Lewis acid catalysts. Some trade names for epoxies include Araldite (Ciba-Geigy), DEN, and DER (Dow Chemical).

Bismaleimides (BMI). Bismaleimides belong to the addition-type reaction in the polyimide family of high-temperature materials. The imide oligomers or bisimide monomers are generally derived from maleic anhydride and aromatic diamines. These monomers are easy to make or modify and are cured by thermally induced addition reactions to yield highly cross-linked polymers having higher thermal stability and lower water absorption than epoxies [11]. Reference 11 also identifies 6 commercial sources of bismaleimide resins and 11 BMI prepregs. Some trade names include Kerimid (Rhone-Poulenc), Compimide (Shell Chemical), and Matrimid (Ciba-Geigy).

Phenolics (PH). Phenolic resins are the reaction product of one or more of the phenols with one or more of the aldehydes. Phenolic resins have the inherent characteristics of low flammability; they produce little smoke on burning, and have good thermal stability. In addition, the phenolic polymer structure facilitates the formation of a high carbon foam structure or char that functions as an insulator. However, phenolic resins cure with the evolution of water, which results in a large void content in the final composite. Hence, current fire-resistant, fiber-reinforced phenolic composites are structurally unsuitable for primary applications, but suitable for secondary or auxiliary applications. Some trade names include Durez (Durez Plastics Division), Bakelite (Union Carbide), and Plenco (Plastics Engineering).

In an extensive thermal and flammability study conducted by NASA researchers [12], modified phenolic, bismaleimide, and polyimide composites were studied as alternatives for the fire-retardant epoxy used as face sheet/adhesive resin in aircraft interior composite panels. The modified phenolic composite had the lowest toxicity of all the materials tested using NASA-Ames pyrolysis toxicity apparatus [13].

Polyimides (PI). The polyimides are characterized as having the imide structure in the polymer backbone, which has exceptional thermal and oxidative stability. A novel class of addition polyimides, known as in situ *polymerization of monomer reactant* polyimide (PMR), was developed by NASA Lewis. The PMR system is characterized by good solubility of the reacting monomers, good reactivity at elevated temperatures, and no volatiles produced during the cure cycle. Various modifications of PMR polyimides have been prepared. The first generation of PMR resin was designated PMR-15.

Polphenylene sulfide (PPS). Polyphenylene sulfide is a partially crystalline high-performance engineering thermoplastic material produced by the reaction of *p*-dichlorobenzene and sodium sulfide. Molecular weight can be controlled and is used to produce two types of polymers. Type 1 is a material of relatively low molecular weight (less than 200) and is the basis for most injection-molding compounds. By using a catalyst, and in some cases, a comonomer, Type 2 polymer with a much higher molecular weight is produced [14]. PPS resins have broad chemical resistance, excellent stability at high temperatures, and inherent flame retardancy. PPS resins are marketed under the trade name of Ryton (Phillips Petroleum Company) and Fortron (Hoechst Celanese Corporation).

Polyaryl sulfone (PAS). PAS is a thermoplastic resin offering good solvent resistance, mechanical properties, and excellent processability. The chemistry of polyaryl sulfone centers on ether oxygen flexibilized phenyl rings with phenyl and sulfone linkages to provide rigidity and high service temperatures [15]. PAS resin is marketed under the trade name of Radel by Amoco Performance Products.

Polyether ether ketone (PEEK). PEEK is a high-temperature, crystalline thermoplastic offering excellent combination of thermal and combustion characteristics. The wholly aromatic structure of PEEK contributes to its high-temperature performance, and its crystalline character gives it resistance to organic solvents, resistance to dynamic fatigue, and retention of ductility on short-term heat aging. PEEK resins are marketed by ICI Advanced Materials.

9.3.2 Fabrication Techniques

Composite components and structures are fabricated by impregnating the fibrous reinforcement with liquid resin using various processes such as hand lay-up, *vacuum-assisted resin transfer molding* (VARTM), filament winding, pultrusion, autoclave curing, etc.

The selection of matrix resin plays a dominant role in the determination of fire performance and chemical resistance of composites. The selection of fiber plays a dominant role in the final mechanical properties of the composite. Similarly, the choice of fabrication technique plays a dominant role in the cost of the composite structures. Nowhere is it truer than in large and complex military structures.

The simplest composite fabrication method is hand lay-up, which involves manual placement of fiber-reinforcement mats or fabric in an open mold. Resin can be applied by pouring, brushing, or spraying, or the mat or fabric can be preimpregnated. Room temperature cure polyesters, vinyl esters, and epoxies are the most common matrix resins. Glass is the most common reinforcement for structural composites by hand lay-up either in the form of woven roving or glass fabric. This technique is now emerging as a semiautomated tape laying, which uses preimpregnated ribbons and is especially desirable for contour matching components. The drawback of this technique includes higher voids, trapped air, and higher resin-to-fiber ratios [16].

To obtain lower voids and higher glass concentrations, vacuum bag molding is employed whereby a flexible film is placed over the resin and fibers in the mold, and vacuum is applied. The resulting pressure reduces voids, trapped air, excess resin, and produces composite structures with higher mechanical performance. Pressure bag molding is analogous to vacuum bag molding except that air pressure (up to 50 lb/in^2) is applied directly to the rubber sheet to further lower voids.

VARTM is often utilized for large and complex structures with low-temperature curing resins. In this technique, vacuum is used to impregnate the fiber preform with resin. The *Seemann composites resin infusion molding process* (SCRIMP), which has been utilized by the U.S. Navy, is shown in Fig. 9.8 [3].

Autoclave curing is used to apply additional heat and pressure (up to 100 lb/in^2) to accelerate the cure and use higher temperature curing resins. This technique produces very low to void-free composite structures with fiber loadings up to 65 percent. The drawback of this technique is that the size of the autoclave limits the size of the structure. However, this is the favored method utilized by the military aircraft industry for manufacturing high-performance, high-service-temperature composite structures.

Thermosetting bulk and sheet molding compounds have been in increasing demand from automotive, transportation, construction, and appliance markets. They are cost effective, and can provide good finish. These formulations are processed by compression molding. Heat (225 to 320°F) and pressure (150 to 1000 lb/in^2) are applied for cure times varying from 30 s to 5 min. Polyesters are the predominant matrix resins, but phenolics and epoxies are also compression molded.

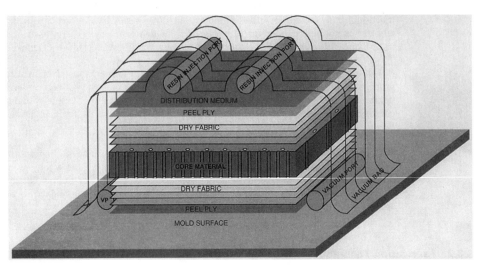

FIGURE 9.8 Seemann composites resin infusion molding process (SCRIMP). (From Ref. 3)

9.3.3 Cost of Composite Systems

For a fiber, the higher the modulus, the higher the price per pound. In general, for a resin matrix, the higher the service temperature, the higher the cost [17]. Many resins are especially compounded for a specific application, and their prices vary all over the spectrum. However, material costs for a large and complex military application can be a relatively small portion of the total cost with fabrication technique, NDE, and rework/repair being the major cost items.

Fiber-reinforced composites are manufactured by a variety of techniques. Consumer thermoplas-

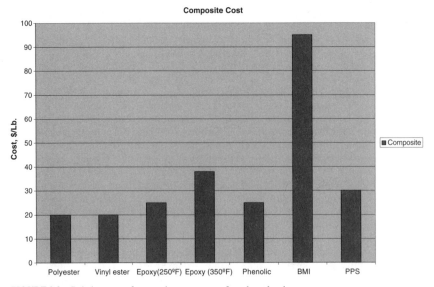

FIGURE 9.9 Relative costs of composite components for selected resins.

tics, such as appliance housings, are made by injection molding. Most thermosets are processed by resin transfer molding. High-performance composites for the aircraft industry are fabricated by autoclave curing techniques. For large complex structures, such as AEM/S system, the U.S. Navy prefers the use of VARTM shown in Fig. 9.8. Almost 80 percent of the total cost of the manufactured component is derived from the processing technique and part inspection.

Both chemical and aerospace industries use a certain multiple of material costs to arrive at an estimate of fabricated costs. For aerospace industry, this multiple varies from 3 to 10 based on prepreg costs and complexity of parts under production. Figure 9.9 presents the relative cost picture for fabricated glass-reinforced composites using selected resin matrices. In this comparison, it was assumed that fabricated components are flat sheets, 4 × 8 ft in dimension.

9.4 FIRE THREAT

A significant concern in any application of organic-matrix-based composites in occupied spaces is the possibility that an accidental (or deliberate) fire may impinge on the structure. In a superstructure of a warship, fire can start due to many reasons. The peacetime external fire threats include helicopter crash, flight deck accidents, collision, fuel spill, weapon-handling accident, etc.

The wartime fire threat includes missile unspent fuel, or shrapnel-damage-initiated fire. The experience of the USS *Stark,* hit with an Exocet missile, is an example. Mishaps, rather than hostile action, have caused all carrier fires since World War II. Three of the most serious fires aboard aircraft carriers include the *Oriskany* in 1966, the *Forrestal* in 1967, and the *Enterprise* in 1969. The *Forrestal* fire was the worst of the three. It began with an accidentally fired air-to-ground rocket, and hit a fully fueled A-4 Skyhawk, spilling blazing JP-5 fuel onto the flight deck. A photo of this mishap published in a national newspaper is shown in Fig. 9.10. A wild fire on the crowded flight deck of the USS *Enterprise* cooked off a 500 pound bomb carried by an F-4 Phantom in 1969. A photo of this fire is shown in Fig. 9.11. The aircraft crew was able to control the blaze in less than an hour. Another fire incident of consequence was the 1975 collision involving the USS *Kennedy* (CV-67) and the USS *Belknap* (CG 26). The resultant fire influenced the U.S. Navy to improve the survivability of aluminum structures through mineral wool fire insulation. Principal use of steel instead of aluminum in the deckhouse design started with the DDG 51 class ship. Today, steel is the preferred choice as material of construction for U.S. Navy surface combatants.

During the Integrated Technology Deckhouse Program (1988–1993) in the U.S. Navy, JAG reports for the 1980–1986 and Navy Safety Center Database for the 1983–1987 time periods were reviewed. The JAG data reviewed indicated that the major fire source is electrical, followed by

FIGURE 9.10 Fire aboard the aircraft carrier *Forrestal* in 1967. (NAVSEA)

FIGURE 9.11 Fire aboard the aircraft carrier USS *Enterprise* in 1969. (NAVSEA)

matches/smoking, open flame/welding/hot work, and arson. Both the JAG and Navy Safety Center fire data also show that the majority of fires occur in the engineering spaces, followed by storerooms, and crew living spaces. The data further indicate that the majority of fires occurred at shipyard followed by at sea and in port.

9.5 POLYMER COMPOSITES AND FIRE

Since World War I, U.S. Navy ship design had evolved from all-steel construction to include combinations of steel hulls and aluminum deckhouses. In the late 1980s, the Navy reversed this trend towards aluminum deckhouses with a move back to all-steel deckhouses for DDG-51. This has made steel the current baseline for the Navy.

Composite materials offer benefits in many different areas for military applications. Some of these benefits include structures with reduced weight, reduced corrosion, and reduced signature visibility or radar cross section. The benefits of using composites in the fire safety area include the lower thermal conductivity of composites (in comparison to metals), which reduces the fire spread by conduction of heat through structural elements such as decks and bulkheads. At the same time, polymer composites raise many fire safety issues.

Organic matrix composite materials will ignite if sufficiently heated. Ignition may occur by spark, a heated surface, or welding/hot work. After ignition, the fire may grow at a slow or fast rate depending upon the type of composite material, interaction with the surroundings, and access to oxygen. When it grows, it produces increasing amounts of energy, mostly due to flame spread along large surfaces. Besides releasing energy, a variety of toxic and nontoxic gases and solids are produced. For a wide range of compartment fires, the room becomes divided into two distinct layers: a hot upper layer and a cold lower layer. The fire may continue to grow, either by increased burning rate, by flame spread, or by ignition of secondary fuel packages, and the upper layer may become very hot. When the temperature in the hot upper layer reaches 500 to 600°C, the radiation from the hot layer toward other combustible materials in the enclosure sets up a condition in which all the combustible material in the enclosure is ignited with a very rapid increase in energy release rates.

This very rapid and sudden transition from a growing fire to a fully developed fire is called *flashover*. This is a stage where fire suddenly jumps from a relatively controllable state to a state of awesome power and destruction, and where human survivability in the enclosure is no longer possible [18–21]. The fully developed fire can burn for a number of hours if no firefighting action is initiated as long as sufficient fuel and oxygen are available for combustion.

Besides fire growth, another potential problem with polymer composites is that of structural integrity during fire. First, heat weakens the polymer binder. Thermoplastic binders begin to creep and then to flow as the impinging flames raise their local temperature past the glass transition temperature. Thermoset binders degrade to a char or gasify or both. The functioning of the binder is thus diminished and the composite loses strength. If the structure is one in which the composite forms only a reinforcing or repair role, the consequences of a local, heat-induced composite failure are not likely to be serious; time is available to repair the damaged material. However, if the affected composite component is part of a primary critical structure such as the wing of an aircraft, the structure may collapse.

In many cases, especially for the marine industry, the composites are made out of glass fiber reinforcement which is incombustible. This glass fiber content is as high as 70 percent by weight in some cases. When the outermost layers of a composite lose their resin due to heat induced gasification, the glass fibers act as an insulating layer, slowing heat penetration into and evolution of gases from the depth of the composite.

Historically, there have not been any large fires involving polymer composites in military applications. In 1983, a British minesweeper constructed of GRP (polyester resin and woven roving glass) sustained a fire in the exhaust uptake of a diesel engine. The fire continued for 4 to 6 h and reflashed numerous times following seawater hose line application due to leaking hydraulic fluid and heat contained by the GRP structure. The fire was basically contained within the machinery space. However, the only area of boundary penetration occurred on the overhead where the fire had been most intense. The bulkhead separating the machinery space from an adjacent machinery space was involved in the fire but did not allow fire propagation to the adjacent space. Difficulty of extinguishing the burning GRP within the machinery space was due to a combination of heavy smoke, obstructions (thermal and acoustical insulation), leaking hydraulic fluid, and heat retention by the GRP structure.

The inherent chemical nature and complexity of polymer matrix composite materials do not lend themselves to easy analytical prediction of their behavior when exposed to a high heat flux from a fire source. Composites exhibit anisotropic heat transfer. They selectively burn, produce smoke, release heat, chemically degrade, produce char, and delaminate.

Assessment of the fire hazard of combustible composites and plastics has evolved over the past three decades to include measurement of flammability characteristics such as ignitability, flame spread, rate of heat release, and smoke and gas production during exposure to heat or fire. The gas phase and aerosol combustion have been studied extensively [22, 23] because of their commercial importance (e.g., furnaces, internal combustion engines, etc.). However, very little is known about the solid-state chemical kinetic processes of flaming combustion that generate the gaseous fuel. In particular, the material property or combination of properties that governs the flammability of polymers and composites is not readily quantified for complex commercial polymers.

9.6 FIRE REQUIREMENTS AND REGULATIONS FOR POLYMER COMPOSITES

Nearly all fire test methods have been designed to represent some realistic set of fire conditions by simulating an expected fire scenario or by reproducing the heat exposure conditions. The current regime of fire tests mandated for regulatory purposes provide numerical results for ranking of materials. The majority of guidelines developed by regulatory agencies focus on material and product performance testing designed to control flame spread and ignitability of combustible materials. The diverse applications of composites for both military and civilian sectors involve different fire safety concerns owing to the individual geometry and configuration of an enclosure, fire load, fire scenario, mission requirements, ease of escape, and the extent of potential human and property loss.

Table 9.4 provides a summary of fire requirements for composite materials in the United States. This table includes fire requirements for use of composites in infrastructure (highways and bridges), ground transportation (cars, trucks, and buses), air transportation (small and large), commercial

TABLE 9.4 Summary of Fire Requirements for Composite Materials in the United States [24]

Sector	Component	Property	Test procedure	Criteria
Infrastructure		Fire requirements not well defined yet		
Surface (cars, trucks, and buses)	Panels	Flame spread	FMVSS 302	Rate of flame spread, 4 in/min
Surface mass transit vehicles (buses, light rails, and passenger trains)	Seat materials	Flame resistance	FAR 25.853	Flame time 10s Burn length 6 in
	Panel, partition, wall, ceiling	Flame spread	ASTM E162	Flame spread index 35
	Floor structure	Fire endurance	ASTM E119	Nominal evacuation time 15 min
	Seat, panels, walls, partitions, ceiling, floor	Smoke emission	ASTM E662	Ds (1.5) 100 Ds (4.0) 20
	All materials	Toxicity	NBSIR-82-2532	As appropriate
Air (commercial aviation aircraft)	Cabin and cargo compartment materials: seat, panel, liner, ducting	Flame resistance -Vertical -Horizontal -45 degree	FAR 25.853 (a-b) FAR 25.853 (b2-b3) FAR 25.855	Flame time: 15s Flame time of drippings: 3s Burn length: 6 and 8 in. Rate of flame spread: -Class b-2: 2.5 in/min -Class b-3: 4 in/min
	Cargo compartment liners, seats	Fire endurance	FAR 25.855 FAR 25.853	No flame penetration of liner Peak temperature 102 mm above specimen: 204°C; mass loss and flame spread criteria for seats
	All large area cabin interior materials	Smoke emission Heat release rate	FAR 25.853 (a-1) FAR 25.853 (a-1) ASTM E906	Ds (4.0) 200 Peak HRR in 5 min: 65 kW/m^2 Total HRR in 2 min: 65 kW-min m^2

Application	Component	Property	Standard	Criteria
Marine (life boats, rescue boats, and small passenger vessels)	Main structure, hull	Resin/laminate flammability (fire retardancy)	MIL-R-21607 or ASTM E-84	Resin qualified under MIL-R-21607 or laminate tested to ASTM E84, flame spread index 100
Marine (high speed craft, fire restricting materials)	Bulkheads, wall and ceiling linings (surface materials)	Heat release rate, surface flammability, smoke production	ISO 9705 room/corner test	-ave HRR 100 kW -max HRR 500 kW -ave smoke prod. 1.4 m^2/s -max smoke prod. 8.3 m^2/s -flame spread 0.5 m from floor -no flaming drops or debris
Marine (high speed craft, fire restricting materials)	Furniture frames, case furniture, other components	Ignitability, heat release, smoke production	ISO 5660 Cone calorimeter	Criteria are currently under development
Military (US Navy submarines)	Structural composites inside the pressure hull	Heat release, smoke, etc.	MIL-STD-2031(SH)	-flame spread index 20 -max smoke Dm 200 -25 kW/m^2: PHR 50 kW/m^2 Tign 300s -50 kW/m^2: PHR 65 kW/m^2 Tign 150s -75 kW/m^2: PHR 100 kW/m^2 Tign 90s -100 kW/m^2: PHR 150 kW/m^2 Tign 60s

TABLE 9.5 MIL-STD-2031 Submarine Composites Fire Performance Acceptance Criteria [25]

Fire test/characteristic	Requirement	Test method
Oxygen-Temperature Index (%)	Minimum	ASTM D-2863 (modified)
% oxygen at 25°C	35	
% oxygen at 75°C	30	
% oxygen at 300°C	21	
Flame spread index	Maximum 20	ASTM E-162
Ignitability (sec)	Minimum	ASTM E-1354
100 kW/m^2 irradiance	60	
75 kW/m^2 irradiance	90	
50 kW/m^2 irradiance	150	
25 kW/m^2 irradiance	300	
Heat release (kW/m^2)	Maximum	ASTM E-1354
100 kW/m^2 irradiance, peak	150	
Average for 300 sec	120	
75 kW/m^2 irradiance, peak	100	
Average for 300 sec	100	
50 kW/m^2 irradiance, peak	65	
Average for 300 sec	50	
25 kW/m^2 irradiance, peak	50	
Average for 300 sec	50	
Smoke Obscuration	Maximum	ASTM E-662
Ds during 300 secs	100	
Dmax	200	
Combustion gas generation (25 kW/m^2)	CO = 200 ppm	ASTM E-1354
	CO_2 = 4%v	
	HCN = 30 ppm	
	HCL = 100 ppm	
Burn-through fire test	No burn-through in 30 minutes	Appendix B
Quarter-scale fire test	No flashover in 10 minutes	Appendix C
Large scale open environment test	Pass	Appendix D
Large scale pressurizable fire test	Pass	Appendix E
N-Gas model smoke toxicity screening test	No deaths / Pass	Appendix F: Modified NBSTTM

marine transportation (cargo and passenger vessels), and military applications [24]. Table 9.5 provides a summary of acceptance criteria for the use of composite materials in U.S. Navy submarines for structural applications [25]. Table 9.6 provides a summary of fire performance goals for the use of composites in U.S. Navy surface ship applications [26]. These are briefly discussed in the following sections.

9.6.1 Infrastructure and Fire Regulations

The use of composites in transportation infrastructure such as bridge and highway repair and seismic retrofit is expected to be a growing market. Advantages of advanced composites for new construction include tailorable mechanical properties, high strength-to-weight ratios, and chemical inertness, which significantly exceed those of conventional engineering materials such as concrete and steel. The *California Transportation Department* (CALTRANS) is spending several billion dollars

TABLE 9.6 Fire Performance Goals and Standard Fire Test Methods for Composite Structures in Surface Ships [26]

Category	Test method	Criteria
\multicolumn{3}{c}{Surface flammability}		

Category	Test method	Criteria
	Surface flammability	

Surface flammability	ASTM E-84	• Interior applications:
		  • Flame spread index ≤ 25
		  • Smoke developed index ≤ 15
		• Exterior applications:
		  • Flame spread index ≤ 25
		  • Smoke data for review by NAVSEA 05P6

	Fire growth	
Fire growth	ISO 9705 "Fire tests-full-scale room test for surface products"; Annex A standard ignition source fire of 100 kW for 10 minutes and 300 kW for 10 minutes	• Net peak heat release rate over any 30 second period less than 500 kW • Net average heat release rate for test less than 100 kW • Flame spread must not reach 0.5 m above the floor excluding the area 1.2 m from the corner with the ignition source • In addition to ISO 9705 requirements for No flaming drops or debris, this requirement is further restricted by USN to No flaming droplets at any location.
	Tenability	
Smoke production	ISO 9705; 100 kW for 10 minutes and 300 kW for 10 minutes	• Peak smoke production rate less than 8.3 m^2/s over any 60 second period of test • Test average smoke production rate less than 1.4 m^2/s
Smoke toxicity	ISO 9705	• Total mass of CO generated during the test less than 1.9 kg
	ASTM E662	• CO: 350 ppm (max); CO_2: 4% (max); HCl: 30 ppm (max); HCN: 30 ppm (max) [criteria from MIL-STD-2031 revised]
	NFPA 269–1996 Ed.	• Data for review by NAVSEA
	Fire resistance	
Bulkheads/ overheads/decks/ doors/hatches/ penetrations	• Navy modified: UL 1709 fire curve using IMO A.754(18) test procedures; • Total number of thermocouples and their placement on unexposed side IAW MIL-PRF-XX 381; • IMO App. A. III & A.IV apply; • Total number of thermocouples and their placement on unexposed side IAW MIL-PRF-XX 381; • Hose stream test • Maximum load[a]	• Peak temperature rise on the unexposed surface not more than 325°F (180°C); • Average temperature rise on the unexposed surface not more than 250°F (139°C); • There should be no passage of flames, smoke, or hot gases on the unexposed face; • Cotton-wool pad: there should be no ignition, i.e., flaming or glowing; • Gap gauges: not possible to enter the gap gauges into any openings in the specimen; • Approval of constructions restricted to the orientation in which they have been tested.

TABLE 9.6 (*Continued*)

Category	Test method	Criteria
	Fire resistance	

Composite joints	UL 2079 "Tests for fire resistance of building joint systems"	• 30 minute rating requiring:
		  • UL-1709 fire curve for a period of 30 minutes
		  • Average temperature rise on the unexposed surface not more than 250°F (139°C)
		  • Peak temperature rise on the unexposed surface not more than 325°F (180°C)
		  • There should be no passage of flames, smoke, or hot gases on the unexposed face
		  • No ignition of cotton wool pad by gases transmitted through assembly
		  • No through openings

	Structural integrity under fire	
Bulkheads/ overheads/decks	UL 1709 fire curve using IMO A.754(18)	• Tests should be performed with a dead load (maximum load) plus live load on the bulkhead or overhead. • No collapse or rupture of the structure for 30 minutes • The maximum average temperature on the unexposed side should not exceed the critical temperature of the composite where structural properties degrade rapidly. This applies if the critical temperature is less than the average temperature rise of 250°F[b]
Composite joints	UL 2079 with UL 1709 fire curve and maximum design load	• No collapse or catastrophic joint failure after a 30 minute exposure
Attachments/ hangers	Furnace testing with UL 1709 fire curve and maximum design load	• Remain intact after 30 minute exposure
	Passive fire protection	
Fire insulation	UL 1709 fire curve; MIL-PRF-XX381 applies.	• 30 minute rating with peak temperature rise less than 325°F (180°C) and average temperature rise less than 250°F (139°C) • Medium weight shock test, Grade A shock qualified
	Active fire protection	
Detection, suppression, and firefighting		• Material fire performance criteria are based in conjunction with existing or additional detection, suppression, and fire-fighting systems.

[a] Max load is dead load plus live load. Live load for firefighters is 50 psf.
[b] Critical temperature is defined as the temperature at which the rapid loss of modulus occurs when determined in accordance with DMTA.

for repair of bridges and highway structures. Candidate repair systems include continuous fiber-reinforced polymer composites of carbon and glass with epoxy and polyester resins. There are currently no requirements for flammability or fire endurance of infrastructure materials because these are exterior applications.

9.6.2 Ground Transportation and Fire Regulations

Over 400,000 motor vehicle fires occur yearly in the United States, claiming over 700 lives and causing nearly 3000 civilian injuries [27]. Most of these fires originate in the engine compartment or as a result of impact, with cigarette ignition of interior materials being a minor cause of vehicle fires. The first and only U.S. requirement for interior materials and components used in cars, trucks, and buses was developed by the *National Highway Transportation Safety Administration* (NHTSA) and established as *Motor Vehicle Safety Standard* (FMVSS) 302 [28]. This requirement is directed at reducing the hazards of interior fires caused by smoking and matches. A bunsen burner flame test is used to measure the rate of flame spread on a 254-mm (10-in.) horizontal specimen. The test procedure has been adopted by the automotive industry in several other countries and incorporated as an *International Organization for Standardization* (ISO) Standard 3795. FMVSS 302 is not a severe fire test due to the relative ease and speed of passenger egress from a motor vehicle in the cigarette ignition scenario [24].

Fire safety requirements specific to passenger railcar interiors are mandated by the *Federal Railroad Administration* (FRA), the regulatory agency responsible for U.S. passenger train safety. The FRA guidelines [29] for the flammability and smoke properties of materials apply to passenger cars in inter-city and Amtrak trains. Similar guidelines have been issued by the *Federal Transit Administration* (FTA), formerly known as the *Urban Mass Transportation Administration* (UMTA) [30], for materials used in light rail, subway cars, and urban mass transit buses. FRA guidelines for selection of materials, summarized in Table 9.4, consist of prescribed limits based on ignition resistance, flame spread, smoke density, and fire endurance tests to ensure the structural integrity of passenger cars. Identical guidelines comprise the *National Fire Protection Association* (NFPA) Standard for Fixed Guideway Systems (NFPA 130). The FRA criteria for the amount of smoke generated at 1.5 and 4.0 min into exposure are $D_s = 100$ and 200, respectively.

Amtrak has expanded on the FRA guidelines by stipulating that exterior and interior railcar components be tested as complete assemblies, i.e., in a finished product form, rather than as separate materials. Amended Amtrak specifications require a toxicological screening of all new materials using the NBS test method for determination of acute inhalation toxicity due to combustion products. In addition, a fire hazard analysis is required to take into account the complete fire load, configuration, and structural design in combination with the material test data providing a systematic approach to the evaluation of material performance.

9.6.3 Air Transportation and Fire Regulations

Flammability requirements for materials used in commercial aircraft cabins have become highly stringent following new regulations based on heat release measurements enacted in 1990. The baseline performance requirements stipulated in *Federal Aviation Regulations* (FAR) resulted from full-scale fire tests carried out at the FAA Technical Center in Atlantic City, NJ. These tests simulated a postcrash external fuel fire penetrating an intact fuselage. The test results indicated that occupant survival is possible until the burning interior cabin materials cause the cabin to flashover [31]. Therefore, it was deemed essential to control the heat release contribution of cabin materials used in large area applications such as sidewalls, ceiling, stowage bins, and partitions. Subsequent testing showed that heat release rate measured in bench-scale fire tests correlated with cabin flashover time. The *Federal Aviation Administration* (FAA) incorporated limits on the total heat release, heat release rate, and smoke emission from materials used in aircraft cabins are contained in FAR 25.853 and are shown in Table 9.4. Recognizing the growing emphasis on fire safety of aircraft interior materials, the *Suppliers of Advanced Composite Materials Association* (SACMA) organized a Flammability Task Force in 1990 to address these new fire regulations [32].

The FAA heat release standard requires all cabin materials to be tested in a modified Ohio State University heat release test apparatus as described in ASTM E 906. The materials are required to have less than 65 kW/m² min total heat release over 2 min and a peak heat release rate of 65 kW/m² over the 5-min duration of the test. The regulations also limit the smoke density of large-area interior materials to D_s 200 at 4 min using ASTM E 662. Other bench-scale tests are required for ignition resistance and flame propagation using a bunsen burner with 12 and 60 s exposure. Bunsen burner tests are also required for interior cabin materials. An oil burner exposure test has been required for aircraft seating since 1987, and for cargo liners since 1991. The specified acceptance criteria are reported in Table 9.4.

9.6.4 Commercial Marine Transportation and Fire Regulations

U.S. regulations for commercial shipping are found in the *Code of Federal Regulations* (CFR), Title 46, which covers nearly every aspect of design and construction of small and large passenger vessels, cargo vessels, tank vessels, mobile offshore drilling units, and shipbuilding materials. CFR Title 46 currently limits the use of composites to small passenger vessels, lifeboats, and various minor components. For most vessels, regulations require the main structure to be steel or equivalent noncombustible material, and most of the ship interior and outfit to be noncombustible. The breakpoint for this is certain small passenger vessels [33] that can be built completely of composite materials provided that the Coast Guard approved fire-retardant resins (MIL-R-21607) are used. There are provisions to allow a general-purpose resin to be used in lieu of fire-retardant resins, such as installing fire-rated boundaries surrounding galleys, limiting ignition sources, fire detection and extinguishing systems in certain spaces, machinery space boundaries lined with noncombustible materials, and restrictions on furnishings [46 CFR 177.410].

The *International Maritime Organization* (IMO), a specialized agency of the United Nations, is responsible for maintaining the *International Convention for the Safety of Life at Sea* (SOLAS). Enforcement of the IMO conventions and standards is the responsibility of the flag state; the IMO has no direct enforcement mechanism. Two significant recent IMO efforts affecting the use of composites and fire testing in general are the adoption of (1) the High Speed Craft (HSC) Code and (2) the *Fire Test Procedures* (FTP) Code [34]. The adoption of the FTP code makes the use of IMO test procedures mandatory for materials and products used for vessels engaged in international voyages. Before the FTP code, each country could use its own national standards, or the IMO recommendations.

The SOLAS regulations are very similar to U.S. domestic regulations in that they require steel or noncombustible vessel construction. In order for the composites to be used in ship construction, other than for high-speed craft, they must be considered "equivalent to steel" as determined by "interim" guidelines by the IMO Maritime Safety Committee [35]. These guidelines include the following criteria:

noncombustible (IMO FTP Code, Part 1);

fire resistant compartment boundaries (IMO FTP Code Part 3);

low smoke/toxicity (IMO FTP Code, Part 2);

determination of structural properties and critical temperature of the composite, in accordance with given guidelines.

SOLAS classifies materials as noncombustible if they do not ignite or evolve combustible gases when heated to 750°C in a vertical cylindrical chamber (ISO 1182). For determining the flammability of surface finishes, the IMO specifies lateral flame spread apparatus (ASTM E 1317). Some IMO requirements for fire safety are summarized in Table 9.4. The U.S. Coast Guard (USCG) is responsible for enforcing compliance with SOLAS requirements for all U.S. ships engaged on international voyages, and for foreign ships entering U.S. ports.

A new regulatory effort in recent years was the adoption of the IMO HSC Code [36]. This code is intended to be a stand-alone document, with a philosophy based on the management and reduction of risk as well as the traditional philosophy of passive fire protection. It encompasses all aspects of the design, construction, and operation of high-speed passenger or cargo craft, and is intended to be used in its entirety. As with nearly all maritime regulations, the code does not specifically allow or restrict composite materials. It uses performance-based criteria, and introduces a new regulatory class of material: *fire-restricting materials,* defined as having low flame spread characteristics, limited rate of heat release, and limited smoke and toxic products emission. Table 9.4 lists the related fire test standards for fire-restricting materials [37]. The definition of this new class of construction materials represents an improvement in the standards and incorporates modern fire test methods.

9.6.5 Military Use of Composites and Fire Regulations

The use of composites in high-technology military applications represents the largest market for advanced materials. The new acquisition reform is leading the military to develop performance-based standards using commercially available test methods. In some cases, waivers may be granted due to mission requirements if materials cannot meet fire requirements.

The use of composites inside naval submarines is now covered by MIL-STD-2031 (SH), "Fire and Toxicity Test Methods and Qualification Procedure for Composite Material Systems Used in Hull, Machinery, and Structural Applications inside Naval Submarines" [25]. This military standard, summarized in Table 9.5, contains test methods and requirements for flammability characteristics such as flame spread index (FSI), specific optical density, heat release rate and ignitability, oxygen-temperature index, combustion gas generation, long-term outgassing, etc. Two guiding criteria [38] were established for the use of composite systems aboard Navy vessels. The first is that the composite system will not be the fire source, i.e., it will be sufficiently fire resistant not to be a source of spontaneous combustion. The second is that ignition of the composite system will be delayed until the crew can respond to the primary fire source, i.e., the composite system will not result in rapid spreading of the fire.

There are no official performance requirements promulgated by the U.S. Navy for the use of composites in surface ship structural applications. However, performance goals have been established for the use of composite materials in the next generation of surface combatants. These fire performance goals are summarized in Table 9.6 [26].

Both the USAF and U.S. Army have military standards regulating the use of composites in military aircraft and fighting vehicles. However, these standards are application specific and designed for specific components.

9.7 FIRE PERFORMANCE AND TEST METHODS FOR COMPOSITES

There are several fire performance characteristics that evaluate the fire safety of composites in military applications. Fire performance characteristics of composite systems may be divided into several categories, some of which are discussed below. It is the combination and totality of such performance characteristics that ensure the fire safety of composite systems that must perform in most hostile fire threat environments.

1. Surface flammability (flame spread, minimize the ignition and spread of fire within the compartment)
2. Tenability (smoke production, toxicity)
3. Fire growth (flashover, minimize the hazard to personnel escaping the fire or their ability to fight the fire)
4. Fire resistance (contain the fire to designated spaces and/or zones)
5. Structural integrity under fire (reduce the risk of structural collapse)
6. Passive fire protection (fire insulation, delay fire spread, flashover, and structural collapse)
7. Active fire protection (detection, suppression, fire fighting)

9.7.1 Surface Flammability

The surface flammability determines the relative burning behavior of construction materials in terms of flame spread along the exposed surface. ASTM E-84, "Standard Test Method for Surface Burning Characteristics of Building Materials" [39] is often the test method of choice in most national codes such as NFPA 301, Code for Safety to Life from Fire on Merchant Vessels [40]. This is also the test method of choice in U.S. Navy MIL-STD-1623D, "Fire Performance Requirements and Approved Specifications for Interior Finish Materials and Furnishings" [41].

ASTM E-84, "Test Method for Surface Burning Characteristics of Building Materials," uses specimen size of 24 ft $\times$ 20.5 in. $\times$ 4 in. maximum thickness in a Steiner tunnel to obtain material data on flame spread index and smoke developed index. Briefly, the calibration procedure sets the flame spread index of cement board and red oak wood at 0 and 100, respectively. However, this test has been found to be misleading for materials that do not remain attached to the ceiling of the Steiner tunnel during the fire test.

A small-scale version of surface flammability fire test method is ASTM E-162, "Surface Flammability of Materials Using a Radiant Heat Energy Source" [42]. A specimen 6 in. wide by 18 in. long and no greater than 1.0-in. thick is placed in the sample holder. This is located in front of a 12 $\times$ 18-in. radiant panel using air and gas as the fuel supply. The radiant panel consists of a porous refractory material and shall be capable of operating up to 1500°F (815°C). A small pilot flame about 2 in. long is applied to the top center of the specimen at the start of the test. The test is completed when the flame front has traveled 15 in. or after an exposure time of 15 min. A factor derived from the rate of progress of the flame front and another relating to the rate of heat liberation by the material under test is combined to provide flame spread index (FSI). Table 9.7 lists the selected composite materials evaluated under a variety of programs for flammability properties at NSWCCD. Table 9.8 lists the flame spread index of several composite materials tested in ASTM E-162.

9.7.2 Smoke and Combustion Gas Generation

Fiber-reinforced polymer composite materials give off smoke when they burn. Smoke is defined as the airborne solid and liquid particulates and gases that evolve when a material undergoes pyrolysis or combustion. Smoke affects visibility and hinders the ability of the occupants to escape and firefighters to locate and suppress the fire.

Smoke production of construction and combustible materials is typically measured either in small-scale tests, such as ASTM E 662 [43] or ASTM E 1354 [44], or full-scale tests, such as ASTM E-84 [39] or ISO 9705 [45]. The full-scale ASTM E-84 test is widely used for building and construction materials. The small-scale smoke density chamber ASTM E 662 test is also widely used for characterization of smoke density of materials as it relates to vision obscuration due to combustion products from flaming and nonflaming modes. The ASTM E 662 test is conducted in a closed chamber of fixed volume and the light attenuation is recorded over a known optical path length. A 3 $\times$ 3-in. sample is subjected to a radiant heat flux of 25 kW/m^2 under piloted ignition (flaming) or nonflaming mode and the corresponding light transmission provides specific optical density (D_s). Visibility through smoke is inversely related to specific optical density. Measurements made with the test relate to light transmission through smoke and is similar to the optical density scale for human vision. In simplified terms, the chamber is calibrated to initial light transmission of 100 percent, meaning no smoke. As the sample is heated by a radiant flux of 25 kW/m^2, either in nonflaming or flaming mode, the amount of light transmitted as a fraction of initial light is used to calculate the specific optical density. The specific optical density of 100 corresponds to the light transmission of 17 percent. The maximum optical density (D_m) over the duration of the test is used to identify materials with relatively high smoke production.

Combustion gas generation is defined as the gases evolved from materials during the process of combustion. Organic materials, when decomposing and burning, can evolve a variety of toxic gases, the most common of which is *carbon monoxide* (CO). In addition, nitrogen-containing materials may evolve *hydrogen cyanide* (HCN) and nitrogen oxides, sulfur-containing materials may evolve sulfur oxides and sulfides, chlorine-containing materials may evolve *hydrogen chloride* (HCl), and other gases may be generated depending upon the chemistry of the matrix resin of a given composite

TABLE 9.7 Selected List of Composite Materials Evaluated at CARDEROCKDIV, NSWCCD

Composite	Identification
Gl/VE (1031)	Glass/vinyl ester, fire retardant, brominated
Gl/VE (1087)	Glass/vinyl ester, non–fire-retardant
Gl/VE (1167)	Nonbrominated vinyl ester resin
Gl/VE (1168)	Brominated bisphenol A epoxy vinyl ester resin
Gl/VE (1169)	Epoxy novolac vinyl ester resin
Gl/VE (1170)	Bisphenol A epoxy vinyl ester resin
Gl/VE (1257)	Sandwich composite, 3.0-in.-thick balsa core and 0.25-in.-thick Gl/VE composite skins
Gl/Modar (1161)	Glass/modified acrylics, fire retarded
Gl/EP (1089)	Glass/epoxy, S2/3501-6, $(0/90)_s$
Gl/EP (1066)	Glass/epoxy, 105/206, RT cure, post cured
Gl/EP (1067)	Glass/epoxy, 125/226, RT cure, post cured
Gl/EP (1040)	Glass/epoxy, E-Glass/F155
Gl/EP (1071)	Glass/epoxy, S2/F155
Gl/EP (1006)	Glass/Epoxy, 7701/7781
Gl/EP (1070)	Sonar bow dome, MXB7780/3783
Gl/EP (1003)	RTM, 9405/9470
Gl/EP (1090)	SL-851-H4, SMC, 50% glass
Gr/EP (1091)	T-300/5208, $(0/90)_s$, 350°F
Gr/EP (1092)	AS4/LC1, anhydride cured
Gr/EP (1093)	Graphite/epoxy: AS4/3501-6
Gr/EP (1094)	P55/ERLX, toughened epoxy
Gl/CE (1046)	Glass/cyanate ester
Gr/M.BMI (1095)	T300/5245C, modified bismaleimide
Gl/BMI (1096)	T2E225/F650
Gr/BMI (1097)	T6T145/F650, $(0/90)_s$, PC 475°F for 4 hours
Gr/BMI (1106)	T6T145/F655, toughened
Gr/BMI (1098)	HTA-7/65FWR
Gl/PH (1014)	J2027/Phencat 10, RT cure, PC at 140°F/6 hrs
Gl/PH (1015)	Mark IV, RT cure, PC at 140°F/6 hrs
Gl/PH (1017)	Fire PRF2, RT cure, PC at 140°F/6 hrs
Gl/PH (1018)	350D66 RT cure, PC at 140°F/6 hrs
Gl/PH (1099)	Q6399, developmental RT curing phenolic system
Gl/PH (1100)	CPH 2265/7781, cures at 250°F
Gl/PH (1101)	CPH 2265/7781, post cured at 350°F
Gr/PH (1102)	3C584/F453-1, structural, heat resistant
Gr/PH (1103)	R1620, toughened, structural
Gr/PH (1104)	402/7781
PE/PH (1073)	Polyethylene 1000, 985PT/Mark IV
Aramid/PH (1074)	Aramid 49, 900-F1000/Mark IV
Gl/PMR (1105)	CPI 2237/6781, PMR-15 polyimide system
Gl/PN (1136)	Glass/Phthalonitrile
Gr/PN (1080)	Graphite/Phthalonitrile, NRL
Gl/SI (1116)	Glass/Silicone
Gl/PP (1082)	Glass/polypropylene
Gl/J-2 (1077)	Glass/Nylon
Gl/PPS (1069)	AG 40-70, polyphenylene sulfide
Gr/PPS (1083)	AC 40-60, polyphenylene sulfide
Gl/PPS (1084)	LG 40-70, polyphenylene sulfide
Gr/PPS (1085)	LC 31-60, T300, polyphenylene sulfide
Gr/PAS (1081)	T650-42/polyaryl sulfone
Gr/PES (1078)	4084/PES-1, IM8/ITA
Gr/PEEK (1086)	APC-2/AS4, poly ether ether ketone
Gl/PEKK (1079)	S2/PEKK, polyether ketone ketone

TABLE 9.8 Flame Spread Index for Various Fiber Reinforced Composites (ASTM E-162)

Composite	ASTM E-162, FSI	Composite	ASTM E-162, FSI
Red oak wood	100	Gl/Phenolic, 1017	6
Gl/Vinyl ester, NFR, 1087	156	Gl/Phenolic, 1018	4
Gl/Vinyl ester, FR, 1031	27	Gr/Phenolic, 1102	6
Gl/VE Sand. Composite, 1257	24	Gr/Phenolic, 1103	20
Gl/Epoxy, 1066	43	Gr/Phenolic, 1104	3
Gl/Epoxy, 1067	12	PE/Phenolic, 1073	48
Gl/Epoxy, 1089	11	Aramid/Phenolic, 1074	30
Gl/Epoxy, 1091	11	Gl/PMR-15, 1105	2
Gl/Epoxy, 1092	23	Gl/SI	2
Gr/M. BMI, 1095	13	Gl/J-2, 1077	13
Gl/BMI, 1096	17	Gl/PPS, 1069	7
Gr/BMI, 1097	12	Gr/PPS, 1083	3
Gr/BMI, 1098	3	Gl/PPS, 1084	8
Gl/Phenolic, 1099	1	Gr/PPS, 1085	3
Gl/Phenolic, 1100	5	Gr/PAS, 1081	9
Gl/Phenolic, 1101	4	Gr/PEEK, 1086	3
Gl/Phenolic, 1014	4	Gl/PEKK, 1079	3
Gl/Phenolic, 1015	4		

material. The Committee on Fire Toxicology of the National Academy of Science has concluded that as a basis for judging or regulating materials performance in a fire, combustion product toxicity data must be used only within the context of fire hazard assessment. The committee believes that required smoke toxicity is currently best obtained with animal exposure methods for purposes of predicting the fire hazard of different materials [46].

Table 9.9 presents smoke density and the relative concentrations of combustion gas generation (Draeger colorimetric tube) in flaming mode during smoke-obscuration tests (ASTM E-662) for several composite materials. With the exception of vinyl ester, all other composite systems had the specific optical density at 300 s of less than 100. Glass- or graphite-reinforced phenolic composites have very low smoke. This is also true for all advanced thermoplastics that also have low maximum smoke density. In general, thermoset composite materials give off more carbon monoxide than thermoplastic composites. The thermoplastic panels evaluated in this study slightly expanded or foamed up in the middle during smoke density tests, presumably due to the gases escaping through the softened front face during fire exposure.

9.7.3 Fire Growth

The growth of a compartment fire depends on the rate at which the initiating source fire ignites materials and other items in the compartment and the heat release rate of the ignited items. To effectively fight the fire, the fire must not reach a flashover condition before fire extinguishment procedures are initiated. Flashover is the condition in which all gases in the upper part of a compartment spontaneously ignite. This produces a thermal condition that will inevitably ignite all combustible items within the compartment, and engulf the compartment with fire. When flashover is reached, the ability of firefighters to fight the fire will be reduced.

For most interior applications of composites with large surfaces, fire growth potential should be the first issue addressed and overcome for habitable environments. In small scale, fire growth potential is measured by heat release rates in ASTM E 1354 [44], "Standard Test Method for Heat and Visible Smoke Release Rate for Materials and Products Using an Oxygen Consumption Calorimeter," commonly referred to as the *cone calorimeter*. In full scale, fire growth potential is measured by ISO 9705 [45], "Full-Scale Room Test for Surface Products."

TABLE 9.9 Smoke and Combustion Gas Generation (ASTM E-662)

Composite	D_s (300s)	D_m	CO ppm	CO_2 % v	HCN ppm	HCL ppm
Gl/Vinyl ester, 1031	463	576	230	0.3	ND*	ND
Gl/Vinyl ester, 1087	310	325	298	1.5	1	0.5
Gl/Vinyl ester, 1167	103	173	300	2	2	ND
Gl/Vinyl ester, 1168	503	593	800	0.5	2	TR**
Gl/Vinyl ester, 1169	154	217	200	2	TR	ND
Gl/Vinyl ester, 1170	185	197	300	2	TR	ND
Gl/VE Sandwich Composite, 1257	550	900	2900	0.6	2	TR
Gl/Modar	11	109	400	0.5	5	ND
Gl/Epoxy, 1089	56	165	283	1.5	5	ND
Gl/Epoxy, 1066	2	408	200	2	5	2
Gl/Epoxy, 1067	16	456	250	1	2	ND
Gl/Epoxy, 1071	17	348	80	0.5	3	1
Gl/Epoxy, 1090	96	155	50	0.2	ND	ND
Gr/Epoxy, 1091	75	191	115	0.9	15	TR
Gr/Epoxy, 1092	66	210	313	2	1	0.5
Gr/Epoxy, 1093	3	353	160	0.5	2	1.5
Gr/Epoxy, 1094	1	301	300	0.6	2	1
Gl/CE, 1046	4	84	—	—	—	—
Gr/M. BMI, 1095	24	158	30	0.3	1	ND
Gr/BMI, 1097	6	171	175	0.8	3	ND
Gr/BMI, 1098	9	117	10	TR	TR	1
Gl/BMI, 1096	34	127	300	0.1	7	TR
Gl/Phenolic, 1100	4	18	300	1	1	1
Gl/Phenolic, 1099	4	43	50	TR	TR	ND
Gl/Phenolic, 1101	1	23	300	1	1	1
Gl/Phenolic, 1014	1	1	300	1	TR	ND
Gl/Phenolic, 1015	1	3	190	1	TR	TR
Gl/Phenolic, 1017	1	4	200	1	ND	ND
Gl/Phenolic, 1018	1	1	200	1	TR	ND
Gr/Phenolic, 1102	1	24	115	0.5	1	1
Gr/Phenolic, 1103	40	138	100	0.1	1	ND
Gr/Phenolic, 1104	1	4	600	1	2	ND
PE/Phenolic, 1073	1	241	700	2	2	ND
Aramid/Phenolic, 1074	2	62	700	1.5	2	ND
Gl/PMR-15, 1105	1	16	200	1	TR	2
Gl/PN (1136)	1	5	60	0.5	TR	ND
Gl/SI (1116)	1	2	50	0.5	TR	TR
Gl/J-2, 1077	—	328	180	1	10	ND
Gl/PPS, 1069	8	87	70	0.5	2	0.5
Gr/PPS, 1083	2	32	100	0.5	1	ND
Gl/PPS, 1084	4	54	100	1	TR	TR
Gr/PPS, 1085	1	26	100	1	TR	TR
Gr/PAS, 1081	2	3	55	0.1	TR	ND
Gr/PES, 1078	1	5	110	1	1	1
Gr/PEEK, 1086	1	1	TR	TR	ND	ND
Gl/PEKK, 1079	1	4	200	1	ND	TR

* ND stands for not defined or not detected.
** TR stands for trace.

9.7.3.1 Heat Release Rate and Ignitability (ASTM E 1354)

Heat release is defined as the heat generated in a fire due to various chemical reactions occurring within a given weight or volume of material. The major contributors are those reactions where CO and CO_2 are generated and oxygen is consumed [47]. Different levels of radiant heat flux simulate fire scenarios in which the composite material is itself burning or in which it may be near another burning material. Heat release rate data provide a relative fire hazard assessment for materials. Materials with low heat release rate per unit weight or volume will do less damage to the surroundings than the material with high release rate. The rate of heat release, especially the peak amount, is the primary characteristic determining the size, growth, and suppression requirements of a fire environment [47].

ASTM E 1354 (oxygen consumption cone calorimeter) covers the measurement of the response of materials exposed to controlled levels of radiant heating and is used to determine the heat release rates, ignitability, mass loss rates, effective heat of combustion, and visible smoke development. These values are becoming increasingly important in determining fire growth and are needed in the various fire models that are being developed. Specific heat flux of 25, 50, 75, and 100 kw/m^2 are required. These thermal insults correspond to a small Class A fire, a large trash can fire, a significant fire, and a pool oil fire. A 100 × 100-mm sample is placed beneath the conical-shaped heater that provides a uniform irradiance on the sample surface. The sample mass is constantly monitored using a load cell, and the effluent from the sample is collected in the exhaust hood above the heater. A spark igniter 12.5 mm from the sample surface is used to initiate the burning of any combustible gas mixture produced by the sample. In the duct downstream of the hood, the flow rate, smoke obscuration, and O_2, CO_2, and CO concentrations are continuously measured. Once the sample ignites, the burning of the sample causes a reduction in the oxygen concentration within the effluent collected by the hood. This reduction in oxygen concentration has been shown to correlate with the heat release rate of the material, 13.1 MJ/kg of O_2 consumed. This is known as the oxygen consumption principle. Using this principle, the heat release rate per unit area of the sample is determined with time using measurements of mass flow rate and oxygen depletion in the gas flow. A composite material during test in cone calorimeter apparatus is shown in Fig. 9.12.

Smoke is also measured during the ASTM E 1354 rate of heat release test method. Smoke obscuration is measured in the flow system by means of a laser photometer. Results are computed in the form of specific extinction area. In addition, data from the cone calorimeter test include mass loss

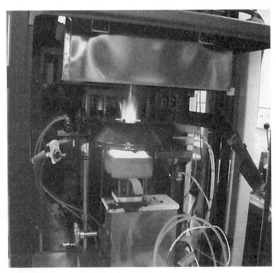

FIGURE 9.12 Cone calorimeter (ASTM E 1354) heat release test.

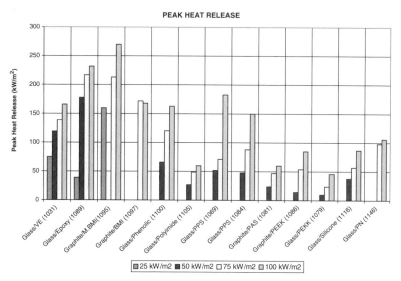

FIGURE 9.13 Peak heat release rates (ASTM E 1354) from several composites.

rate, average heat release rates at various time intervals, total heat release, heat of combustion, etc. For this reason, cone calorimeter has now become a very widely used small-scale test method to determine the fire-growth potential of combustible materials. Table 9.10 gives the peak and average heat release rates, time to ignition, mass loss, total heat release, and extinction area for various composite materials at different heat fluxes [48–50]. Figures 9.13 and 9.14 show the peak heat release rates and time to ignition for selected composites, respectively.

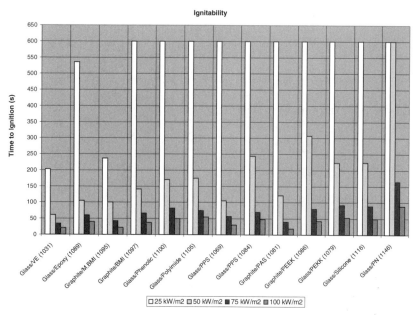

FIGURE 9.14 Time to ignition (ASTM E 1354) for several composites.

TABLE 9.10 Heat Release and Ignitability of Composite Materials

Material system	Irradiance (kW/m^2)	Weight loss (%)	Ignitability (s)	Peak heat release (kW/m^2)	Average heat release 300 s (kW/m^2)	Total heat release (MJ/m^2)	Extinction area (m^2/kg)
			Thermosets				
Douglas fir plywood	25	74	306	188	90	69.2	98
	50	82	22	314	98	76.5	75
	75	87	8	335	157	84.8	141
	100	—	—	—	—	—	—
Glass/VE (1031)	25	14	278	75	29	11	1185
	50	26	74	119	78	25	1721
	75	29	34	139	80	27	1791
	100	28	18	166	—	22	1899
Glass/VE (1087)	25	36	281	377	180	55	1188
	50	—	—	—	—	—	—
	75	34	22	499	220	68	1218
	100	33	11	557	—	64	1466
Glass/VE (1167)	25	—	320	308	180	64	836
	50	—	85	276	184	59	999
	75	—	42	281	190	59	986
	100	—	—	—	—	—	—
Glass/VE (1168)	25	—	214	147	92	29	1341
	50	—	52	152	86	28	1524
	75	—	29	217	108	33	1569
	100	—	—	—	—	—	—
Glass/VE (1169)	25	—	302	342	211	68	796
	50	—	85	302	198	62	815
	75	—	42	303	203	63	872
	100	—	—	—	—	—	—
Glass/VE (1170)	25	—	259	356	190	58	914
	50	—	75	348	179	55	1027
	75	—	36	432	202	61	1050
	100	—	—	—	—	—	—
Gl/VE Sandwich Composite (1257)	25	15	306	121	58	121.1	933
	50	18	70	126	93	126.0	1063
	75	23	28	150	99	149.7	986
	100	—	—	—	—	—	—
Glass/Modar (1161)	25	—	421	149	85	58.2	100
	50	—	119	160	91	64.3	126
	75	—	61	181	105	67.0	161
	100	—	—	—	—	—	—
Glass/epoxy (1089)	25	—	535	39	30	10	470
	50	—	105	178	98	30	580
	75	—	60	217	93	28	728
	100	—	40	232	93	24	541
Glass/epoxy (1066)	25	20	140	231	158	52	1096
	50	23	48	266	154	48	1055
	75	24	14	271	157	48	1169
	100	24	9	489	—	46	1235

TABLE 9.10 (*Continued*)

Material system	Irradiance (kW/m²)	Weight loss (%)	Ignitability (s)	Peak heat release (kW/m²)	Average heat release 300 s (kW/m²)	Total heat release (MJ/m²)	Extinction area (m²/kg)
			Thermosets				
Glass/epoxy (1067)	25	18	209	230	120	41	1148
	50	20	63	213	127	39	1061
	75	21	24	300	138	43	1109
	100	20	18	279	—	32	1293
Glass/epoxy (1040)	25	—	—	—	—	—	—
	50	19	18	40	2	29	566
	75	21	13	246	1	38	605
	100	23	9	232	5	47	592
Glass/epoxy (1071)	25	7	128	20	4	1	1356
	50	5	34	93	—	3	1757
	75	23	18	141	99	30	1553
	100	25	10	202	108	34	1310
Glass/epoxy (1006)	25	14	159	81	63	28	2690
	50	28	49	181	108	39	1753
	75	24	23	182	—	35	1917
	100	29	14	229	131	41	1954
Glass/epoxy (1070)	25	23	229	175	95	45	1119
	50	28	63	196	143	49	1539
	75	27	30	262	133	43	1440
	100	30	23	284	—	36	1640
Glass/epoxy (1003)	25	17	198	159	93	36	1162
	50	22	50	294	135	43	1683
	75	22	73	191	121	41	1341
	100	22	19	335	122	37	1535
Glass/epoxy (1090)	25	19	479	118	67	38	643
	50	28	120	114	90	55	803
	75	34	54	144	115	64	821
	100	34	34	73	150	71	1197
Graphite/epoxy (1091)	25	7	NI	NI	NI	NI	601
	50	—	—	—	—	—	—
	75	25	53	197	90	30	891
	100	38	28	241	—	28	997
Graphite/epoxy (1092)	25	—	275	164	99	32	525
	50	—	76	189	116	37	593
	75	—	32	242	112	37	363
	100	—	23	242	113	71	235
Graphite/epoxy (1093)	25	13	338	105	69	—	—
	50	24	94	171	93	—	—
	75	23	44	244	147	—	—
	100	22	28	202	115	—	—
Glass/CE (1046)	25	8	199	121	74	30	794
	50	22	58	130	71	49	898
	75	23	20	196	116	58	1023
	100	24	10	226	141	47	1199

TABLE 9.10 (*Continued*)

Material system	Irradiance (kW/m²)	Weight loss (%)	Ignitability (s)	Peak heat release (kW/m²)	Average heat release 300 s (kW/m²)	Total heat release (MJ/m²)	Extinction area (m²/kg)
			Thermosets				
Graphite/M.BMI (1095)	25	19	237	160	103	32	645
	50	—	—	—	—	—	—
	75	24	42	213	115	36	685
	100	26	22	270	124	38	706
Graphite/BMI (1097)	25	5	NI	NI	NI	NI	238
	50	—	—	—	—	—	—
	75	30	66	172	130	45	933
	100	31	37	168	130	41	971
Graphite/BMI (1098)	25	—	NI	NI	NI	NI	NI
	50	13	110	74	51	14	228
	75	15	32	91	65	17	370
	100	16	27	146	75	22	383
Glass/BMI (1096)	25	17	503	128	105	40	324
	50	25	141	176	161	60	546
	75	30	60	245	199	76	604
	100	30	36	285	219	73	816
Glass/phenolic (1099)	25	—	NI	NI	NI	NI	NI
	50	—	121	66	43	18	4
	75	—	33	102	86	33	85
	100	—	22	122	95	40	—
Glass/phenolic (1100)	25	—	NI	NI	NI	NI	—
	50	—	125	66	48	17	308
	75	—	20	120	63	21	365
	100	—	40	163	74	21	441
Glass/phenolic (1101)	25	—	NI	NI	NI	NI	NI
	50	—	210	47	38	14	176
	75	—	55	57	40	16	161
	100	—	25	96	70	22	620
Glass/phenolic (1014)	25	—	NI	NI	NI	NI	NI
	50	12	214	81	40	17	83
	75	16	73	97	54	20	246
	100	16	54	133	78	21	378
Glass/phenolic (1015)	25	—	NI	NI	NI	NI	NI
	50	6	238	82	73	15	75
	75	8	113	76	37	7	98
	100	13	59	80	62	12	58
Glass/phenolic (1017)	25	—	NI	NI	NI	NI	NI
	50	10	180	190	139	43	71
	75	14	83	115	84	17	161
	100	18	43	141	73	19	133
Glass/phenolic (1018)	25	—	NI	NI	NI	NI	NI
	50	3	313	132	22	12	143
	75	11	140	56	44	11	74
	100	13	88	68	58	13	66

TABLE 9.10 (*Continued*)

Material system	Irradiance (kW/m²)	Weight loss (%)	Ignitability (s)	Peak heat release (kW/m²)	Average heat release 300 s (kW/m²)	Total heat release (MJ/m²)	Extinction area (m²/kg)
			Thermosets				
Graphite/phenolic (1102)	25	4	NI	NI	NI	NI	NI
	50	—	—	—	—	—	—
	75	28	79	159	80	28	261
	100	—	4	196	—	—	—
Graphite/phenolic (1103)	25	—	NI	NI	NI	NI	N
	50	28	104	177	112	50	253
	75	27	34	183	132	50	495
	100	29	20	189	142	51	493
Graphite/phenolic (1104)	25	—	NI	NI	NI	NI	NI
	50	9	187	71	41	14	194
	75	11	88	87	—	11	194
	100	11	65	101	—	11	232
PE/phenolic (1073)	25	30	714	NI	NI	NI	NI
	50	61	129	98	83	107	294
	75	60	28	141	92	104	500
	100	67	10	234	131	96	580
Aramid/phenolic (1074)	25	4	1110	NI	NI	NI	NI
	50	43	163	51	40	57	156
	75	40	33	93	54	45	240
	100	65	15	104	72	95	333
Glass/polyimide (1105)	25	—	NI	NI	NI	NI	NI
	50	11	175	40	27	21	170
	75	13	75	78	49	22	131
	100	14	55	85	60	20	113
Glass/PN (1273)	25	—	NI	0	0	0	0
	50	—	437	35	24	10.9	157
	75	—	165	83	49	18.5	75
	100	—	88	109	57	22.3	58
Graphite/PN (1080)	25	—	—	—	—	—	—
	50	—	—	—	—	—	—
	75	—	—	—	—	—	—
	100	13	75	119	36	12	610
Glass/PP (1082)	25	37	168	187	153	88	702
	50	36	47	361	248	82	959
	75	37	23	484	265	82	1077
	100	36	13	432	—	82	1120
Glass/J-2 (1077)	25	—	193	67	38	—	803
	50	—	53	96	49	—	911
	75	—	21	116	48	—	866
	100	—	13	135	76	—	1011
Glass/PPS (1069)	25	—	NI	NI	NI	NI	NI
	50	12	105	52	25	32	585
	75	12	57	71	56	24	575
	100	14	30	183	106	41	749

TABLE 9.10 (Continued)

Material system	Irradiance (kW/m²)	Weight loss (%)	Ignitability (s)	Peak heat release (kW/m²)	Average heat release 300 s (kW/m²)	Total heat release (MJ/m²)	Extinction area (m²/kg)
			Thermosets				
Graphite/PPS (1083)	25	—	NI	NI	NI	NI	NI
	50	—	—	—	—	—	—
	75	34	69	81	60	37	431
	100	23	26	141	80	37	752
Glass/PPS (1084)	25	—	NI	NI	—	—	—
	50	13	244	48	28	39	690
	75	15	70	88	67	35	954
	100	16	48	150	94	35	613
Graphite/PPS (1085)	25	—	NI	NI	—	—	—
	50	16	173	94	70	26	604
	75	17	59	66	50	23	—
	100	26	33	126	88	33	559
Graphite/PAS (1081)	25	—	NI	NI	NI	NI	NI
	50	3	122	24	8	1	79
	75	18	40	47	32	14	211
	100	18	19	60	44	14	173
Graphite/PES (1078)	25	—	NI	NI	NI	NI	NI
	50	—	172	11	6	3	145
	75	—	47	41	23	22	88
	100	—	21	65	39	23	189
Graphite/PEEK (1086)	25	—	NI	NI	NI	NI	NI
	50	2	307	14	8	3	69
	75	18	80	54	30	35	134
	100	16	42	85	56	28	252
Graphite/PEKK (1079)	25	—	NI	NI	NI	NI	NI
	50	6	223	21	10	15	274
	75	10	92	45	24	20	—
	100	6	53	74	46	24	891

9.7.3.2 Full-Scale Room Test (ISO 9705)

In 1996, the HSC entered into force as part of the SOLAS convention [36]. This code deals with many aspects of the construction and operation of high-speed craft. The most common type of ship that is regulated by the code is the passenger and vehicle ferry that operates within 4 h from the shore. The code permits that a high-speed craft be constructed of combustible materials, provided certain fire performance criteria are met. Materials that meet these criteria are referred to as "fire-restricting materials." The determination of fire-restricting materials is based primarily on one of two tests. Bulkhead linings and ceiling materials are tested using the ISO 9705 room/corner test. Acceptance criteria for ISO 9705 are published in resolution MSC.40 (64) of the IMO, "Standards for Qualifying Marine Materials for High-Speed Craft as Fire-Restricting Materials" [37]. Furniture components (other than fabrics, upholstery, or bedding) and other components are tested using the ISO 5660 cone calorimeter. No acceptance criteria are published for ISO 5660.

From the mid-1970s to the mid-1980s, both ASTM and ISO had committees working on developing a standard test method for measuring fire growth. This effort resulted in ASTM E603-98,

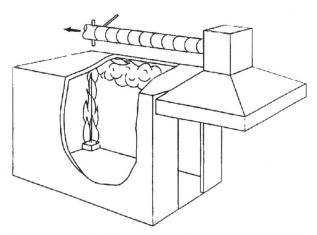

FIGURE 9.15 Schematics of the ISO 9705 room test.

"Standard Guide for Room Fire Experiments," [51] and ISO 9705, "Fire Tests—Full-Scale Room Test for Surface Products" [45]. ASTM E 603 is meant to provide guidance for developing a full-scale fire test, while ISO 9705 is a standard fire test and has more prescribed test procedures. As a result, the methods prescribed in ISO 9705 are used by the fire safety community to determine the fire growth of combustible boundaries.

ISO 9705 [45] is a full room test for assessing composite fire growth for interior applications or enclosed spaces such as a deckhouse on a ship. The enclosure provides an enhanced heat feedback effect, due to accumulating hot smoke, which is not present in an open fire exposure. The ISO 9705 room/corner test consists of an 8-ft. (2.44 m) wide, 12-ft. (3.66 m) deep, and 8-ft. (2.44 m) high room constructed of noncombustible material with a 6.5-ft. (2.0 m) high and 2.6-ft. (0.8 m) wide door. A schematic of the ISO 9705 test arrangement is shown in Fig. 9.15. Instrumentation for measuring the heat release rate of the room, smoke-production rate, and species concentration is installed in the exhaust stack. The room also contains a single heat flux gauge located in the center of the floor. The material being tested is applied to both side walls, the back wall, and the ceiling. A propane burner is located on the floor in either back corner of the room.

An ignition source, which is a propane sand burner, is placed flush against a corner in the room lined with the combustible material. There are three heat release rate curves suggested for the ignition source in the Annexes A and B of ISO 9705. They are:

- 100 kW for 10 min followed by 300 kW for 10 min using a 0.17-m square burner
- 176 kW for 20 min using a 0.30-m square burner
- 40 kW for 5 min followed by 160 kW for 10 min using a 0.30-m square burner.

IMO Resolution MSC.40 (64) requires the 100 kW for 10 min followed by the 300-kW fire for 10 min as an ignition source to evaluate combustible boundary materials for high-speed craft [37]. Figure 9.16 shows a full-scale room/corner test at 100-kW fire exposure.

The 100-kW for 10 min, 300-kW for 10 min fire curve can be thought of as a representation of a growing fire inside a compartment. The 100-kW portion of this curve represents the heat released by an initiating source fire such as a medium-size trash can filled with trash. The 300-kW portion of the fire curve represents the fire spreading to an adjacent object(s). Such a heat release rate could be produced by one large sea bag with dimensions of 3.3 ft (1.0 m) by 3.3 ft (1.0 m), or two medium-size sea bags with dimensions of 1.65 ft (0.5 m) by 3.3 ft (1.0 m). The duration of the ISO 9705 test (20 min) is consistent with the time required for a firefighting team to detect and respond to a fire.

IMO MSC.40(64) sets forth acceptance criteria to qualify marine materials as "fire-restricting materials" [37] and the same acceptance criteria have also been adopted by the U.S. Navy for the use

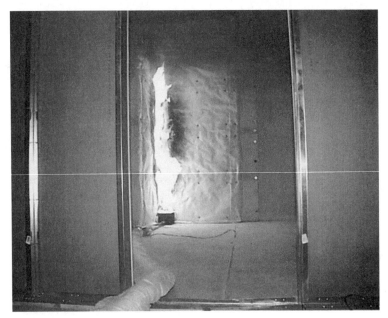

FIGURE 9.16 ISO 9705 fire test during 100-kW fire exposure.

of organic-matrix-based composite materials in surface ships. These fire performance requirements are:

- The time average of heat release rate (HRR), excluding the HRR from the ignition source, does not exceed 100 kW.
- The maximum HRR, excluding the HRR from the ignition source, does not exceed 500 kW averaged over any 30-s period of time during the test.
- Flame spread must not reach any further down the walls of the test room than 0.5 m from the floor excluding the area which is within 1.2 m from the corner where the ignition source is located.
- No flaming drops or debris of the test sample may reach the floor of the test room outside the area that is within 1.2 m from the corner where the ignition source is located.
- The time average of the smoke production rate does not exceed 1.4 m^2/s.
- The maximum value of the smoke production rate does not exceed 8.3 m^2/s averaged over any period of 60 s during the test.

The USCG conducted ISO 9705 room/corner fire tests on seven different composite materials in collaboration with Southwest Research Institute in 1997 [52–54]. The NSWCCD also conducted ISO 9705 fire tests on selected passive fire protection systems in 2001. The test results are summarized below.

In the USCG study in collaboration with Southwest Research, eight glass-fiber-reinforced composite materials and one textile wall covering were tested in full scale in the ISO 9705 room. The same materials were also evaluated in small scale according to the test procedures of the cone calorimeter, the IMO surface flammability test (Part 5 of the IMO Fire Test Procedures of FTP Code), and the IMO smoke and toxicity test (Part 2 of the FTP Code) [52–54]. The materials that were utilized in this study are summarized in Table 9.11.

The results of the ISO 9705 tests are summarized in Table 9.12. Material Nos. 1 and 6 slightly exceeded the ISO 9705 smoke production limits for fire-restricting materials. Material No. 6 is iden-

TABLE 9.11 Identification of Materials Evaluated by USCG [52]

USCG ID No.	Identification
1	FR Phenolic
2	Fire restricting material
3	FR Polytester
4	FR Vinyl ester
5	FR Epoxy
6	Coated FR epoxy
7	Textile wall covering
8	Polyester
9	FR modified acrylic

tical to Material No. 5, but painted with an intumescent coating. Material No. 7 did not exceed the ISO 9705 criteria for heat release and smoke production, but failed because flaming debris fell to the floor during the test. However, flaming persisted for only a few seconds. Furthermore, this phenomenon occurred only once during the test. A photograph of an ISO 9705 test just prior to flashover (Material 5) is included in Fig. 9.17.

For both load-bearing and non-load-bearing composite structures, the U.S. Navy accepts the use of fire insulation as a protective cover over combustible composite structures, such as brominated vinyl esters, which do not meet the acceptance criteria for fire-restricting materials. The use of a protective cover is not allowed by IMO [37]. The U.S. Navy accepts this practice because it is expected that naval shipboard configuration control will assure the presence of the protective cover over the life of the ship.

Naval Surface Warfare Center, Carderock Division, and Omega Point Laboratories conducted the room/corner fire tests on composite corners protected with fire insulation, such as Structo-Gard® (manufactured by Thermal Ceramics), Superwool®, and intumescent mat. The test corner was made from a prefabricated composite corner with 4 × 8-ft. sides and 4 × 4-ft. overhead. The composite was a sandwich with balsa core (3.0 in. thick) and glass-reinforced fire-retarded vinyl ester skins (0.25 in. thick). The composite corner, as well as all edges, was protected with two layers of fire insulation to prevent burning of composites through the seams. Fire insulation was attached with self-drilling screws with studs and mushroom caps using a modified 18-in. (12-in. spacing is Navy standard) spacing scheme. All seams were sealed with glass fabric tape bonded with adhesive.

The summary of results from these tests is given in Table 9.13. All three systems passed the criteria set for ISO 9705 in HSC Code. Figures 9.18 to 9.20 show the GRP/balsa core composite corner with StructoGard before, during, and after the ISO 9705 fire test.

Gas temperatures from the thermocouple tree in the room corner opposite the fire source were also measured. The peak measured gas temperature

FIGURE 9.17 Material No. 5 just prior to flashover.

TABLE 9.12 USCG Test Data from ISO 9705 Room Test [53]

ID	PHR from cone at 50 kW/m² heat flux	Critical heat flux for ignition, kW/m²	Measured ignition temp from LIFT, °C	Thermal inertia from LIFT κρc kW²·sec/m⁴·K	Time to flashover from full scale ISO 9705 test (secs)	PHRR from ISO 9705, kW	AHRR from ISO 9705, kW	Peak SPR from ISO 9705, m²/s	Avg. SPR from ISO 9705, m²/s
1	33.9	42	—	—	No flashover	159	62	5.41	1.50
2	NI	69	—	—	No flashover	129	31	0.47	0.15
3	116.25	15	375	1.65	372	677	191	21.7	10.00
4	135.25	18	370	1.89	318	463	190	32.1	9.08
5	73.0	42	453	1.73	990	421	115	26.4	6.39
6	42.4	40	643	8.00	No flashover	134	28	3.46	1.45
7	68.6	19	647	0.27	No flashover	131	17	0.16	0.10
8	361.50	23	337	0.74	108	568	170	4.10	2.28
9	129.33	22	385	1.72	666	542	109	3.81	0.42

TABLE 9.13 NSWCCD Test Data from ISO 9705 Test

ISO 9705 Criteria for GRP usage in interior spaces	Heat release rate (excluding source)		Smoke production rate, SPR		Total CO mass (kg)
	Peak (kW)	Test average (kW)	Peak (m²/sec)	Test average (m²/sec)	
Criteria	500	100	8.300	1.400	1.90
GRP/StructoGard®	57	25	0.500	0.348	0.34
GRP/Super Wool®	42	12	0.130	0.058	NA
GRP/intumescent mat	66	16	0.153	0.077	NA

was 724°F, which occurred at the end of the 20-min fire exposure. Graphical presentation of the temperatures from the three thermocouples behind the StructoGard is given in Fig. 9.21. The peak measured temperature was 303°F from the thermocouple in the middle, 24 in. from the floor. There was no apparent damage or indications of composite layer delamination.

9.7.4 Fire Resistance

The fire resistance of bulkheads, decks, and overheads is their ability to prevent ignition of items on the nonfire side of the bulkhead (backside). These physical boundaries retard the passage of flame and smoke and stop the spread of fire to adjoining spaces. The 30- to 60-min protection period is important because it enables personnel to escape from the fire area and provides time for a concerted firefighting effort by the crew.

The word "fire resistance" is sometimes misused in the context of GRP (composite) applications. Historically, this word is used for expressing the ability of building structures to limit the fire spread

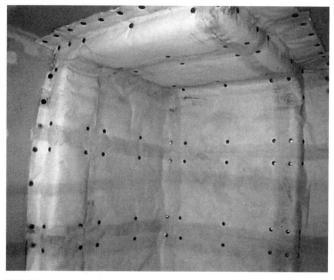

FIGURE 9.18 Composite corner prior to testing.

FIGURE 9.19 300-kW portion of fire test.

from room of fire origin to adjoining spaces, such as bulkheads and overheads. However, this word has been sometimes used in the context of GRP applications by suggesting that composite is fire resistant to imply that it has limited flame spread, fire growth, and smoke production. The more appropriate terminology for such material characterization is "fire-restricting," which is the IMO preferred characterization to imply that materials have low surface flammability, heat release rates, and smoke production. The term "fire resistance" is used in this section in the context of fire spread to adjoining spaces as measured by the temperatures on the backside or the unexposed side.

FIGURE 9.20 Composite corner with StructoGard after testing.

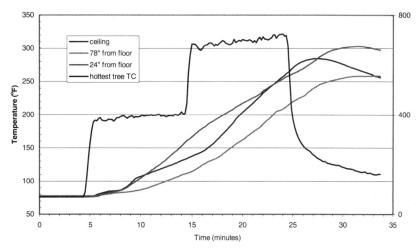

FIGURE 9.21 Thermocouples behind StructoGard.

9.7.4.1 Fire Resistance Test Method

Fire resistance rating of walls is perhaps one of the oldest fire test standards that is still in use. Underwriters Laboratory utilizes this test method to provide fire ratings for all assemblies used in building construction. IMO A.754(18), "Recommendation on Fire Resistance Tests for "A," "B," and "F" Class Divisions" [55] is applicable to merchant vessels with products such as bulkheads, decks, overheads, doors, ceilings, linings, windows, fire dampers, pipe penetrations, and cable transits.

Historically, ASTM E 119, "Standard Test Methods for Fire Tests of Building Construction and Materials," [56] has been used to characterize the fire resistance of building construction. A similar fire curve is also used by IMO Resolution A.754 (18) [55]. The U.S. Navy previously used fire resistance in accordance with ASTM E 119 test method for all fire zone bulkheads using the fire curve provided in ASTM E 119 (standard time-temperature curve). After a large weapon-effects-induced fire aboard the USS *Stark* in 1992, the Naval Research Laboratory performed postflashover fire tests in ex-USS *Shadwell* [57]. The results indicated that UL-1709 fire curve [58] more closely approximated the thermal conditions in the ship compartment during the postflashover fire. As such, the U.S. Navy is now using the fire exposure of UL 1709 fire curve as a benchmark for evaluating fire resistance.

ASTM E 119 fire curve, shown in Fig. 9.22, specifies a furnace temperature of 1000°F in 5 min, 1300°F in 10 min, 1550°F in 30 min, and 1700°F after 60 min. This fire curve is also known as the slow-rise curve typical of Class A fire from common combustibles such as wood. As shown in Fig. 9.22, the UL 1709 fire curve rises to 2000°F within the first 5 min of the test and is between 1800 and 2200°F at all times after the first 5 min of the test. The heat flux to the sample during the 2000°F part of the UL 1709 exposure must be 204 ± 16 kW/m². This fire curve is also known as the "rapid rise" curve typical of Class B fire from flammable liquids such as hydrocarbon pool fire. One of the distinguishing features of a UL 1709 (postflashover fire, hydrocarbon pool fire) is the rapid development of high temperatures and heat fluxes that can subject exposed structural members to thermal shock much more rapidly than ASTM E 119. The exposure scenario that is simulated by the UL 1709 fire curve is the condition of total, continuous engulfment of a member or assembly in the luminous flame area of a large, free-burning-fluid-hydrocarbon pool fire.

IMO Resolution A.754 (18) [55] uses a standard time-temperature fire curve which is similar to the ASTM E 119 [56] for characterizing the A, B, and F class divisions in merchant vessels in compliance with the provisions of the International Convention for the Safety of Life at Sea (SOLAS), 1974. This test basically involves subjecting the structural component to a heated furnace environment for the desired duration. The furnace is used to simulate the conditions of a room fire. If the

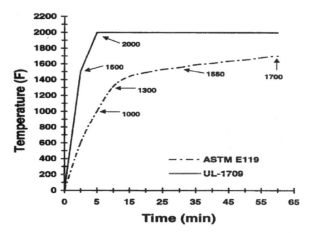

FIGURE 9.22 Temperature versus time profiles for ASTM E 119 and UL 1709 fire curves.

endpoint criteria are not reached prior to the end of the test period, the assembly is rated as acceptable for the test period, e.g., 30 or 60 min. The sample material being tested is mounted to the furnace and acts as one side of the furnace. The sample may be mounted in a vertical (bulkhead) or horizontal (deck) orientation. The sample can be tested structurally loaded or nonloaded. Fire resistance is measured by the heat transmitted through the sample (measured using thermocouples mounted on the unexposed side of the sample) and the transmission of hot gases through the assembly (sufficient to ignite cotton pad by hot gases leaking through the assembly).

In general, IMO classifies "A" divisions as those formed by bulkheads and decks that comply with the following:

- Constructed of steel or other equivalent material; other equivalent material means any noncombustible material, which, by itself or due to insulation provided, has structural and integrity properties equivalent to steel at the end of the applicable exposure to the standard fire test (e.g., aluminum alloy with appropriate insulation)
- Prevent passage of smoke and flame to the end of the 1-h standard fire test
- Shall be insulated with noncombustible materials such that the average temperature of the unexposed side will not rise more than 139°C above the original temperature, within the time listed below:
  - Class A-60   60 min
  - Class A-30   30 min
  - Class A-0   0 min

"B" Class divisions are those divisions formed by bulkheads, decks, ceilings, or lining that comply with the following:

- Prevent passage of flame to the end of the 30 min of the standard fire test
- Shall be constructed of approved noncombustible materials
- Shall have an insulation value such that the average temperature of the unexposed side will not rise more than 139°C above the original temperature, within the time listed below:
  - Class B-30   30 min
  - Class B-15   15 min
  - Class B-0   0 min

"F" Class divisions are applicable only to fishing vessels, and are not discussed further.

The acceptance criteria for fire resistance are the unexposed side temperature rise in combination with fire integrity. In general, this includes the following:

- Peak temperature rise on the unexposed surface not more than 325°F (180°C)
- Average temperature rise on the unexposed surface not more than 250°F (139°C)
- No passage of flames, smoke, or hot gases on the unexposed face
- Cotton-wool pad: there should be no ignition, i.e., flaming or glowing
- Gap gauges: not possible to enter the gap gauges into any openings in the specimen

9.7.4.2 Composite Fire Performance in ASTM E 119 Fire Tests

In 1991, NSWCCD conducted four large-scale fire resistance tests at Southwest Research Institute using the ASTM E 119 fire curve. The purpose of this series of tests was to determine the response of the full-scale panel system to fire, to determine the effects of "fire hardening" a conventional panel, and to determine several parameters for fire-exposed composite structures. The results of these tests are summarized in Table 9.14 [59]. The standard composite consisted of solid glass-reinforced fire-retarded vinyl ester composite with foam core hat stiffeners. The fire-hardened composite panels consisted of solid glass-reinforced fire-retarded vinyl ester composite with hat stiffeners (with a middle skin) and 1.25-in.-thick fire insulation (StructoGard). The unprotected bulkhead and deck panels (the baseline for performance) did not provide sufficient resistance to the standard fire exposure

TABLE 9.14 Composite Panel Fire Test Performance Under ASTM E 119 [59]

Test number		Test 1	Test 2	Test 3	Test 4
Panel type		Gl/VE standard composite	Gl/VE standard composite	Fire hardened (insulated) composite	Fire hardened (insulated) composite
Configuration		Vertical (bulkhead)	Horizontal (deck)	Horizontal (deck)	Vertical (bulkhead)
Static load		Line 28,000 lbs*	Distributed 8,673 lbs	Distributed 8,673 lbs	Line 28,000 lbs
Test duration (min)		45	60	120	60
Structural response					
Maximum deflection (in.)		9.2	5.6	2.8	0.7
Deflection time (min.)		45	65	126	60
Time to exceed 2″ def. (min:sec)		7:15	5:45	97	>60
Thermal response (min:sec)					
Hot side temperature reaches 250°F		1:15	1:15	20:15	20:00
Cold side temperature reaches 250°F	1 TC	7:45	6:45	40:30	46:00
	AV.	9:45	8:30	54:15	55:00
Cold side temperature reaches 450°F	1 TC	13:15	12:45	78:45	>60
	AV.	16:30	16:45	103:15	>60

* The load was reduced to 1000 pounds/lin ft at 7 minutes to prevent total collapse of the panel.

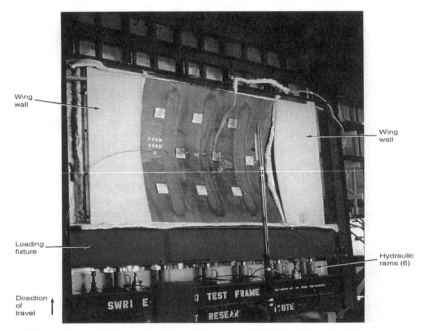

FIGURE 9.23 Standard composite panel (bulkhead) during ASTM E 119 fire test under load.

conditions (ASTM E 119) for the required time period. They exhibited excessive deflections and cold face temperature rise. Figure 9.23 shows the bulkhead test as the unprotected panel buckled. However, no flames or hot gases passed through these conventional panels during the fire exposures.

The fire-hardened panels readily met all of the goals established for these structures. The fire-hardening techniques evaluated in this study were responsible for increasing the survivability of the composite deck and bulkhead systems by a factor of 10 over the survival times of the nonprotected systems.

9.7.4.3 Composite Fire Performance in Module Fire Tests

A cooperative effort between the United Kingdom, Canada, and the United States led to a series of composite module full-scale fire tests. These modules were approximately $8 \times 8 \times 16$ ft. The purpose of these fire tests was to understand how large composite structures would perform aboard surface combatant ships during a significant fire event, determine the response of the composite module to fires of varying sizes, the effects of fire on structural strength, correlation of large scale tests with small-scale tests, and extinguishability of large surface areas of combustible materials using shipboard firefighting equipment and tactics. The effectiveness of insulation as a protective measure for composite substrates was also demonstrated. The United Kingdom provided two composite modules for fire testing in 1989. These modules consisted of glass/polyester and glass/polyester with phenolic cladding. The United States manufactured a third module using glass-reinforced vinyl ester composite and also evaluated fire protection concepts with a barrier insulation and with fire-hardened structural components.

The Carderock Division of the Naval Surface Warfare Center and the Naval Research Laboratory performed a series of fire tests. The fire sizes ranged from very small, such as would be encountered in wastebaskets, to very large postflashover fires. Figure 9.24 shows the U.S.-manufactured composite module, and Fig. 9.25 shows the locations of the fires. Figure 9.26 shows the exterior JP-5 pool fire test. Table 9.15 presents the summary of the fire test results.

The composite system tested showed resistance to heat and flame penetration, substantially increasing the time necessary for fire to spread between compartments. Flammability of the base

FIGURE 9.24 Photograph of the U.S. module on its steel deck plate.

GRP material was greatly improved by covering it with insulation material. Structural stability of the module was enhanced by incorporating a secondary composite structure within the hat-stiffened beams and by adding insulation material. The fire performance of the glass vinyl ester composite resembled that of the polyester GRP used in the 1989 testing. With sufficient imposed heat flux, flame will spread on all three materials: vinyl/ester, polyester, and phenolic. Unless there were unusual or severe conditions (e.g., hot gas layer), fires did not appear to propagate on the materials tested without an additional external source of heat.

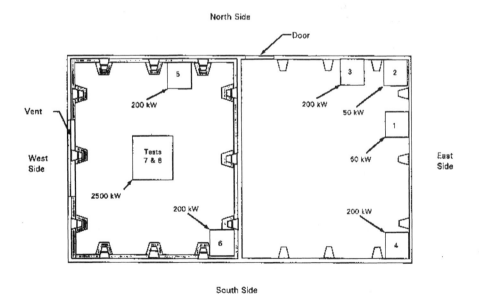

FIGURE 9.25 U.S. composite module, sizes and locations of the fire tests.

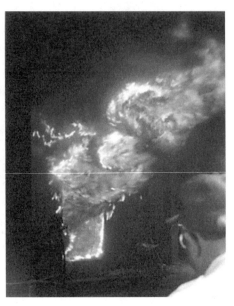

FIGURE 9.26 Postflashover fire in the U.S. module, west end, near the end of the test.

9.7.4.4 Composite Fire Performance Using UL 1709 Fire Curve (Solid Composites)

The U.S. Navy has previously conducted ASTM E 119 fire tests on solid composite panels in 1991 [59]. However, the exposure specified in ASTM E 119 does not adequately characterize postflashover fires. Subsequent to the flashover tests conducted by NRL and NAVSEA in ex-USS *Shadwell* in 1999 [57], UL 1709 is now the U.S. Navy benchmark for measuring fire resistance of bulkheads and decks to protect against rapidly developing fires. One of the most distinguishing features of a postflashover fire is the rapid development of high temperatures and heat fluxes that can subject exposed structural members to a thermal shock much greater than ASTM E 119. The ASTM E 119 and UL 1709 fire curves are shown in Fig. 9.22.

The U.S. Navy conducted fire resistance tests on both solid and sandwich composite panels using UL 1709 fire curve [60]. These tests were conducted at South West Research Institute (SwRI) in September 1998. SwRI's large horizontal furnace was first calibrated prior to conducting fire tests. This calibration was then used to determine the range of temperatures required to provide the heat flux specified by the UL 1709 standard. The standard requires that the average temperature within the exposure furnace shall be 2000 ± 200°F and provide an incident heat flux to the sample of 204 ± 16 kW/m^2 within 5 min and hold steady for the duration of the test. The calibration indicates that this flux corresponds to an average furnace temperature of 1967°F. Similarly, the upper limit of 220 kW/m^2 is associated with a furnace temperature of 2016°F while the lower limit of 188 kW/m^2 corresponds to a temperature of 1918°F.

Four glass/vinyl ester solid composite panel test assemblies were fire tested under the conditions of UL 1709 for a period of 30 min. These test assemblies were (1) 4 ft × 4 ft × 0.5 in. thick solid composite panel fire insulated with 1.25-in.-thick StructoGard; (2) 4 ft × 4 ft × 0.5 in. thick solid composite panel with six cables running through a steel multicable transit and fire insulated with 1.25-in.-thick StructoGard; (3) 4 ft × 4 ft × 0.5 in. thick solid composite panel fire protected with 0.3-in.-thick adhesively bonded intumescent mat; and (4) 8 ft × 8 ft solid composite panel with fire-hardened hat stiffeners (balsa wood core with GRP layer) and fire insulated with 1.25-in.-thick StructoGard and subjected to structural loading.

At the end of 30-min exposure under UL 1709 fire curve, the test results show that the maximum average unexposed surface temperature and maximum single thermocouple temperature exhibited by (1) 4 ft × 4 ft × 0.5 in. thick composite bulkhead panel, fire insulated with 1.25-in. StructoGard,

TABLE 9.15 Summary of Composite Module Fire Tests

Fire type	Origin → Base material → ↓ Fire size/duration ↓	UK 1 Glass/ Polyester	UK 2 Phenolic over Glass/Polyester	US Glass/ Vinyl Ester
Minor fires	50 kW wall/5 min.	Some charring and blistering, fire self-extinguished on flame removal.	Some charring, fire self-extinguished on flame removal.	Some charring, fire self-extinguished on flame removal.
	50 kW corner/5 min.	Some charring and blistering, fire self-extinguished on flame removal.	Some charring, fire self-extinguished on flame removal.	Some charring, fire self-extinguished on flame removal.
Small fires	200 kW center/20 min.	Blistering on roof and sides, no ignition of composite.	Flashover at approximately 16 minutes. The source was closer to the module walls than in previous test.	Test was not performed.
	200 kW wall/5 min.	Composite became involved in fire. Flames reached roof and spread across ceiling. Extinguished with water spray.	Composite became involved in the fire, but only in immediate area of source.	Composite became involved in fire. Flames reached roof. Extinguished with water.
	200 kW corner/5 min.	Composite became involved in fire. Fire spread throughout corner, and flashed-over at 5 minutes.	Composite became involved in fire. Fire spread in corner, but did not flash-over.	Composite became involved in fire. Fire spread throughout corner, and flashed-over at 5 minutes.
	200 kW (insulated) wall/30 min.	Test was not performed.	Test was not performed.	Some smoking from propane and insulation binder. No composite involvement in fire.
	200 kW (insulated) corner/30 min.	Test was not performed.	Test was not performed.	Some smoking from propane and insulation binder. No composite involvement in fire.
Large interior fires (30 min.)	2500 kW post flashover	Extremely intense fire. Composite involved. After fuel shut-down, hot steel keeps nearby composite burning. Extinguished with water.	Extremely intense fire. Composite involved. Major debonding between Phenolic and Polyester composites. After fuel shut-down, hot steel keeps nearby composite burning. Extinguished with water.	Extremely intense fire. Composite involved. Some flaming transmitted to outside by steel curtain plate. After fuel shut-down, and 5 min. wait, extinguished with water.
	2500 kW (insulated) post flashover	Test was not performed.	Test was not performed.	While there were malfunctions with the fuel system, this test showed that actual composite participation in the fire was significantly delayed.
Exterior fires	6900 kW exterior/30 min.	Composite became involved as fire touched module. Appeared to die out as wind moved flames away. Extinguished with AFFF.	Composite became involved as fire touched module. Appeared to die out as wind moved flames away. Extinguished with AFFF.	Test was not performed.

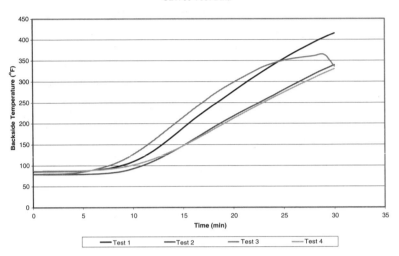

FIGURE 9.27 Fire resistance data (unexposed side average temperatures) for four solid composite panels using UL 1709 fire curve.

was 415 and 457°F, respectively; (2) 4 ft × 4 ft × 0.5 in. thick solid composite deck panel with a multicable transit, fire-insulated with 1.25-in. StructoGard, was 340 and 360°F, respectively; (3) 4 ft × 4 ft × 0.5 in. thick solid composite bulkhead panel, fire-protected with 0.3-in.-thick intumescent mat, was 365°F and 377°F, respectively; (4) 8 × 8-ft solid composite deck panel with fire-hardened hat stiffeners (balsa wood core with GRP layer), fire-insulated with 1.25-in. StructoGard, and subjected to a structural load of 8646 lb, was 331 and 403°F, respectively. The maximum deflection in the center of the deck was 1.4 in., which is less than $L/50$.

The backside temperature for each test is shown in Fig. 9.27. Figure 9.28 shows the temperature

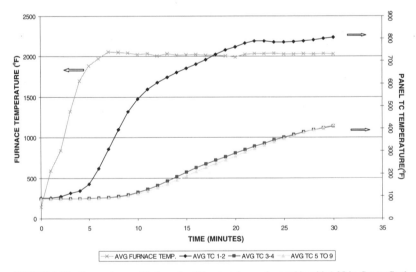

FIGURE 9.28 Temperature profile through solid composite panel assembly with 1.25-in. StructoGard fire insulation. Average of thermocouples 1–2, 3–4, and 5–9 represent temperatures under the insulation on the hot side, midsection of the solid composite panel, and the backside of the composite panel, respectively.

FIGURE 9.29 Intumescent mat bulkhead test (Test 3) at the end of 30-min fire test.

profile through the thickness of the solid composite assembly with fire insulation (Test 1). This thermal profile shows that the temperature under the insulation, at the interface with composite surface, is approximately 800°F. The ignition temperature and minimum temperature for flame spread for solid composite (glass/FR VE, 12.5 mm thick, 1920 kg/m^3) is 675 K (755°F) and 625 K (665°F), respectively [61]. Figure 9.29 shows a photo of solid composite panel with intumescent mat (Test 3) at the end of the 30-min fire test. The seams of the protective cover (intumescent mat) had opened up during the fire test, and the composite panel under the protective cover had ignited. Figures 9.30, 9.31, and 9.32 show the deck panel before the attachment of the fire insulation, in the furnace with the fire insulation before the fire test, and at the end of the 30-min fire test, respectively. In all cases, the hot side of the composite panels under the insulation was blackened and charred. There was no significant damage to the unexposed face other than discoloration and blistering within the composite. In all cases, there was no holing, and no passage of flames or hot gases.

9.7.4.5 Composite Fire Performance Using UL 1709 Fire Curve (Sandwich Composites)

In September 1999, NSWCCD conducted UL 1709 fire exposure evaluations of glass/vinyl ester/balsa wood core sandwich composite panels [62]. A typical U.S. Navy sandwich composite panel consisted of 0.25-in.-thick glass/vinyl ester skins with 3.0-in.-thick balsa wood core. These fire

FIGURE 9.30 Solid composite deck panel (Test 4) before the fire insulation.

FIGURE 9.31 Solid composite deck panel with fire insulation (Test 4) in the furnace before the fire test.

tests were conducted at SwRI in September 1999. A total of five glass/vinyl ester composite sandwich panel test assemblies were fire tested under the conditions of UL 1709 for a period of 30 min. These test assemblies were (1) a bare 4 ft × 4 ft × 3.5 in. thick composite sandwich panel (1257); (2) a 4 ft × 4 ft × 3.5 in. thick composite sandwich panel with six cables running through a composite multicable transit and fire insulated with 1.25-in.-thick StructoGard (1292); (3) a 4 ft × 4 ft × 3.5 in. thick composite sandwich panel fire protected with 0.3-in.-thick I-10A (3M Company) intumescent mat scrimped with phenolic resin (1274); (4) a 4 ft × 4 ft × 3.5 in. thick composite

FIGURE 9.32 Solid composite deck panel after the UL-1709 fire resistance test (Test 4).

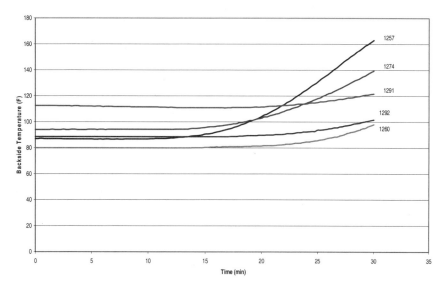

FIGURE 9.33 Backside temperature profiles from sandwich composite panels during UL 1709 fire exposure tests.

sandwich panel fire-insulated with 1.25-in.-thick adhesively bonded StructoGard (1291); and (5) a 4 ft × 4 ft × 3.5 in. thick composite sandwich panel fire protected with 0.3-in.-thick adhesively bonded intumescent mat (1260).

The test results show that all five of the sandwich composite deck panels have average backside surface temperatures and maximum single-point temperature measurements well below the NAVSEA criteria for a 30-min fire exposure. The average backside temperature profiles for all five tests are given in Fig. 9.33. The 4 ft × 4 ft × 3.5 in. thick bare composite panel (Test 1, 1257) had an average unexposed surface temperature of 163°F, and the maximum single thermocouple temperature of 186°F. The 4 ft × 4 ft × 3.5 in. thick composite sandwich panel with six cables running through a composite multicable transit and fire insulated with 1.25-in.-thick StructoGard (Test 2, 1292) had an average unexposed surface temperature of 102°F, with a maximum single thermocouple temperature of 108°F. The 4 ft × 4 ft × 3.5 in. thick solid composite panel fire protected with 0.3-in.-thick intumescent mat scrimped with phenolic resin (Test 3, 1274) had an average unexposed surface temperature of 139°F, with a maximum single thermocouple temperature of 163°F. The 4 ft × 4 ft × 3.5 in. thick composite sandwich panel fire insulated with 1.25-in.-thick adhesively bonded StructoGard (Test 4, 1291) had an average unexposed surface temperature of 122°F, with a maximum single thermocouple temperature of 123°F. The 4 ft × 4 ft × 3.5 in. thick composite sandwich panel fire protected with 0.3-in.-thick adhesively bonded intumescent mat (Test 5, 1260) had an average unexposed surface temperature of 98°F with a maximum single thermocouple temperature of 107°F.

The panels with StructoGard insulation showed signs of flaming and combustion under the insulation. This was due to opening at the seams in the insulation. The surface under the insulation was charred. The bare composite panel and the panels with intumescent mat insulation had severe damage to the exposed side of the composite. The composite skin on the exposed side separated from the balsa wood core, allowing the balsa to be directly exposed to the fire. In all cases, there was no deformation or discoloration to the unexposed composite skin of the sandwich panels.

Figures 9.34 and 9.35 show the sandwich composite panel (Test 4) before and after the UL 1709 fire exposure test. Figure 9.36 shows the thermal gradient through the thickness for a sandwich composite panel (Test 4) protected with 1.25-in.-thick StructoGard fire insulation. In this figure, the y axis on the left shows the furnace temperature, and the y axis on the right shows the temperature profile through the thickness of the composite. This figure shows that the maximum temperature under

FIGURE 9.34 Sandwich composite panel with StructoGard fire insulation before the UL 1709 fire exposure test.

FIGURE 9.35 Sandwich composite panel with StructoGard fire insulation after the UL 1709 fire exposure test.

the insulation, at the interface with the composite surface, is approximately 700°F. The ignition temperature and minimum temperature for flame spread for a sandwich composite (87.5 mm thick, 466 kg/m^3) is 657 K (723°F) and 607 K (633°F), respectively [61]. The maximum temperature under the composite skin on the exposed side, at the interface with balsa wood core, is slightly over 400°F. However, the maximum temperature at the midsection of the balsa wood core is under 200°F. This large thermal gradient due to balsa wood core is primarily responsible for low backside temperatures during the UL 1709 fire exposure tests in a sandwich composite panel.

This thermal gradient is further demonstrated in Fig. 9.37 for a sandwich composite (3.5 in. thick with 3.0-in. balsa core) during cone calorimeter tests at 25, 50, 75, and 86 kW/m^2 heat fluxes. Figure 9.37 shows the temperature drop through thickness (3.5 in.) of a sandwich composite at the end of a 30-min test at a given heat flux. At 75 kw/m^2 heat flux exposure, the temperature at the end of a 30-min test drops from a surface temperature of 1358°F to a temperature of 1253, 861, 403, and 255°F at the sandwich composite thickness of 0.5, 2.5, 4.3, and 4.7 cm, respectively.

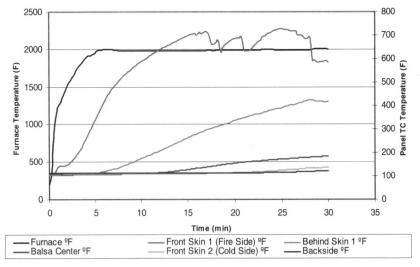

FIGURE 9.36 Thermal gradient in a sandwich composite panel with StructoGard fire insulation (Test 4, 1921) during UL 1709 fire exposure.

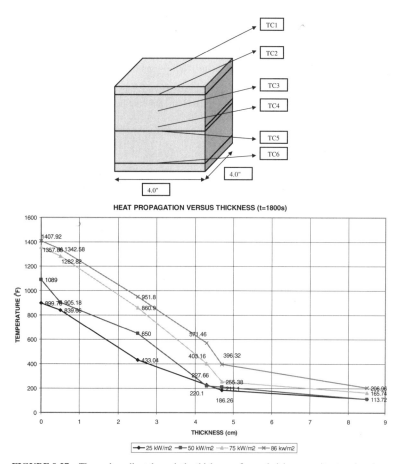

FIGURE 9.37 Thermal gradient through the thickness of a sandwich composite at various heat fluxes (cone calorimeter).

9.7.5 Structural Integrity Under Fire

The mechanical properties of polymer composite materials degrade at elevated temperatures. This degradation is more pronounced for resin-dominated properties, such as compression and shear, than for fiber-dominated properties, such as tension. Because of this phenomenon, the structural integrity of the composite structures should be evaluated at elevated temperatures due to fire.

The mechanical functions of the structure must not be compromised during the fire scenario. A goal of 30 min without collapse or catastrophic failure will allow firefighters adequate time to extinguish the fire before decks or bulkheads become unsafe. Analytical evaluation of this phenomenon is not currently well defined. Material properties are highly nonlinear, as is the structural response. In most cases, full-scale testing is the best option for evaluation. Care must be taken to ensure boundary conditions, loads, and design details are well represented in the test set-up. Full-scale geometry should be tested as much as possible as response may not be easily scalable.

The U.S. Navy fire performance goal for structural integrity under fire is as follows:

- UL-1709 fire curve for a period of 30 min.
- Test procedures in accordance with IMO A.754 (18).

- For composite joints, test procedures in accordance with UL 2079.
- Tests should be performed with the maximum dead and live loads and an additional live load simulating the load from firefighters (50 lb/ft²).
- A load factor should be applied to ensure a margin of safety is retained.
- No collapse or rupture of the structure for 30 min.
- The maximum average temperature on the unexposed side should not exceed the critical temperature of the composite where structural properties degrade rapidly. This applies if the critical temperature is less than the average temperature rise of 250°F.

Unlike flame spread index, smoke obscuration, or heat release rate, there is no small- or bench-scale test that can be applied to assess the residual strength or strength remaining after fire exposure to carry load. Because of the nature of their construction, composite materials do not lend themselves to easy analytical calculation of their behavior when exposed to a high heat flux. Composites exhibit anisotropic heat transfer, they burn, give off smoke, and release heat, char, and delaminate.

9.7.5.1 Residual Flexural Strength (ASTM D-790) After Fire Exposure

Flexural strength was selected to characterize the residual mechanical integrity of selected composite panels after fire exposure. As part of the testing protocol, all specimens (3 × 3 × 0.25 in.) were exposed at radiant heat source of 25 kW/m² for a period of 20 min during ASTM E 662 smoke-generation test in a flaming mode. The specimens were reclaimed and cut into ½ × 3-in. coupons, each specimen yielding five coupons. These coupons were tested in accordance with ASTM D-790 using a universal testing machine. Specimens were tested for flexural strength before and after the fire test. The percent residual strength retained (%RSR) after the fire test for selected thermoset and thermoplastic composite materials is given in Fig. 9.38. Table 9.16 gives the flexural strength values before and after the fire test. The values represent an average for all five coupons.

Graphite/PEEK retained the maximum flexural strength (75 percent) of all composites evaluated at this level of fire exposure followed by graphite/phenolic (53 percent). Glass/epoxy delaminated during the fire exposure due to resin charring resulting in loss of interlaminar strength. One inter-

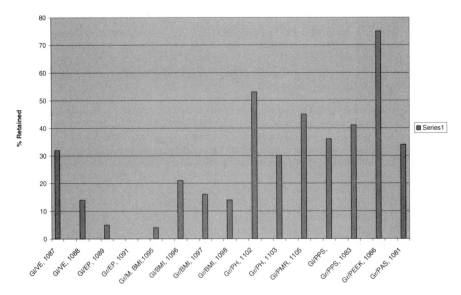

FIGURE 9.38 Percent residual modulus for selected composites after fire test.

TABLE 9.16 Residual Flexural Strength of Selected Composites (ASTM D-790)

Composite	Before, Ksi	After, Ksi	% Retained
Gl/VE (1)	53.9	17.3	32
Gl/VE (2)	58.9	8.4	14.2
Gl/Epoxy (3)	168	9	5.1
Gr/Epoxy (5)	104	0	0
Gr/M. BMI (9)	124.6	5.1	4.2
Gl/BMI (10)	148	31.1	21
Gr/BMI (11)	115.1	18.4	16
Gr/BMI (13)	175.4	24.1	13.7
Gr/Ph (20)	53.9	28.7	53
Gr/Ph (21)	40.9	12.2	29.9
Gl/PMR (23)	113.8	51.2	45.0
Gl/PPS (24)	46.5	16.7	35.9
Gr/PPS (25)	71.8	29.4	40.9
Gr/PEEK (926)	144.0	108.5	75.3
Gr/PAS (27)	116.6	39.1	33.5

esting observation was that all thermoplastic panels evaluated in this study, with the exception of graphite/PEEK, had slightly expanded or foamed up in the middle presumably because of the gases escaping through or from the softened front face during fire exposure. Thermoplastics soften during heating and become solid again after cooling. As such, measurements of flexural strength retained after fire test may not give true or accurate flexural properties during fire.

9.7.5.2 Residual Strength of Composites During Fire

The methodology for the assessment of residual strength of composites during fire involves three basic steps. These steps are described below.

1. Structural performance at elevated temperatures
2. Temperature/Time/Thickness profile during fire exposure
3. Assessment of residual strength via interrelationship of temperature, mechanical property, and time

9.7.5.2.1 Structural Performance of Composites at Elevated Temperatures. Composites retain most of their load-bearing characteristics below a certain "critical" temperature. Above this critical temperature, composites begin to lose their mechanical properties rapidly and, in some cases, catastrophically. The aircraft industry often uses glass transition temperature as the guidance for critical temperature for composite performance at elevated temperatures. As a guidance for elevated temperature performance, the plastics industry routinely uses heat distortion point, which is the temperature at which a standard static bar deflects 0.010 in. under a static load of 66 or 264 lb/in^2.

Table 9.17 gives the elevated temperature properties for glass/vinyl ester composites. These composites were fabricated with 8 plies, 24 oz/yd^2 woven roving/Derakane 510A vinyl ester and had a thickness of 0.2-in. All plies were oriented in the 0-degree direction. The panel was postcured at 150°F for 8 h. For tensile properties, ASTM 695 was used. For compressive properties, ASTM 638 was used. For shear properties, V-notch Beam method, ASTM D5379 was used.

Testing at RT was performed on a Satec Baldwin 60,000-lb screw-driven test frame. Testing at elevated temperature was performed using MTS testing machine fitted with an environmental chamber (MTS systems, Model # P-2-CFCO20). Test samples were placed in a circulating oven at 150°F for 30 min. When soaked, they were transferred to the machine in a fixture preheated to 150°F.

TABLE 9.17 Properties of Gl/VE at Elevated Temperature

Property	Warp (RT)	Fill (RT)	Warp (150°F)	Fill (150°F)
Tensile modulus, Msi	3.5	3.8	2.7	NA
Tensile strength, Ksi	56.1	46.0	46.9	NA
Compressive modulus, Msi	4.3	4.3	3.4	3.7
Compressive strengths, Ksi	48.6	45.9	37.0	33.7
Shear modulus, G12, Msi	0.65		0.31	
Shear strength, Ksi	11.3		4.3	

9.7.5.2.2 Structural Performance at Elevated Temperatures: DMTA. Organic polymer-matrix-based, fiber-reinforced composite materials undergo viscoelastic transitions followed by reversible and irreversible thermal damage when exposed to elevated temperatures due to shipboard fires. A characteristic property of all polymeric materials is the glass transition temperature. This is the temperature at which the material properties change from those of a plastic (glassy state, hard, brittle, high strength) to those of a rubber (leathery, soft, low modulus).

Dynamic mechanical thermal analysis (DMTA), also referred to as *dynamic mechanical analysis* (DMA), is a small-scale technique that is often used in polymer laboratories to determine the flexural modulus of resins as a function of temperature, and by extension, that of fiber-reinforced composite materials. This method has great sensitivity in detecting changes in internal molecular mobility, probing phase structure and morphology, assessing the glass transition temperature (T_g), and determination of isothermal aging effects on load-bearing characteristics. Dynamic mechanical thermal analysis produces quantitative information on the viscoelastic and rheological properties of a material by measuring the mechanical response of a sample as it is deformed under periodic stress.

The property measured by DMTA that is of greatest interest in determining the load-bearing capabilities of composite materials is the bending storage modulus E', which agrees closely with the flexural modulus as measured by three-point bending method in accordance with ASTM D790. The storage modulus E' is the elastic response and corresponds to completely recoverable energy. The loss modulus E'' is the viscous response corresponding to the energy lost (dissipated as heat) through internal motion. Also important is the ratio of loss modulus to storage modulus E''/E' as a material passes through the glass transition point. This ratio E''/E' is also defined as the loss factor (tan δ), and indicates the balance between the elastic phase and the viscous phase in a polymer. It can influence impact properties, and is an essential evaluation factor in determining the effects of post curing and heat aging.

The dynamic mechanical testing for glass-reinforced vinyl ester (brominated) was conducted in two distinct steps [63]. In the first step, glass-reinforced vinyl ester composite panels were isothermally aged and tested in DMTA for storage (E') and loss (E'') moduli for a period of 8 h at 77, 150, 200, 250, 300, 400, 500, and 600°F, respectively. Table 9.18 summarizes the modulus data and also presents the percent residual bending modulus retained at various temperatures during isothermal aging. Data show that E' decreases at subsequently increasing temperatures until 200°F. A significant drop in E' takes place between 200 and 250°F followed by a catastrophic drop between 250 and 300°F.

In the second step, all samples were cooled to room temperature (RT) and retested under ambient conditions. Storage modulus data obtained from this phase of testing represented recovery of structural performance or residual strength of composites after having been subjected to elevated temperatures. Loss factor (tan δ) values obtained from this phase of testing provided valuable insight into the threshold temperature limit of reversible versus irreversible thermal damage for glass/vinyl ester composites. Table 9.19 presents the data obtained for storage modulus of samples at room temperature, which were previously isothermally aged at different temperatures for a period of 8 h. Data show that samples previously isothermally aged up to 150°F for a period of 8 h, and subsequently cooled to room temperature, do not exhibit thermal damage and recover all of its original structural

TABLE 9.18 Storage Modulus (E′) During Isothermal Aging For 8 Hours at Different Temperatures

Isothermal aging temperature, °F	Storage modulus, E′, GPa	Storage modulus, E′, Msi	% Residual modulus retained
77	20.89	3.03	100
150	17.78	2.58	85
200	16.59	2.40	79
250	12.02	1.74	58
300	1.86	0.27	9
400	1.32	0.19	6
500	0.93	0.13	4
600	0.14	0.02	0.7

performance. Beyond 150°F and up to 400°F, the glass/vinyl ester samples begin to exhibit thermal damage. However, load-bearing structures exposed to these temperatures up to 400°F still retain up to 70 percent of original flexural properties.

Beyond 400°F, the glass vinyl ester samples suffer significant thermal damage and begin to lose load-bearing viability as a composite structure. This is shown in Fig. 9.39. This can be further observed in dynamic scans for loss factor (tan δ) obtained from DMTA testing of previously isothermally aged samples and shown in Fig. 9.40. At temperatures of isothermal aging beyond 400°F, vinyl ester resin exhibits chemical breakdown as evidenced by the loss of matrix resin viscoelasticity, and is no longer capable of transferring the load to the fiber.

9.7.5.2.3 Temperature/Time/Thickness Profile during Fire Exposure. Thermal profiles through the thickness of composites were obtained by using cone calorimeter (ASTM E-1354) in horizontal orientation [64–66]. K-type thermocouples were embedded at different thicknesses during composite fabrication. Figure 9.41 shows the thermal profiles for a room-temperature cure glass/vinyl ester with and without intumescent coating (30 mils). Figure 9.42 shows the thermal profiles for a 250°F autoclave cured glass/epoxy (7781/7701) at 25, 50, 75, and 100 kW/m². The thermocouples were embedded at a thickness of 0 (top), 0.17, 0.34, and 0.51 (bottom) in. Thermal profiles for composites occasionally show spikes due to sudden bursts of hot gases.

9.7.5.2.4 Effect of Elevated Temperatures on Mechanical Response of Glass/Vinyl Ester Composite Beams. The NSWCCD 8 × 8-ft "Structural Survivability Test Chamber (SSTC)" was

TABLE 9.19 Room Temperature Storage Modulus (After Cooling) of Previously Isothermally Aged Samples at Various Temperatures for 8 Hours

Isotherm temp., °F	E′ (GPa)	E′ (Msi)	% Recovery
77	22.39	3.25	100
150	21.72	3.15	97
200	19.95	2.89	89
250	17.78	2.58	79
300	17.58	2.55	78
400	15.85	2.29	70
500	11.22	1.63	50
600	6.31	0.91	28

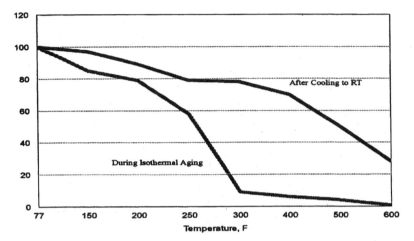

FIGURE 9.39 Effect of elevated-temperature isothermal aging on residual modulus.

used in a scaled-up version of the four-point bending method (ASTM D790) for mechanical testing of composites at elevated temperatures due to shipboard fires. The facility is capable of loading beams up to 6 ft long, and prototype structures up to $4 \times 4 \times 4$ ft. It is capable of operating at temperatures up to 600°F. It has also been adapted to provide a radiant heat flux of 0–100 kW/m^2 on a structure's surface (much like cone calorimeter) combined with static loading. This chamber is illustrated in Fig. 9.43.

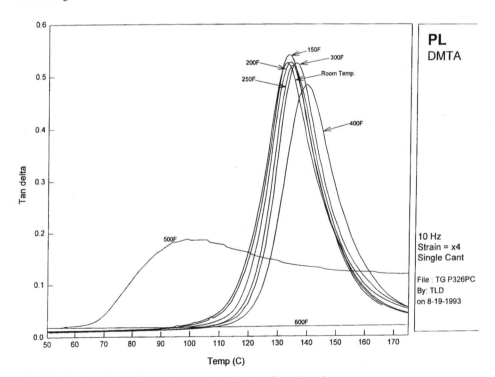

FIGURE 9.40 Tan delta for RT-cooled but previously isothermally aged panels.

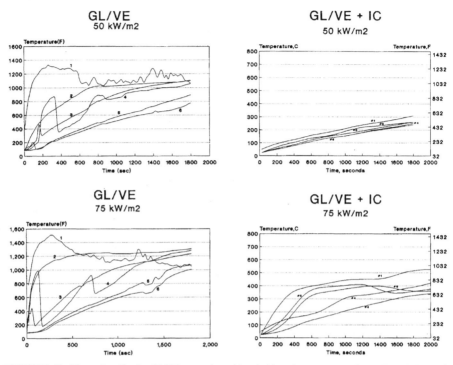

FIGURE 9.41 Thermal profiles for GL/VE composites with and without intumescent coatings at 50 and 75 kW/m^2 heat flux.

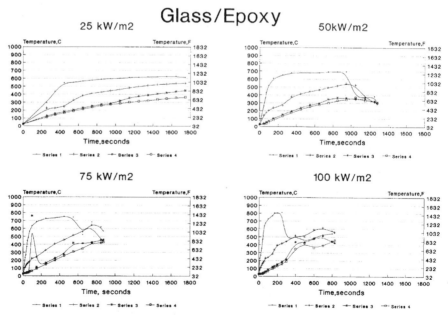

FIGURE 9.42 Thermal profiles of glass/epoxy composites at various heat fluxes.

FIGURE 9.43 Structural survivability test chamber.

Isothermal testing of several solid glass/vinyl ester beams was completed at room temperature and at elevated temperatures of 180, 235, 300, and 350°F with incremental loading up to 815 lb [67]. In this isothermal testing mode, the beams were installed in the loading device and the cable transducer was attached. The witness beam was placed in the chamber next to the beam under test. The chamber was then preheated to the isothermal condition under test for about 60 min, then the weight tray was put on top of the beam, and the weights were placed in the tray in approximately 50-lb increments. The deflections for the 1.25-in.-thick, 6-ft-long, and 6-in.-wide solid composite beams are shown in Fig. 9.44. These data show that the temperature to 50 percent modulus for solid composite beams appears to be in the vicinity of 300°F.

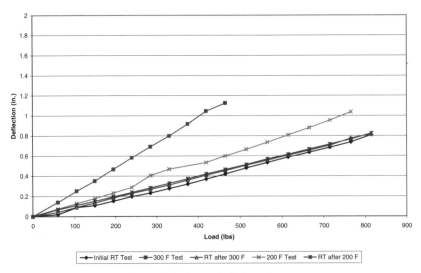

FIGURE 9.44 Solid composite beam deflections under load at various temperatures.

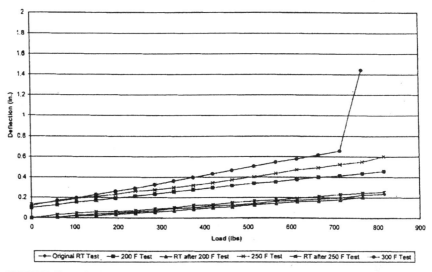

FIGURE 9.45 Deflections in sandwich composite beams at elevated temperatures.

Isothermal testing of sandwich beams (balsa wood and PVC foam core) at temperatures up to 300°F was also completed. The sandwich (balsa core) test beams were 6 ft long, 6 in. wide, and 2.5 in. thick (0.25-in. glass/vinyl ester skins with 2.0-in. thick D100 balsa core). The deflections for the sandwich composite beams are shown in Fig. 9.45. Data show that the 2.5-in.-thick sandwich beam (balsa core) snapped (broke) near 300°F at a loading of 775 lb. The deflection in the beam at break was $L/90$ (0.663 in.). The failure mechanism appeared to be the kink in the top skin, which is shown in Fig. 9.46.

FIGURE 9.46 Sandwich beam after testing (side view of kink).

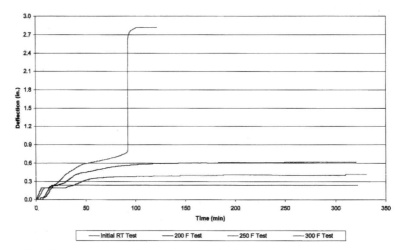

FIGURE 9.47 Nonisothermal tests under constant load at various temperatures.

The isothermal tests for PVC foam core sandwich beams at 200 and 250°F indicate marked increase in deflections when compared to room temperature. At the end of the testing, the PVC foam core beam had taken a permanent set in the center.

Finally, the elastic and creep properties of the 6-ft-long, 6-in.-wide solid (1 in. thick) and sandwich (2.5 in. thick) beams were examined by conducting nonisothermal tests with static loads at temperatures up to 350°F. Figure 9.47 shows the deflections from 6-ft.-long, 6-in.-wide, 2.5-in.-thick sandwich beam with balsa core during nonisothermal tests at RT, 200, 250, and 300°F. The tests were conducted at constant load of 700 lb for 5 h. The beam 4 (300°F) snapped at about 90 min into the test. The failure mechanism appeared to be the kink (wrinkle) in the top skin, which was consistent with the observations from isothermal testing of similar sandwich beams with balsa core. The deflection just prior to snapping was $L/73$ (0.817 in.).

Results from composite solid and sandwich beam tests have been used to develop predictive and modeling techniques by D. Palmer at NSWCCD and G. Petrie at UNO [67]. A theoretical analysis of the test data with *Boeing Non Linear Visco-elastic Composite Analysis* (BONVICA) was performed by Dr. Hugh MacManus at MIT. Results of this study showed that a simple elastic bending analysis accurately predicted deflections and strains for both solid and sandwich beams at temperatures up to 235°F. Above this temperature the deflections were overpredicted.

9.7.6 Passive Fire Protection

9.7.6.1 Passive Fire Protection—Metallic Structures

Historically, the U.S. Navy shipboard structures have been metallic, using steel or aluminum. In many cases, the critical structures in the aircraft industry, such as fuselage, are also made of steel or aluminum. These metallic structures transmit heat through the boundary to the other side, and even permit burn-through. Fire tests have shown that temperatures in excess of 450°F can significantly reduce the structural properties of aluminum or ignite the common combustibles on the other side, commonly referred to as the back side, cold side, or unexposed side.

Shipboard compartment fires tend to proceed at a progressive rate, increasing the temperature in the overhead as more combustibles become involved. Under ideal conditions of fuel type and availability of air, the compartment may suddenly flashover, igniting all of the combustibles in the space and producing a sharp rise in the overhead temperature. The temperature at which flashover occurs is generally considered to be 500 to 600°C and represents the transition to a fully involved fire.

TABLE 9.20 Effect of Insulation on Deck Temperature in Overhead Compartment

Condition	Temperature in	5 min	10 min	20 min
Bare steel	°C	475	700	825
	°F	887	1292	1517
2.5 cm of mineral wool on the fire side of the deck	°C	70	379	671
	°F	158	714	1240

Because of the large air demand to achieve and sustain flashover, fires of this magnitude seldom occur, but when they do, the results can be devastating. As metal bulkheads and decks heat up, there is a rapid rise in temperature in adjacent and overhead compartments.

The U.S. Navy has conducted tests to evaluate the tendency for fire spread from ship compartments after flashover in the U.S. Navy's fire test ship in Mobile, Alabama, the ex-USS *Shadwell* LSD-15 [57, 68]. The fire compartment contained a volume of approximately 81 m^3. The compartment was uninsulated, consisting of surrounding bulkheads and decks of 0.48 cm steel. Thermocouples and radiometers were installed to measure the temperatures and heat fluxes. In order to simulate a worse case flashover condition, a 17.4 L/min diesel spray fire, emitting approximately 10 MW of energy, was ignited in the fire compartment.

Table 9.20 [57, 68] shows deck temperatures in overhead compartment with and without fire insulation (1-in.-thick mineral wool) at three particular times: 5, 10, and 20 min after flashover of fire compartment. The estimated radiant heat flux at a 0.3-m standoff above the deck in the overhead compartment and 0.3 m forward of the bulkhead in the adjacent compartment was calculated as well. The empirical results were compared to analytical values predicted from standard thermodynamic formulae, energy balance calculations, and heat transfer coefficients. The measured temperature values agreed well with the calculated predictions. For engineering hazard analysis, these estimates provide reasonable approximations for determining times to critical events.

By comparing the general thermal effects with the ignition data for common combustible materials, it is possible to predict when fires would start in the surrounding compartments due to ignition of normal combustibles located in those compartments. Figure 9.48 shows the estimated ignition times for four configurations of materials in the overhead compartment. Also shown are the ignition times

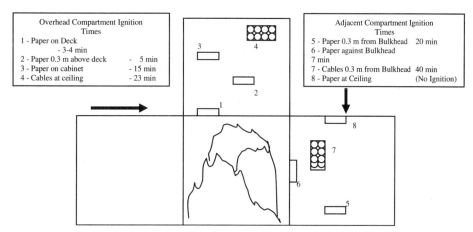

FIGURE 9.48 Estimated ignition times for combustibles at various locations.

for materials in the compartment adjacent to the fire compartment. The report also addresses the effect of thermal conditions on people or delicate electronics equipment.

As part of this same program, limited testing was also conducted to study the ability of insulation to retard fire spread. Results show that insulation (mineral wool) failed relative to what its calculated performance should have been. In fact, after a 20-min exposure, the temperature was over 400°C above what its K factor calculations would indicate. Fire insulation is typically tested against the time/temperature curve of ASTM E 119, which is shown in Fig. 9.22. The fire threat used to simulate flashover in the fire compartment is much more severe. The higher temperature exposure defeated the insulating properties of 2.5-cm-thick mineral wool by melting the fibers and destroying the binder. The actual fire exposure that was simulated for flashover is very close to the time/temperature curves used in UL 1709, which is a rapid rise fire test and is also shown in Fig. 9.22. Accordingly, the U.S. Navy is now using the fire exposure of UL 1709 as a benchmark for measuring insulation performance to protect against flashover fires. MIL-PRF-XX381 [69], which is a military performance specification for high-temperature fire insulation, reflects this change in the Navy's preference.

9.7.6.2 Passive Fire Protection—Composite Structures

The major difference between the metallic and organic-matrix-based composite structures is that enclosed spaces consisting of conventional matrix-based composite structures may be driven to flashover by trash-can-size fires. Suppression of fire growth potential calls for measures that either preclude the heat from an external fire getting to the surface of a composite or which dampen the inherent response of the resin to this heat. One approach is the fire insulation of the composite. This has been suggested as a solution for both the hazard of fire involvement on the part of composite (combustible) structure and for the threat of structural collapse. A sufficiently thick layer (e.g., 1.25 in.) of fire insulation can keep the temperature of the sandwich composite (exposed side) below its ignition temperature (reducing hazard of fire involvement) and also below its glass transition temperature on the backside for periods of 30 min (reducing threat of structural collapse).

For military applications, fire insulation attachment methods for composite structures should be robust enough to withstand the effects of blast and shock in addition to rigorous wear and tear of use in hostile environments. The U.S. Navy fire insulation attachment method for composite structures is shown in Fig. 9.49. Fire insulation is attached in accordance with NAVSEA Drawing 5184182. The insulation is shock qualified as Grade A, meaning it must remain intact and functional. An example of such an installation in interior spaces is shown in Fig. 9.50.

The NAVSEA-proposed MIL-PRF-XX381, "Performance Specification, Insulation, High Temperature Fire Protection, Thermal and Acoustic," covers the requirements for fire-insulation materials [69]. StructoGard® and Solimide® Firesafe insulation are currently approved fire-insulation materials for shipboard use. Solimide Firesafe is a polyimide foam adhered to an amorphous silica insulation material that is mainly used for thermal and acoustical insulation. StructoGard is a high-temperature, soluble, amorphous, man-made mineral fiber fire-insulation blanket that is mechanically attached to composite panels, as shown in Fig. 9.49, by self-tapping screws with metallic studs. Asbestos and ceramic fibers and components containing asbestos and ceramic fibers are prohibited.

In UL 1709 fire exposure testing conducted in 1999 [62] with a single layer of $1^1/_4$-in.-thick StructoGard® covering sandwich composite panel, burning or flaming was observed between the seams of the insulation. In sandwich composite construction, one of the fire safety goals is to prevent the sandwich composite from becoming involved during an incipient fire, such as the fire in the ISO 9705 (room/corner fire test). However, seam burning was also observed when exposed to the full-scale ISO 9705 Room Fire Test [70]. NAVSEA is now requiring that all load- and non-load-bearing composite structures should be protected with two layers of fire insulation with seams of the top layer overlapping the seams of the bottom layer by at least 6 in.

9.7.6.3 Passive Fire Protection—Thickness

The level of passive fire protection is dependent on the thickness of fire insulation and its heat transfer characteristics. As part of the Lightweight Insulation and Passive Fire Protection program, stan-

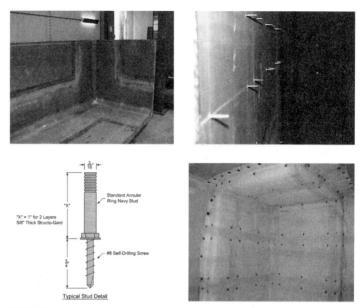

FIGURE 9.49 Passive fire protection (fire insulation) attachment method.

dard mineral wool and a candidate replacement insulation (StructoGard) were tested using the rapid temperature rise fire exposure described by UL Standard 1709. The mineral wool insulation samples tested were 2.5 cm (1 in.) and 5.1 cm (2 in.) thick. The replacement material, StructoGard, is a small-diameter, soluble, amorphous, man-made mineral fiber insulation supplied by Manville Corporation [71]. The StructoGard was tested at 1.8 cm (0.7 in.) and 3.9 cm (1.5 in.) thickness. The materials were tested in a steel deck configuration with the insulation exposed to the fire. Table 9.21 gives the average weight of the materials (including facing) and the time to reach critical backside temperature of 232°C (450°F). While there is a rank ordering correlation between these tests, the UL 1709

FIGURE 9.50 Fire insulation installation on composite interior spaces of a prototype deckhouse.

TABLE 9.21 Comparison of Mineral Wool Versus StructoGard under UL 1709 Fire Exposure [71]

	Weight $Kg/m^2 (lbs/ft^2)$	Time to critical temperature (minutes) 232°C (450°F)
Navy mineral wool		
2.5 cm (1 in.)	2.25 (0.46)	9.7
5.1 cm (2 in.)	3.71 (0.76)	11.3
Manville StructoGard		
1.8 cm (0.7 in.)	2.25 (0.46)	12.0
3.9 cm (1.5 in.)	4.00 (0.82)	29.5

test caused the fairly rapid degradation of the mineral wool material. Until recently, this critical backside temperature of 450°F was the U.S. Navy acceptance criterion for fire resistance. However, with the new proposed MIL-PRF-XX 381 [69], this acceptance criterion for fire resistance has now changed to average backside temperature rise of 250°F. Figures 9.51 to 9.53 provide the estimated fire resistance performance (backside temperatures) of StructoGard FA (0.625 in.), StructoGard FB (1.25 in.), and StructoGard FC (1.875 in.), respectively, in conjunction with 0.25-in. carbon steel under UL 1709 fire exposure [72, 73].

9.7.6.4 Passive Fire Protection—Unrestricted versus Restricted

Passive fire protection, such as fire insulation, may be attached either on the fire- or backside. However, the fire resistance performance of a division is significantly reduced if placed on the backside. Under IMO guidance, in tests for "A" class bulkheads for "general application," it may be possible for approval to be granted on the basis of a single test only, provided that the bulkhead has been tested in the most onerous manner, which is considered to be with the insulation on the unexposed face and the stiffeners on that side.

Effect of fire insulation placement on fire resistance performance is shown in Fig. 9.54 [73]. As

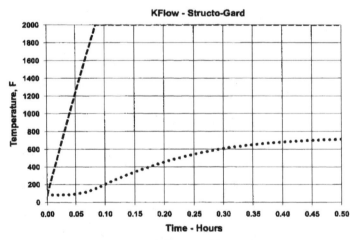

* KFlow is for comparison purposes only, not for design specifications.

FIGURE 9.51 Estimated performance for StructoGard FA (0.625 in.) under UL 1709 fire exposure [72].

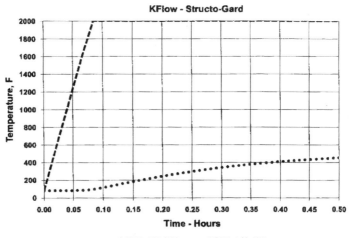

FIGURE 9.52 Estimated performance for StructoGard FB (1.250 in.) under UL 1709 fire exposure [72].

shown in this figure, the backside temperatures are increased at 5-min period during the UL 1709 fire test from 260°F (insulation on fire side, 1 in. thick) to 557°F (insulation on backside, 1 in. thick). The temperature on the backside at 5-min period was almost ambient when the 1-in.-thick insulation was placed on both fire- and backside.

9.7.6.5 Passive Fire Protection—Intumescent Coatings

Intumescent coatings function on their ability to swell or expand (intumesce) to produce char or foam, which insulates and protects the substrate underneath from direct exposure to fire. These materials

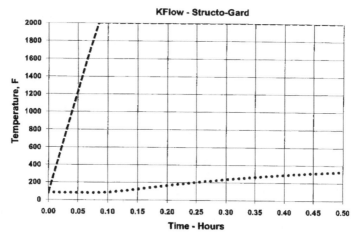

FIGURE 9.53 Estimated performance for StructoGard FC (1.875 in.) under UL 1709 fire exposure [72].

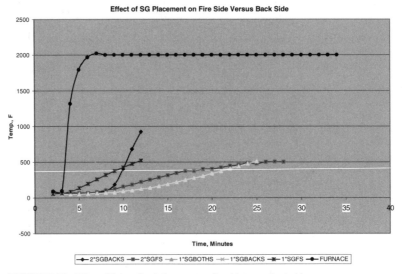

FIGURE 9.54 Effect of StructoGard placement on fire side versus back side.

may be water or solvent based. Water-based intumescent materials are nontoxic and environmentally benign, which is vital to remain consistent with normal military construction processes. Some shipbuilding contractors have also proposed combinations of currently used fire insulation with intumescent coatings to reduce the thickness, and thus the weight of the passive fire protection systems.

In the U.S. Navy, passive fire protection systems, which are designed to replace fire insulation, are now evaluated under proposed MIL-PRF-XX 381, "Performance Specification, Insulation, High Temperature Fire Protection, Thermal and Acoustic" [69]. They should be capable of withstanding a fire test using UL-1709 fire curve, which simulates a postflashover fire, for a 30-min minimum time while holding the backside average temperature rise of the substrate to a maximum of 250°F. The passive fire protection system shall be capable of withstanding shipboard environment, including the conditions of humidity, heat and cold, exposure to light, vibration, and shock as specified.

Previous work by the U.S. Navy included the study of intumescent coatings by NRL and NSWCCD. The objective of the NRL [74] study was to identify a fire-protective coating that would adequately protect the antisweat hull insulation (PVC nitrile rubber). The coating currently used is Ocean 9788 (an intumescent type). The objective of the NSWCCD [75] study was to identify suitable fire-protective coating candidates to protect glass-reinforced vinyl ester GRP structures. Other noteworthy studies of intumescent coatings include "Innovative Fire Resistant Coatings for Use on Composite Products Aboard U.S. Commercial Ships" [76], MARITECH Project DTMA91-95-H-00091 [77], and MANTECH GRP Forward Director Room [78]. When Northrop Grumman was required to protect their innovative composite topside demonstration module [78], an epoxy-based intumescent coating applied at a thickness of over 0.5 in. was used. This intumescent coating provided excellent passive fire protection to the GRP test article during full-scale fire resistance tests using UL 1709 fire exposure. However, when this material was tested under the ISO 9705 room/corner fire test, smoke production exceeded allowable limits. Epoxy-based intumescent coating suppliers do not recommend using their products for interior spaces for this reason.

In FY 01, the U.S. Navy initiated the study of intumescent coatings to determine if commercially available intumescent and *passive fire protection* (PFP) coatings:

- Can be used as a replacement for fire insulation in shipboard interior applications
- Can be used as an adjunct to fire insulation
- Can be used to prevent flashover

- Can withstand the wear and tear of shipboard environment and rigors of deployment without degradation from cleaning and top coating

The objective of this program was to identify PFP coatings for shipboard interior applications capable of meeting USN fire resistance requirements of 30-min rating with backside average temperature rise less than 250°F using the UL 1709 fire curve which simulates a postflashover fire. In addition to meeting fire performance requirements, PFP coatings were required to meet durability (adhesion, impact), environmental (washability, humidity, fluid, and chemical resistance), and health (offgassing, fibers) requirements.

9.7.6.5.1 Intumescent Coatings—Small-Scale Tests. A total of 19 coatings were selected for evaluation. The coating candidates, identified with letter designations, are listed in Table 9.22 [79–81]. For control and comparison purposes, currently used Navy paint systems were also included in this task. These paints included: Coating G (fire-protective coating for antisweat PVC nitrile rubber hull insulation); Coating H (MIL-PRF-46081, coating compound, thermal insulating, intumescent); and Coating Y (topcoat, MIL-DTL-24607 enamel, interior, nonflaming (dry), chlorinated alkyd resin, semigloss). Coating Y is a general-use coating for interior ship bulkhead and overhead applications. All coatings were applied with manufacturer's recommended primer. Where the supplier did not have a recommendation, Navy standard F-150 primer was used. All coatings were evaluated at the manufacturer's recommended thickness. Some coatings were applied at multiple thickness. Three different material substrates were chosen to evaluate the candidate coatings. These included 0.25-in.-thick mild steel, 1.25-in.-thick StructoGard fire insulation, and 3.5-in.-thick GRP sandwich composite (3.0-in. balsa core with 0.25-in. glass/vinyl ester composite skins [79–81].

Table 9.23 shows the small-scale screening test acceptance criteria developed for down selection of intumescent coatings. The screening tests consisted of the areal density (weight per unit area); adhesion, and impact resistance tests to evaluate the effects of wear, tear, and rigors of long-term shipboard deployment; ASTM E 162 Radiant Panel Test [42], ASTM E 662 Smoke Density Test

TABLE 9.22 Selection of PFP Coatings for Small Scale Testing

Coating ID*	Base	Generic coating	Primer system	DFT (mils)	Areal density lbs/ft^2	Comments
A	Water	Vinyl acetate	F150	50	0.36	
E	Water	Acrylic latex	Primer 490	55	0.37	
F	Solvent	Chlorinated rubber	Primer 490	200	1.05	
G50	Solvent	Alkyd	634 Primer	50	0.41	Control
G10	Solvent	Alkyd	634 Primer	10 mils	0.082	Control
H50	Solvent	Polyamide epoxy	F 150	50 mils	0.36	Control
H10	Solvent	Polyamide epoxy	F 150	10 mils	0.071	Control
I30	Water	Latex	F 150	30 mils	—	
I50	Water	Latex	F 150	50 mils	0.37	
J	100% solids	Epoxy	251	195 mils	1.08	
K	100% solids	Epoxy	F 150	50 mils (4)	0.31	
L	Water	Latex	F 150	50 mils (4)	0.29	
N	100% solids	Epoxy	Proprietary	50 mils	0.35	
O	Water	Acrylic	Proprietary	50 mils	0.38	
P	100% solids	2 part silicone	F 150	1000 mils	1.63	
Q	Water	Acrylic	F 150	62.5 mils	0.27	
Q2	Solvent	Vinyl toluene butadiene	F 150	62.5 mils	0.50	
R	Water	Silicone	F 150	1000 mils	1.96	
Y	Solvent	Chlorinated alkyd	F 150	10 mils	—	Control
SG	Fibrous sheet	HT glass fibers	—	1.25 inch	1.0	Control

* Number next to coating ID represents the coating thickness in mils.

TABLE 9.23 Acceptance Criteria for Small Scale Screening Tests

Test	Acceptance criteria			Source
Areal density	Maximum areal density of 1.02 lbs/ft^2			MIL-PRF-XX 381 [69]
Adhesion tests ASTM D 4541 pull-off test	270 psi			Equal to or better than coating G at 50 mils
Knife test MIL-PRF 24596	Difficult to furrow; no flaking or chipping			MIL-PRF 24596
Impact tests ASTM D2794	50 in-lb			Performance equivalent or better than coatings G and H
Flame spread ASTM E 162 [82]	Not greater than 25			MIL-PRF-XX 381 [69]
Smoke generation ASTM E 662 [42]	Not greater than 200			MIL-STD 2031 [25]
Heat release, kW/m^2 ASTM E 1354 [43]	Heat flux	Peak	Avg. 300s	MIL-STD 2031 [25]
	25	50	50	
	50	65	50	
	75	100	100	
Ignitability ASTM E 1354	Heat flux	Time to Ign, secs		MIL-STD 2031 [25]
	25	300		
	50	150		
	75	90		
Simulated UL-1709 fire resistance test	Duration: 30 minutes; unexposed avg. temp. rise: 250°F			MIL-PRF-XX 381 [69]

FIGURE 9.55 Simulated UL 1709 fire resistance test apparatus. (From Refs. 25 and 82)

[43], and ASTM E 1354 Cone Calorimeter Heat Release Test [44] to determine the combustibility of coatings for interior applications; and simulated UL 1709 fire resistance test using 2 × 2-ft panels to determine the backside (unexposed) temperatures. For the purposes of evaluating intumescent coatings as stand-alone replacement for currently used fire insulation, the simulated UL 1709 fire resistance test is the most meaningful small-scale screening test [25, 82]. In this test, a 0.6-m (2-ft) square specimen in the vertical orientation (bulkhead) is placed in front of a 147-kW propane jet burner that produces temperatures of 2000°F at the specimen surface. The fire resistance test generates heat flux of approximately 180 kW/m^2 on the exposed surface of the substrate. This heat flux corresponds to hydrocarbon pool fire and the fire curve in the UL 1709. In accordance with proposed MIL-PRF-XX 381 [69], the pass/fail criterion for a 30-min rating is backside peak temperature rise less than 325°F, and more critically, average backside temperature rise less than 250°F using UL 1709 fire curve that simulates a postflashover fire. The results from this test can be used to compare material performance in containing and preventing the spread of fire, smoke, and fire gases between compartments or spaces. Because of direct flame impingement from the burner on the exposed surface, this test is capable of discriminating between coatings that produce soft or fragile char during heating. The test apparatus is shown in Fig. 9.55.

Tables 9.24, 9.25, and 9.26 present the summary of small-scale screening test results for several intumescent coatings in

TABLE 9.24 Summary of Screening Tests on Steel Substrate

Coating	Mil thickness	Density	Adhesion	Impact	Flame spread	Smoke	Heat release	Simulated UL-1709	Simulated UL-1709; failure time	Comments
A	50	P	P	P	P	P	P	F; 798	02:20	Char layer does not bond to the panel.
E	55	P	P	P	P	P	F	F; 504	05:00	Expands well, some cracks in middle.
F	200	F	P	P	F	P	F	F; 512	13:20	Some expansion, some burning noticed.
G50	50	P	P	P	P	P	F	F; 880	02:40	Good char layer, adheres well.
G10	10	P	P	P	P	P	P	F; 1032	02:00	Little intumescence, paint separated in center.
H50	50	P	P	P	P	P	F	F; 779	01:00	Some expansion in center, bare steel on sides.
H10	10	P	P	P	P	F	F	F; 962	02:20	Some expansion in center, bare steel on sides.
I30	30	P	NA	NA	NA	NA	NA	F; 596	05:40	Expands rapidly with good adhesion, little smoke.
I50	50	P	P	F	P	P	F	F; 462	11:20	Good char formation, some cracks near bottom.
J	195	F	P	P	P	F	F	F; 445	11:20	Little intumescence, good adhesion to the panel.
K	50	P	P	F	P	P	F	F; 1324	01:20	Char layer does not bond to steel, fell off.
L	50	P	P	P	P	P	P	F; 1216	03:20	Very little intumescence, chunks of char fell off.
N	50	P	P	P	P	F	F	F; 787	02:00	Large section of char peeled away, test terminated.
O	50	P	P	P	P	P	P	F; 774	02:40	Coating expands and adheres to the substrate.
P	1000	F	NA	Bounces	F	F	F	F; 1306	12:00	Lots of smoke initially, flaming chunks fell off.
Q	62.5	P	P	P	P	F	F	F; 1292	02:40	Char layer did not adhere, steel exposed in places.
Q2	62.5	P	F	P	P	F	F	F; 1273	02:40	Coating peeled away from panel in sections.
R	1000	F	NA	F	P	P	NA	NA	NA	Large chunks fell off during handling prior to test.
Y	10	NA	NA	NA	P	P	F	NA	NA	Standard Navy interior paint, not an int. ctg.
Steel (B)	0	—	—	—	—	—	—	F; 1342	<1:00	Steel glowing red on the back side, buckling.
SG	1.25 in	1.0	NA	NA	—	—	—	P; 286	NA	Standard StructoGard performed very well.

9.69

TABLE 9.25 Summary of Screening Test Results on StructoGard

CTG ID	DFT mils	Flame spread index	Smoke density	Heat release	Simulated UL-1709	Comments
SG(1)	NA	P	P	F	P	Unpainted control
SG + A	10–15	P	P	F	P	Large pieces of coating sagged and came off.
SG + E	10–15	F	P	F	P	The coating adhered during the test.
SG + G	10–15	P	P	F	P	Some fissures formed but the coating did not fall off.
SG + I	10–15	P	P	F	P	Adhesion to SG substrate was poor.
SG + O	10–15	P	P	F	F	Some coating came off during the tests.
SG + Y	10–15	F	P	F	F	Paint peeled off during the fire resistance test.
5/8″ SG	NA	NA	NA	NA	F	Does not meet the acceptance criteria for fire resistance test.
5/8″ SG + 50 mils A	50	NA	NA	NA	P	The char survived the test intact, but was only held up by the test frame. It had not bonded to the SG.

(1): StructoGard, 1.25 inch thick.

conjunction with steel (0.25-in.), StructoGard (1.25-in.), and GRP (3.5-in.-thick sandwich composite with glass/VE and balsa core) substrates, respectively. Figures 9.56 and 9.57 show the average backside temperatures on steel and StructoGard substrates from simulated UL 1709 fire resistance test for selected intumescent coatings at the thickness shown in Tables 9.24 and 9.25, respectively.

None of the coatings, tested at manufacturer's recommended coating thickness on steel substrate,

TABLE 9.26 Summary of Screening Test Results on GRP Substrate (Sandwich Composite)

Coating ID	DFT mils	Flame spread index	Smoke density	Heat release	Simulated UL-1709	Comments
GRP (1)	NA	P	F	F	P	GRP face flaming and smoking in the simulated UL-1709 test.
GRP + A	50	P	P	F	P	Pieces of coating fell off during simulated UL-1709 test. GRP face burning. Test terminated early.
GRP + E	50	P	P	F	P	Severe flaming on the edge during the simulated UL-1709 test.
GRP + G	50	P	P	F	P	Simulated UL-1709 test terminated after 7 minutes due to sever cracks in the coating and GRP involvement.
GRP + I	50	P	P	F	P	Some burning of GRP at the edges.
GRP + J	195	P	F	F	P	Black smoke and heavy flames during simulated UL-1709 test. Coating adherence to the substrate was excellent.
GRP + O	50	P	F	F	P	Severe cracking of the coating during simulated UL-1709 test. Test terminated early.

(1): 3.5 thick GRP (0.25 inch thick glass/vinyl ester skins with 3.0 inch thick balsa wood core).

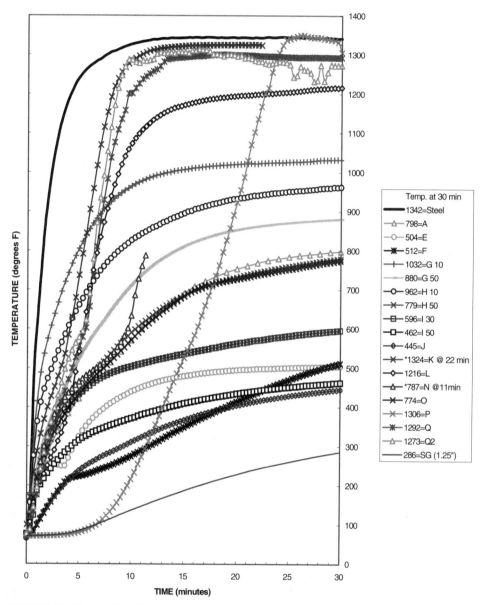

FIGURE 9.56 Simulated UL 1709 fire resistance test results for steel substrate.

met the criterion established in proposed MIL-PRF-XX 381 [69] of a maximum average backside temperature rise of 250°F. The most common failure mode observed during simulated UL 1709 fire resistance tests was the poor substrate adherence of the intumescent coatings and the fragility of the intumesced char, which resulted in the falling off of large chunks of intumescent coatings during these fire tests. An example of this behavior is shown in Fig. 9.58. Several of the coatings did demonstrate the ability to slow heat transfer through the panel considerably. Coatings E, F, I50, and J had average backside temperatures under 515°F after 30 min in conjunction with steel substrate.

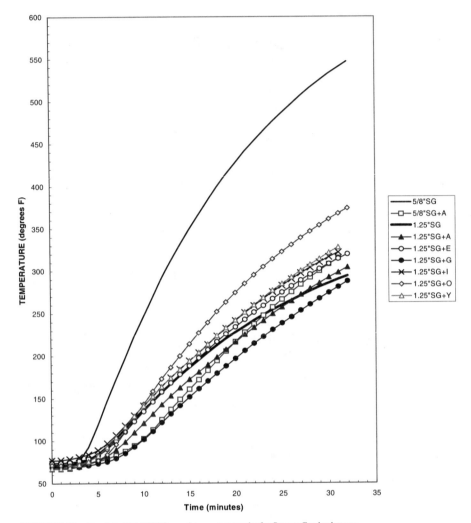

FIGURE 9.57 Simulated UL 1709 fire resistance test results for StructoGard substrate.

9.7.6.5.2 Intumescent Coatings—Room/Corner Fire Tests. The small-scale screening tests resulted in the selection of three intumescent coatings, namely Coatings A, E, and I, for further investigations in intermediate room/corner and full-scale fire tests in U.S. Navy research ship, ex-USS *Shadwell*.

Coating candidates A, E, and I were tested in the room corner configuration along with 1.25-in. StructoGard and bare steel corner for the purpose of comparison. In this study, an open steel corner (two sides and the ceiling) was constructed and used instead of an insulated room corner test. The 4 × 8-ft open steel corner was subjected to the ISO 9705 fire curve (100 kW for the first 10 min, 300 kW for the last 10 min). The open corner consisted of five 4 × 4-ft panels, which were bolted to a steel frame. The panels were coated with recommended primer and selected coating to a thickness of 50 mils prior to installation in the corner test assembly. After all panels were installed, a final skim coat of selected coating was applied to create a smooth transition from one panel to the next. The

FIGURE 9.58 Coating K in the simulated UL 1709 with char pieces that fell off.

thickness of this skim coat averaged approximately 2 to 3 mils. The tests were conducted underneath the exhaust hood. Data were collected on the flame spread, heat release, gas species production, and smoke production rates as well as surface and backside temperatures. The ability of coatings to adhere to the substrate during the fire tests was also evaluated qualitatively. Coating E applied on the steel corner during 300-kW fire exposure is shown in Fig. 9.59. The heat release rates for the intumescent coatings observed in these tests are very low. The net peak heat release rate for all coatings was less than 100 kW. This suggests that intumescent coatings (A, E, or I) may not, by themselves, cause a flashover when applied (50 mils or less) on substrate such as steel. Some char was observed

FIGURE 9.59 Coating E with steel corner during 300-kW fire exposure test.

FIGURE 9.60 View of test bulkhead and insulated, non-test surfaces.

to have fallen off from the ceiling during all corner tests. Some visible flaming of Coating A was observed during the test directly in the corner. However, no visible flame spread was observed during the test for Coating A. The post fire inspection of the corner for Coating A revealed that large fissures were present in the coating. Coating I demonstrated very poor adhesion to the overhead panel during the 300-kW portion of the fire test. In two instances, large sections of the overhead char layer fell from the panel.

9.7.6.5.3 Intumescent Coatings—Full-Scale Fire Tests in Ex-USS Shadwell. Large-scale experimentation was conducted in compartment 3-81-2 aboard the ex-USS *Shadwell* [80]. Selected intumescent and baseline coatings were applied to both sides of a 6.1-m (20-ft)-long, steel bulkhead separating the fire compartment from an instrumented boundary compartment. A view of the test-ready bulkhead, seen on the left from the fire compartment side, is shown in Fig. 9.60.

All fire compartment surfaces, except the deck and test bulkhead itself, were permanently covered with 1.25-in.-thick fire insulation (StructoGard FB). This layer of insulation insured the test bulkhead was exposed to the maximum insult possible. The insulation also served to prevent widespread, heat-related damage to the nontest surfaces of the fire compartment. To insure sufficient air was available for maximum burning efficiency and maximum sustained burning temperatures, the forward and after archways of the fire compartment remained open during execution of all fire testing.

Three test insults were generated to evaluate performance of test coatings against the proposed MIL-PRF-XX381 performance criteria. The radiant and wood crib (incipient) tests were performed to expose all bulkhead treatments to the widest range of potential fire threats possible. A pair of propane-fueled, exposed-element, resistance heaters produced a measured, nonflaming, radiant insult of 5 to 8 kW/m^2. The growing, incipient fire, fueled by two wood cribs constructed of kiln-dried red oak, generated an insult computed to be 1.5 MW. A hydrocarbon-fueled spray fire, *n*-heptane pressurized to 40 lb/in^2 and released through a pair of BETE Model FF052, extra-wide-angle nozzles generated an insult computed at 2 MW. Test duration was 30 min.

Figures 9.61 and 9.62 show the performance of selected intumescent coatings against spray fire (2 MW) and wood crib (1.5 MW) fire tests, respectively. All coating combinations failed the acceptance criteria for average and peak farside temperature rise of 250 and 325°F, respectively. More specifically, any coating identified as having PFP qualities that is applied to any surface aboard ship must be capable of withstanding the rigors of a fire and be capable of mitigating temperature rise

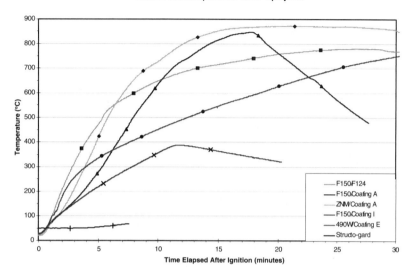

FIGURE 9.61 Performance of test coatings in 2 MW heptane spray fire.

within or across the component being coated. None of the subject intumescent coatings showed the ability to do this during large-scale fire testing. Test coatings E and I formed char layers on the fire-side of the test bulkhead. The char layers formed by exposure to the incipient (wood crib) fire tended to be extremely fragile, showed a propensity to crack during the fire and were highly variable in their thickness. Char layers formed by Coatings E and I developed an armorlike shell when exposed to the higher temperatures generated by the spray fire. The spray-fire-generated char layers were also

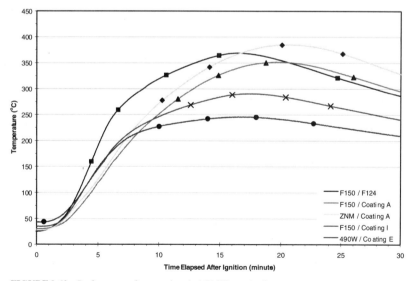

FIGURE 9.62 Performance of test coatings in 1.5 MW wood crib.

highly variable in thickness and showed a tendency to crack. Test coating A delaminated from more than 90 percent of the bulkhead surface on both the fire and nonfire sides when exposed to both the incipient and spray fire insults [80]. Test coating A was observed to support flaming combustion after the product delaminated and fell to the test compartment deck.

9.7.7 Active Fire Protection

There is a trend developing in military applications where more organic composites and plastics are expected to be used in future aircraft, tanks, ships, and submarines. The purpose of this section is to emphasize the fullest practical degree of fire protection, fire detection, and fire extinguishment in military applications. The designers of the next generation of military platforms should bear in mind that the overall fire safety is the combination of material fire safety, passive and active fire protection, and personnel safety. The approach should build on the current practice and integrate passive and active firefighting systems, state-of-the-art heat and smoke alarm systems to reduce detection time of any fire in the zone of origin, more efficient smoke removal and sprinkler systems for the containment and extinction of any fire in the space of origin, and crew monitoring and locator beacons to facilitate fire fighting and escape from fire zones.

REFERENCES

1. Gagorik, J.E., Corrado, J.A., and Kornbau, R.W., "An Overview of Composite Developments for Naval Surface Combatants," *36th International SAMPE Symposium and Exhibition,* Vol. 36, Covina, CA, April 1991.
2. Beach, J.E., and Cavallaro, J.L., "Structures and Materials: An Overview," *Naval Surface Warfare Center Carderock Division Technical Digest,* December 2001.
3. Camponeschi, E.T., Jr., and Wilson, K.M., "The Advanced Enclosed Mast Sensor System: Changing U.S. Navy Ship Topsides for the 21st Century," Naval Surface Warfare Center Carderock Division Technical Digest, December 2001.
4. Frame, B.J., and Hansen, J.G.R., "Fire Resistant Composite Materials for Energy Absorption Applications," *Proceedings of the 47th International SAMPE Symposium and Exhibition,* Vol. 47, May 12–16, 2002.
5. *Marine Composites,* 2nd edition, Eric Greene Associates, Annapolis, MD, ISBN 0-9673692-0-7.
6. Wigotsky, V., *Plastics Engineering,* 44(7), July 1988.
7. Goodman, S., *Handbook of Thermoset Plastics,* Noyes Publications, 1986.
8. Sorathia, U., Dapp, T., and Beck, C., "Fire Performance of Composites," *Materials Engineering,* 109(9), September 1992.
9. *Engineered Materials Handbook, Composites,* Volume 1, ASM International, Metals Park, Ohio.
10. Sorathia, U., Ness, J., and Blum, M., "Fire Safety of Composites in the U.S. Navy," *Composites: Part A, Applied Science and Manufacturing,* 30: 707–713, 1999.
11. Loszewski, R., and Kirshenbaum, S, "Improved Graphite Fiber Prepreg for 350–450°F Service Composites," NASA Report, Washington, DC, June 1986.
12. Kourtides, D.A., Gilwee, W.J., and Parker, J.A., *Polymer Engineering & Science,* 19(1): 24, 1979 and 19(3): 226, 1979.
13. Knop, A., and Pilato, L.A., *Phenolic Resins,* Springer-Verlag, Germany, 1985.
14. Dix, J.S., "PPS: The Versatile Engineering Plastic," *Chemical Engineering Progress,* January 1985.
15. Cole, B.W., and Crowe, W.G., "High Performance Thermoplastic Composites," SPE/APC 1988 Technical Conference, Society of Plastic Engineers, Brookfield, CT, 1988.
16. English, L., ed., "Part I, The Basics; Part II, Reinforcements; Part III, Matrix Resins," *Materials Engineering,* 4(9), September 1987.
17. Shumay, W.C., Jr., Senior Editor, *Advanced Materials & Processes,* 133(6), June 1988.

18. Karlsson, B., and Quintiere, J.G., *Enclosure Fire Dynamics*, CRC Press, ISBN 0-8493-1300-7.
19. Ohlemiller, T., and Cleary, T., "Upward Flame Spread on Composite Materials," *Fire and Polymers II—Materials And Tests for Hazard Prevention*, G.L. Nelson, ed., American Chemical Society, Washington, DC, 1995, Chap. 28.
20. Sorathia, U., Lyon, R., Gann, R.G., and Gritzo, L., "Materials and Fire Threat," *Fire Technology*, 33(3), 1997. (Reproduced from *SAMPE Journal*, 32(3), May/June 1996).
21. Proceedings, "Fire Risk and Hazard Assessment Research Application Symposium," Fire Protection Research Foundation, Baltimore, MD, June 20–22, 2001.
22. Linan, A., and Williams, F.A., *Fundamental Aspects of Combustion*, Oxford University Press, New York, 1993.
23. Strehlow, R.A., *Combustion Fundamentals*, McGraw-Hill, New York, 1984.
24. Sorathia, U., Lyon, R., Ohlemiller, T., and Grenier, A., "A Review of Fire Test Methods and Criteria for Composites," *SAMPE Journal*, 33(4), July/August 1997.
25. MIL-STD-2031 (SH), "Fire and Toxicity Test Methods and Qualification Procedure for Composite Material Systems Used in Hull, Machinery, and Structural Applications Inside Naval Submarines," February 1991.
26. Sorathia, U., Long, G., Scholl, B., Blum, M., Ness, J., and Gracik, T., "Performance Requirements for Fire Safety of Materials in U.S. Navy Ships and Submarines," *Proceedings of the 46th International SAMPE Symposium and Exhibition*, Vol. 46, May 2001.
27. Conley, C.J., "U.S. Vehicle Fire Trends and Patterns Through 1992," NFPA Report, National Fire Protection Association, Quincy, MA, October 1994.
28. Title 49—"Transportation 571-302, Standard 302, Flammability of Interior Materials," *Federal Register*, 36: 232, 1972.
29. Federal Railroad Administration, "Guidelines for Selecting Materials to Improve Their Fire Safety Characteristics," *Federal Register*, 49: 162, 1984.
30. Urban Mass Transportation Administration, "Recommended Fire Safety Practices for Rail Transit Materials Selection," *Federal Register*, 49: 158, 1984.
31. Sarkos, C.P., and Hill, R.G., "Heat Exposure and Burning Behavior of Cabin Materials During an Aircraft Post-Crash Fuel Fire," NMAB Report 477-2, National Materials Advisory Board, National Research Council, National Academy Press, 1995, p. 25.
32. Nollen, D.A., "Flammability Regulations Affecting Advanced Composite Materials," *Journal of Fire Sciences*, 8: 227–238, 1990.
33. *Code of Federal Regulations*, Title 46, Shipping, Chapter 1, Part 175–185, Subchapter T, Small Passenger Vessels (Under 100 Gross Tons), 46 CFR § 175–185.
34. IMO Resolution MSC.61 (67), "International Code for the Application of Fire Test Procedures," International Maritime Organization, London, UK, December 1996.
35. IMO, MSC Circular 732, "Interim Guidelines on the Test Procedure for Demonstrating the Equivalence of Composite Materials to Steel Under the Provisions of the 1974 SOLAS Convention," June 28, 1996.
36. *The International Code of Safety for High Speed Craft*, International Maritime Organization, Maritime Safety Committee, Adopted as Chapter 10 to SOLAS, May 1994.
37. IMO Resolution MSC.40 (64), "Standard for Qualifying Marine Materials for High Speed Craft as Fire-Restricting Materials," International Maritime Organization, London, December 1994.
38. DeMarco, Ronald A., "Composite Applications at Sea: Fire Related Issues," *36th International SAMPE Symposium*, April 15–18, 1991.
39. ASTM E84-98, 2000, "Standard Test Method for Surface Burning Characteristics of Building Materials," American Society for Testing and Materials, Vol. 4.07 Building Seals and Sealants, Fire Standards, Dimension Stone, American Society for Testing and Materials, Conshohocken, PA.
40. NFPA 301, "Safety to Life from Fire on Merchant Vessels," National Fire Protection Association, Quincy, MA, 2001.
41. MIL-STD-1623D (SH), "Fire Performance Requirements and Approved Specifications for Interior Finish Materials and Furnishings (Naval Shipboard Use)," Naval Sea Systems Command, Washington, DC, 1985.
42. ASTM E 162, "Standard Test Method for Surface Flammability of Materials Using a Radiant Heat Energy Source," 1997 *Annual Book of ASTM Standards*, Vol. 04.01, 1997.

43. ASTM E 662-93, "Standard Test Method for Specific Optical Density of Smoke Generated by Solid Materials," Fire Test Standards, Dimension Stone, American Society for Testing and Materials, Conshohocken, PA.
44. ASTM E 1354-94, "Standard Test Method for Heat and Visible Smoke Release Rate for Materials and Products Using an Oxygen Consumption Calorimeter," 1997 *Annual Book of ASTM Standards*, Vol. 04.01: Building Seals and Sealants, Fire Standards, Dimension Stone, American Society for Testing and Materials, Conshohocken, PA.
45. ISO 9705, "Fire Tests—Full-Scale Room Test for Surface Products," International Organization for Standards, Geneva, Switzerland, 1993.
46. Committee on Fire Toxicology, National Research Council, *Fire & Smoke, Understanding the Hazards*, National Academy Press, Washington, DC, 1986.
47. Tewarson, A., "Generation of Heat and Chemical Compounds in Fires," *Fire Protection Handbook*, J.P. Dinenno, ed., Society of Fire Protection Engineering, 1988.
48. Sorathia, U., and Beck, C.P., "Fire-Screening Results of Polymers and Composites," *Proceedings of Improved Fire and Smoke Resistant Materials For Commercial Aircraft Interiors*, Publication NMAB-477-2, National Research Council, National Academy Press, Washington, DC, 1995.
49. Macaione, D.P., and Tewarson, A., "Flammability Characteristics of Composite Materials," *ACS Symposium Series*, 1990.
50. Brown, J.E., Braun, E., and Twilley, W.H., "Cone Calorimeter Evaluation of the Flammability of Composite Materials," NBSIR 88-3733, NIST, Gaithersburg, MD, March 1988.
51. ASTM E603-98, "Standard Guide for Room Fire Experiments," American Society for Testing and Materials, Vol. 4.07 Building Seals and Sealants, Fire Standards, Dimension Stone, American Society for Testing and Materials, Conshohocken, PA, 1999.
52. Janssens, M., Garabedian, A., and Gray, W., "Establishment of International Standards Organization (ISO) 5660 Acceptance Criteria for Fire Restricting Materials on High Speed Craft," United States Coast Guard Report, CG-D-22-98.
53. Janssens, M., et al., "Fire, Smoke, and Toxicity of Composites on High Speed Craft," United States Coast Guard Report, CG-D-27-98.
54. Beyler, C.L., Hunt, S.P., Lattimer, B.Y, Iqbal, N., Lautenberger, C., Dembsey, N., Barnett, J., Janssens, M., Dillon, S., and Grenier, A., "Prediction of ISO 9705 Room/Corner Test Results," USCG Report, CG-D-22-99, Vols. I and II, November 1999.
55. IMO Resolution A.754 (18), "Recommendation on Fire Resistance Tests for 'A,' 'B' and 'F' Class Division," FTP Code—International Code for Application of Fire Test Procedures, International Maritime Organization, London, 1998.
56. ASTM E119-98, "Standard Test Methods for Fire Tests of Building Construction and Materials," Vol. 4.07, Building Seals and Sealants, Fire Standards, Dimension Stone, American Society for Testing and Materials, Conshohocken, PA, 1999.
57. Leonard, J.T., et al., "Post Flashover Fires in Simulation Shipboard Compartment," NRL Memo Report 6886, Naval Research Laboratory, Washington, DC, September 3, 1991.
58. UL 1709, "Rapid Rise Fire Tests of Protection Materials for Structural Steel," Underwriters Laboratories, Inc., Northbrook, IL, 1991.
59. Rollhauser, C.M., Beck, C.P., Purcell, R., Griffith, J., Wenzel, A., and Douglas, H.E., "Fire Testing of Structural Composite Panels for the Integrated Technology Deckhouse," Fire Test Report, Southwest Research Institute, San Antonio, TX, June 1996.
60. Sorathia, U., Lugar, J., and Beck, C., "UL 1709, Fire Testing of Glass/Vinyl Ester Composite Panels," presented at *Fire and Materials Conference*, 1999.
61. Latimer, B.Y., and Sorathia, U., "Corner Fires—Part 1–4, *Fire Safety Journal*, 2002, submitted for publication.
62. Sorathia, U., Blum, M., McFarland, M., and Coffin, P., "UL 1709, Fire Testing of Glass/Vinyl Ester and Balsa Core Sandwich Composite Panels," presented at *Fire and Materials Conference*, 2000.
63. Sorathia, U., and Dapp, T., "Structural Performance of Glass/Vinyl Ester Composites at Elevated Temperatures," *SAMPE Journal*, 33(4), July/August 1997.
64. Sorathia, U., Beck, C., and Dapp, T., "Residual Strength of Composites during and after Fire Exposure," *Journal of Fire Sciences*, 11(3), May/June 1993.

65. Ritter, G.A., Corlett, R., et al., "Real-Time Measurement of Flexural Rigidity in Fire-Impacted Composite Plates," *Fire and Materials—1st International Conference and Exhibition,* Washington, DC, September 1992.
66. Milke, J.A., and Vizzini, A.J., "Thermal Response of Fire Exposed Composites," *Journal of Composites Technology & Research,* 13(3): 145–151, Fall 1991.
67. Petrie, G.L., Sorathia, U., and Warren, L.W., "Testing and Analysis of Marine Composite Structures in Elevated Temperature Conditions," *Proceedings of the 44th International SAMPE Symposium,* Vol. 44, May 23–27, 1999.
68. William, F., Darwin, R., Scheffey, J., "Fire Spread By Heat Transmission Through Steel Bulkheads and Decks," *Fire Safety Journal,* Paper 6, Fire Safety on Ships, 1994.
69. MIL-PRF-XX381, "Performance Specification, Insulation, High Temperature Fire Protection, Thermal and Acoustic," U.S. Navy.
70. Sorathia, U., "Fire Performance of Composites," Tutorial presented at *47th International SAMPE Symposium,* May 12–26, 2002.
71. NRL Ltr 3900, Ser 6180/312, Fire Resistance Tests of Insulation Subjected to UL 1709 Fire Exposure," August 1993.
72. Meier, J., "Heat Flows for StructoGard," *Thermal Ceramics,* Augusta, GA, May 1997.
73. Rollhauser, C.M., Beck, C.P., and Douglas, H.E., "Development of a Fire Protective Insulation System for Vinyl Ester Composite Substrates," presented at *ABC Conference,* 1996.
74. Williams, F.W., et al, "Evaluation of Intumescent and Other Fire Protective Coatings for Submarine Applications," NRL Report, 3900, Ser 6180/284, 1993.
75. Sorathia, U., Rollhauser, C.M., and Hughes, W.A., "Improved Fire Safety of Composites for Naval Applications," *Fire and Materials,* 16: 119–125, 1992.
76. Grand, A.F., Priest, D.N., and Deggary, N., "Innovative Fire Resistant Coatings for Use on Composite Products Aboard U.S. Commercial Ships," Final Report, Omega Point Laboratory Project No. 13260-94700, DOT Contract No. DTRS-57-92-C-00111, June 1993.
77. Assaro, R., MARITECH Project DTMA91-95-H-00091, March 2001.
78. Program Review, Northrop Grumman, Composite Forward Director Room, 2000.
79. Sorathia, U., Long, G., Blum, M., Gracik, T., Ness, J., Le, A., Lindsey, N., McFarland, M., and Dallek, S., "Evaluation of Intumescent Coatings as Passive Fire Protection System for Shipboard Interior Applications," presented at *BCC Conference,* 2002.
80. Durkin, A., Williams, F., and Pham, H., "The Performance of Water Base Intumescent Coatings as Passive Fire Protection Components on an Internal Ship Bulkhead," NRL Report, 6180/0039, March 15, 2002.
81. Sorathia, U., Gracik, T., Ness, J., Durkin, A., Williams, F., Hunstad, M., and Berry, F., "Evaluation of Intumescent Coatings for Shipboard Fire Protection, *Journal of Fire Science,* submitted for publication.
82. Sorathia, U., Long, G., Gracik, T., Blum, M., and Ness, J., "Screening Tests for Fire Safety of Composites for Marine Applications," *Fire and Materials,* 25: 215–222, 2001.

INDEX

The letter t following an entry indicates a table. The letter f indicates a figure.

aircraft, 5.46–5.47
ASTM E 119, 9.41–9.42
ASTM E-1354, 9.22, 9.26
ASTM E-162, 9.22
ASTM E-662, 9.22
ASTM E-84, 9.22
automobiles, 5.47

burn injury, 5.29f

chemicals (see liquids and chemicals)
combustion
    flaming, 1.2
    pyrolysis, 1.3, 1.27
    smoldering, 1.2
composites, fabrication techniques (see also composites, polymer; composites, requirements and regulations for; composites, fire resistance)
    autoclave curing, 9.9
    compression molding, 9.9
    hand lay-up, 9.9
    VARTM, 9.9
composites, fire performance and test methods for (see also composites, polymer; composites, requirements and regulations for; composites, fabrication techniques; composites, fire resistance; composites, passive fire protection)
    fire growth, 9.24–9.31
    full-scale room test, 9.32–9.37
    heat release rate, 9.26–9.27
    ignitability, 9.26
    smoke and gas generation, 9.22–9.24
    surface flammability, 9.22
composites, fire resistance, 9.37–9.51 (see also composites, fabrication techniques; composites, polymer; composites, requirements and regulations for; composites, regulations for passive fire protection)
    "A" class divisions, 9.40
    "B" class divisions, 9.40
    ASTM D-790, 9.52
    ASTM E 119, 9.41–9.42
    fire resistance test method, 9.39–9.41
    fire-restricting, 9.38
    IMO A 754, 9.39
    module fire tests, 9.42–9.44
    residual strength, 9.52–9.60
    sandwich composite, 9.47–9.51
    structural integrity, 9.51–9.52
    thermal gradient, 9.50
    UL-1709, 9.39, 9.44–9.47
composites, military applications of, 9.2
composites, passive fire protection, 9.60–9.76 (see also composites, fire resistance; composites, fabrication techniques; composites, polymer; composites, requirements and regulations for)
    composite structures, 9.62
    intumescent coatings, 9.65–9.76
    metallic structures, 9.60–9.62
    thickness, 9.62–9.64
composites, polymer (see also composites, fabrication techniques; composites, requirements and regulations for; composites, fire resistance; composites, passive fire protection; polymers, properties; materials)
    bismaleimides, 9.8
    cost, 9.10–9.11
    epoxy, 9.8
    fibers, 9.6
    fire threat, 9.11–9.12
    flashover, 9.13
    mechanical properties of, 9.6–9.7
    phenolics, 9.8
    polyaryl cellulose, 9.9
    polyester, 9.7
    polyether ether ketone, 9.9
    polyimides, 9.8
    polyphenylene sulfide, 9.8
    structural integrity, 9.13
    thermoplastics, 9.6
    thermosets, 9.6
    vinyl ester, 9.7–9.8
composites, requirements and regulations for, 9.13–9.21 (see also composites, polymer; composites, fabrication techniques; composites, requirements; composites, fire resistance; composites, passive fire protection)

composites, requirements and regulations for, *(cont.)*
  air transportation, 9.19–9.20
  ASTM E-1354, 9.22, 9.26
  ASTM E-162, 9.22
  ASTM E-662, 9.22
  ASTM E-84, 9.22
  ground transportation, 9.19
  ISO 9705, 9.24, 9.33
  marine transportation, 9.20–9.21
  military use, 9.21

dust (see gas)

enclosure effects, 1.46–1.48
  CFD models, 1.48
  radiant heat flux, 1.47
  smoke layer interface, 1.46
  smoke layer, 1.46, 1.47, 1.48
  zone models, 1.48

fiber properties
  heat release rate, 5.2
  limiting oxygen index (LOI), 5.2
  LOI (see limiting oxygen index)
fiber-reinforced polymer, 9.1
fire retardants (see flame retardants)
flame (also see flame temperature)
  diffusion, 1.3
  premixed, 1.3
flame retardants (see also plastics, flame retardants for)
  alumina trihydrate, 1.42
  combustibility ratio, 1.42
  heat of combustion, 1.42
  heat of gasification, 1.42
  heat of vaporization, 1.42
  pyrolysis, 1.42
flame spread (also see solids)
  concurrent, 1.38
  critical heat flux for flame spread, 1.41
  flame length, 1.39
  heat release rate/unit area, 1.38
  opposed flow, 1.38
  wind-aided, 1.38
flame temperature (also see flame)
  actual, 1.16
  adiabatic flame temperature, 1.12
flammability
  gas, 1.16–1.20
  liquid, 1.21–1.26
  solid, 1.26–1.42
flammability limits (also see flammability)
  critical adiabatic flame temperature, 1.19
  critical energy density, 1.18
  fuel lean, 1.16
  fuel rich, 1.16
  LeChatelier's rule, 1.19
  lower flammability limits, 1.16, 1.18
  upper flammability limits, 1.16, 1.18, 1.19
flammability of furnishings, 5.16–5.23, 5.37–5.45
  aircraft, 5.46–5.47
  ASTM E 1357, 5.22
  automobiles, 5.47
  California TB 116, 5.22
  California Technical Bulletins, 5.22
  fire blocking layers, 5.40
  flooring radiant panel test, 5.16–5.17
  mattress and furniture tests, 5.17t
  NFPA 2113, 5.19
  OSU test, 5.16
  oxygen consumption calorimeter test, 5.16
  passenger trains, 5.23t, 5.45–5.46

gas (also see flammability limits)
  autoignition temperature, 1.19
  flame speed, 1.20
  flammability limits, 1.16–1.19

heat of combustion, 1.5t, 1.10–1.12 (also see thermochemistry)
  air heat of combustion, 1.12
  gross heat of combustion (high heating value), 1.10
  net heat of combustion (low heating value), 1.10
  oxygen consumption calorimetry, 1.11
  oxygen heats of combustion, 1.11
  ventilation limit, 1.11

IMO A. 754, 9.39
ISO 9705, 9.24, 9.33

liquid (see also liquids and chemicals)
  burning rate, 1.24
  Clapeyron-Clausius equation, 1.22
  combustibility ratio, 1.26
  combustible liquid, 1.21
  effective heat of combustion, 1.25
  fire point, 1.21
  flammable liquid, 1.21
  flash point, 1.21
  heat flux, 1.24
  heat of gasification, 1.24
  heat release parameter, 1.26
  mass-burning ratio, 1.24
  mean beam length, 1.24
  Raoult's law, 1.23
  vapor pressure, 1.21, 1.22
liquids and chemicals (see also liquid)
  autoignition temperature, 8.13–8.14

boiling, 8.9
Chemical Abstract Service (CAS), 8.1
Chemtrec, 8.1
combustion characteristics, 8.16–8.37
European Chemicals Bureau, 8.1
fireballs, 8.4
flammability, 8.6t, 8.10, 8.16–8.19
flash points, 8.11–8.12
hazard classification, 8.13–8.16
heat release rate, 8.24–8.37
ignition characteristics, 8.10–8.16
oils, mesoscale burning of, 8.31, 8.36
pool fires, large-scale, 8.28–8.31, 8.36
pool fires, small-scale, 8.25–8.28, 8.36
release rate of fluid vapors, 8.20–8.24
release rates, 8.35–8.37
smoke point, 8.37–8.38
spray fires, large scale, 8.32–8.35
vaporization, 8.5–8.10

materials (see polymers, properties, wood, physical structure and properties of, composites, polymer)
  charring behavior, 2.5–2.8
  decomposition behavior, 2.5–2.8
  fire growth behavior, 2.13–2.15
  flame spread behavior, .13–2.15
  ignition behavior, 2.8–2.13
  melting behavior, 2.4–2.5
  softening behavior, 2.4–2.5
  vaporization behavior, 2.5–2.8
materials, structural
  "high rise" curves, 6.10–6.11
  beams and girders, 6.34–6.36, 6.58–6.60
  columns, 6.32–6.34, 6.53–6.58
  concrete, 6.14–6.18, 6.42–6.43
  connections, 6.21–6.22, 6.60
  fire resistance, calculation, 6.43–6.61
  flame shielding, 6.30–6.32
  floor/ceiling and roof/ceiling assemblies, 6.36–6.39
  girders (see beams and girders)
  gypsum, 6.23–6.24
  roof/ceiling assemblies (see floor/ceiling and roof/ceiling assemblies)
  spray-applied materials, 6.24–6.28
  standard tests for, 6.4–6.7
  steel, 6.12–6.14, 6.44–6.52
  structural fire response, 6.11
  structural integrity, 6.7
  superimposed loading, 6.7–6.9
  trusses, 6.22, 6.39–6.42, 6.60
  walls, 6.36
  water-filled columns, 6.28–6.30
  wood, 6.19–6.21

National Fire Protection Association (NFPA), 5.19, 5.21
National Institute for Standards and Technology (NIST), 6.3
  natural materials (see materials)

Occupational Safety and Health Association (OSHA), 5.19

passenger trains, 5.23t, 5.45–5.46
performance standards, 5.9t
  electric arc tests, 5.13–5.14, 5.32–5.35
  instrumented Thermal Mannequin test, 5.12, 5.19, 5.28t, 5.32t
  molten metal test, 5.14–5.15
  radiant protective performance test (RPP test), 5.11
  RPP test (see radiant protective performance test
plastics, fire behavior of
  charring, 3.36
  combustion efficiency, 3.36
  combustion efficiency, 3.37
  critical heat flux, 3.31
  diffusion barrier, 3.37
  extinction, 3.28
  fire calorimetry, 3.36
  fire point temperature, 3.28
  flame spread rate, 3.31
  flash point temperature, 3.28
  heat of fusion, 3.35
  heat release rate, 3.36
  heat release rate, 3.38, 3.39, 3.41
  ignition criteria, 3.28, 3.29f
  ignition resistance, 3.40
  ignition temperature, 3.31
  ignition time, 3.30
  incipient burning, 3.42
  latent heat of vaporization, 3.35
  mass loss rate, 3.35
  piloted ignition, 3.28
  recession velocity, 3.34
  self-extinguish, 3.41
  stationary state, 3.34
  thermal diffusivity, 3.35
  thermal inertia, 3.33, 3.34t
  thermal response parameter (TRP), 3.31
  unsteady heat conduction, 3.30
plastics, flame retardants for
  ABS, 4.19, 4.48–4.58
  alumina trihydrate (ATH), 4.1, 4.2, 4.7
  ammonium phosphate, 4.31
  antagonism, 4.22
  antimony pentoxide, 4.22
  antimony trioxide, 4.2, 4.9, 4.21–4.23, 4.70–4.71
  ATH, 4.1, 4.2, 4.7

plastics, flame retardants for *(cont.)*
   brominated diphenyl oxides (DPO), 4.19
   brominated oligomers, 4.18
   bromine, 4.10
   chlorinated paraffins, 4.9–4.10
   chlorophosphates, 4.22
   dibromostyrene, 4.20
   dimethyl methyl phosphonate (DMMP), 4.32
   elastomer/rubber, 4.73–4.76
   Emmons fire triangle, 4.5
   engineering thermoplastics (ETP), 4.19, 4.66
   epoxy, 4.72–4.73
   ethylene diamine phosphate (EDAP), 4.33
   ethylene-vinyl acetate (EVA), 4.8
   EVA, 4.8
   exfoliation, 4.76, 4.77
   HBCD, 4.11, 4.21, 4.45–4.46
   hexabromocyclododecane, 4.10
   high-impact polystyrene (HIPS), 4.9, 4.19
   HIPS, 4.9, 4.19, 4.46–4.48
   intercalated structures, 4.77
   intumescence, 4.7, 4.31, 4.36
   layered silicates, 4.77
   magnesium carbonate, 4.8
   magnesium hydroxide, 4.1, 4.8
   melamine, 4.33–4.35
   nanoclays, 4.1
   nanocomposites, 4.76–4.84
   nylon, 4.9, 4.58–4.66
   octabromodiphenyl oxide, 4.19
   oxygen index, 4.22
   PC/ABS, 4.50–4.58
   PDMS, 4.79
   pentabromobenzyl acrylate, 4.20
   phenolic resins, 4.73, 9.8
   phosphate esters, 4.31, 4.70
   phosphine esters, 4.31
   phosphonates, 4.31–4.32
   PMMA, 4.79
   polybutylene terephthalate (PBT), 4.66–4.68
   polycarbonate (PC), 4.49–4.58
   polyesters, thermoplastic, 4.66, 9.7
   polyethylene oxide (PEO), 4.78
   polyethylene terephthalate (PET), 4.68–4.69, 5.6
   polyethylene, 4.9, 4.19, 4.44
   polyolefins, 4.7, 4.9
   polypropylene, 4.19, 4.38–4.44
   polystyrene, 4.44–4.46, 4.80
   polyurethane, 4.9, 4.31, 4.73
   polyvinyl chloride (PVC), 4.7, 4.69–4.72
   PVC, 4.7, 4.69–4.72
   red phosphorous, 4.31
   scorch performance, 4.22
   sodium antimonate, 4.22
   styrenics, 4.44
   tetrabromobisphenol, 4.18
   tetrabromopthalic anhydride, 4.10
   tribromoneopentyl alcohol, 4.20
   tribromophenol, 4.18
   tribromophenol, 4.19
   triphenol phosphate (TPP), 4.6
   UL-94 V-0, 4.19, 4.20, 4.31, 4.82
   zinc stearate, 4.11
polymers, burning process of
   activation energy, 3.17
   char yield, 3.19, 3.26t
   charring, 3.20, 3.36
   depolymerization, 3.18
   flaming combustion, 3.15
   flammability parameter, 3.26
   frequency factor, 3.19
   fuel-generation model, 3.17
   heat of gasification, 3.16, 3.35
   heat release capacity, 3.26
   ignition, 3.15, 3.30
   mass loss rate, 3.17, 3.22
   rate constants, 3.18
   side-chain scission, 3.16
   specific heat release rate, 3.23
   stationary-state hypothesis, 3.18
   surface temperature, 3.15
   thermal decomposition, 3.15
   thermal degradation, 3.16
   thermolysis, 3.15
   Van Krevelen formula, 3.20
polymers, properties (see materials)
   addition polymer, 3.2
   amorphous, 3.3
   condensation polymer, 3.2
   crystalline, 3.3
   density, 3.3
   elastomer, 3.1, 3.7
   glass transition temperature, 3.3
   heat capacity, 3.3
   heat of combustion, 3.9, 3.26t
   heat of gasification, 3.9–3.10, 3.16, 3.35
   ignition temperature, 3.7
   mass loss rate, 3.14
   melting temperature, 3.4
   modulus of elasticity, 3.4
   molar heat, 3.14
   monomer, 3.1
   onset degradation temperature, 3.14
   plastic, 3.1
   polymerization, 3.2
   pyrolysis, 3.14, 3.21, 3.28
   thermal conductivity, 3.3
   thermal diffusivity, 3.9

thermal inertia, 3.8, 3.33
thermal properties, 3.8–3.9
thermoplastic, 3.3, 3.7, 9.6
thermoset, 3.3, 3.7, 9.6
protective clothing
   applications, 5.23
   ASTM, 5.21
   electric arc protection, 5.33–5.35
   flash fire protection, 5.25–5.27, 5.30–5.32
   molten metal protection, 5.35–5.36
   National Fire Protection Association (NFPA), 5.19, 5.21
   NFPA 2112 (2001), 5.20
   Occupational Safety and Health Association (OSHA), 5.19
   OSHA 29 CFR 1910.269, 5.20t
   performance standards, 5.19–5.22
   thermal protective performance test (TPP test), 5.8, 5.19, 5.32t
   TPP test (see thermal protective performance test)
   vertical flammability test, 5.8, 5.19, 5.28t

self-heating (also see solids)
   smolder, 1.41
   spontaneous combustion, 1.41
smoke 1.42–1.46
   Bouger's law, 1.43
   carbon monoxide, 1.45, 1.48
   combustion products, 1.45
   enclosure fires, 1.46
   entrainment, 1.43
   exposure duration (dose), 1.45
   extinction area, 1.44
   extinction coefficient, 1.43
   flashover, 1.46, 1.47
   fuel mass loss rate, 1.45
   global equivalence ratio, 1.45, 1.46, 1.48
   nonthermal damage, 1.46
   optical density, 1.43, 1.44
   production, 1.43
   toxicity, 1.45
   visibility, 1.44
   yield factor, 1.43, 1.45
solids
   Biot number, 1.29
   burnout, 1.38
   characteristic temperature rise, 1.32, 1.34, 1.36
   characteristic thickness, 1.29
   characteristic time, 1.32, 1.34
   char-forming materials, 1.28
   combustibility ratio, 1.38
   combustibility ratio, 1.42
   convection, 1.31
   convective heat-transfer coefficient, 1.31
   convective-radiative boundary condition, 1.30
   critical heat flux for flame spread, 1.41
   density, 1.30
   effective ignition temperature, 1.27
   fire retardants, 1.42
   flame spread, 1.38–1.41
   flaming ignition, 1.29
   heat of gasification, 1.27, 1.42, 3.9–3.10, 3.35
   heat of combustion, 3.9
   heat transfer coefficient, 1.29, 1.36, 1.37
   ignition temperature, 1.30, 1.36, 1.37, 1.38, 3.7
   imposed heat flux, 1.30, 1.35
   incident radiant heat flux, 1.30, 1.47
   net heat flux, 1.37
   piloted ignition, 1.27, 1.35, 3.28
   pyrolysis, 1.27, 1.37, 1.42, 3.14, 3.16, 3.21, 3.28
   reradiation, 1.31
   self-heating, 1.41–1.42
   semi-infinite solid, 1.33
   specific heat, 1.30
   surface density, 1.33
   thermal conductivity, 1.29, 1.30
   thermal inertia, 1.33, 1.34, 3.33
   thermally thick, 1.28, 1.33, 1.34, 1.37
   thermally thin, 1.28, 1.29, 1.30, 1.36
   thermoplastics, 1.28
   time to ignition, 1.30, 1.35
   unpiloted ignition, 1.27, 1.37
stoichiometry
   air stoichiometric ratio, 1.4
   fuel equivalence ratio, 1.5
   mixture fraction, 1.5, 1.7
   oxygen stoichoiometric ratio, 1.4
   stoichiometric reaction, 1.3
   yields, 1.4
structural materials (see materials, structural)
synthetic materials (see materials, polymers)

test methods, 2.1, 2.2
   burning behavior, 2.15–2.26
   flame spread and fire growth, 2.26–2.49
   International Standards Organization, 3.7
thermochemistry
   conservation of energy, 1.8
   Dalton's law, 1.8
   enthalpy of reaction, 1.7
   heat of combustion, 1.5t, 1.10, 1.11
   heat of formation, 1.8, 1.9t
   heat of reaction, 1.7, 1.8, 1.10
   internal energy, 1.8
   thermodynamics, first law, 1.8
thermodynamics
   conservation of energy, 1.8
   first law, 1.8

UL Fire Resistance Directory, 6.32
UL-1709, 9.39, 9.44–9.47

wood (see wood products; wood, protected frame construction; wood, physical structure; wood, thermal properties; wood members, exposed)
wood members, exposed, 7.40–7.48 (see also wood, physical structure and properties of; wood products; wood, thermal properties; wood, protected frame construction)
  beams, 7.41–7.42
  columns, 7.42–7.43
  connections, 7.48
  empirical design method, 7.41
  heavy timber, 7.40–7.41
  mechanics-based design method, 7.44–7.48
  member design, 7.46
  member strength, 7.45
  timber decks, 7.46
wood products (see also wood, physical structure and properties of; wood products; wood thermal properties of; wood members, exposed; wood, protected frame construction)
  beams and stringers, 7.3
  decking, 7.3
  dimension lumber, 7.2
  engineered wood products, 7.4–7.5, 7.23
  fiberboard, 7.4
  fire-retardant treatments, 7.9
  hardboard, 7.4
  panel products, 7.3-7-4, 7.23
  particle board panels, 7.3–7.4
  posts and timbers, 7.3
  sawn timber, 7.2–7.3
  wood structured panels, 7.3
wood, physical structure and properties of (see also wood; wood products; wood, thermal properties; wood members, exposed; wood, protected frame construction; materials)
  absorptivity, 7.10
  bending strength, 7.26
  cellulose, 7.7, 7.31
  compression strength, 7.26
  density, 7.10, 7.15
  design values, 7.24
  emissivity, absorptivity, 7.10
  heartwood, 7.6
  hemicellulose, 7.7, 7.31
  lignin, 7.7, 7.31
  modulus of elasticity, 7.25
  moisture content, 7.7, 7.25
  specific heat, 7.16–7.17
  steam softening, 7.25
  tension strength, 7.26
  thermal conductivity, 7.17–7.22
wood, properties of, see wood, physical structure and properties
wood, protected frame construction (see also, wood; wood members, exposed; wood, thermal properties; wood, physical structure and properties of; wood products)
  fire-rated assemblies, listed, 7.48
  fire-rated assemblies, non-listed, 7.49–7.50
  roof and floor ceiling assemblies, 7.51
  wall assemblies, 7.50
wood, reaction to fire (see wood, thermal properties)
wood, thermal properties
  cellulose, 7.8
  charring rate, 7.29, 7.32–7.34, 7.46
  fire growth, room/corner, 7.39–7.40
  flame spread, opposed-flow, 7.38–7.39
  flame spread, wind-aided, 7.36
  heat of combustion, 7.31
  heat of gasification, 7.31
  hemicellulose, 7.8
  ignition temperature, 7.28
  lignin, 7.8
  piloted ignition, 7.27–7.29
  pyrolysis models, 7.35
  smoke, 7.35
  smoldering combustion, 7.35
  thermal degradation, 7.11
  toxicity, 7.36